LIVERPOOL JMU LIBRARY

3 1111 00717 4467

Handbook of Derivatives
for Chromatography

Handbook of Derivatives for Chromatography

Second Edition

Edited by

KARL BLAU and JOHN M. HALKET

Bernhard Baron Memorial Research Laboratories, Queen Charlotte's and Chelsea Hospital, London, UK

JOHN WILEY & SONS
Chichester · New York · Brisbane · Toronto · Singapore

First edition © 1977 by Heyden & Sons Ltd

Second edition © 1993 by John Wiley & Sons Ltd,
　　　　　　　　　　Baffins Lane, Chichester,
　　　　　　　　　　West Sussex PO19 1UD, England
　　　　　　　　　　Telephone　National: Chichester (0243) 779777
　　　　　　　　　　　　　　　　International: +44 243 779777

Reprinted with corrections July 1994
Reprinted October 1995, August 1996

All rights reserved.

No part of this book may be reproduced by any means,
or transmitted, or translated into a machine language
without the written permission of the publisher.

Other Wiley Editorial Offices

John Wiley & Sons, Inc., 605 Third Avenue,
New York, NY 10158-0012, USA

Jacaranda Wiley Ltd, G.P.O. Box 859, Brisbane,
Queensland 4001, Australia

John Wiley & Sons (Canada) Ltd, 22 Worcester Road,
Rexdale, Ontario M9W 1L1, Canada

John Wiley & Sons (SEA) Pte Ltd, 37 Jalan Pemimpin #05-04,
Block B, Union Industrial Building, Singapore 2057

Library of Congress Cataloging-in-Publication Data

Handbook of Derivatives for chromatography.—2nd ed. / edited by
　　Karl Blau and John Halket.
　　　　　p.　　cm.
　　Includes bibliographical references and index.
　　ISBN 0 471 92699 X
　　　1. Chromatographic analysis—Handbooks, manuals, etc.　I. Blau,
　　Karl.　II. Halket, John.
　　QD79.C4H35　　1993　　　　　　　　　　　　92-35141
　　543'.089—dc20　　　　　　　　　　　　　　　　CIP

British Library Cataloguing in Publication Data

A catalogue record for this book is available from the British Library

ISBN　0　471　92699　X

Typeset in 10/12pt Palatino by Thomson Press (India) Ltd., New Delhi
Printed and bound by Bookcraft, Midsomer Norton, Avon

Contents

Preface		xi
Acknowledgements		xv
Contributors		xvii
Abbreviations		xix

1. A Guide to the Handbook: The Selection of Derivatives — 1
The Editors

    1 Introduction 1
    2 What do you want to analyse? 1
    3 The choice of chromatographic methods 3
    4 The synoptic tables of this chapter 3
    5 Where to find help with thin-layer chromatography (TLC) 3
    6 If you want to use column chromatography (HPLC) 3
    7 For gas chromatography and GC–MS 3
    8 Applications for mass spectrometry 9
    9 Conclusions 9
   10 References 10

2. Esterification — 11
Karl Blau and *André Darbre*

    1 Theoretical basis 12
        1.1 Introduction 12
        1.2 Reaction mechanisms of esterification 12
    2 Practical methods 14
        2.1 The use of an alcohol and hydrogen chloride 14
        2.2 Esterifications involving other acid catalysts 15
        2.3 The use of diazomethane and analogues 17
        2.4 'Silver salt', 'potassium bromide' and similar methods 19
        2.5 Silylation 21
        2.6 Coupling reactions using carbodiimides 22
        2.7 Esterifications using other coupling reactions 23
        2.8 Pyrolysis of quaternary ammonium salts 24

		2.9 The use of dialkyl acetals	25
		2.10 Transesterification of triglycerides and amino acid esters	25
		2.11 Miscellaneous esterification reactions	26
		2.12 Conclusions	27
	3	References	28

3. Acylation — 31
Karl Blau

	1	Introduction	31
	2	Reaction mechanisms	33
		2.1 General	33
		2.2 The reactivity of acylimidazoles	33
	3	Types of acylation reactions	34
		3.1 General	34
		3.2 Detector-oriented acylation reactions	34
	4	Practical applications	36
		4.1 General	36
		4.2 Acetylation reactions	36
		4.3 Methods for making propionyl derivatives	39
		4.4 Formyl derivatives	39
		4.5 Perfluoroacyl derivatives	40
		4.6 Alternative perfluoroacylation procedures	43
		4.7 Miscellaneous acylations	45
	5	Conclusions	47
	6	References	47

4. Advances in Silylation — 51
R. P. Evershed

	1	Introduction	52
	2	General features of silylation reactions	52
		2.1 Mechanistic considerations	52
		2.2 Practical considerations	53
	3	Gas chromatography of silyl derivatives	54
	4	Preparation of silyl derivatives and applications to the gas chromatography of organic compounds	55
		4.1 Trimethylsilyl (TMS) derivatives	55
		4.2 Applications of trimethylsilylation to the analysis of organic compounds	60
		4.3 Silyl derivatives other than TMS	75
		4.4 Application of t-butyldimethylsilylation to the GC analysis of organic compounds	76
		4.5 Other alkylsilyl and aryldimethylsilyl derivatives	85
		4.6 Halogenated silyl derivatives	93
		4.7 Nitrogen-containing silyl derivatives	96
	5	Conclusions	99
	6	References	100

5. Alkylation — 109
Pavol Kováč

	1	Introduction	109
	2	Alkylation procedures	110
		2.1–2.18 Alkylations for special applications	111–126
	3	Conclusions	126
	4	References	127

6. Derivative Formation by Ketone–Base Condensation — 131
Roberta H. Brandenberger and Hans Brandenberger

    1 Analysis of carbonyl compounds as Schiff bases   131
    2 Analysis of primary amines as Schiff bases   133
    3 Analysis of primary amines as mustard oils   134
    4 Selected procedures   136
        4.1 Schiff bases from carbonyl compounds   136
        4.2 Schiff bases from amines   137
        4.3 Isothiocyanates from primary amines   138
    5 References   139

7. Formation of Cyclic Derivatives — 141
Karl Blau and André Darbre

    1 Introduction   142
    2 Acetals and ketals   142
    3 Cyclic siliconides   145
    4 Cyclic boronates   146
    5 Quinoxalinols from α-keto acids   147
    6 Cyclization reactions of guanidino groups   148
    7 Other miscellaneous heterocyclic derivatives   149
    8 Cyclic derivatives of amino acids   150
    9 Sequencing of peptides and polypeptides   151
    10 Conclusions   153
    11 References   153

8. Coloured and UV-absorbing Derivatives — 157
Famei Li and C. K. Lim

    1 Reagents for amines and amino acids   158
    2 Reagents for carboxylic acids   164
    3 Reagents for hydroxy compounds   168
    4 Reagents for carbonyl compounds   170
    5 Conclusions   172
    6 References   172

9. Fluorescent Derivatives — 175
Nikolaus Seiler

    1 Introduction   176
    2 Sulphonyl chlorides   177
    3 Carbonyl chlorides   186
    4 Halogenonitrobenzofurazans   187
    5 Isocyanates and isothiocyanates   190
    6 Fluorescamine   191
    7 2-Methoxy-2,4-diphenyl-3(2H)-furanone (MDPF)   192
    8 Schiff base-forming and related reagents   192
    9 Further reagents for the labelling of amino compounds   196
    10 Reagents for carbonyl compounds   196
    11 Condensation of aliphatic aldehydes with 1,3-diketones and ammonia to lutidine derivatives   199
    12 Sulphhydryl reagents   199
    13 Fluorescence labelling of acidic functions   201
    14 Fluorescence labelling of alcohols   203
    15 Fluorescence labelling of adenine and its derivatives by chloro- or bromoacetaldehyde   203

16	Conclusions	204
17	References	204

10. Derivatization for Chromatographic Resolution of Optically Active Compounds — 215
M. W. Skidmore

1	Introduction	215
2	Indirect versus direct methods	217
3	Selection of chiral derivatizing reagents for chromatographic resolution of enantiomers	218
4	Reactions of individual functionalities	219
	4.1 Hydroxyl groups (alcohols and phenols)	219
	4.2 Aliphatic and acyclic carboxylic acids	225
	4.3 Amines	230
	4.4 Aldehydes and ketones	237
	4.5 Lactones	240
	4.6 Amino acids	241
	4.7 Amino alcohols	247
	4.8 Synonyms for some of the reagents described in the text	249
5	Conclusions and future trends	249
6	References	249

11. Ion-pair Extraction in Chromatography — 253
Bengt-Arne Persson and Göran Schill

1	Application of the ion-pair concept	253
2	Ion-pair extraction and side reactions	254
3	Ion-pair distribution in liquid chromatography	256
4	Detection by ion-pairing agent	260
5	Ion-pair mediated derivatization	263
6	Theoretical background for the distribution of ion-pairs	263
7	Conclusions	264
8	References	264

12. Derivatization for Fast Atom/Ion Bombardment Mass Spectrometry — 267
Russell C. Spreen

1	Introduction	267
2	The FAB matrix: performance considerations	269
3	Derivatives	270
	3.1 Non-specific methods: sample preparation	271
	3.2 Specific derivatization methods: reactions to alter structure	273
	3.3 Derivatization for soft ionization MS–MS	280
	3.4 Creation of a charged site	284
4	Derivatization and continuous flow FAB: future prospects	286
5	Conclusions	286
6	References	286

13. Derivatives for Supercritical Fluid Chromatography — 289
Keith D. Bartle, Anthony A. Clifford and Naila Malak

1	Introduction	289
2	Derivatization to increase solubility	290
3	Derivatization for selective and sensitive detection	293
4	Conclusions	294
5	Acknowledgements	294
6	References	294

14. Derivatives for Gas Chromatography–Mass Spectrometry — 297
John M. Halket

　　1　Introduction　297
　　2　The mass spectrometric detector　298
　　3　Silyl derivatives　303
　　4　Perfluoroacylation　313
　　5　Esterification　317
　　6　Cyclization　318
　　7　Conclusions　321
　　8　References　322

15. Post-chromatographic Derivatization — 327
Karl Blau

　　1　The rationale and scope of post-chromatographic derivatization　327
　　2　Revelation of spots on paper and thin-layer chromatograms　329
　　3　Post-chromatographic derivatization in liquid chromatography　329
　　4　Post-chromatographic derivatization in gas chromatography　336
　　5　Radioactive derivatization　338
　　6　Miscellaneous applications　338
　　7　Conclusions: future developments　343
　　8　References　344

16. Practical Considerations — 349
Karl Blau

　　1　Introduction　349
　　2　Apparatus and equipment　350
　　3　Solvents and reagents　355
　　4　Safety　355
　　5　Conclusions　356
　　6　References　356

Index — 357

Preface

Welcome to this new edition of *Handbook of Derivatives for Chromatography*! It reflects developments in chemical derivatization that have gone hand-in-hand with the unstoppable advance of the science (and art!) of chromatography. In the first edition we could already see consolidation occurring in many areas of derivatization, and in fact there has been a modest expansion in the scope and variety of derivatization approaches; indeed, in some areas derivatization may even have declined in significance as a result of developments elsewhere; the burgeoning of ion chromatography where derivatization may be unnecessary is an example [1]. This edition describes new applications of derivatization chemistry and also approaches which exploit advances in chromatographic methods and improved instrumentation.

There has, in particular, been the welcome spread of 'bench-top' GC–MS systems, which have become cheap enough to bring mass spectrometry to many who at one time regarded it as an enviable but unaffordable luxury, and who are now enthusiastically exploiting the advantages of positive chemical identification which is its hallmark. Detector-oriented derivatization schemes are nothing new, but it is a particular feature of our new edition that mass spectrometry-oriented derivatization approaches are emphasized, because we want to bring the whole area of derivatization to the attention of this widening readership in the hope of making their work easier.

Much of the new handbook is relatively unchanged. Any enthusiastic cook is bound to agree with Shakespeare that '...age cannot wither...nor custom stale...' a good recipe. Some sections have disappeared: no mention is made of the advances that have been made in the chromatography of inorganic anions and cations. We now have two books specifically devoted to inorganic chromatography [2, 3], which cover this subject in greater detail than was possible in just a couple of chapters. There is also no chapter on reaction chromatography; again, a whole book devoted to this has appeared in the meantime [4]; see also some of the other books quoted below.

A brand new chapter deals with derivatization to produce coloured derivatives, the one major area of derivatization that was not dealt with adequately in the first edition, and there are other new chapters. But, precisely because recipes do not get outdated, the two editions of the handbook may be used side by side.

The first edition of the handbook was firmly based on the philosophy, adopted from Alan Pierce's pioneering work on silylation [5], that derivatization reactions should be presented on the basis of unit processes such as esterification, alkylation, silylation etc., so that the chemical groupings of the substance to be analysed would suggest the derivatives to be used. The present edition no longer entirely follows this approach. It seemed unnecessary to restrict ourselves too rigidly, and whenever we have encountered an area where a modification of the original philosophy looked more helpful, we cheerfully describe other ways of approaching derivatization. Our main concern is the convenience of the reader. It is worth noting that, soon after our first edition appeared, Daniel Knapp published his excellent *Handbook of Analytical Derivatization Reactions* [6], which was firmly based on the alternative philosophy of the *kind of sample* being the basis for the selection of the most appropriate derivatization process. The success of that approach shows that there is more than one solution. His handbook is complementary to ours, and can usefully be consulted side by side with our two volumes. The same applies to the more specialized *Chemical Derivatization in Liquid Chromatography* by Lawrence and Frei [7], and the book on *Chemical Derivatization in Gas Chromatography* by Drozd [8]. Useful works like these and the others listed below [9–29] fortunately do not go out of fashion and have a lasting value.

Isolation and 'clean-up' procedures, we suggested earlier, merited a separate handbook. Because this area had lagged behind other aspects of chromatography, a tremendous volume of development has occurred in the last decade, with the wholesale adoption of solid-phase extraction, particularly using small columns, and latterly, specific filter discs. This has made the isolation of many compounds much easier, more quantitative and more reproducible, and has transformed the analytical scene in many fields. Another development is the use of microwaves to accelerate chemical reactions. Many microwave ovens are now commercially available which are specifically designed to promote chemical reactions, and they will be found useful for derivatization, although the small scale of most derivatization reactions permits even domestic-type microwave ovens to be used successfully as long as they are not returned to the kitchen for cooking afterwards....

We foresee automation of derivatization, mechanization of sample handling, computer-aided data reduction and 'expert systems' being increasingly taken for granted in chromatographic as well as mass spectrometric analysis. The era of the 'drive-in lab' is almost here: drive up, inject your sample, and drive round to the office for the report....

We sadly have to record the death, during the production of this Handbook, of our contributor Professor Göran Schill. He was a pioneer in the field of analytical pharmaceutical chemistry, and an outstanding scientist. His scientific achievements are his lasting memorial, but he will be greatly missed.

Our writing team of both original and new contributors dedicates this edition to chromatographers and mass spectrometrists everywhere, in the hope that it will give them practical help with chemical derivatization for chromatography.

The Editors

References

[1] D. T. Gjerde and J. S. Fritz, *Ion Chromatography*, Hüthig, Heidelberg (1987).

[2] J. C. MacDonald, *Inorganic Chromatographic Analysis*, Wiley, New York (1985).

[3] G. Schwedt, *Chromatographic Methods in Inorganic Analysis*, Hüthig, Heidelberg (1981).

[4] V. G. Berezkin, *Chemical Methods in Gas Chromatography*, Elsevier, Amsterdam (1983).

[5] A. E. Pierce, *Silylation of Organic Compounds*, Pierce, Rockford (1979).

[6] D. R. Knapp, *Handbook of Analytical Derivatization Reactions*, Wiley, New York (1979).

[7] J. F. Lawrence and R. W. Frei, *Chemical Derivatization in Liquid Chromatography*, Elsevier, Amsterdam (1976).

[8] J. Drozd, *Chemical Derivatization in Gas Chromatography*, Elsevier, Amsterdam (1981).

[9] R. W. Frei and J. F. Lawrence, *Chemical Derivatization in Analytical Chemistry. Volume 1: Chromatography*, Elsevier, Amsterdam (1981).

[10] R. Ikan, *Chromatography in Organic Microanalysis*, Academic Press, New York (1983).

[11] C. K. Lim (Editor), *HPLC of Small Molecules*, IRL Press, Oxford (1986).

[12] R. W. Frei, *Handling of Samples in Chromatography*, Gordon & Breach, New York (1986).

[13] G. Zweig and J. Sherma (Editors), *Handbook of Chromatography, General Data and Principles*, CRC Press, Boca Raton (1985).

[14] E. Heftmann (Editor), *Chromatography. Fundamentals and Applications of Chromatographic and Electrophoretic Methods, Part B: Applications*, Elsevier Amsterdam/New York (1983).

[15] D. B. Jack, *Drug Analysis by Gas Chromatography*, Academic Press, New York (1984).

[16] S. H. Y. Wong (Editor), *Therapeutic Drug Monitoring and Toxicology by Liquid Chromatography*, Marcel Dekker, New York (1985).

[17] A. C. Moffat, J. V. Jackson, M. S. Moss and B. Widdop (Editors), *Clarke's Isolation and Identification of Drugs*, Pharmaceutical Press, London (1986).

[18] E. Heftmann, *Chromatography of Steroids*, Elsevier, Amsterdam (1976).

[19] J. C. Touchstone, *CRC Handbook of Chromatography: Steroids*, CRC Press, Boca Raton (1986).

[20] A. Baerheim Svendsen and R. Verpoorte, *Chromatography of Alkaloids, Part A: Thin-Layer Chromatography*, Elsevier, Amsterdam (1983).

[21] R. Verpoorte and A. Baerheim Svendsen, *Chromatography of Alkaloids, Part B: Gas–Liquid Chromatography and High-Performance Liquid Chromatography*, Elsevier, Amsterdam (1984).

[22] M. Popl, J. Fähnrich, V. Tatar and J. Cazes (Editors), *Chromatographic Analysis of Alkaloids*, Marcel Dekker, New York (1990).

[23] A. Kuksis, *Chromatography of Lipids in Biomedical Research and Clinical Diagnosis*, Elsevier, Amsterdam (1987).

[24] W. W. Christie, *High Performance Liquid Chromatography and Lipids*, Pergamon, Oxford (1987).

[25] W. W. Christie, *Gas Chromatography and Lipids: A Practical Guide*, Oily Press, Ayr (1989).

[26] G. H. Wagman and M. J. Weinstein, *Chromatography of Antibiotics*, 2nd edn., Elsevier, Amsterdam (1984).

[27] C. J. Biemann and G. D. McGinnis (Editors), *Analysis of Carbohydrates by Gas Chromatography and Mass Spectrometry*, CRC Press, Boca Raton (1989).

[28] D. Stevenson and I. D. Wilson, *Recent Advances in Chiral Separations*, Plenum, New York (1991).

[29] E. J. Lough (Editor), *Chiral Liquid Chromatography*, Blackie, Glasgow (1989).

Acknowledgements

Many thanks to all who have written to us about the *Handbook of Derivatives for Chromatography* with comments, corrections and suggestions for the new edition.

Particular thanks are due to Georgina Going, librarian at the Institute of Obstetrics and Gynaecology at Queen Charlotte's and Chelsea Hospital. We would like to thank Samantha Baker of Fisons Scientific Equipment, Loughborough, Heinz Engelhardt of the University of the Saarland, P. Knight of SGE (UK) Ltd, Milton Keynes, Ira Krull of Northeastern University, Boston, Franz Morgenbesser of ICT Handels GmbH, Frankfurt, Tom Rendl and Colin Rix of Alltech Associates, Deerfield, IL and Carnforth, Lancs., Crystal Ridgway of Wheaton Science Products, Inc., Millville, NJ, Tim Warriner of Pierce & Warriner (UK) Ltd, Chester, and Claire Wittekind of V. A. Howe & Co. Ltd, Banbury, for their help with illustrations.

And of course we are especially grateful to first Jenny Cossham and then Mary Wong, Irene Cooper and James Deeny for their encouragement and patience, and to the whole production team at John Wiley & Sons.

Contributors

Keith D. Bartle, *University of Leeds, School of Chemistry, Leeds LS2 9JT, UK.*

Karl Blau, *Department of Chemical Pathology, Queen Charlotte's and Chelsea Hospital, London W6 0XG, UK.*

Hans Brandenberger, *Branson Research, CH-8708 Männedorf, Switzerland.*

Roberta H. Brandenberger, *Branson Research, CH-8708 Männedorf, Switzerland.*

Anthony A. Clifford, *University of Leeds, School of Chemistry, Leeds LS2 9JT, UK.*

André Darbre, *formerly Department of Biochemistry, King's College, Strand, London WC2R 2LS, UK.*

R. P. Evershed, *Department of Biochemistry, University of Liverpool, Liverpool L69 3BX, UK.*

John M. Halket, *Trace Analysis Unit, Bernhard Baron Memorial Research Laboratories, Queen Charlotte's and Chelsea Hospital, London W6 0XG, UK.*

Pavol Kováč, *National Institute of Diabetes and Digestive and Kidney Diseases, National Institutes of Health, Bethesda, MD 20892, USA.*

Famei Li, *Department of Clinical Biochemistry, King's College School of Medicine and Dentistry, London SE5 9PJ, UK.*

C. K. Lim, *MRC Toxicology Unit, Medical Research Council Laboratories, Woodmansterne Road, Carshalton, Surrey SM5 4EF, UK.*

Naila Malak, *University of Leeds, School of Chemistry, Leeds LS2 9JT, UK.*

Bengt-Arne Persson, *Department of Bioanalytical Chemistry, Astra Hässle AB, S-431 83 Mölndal, Sweden.*

Göran Schill, *Department of Analytical Pharmaceutical Chemistry, University of Uppsala Medical Centre, S-751 23 Uppsala, Sweden.*

Nikolaus Seiler, *Marion Merrell Dow Research Institute, 16, rue d'Ankara, 67084 Strasbourg Cédex, France.*

M. W. Skidmore, *Residue and Environmental Chemistry, ICI Agrochemicals, Jealott's Hill Research Station, Bracknell, Berkshire RG12 6EY, UK.*

Russel C. Spreen, *Structure Chemistry, Mass Spectrometry Facility, Zeneca Pharmaceuticals Group, Wilmington, DE 19897, USA.*

Abbreviations

These are the main abbreviations used in the Handbook, but others appear at various places and are explained whenever they are first mentioned. All other abbreviations such as units and dimensions are assumed to be too widely known to need inclusion.

AcOH	Acetic acid
ADMS	Allyldimethylsilyl
AITC	2,3,4-Tri-O-acetyl-α-D-arabinopyranosyl isothiocyanate
AP	S-acetoxypropionyl
BASTFA	N,O-Bis(allyldimethylsilyl)trifluoroacetamide
BMDS	Bromomethyldimethylsilyl
Bns	5-Di-n-butylaminonaphthalene-1-sulphonyl
t-BOC	t-Butyloxycarbonyl
b.p.	Boiling point
BSA	N,O-Bistrimethylsilylacetamide
BSCl	Benzenesulphonyl chloride
BSTFA	N,O-Bistrimethylsilyltrifluoroacetamide
BtDMS	tert-Butyldimethylsilyl
Bu (n,iso,tert)	Butyl (normal, iso or tertiary)
CEDMS	2-Cyanoethyldimethylsilyl
CHD	1,3-Cyclohexanedione, dihydroresorcinol
CDI	N,N'-Carbonyldiimidazole
CI	Chemical ionization (mass spectrometry)
CMDMS	Chloromethyldimethylsilyl
DADS	N,N-Diethylaminodimethylsilyl
Dabsyl	Dimethylaminoazobenzenesulphonyl
DABSCl	Dimethylaminoazobenzenesulphonyl chloride
Dansyl, Dns	5-Dimethylaminonaphthalene-1-sulphonyl
DCCI	Dicyclohexylcarbodiimide
DCMMS	Di-(Chloromethyl)methylsilyl
DEHS	Diethylhydrogensilyl
Dis-Cl	2-p-Chlorosulphophenyl-3-phenylindene
DMAA	N,N-Dimethylacetamide
DMAM	Dimethylaminomethyl

DMCS	Dimethylchlorosilane
DMDAS	Dimethyldiacetoxysilane
DMF	N,N-Dimethylformamide, formdimethylamide
DMNFHS	Nonafluorohexyldimethylsilyl
DMIPS	Isopropyldimethylsilyl
DMS	Dimethylsilyl
DMSO	Dimethyl sulphoxide
DNFB	1-Fluoro-2,4-dinitrobenzene
DNP	2,4-Dinitrophenyl
2,4-DNPH	2,4-Dinitrophenylhydrazine
Dns, dansyl	5-Dimethylaminonaphthalene-1-sulphonyl
DP	Degree of polymerization
DPE	1,2-Diphenylethylenediamine
DTBS	Di-tert-butylsilyl
EC	Electron capture
ECD	Electron capture detector
EDCI	1-Ethyl-3-(dimethylaminopropyl)carbodiimide
EDMS	Ethyldimethylsilyl
EDTA	Ethylenediaminetetraacetic acid
EI	Electron impact (mass spectrometry)
EO–TMS	Ethoxime and trimethylsilyl
Et	Ethyl
FAB	Fast atom bombardment (mass spectrometry)
FD	Field desorption (mass spectrometry)
FDNB	1-Fluoro-2,4-dinitrobenzene
FID	Flame ionization detector
FLEC	(+)-1-(9-Fluorenyl)ethyl chloroformate
FLOP	S-(+)-Flunoxaprofen
Flophemesyl	Pentafluorophenyldimethylsilyl
FMOC	9-Fluorenylmethyl chloroformate
FNBT	4-Fluoro-3-nitrobenzotrifluoride
GC, GLC	Gas chromatography
GC–MS	Combined gas chromatography and mass spectrometry
GITC	2,3,4,6-Tetra-O-acetyl-β-D-glucopyranosyl isothiocyanate
HFB	Heptafluorobutyryl
HFBA	Heptafluorobutyric anhydride
HMDS	Hexamethyldisilazane
HPLC, LC	High-performance liquid chromatography
IMDMS	Iodomethyldimethylsilyl
IPDMS	Isopropyldimethylsilyl
IR	Infrared
KRI	Kovats retention index
LC–MS	Combined HPLC and mass spectrometry
MDiPS	Di-isopropylmethylsilyl
MBTFA	N-Methylbis(trifluoroacetamide)
MDPF	2-Methoxy-2,4-diphenyl-3-furanone
MEK	Methyl ethyl ketone
Me	Methyl
Mns	6-Methylanilinonaphthalene-2-sulphonyl
MO–TMS	Methoxime and trimethylsilyl
m.p.	Melting point
MPTA	S-(−)-α-Methoxy-α-(trifluoromethyl)phenylacetyl

MSTFA	N-Methyl-N-trimethylsilyltrifluoroacetamide
MS	Mass spectrometry
MTBSTFA	N-Methyltert-butylsilyltrifluoroacetamide
MW	Molecular weight
NAC	N-Acetyl-L-cysteine
NBD-Cl	4-Chloro-7-nitrobenzo-2,1,3-oxadiazole, 4-chloro-7-nitrobenzofurazan
NCF	2-Naphthyl chloroformate
NIC	Naphthyl isocyanate
NICI	Negative-ion chemical ionization (mass spectrometry)
NITC	Naphthyl isothiocyanate
NMIM	N-Methylimidazole
OPA	ortho-Phthaldialdehyde
n-PDMS	n-Propyldimethylsilyl
PEIC	R-(+)-1-Phenylethyl isocyanate
PFB	Pentafluorobenzyl
PFBzO	Pentafluorobenzoyl
PFBzO-Cl	Pentafluorobenzoyl chloride
PFP	Pentafluoropropionyl
PFPA	Pentafluoropropionic anhydride
PhDMS	Phenyldimethylsilyl
PIC	Phenyl isocyanate
PICSI	Picolinyldimethylsilyl
PITC	Phenyl isothiocyanate
PMP	1-Phenyl-3-methyl-5-pyrazolone
Pr (n or iso)	Propyl (normal or iso)
PTC	Phenythiocarbamoyl
PTFE	Polytetrafluoroethylene (e.g. Teflon)
PTH	Phenylthiohydantoin
RIA	Radioimmunoassay
RP	Reversed phase (as in RP-HPLC)
SFC	Supercritical fluid chromatography
SIM	Selected ion monitoring
SPE	Solid-phase extraction
TBDMCS	t-Butyldimethylchlorosilane
TBDMS	t-Butyldimethylsilyl
TBDMSIM	t-Butyldimethylsilylimidazole
TBMPS	t-Butylmethoxyphenylsilyl
TCEP	Tris-(2-Cyanoethoxy)propane
TES	Triethylsilyl
TFA	Trifluoroacetyl
TFAA	Trifluoroacetic anhydride
THF	Tetrahydrofuran
TIC	Total ion current
TLC	Thin-layer chromatography
TMAH	Trimethylanilinium hydroxide
TMCS	Trimethylchlorosilane
TMS	Trimethylsilyl
TMSDEA	Trimethylsilyldiethylamine
TMSIM	Trimethylsilylimidazole
TMIPS	Cyclotetramethylene-isopropylsilyl
TMTBS	Cyclotetramethylene-t-butylsilyl
TMTFTH	Trimethyl(−)(α,α,α-trifluoro-m-tolyl)ammonium hydroxide

TnBS	Tri-n-butylsilyl
TNBS	Trinitrobenzenesulphonic acid
TnHS	Tri-n-hexylsilyl
TnPS	Tri-n-propylsilyl
TNP	Trinitrophenyl
TOF	*S*-Tetrahydro-5-oxo-2-furancarboxyl
TSCl	Toluenesulphonyl chloride
UDPGA	Uridine-5′-diphosphoglucuronic acid
UV	Ultraviolet

A Guide to the Handbook: The Selection of Derivatives

Karl Blau and **John M. Halket**
Department of Chemical Pathology and Trace Analysis Unit, Queen Charlotte's and Chelsea Hospital, London

1. INTRODUCTION   1
2. WHAT DO YOU WANT TO ANALYSE   1
3. THE CHOICE OF CHROMATOGRAPHIC
   METHODS   3
4. THE SYNOPTIC TABLES OF THIS CHAPTER   3
5. WHERE TO FIND HELP WITH THIN-LAYER
   CHROMATOGRAPHY (TLC)   3
6. IF YOU WANT TO USE COLUMN
   CHROMATOGRAPHY (HPLC)   3
7. FOR GAS CHROMATOGRAPHY AND GC–MS   3
8. APPLICATIONS FOR MASS SPECTROMETRY   9
9. CONCLUSIONS   9
10. REFERENCES   9

1. Introduction

It might at first seem unnecessary to offer a guide to a handbook which is in any case already organized into clearly defined chapters each with its table of contents. This is probably true for experienced analysts with a strong background in synthetic organic and analytical chemistry: they will know where to find what they need without any guide. Many other readers, however, are not organic or analytical chemists. They are using chromatographic methods as a means to an end, and want to analyse their compounds without having to go too deeply into the theory and practice of analytical chromatographic instrumentation and its application to solving various problems in a wide variety of disciplines and areas of research. These researchers just want to get on with running their samples, and we hope that they will find this guide a helpful way into the handbook.

2. The most important question is: What do you want to analyse?

The nature of the compounds to be analysed and their chemical properties, more than any other factors, govern the choice of derivatives and of the methods that are going to be used for their analysis. A good look at the chemical structure will immediately suggest the strategy needed to prepare a derivative that will be suitable for analysis. If the compounds come into a widely analysed category, Table 1 will show the most widely applied approach. This is because the organization of the handbook is based on the derivatization procedures appropriate to the different types of chemical groups occurring in the compounds that need to be analysed. Many compounds may be analysed without separating them out of wherever they are found, the 'matrix'. There may be a highly specific colour reaction, for example. But very often one is not that lucky. Matrix effects frequently affect the analytical results; or interfering substances may be present, making results unreliable. In such situations one has a choice: one can either get the desired substance out of the matrix on its own by a specific isolation procedure, or else one can use a preliminary clean-up followed by chromatography to separate out the compound of interest. It is then the chromatographic spot (in TLC) or the separated peak (in HPLC or GC) which is measured for quantitation, and in GC–MS for identification as well.

Table 1 Popular derivatization approaches by analytical technique and analyte class. Figures in brackets refer to the chapters in the handbook

	GC (GC–MS)	HPLC (TLC)
Aflatoxins	Use HPLC or TLC	Post-column derivatization (15)
Amino acids	n-Butyl ester, TFA (3, 14)	OPT (8, 9), Dansyl (9)
Aminoglycosides (Antibiotics)	Use HPLC or TLC	FDNB (8), OPT (8, 9)
Bile acids	Me ester–TMS	p-Bromophenacyl esters (2, 8)
Biogenic amines	Perfluoroacyl (3)	Not usually necessary Post-column (15)
Carbohydrates	TFA (3), TMS (4), boronate (7)	Phenyl isocyanate (8) Post-column (15)
Drugs	Multifunctional. See Tables 1 and 2	Often not necessary
Fatty acids	Me ester (2)	p-Bromophenacyl ester (2, 8)
Keto acids	Oxime (6)–TMS (4, 14)	2,4-DNP (6, 8)
Mycotoxins	See Aflatoxins	
Nitrosamines	Chromatograph without derivatization	Chromatograph without derivatization
Nucleotides, nucleosides and bases	TMS (4)	Depends on specific analyte (8, 9)
Organic acids	TMS (4, 14)	p-Bromophenacyl ester (2, 8)
keto	Oxime (6)–TMS (4, 14)	2,4-DNP (6, 8)
α-keto	Quinoxalinol–silyl (4, 7, 14)	Quinoxalinol (7, 9)
Peptides	Me ester (2)–TFA (3)	Dansyl (9)
Polychlorinated biphenyls (PCB)	No derivatizable groups	No derivatizable groups
Polyamines	TFA (3)	Benzoyl (8), dansyl (9)
Polycyclic aromatic hydrocarbons (PAH)	(TMS for metabolite–OH groups)	Chromatograph without derivatization
Porphyrins	TMS (4)	Chromatograph without derivatization
Prostaglandins (thromboxanes)	Me ester–oxime–TMS (4, 6, 14) PFB ester–oxime–TBDMS (2, 4, 6, 14)	p-Bromophenacyl ester (2, 8)
Steroids	MO–TMS (4, 6, 14) HFB (3, 14)	Keto: 2,4-DNP (8), dansyl (9) benzoyl (8)
Sugars	Reduce to alditols, then TFA (3) or TMS (4)	Phenyl isocyanate (8) Post-column (15)
Vitamins:		
A (retinol)	TMS (4)	Chromatograph without derivatization
B$_1$ (thiamine)	Acyl (3)	Chromatograph without derivatization, or post-column (15)
B$_6$	Use HPLC	Chromatograph without derivatization, or post-column (15)
C	TMS (4)	Chromatograph without derivatization, or oxidize to dehydro-form and prepare 2,4-DNP (8) or quinoxalinol (7, 9) derivatives
D and metabolites	TMS (4), boronate–TMS (4, 7, 14)	Chromatograph without derivatization
E (tocopherol)	TMS (4)	Chromatograph without derivatization
K	Chromatograph without derivatization (HPLC preferred)	Chromatograph without derivatization

3. The choice of chromatographic methods

The nature of the compounds to be analysed affects the way they can be chromatographed: for some compounds thin-layer chromatography is the best, and very often this may not require derivatization until after the separation, when the separated spots have to be visualized by colorimetric or fluorimetric detection methods (see Chapter 15 for further information on post-chromatographic derivatization). Other compounds are best analysed by gas chromatography or GC–MS, and yet others by HPLC. The choice may also be affected by what is available in the laboratory: almost everybody has the facilities for TLC, but GC and HPLC are now widely available, and GC–MS is also becoming more accessible. HPLC very often requires pre-chromatographic derivatization, but post-chromatographic procedures, such as photochemical or electrochemical reactions, are increasingly being used, either instead of or as well as pre-chromatographic derivatization (again, see Chapter 15). GC and GC–MS usually require pre-chromatographic derivatization reactions, and only rarely need post-chromatographic procedures.

4. The synoptic tables of this chapter

These indicate the chapters in the handbook where the derivatizations of various substances by compound type are presented. Table 1 shows the derivatization procedures commonly used for widely analysed compound classes. Tables 2 and 3 list derivatizations applicable to oxygen- and nitrogen-containing functional groups, respectively. In addition, the following paragraphs briefly point to the chapters that are appropriate to the different types of chromatographic separation techniques. For the rest, the tables of contents of the individual chapters, together with the index at the back of the handbook, complete the various signposts to the selection of suitable procedures.

Finally, we would stress that there is no ideal or best or perfect solution: there will usually be more than one solution to a derivatization or chromatographic problem. Different analysts may have different approaches, may have limitations of equipment availability or may have definite personal preferences about what kind of chromatography they like best. We hope, and indeed are fairly confident, that somewhere within this handbook there will be found the solutions to most chromatographic problems.

5. Where to find help with thin-layer chromatography (TLC)

The majority of compounds can be chromatographed by TLC without derivatization [1–5]. On the other hand, it is much easier to devise and follow separations if the compounds are already coloured, and here Chapter 8 will be helpful. Coloured derivatives are particularly useful for compounds that do not themselves chromatograph well, such as amines. The advantage of TLC is that, when all else fails, separated organic compounds can be revealed by charring. Fluorescent derivatives (Chapter 9) may also be useful. For optically active compounds special TLC plates can also effect resolution (Chapter 10).

Where TLC scores particularly well, however, is in following the course of derivatization reactions. There is no better, quicker or more convenient way of monitoring the completeness of the conversion of a compound to its derivative and making sure that no side reactions have occurred, and this is of the greatest help in optimizing the derivatization procedure.

6. If you want to use column chromatography (HPLC)

Derivatization for liquid chromatography—column chromatography, HPLC—has received special attention [6–8]. It has found particular application in the analysis of pharmaceuticals, drugs, vitamins and hormones and related compounds [9, 10]. In this handbook applications in which HPLC is involved will be found in most chapters, but the chapters that are particularly useful are those on coloured, UV-absorbing and fluorescent derivatives (Chapters 8 and 9); optical resolution, because of the development of chiral separation columns for HPLC (Chapter 10); ion-pair chromatography—essentially an HPLC methodology; and post-chromatographic derivatization, where the majority of the techniques are designed to follow HPLC separation (Chapter 15).

7. For gas chromatography and GC–MS

Gas chromatographic procedures are the ones that use pre-chromatographic derivatization the most, because analysis in the vapour phase requires volatile derivatives [11]. Probably well over half the recipes in the various chapters of this Handbook are designed to make substances into volatile derivatives for gas chromatography

Table 2 Synoptic chart on derivative formation with oxygenated groupings

Functional group	Procedure	Examples of products	Chapter	Comments
–OH (Primary, secondary and tertiary alcohols; phenols; carbohydrates)	Silylation	–O–Si(CH$_3$)$_3$	4	
	Acylation	–O–CO–CH$_3$; –O–CO–CF$_3$	3	
	Benzoylation	–O–CO–C$_6$H$_5$; –O–CO–C$_6$F$_5$	3	
	Alkylation	–O–CH$_3$; –O–CH$_2$–C$_6$F$_5$	5	
	Dansylation	Ar–O–Dns	9	Fluorescent derivative (phenols)
	Reaction with Dis-Cl	–O–Dis	9	Fluorescent derivatives of phenols and alcohols
	Reaction with FDNB	⟨structure: –O–C$_6$H$_3$(NO$_2$)$_2$⟩	8	For GLC of phenols with ECD
	Reaction with NBD-Cl	7-Nitrobenzofurazan	9	Fluorescent derivative (phenols)
	Ion-pair formation	Ar–O$^-$ M$^+$	14	For phenols; 'M' can be a variety of counter-ions
⟩C=O (Aldehydes and ketones)	Oxime formation	⟩C=N–OH; ⟩C=N–O–CH$_3$	6	May form *syn* and *anti* isomers
	Oxime formation and silylation	⟩C=N–O–Si(CH$_3$)$_3$	6	May form *syn* and *anti* isomers
	Ketal/acetal formation	⟨1,3-dioxolane structure⟩	7	
	Hydrazone formation	⟩C=N–NH–C$_6$H$_5$	6,8,9	Fluorescent and electron capturing derivatives available
	Schiff's base formation	⟩C=N–R	6	

	Silylation	=C–CO–O–Si(CH$_3$)$_3$
		  |
		  O–Si(CH$_3$)$_3$
			4	Only when enol formation is favoured, e.g. pyruvate

	Oxidation	–COOH	8	For aldehydes and methyl ketones (iodoform reaction); derivatized as carboxylic acids

Group	Reaction	Product	Ref.	Notes
−COOH (Carboxylic acids)	Esterification (alkyl)	−CO$_2$−CH$_3$; −CO$_2$−CH$_2$CF$_3$	2	
	Esterification (aryl)	−CO$_2$−CH$_2$C$_6$H$_5$; −CO$_2$−CH$_2$C$_6$F$_5$	2	
	Silylation	−CO$_2$−Si(CH$_3$)$_3$	4	
	Ion-pair formation	R−COO$^-$ M$^+$	11	'M' may be a variety of counter-ions
−C−C− | | OH OH (Glycols)	As for −OH, but also: Cyclic boronate formation	[cyclic boronate structure]	7	R = alkyl (most often butyl), or phenyl
	Acetal or ketal formation	[cyclic acetal/ketal structure with R, R']	7	
−CH−COOH | OH (α-Hydroxy acids)	As for the individual groupings, but also: Boronation	[cyclic boronate with C$_4$H$_9$]	7	
	Simultaneous acylation and esterification	−CH−COO−C$_2$H$_5$ | O−CO−C$_3$F$_7$	2, 3	A number of variations of this approach are available
−CO−COOH (α-Keto acids)	As for the individual groupings, but also: Cyclization with 1,2-diaminobenzene followed by silylation	[quinoxalinone with Si(CH$_3$)$_3$]	7, 9	
R−CO−OR' (Esters)	Esters may be analysed chromatographically without derivatization, but where R' is involatile: Ester interchange (transesterification)	R−CO−OCH$_3$	2	
R−CO−O−CO−R' (Acid anhydrides)	Esterification	R−CO−OR'' + R'−CO−OR''	2	Treat exactly the same as an acid, using e.g. EtOH/HCl

Table 3 Synoptic chart on the formation of derivatives from nitrogen-containing groupings

Functional group	Procedure	Examples of product	Chapter	Comments
$-NH_2$ (Primary amines; amino acids; amino sugars)	Acylation	$-NH-CO-CH_3$; $-NH-CO-CF_3$ $-NH-CO-C_6H_5$; $-NH-CO-C_6F_5$	3	
	Benzoylation		3	
	Silylation (mild)	$-NH-Si(CH_3)_3$	4	Mixtures may be obtained
	Silylation (vigorous)	$-N[Si(CH_3)_3]_2$	4	
	Treatment with CS_2	$-N=C=S$	6	Volatile, for use with GLC
	Thiourea formation	(acridine-9-yl thiourea derivative)	9	Fluorescent derivative
	Schiff's base formation	$-N=CH-C_6F_5$	9	
	2,4-Dinitrophenylation	$-NH-C_6H_3(NO_2)_2$	8	
	Sulphonamide formation	$-NH-SO_2-$ (dinitrophenyl sulphonyl)	8	
		$-NH-Dns$	9	Fluorescent derivative (several variants available)
	Carbamate formation	$-NH-CO_2-CH_3$	3, 10	
	Treatment with fluorescamine	N-substituted 2-phenyl-pyrrolin-4-ones	9	Specific fluorogenic reagent for primary amino groups
	Treatment with pyridoxal	Pyridoxylidine derivative	9	Semi-specific fluorogenic reagent for primary amines
	Treatment with NBD-Cl	7-Nitrobenzofurazan	9	Fluorescent product
	Alkylation	$-N{<}^{CH_3}_{CH_3}$	5	
	Ion-pair formation	$R-NH_3^+ X^-$	14	'X' may be a variety of counter-ions
$-NH-R$ (Secondary amines, imino acids, substituted amino sugars)	Acylation	$-N(R)-CO-CH_3$; $-N(R)-CO-CF_3$ $-N(R)-CO-C_6H_5$; $-N(R)-CO-C_6F_5$	3	As with primary amines
	Benzoylation		3	As with primary amines
	Silylation	$-N(R)-Si(CH_3)_3$	4	May need 'forcing' conditions

Functional group	Reaction	Product		Notes
R–N–R' | R" (Tertiary amines)	2,4-Dinitrophenylation	(structure: R–N(R')–C₆H₃(NO₂)₂)	8	As with primary amines
	Sulphonamide formation		8	As with primary amines (more probability of side reactions)
	Treatment with NBD-Cl		9	As with primary amines
	Ion-pair formation	$R_2NH_2^+ X^-$	11	'X' may be a variety of counter-ions
	Carbamate formation	$R'\!\!>\!\!N\text{-}CO_2\text{-}CH_2\text{-}C_6F_5$	3	
Quaternary ammonium salts	Thermal decomposition	Tertiary amines		HPLC directly; Chapter 8 of first edition
	Ion-pair formation	$R_4N^+ X^-$	11	'X' may be a variety of counter-ions
–CO–NH₂ (Amides)	Silylation (vigorous)	$-C=N-Si(CH_3)_3$ | $O-Si(CH_3)_3$	4	Silylamides are themselves powerful silylating reagents
	Acylation (vigorous)	$-CO-NH-CO-C_6F_5$	3	
	Alkylation	$-CO-N(CH_3)_2$ | R'	5	e.g. CH₃I/NaH/DMSO
R–NH–CO–NH–R' (substituted ureas; carbamides)	Acylation	R–N–CO–N–CO–CF₃ | | CO–CF₃ R	3	
–CO–NH–R (Alkylamides)	Acylation (vigorous)	$-CO-N-CO-C_6F_5$ | R	3	
	Silylation	$-CO-N-R$ | $Si(CH_3)_3$	4	Silylamides are themselves powerful silylation reagents
	Alkylation	$-CO-N-R$ | CH_3	5	e.g. CH₃I/NaH/DMSO

(Continued overleaf)

Table 3 (continued)

Functional group	Procedure	Examples of product	Chapter	Comments
R–NH–C–NH–R' ‖ NH (Substituted guanidines)	Acylation (vigorous)	N–CO–CF$_3$ ‖ R–N–C–NR'–CO–CF$_3$ | CO–CF$_3$	3	Acylation is difficult, and stability of the products is poor
	Cyclic derivatives		7	
R–CH–NH$_2$ | COOH (Amino acids)	Silylation	R–CH–NH–Si(CH$_3$)$_3$ | COOSi(CH$_3$)$_3$	4, 14	
	Esterification + Acylation	R–CH–NH–CO–CF$_3$ | COOC$_4$H$_9$	2, 3, 14	
–CH—CH– | | OH NH$_2$ (Amino alcohols)	As for individual groups, but see also:			

	Cyclic boronate formation	–CH—CH–
		| |
		O NH
		  \\ /
		   B
		   |
		   C$_4$H$_9$
			7	

	Simultaneous acylation and silylation	| –CH–CH–NH–CO–CF$_3$ | O–Si(CH$_3$)$_3$	3	Using N-methyl-N-trimethylsilyl-trifluoroacetamide for example
–NO$_2$ (Nitro compounds)	Chromatograph without derivatization		8	Electron-capturing; coloured

or GC–MS. The chapters on esterification, acylation, silylation, alkylation, ketone–base condensation and cyclic derivative formation (Chapters 2–7) abound with such procedures, and more will be found in the chapter on GC–MS (Chapter 14). The best way to pick a suitable approach is to look at the reactive groupings present in the substances to be analysed and to use the tables in this guiding chapter as signposts to the right derivative or derivatives for those groupings.

8. Applications for mass spectrometry

Obviously GC–MS methods will predominantly use the approaches indicated for GC and GC–MS in the previous section. Additionally, the chapter on analysis by soft ionization techniques (Chapter 12) will contain procedures additional to these, and also to those specifically referred to in Chapter 14.

9. Conclusions

At this point we would like to conclude by suggesting very strongly that, once it has been decided to make a derivative, once a suitable one has been chosen, and once the approach to derivatization is clear, then it is a good idea to spend a little time in optimizing the derivatization. As many examples in this handbook demonstrate, different workers very often quote different conditions for preparing the same derivative. These conditions are sometimes not ideal: they often involve temperatures that are unnecessarily high and times that are unnecessarily long to achieve the desired derivatization, and this may even adversely affect the yield!

Each compound has an optimal reaction path to complete derivatization, although where a mixture is concerned it may be hard to reconcile the various different optimal reaction paths, and compromise conditions will have to be chosen. But time spent in playing about with the parameters of temperature, time, reagent concentrations, solvents and catalysts is time well spent, because it will eventually save time in running the subsequent analyses—over and over again. Besides, it also makes the final report more professional-looking and convincing!

10. References

[1] E. Stahl, *Thin-Layer Chromatography, a Laboratory Handbook*, 2nd fully revised and expanded edition, Springer, Berlin–Heidelberg–New York (1969).
[2] I. Smith, *Chromatographic and Electrophoretic Techniques*, 4th ed., Heinemann, London (1976).
[3] J. G. Kirchner, *Thin-Layer Chromatography*, 2nd ed., Wiley, New York (1978).
[4] J. C. Touchstone and M. F. Dobbins, *Practice of Thin-Layer Chromatography*, Wiley, New York (1978).
[5] Anon., *Dyeing Reagents for Thin Layer and Paper Chromatography*, E. Merck, Darmstadt (1980).
[6] J. F. Lawrence and R. W. Frei, *Chemical Derivatization in Liquid Chromatography*, Elsevier,, Amsterdam (1976).
[7] R. W. Frei and J. F. Lawrence, *Chemical Derivatization in Analytical Chemistry. Volume 1. Chromatography*, Elsevier, Amsterdam (1981).
[8] S. H. Y. Wong (Editor), *Therapeutic Drug Monitoring and Toxicology by Liquid Chromatography*, Marcel Dekker, New York (1985).
[9] C. K. Lim (Editor), *HPLC of Small Molecules*, IRL Press, Oxford (1986).
[10] H. L. J. Makin and R. Newton, *High Performance Liquid Chromatography in Endocrinology* (Monographs in Endocrinology Series, Vol. 30), Springer, Berlin (1988).
[11] J. Drozd, *Chemical Derivatization in Gas Chromatography*, Elsevier, Amsterdam (1981).

2

Esterification

Karl Blau
Department of Chemical Pathology, Bernhard Baron Memorial Research Laboratories, Queen Charlotte's and Chelsea Hospital, London
and
André Darbre
Formerly, Department of Biochemistry, King's College, London

1. THEORETICAL BASIS	12	
1.1 Introduction	12	
1.2 Reaction mechanisms of esterification	12	
1.2.1 A_{AC^2} mechanism	12	
1.2.2 A_{AL^1} mechanism	13	
1.2.3 A_{AC^1} mechanism	13	
1.2.4 Catalysis for esterification	13	
1.2.5 Removal of water	13	
1.2.6 Transesterification	14	
2. PRACTICAL METHODS	14	
2.1 The use of an alcohol and hydrogen chloride	14	
2.1.1 Esterification with alcoholic HCl (general method)	14	
2.1.2 Esterification with alcoholic HCl (micro method)	14	
2.1.3 Methyl esters of amino acids	15	
2.1.4 Higher esters	15	
2.2 Esterifications involving other acid catalysts	15	
2.2.1 Fatty acid methyl or ethyl esters	15	
2.2.2 Making 2-chloroethyl esters	15	
2.2.3 Methyl esters with boron trifluoride/methanol	15	
2.2.4 Higher esters with boron trifluoride	15	
2.2.5 Chloroethyl esters with boron trichloride	16	
2.2.6 Thionyl chloride in ester formation	16	
2.2.7 Esterification with an alcohol and an anhydride (general method)	16	
2.2.8 Trichloroethyl esters	16	
2.2.9 Double derivatizations with an alcohol and an anhydride	16	
2.3 The use of diazomethane and analogues	17	
2.3.1 General diazomethane procedure for methyl esters	17	
2.3.2 Methyl esters with trimethylsilyldiazomethane	18	
2.3.3 Ethyl ester formation with diazoethane	18	
2.4 'Silver salt', 'potassium bromide' and similar methods	19	
2.4.1 Methyl esters via the silver salts and methyl iodide	19	
2.4.2 Benzyl esters via the silver salts and benzyl chloride	20	
2.4.3 Methyl esters: potassium salts and methyl iodide	20	
2.4.4 Methyl esters: potassium salts with dimethyl sulphate	20	
2.4.5 Butyl esters with butyl iodide	20	
2.4.6 Phenacyl esters: phenacyl bromide and trimethylamine	20	
2.4.7 *p*-Bromo, *p*-methyl and *p*-nitrobenzyl esters	20	
2.4.8 *p*-Bromophenacyl esters	20	
2.4.9 *p*-Nitrophenacyl esters	21	
2.4.10 Pentafluorobenzyl esters	21	
2.5 Silylation	21	
2.5.1 Simple general silylation procedure for acids	22	
2.5.2 *tert*-Butyldimethylsilyl esters	22	
2.5.3 Halogenated silyl esters	22	
2.6 Coupling reactions using carbodiimides	22	
2.6.1 Coupling with DCCI	23	
2.6.2 Coupling with N,N'-Carbonyldiimidazole (CDI)	23	
2.7 Esterification using other coupling reactions	23	
2.7.1 Coupling with 6-chloro-1-*p*-chlorobenzenesulphonyloxy-benzotriazole (CCBBT)	23	

Handbook of Derivatives for Chromatography
Edited by K. Blau and J. M. Halket © 1993 John Wiley & Sons Ltd

2.7.2 Benzyl esters with N,N'-dicyclohexyl-O-benzylisourea ... 24
2.8 Pyrolysis of quaternary ammonium salts ... 24
  2.8.1 Methyl esters with tetramethylammonium hydroxide ... 25
  2.8.2 Methyl esters with trimethylanilinium hydroxide (TMAH) ... 25
2.9 The use of dialkyl acetals ... 25
  2.9.1 Methyl esters with DMF-DMA ... 25
  2.9.2 Retinoic acid methyl ester ... 25
2.10 Transesterification of triglycerides and amino acid esters ... 25
  2.10.1 Sodium methoxide transmethylation ... 26
  2.10.2 Transmethylation with methanolic HCl ... 26
  2.10.3 Transesterification of lipids using $AlCl_3$ ... 26
  2.10.4 Transmethylation with TMTFTH ... 26
  2.10.5 Formation of isopropyl esters from triglycerides ... 26
  2.10.6 Formation of n-butyl esters from triglycerides ... 26
  2.10.7 Alternative transbutylation procedure for lipids ... 26
  2.10.8 Amino acid isobutyl esters from the methyl esters ... 26
2.11 Miscellaneous esterification reactions ... 26
  2.11.1 Methyl esters with trimethyloxonium tetrafluoroborate ... 27
  2.11.2 Benzyl esters with benzyl-3-p-tolyltriazenes ... 27
2.12 Conclusions ... 27
3. REFERENCES ... 28

1. Theoretical basis

1.1 Introduction

Acids are reactive substances, and many are too polar to chromatograph well. This is particularly true in gas chromatography: underivatized acids tend to tail because of non-specific interaction with the 'inert' support. Because of associative effects, they are also not as volatile as their molecular weights might indicate. Esters, on the other hand, are non-polar, more volatile, and do chromatograph well. Esterification is therefore the obvious first choice for derivatization of acids, particularly for gas chromatography. General organic chemistry textbooks provide an introduction to esterification, and a comprehensive monograph describes the chemistry of carboxylic acids and esters [1]. Also available is a useful list of esterification reagents [2], and books devoted to methods for preparing esters [3, 4]. Because esterification is a widely used general procedure, most chemical companies specializing in chromatography have a section on esterification reagents in their catalogues, often with application notes and references.

Esterification involves the condensation of the carboxyl group of an acid and the hydroxyl group of an alcohol, with the elimination of water.

$$R-COOH + HO-R' \longrightarrow R-COO-R' + H_2O$$

There are many possible ways in which this can be achieved, and reaction schemes have been worked out for particular purposes, including derivatization for chromatography.

1.2 Reaction mechanisms of esterification

In the esterification of a carboxylic acid with an alcohol, reaction occurs by acyl-oxygen (1) or alkyl-oxygen (2) heterolysis:

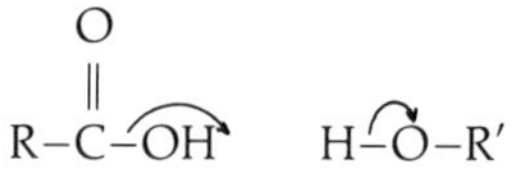

$$R-\overset{O}{\underset{\|}{C}}-OH \quad H-O-R' \quad (1)$$

$$R-\overset{O}{\underset{\|}{C}}-O-H \quad HO-R' \quad (2)$$

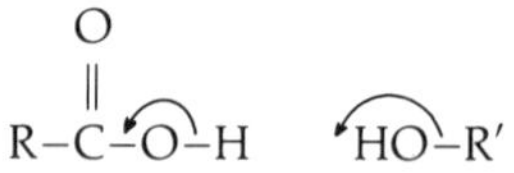

Ingold [5] showed that acid-catalysed esterification involves three possible mechanisms, denoted as A_{AC^1}, A_{AC^2} and A_{AL^1}, where A is the substrate (conjugate acid of the carboxylic acid), AC and AL denote acyl and alkyl bond heterolysis, and the numbers represent the molecularity of the rate-determining step.

1.2.1 A_{AC^2} Mechanism

The esterification reaction is mainly bimolecular, and goes via acyl–oxygen fission, where the rate-limiting step is the attack of alcohol on the protonated carboxylic acid. Ingold proposed the following schemes:

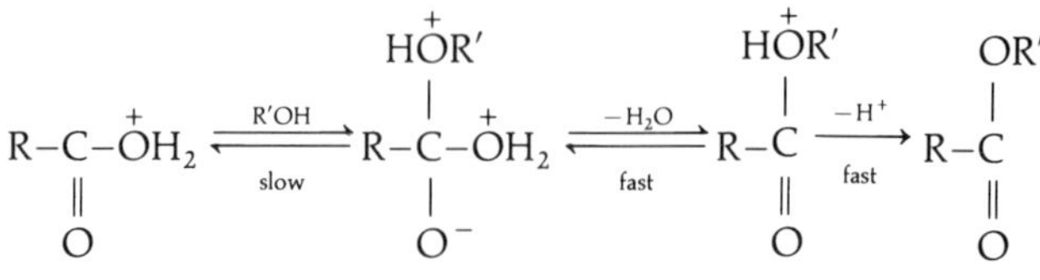

$$R-\overset{+}{\underset{\|}{\underset{O}{C}}}-OH_2 \underset{slow}{\overset{R'OH}{\rightleftharpoons}} R-\overset{+}{\underset{\|}{\underset{O^-}{C}}}-\overset{\overset{+}{H O R'}}{\underset{|}{O} H_2} \overset{-H_2O}{\underset{fast}{\longrightarrow}} R-\overset{\overset{+}{H O R'}}{\underset{\|}{\underset{O}{C}}} \overset{-H^+}{\underset{fast}{\longrightarrow}} R-\overset{OR'}{\underset{\|}{\underset{O}{C}}}$$

$$R-\overset{+}{C}-OH \underset{slow}{\overset{R'OH}{\rightleftarrows}} R-\overset{H\overset{+}{O}R'}{\underset{OH}{C}}-OH \rightleftarrows R-\overset{OR'}{\underset{OH}{C}}-\overset{+}{O}H_2 \xrightarrow[fast]{-H_2O} \overset{OR'}{\underset{O-H}{C^+}} \underset{fast}{\overset{-H^+}{\rightleftarrows}} R-\overset{OR'}{\underset{O}{C}}$$
|
OH

The carbon atom becomes tetrahedrally bonded, and for this reason retardation of the reaction rate may occur where —R and —R' are bulky groups which may interact sterically.

1.2.2 A_{AL^1} Mechanism

This mechanism is common for tertiary alcohols, and explains the formation of racemized esters when esterifying with an optically active alcohol. This reaction is slow for primary alcohols. The alcohol is first protonated and then loses water to form a carbonium ion, which reacts rapidly with the acid.

$$R'-OH \overset{H^+}{\rightleftarrows} R'-\overset{+}{O}H_2 \underset{slow}{\overset{-H_2O}{\rightleftarrows}} R'^+$$

$$R-\overset{O}{\underset{OH}{\overset{\|}{C}}} + R'^+ \rightleftarrows R-\overset{OR'}{\underset{O-H}{C^+}} \overset{-H^+}{\rightleftarrows} R-\overset{OR'}{\underset{O}{C}}$$

However, the carbonium ion can also attack the alcohol or anions to form by-products.

1.2.3 A_{AC^1} Mechanism

This occurs with sterically hindered acids, e.g. 2,4,6-trimethylbenzoic acid, using sulphuric acid catalysis, where the acylium ion intermediate reacts with the alcohol.

$$R-\overset{O}{\overset{\|}{C}}-OH + H_2SO_4 \rightleftarrows R-\overset{+}{C}\overset{OH}{\underset{OH}{\diagdown}} + HSO_4^-$$

$$R-\overset{+}{C}\overset{OH}{\underset{OH}{\diagdown}} + H_2SO_4 \rightleftarrows R-\overset{+}{C}=O + HSO_4^- + H_3O^+$$

$$R-\overset{+}{C}=O + R'OH \rightleftarrows R-\overset{\|}{\underset{O}{C}}-OR' + H^+$$

These mechanisms have in common that they are acid-catalysed and that they involve the removal of water.

1.2.4 Catalysis for esterification

Initially hydrogen chloride was the favoured catalyst, because of its acid strength and because it is readily removed. Insofar as sulphuric acid is less easily removed and has other drawbacks, such as the risk of charring and other dehydrating reactions, it has been less popular as an acidic esterification catalyst. Other acids, both conventional and 'Lewis-type', have been used as esterification catalysts, including trifluoroacetic and dichloroacetic acids, benzene- and p-toluene-sulphonic acids, sulphuryl and thionyl chlorides, phosphorus trichloride and oxychloride, polyphosphoric acids, and strongly acidic cation exchange resins, to name but a few in a by no means exhaustive roll-call. For analytical work esterification is best done in the presence of a volatile catalyst such as hydrogen chloride or thionyl chloride, which can be removed along with excess of the alcohol.

1.2.5 Removal of water

A good way of driving esterification reactions is to remove the water as it is formed. Although the usual way is to use a chemical reagent, other ways have been devised. Thus, when the reaction mixture boils above 100 °C, water may distil off as soon as it is formed, so esters of n-butanol (b.p. 117 °C) or higher homologues may be chosen. However, higher alcohols tend to be less reactive and are also less effective solvents for the acid being esterified. An alternative method is to remove the water as it appears by using a Dean and Stark trap.

An effective means of eliminating water is by azeotropic distillation: this promotes rapid esterification and high yields [6] and comes into its own when dry reagents are not available, but the method is not readily applicable to small-scale work. Similarly, a volatile alcohol may be allowed to distil away from the reaction mixture, entraining the water with it [7]. Water may be eliminated from a refluxing azeotropic distillate by allowing it to pass through a Soxhlet unit with a filter thimble containing a desiccant such as calcium carbide [8], anhydrous magnesium sulphate [9, 10] or Linde molecular sieve type 4A or 5A [11].

Chemical methods are very effective for getting rid of water. Lorette and Brown [12] introduced 2,2-dimethoxypropane as a water scavenger. It reacts in acid solution to yield acetone:

$$\underset{\underset{CH_3}{|}}{\overset{\overset{CH_3}{|}}{C}}(OCH_3)_2 + H_2O \longrightarrow \underset{\underset{CH_3}{|}}{\overset{\overset{CH_3}{|}}{C}}=O + 2CH_3OH$$

Although methanol is produced, the procedure is not restricted to the preparation of methyl esters: in the presence of excess of another alcohol the reaction will lead to the formation of its ester, and if necessary things can be so arranged that the methanol produced volatilizes away, together of course with the acetone. Darkening may be prevented by the use of dimethyl sulphoxide [13]. Other water scavengers which have been used include 100% sulphuric acid [14] and graphite bisulphate [15].

1.2.6 Transesterification

Esters may be solvolysed by alcohols [16] and, if the alcohol is present in excess, the original ester grouping may be replaced by an ester of the new alcohol: this is called transesterification, interesterification or ester interchange. It has been widely used for making esters of higher alcohols from those of lower alcohols, particularly from methyl esters (among the easiest to prepare), where the transesterification can be driven by volatilization of the methanol; alternatively, methanol can be removed with Linde sieve type 3A [17]. As well as anhydrous hydrogen chloride, other acidic catalysts that may be used include boron trifluoride or tribromide, sulphuric acid and perchloric acid. Transesterification can also be done with basic catalysts such as sodium or potassium methoxides in aqueous methanol, and such methanolysis reactions are a popular way of making methyl esters from triglyceride and other similar fats, enabling their fatty acid composition to be determined [18–24]. Transesterification using sodium methoxide, ethoxide or even hydroxide in isopropanol was equally effective in making isopropyl esters [25].

2. Practical methods

2.1 The use of an alcohol and hydrogen chloride

This is the classical Fischer–Speier procedure. Originally the acid was dissolved in the alcohol and boiled under reflux while anhydrous hydrogen chloride was bubbled into the solution from a generator. In early generators, concentrated sulphuric acid dripped into concentrated hydrochloric acid from a separating funnel, with a calcium chloride tube to dry the HCl gas. Later, Kipp-type generators were used, with sulphuric acid dripping onto solid ammonium chloride. The availability of bottled hydrogen chloride in cylinders replaced these arrangements. Nowadays, particularly when working on a small scale, methanolic or ethanolic HCl is readily made by the dropwise addition of acetyl chloride, in an amount required for achieving the desired HCl concentration, to the stirred alcohol, with cooling. The small proportion of the acetate ester formed is evaporated at the end of the esterification, along with the excess of the alcohol. Alternatively, thionyl chloride can be used instead of acetyl chloride, and the only side product is then the very volatile SO_2 [26]. Because so many esterifications are for making methyl or ethyl esters, it is easy to evaporate off the surplus methanol or ethanol at the completion of the reaction, and at the same time the catalyst acid (or acids if thionyl chloride was used) disappears and the water formed during the reaction is entrained with the alcohol. With higher alcohols, HCl from a cylinder is the best way of achieving the desired concentration, usually about 3 M, which may be checked before use by titration. After esterification has been completed, the reaction mixture may be poured into excess water; the acid is then neutralized and the ester is extracted into a non-polar solvent.

2.1.1 Esterification with alcoholic HCl (general method)

A few milligrams of the acid are treated with 1 ml of the alcohol containing 3 M HCl, and heated at 60 °C for methanol or at higher temperature for the higher alcohols. Reaction times for maximum yields need to be established experimentally. For methyl or ethyl esters, the alcohol is evaporated off, leaving the ester as residue. Addition of 5% by volume of benzene to the alcohol may help to entrain the water formed in the reaction. Alternatively, the reaction mixture is poured into 5 ml of water, the mixture is made alkaline with 3 M potassium bicarbonate and the esters are extracted into 3 × 1 ml portions of hexane. The hexane solution may be dried over anhydrous sodium sulphate or, where absorption losses on the desiccant are undesirable, concentrated into a small volume for analysis.

2.1.2 Esterification with alcoholic HCl (micro method)

One or two milligrams of the acid in a small vial are heated with 100 µl of the alcohol containing 3 M HCl for 30 minutes at 70 °C (methanol or ethanol). Higher alcohols may need higher temperatures and/or longer heating times. The alcohol is evaporated in a stream of nitrogen or *in vacuo*, leaving the ester as residue. Yields may if necessary be improved by repeating the procedure once more on the product of the first reaction.

2.1.3 Methyl esters of amino acids

Amino acid(s) at the milligram level are treated with 2 ml of 4 M methanolic HCl at 70 °C for two hours. The excess methanol is evaporated to dryness is a stream of nitrogen. Traces of water may be entrained from the dried residue by addition of 100 μl of methylene chloride and re-evaporation. The residue consists of the hydrochlorides of the methyl esters of the amino acid(s) [27, 28].

2.1.4 Higher esters

Oxepinac in biological fluids was esterified in n-propanol by a classical procedure. The extract from the fluid was dissolved in 1 ml of n-propanol and saturated with gaseous HCl at 60 °C for an hour. After evaporation under reduced pressure, the residual esters were treated with 1 ml of 3% sodium bicarbonate and 5 ml of chloroform for 2 minutes. The chloroform layer was evaporated to dryness and the residue was dissolved in 200 μl of n-propanol for analysis [29].

Sioufi *et al.* made isobutyl esters of baclofen enantiomers by treating the acid from plasma or urine eluted from a solid-phase (C_{18}) column, with 0.5 ml of HCl in isobutyl alcohol made by adding 250 μl of acetyl chloride to 5 ml of the alcohol. The reaction was effected at 100 °C for 15 minutes, after which the excess alcoholic HCl was evaporated to dryness in a stream of nitrogen [30].

2.2 Esterifications involving other acid catalysts

Sulphuric acid has also been used in esterification but, because it is less easily eliminated at the end of the reaction than hydrochloric acid, the isolation of the esters usually requires their extraction into a non-polar solvent such as hexane. Also, with sulphuric acid there is sometimes a risk of dehydration reactions, charring and/or oxidative side reactions, which is why it is not so popular.

2.2.1 Fatty acid methyl or ethyl esters

Free fatty acids are dissolved in a 4:1 mixture of the alcohol with light petroleum (b.p. 40–60 °C) containing 2% sulphuric acid at 70 °C for an hour. Triglyceride fats can be transesterified in the same way but this takes four hours [31]. Similar transesterification procedures have been described for triglyceride fats using chloroform [32] or ether [33] as solvent: the use of acid conditions rather than the usual base-catalysed transesterification conditions was to avoid isomerization.

2.2.2 Making 2-chloroethyl esters

These esters may be made from the free acids or by transesterification of the methyl esters using 2% sulphuric acid in 2-chloroethanol at 60 °C for two hours. The mixture is poured into water and the esters are extracted into petroleum ether [34]. These esters may also be made using the commercially available 10% boron trichloride in 2-chloroethanol (see Section 2.2.5).

2.2.3 Methyl esters with boron trifluoride/methanol

The next most popular esterification catalyst is the Lewis acid boron trifluoride (BF_3), which is commercially available as a 14% solution in methanol for making methyl esters. The analogous BCl_3 reagents are also available and almost equally useful. BCl_3 is also available as the 2-chloroethanol complex, as mentioned above, and BF_3 as the n-propanol and n-butanol complexes.

In use, BF_3-catalysed esterifications are rapid: the reactions are complete after a few minutes on a boiling water bath. Bannon *et al.* have studied the reaction in detail, and point to vigorous shaking during extraction of the esters into a non-polar solvent as a crucial factor [35].

To the acid (1–20 mg) in a ground-glass stoppered test tube or screw-capped vial is added 1 ml of 14% BF_3–methanol or 10% BCl_3–methanol, and the solution is heated on a boiling water bath for 2 minutes. The esters are extracted into n-heptane [36–38].

2.2.4 Higher esters with boron trifluoride

Acids (up to 100 μg) are taken up in 150 μl of a 14% solution of BF_3 in the selected alcohol and heated at 100 °C for up to 20 minutes, although shorter times may be sufficient: this may readily be established experimentally. The mixture is cooled in an ice bath and a mixture of 200 μl each of ether and n-pentane is added. The excess catalyst is neutralized with 150 μl of a 50–50 mixture of n-pentane and pyridine, with cooling to get rid of the heat of neutralization. Centrifugation separates the layers: the upper layer contains the esters [39].

2.2.5 Chloroethyl esters with boron trichloride

Short-chain fatty acids, about 10 μmol, are heated with 1–2 ml of 10% BCl_3–2-chloroethanol at 100 °C for 30 minutes. The mixture is cooled in ice and the esters are extracted into n-hexane or petroleum ether (b.p. 30–60 °C). The organic layer may be dried with anhydrous sodium sulphate, and may be concentrated in a stream of nitrogen for analysis by GC with electron capture detection [see refs 40 and 41 for similar procedures].

2.2.6 Thionyl chloride in ester formation

Thionyl chloride has already been mentioned, and has been used in esterification procedures directly, not just to make alcoholic solutions of HCl.

The acid (up to 50 mg) is placed in a ground-glass stoppered test tube or a screw-cap vial with 200 μl of dry mthanol (or other alcohol) and cooled in a solid CO_2–acetone cooling bath. Thionyl chloride (20 μl) is added, with shaking, and the flask is allowed to warm up and then left at 40 °C for two hours. The solution is evaporated to dryness, either in a stream of nitrogen for volatile alcohols or in a vacuum desiccator over a dish of flake NaOH for higher alcohols. This procedure is loosely based on the method of Brenner et al. [42]. Thionyl chloride should be distilled from linseed oil before use [43].

2.2.7 Esterification with an alcohol and an anhydride (general method)

Chemical removal of water has already been mentioned, and an obvious way of doing this is with an anhydride. Although there is competition for the anhydride between the water formed in the esterification reaction and the excess of esterifying alcohol, under the right conditions the anhydride reacts preferentially with the water. This approach is particularly useful in circumstances where the acid to be esterified contains another group that is capable of being acylated because, again under the right conditions, it is possible to esterify and acylate simultaneously. These conditions, with the use of anhydrides, have been used particularly to make strongly electron capturing derivatives for gas chromatographic analysis.

The acid (up to ca. 100 mg) is allowed to react with the alcohol (one molar equivalent) in the presence of a slight excess (1.2–1.5 molar equivalent) of trifluoroacetic anhydride (TFAA). The reaction is rapid; 10 minutes is usually long enough, but gentle warming may be necessary. Less reactive primary, secondary, tertiary and polyhydroxy alcohols, phenols and thiophenols give higher yields than methanol or ethanol, which tend to react with the anhydride to form trifluoroacetate esters [44].

2.2.8 Trichloroethyl esters

The carboxylic acids (10 mg total) are refluxed on a steam bath with 1 ml of a freshly prepared 1:9 mixture of 2,2,2-trichloroethanol and TFAA for 10 minutes. The reaction mixture is diluted with 100 μl of ethyl acetate and passed through a short column of silica gel to retain the reaction mixture and avoid it causing a huge solvent peak [45].

The silica gel may adsorb the esters of the short-chain fatty acids. For the esterification of those acids, a chloroform solution of the C_2–C_8 aliphatic acids (50 μl) is treated with an equal volume of a freshly prepared 1% solution of 2,2,2-trichloroethanol in chloroform, together with 150 μl of heptafluorobutyric anhydride (HFBA) in a closed tube at room temperature for 30 minutes. Excess trichloroethanol is removed by reaction with 50 μl of a 25% solution of palmitic acid in chloroform, added subsequently and left to react at room temperature for 15 minutes. (The trichloroethyl palmitate peak comes off the GC column long after the shorter-chain fatty acid ester peaks). Further purification is achieved by shaking with 100 μl of chloroform and 100 μl of 0.1 M HCl; the aqueous layer is discarded, and the organic layer is washed with 100 μl of 0.1 M NaOH and evaporated to dryness. The residue is dissolved in 100 μl of diethyl ether for analysis by GC with electron capture detection [46].

2.2.9 Double derivatizations with alcohol and an anhydride

Hydroxy-acids (up to 10 μmol) were dissolved in 100 μl of chloroform and treated with 10 μl of 1:4 ethanol–chloroform and, after mixing, 10 μl of 1:3 pyridine–chloroform. Then 30 μl of HFBA are added, the solution is mixed and concentrated to about half the volume in a stream of nitrogen, followed by heating in a sealed tube at 100 °C for 4 minutes. After cooling, 140 μl of chloroform and 90 μl of 0.1 M HCl are added, vortexed and allowed to separate. A similar washing step is done using 0.25 M NaOH, the lower layer is then concentrated to dryness, and the residue is taken up for analysis by GC [47].

Amino acids (again, up to 10 μmol) are treated with 250 μl of a 1:4 mixture of 2,2,3,3,3-pentafluoropropan-1-ol and pentafluoropropionic anhydride (PFPA) at 75 °C for 15 minutes. After cooling and evaporation to dryness in a stream of nitrogen, the reaction is completed by heating with 100 μl of PFPA at 75 °C for 5 minutes.

After again evaporating to dryness, the residue is taken up in ethyl acetate for analysis by gas chromatography with electron capture detection [48].

2.3 The use of diazomethane and analogues

One of the quickest esterification reactions is to make the methyl ester of an acid by reaction with diazomethane. The reason for using this method rather than any of the other ways of making methyl esters is not only because the reaction is quick, but also because the yield is high, side reactions are minimal and, once the reagent is made, the process is convenient, but above all because the conditions are mild.

Diazomethane is a yellow gas which is usually used as an ethereal solution, often with some methanol present, and what drives the reaction is the elimination of gaseous nitrogen.

$$R-COOH + CH_2N_2 \longrightarrow R-COO-CH_3 + N_2$$

Because of its yellow colour, diazomethane acts as its own indicator: when it is used up the yellow colour disappears, and more diazomethane is added until the yellow colour just persists.

CAUTION: Diazomethane is carcinogenic and unstable, and should be prepared only in small quantities. The ethereal solution should be handled in a well ventilated hood or fume cupboard. Small amounts of the ethereal solution (less than 100 ml) may be kept for some days in a refrigerator, but ground-glass stoppers must be avoided. Safe handling requires the avoidance of overheating to minimize the risk of explosion, but in any case derivatization should be restricted to the micro-scale, and this more than anything protects the user against all these hazards.

Diazomethane is best used on a small scale, and a suitable kit designed by Fales et al. [49] is commercially available, use of which avoids the need to distil the diazomethane formed. Although many procedures involve N-methyl-N'-nitro-N-nitrosoguanidine (MNNG) as a starting material, this is a potent carcinogen, and the De Boer and Backer reagent Diazald (N-methyl-N-nitroso-p toluenesulphonamide) is preferable [50]. The most recent method describes the use of this reagent in the small-scale apparatus [51]. An alternative procedure is to use the apparatus assembled from side-arm test tubes designed by Schlenk and Gellerman [52] where the diazomethane is carried through to the ether or directly to the reaction solution in a stream of nitrogen, and the need for distillation is again avoided (Figure 1).

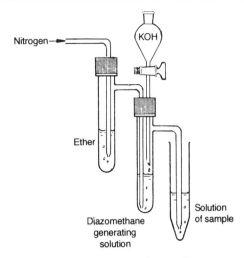

Figure 1. Apparatus for making diazomethane on a small scale without the need for distillation. (After Schlenk and Gellerman [52]).

2.3.1 General diazomethane procedure for methyl esters

The acids (50–30 mg) are dissolved in 2 ml of ether containing 10% of methanol, and the gas generated from 2 mmol of Diazald in the apparatus of Figure 1 is bubbled into the solution in a stream of nitrogen (at a rate of about 5 ml/min of nitrogen). After a few minutes the yellow colour persists, and the nitrogen in the diazomethane generator can be turned off and the apparatus disconnected. The ether/MeOH is then evaporated in a stream of nitrogen, yielding the esters as a residue. Some methods call for the prior elimination of excess diazomethane with a few drops of dilute ethereal acetic acid. This will of course contribute some methyl acetate, but this is volatile enough to evaporate with the ether. In a micro-modification of this method, the acids are adsorbed on a short column of Celite 545 in a small capillary tube. The diazomethane is generated in a stirred 1 ml serum vial fitted with a septum into which the capillary column in inserted. The gas is generated in the vial and passes through the capillary column, esterifying the acids in situ, and the esters are then eluted with a little ether [53].

Diazomethane is not ideal for esterification of phenolic acids because the phenolic hydroxyl groups are also methylated, albeit at a slower rate, which may lead to mixtures of partially methylated products. However, esterification can usually be achieved without O-methyl ether formation by cooling the reaction mixture to below 0 °C [54, 55].

2.3.2 Methyl esters with trimethylsilyl-diazomethane

An interesting variant is the use of the reagent trimethylsilyldiazomethane [56], which is a safe and stable substitute for diazomethane and is used in the same way [57].

The acid (1 mM) in 2 ml of methanol and 7 ml of benzene is mixed with trimethylsilyldiazomethane (30% molar excess) in 1 ml of benzene, and the reaction mixture is stirred at room temperature for 30 minutes and then concentrated [57].

2.3.3 Ethyl ester formation with diazoethane

Diazoethane and higher diazo-compounds may be used in much the same way as diazomethane. Diazoethane can be made from N-nitrosourea [54] or N-ethyl-3-nitro-N-nitrosoguanidine [58] using the same apparatus [49, 52]. There is not a great deal to choose between methyl and ethyl esters made this way, but the following example indicates one rationale for making the distinction.

Indomethacin is metabolized to the O-desmethyl derivative; because diazomethane would remethylate this back to the initial drug as well as making the methyl ester, diazoethane was used, so that both the drug and its metabolite could be determined as the ethyl ester (Figure 2). However, in the case mentioned, silylation would probably do the job more effectively.

The drug and its metabolite were extracted with ethylene dichloride from 0.5 ml of serum or urine buffered at pH 5 with an equal volume of citrate buffer. The layers were separated and the organic layer was blown to dryness with nitrogen.

Diazoethane made from N-ethyl-N-nitrosourea [59] in ethereal solution (0.5 ml) was added and the solution was allowed to stand at 30 °C for 30 minutes. The solution was again blown down with nitrogen, and the residue was taken up in benzene for analysis by GC–ECD [60].

Diazopropane is made from the analogous N-n-propyl-3-nitro-N-nitrosoguanidine [61, 62], and Wilcox has made a series of diazoalkanes from the N-alkyl-N-nitrosoureas [63]. The higher diazoalkanes have not been widely used, but their application is the same as described for diazomethane and diazoethane, and the advantage again lies in the mild conditions of esterification and the convenience in obtaining the products.

Phenyldiazomethane has been used to make benzyl esters, specifically for avoiding the risk of evaporative losses where lower esters might be too volatile. It can be made from azibenzil [64] or from N-benzyl-N-nitroso-p-toluenesulphonamide [65–69]. It is kept as a red solution, 50% in either ethyl ether or petroleum ether, and as with diazomethane, disappearance of the red colour is used to indicate the completeness of the reaction. The analogous diazobenzene can be made in much the same way from N-phenyl-N-nitrosourea [68]. It is used to make phenyl esters in the same way as benzyl esters are made with phenyldiazomethane. Finally, pentafluoro-

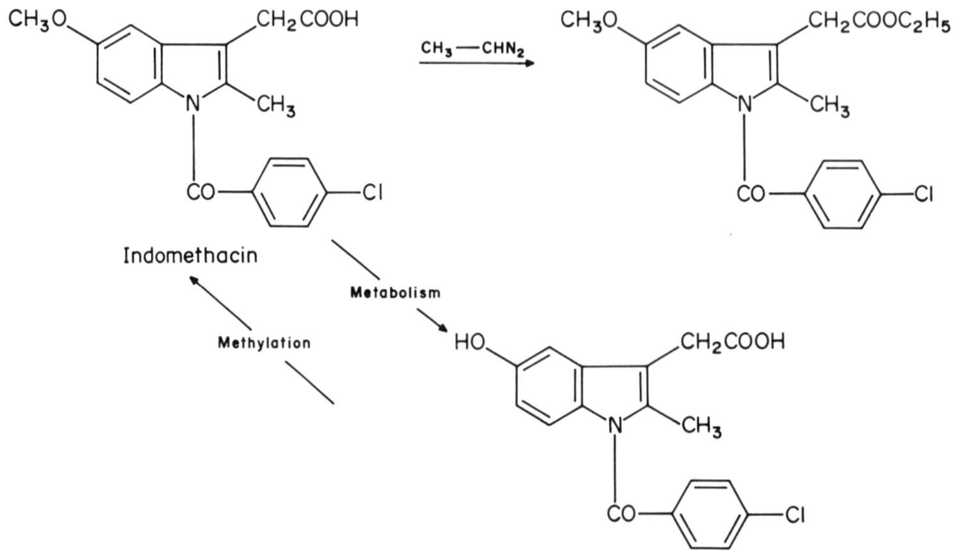

Figure 2. Indomethacin, its ethyl ester and its metabolite.

phenyldiazoalkanes have also been described for rapidly making esters with electron capturing properties for very sensitive detection in gas chromatography [70, 71].

2.4 'Silver salt', 'potassium bromide' and similar methods

Another rapid procedure was first achieved by making the silver salt of the acid and treating it with an alkyl halide.

$$R-COOAg + Cl-R' \longrightarrow R-COO-R' + AgCl$$

Although this is not quite as fast as an aqueous ionic reaction, the insolubility of silver chloride is a strong driving force, and yields tend to be good. However, although the reaction itself is rapid, essentially quantitative and fairly free of side reactions, making the silver salt can be involved and time consuming—and expensive [72].

Formally this approach also covers a host of similar reactions where other salts of the acid (most often the potassium salts, but alternatively various quaternary ammonium type salts and even some tertiary amine salts) are involved, and then reacted with (usually) the *bromo*-derivative of the alkyl or aryl group that is to be esterified with the acid or acids in question. Because of the versatility and convenience of this approach, it has largely displaced the silver salt method, but formally what drives all these reactions is the production *within an organic matrix* of an insoluble or at least a very sparingly soluble salt.

A valuable variant of this approach was first described by Pedersen [73]. He used cyclic polyethers, the so-called 'crown' ethers (from their conformational shape), for complexing many cations, particularly potassium. Crown ethers hold the cation in the centre of the ring by mainly electrostatic forces. Reaction of a crown ether in an organic solvent with the potassium salt of an acid results in the potassium ion being complexed into the crown (see, for example, Figure 3), and this makes the anion very reactive to attack by an alkyl halide (it has been referred to as a 'naked' anion [74, 75]), leading to esterification under mild conditions. Rigorous exclusion of water is not necessary and the products are obtained in higher yield and free of by-products. The great advantage of this approach lies in the scope and variety of the ester groups that can be introduced in this way. (For reviews see references 76 and 77).

2.4.1 *Methyl esters via the silver salts and methyl iodide*

This is one of the few silver salt procedures which is on a small scale and designed as a derivatization scheme for chromatography. Non-esterified fatty acids are separated by GC and trapped at the column outlet in

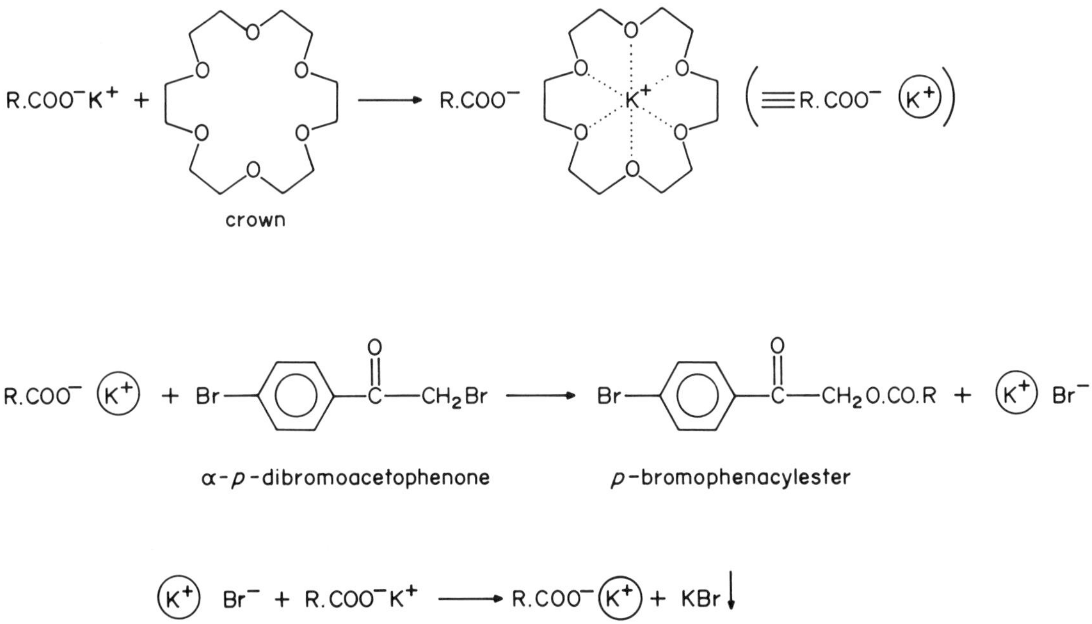

Figure 3. Formation of esters with 'crown' ethers, alkyl or aryl halides and the potassium salts of the acids.

a narrow tube filled with silver oxide. The trapped silver salts are reacted *in situ* with methyl iodide for 2 minutes at 100 °C, and the tube is then inserted into the injection port of a gas chromatograph for analysis of the fatty acid methyl esters. A mixture of fatty acids can be similarly applied to a tube packed with silver oxide [78].

2.4.2 Benzyl esters via the silver salts and benzyl chloride

The ethyl and methyl esters of short-chain fatty acids are very volatile, so to avoid evaporation losses the *benzyl* esters of short-chain fatty acids were made, in a narrow tube filled with silver oxide, using 4% benzyl chloride in ether [79].

2.4.3 Methyl esters: potassium salts and methyl iodide

To the dried mixture of acids (up to 20 mg) are added 2 ml of dried acetone, 2 ml of methyl iodide and 30–50 mg of anhydrous K_2CO_3. After refluxing at 60–70 °C under a calcium chloride drying tube for 20–30 minutes, the solution is made up to 5 ml with acetone and analysed by GC [80]. Serum free fatty acids have been derivatized for analysis by very similar versions of this procedure [81, 82].

2.4.4 Methyl esters: potassium salt with dimethyl sulphate

This procedure was worked out for the drug probenecid extracted from biological fluids (0.2 ml). To the extract were added 1 ml of 0.5% K_2CO_3 in methanol (10 ml of 5% aqueous K_2CO_3 diluted to 100 ml with MeOH) and 0.1 ml of Me_2SO_4. The mixture was heated at 70 °C for 5 minutes and evaporated to dryness. The residue was mixed with 1 ml of pH 5.6 acetate buffer and 5 ml of chloroform, the chloroform was separated and evaporated to dryness, and the residue was dissolved in 100 μl of chloroform for analysis by gas chromatography [83].

2.4.5 Butyl esters with butyl iodide

Instead of a potassium salt, the trimethylphenyl-ammonium (trimethylanilinium) salt is used to make fatty acid butyl esters. To 0.05 mmol of the acid(s) is added 1 ml of 0.1 M trimethylanilinium hydroxide in methanol. Then 100 μl of butyl iodide are added and the mixture is shaken vigorously and left at room temperature for 10 minutes. The precipitate may be spun down and the supernatant analysed by gas chromatography. This procedure is very versatile, and a variety of other iodo-compounds may be substituted for butyl iodide, allowing esterification under very mild conditions [84].

2.4.6 Phenacyl esters: phenacyl bromide and trimethylamine

The esterification of acids with alcohols containing aromatic groups improves detectability in HPLC (see also Chapter 8). Here, fatty acids (100 μg) were treated with 10 μl of a 1.2% solution of phenacyl bromide in acetone at room temperature overnight, or at 50 °C for 2 hours. The supernatant was injected directly for HPLC with UV detection [85].

2.4.7 p-Bromo, p-methyl and p-nitrobenzyl esters

The short-chain fatty acids are dissolved in ethanol (ca. 1 mg/ml). The reaction is done in a melting-point capillary. The acid solution (10 μl) is mixed with 3 μl of 0.3% ethanolic KOH and 5 μl of the selected *para*-substituted benzyl bromide (3 mg/ml in ethanol). The tube is flame sealed and the reaction is carried out at 110 °C for an hour. After cooling, the tube is opened and the solution is analysed by GC [86].

2.4.8 p-Bromophenacyl esters

Short-chain fatty acids were esterified in this procedure, where exclusion of water is clearly unnecessary. The acids (at about 0.1 M in ethanol, 2.5 ml) are mixed with 5 ml of water, and the solution is neutralized with 5 M KOH to phenolphthalein and then made just acid again with 3 drops of 1 M HCl. *p*-Bromophenacyl bromide (0.63 g) is added and the solution is refluxed for 10 minutes. If it goes cloudy, a little more ethanol is added until it clears. After cooling, the esters are extracted into 25 ml of ethyl acetate, with centrifuging to separate the layers. The upper layer is taken off, dried over sodium sulphate and analysed by GC [87].

Extractive alkylation, or 'liquid–liquid phase transfer catalysis', has been described for making these esters. The acid(s) (0.1–1 mM) were treated with 8 ml of pH 5 buffer containing 3 mM tetrahexylammonium hydrogen sulphate and 1 ml of methylene dichloride containing 36 mM *p*-bromophenacyl bromide and the internal standard n-hexacosane. The reaction was done at 42 °C with occasional shaking [88].

Crown ether procedures too have been used to make these esters, again mostly from the short-chain fatty acids. The ether used was '18-crown-6', 1,4,7,10,13,16-

hexaoxacyclo-octadecane, dissolved with p-bromophenacyl bromide in acetonitrile in a molar ratio of 1:10. The acids are placed in a reaction vial and converted to the potassium salts with a 3–5 times molar excess of potassium bicarbonate, and then taken to dryness. To this residue the acetonitrile solution of alkylating agent and crown ether is added, and the vial is closed, followed by heating at 80 °C for 30 minutes. For HPLC analysis the cooled solution may be used directly; alternatively, the reaction mixture may be taken to dryness and the esters dissolved in a suitable solvent for analysis [89, 90].

2.4.9 p-Nitrophenacyl esters

This type of derivative was used with prostaglandins, of which up to 5 mg in 1 ml of dried acetonitrile was treated with a threefold molar excess of p-nitrophenacyl bromide followed by 2 µl of N,N-di-isopropylethylamine. Esterification was for an hour at room temperature, and the reaction mixture was used for HPLC analysis with UV detection [91, 92].

A crown ether method has also been described for these esters. The acid(s) are again neutralized with methanolic KOH to phenolphthalein and the solution is evaporated to dryness. To the potassium salts is added 1.2 ml of a solution, in acetonitrile, of 43 mg of p-nitrophenacyl bromide and 56 mg of dicyclohexyl-18-crown-6. The mixture is refluxed for 1.5 hours, and the product is analysed by HPLC [93].

2.4.10 Pentafluorobenzyl esters

These electron-capturing esters are popular for the GC analysis of short-chain fatty acids because their use avoids evaporative losses of the esters of the lower acids, but many other acids have also been esterified to these derivatives. Conventional methods have been widely used, but a crown ether method has also been described.

The acids (5 µmol of each) are dissolved in 100 ml of acetone and 250 mg of pentafluorobenzyl bromide (NB: this is a strong lachrymator, so appropriate precautions should be taken), and 50 mg of potassium bicarbonate are added. The mixture is refluxed for 3 hours and then 500 ml of ether and 200 ml of ethyl acetate are added. The combined solution is briefly washed with 10 ml of ether-saturated water and dried over sodium sulphate before evaporation to dryness. The residue is taken up in 100 ml of hexane containing 1% each of acetone and ether [94]. This widely quoted procedure is readily scaled down: a much smaller-scale version has been described for the drug flurbiprofen [95].

As an alternative to the potassium salt, tertiary amines and tetra-alkylammonium salts have been used instead. The acids, e.g. the dried residue from a serum extraction, are treated with an acetonitrile solution of pentafluorobenzyl chloride and di-isopropylethylamine, in a 3:1 molar ratio, at 40 °C for 5 minutes. The reaction mixture is blown dry with nitrogen; if necessary the esterification may be repeated [96]. Indole-3-acetic acid was similarly esterified, but using acetone as solvent and N-ethylpiperidine as the cation and reaction for 45 minutes at 60 °C [97]. Tetrabutylammonium hydrogen sulphate (0.1 mmol) was converted to the hydroxide with sodium hydroxide (0.2 mmol) and added to the dried acids in a serum extract with 1 ml of methylene chloride and 20 µl of pentafluorobenzyl bromide. The mixture was vigorously shaken for 30 minutes and blown down to dryness with nitrogen, and the esters were extracted into hexane for HPLC [98, 99].

Extractive pentafluorobenzylation was described by Jacobsson et al. They extracted 9 mmol of acid with 50 ml of acetone containing 2.8 mmol of pentafluorobenzyl chloride, and added 6 mmol of K_2CO_3. The reaction was done under reflux for 6 hours, and the mixture was evaporated to dryness in a rotary evaporator. The esters were dissolved in 10 ml of hexane, the solution was washed with 10 ml of water, and further clean-up, if required, was done on a silica gel column [100].

The crown ether procedure was used on short-chain fatty acids. About 5 nmol of the acids (plus internal standard, either n-butyric or isovaleric acid) were treated with 20 µl of 1 M HBr and 2 µl of 1 M $NaHCO_3$ and evaporated to dryness. The residue was treated with 10 µl of acetonitrile containing 5 µl of pentafluorobenzyl bromide and 11 mg of pentaoxapentadecane (15-crown-5). The reaction mixture was heated at 80 °C for two hours with occasional shaking, and then diluted with 1–2 µl of toluene before analysis by gas chromatography [101].

A wide variety of other esters has been made, or may potentially be accessible, by the type of procedure outlined in this section. To give a couple of examples, the naphthacyl esters were made for improved detectability in HPLC analysis in the same way as the phenacyl esters mentioned earlier [102], the fluorescent esters of 4-methyl-7-methoxycoumarin were made from the 4-bromomethyl-7-methoxycoumarin by the potassium salt procedure [103] and anthrylmethyl esters were obtained by an extractive alkylation procedure [104].

2.5 Silylation

Just as alcohols readily form trimethylsilyl (TMS) ethers, so acids form trimethylsilyl *esters*, and just as readily.

The commercial availability of a wide range of highly reactive and 'user-friendly' reagents for what is generally referred to as 'silylation' has made it very easy to make TMS and other silyl esters, and this is therefore one of the most popular ways of esterifying acids for chromatography, particularly gas chromatography. Much of the impetus for this came initially from Pierce's classic review on silylation [105], and more recent developments in derivatization by silylation will be found in Chapter 4 of this handbook. However, if one is looking for a quick and easy way of esterifying an acid in high yield for analysis by gas chromatography, then N,N-bis(trimethylsilyl)trifluoroacetamide (BSTFA)—or bis(trimethylsilyl)-acetamide (BSA) if using an electron capture detector—is the reagent of choice. Below are given three useful silylation procedures. For the rest, look at Chapter 4.

2.5.1 Simple general silylation procedure for acids

The acids (a few mg in a reaction vial) are treated with 50 μl of BSA or BSTFA. Silylation is usually complete as soon as the acids have dissolved, and GC analysis can be done by direct injection of the reaction mixture, but many recipes recommend for 5–15 minutes at 60 °C. Where there are problems due to steric effects, the addition of 1% of trimethylchlorosilane (TMCS) to the BSA or BSTFA works wonders. BSTFA is usually preferred for gas chromatography or GC–MS because excess reagent and reaction products are very volatile. The reactions may also be done in pyridine, dimethylformamide or acetonitrile solution. For use with electron capture detectors, BSTFA cannot be used; BSA must be substituted. Many variants of this general method have been described. Any particular application can readily be optimized by studying the effects of solvent, time and temperature on the yield. Any but the most hindered hydroxy groups will also be silylated at the same time; very occasionally molecular rearrangements have been encountered on silylation. Amino acids can be doubly derivatized by silylation, but require higher temperatures and forcing conditions for maximum yields [106], and silylated amino groups are very much more sensitive to moisture than silyl ethers or silyl esters.

Short-chain fatty acids from blood or urine were silylated with TMS-imidazole in a reaction where the extracted acids in ether were heated with the reagent at 60 °C for 15 minutes and the reaction mixture was injected directly into the gas chromatograph. The first 16 esters came off before the huge reagent peak. It is of course more usual to have the reagent peak come off well before the compounds of interest, but if it works the other way.... [107].

2.5.2 tert-Butyldimethylsilyl esters

For some applications, and particularly for GC–MS (see also Chapter 14), there are advantages to using tert-butyldimethylsilyl (TBDMS) esters, and these are made from up to 5 mg of the acid(s) in 100 μl of dimethylformamide containing 20 μmol of imidazole and 10 μmol of TBDMCS. The reaction mixture is heated at 60 °C for 15 minutes, an equal volume of 5% NaCl is added and the esters are extracted into 1 ml of ether for analysis by GC or GC–MS [108]. An improved reagent for making TBDMS esters is N-TBDMS-N-methyltrifluoroacetamide, or MTBSTFA, used much as in 2.5.1 above. For most acids room temperature is probably sufficient for complete esterification in 5–15 minutes; yields are good and the derivatives are stable [109].

2.5.3 Halogenated silyl esters

Chloromethyldimethylsilyl (CMDMS) and bromomethyldimethylsilyl (BMDS) esters are useful for esterifying acids with all the advantages of silylation, but with improved sensitivity in GC with electron capture detection [110]. The BMDS produced a significant improvement in sensitivity, and of course the iodo analogue even more so, but the latter had to be made by halide ion exchange. A standard procedure is to dissolve the acids (a few mg) in 600 μl of pyridine, add 200 μl of di(chloromethyl)tetramethyldisilazane and 100 μl CMDMS-Cl and leave the reaction mixture at room temperature for 30 minutes [111]. Attempts to make the BMDS esters from carboxylic acids led to the formation of bromomethyltetramethyldisiloxane esters, i.e. introduced an unexpected extra dimethylsiloxane grouping, but the products nevertheless gave good peaks with sensitive detection [112].

2.6 Coupling reactions using carbodiimides

Since esterification is essentially a condensation reaction between a carboxylic acid and an alcohol with the elimination of water, coupling reagents such as the carbodiimides [113–115] also work well for esterification. Figure 4 shows a typical reaction using dicyclohexylcarbodiimide (DCCI) [116]. Dicyclohexylurea is formed as a by-product.

CAUTION! Take care when handling DCCI: use rubber gloves, protective eye wear and a good fume hood.

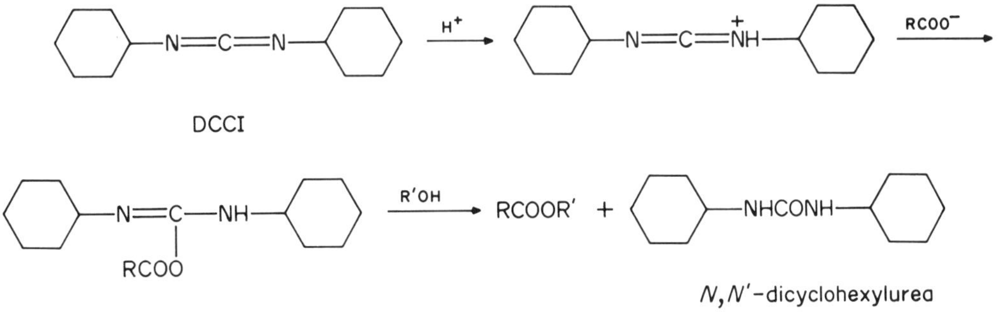

Figure 4. Esterification using DCCI.

2.6.1 Coupling with DCCI

About 10 μmol of the acid or acids in 25 μl of the alcohol and 5 μl of pyridine are treated with a slight molar excess of DCCI and shaken gently at intervals over 30–120 minutes at room temperature. N,N'-Dicyclohexylurea precipitates out and is spun down. The supernatant can then be analysed directly by gas chromatography. With solid alcohols, more pyridine and heating at 40–80 °C may be necessary [116].

2.6.2 Coupling with N,N'-carbonyldiimidazole (CDI)

The general scheme is shown in Figure 5.

An extract of plasma acids was made by acid extraction with heptane–isopropyl alcohol (3:7). The residue from this extraction, in a reaction vial, was dissolved in 0.1 ml of 0.325 M CDI in freshly distilled ethanol or hydrocarbon-stabilized chloroform. 10% Triethylamine in dry methanol (1 ml) was added. After one minute or more, 3 ml of 1 M NaOH saturated with carbon tetrachloride was added, with vortex mixing for 2 minutes. After releasing the pressure in the vial by puncturing the septum, the vial was centrifuged to separate the layers and the lower (chloroform) layer was sampled with a microlitre syringe for analysis by gas chromatography [117].

2.7 Esterification using other coupling reactions

2.7.1 Coupling with 6-chloro-1-p-chlorobenzenesulphonyloxybenzotriazole (CCBBT)

This is one of a series of novel coupling reagents. The general scheme is shown in Figure 6.

The acid (10 μmol) and a molar equivalent of triethylamine are dissolved in 20 μl of acetonitrile, and a molar equivalent of CCBBT is added. The reaction mixture is cooled in ice and is then allowed to warm up to room temperature, with intermittent shaking, over 1.5 hours. Another equivalent of triethylamine together with 1.5 μl of MeOH is added, again with ice-cooling, and the mixture is again allowed to warm up to room temperature with intermittent shaking over 1.5 hours. The mixture is taken to dryness and the methyl ester(s) are extracted into ether. The ether may be briefly washed with sodium bicarbonate and with dilute HCl and water, dried over

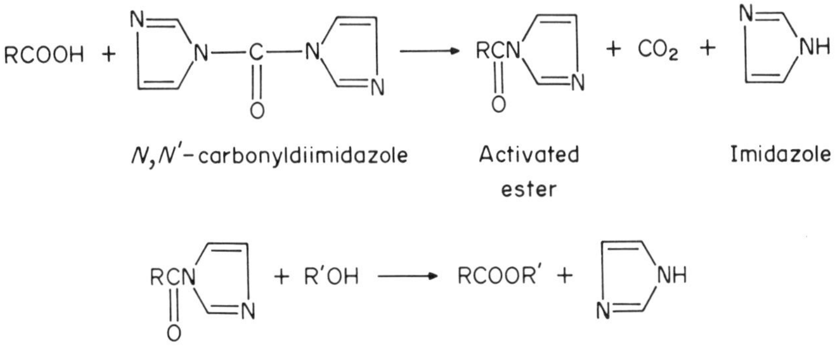

Figure 5. Esterification using CDI.

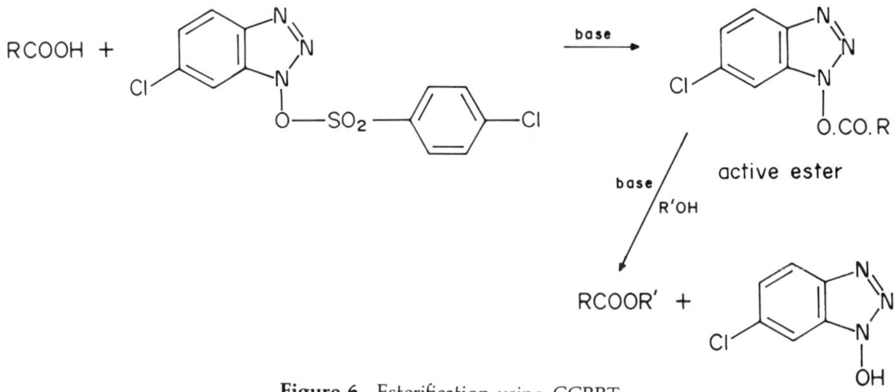

Figure 6. Esterification using CCBBT.

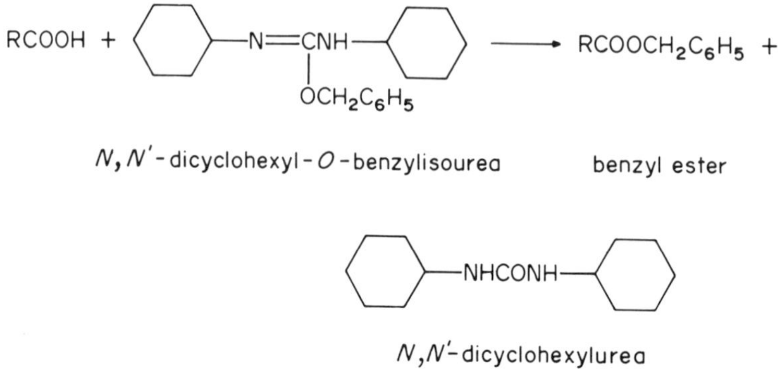

N,N′-dicyclohexyl-*O*-benzylisourea   benzyl ester

N,N′-dicyclohexylurea

Figure 7. Esterification with dicyclohexyl-*O*-benzylurea.

anhydrous magnesium sulphate and concentrated for analysis [118].

2.7.2 Benzyl esters with N,N′-dicyclohexyl-O-benzylisourea

The reaction is shown in Figure 7. Dicyclohexylurea is again formed as a by-product.

The acids (or the residues from an extract) are dissolved in a suitable solvent (benzene, dioxan, tetrahydrofuran or CCl_4) and refluxed with a slight molar excess of N,N′-dicyclohexyl-O-benzylisourea for an hour. The precipitate of dicyclohexylurea is centrifuged down, and the supernatant may be used directly for gas chromatographic analysis [119, 65].

2.8 Pyrolysis of quaternary ammonium salts

These reactions are done by injecting reagent together with the acid(s) into the heated inlet of a gas chromatograph, where rapid pyrolysis occurs, and the products are immediately swept onto the analytical column. The beauty of this approach is that it is all achieved with the microlitre syringe used for the injection: if nothing else, this saves washing glassware. The precise location of sample and reagent within the syringe does not appear to be critical. Some recipes call for sample to be drawn up first, followed by reagent, others reverse this, and others still sandwich the sample between two slugs of reagent, but, once inside the flash heater, the reaction appears to go regardless [120].

Initially, tetramethylammonium hydroxide was used [121–124], but tetrabutylammonium hydroxide was better for very volatile acids [125], and subsequently trimethylanilinium hydroxide (TMAH) was found to be preferable to the tetramethyl compound, and has been widely used as a methylating agent for a variety of different classes of compounds, particularly drugs, but also including acids [120, 126–129]. The *m*-trifluoromethyl substituted analogue of TMAH has also been used for free fatty acids and lipid components in biological material, e.g. needle biopsies. This reagent, trimethyl-(α,α,α-trifluoro-*m*-tolyl)ammonium hydroxide or TMTFTH, was chosen because it could be used at a lower injector temperature and because it was less destructive of labile unsaturated fatty acids [130].

2.8.1 Methyl esters with tetramethylammonium hydroxide

The acid is dissolved in methanol and titrated with 24% tetramethylammonium hydroxide in methanol to phenolphthalein. The sample (1–5 μl) is injected for gas chromatography with a flash heater, loosely packed with glass wool, at a temperature of 350 °C [122]. A similar procedure, but with the trimethylammonium salt of the acid made first in a capillary probe, and the probe then quickly inserted into the heated inlet zone of the gas chromatograph, is claimed to give a higher yield [124]. For the volatile formic and for lactic acid, the tetrabutylammonium salts were made instead and pyrolysed in the same way [125].

More recently, the reagent of choice for the pyrolytic methylation reactions has been trimethylanilinium hydroxide, because the N,N-dimethylaniline is a better leaving group and thus a lower inlet temperature may be used [126–130]. An even better leaving group is obtained with the m-trifluoromethyl derivative of TMAH, TMTFTH [130].

2.8.2 Methyl esters with trimethylanilinium hydroxide (TMAH)

A 10 μl syringe is used to draw up 1 μl of trimethylanilinium hydroxide, 0.2–1 μl of the sample and a further 1 μl of trimethylanilinium hydroxide. Analysis is by gas chromatography with a flash heater set at 250 °C. The contents of the syringe are quickly injected. Some acids may require a higher temperature for maximum yield [120].

2.9 The use of dialkyl acetals

Dimethylformamide dimethylacetal (DMF–DMA, see below)

$$\begin{array}{c} CH_3 \\ \\ CH_3 \end{array} \!\!\!\!\! N\!-\!CH \!\!\!\!\! \begin{array}{c} OCH_3 \\ \\ OCH_3 \end{array}$$

is another useful methylating reagent which may be used to esterify acids to their methyl esters by the syringe injection technique. With this reagent hydroxyl groups are *not* methylated. Ethyl, propyl, n-butyl and t-butyl acetals may be used analogously [131]. However, with amino acids this reagent gives the dimethylaminomethylene amino acid methyl esters, which is an original and convenient way of making volatile amino acid derivatives [132]. It would be interesting to use the nitrogen–phosphorus rubidium bead detector with these derivatives, because one should get excellent discrimination in their favour, leading to sensitive analyses with little interference.

2.9.1 Methyl esters with DMF–DMA

The fatty acid sample or biological acid extract is dissolved in DMF–DMA and injected into the gas chromatograph. Alternatively, the syringe is charged with 1 μl of DMF–DMA and 2 μl of pyridine containing the acids (up to about 2 μg) and again 1 μl of DMF–DMA. The contents are injected directly: an inlet temperature of at least 190 °C should be used.

2.9.2 Retinoic acid methyl ester

Retinoic acid was extracted from 30 ml of plasma, the extract was dried down and the residue was treated with 50 μl of DMF–DMA by essentially the same procedure as above. The product was analysed by GC or by GC–MS [133].

2.10 Transesterification of triglycerides and amino acid esters

The neutral lipid fraction extracted from biological materials contains among other components all the triglyceride fats. The fatty acid composition of these is sometimes of great significance, and methods for the establishment of this composition are based on the quantitative analysis of fatty acid methyl esters by gas chromatography. Initially, alkaline hydrolysis released the fatty acids, which were then extracted, concentrated and esterified, some kind of sulphuric acid/methanol procedure being widely used. However, this sequence of separate procedures has been superseded because losses may occur due to incomplete hydrolysis or side reactions such as polymerization or the alteration of unsaturated fatty acids. The preferred method is transesterification in either alkaline or acid media [134]. The most recent methods use methanolic HCl, because in alkaline conditions some hydrolysis may occur as a side reaction and cause lowered yields [19, 135, 136]. There are of course effective quantitative methods for the separation of the fatty acid methyl esters by gas chromatography, both on packed and on capillary columns.

2.10.1 Sodium methoxide transmethylation

The lipid material (20 mg) in 1 ml of dry benzene, in a reaction vial, is treated with 2 ml of 0.5 M methanolic sodium methoxide, and the vial is capped and heated at 80 °C for 20 minutes. After cooling, the mixture is poured into 10 ml of water and the methyl esters are extracted twice with 1 ml portions of hexane or heptane. The combined extracts are dried with anhydrous sodium sulphate and concentrated for gas chromatographic analysis.

2.10.2 Transmethylation with methanolic HCl

The lipid material (up to about 10 mg) in a reaction tube is treated with 1 ml of 3 M HCl/MeOH, then purged with nitrogen, and the tube is capped and heated at 80 °C for an hour. After cooling, 1 ml of 0.9% NaCl and 0.3 ml of hexane are added and the tube is vortexed vigorously and centrifuged. The upper, hexane, layer can be sampled directly for gas chromatography. This procedure can be used directly on plant tissues (up to 50 mg) without prior extraction of the lipids, although the presence of water does affect the yield somewhat. This may to some extent be overcome by the addition of dimethoxypropane [12] as a water scavenger [136], but in that case the use of dimethyl sulphoxide to inhibit by-product formation is recommended [137].

2.10.3 Transesterification of lipids using $AlCl_3$

Lipids (2–10 mg) are treated with 4 ml of 10% $AlCl_3$ in methanol or other appropriate alcohol on a boiling water bath for an hour. The esters are extracted into 5 ml of light petroleum ether followed by 2 ml of saturated NaCl. The upper layer is concentrated and the residue is dissolved in 0.2–1 ml of CS_2 for analysis [138].

2.10.4 Transmethylation with TMTFTH

As with the methylation of free fatty acids [130], TMTFTH may also be used for transmethylation. To 10 mg of the triglyceride extract in a reaction vial are added 0.5 ml of benzene and 200 μl of 0.2 M methanolic TMTFTH. The vial is capped and shaken and the reaction proceeds at room temperature for 30 minutes. The solution may be sampled directly for gas chromatography [139], with the same advantages over the tetramethylammonium procedure [140].

2.10.5 Formation of isopropyl esters from triglycerides

This procedure is essentially similar to that given in 2.10.1 above, but uses isopropyl alcohol and sodium methoxide [25]. The intention is to prevent losses of volatile short-chain fatty acid esters.

2.10.6 Formation of n-butyl esters from triglycerides

The lipid extract (25 mg) is refluxed for 2 minutes in 1 ml of 0.5 M NaOH in butanol, and then 1.5 ml of 12.5% BF_3 in butanol and 1 ml of hexane are added and the solution is again refluxed, this time for 3 minutes. After cooling, 20 ml of water is added and the hexane layer is separated, washed twice with an equal volume of water, dried and injected for gas chromatography [141].

2.10.7 Alternative transbutylation procedure for lipids

Fatty acids (in milk fat) to a total of up to 10 mg were treated with 250 μl of di-n-butyl carbonate and 0.5 ml of 0.2 M sodium butoxide. The tube was capped and the mixture was heated for 2 minutes on a steam bath. The contents of the tube were transferred to a Babcock milk fat test bottle and rinsed in with 0.5 ml of water, 0.5 ml of acetone and a further 0.5 ml of water. Saturated NaCl (10 ml) was added and the volume was adjusted to the mark with water. The layers were separated by centrifugation and the upper layer was taken directly for analysis by gas chromatography [142].

2.10.8 Amino acid isobutyl esters from the methyl esters

The residue of methyl esters (e.g. from esterification according to procedure 2.1.1 above) is treated with 200 μl of methylene chloride, which is evaporated off to entrain any traces of water. Transesterification is with 200 μl of 1.25 M HCl in isobutyl alcohol at 110 °C for 150 minutes [143]. An essentially similar procedure for making isoamyl esters has also been described [144]. The similar method of transbutylation developed by the Missouri group [145] has been replaced by the direct butyl esterification procedure [146, 147].

2.11 Miscellaneous esterification reactions

Alkyloxonium fluoroborates are powerful alkylating reagents, especially for esterifying sterically hindered acids.

2.11.1 Methyl esters with trimethyloxonium tetrafluoroborate

The acid (up to 5 mg) in 50 μl of di-isopropylethylamine is treated with a 10% molar excess of Me$_3$OBF$_4$ in a 7.5% suspension in methylene chloride at room temperature overnight. The methylene chloride layer is washed with successive 10 ml portions of 1 M HCl, 1 M KHCO$_3$ and saturated NaCl, dried over anhydrous sodium sulphate and taken for chromatographic analysis [148]. Several similar recipes have been published for sterically hindered acids [149, 150].

Benzyl and *p*-nitrobenzyl esters have been made for HPLC using the *p*-tolyltriazene reagents (see Figure 8). The evolution of nitrogen (as with the diazoalkanes) drives the reaction to completion.

2.11.2 Benzyl esters with benzyl-3-p-tolyltriazenes

The acids (up to 10 mg) in 5 ml of ether or alcohol in a reaction tube are treated with 50 mg of the triazene reagent, and the tube is sealed and heated at 40 °C for 2–3 hours. The solution may be directly analysed, or concentrated to dryness with N$_2$, taken up in ether, washed with successive equal portions of 1 M HCl and water, dried and concentrated for HPLC [151].

2.12 Conclusions

It is confusing to have this wealth of esterification methods: why are there so many? Of course there are no bounds to the ingenuity and skill of organic chemists, but as far as derivatization for chromatography goes, clearly some procedures have been developed in response to particular problems: perhaps the acid is sensitive to particular environments (strongly acidic or basic conditions, moisture), or thermolabile or sterically hindered. Other methods may have been chosen for speed, for convenience or for quantitative yield. At this point perhaps some guidance would be helpful.

If there is nothing particularly unusual about the acids to be derivatized, the easiest thing is to dissolve them up in the silylating agents BSA or BSTFA (method 2.5.1), and this has the advantage that hydroxyls will also be silylated. For making methyl esters, the use of diazomethane is quick and more or less quantitative, but hydroxyls may be partially methylated (method 2.3.1). Methyl esters can also be made very simply without the need for special reagents using methanolic HCl made with acetyl chloride (method 2.1.1. or 2.1.2), and only slightly less simply with BF$_3$/methanol (method 2.2.1). After these straightforward procedures, methylation procedures such as the use of DMF-dimethyl acetal reagents are very easy, requiring only a microlitre syringe (method 2.9.1), as is on-column pyrolysis of trimethylanilinium salts (method 2.8.2).

Although this covers the majority of common applications, for alcohols other than methanol many of the recipes mentioned in this chapter will be found applicable, particularly the wide variety of procedures described in Section 4; silver salt, potassium salt and crown ether methods. A useful short cut is to consult chemical catalogues, especially those of companies specializing in chromatography. Although these companies obviously promote their own products, they offer derivatization reagents especially formulated and conveniently packaged in small lots, often in ampoules and sealed under nitrogen for prolonged shelf-life. Often the companies may include helpful comments or application leaflets describing the scope and detailed instructions for the use of each reagent. A wide selection of esterification recipes, together with a host of various specific applications, are collected in Chapter 3 of Knapp's excellent handbook [152].

In the final analysis, one need not be too concerned about hitting on 'the perfect choice' of an esterification procedure for any given application. It must by now be evident, from the variety presented, that there are any number of procedures which will give essentially equivalent results. The best thing is to use the one that comes readiest to hand: if it should prove less than ideal, one can then try another and hopefully better approach.

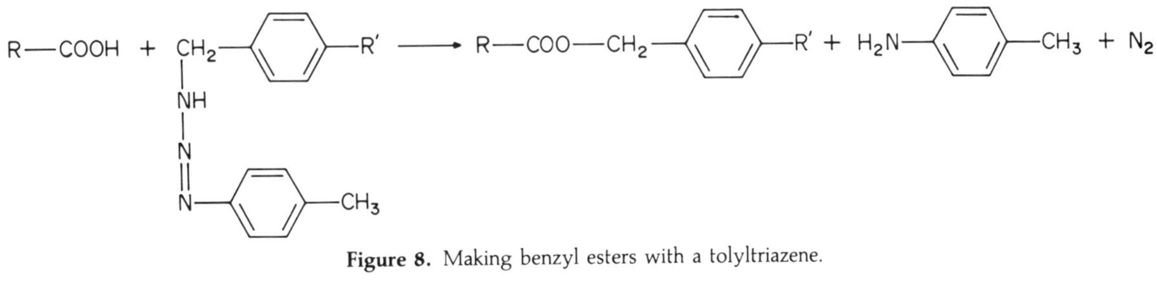

Figure 8. Making benzyl esters with a tolyltriazene.

3. References

[1] S. Patai (Editor), *The Chemistry of Carboxylic Acids and Esters*, Interscience, London (1969).
[2] G. Zweig and J. Sherma (Editors), *Handbook of Chromatography*, Vol. II, CRC Press, Cleveland (1972), p. 226.
[3] I.T. Harrison and S. Harrison, *Compendium of Organic Synthetic Methods*, Vol. 1, Wiley, New York (1971), p. 271.
[4] I. T. Harrison and S. Harrison, *Compendium of Organic Synthetic Methods*, Vol. II, Wiley, New York (1974), p. 108.
[5] C. K. Ingold, *Structure and Mechanism in Organic Chemistry*, Cornell University Press, Ithaca (1969), p. 1128.
[6] M. Dymycky, E. F. Mellon and J. Naghski, *Anal. Biochem.*, **41**, 487 (1971).
[7] D. E. Johnson, S. J. Scott and A. Meister, *Anal. Chem.*, **33**, 669 (1961).
[8] E. Thielepape, *Ber. Dtsch. Chem. Ges.*, **66**, 1454 (1933).
[9] B. R. Baker, *J. Am. Chem. Soc.*, **66**, 1454 (1933).
[10] B. R. Baker, M. V. Querry, S. R. Safir and S. Bernstein, *J. Org. Chem.*, **12**, 138 (1947).
[11] R. L. Stern and E. N. Bolan, *Chem. Ind. (London)*, 825 (1967).
[12] N. B. Lorette and J. H. Brown, Jr., *J. Org. Chem.*, **24**, 261 (1959).
[13] P. G. Simmonds and A. Zlatkis, *Anal. Chem.*, **37**, 302 (1965).
[14] M. S. Newman, *J. Am. Chem. Soc.*, **63**, 2431 (1941).
[15] J. Bertin, H. B. Kagan, J.-L. Luche and R. Setton, *J. Am. Chem. Soc.*, **96**, 8113 (1974).
[16] R. S. Juvet, Jr. and F. M. Wachi, *J. Am. Chem. Soc.*, **81**, 6110 (1959).
[17] D. P. Roelofsen, J. A. Hagendoorn and H. van Bekkum, *Chem. Ind. (London)*, 1622 (1966).
[18] V. L. Davison and M. J. D. Dutton, *J. Lipid Res.*, **8**, 147 (1968).
[19] S. W. Christopherson and R. L. Glass, *J. Dairy Sci.*, **52**, 1289 (1969).
[20] C. R. Scholfield, *Anal. Chem.*, **47**, 1417 (1975).
[21] K. Oette and M. Doss, *J. Chromatogr.*, **32**, 439 (1968).
[22] K. Oette, M. Doss and M. Winterfeld, *Z. Klin. Chem. Klin. Biochem.*, **8**, 525 (1970).
[23] B. J. Holub, *Biochim. Biophys. Acta*, **369**, 111 (1974).
[24] S.-N. Lin and E. C. Horning, *J. Chromatogr.*, **112**, 483 (1975).
[25] J. L. Giegel, A. B. Ham and W. Clema, *Clin. Chem.*, **21**, 1575 (1975).
[26] S. P. Campana and G. Goisiis, *J. Chromatogr.*, **236**, 1978 (1982).
[27] A. Darbre and A. Islam, *Biochem. J.*, **106**, 923 (1968).
[28] A. Islam and A. Darbre, *J. Chromatogr.*, **71**, 223 (1972).
[29] H. Hakusui, W. Suzuki and M. Sano, *J. Chromatogr.*, **182**, 47 (1980).
[30] H. Sioufi, G. Kaiser, F. Leroux and J. P. Dubois, *J. Chromatogr.*, **450**, 221 (1988).
[31] L. Gosselin and J. de Graeve, *J. Chromatogr.*, **110**, 117 (1975).
[32] K. V. Peisker, *J. Am. Oil. Chem. Soc.*, **41**, 87 (1964).
[33] G. W. McGinnis and L. R. Dugan, *J. Am. Oil Chem. Soc.*, **42**, 305 (1965).
[34] K. Oette and E. H. Ahrens, Jr., *Anal. Chem.*, **33**, 1847 (1961).
[35] C. D. Bannon, J. D. Craske, N. T. Hai, N. L. Harper and K. L. O'Rourke, *J. Chromatogr.*, **247**, 63 (1982).
[36] L. D. Metcalfe and A. A. Schmitz, *Anal. Chem.*, **33**, 363 (1961).
[37] L. D. Metcalfe, A. A. Schmitz and J. R. Pelka, *Anal. Chem.*, **38**, 514 (1966).
[38] F. D. Gunstone and I. Ismael, *Chem. Phys. Lipids*, **1**, 209 (1967).
[39] P. A. Biondi and M. Cagnasso, *J. Chromatogr.*, **109**, 389 (1975).
[40] W. H. Gutenmann and D. J. Lisk, *J. Assoc. Off. Anal. Chem.*, **47**, 353 (1964).
[41] D. W. Woodham, C. W. Collier, C. D. Loftis and W. G. Mitchell. *J. Agric. Food Chem.*, **19**, 186 (1971).
[42] M. Brenner, H. R. Müller and R. W. Pfister, *Helv. Chim. Acta*, **33**, 568 (1950).
[43] P. B. Hagen and W. Black, *Can. J. Biochem.*, **43**, 309 (1965).
[44] R. C. Parish and L. M. Stock, *J. Org. Chem.*, **30**, 927 (1965).
[45] R. V. Smith and S. L. Tsai, *J. Chromatogr.*, **61**, 29 (1971).
[46] C. C. Alley, J. B. Brooks and G. Choudhary, *Anal. Chem.*, **48**, 387 (1976).
[47] J. B. Brooks, C. C. Alley and J. A. Liddle, *Anal. Chem.*, **46**, 1930 (1974).
[48] S. Wilk and M. Orlowski, *Anal. Biochem.*, **69**, 100 (1975).
[49] H. M. Fales, T. M. Jaouni and J. F. Babashek, *Anal. Chem.*, **45**, 2302 (1973).
[50] T. J. de Boer and H. J. Backer, *Recl. Trav. Chim. Pays Bas*, **73**, 229 (1954).
[51] F. Ngan and M. Toofan, *J. Chromatogr. Sci.*, **29**, 8 (1991).
[52] H. Schlenk and J. L. Gellerman, *Anal. Chem.*, **32**, 1412 (1960).
[53] D. P. Schwartz and R. S. Bright, *Anal. Biochem.*, **61**, 271 (1974).
[54] B. Plazonnet and W. J. A. VandenHeuvel, *J. Chromatogr.*, **142**, 587 (1977).
[55] J. R. Watson, P. Crescuolo and F. Matsui, *J. Pharm. Sci.*, **60**, 455 (1971).
[56] D. Seyfert, H. Menzel, A. W. Dow and T. C. Flood, *J. Organomet. Chem.*, **44**, 279 (1972).
[57] N. Hashimoto, T. Aoyama and T. Schioiri, *Chem. Pharm. Bull.*, **29**, 1475 (1981).
[58] M. G. Horning, K. Letratanangkoon, R. N. Stillwell, W. G. Stillwell and T. E. Zion, *J. Chromatogr. Sci.*, **12**, 630 (1974).
[59] F. Arndt, *Org. Synth. Coll. Vol.*, **2**, 165 (1944).
[60] L. Helleberg, *J. Chromatogr.*, **117**, 167 (1976).
[61] T. Bruzzese, M. Cambieri and F. Recusani, *J. Pharm. Sci.*, **64**, 462 (1975).
[62] F. Marcucci and E. Mussini, *J. Chromatogr.*, **25**, 11 (1966).
[63] M. Wilcox, *Anal. Biochem.*, **16**, 253 (1966).
[64] P. Yates and B. L. Shapiro, *J. Org. Chem.*, **23**, 759 (1973).
[65] H.-P. Klemm, U. Hintze and G. Gercken, *J. Chromatogr.*, **75**, 19 (1973).
[66] U. Hintze, H. Röper and G. Gercken, *J. Chromatogr.*, **87**, 481 (1973).
[67] D. L. Corina and P. M. Dunstan, *Anal. Biochem.*, **53**, 571 (1973).
[68] D. L. Corina, *J. Chromatogr.*, **87**, 254 (1973).

[69] E. K. Doms, *J. Chromatogr.*, **105**, 79 (1975); **140**, 29 (1977).
[70] W. W. Christie and V. M. Stepanov, *J. Chromatogr.*, **392**, 259 (1987).
[71] U. Hofmann, S. Holzer and C. Meese, *J. Chromatogr.*, **508**, 349 (1990).
[72] C. W. Gehrke and D. F. Goerlitz, *Anal. Chem.*, **35**, 76 (1963).
[73] C. J. Pedersen, *J. Am. Chem. Soc.*, **89**, 7017 (1967).
[74] C. L. Liotta, H. P. Harris, M. McDermott, T. Gonzalez and K. Smith, *Tetrahedron Lett.*, 2417 (1974).
[75] C. L. Liotta and H. P. Harris, *J. Am. Chem. Soc.*, **96**, 2250 (1974).
[76] C. J. Pedersen and H. K. Frensdorff, *Angew. Chem. Int. Ed. Engl.*, **11**, 16 (1972).
[77] G. W. Gokel, *Crown Ethers and Cryptands*, Royal Society of Chemistry, London (1991).
[78] C. B. Johnson and E. Wong, *J. Chromatogr.*, **109**, 403 (1975).
[79] C. B. Johnson, *Anal. Biochem.*, **71**, 594 (1976).
[80] S. L. Ali, *Chromatographia*, **7**, 655 (1974).
[81] W. Dünges, *Chromatographia*, **6**, 478 (1973).
[82] A. Grünert and K. H. Bassler, *Fresenius' Z. Anal. Chem.*, **267**, 342 (1973).
[83] K. Sabih and C. D. Klaassen, *J. Pharm. Sci.*, **60**, 745 (1971).
[84] R. H. Greeley, *J. Chromatogr.*, **88**, 229 (1974).
[85] R. F. Borch, *Anal. Chem.*, **47**, 2437 (1975).
[86] J. R. Watson and P. Crescuolo, *J. Chromatogr.*, **52**, 63 (1970).
[87] E. O. Umeh, *J. Chromatogr.*, **56**, 29 (1971).
[88] Y. L'Emaillat, J. F. Menez, F. Berthou and L. Bardou, *J. Chromatogr.*, **206**, 89 (1981).
[89] H. D. Durst, M. Milano, E. J. Kikta, S. A. Connelly and E. Grushka, *Anal. Chem.*, **47**, 1797 (1975).
[90] R. L. Patience and J. D. Thomas, *J. Chromatogr.*, **249**, 183 (1982).
[91] W. Morozowich and S. Douglas, *Prostaglandins*, **10**, 19 (1975).
[92] F. A. Fitzpatrick, *Anal. Chem.*, **48**, 499 (1976).
[93] E. Grushka, H. D. Durst and E. J. Kikta, *J. Chromatogr.*, **112**, 673 (1975).
[94] F. Kawahara, *Anal. Chem.*, **40**, 2073 (1968).
[95] D. G. Kaiser, S. R. Shaw and G. W. Vangiessen, *J. Pharm. Sci.*, **63**, 567 (1964).
[96] A. J. F. Wickramasinghe and S. R. Shaw, *Biochem. J.*, **141**, 179 (1974).
[97] E. Epstein and J. D. Cohen, *J. Chromatogr.*, **209**, 413 (1981).
[98] O. Gyllenhaal, P. Hartvig and H. Brötell, *J. Chromatogr.*, **129**, 295 (1976).
[99] A. G. Netting and A. M. Duffield, *J. Chromatogr.*, **257**, 174 (1983).
[100] S. Jacobsson, A. Larsson, A. Arbin and A. Hagman, *J. Chromatogr.*, **447**, 329 (1988).
[101] J. Chauhan and A. Darbre, *J. Chromatogr.*, **240**, 107 (1982).
[102] M. J. Cooper and M. W. Anders, *Anal. Chem.*, **46**, 1849 (1974).
[103] W. Dünges, *Anal. Chem.*, **49**, 442 (1977).
[104] J. D. Baty, S. Pazouki and J. Dolphin, *J. Chromatogr.*, **395**, 403 (1987).
[105] A. E. Pierce, *Silylation of Organic Compounds*, Pierce, Rockford (1979).
[106] C. W. Gehrke and K. Leimer, *J. Chromatogr.*, **57**, 219 (1971).
[107] O. A. Mamer and B. F. Gibbs, *Clin. Chem.*, **19**, 1006 (1973).
[108] G. Phillipou, R. F. Seamark and D. A. Bingham, *Lipids*, **10**, 714 (1975).
[109] T. P. Mawhinney and M. A. Madson, *J. Org. Che.*, **47**, 3336 (1982).
[110] C. Eaborn, C. A. Holder, D. R. M. Walton and B. S. Thomas, *J. Chem. Soc.*, C 2502 (1969).
[111] H. Morita and W. Montgomery, *J. Chromatogr.*, **123**, 454 (1976).
[112] J. B. Brooks, J. A. Liddle and C. C. Alley, *Anal. Chem.*, **41**, 203 (1975).
[113] H. G. Khorana, *Chem. Rev.*, **53**, 145 (1953).
[114] F. Kurzer and K. Dughadi-Zadeh, *Chem. Rev.*, **67**, 107 (1967).
[115] Y. S. Klausner and M. Bodansky, *Synthesis*, 453 (1972).
[116] E. Felder, U. Tiepolo and A. Mengassini, *J. Chromatogr.*, **82**, 291, 390 (1973).
[117] H. Ko and M. E. Royer, *J. Chromatogr.*, **88**, 253 (1974).
[118] M. Itoh, D. Hagiwara and J. Notani, *Synthesis*, 456 (1975).
[119] E. Vowinkel, *Chem. Ber.*, **100**, 16 (1967).
[120] S. Barnes, R. Waldrop and D. G. Pritchard, *J. Chromatogr.*, **231**, 155 (1982).
[121] E. W. Robb and J. J. Westbrook, *Anal. Chem.*, **35**, 1644 (1963).
[122] J. J. Bailey, *Anal. Chem.*, **39**, 1485 (1967).
[123] D. T. Downing and R. S. Greene, *Lipids*, **3**, 96 (1968); *Anal. Chem.*, **40**, 827 (1968).
[124] D. T. Downing, *Anal. Chem.*, **39**, 218 (1967).
[125] J. W. Schwarze and M. N. Gilmour, *Anal. Chem.*, **41**, 1686 (1969).
[126] B. S. Middleditch and D. M. Desiderio, *Anal. Lett.*, **5**, 605 (1972).
[127] M. S. Roginsky, R. S. Gordon and M. J. Bennett, *Clin. Chim. Acta*, **56**, 261 (1974).
[128] I. Gan, J. Korth and B. Halpern, *J. Chromatogr.*, **92**, 435 (1974).
[129] R. Gugler and C. Jensen, *J. Chromatogr.*, **117**, 175 (1976).
[130] J. MacGee and K. G. Allen, *J. Chromatogr.*, **110**, 35 (1974).
[131] J. P. Thenot, E. C. Horning, M. Stafford and M. G. Horning, *Anal. Lett.*, **5**, 217 (1972).
[132] J. P. Thenot and E. C. Horning, *Anal. Lett.*, **5**, 519 (1972).
[133] T.-C. Chiang, *J. Chromatogr.*, **182**, 335 (1980).
[134] A. J. Sheppard and J. L. Iverson, *J. Chromatogr. Sci.*, **13**, 448 (1975).
[135] W. W. Christie, *J. Lipid Res.*, **23**, 1072 (1982).
[136] J. Browse, P. J. McCourt and C. R. Somerville, *Anal. Biochem.*, **152**, 141 (1986).
[137] P. G. Simmonds and A. Zlatkis, *Anal. Chem.*, **37**, 302 (1965).
[138] R. Segura. *J. Chromatogr.*, **441**, 99 (1988).

[139] D. K. McCreary, W. C. Kossa, S. Ramachandran and R. R. Kurtz, *J. Chromatogr. Sci.*, **16**, 329 (1978).

[140] J. C. West, *Anal. Chem.*, **47**, 1708 (1975).

[141] J. L. Overson and A. J. Sheppard, *J. Assoc. Off. Anal. Chem.*, **60**, 284 (1977).

[142] J. Sampugna, R. E. Pitas and R. G. Jensen, *J. Diary Sci.*, **49**, 1462 (1966).

[143] S. L. MacKenzie and D. Tenaschuk, *J. Chromatogr.*, **97**, 19 (1974).

[144] P. Felker and R. S. Bandurski, *Anal. Biochem.*, **67**, 245 (1975).

[145] C. W. Gehrke, W. M. Lamkin, D. L. Stalling and L. L. Wall, *Quantitative Gas–Liquid Chromatography of Amino Acids in Proteins and Biological Substances*, Analytical Biochemistry Laboratories, Columbia (1968).

[146] D. Roach and C. W. Gehrke, *J. Chromatogr.*, **44**, 269 (1969).

[147] R. W. Zumwalt, K. Kuo and C. W. Gehrke, *J. Chromatogr.*, **57**, 193 (1971).

[148] S. Hünig and M. Kiessel, *Chem. Ber.*, **91**, 380 (1958).

[149] D. J. Raber and P. Gariano, *Tetrahedron Lett.*, 4741 (1971).

[150] R. T. Dean, L. J. DeFilippi and D. E. Hultquist, *Anal. Biochem.*, **76**, 1 (1976).

[151] I. R. Politzer, B. J. Dowty, G. W. Griffin and J. L. Laseter, *Anal. Lett.*, **6**, 539 (1973).

[152] D. R. Knapp, *Handbook of Analytical Derivatization Reactions*, Wiley, New York (1979).

3

Acylation

Karl Blau
Department of Chemical Pathology, Bernhard Baron Memorial Research, Queen Charlotte's and Chelsea Hospital, London

1. INTRODUCTION		31
2. REACTION MECHANISMS		33
2.1 General		33
2.2 The reactivity of acylimidazoles		33
3. TYPES OF ACYLATION REACTIONS		34
3.1 General		34
3.2 Detector-oriented acylation reactions		34
4. PRACTICAL APPLICATIONS		36
4.1 General		36
4.2 Acetylation reactions		36
4.2.1	Acetylation with acetic anhydride and acetic acid	36
4.2.2	Acetylation of urinary sugars with acetic anhydride and sodium acetate	36
4.2.3	Acetylations with acetic anhydride and pyridine	37
4.2.4	Acetylations with other basic catalysts	38
4.2.5	Acetylation in aqueous solution	38
4.2.6	Acetylation with simultaneous silylation	38
4.2.7	Acid-catalysed acetylations	38
4.2.8	Acetolysis procedures	38
4.2.9	Acetylation with acetyl chloride	39
4.2.10	Acetylation with ketene	39
4.3 Methods for making propionyl derivatives		39
4.4 Formyl derivatives		39
4.5 Perfluoroacyl derivatives		40
4.5.1	Acylation with anhydrides on their own	40
4.5.2	Acylation with anhydride in a solvent	41
4.5.3	Acylation with anhydride and a basic catalyst	41
4.5.4	Simultaneous acylation and esterification	43
4.5.5	Acylation of amides	43
4.5.6	Hydrolysis of sulphate esters and glucuronides with simultaneous perfluoroacylation	43
4.5.7	Commentary	43
4.6 Alternative perfluoroacylation procedures		43
4.6.1	The use of perfluoroacylimidazoles	43
4.6.2	Combined acylation and silylation reactions	44
4.6.3	N-Methylbis (trifluoroacetamide) (MBTFA)	44
4.6.4	Pentafluorobenzoyl (PFBzO) derivatives	44
4.6.5	Derivatization of amides to PFBzO derivatives	45
4.7 Miscellaneous acylations		45
4.7.1	General	45
4.7.2	Pivaloyl derivatives	46
4.7.3	Trichloroacetylation	46
4.7.4	Benzoyl and benzenesulphonyl derivatives	46
4.7.5	Preparation of 3,5-dinitrobenzoates	46
4.7.6	Derivatization with chloroformates	47
5. CONCLUSIONS		47
6. REFERENCES		47

Handbook of Derivatives for Chromatography
Edited by K. Blau and J. M. Halket © 1993 John Wiley & Sons Ltd

1. Introduction

Acylation is one of the most widely used derivatization procedures for chromatography. The reasons are, first, that acylation reduces the polarity of amino, hydroxy and thiol groups, and this usually improves their chromatographic properties, reducing non-specific adsorption effects such as streaking in TLC and tailing in gas chromatography and the appearance of 'ghost peaks'. Amides, like esters, usually chromatograph well, and the occasional case of tailing in gas chromatography is usually easily overcome by using only moderately polar stationary phases. Second, acylation may confer improved stability on compounds by protecting unstable groupings, such as for example the neighbouring phenolic groups in the catecholamines, which are very susceptible to oxidation. Third, acylation may confer volatility on substances such as carbohydrates or amino acids which have so many polar groupings that they are involatile and normally decompose on heating, and this makes it possible to analyse these classes of compounds by gas chromatography. Here acylation is an alternative to silylation, and may be preferable in cases where the acylated compound is more stable than the silylated one, as with primary amines, for example. Fourth, acylation may help to effect chromatographic separations which might not be possible with the underivatized compounds. Finally, acylation is one of the ways of introducing detector-oriented groupings, such as the various halogen-containing acyl groups used to make compounds detectable at very low levels with the electron capture detector (ECD) in gas chromatography.

These are the reasons for the development of a great

Table 1 The main types of groupings capable of being acylated

Chemical structure	Description
$-NH_2, -NH-, -N-$	Primary, secondary and tertiary amines
$-NH-CO-R$	Amides
$-CH_2-OH, -CH(OH)-, -C(OH)-$	Primary, secondary and tertiary alcohols
$-SH$	Thiols
Ph–OH	Phenols
$-C(OH)=C-$	Enols
$-C(OH)-C(OH)-$	Glycols
$C=C$	Unsaturated compounds
benzene ring	Aromatic rings

variety of applications of derivatization by acylation. However, the underlying principles and methods are straightforward, and once they are understood it is not difficult to pick one or two of the more promising acylation procedures to try out in a given application, with every expectation of a successful solution to almost any analytical problem.

The main types of chemical groups and structures that may be acylated are shown in Table 1.

2. Reaction mechanisms

2.1 General

Acylation involves the introduction of an acyl group into a molecule with a replaceable hydrogen atom (see also Table 1).

$$R-\overset{O}{\underset{\|}{C}}-X + RY'-H \rightarrow R-\overset{O}{\underset{\|}{C}}-RY' + HX \quad (1)$$

Less commonly, an acyl group may be added across a double bond.

$$R-\overset{O}{\underset{\|}{C}}-X + \overset{}{\underset{}{>}}C=C\overset{}{\underset{}{<}} \longrightarrow R-\overset{O}{\underset{\|}{C}}-\overset{|}{\underset{|}{C}}-\overset{|}{\underset{|}{C}}-X \quad (2)$$

The acylating agent R–C(:O)–X can lose the group –X by (a) electrophilic, (b) nucleophilic or (c) free radical mechanisms, represented in equation (3).

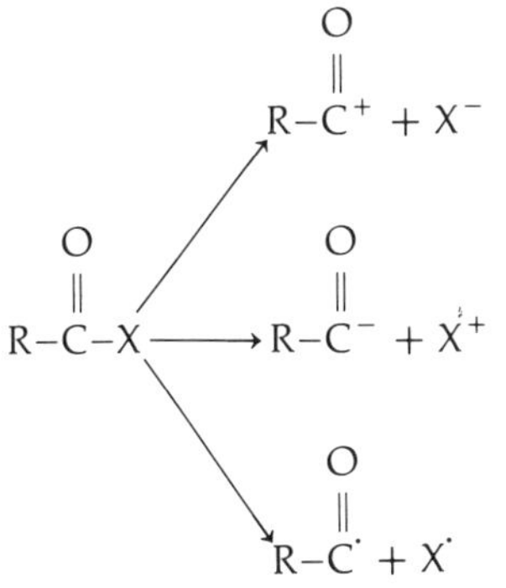

$$R-\overset{O}{\underset{\|}{C}}-X \longrightarrow \begin{cases} R-\overset{O}{\underset{\|}{C}}{}^+ + X^- \\ R-\overset{O}{\underset{\|}{C}}{}^- + X^+ \\ R-\overset{O}{\underset{\|}{C}}{}^\cdot + X^\cdot \end{cases} \quad (3)$$

Direct electrophilic acylations of the first of these types are the most common mode of acylation [1]. The mechanism of acylation, particularly of alcoholic groups, is of course essentially the same as that for esterification, so for further consideration the relevant section of Chapter 2 may be consulted, where there is a more detailed analysis of reaction mechanisms.

2.2 The reactivity of acylimidazoles

Amides are already acylated amines, and have a low reactivity in nucleophilic reactions because the amide carbonyl group is more negatively charged than the carbonyl groups of normal esters (4) and of ketones, because of the delocalization of the lone pair of electrons on the nitrogen into the carbonyl group (equation (5)).

$$R-\overset{O}{\underset{\|}{C}}-O-R' \quad (4)$$

$$R-\overset{O}{\underset{\|}{C}}-NH-R' \longleftrightarrow R-\overset{O^-}{\underset{|}{C}}=\overset{+}{N}H-R' \quad (5)$$

At first sight the reactivity of acylated imidazoles is therefore unexpected but their acylating ability is comparable with that of acid anhydrides or chlorides. This can be seen in a comparison of the susceptibility to hydrolysis of N-acetylpyrrole and N-acetylimidazole (6).

Half-life in water at 25°C and pH 7

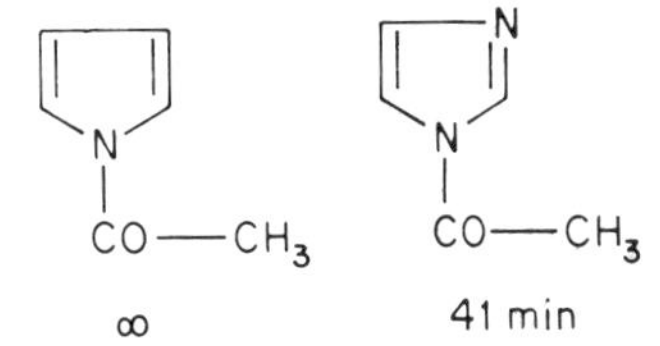

N-acetylpyrrole: ∞

N-acetylimidazole: 41 min (6)

The high reactivity of the latter is due to the delocalization of the nitrogen electrons into the heterocyclic ring. This is promoted by the second nitrogen, which tends to increase the aromaticity of the ring system.

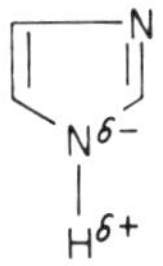

(7)

A number of acylated imidazoles are commercially available as acylating agents. Others are readily prepared by acylating imidazole in tetrahydrofuran with the appropriate acid chloride, or by condensation of the acid with carbonyl diimidazole. The imidazole portion of acylimidazole acylating reagents is, of course, the other product that appears during the acylation reaction besides the desired derivative, but at least in GLC analyses it volatilizes with or at least close to, the solvent peak and does not cause any interference.

3. Types of acylation reactions

3.1 General

In general, acylation reactions for chromatography are carried out with three main types of acylating reagent: acid anhydrides, acid halides and reactive acyl derivatives such as acylated imidazoles, acylated amides or acylated phenols, and the different types are chosen for different reasons. Acyl halides are highly reactive, which may be important with compounds that are difficult to acylate, because of steric factors for instance. A drawback to acyl halides is that a basic acceptor for the halogen acid produced in the reaction is usually required (equation (8)).

$$R-NH_2 + R'COCl + B \longrightarrow$$
$$R-NH-CO-R' + B^+HCl^- \qquad (8)$$

Getting rid of the excess of reagent and of the halide salt formed with the acceptor may be awkward. Anhydrides may therefore be preferred because excess of reagent is easier to remove, leaving a cleaner product (equation (9)).

$$R-NH_2 + R'-CO-O-CO-R' \longrightarrow$$
$$R'-CO-NH-R + R'-COOH \qquad (9)$$

On the other hand, anhydrides tend to be less volatile than acyl halides and are therefore more likely to interfere with the more volatile derivatives in analysis by gas chromatography. Also, the reaction medium is acidic, and so this approach cannot be used with acid-sensitive compounds. For such compounds an acylation reagent which has a basic leaving group on reaction, such as an acylimidazole, may be preferable (equation (10)).

$$R-NH_2 + \underset{\underset{CO-R'}{|}}{\underset{N}{\boxed{}}} \longrightarrow$$

$$R'-CO-NH-R + \underset{\underset{H}{|}}{\underset{N}{\boxed{}}} \qquad (10)$$

However, excess of the reagent may need to be removed before analysis, usually by some kind of washing or extraction procedure. Provided that this is acceptable, small scale acylations for chromatography, in common with other derivatization reactions, work well because they usually involve a large molar excess of acylating agent to drive the reaction to completion.

The nature of the solvent used may have a considerable influence on the yield, and even on the actual course of the reaction, as will be seen when we come to describe specific methods. It has been found that it is generally best if one can remove the excess of the acylating reagents, because their great chemical reactivity may give rise to problems during chromatography, particularly GC. These have included irreversible alteration of the chromatographic column, corrosion or other damage within the chromatographic or GC–MS system, and re-derivatization of non-volatile residues that have accumulated at the top of the column, with the appearance of 'ghost' peaks. For these reasons the acid anhydrides, particularly the more volatile ones which can easily be evaporated off, have proved more popular than the acyl halides.

3.2 Detector-oriented acylation reactions

Detection is a critical factor in chromatography, and the nature of the detection system governs the choices at every stage of an analytical scheme from derivatization to quantification. The design of derivatization reactions is particularly important in acylation, because some derivatization schemes involving acylation are deliberately designed to exploit the advantages associated with particular detectors. This is particularly true of the electron capture detector (ECD), because it is the most sensitive detector in gas chromatograpy and therefore ideally suited for the analysis of substances at very low concentrations.

The ECD responds particularly well to molecules that have the ability to stabilize an attached electron. Such molecules are, above all, those that contain halogen atoms, and to a lesser extent nitro and carbonyl groupings. Because the ECD responds less well than the flame ionization detector (FID) to molecules that lack these groupings, including the solvent, the chromatographic profile usually shows a more rapid return to baseline after an injection. In comparison with the FID the detection sensitivity may be better by one or more orders of magnitude, and, although the linear range of the ECD is not as wide as that of the FID, advances in instrumentation and design have extended the linearity of the ECD response to about four orders of magnitude for many compounds. Thus the ECD discriminates greatly in favour of electron-capturing substances, and is the most sensitive detector for their analysis. One application is the analysis of pesticides, many of which are halogen-containing compounds that may not even need derivatization to be successfully analysed by gas chromatography with electron capture detection.

Table 2 lists a number of acylating reagents which are designed to give electron-capturing derivatives, together

Table 2 Electron-capturing acyl groupings, together with the classes of compounds into which they may be introduced

Electrophore	Functions to be derivatized
Trifluoroacetyl, pentafuoropropionyl or heptafluorobutyryfl (TFA, PFP, HFB)	Hydroxyls; primary and secondary amines; amides; thiols
Trichloroacetyl or pentafluorobenzoyl	Hydroxyls; primary and secondary amines; thiols
Pentafluorobenzyloxycarbonyl or Trichloroethyloxycarbonyl	Tertiary amines

with the groupings that may be derivatized; Table 3 lists analogous reagents which may be similarly introduced, but are not acylating agents and are described in other chapters.

The great sensitivity of analysis that is obtainable with the ECD has been widely applied by analysts interested in measuring compounds present at very low concentrations in various matrices, predominantly biological fluids and tissues and water samples. Much ingenuity has been shown in using the various acylating reagents listed in Tables 2 and 3 to make derivatives which will respond well in the ECD. Examples are catecholamines and other biogenic amines, pesticide residues, drugs and their metabolites, and phenolic substances.

Although these derivatives all show enhanced responses in the ECD, more than one mechanism of electron capture has been described [2], and an understanding of these mechanisms may help in choosing the best derivatization reagent for a particular application [3]. Broadly speaking, the mechanisms are electron attachment to give a stable negative molecular ion:

$$A-B + e^- \rightarrow A-B^- \quad (11)$$

or, alternatively, an ionization and dissociation process:

$$A-B + e^- \rightarrow A + B^- \quad (12)$$

Low detector temperatures apparently encourage the attachment process and high detector temperatures the dissociative process. The presence of molecules capable of either process will diminish the electron flux within the detector and consequently the standing current, and it is this reduction in the standing current which is amplified and measured to produce the detector signal. For practical purposes it is, however, enough to know that the sensitivity of detection increases in the order:

$$F < NO_2 = Cl < Br < I$$

Table 3 Non-acyl electron-capturing groupings and the classes of compounds which may be derivatized with them, together with the appropriate chapters

Electrophore	Classes of compound	Chapter
p-Bromobenzyl	Carboxylic acids	2
Bromomethyldimethylsilyl	Hydroxyls	4
p-Bromophenacyl	Hydroxyls	2
p-Chlorobenzyl	Carboxylic acids	2
2-Chloroethyl	Carboxylic acids	2
Chloromethyldimethylsilyl	Hydroxyls	4
2,4-Dinitrobenzenesulphonyl and 2,4-dinitrophenyl	Primary and secondary amines	8
Iodomethyldimethylsilyl	Hydroxyls	4
p-Nitrobenzyl	Carboxylic acids	2
Pentafluorobenzyl	Hydroxyls, carboxylic acids, thiols and sulphonamides	5
Pentafluorbenzylidene	Primary amines	6
Pentafluorophenyldimethylsilylsilyl (flophemesyl)	Hydroxyls	4
$C_6H_5-NH-N=$, $C_6H_5-O-N=$,	Carbonyls	6
$(CF_2Cl)_2C=O$ forms oxazolidinones of	Primary amines, hydroxyls and thioacids	7
Pentafluoropropyl, trichloroethyl	Carboxylic acids	2

Table 3 gives a list of other electrophores which may also be used to make electron-capturing derivatives as alternatives to acylation reagents, and where they are described in other chapters of the handbook.

Sensitivity of detection with the ECD also increases with the number of electron-capturing groupings in the molecule, although there may be exceptions. However, increased numbers of electron-capturing groups may also increase the retention time of a compound, as the larger groupings such as the nitro group and bromo and iodo groups add significantly to the molecular weight of the derivatives. These groupings also tend to be associated with non-specific adsorption effects on certain columns, and this leads to tailing and asymmetrical peaks. With fluoro groups, however, as one goes from trifluoroacetyl (TFA) to heptafluorbutyryl (HFB) the loss of volatility is not too serious, and with some compounds there may even be a minimum in the retention times of their perfluoroacyl derivatives at the pentafluoropropionyl (PFP) derivatives [4]. In general it has been found better to introduce five fluorine atoms than two chlorines, because the fluoro compounds are likely to be more volatile and can therefore be analysed at lower temperatures. Volatility considerations deserve some thought, and the right choice of volatile derivative may help to optimize the analysis because, although one generally wants to make compounds more volatile by derivatization, if compounds are too volatile evaporation may be a danger. Making a less volatile derivative early on in the analytical scheme may prevent evaporative losses.

Other detectors besides the ECD may lead to the design of detector-oriented derivatization schemes, particularly where, as with the ECD, they improve the sensitivity or specificity of detection. One of these is the thermionic nitrogen/phosphorus detector (NPD), for which specific derivatives have been proposed, either in the nitrogen mode [5–8] or in the phosphorus mode [9,10]. More recently, the availability of the electrochemical detector has occasioned the application of specific derivatives such as the O-acetylsalicyloyl derivatives for the HPLC of amines [11].

4. Practical applications

4.1 General

The previous chapter, on esterification, gave numerous practical applications. With acylation, again, it would be impossible to list the large number of applications that have been published. As one goes through the procedures that have been included, it soon becomes evident that the mechanistic unity described earlier for acylation reactions is reflected in the general resemblance found among the acylation procedures. Once this has been assimilated it is not too hard to work out acylation procedures for oneself, so that any compound can be derivatized, regardless of whether or not an application has been previously described.

As with other derivatization reactions one should be able to work on a very small scale, particularly for gas chromatography, because the sample volume injected will only be a microlitre or two; but even in TLC the applied sample need not be on more than a microlitre scale. Larger scale derivatization is only necessary where a derivative has to be characterized by rigorous classical chemical methods, but these days mass spectrometric structural confirmation is often enough, so that derivatization on anything greater than milligram amounts is rarely called for. This applies particularly to derivatization for electron capture detection, where one may be analysing nanogram or even picogram amounts.

Most reactions for derivatization can be done in small vials of the type used for automatic samplers. Many of these have a tapered form, so that even microlitre volumes can be sampled with a microlitre syringe. Acylations should always be done in all-glass systems, because plastics may not stand up to some of the more aggressive acylating reagents; in addition, plastics contain plasticizers, which may give interfering peaks on chromatograms, and they may physically dissolve derivatives and so affect the reaction yields. Polytetrafluoroethylene (PTE) has been noted as having a physical affinity for perfluoroacylation reagents and their derivatives [12, 13]; see, for example, Figure 1.

4.2 Acetylation reactions

4.2.1 *Acetylation with acetic anhydride and acetic acid*

This is one of the standard acetylation procedures. The sample (up to 5 mg) is dissolved in chloroform (5 ml) and acetylated by warming with acetic anhydride (0.5 ml) and acetic acid (1 ml) for 2–16 hours at 50 °C. Excess reagents may be removed *in vacuo*, the residue being taken up in chloroform for analysis by GC. These volumes may be scaled down if required [14].

4.2.2 *Acetylation of urinary sugars with acetic anhydride and sodium acetate*

The use of sodium acetate as a basic catalyst is another standard acetylation technique. In this application, the

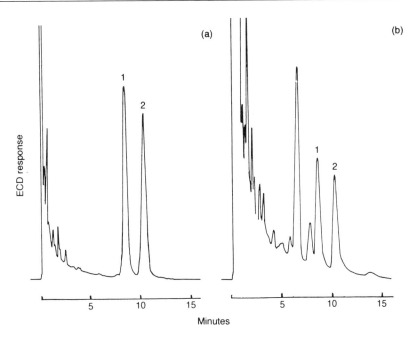

Figure 1. The effect of polytetrafluoroethylene on perfluoroacyl derivatization. The profile in curve A was obtained from derivatization of β-phenylethylamine (PE) with pentafluorobenzoyl chloride (PFBzO-Cl) in an all-glass system, that in curve B from the same reaction in a reaction vial sealed with a polytetrafluoroethylene-lined cap. Peak 1: PE derivative; peak 2: derivative of the internal standard, tolylethylamine (Ref. 13). Note the major peak of pentafluorobenzoic acid, the other by-products, and the reduced yield of derivatives in profile B.

carbohydrates are isolated from urine, and the dried residue from the isolation procedure is first oximated to derivatize carbonyl groups. Hydroxyls are acetylated with acetic anhydride (0.3 ml) in the presence of sodium acetate (12.5 mg) at 100 °C for one hour. Excess anhydride is removed as before prior to dissolution of the products in ethyl acetate for analysis [15].

4.2.3 Acetylations with acetic anhydride and pyridine

Another popular basic catalyst, particularly in carbohydrate chemistry, is pyridine, which has the ability to promote smooth reactions, and additionally has great solvent power and acts as an acceptor for the acetic acid formed in the reaction. However, it may also react with acetic anhydride with the formation of N-acetyl-1,2-dihydro-2-pyridylacetic acid [16], so that care must be taken that this does not interfere in subsequent analyses. Some examples of the use of pyridine follow.

Alditols may be acetylated as such, or after formation by the sodium borohydride reduction of sugars. Derivatization is with a 1:1 mixture of acetic anhydride and pyridine for 20 minutes at 100 °C. The reaction mixture is evaporated to dryness and the residue is dissolved in ethyl acetate for analysis on a column of 3% ECNSS-M. This procedure has the advantage that sugars treated this way give single peaks [17].

Paracetamol (up to a few milligrams), either as such in a pharmaceutical product, or as a dried down ethyl acetate extract isolated from a physiological fluid, is treated with a 3:1 acetic anhydride and pyridine mixture (20 μl) for 20 minutes at 45 °C. The mixture can be injected directly for GC analysis [18].

α-Tocopherol, extracted from up to 1 ml of plasma and after drying to a residue, is dissolved in 20 μl of dry pyridine and 100 μl of acetic anhydride and reacted at 50 °C for 15 min. The solution is evaporated to dryness in a vacuum at the same temperature. The residue is taken up in 1.5 ml of hexane and the solution is washed with 1 ml of water and centrifuged. The upper layer is evaporated to dryness and the residue is dissolved in 200 μl of absolute alcohol for analysis [19]. A very similar procedure was used for acetylation of alkanolamine antihistamine drugs, only omitting the water wash [20].

Acetylation as the second step in derivatization of amino acids for gas chromatography can be done on a milligram scale, following initial preparation of the n-propyl esters, with 1:4 acetic anhydride and pyridine

(0.4 ml) at room temperature for 10 minutes. Excess reagents are removed *in vacuo* and the residue is dissolved in ethyl acetate for further analysis [21, 22].

4.2.4 Acetylations with other basic catalysts

A more basic acetylation catalyst was used by Adams, again after making amino acid n-propyl esters, who used a freshly prepared mixture of acetone, triethylamine and acetic anhydride in the proportions of 5:2:1 at 60 °C for 30 seconds. Excess reagents were blown off cautiously to avoid volatilization losses, and the residue was dissolved in ethyl acetate for injection into the gas chromatograph [23].

Another useful basic catalyst is N-methylimidazole (NMIM), which serves as a more powerful catalyst, base acceptor of the acid and solvent. With hydroxy compounds, on the milligram scale, the equivalent proportions used are hydroxy compound:acetic anhydride:NMIM 1:1.2:2, i.e. a 20% molar excess of anhydride per hydroxy group with an equivalent amount of NMIM as acid acceptor. The hydroxy compound was first dissolved in NMIM and the anhydride was added afterwards. The reactions, which generates its own heat, was complete in less than 10 minutes [24]. The reaction mixture was injected directly for GC analysis.

4.2.5 Acetylation in aqueous solution

Although water hydrolyses acetic anhydride, amino groups and phenolic hydroxyls react much faster, and can therefore be acetylated in aqueous solution. This is the basis of the Schotten–Baumann conditions: the reaction is driven by addition of sodium bicarbonate to remove the acid formed. The stirred aqueous solution of amines (or amine extract from tissues or biological fluids), totalling 4 ml, is treated with acetic anhydride (0.3 ml). A slight molar excess of sodium bicarbonate is added in small portions, letting effervescence subside each time. At the end of the reaction the products are extracted with dichloromethane (2 × 5 ml), and the combined extracts are blown down in a stream of nitrogen [25–27]. Similar procedures have been described for phenols [28, 29].

4.2.6 Acetylation with simultaneous silylation

If one can get two derivatization reactions to take place simultaneously it may save a step. For example, hexosamines (5 mg) are mixed with dimethylformamide (50 µl) containing the internal standard N-acetylglucosaminitol (1 mg/ml), pyridine (40 µl) and acetic anhydride (10 µl) to acetylate the amino groups, and hexamethyldisilazane (30 µl) to silylate the hydroxyls. The ammonia produced helps to remove the acetic acid formed. After vigorous mixing the reaction mixture is left for half an hour at room temperature. The mixture can be injected directly for analysis by GC [30].

4.2.7 Acid-catalysed acetylations

Methanesulphonic acid has been widely used. In the acetylation of glucuronides, the sample is dissolved in acetic anhydride (1 ml) and methanesulphonic acid (50 µl) is added. The reaction mixture is left at room temperature for 24 hours and then poured into ice–water to decompose the anhydride and extracted with chloroform. The chloroform extract is dried over anhydrous sodium sulphate and concentrated *in vacuo*. The dried residue is methylated with diazomethane, evaporated to dryness and taken up in chloroform for GC analysis [31, 32].

In an ingenious procedure, the mixture of acetic anhydride and methanesulphonic acid is adsorbed onto a Celite microcolumn (1.5 cm in length, in a melting point capillary). The alcohols to be acetylated (1–3 mg) are injected dissolved in a suitable non-polar solvent with a microlitre syringe (1–10 µl) and displaced with a bed volume of solvent using gentle air or nitrogen pressure. The syringe is used to take up the eluate at the other end of the column for direct injection into the analytical system. Alternatively, the excess acidic reagents can be removed on another similar microcolumn packed with Celite impregnated with K_2HPO_4 [33].

4.2.8 Acetolysis procedures

Acetolysis is another example of two reactions being done in a single step, here hydrolysis as well as acetylation. Polysaccharides can be broken down to their constituent sugars, which are simultaneously acetylated. The polysaccharide (0.1 g) is stirred with a 1:1:1 mixture of glacial acetic acid, concentrated sulphuric acid and acetic anhydride at 40 °C overnight (or longer if necessary) and the reaction mixture is neutralized with pyridine and ice. The acetylated sugars are extracted from the aqueous solution with chloroform, and the chloroform solution is washed with a little water, dried and evaporated to dryness for chromatography [34].

Sulphate ester conjugates can be converted to acetyl derivatives in much the same way. This is a relatively mild procedure by which esters which are too unstable to acid hydrolysis can be characterized. The sulphate ester (10 mg) is dissolved in a 1:40 solution of methanesulphonic acid in acetic anhydride (2 ml) at 0 °C and the mixture is then heated at 100 °C for 30 min. The mixture is cooled, ice (2 g) is added, and the acetylated products are extracted into benzene (3 × 4 ml). The combined extracts are concentrated for chromatographic analysis. Yields depend on the nature of the sulphate ester, and may be increased by (a) increased reaction times, and (b) increasing the excess of reagents [35].

4.2.9 Acetylation with acetyl chloride

Acetyl chloride is not a widely used acetylation reagent for derivatization because, ideally, a base has to be included to remove the hydrochloric acid formed. Two examples are given, one using vapour phase acetylation.

Hydroxyanthranilic acid, a product of tryptophan metabolism, is derivatized for gas chromatography as the N-acetyl methyl ester. Methylation with diazomethane first makes the O-methyl methyl ester; after evaporation to dryness in a stream of nitrogen, the residue is acetylated in the dark in 50% benzene/acetyl chloride. The solution is again blown dry with nitrogen and dissolved in methanol for chromatography [36].

In the vapour phase method, the substance to be acetylated is supported on a bed of Celite 545 held by a glass wool plug in a melting point capillary tube. Acetyl chloride vapour is sucked through the Celite bed by holding the lower end of the capillary in the headspace over a vial of the reagent and applying a slight vacuum (15–20 cm Hg) for ten minutes. Excess reagent is removed by pulling air through the Celite bed for half a minute, and the derivative is eluted from the Celite with carbon disulphide or methylene chloride, collecting the first few microlitres for analysis. The Celite bed can also first be used to concentrate the compounds to be acetylated from dilute solutions [37].

4.2.10 Acetylation with ketene

Ketene, $CH_2=CO$, is produced by the pyrolysis of acetone.

$$\begin{array}{c} CH_3 \\ \diagdown \\ C=O \longrightarrow CH_2=C=O + CH_4 \quad (13) \\ \diagup \\ CH_3 \end{array}$$

The easiest way of generating it is to pass acetone vapour over a red-hot wire filament [38, 39]. If the gas is bubbled into an aqueous solution of an amine or amino acids, all amino groups are smoothly and quantitatively acetylated without acidic by-products, but none of the hydroxyls [40, 41]. Alcohols can be acetylated by bubbling ketene directly through the alcohol or mixture of alcohols, either neat or dissolved in a non-polar solvent [39].

CAUTION. Ketene is a respiratory tract irritant and MUST be used only in an efficient fume cupboard.

4.3 Methods for making propionyl derivatives

Amines can be directly derivatized with the anhydride. The amine or amine hydrochloride (5 mg) is heated with freshly distilled propionic anhydride (0.5 ml) in a sealed reaction vial at 100 °C for 30 min. The product is poured onto water (5 ml) cautiously made alkaline with 2M NaOH and the amides are extracted into methylene chloride (3 × 1 ml).

Amino acid N-propionyl derivatives were prepared by Youngs, who first made the amino acid isoamyl esters and then treated these with propionyl chloride using anhydrous sodium carbonate to remove the HCl formed. The esterified mixture of amino acids, totalling about 5 mg, was dissolved in carbon tetrachloride (5 ml), and sodium carbonate (100 mg) was added followed by propionyl chloride (0.2 ml). After refluxing for 30 minutes, the sodium salts were filtered off and the solution was concentrated and analysed by gas chromatography [42].

4.4 Formyl derivatives

Making a formyl derivative adds only 28 to the molecular weight of a compound, and for this reason has found particular favour in situations where several groups have to be blocked, such as in steroid chemistry, where the formyl derivatives appeared to be among the most suitable for the analysis of steroids by gas chromatography. However, it must be said that formyl derivatives are not necessarily the most volatile: some of the perfluoroacyl or methoxycarbonyl derivatives may be more volatile than the corresponding formyl ones.

Edwards, Makin and Barratt found that acylation of steroids with 98–100% formic acid could result in artefacts, and adjusted the strength of the acid to 95% by adding water. Although a limit of 30 °C is usually recommended for formylation of steroids, they formylated at 40 °C for 30 minutes and still had average yields of 95% [43]. Alternatively, the 98–100% formic acid can be used at 50 °C for 10 minutes [44] or in the vapour phase at 80 °C in one hour with no appreciable destruction of any steroid [45].

Amines, either as salts or the free bases, can be formylated using sodium formate: the base (or its salt, 1–5 mg) is treated with 98–100% formic acid (20 μl), anhydrous sodium formate (10 mg) and acetic anhydride (200 μl) at 25 °C for 30 minutes in a reaction vial sealed with a cap containing a PTFE liner. Excess reagents are removed *in vacuo* and the residue is taken up in ethyl acetate for analysis [46–48].

Another procedure is to treat the free base dissolved in chloroform (and if necessary liberated from the salts with triethylamine) with a molar equivalent of formic acid together with a molar equivalent of dicyclohexylcarbodiimide. After addition of excess 1 M potassium bicarbonate, the products are extracted into a suitable solvent [47]. Formylations in the peptide field are supposed not to cause any racemization. It may be added that, where there are other groups such as hydroxyls present, e.g. in the amino acid and peptide field, these may need to be blocked by some other protective group, and carboxyls need to be esterified, to yield derivatives intended for GC analysis [47]. The small size of the formyl group recommends it but, as mentioned earlier, other derivatives, although of greater molecular weight, may actually be more volatile than the corresponding formyl derivatives.

4.5 Perfluoroacyl derivatives

The most popular perfluoroacyl groups are trifluoroacetyl (TFA), pentafluoropropionyl (PFP) and heptafluorobutyryl (HFB). TFA derivatives were originally developed in carbohydrate chemistry [49], and extended to peptide chemistry by Weygand [50, 51], but one of the major applications of perfluoroacyl derivatives has been to prepare electron-capturing derivatives for use with the electron capture detector for gas chromatography. This detector gives a greatly enhanced response to halogenated derivatives, and the sensitivity of detection increases in the order F < Cl < Br < I. However, as we ascend the series, volatility and stability tend to decline, so that fluoro and chloro derivatives are the most widely used in practice. In general, the more halogens, the greater the sensitivity of detection; also, O-perfluoroacyl derivatives are more electron-capturing than the corresponding amine derivatives [52]. The precise choice of derivative to use depends on the analytical priorities: the TFA—or sometimes the PFP—derivatives tend to be the most volatile, and HFB derivatives give the greatest sensitivity with electron capture detectors, but chromatographic separation may also need to be considered, and might be optimal for one or other type of derivative [53]. Although the perfluoroacyl anhydrides behave similarly, their reactivities decrease in the order TFA > PFP > HFB [54]. Fashions in research have tended to lead to amino acids being acylated mainly to TFA derivatives, with the aim of merely obtaining volatile derivatives and not for use with the ECD, phenolic amines to PFP derivatives, with analysis by GC with the ECD very much in mind, and steroids to HFB derivatives. In the recipes that follow, although one or the other derivatives may be specified, the three different reagents are sufficiently similar for TFA, PFP and HFB to be essentially interchangeable, except that reaction times and temperatures may need to be adjusted to allow for the different reactivities.

Although the anhydride may be used on its own, the acylation reactions go most smoothly and quickly in a solvent and with a catalyst. The precise nature of the solvent is not important, but it is common to use a solvent such as acetonitrile which can act in both capacities. The following conditions are satisfactory for phenols, amines, phenolic amines and alcohols.

4.5.1 Acylation with anhydrides on their own

Lee, studying the acylation of anilines and chloroanilines, found that reaction with TFAA was complete for most of them in 5 minutes at room temperature, and in 15 minutes for all of them; with PFPA the reaction was complete for most in 15 minutes at room temperature, and after two hours for all; and with HFBA anhydride for most in 18 hours at room temperature and for 30 minutes at 60 °C [54].

Acylation of amino acid esters to render amino acids accessible to gas chromatography is a special problem: the great variation in structures among this group makes it hard to select conditions that will acylate all of the amino acids in good yield. Acylation of arginine and histidine requires vigorous conditions, but too vigorous conditions will destroy some of the more sensitive amino acids such as tryptophan. Two procedures have been developed; both of them depend on esterification of the amino acids first, with evaporation to dryness to remove the alcoholic HCl used for the esterification (for details of the esterification procedures see Chapter 2). In the first procedure, the amino acid ester mixture (about 2 mg) is treated with 200 μl of TFAA for 30 minutes at room temperature or, if arginine is present, for 10 minutes at 140 °C. The final mixture is evaporated to dryness *in vacuo* for just 4 minutes to minimize evaporative losses [55], and the residue is dissolved in methylene chloride for gas chromatography [56].

The reactivity of the compounds to be derivatized clearly plays an equally important part. Thus, for example, ephedrine, pseudoephedrine and analogues were acylated with TFAA for 5 minutes at 60 °C [57], tricyclic antidepressants with HFBA for 10 minutes at 60 °C [58], the methyl esters of pipecolic acid, proline, glutamic acid and γ-aminobutyric acid with HFBA for 20 minutes at 150 °C [59], and catecholamines (after methylation of the β-hydroxyl) with PFPA for 15 minutes at 80 °C [60].

4.5.2 Acylation with anhydride in a solvent

In the second procedure for amino acids (first converted to the n-butyl esters), acylation is done with a 1:3 mixture of TFAA and methylene chloride (1 ml for each 10 mg of amino acid mixture) in a reaction vial with a PTFE-lined screw-cap seal at 150 °C for 5 min. Excess of the reagents is evaporated in a stream of nitrogen, but not to complete dryness, to avoid evaporative losses [61]. Conditions for GC analysis of amino acids as the N-TFA amino acid n-butyl esters have been carefully optimized, and details may be found in the summarizing paper [62].

The acylation procedures as described are equally applicable to PFP and HFB derivatives; however, for amino groups, only the N-HFB derivatives capture really well in the ECD for significantly more sensitive detection than with the flame ionization detector [63, 64], although the derivatives of the hydroxyl amino acids can be detected with much better sensitivity because, as mentioned already, O-perfluoroacyl derivatives are more electron-capturing.

Slightly different conditions were chosen by Gamerith, who used TFAA with acetonitrile as a solvent in the ratio of 1:2 and re-examined the optimal acylation conditions. Although most amino acid n-propyl esters were completely acylated in 6 minutes at 30 °C or even room temperature, serine, tyrosine, lysine and tryptophan needed higher temperatures, up to 150 °C, but losses occurred at the higher temperatures. Gamerith concluded that the best compromise for all amino acids was 5 minutes at 150 °C [65, 66].

The conditions used for amino acid esters with methylene chloride as solvent were found to be equally applicable to the derivatization of nucleic acid bases and hydrolysates of DNA [67].

Amines or amine extracts (1–5 mg) are treated with 500 μl of acetonitrile and 50 μl of anhydride for 5 minutes at room temperature. These conditions are best, particularly with indole compounds, which are acid-sensitive [68]. More usually, a 1:1 mixture of acetonitrile and anhydride is used at 60 °C for up to 20 minutes, depending on the substances to be derivatized. It may be possible to use lower temperatures and shorter reaction times: the best thing is to work out optimal reaction conditions with pure compounds first [67, 70]. Excess solvent and anhydride are completely removed *in vacuo*, or the reaction mixture is blown dry with nitrogen, and the dried residue is dissolved in ethyl acetate (if required, the solvent may contain an injection standard) for analysis [71, 72]. Removal of excess reagent is essential before injection into a GC with an ECD, because if reagent gets into the detector it will saturate it for a long time—perfluoroacyl anhydrides contain a lot of halogen atoms, are very polar and are hard to flush out—far longer indeed than the analysis time of the derivatized compounds.

Other solvents may be used instead of acetonitrile: ether [74] or ether/heptane [75], and ethyl acetate is popular because of its good solvent power for aromatic compounds. The following papers describe the analysis of drugs or metabolites from blood or cerebrospinal fluid, where the substance if first extracted, usually by a solvent extraction step, dried down and then derivatized [76–79].

4.5.3 Acylation with anhydrides and a basic catalyst

Although amines do not have to be in the free-base form for acylation by perfluoroacyl anhydrides, and derivatize smoothly even as their salts, some workers have included bases for catalytic purposes, and also to remove acid formed during the reaction, in addition to a solvent. In one recipe, the amine is dissolved in benzene (500 μl) and trimethylamine is added (100 μl of a 0.05 M solution in benzene) followed by 10 μl of anhydride, and the reaction is allowed to go to completion at room temperature. With the low concentration of anhydride this may take some hours. Excess anhydride is removed by washing with 3 M ammonium hydroxide and the benzene solution is dried before GC analysis [80, 81]. In a similar procedure, the rather uncommon chlorodifluoroacetic anhydride is used. Here the base, extracted from its biological matrix, is dissolved in chloroform (20 μl) and derivatized with triethylamine (1.5 μl) and the anhydride (3 μl) at 50 °C for 30 min. Excess reagent is neutralized with alkali and the derivatives are concentrated to dryness and taken up in toluene for GC analysis [82]. Trichothecenes (250 mg) are treated with TFAA and about 10 mg of solid sodium bicarbonate for 30 minutes at 80 °C [83]. Alternatively, 0.5 ml of 10% acetonitrile in toluene is added followed by 100 μl of triethylamine and 100 μl of PFPA. The reaction is carried out at 60 °C for 15 minutes, and the reaction solution is washed with two 0.5 ml portions of 5% aqueous ammonia and one 0.4 ml portion of water to remove the surplus reagents [84].

In the carbohydrate field, trifluoroacetylation is usually done in a solvent, with or without a basic catalyst. Thus, sugars (1–5 mg) are treated with tetrahydrofuran (THF, 200 μl) and TFAA (50 μl) at 40–50 °C for 10 minutes, with occasional shaking. The solution is blown dry with nitrogen and dissolved in acetonitrile for GC. When the reaction was done in THF each sugar gave only a single peak, but when acetonitrile was the solvent multiple peaks were obtained; however the final derivatives were taken up in acetonitrile because they were more stable in that solvent than in THF [85]. Another procedure for sugars (100 μg) used a mixture of methylene dichloride (65 μl). TFAA (30 μl) and pyridine (5 μl) at 20 °C for 3 hours [86]. The mixture was analysed directly, but as with acetic anhydride, occasional interference may be found from a derivative of pyridine with TFAA, so it is probably better to remove surplus reagents before analysis as in the previous methods.

Sugars are advantageously reduced to their alditols prior to acylation because this produces a single derivative from each sugar, as already mentioned in the recipe for the acetylation of alditols [17]. A similar procedure can also be used with TFAA and acetonitrile. The alditols from the reduction reaction (1–3 mg) are heated in a sealed reaction vial with 100 μl of acetonitrile containing 0.8 M sodium trifluoroacetate as basic catalyst and 100 μl TFAA for 10–15 minutes at 60 °C. Excess reagents are removed *in vacuo* and the residue taken up in methylene chloride for GC analysis [87]. Milder conditions, 10 minutes at room temperature, may be used if dimethylformamide (DMF, 50 μl) is used as solvent with 0.3 mg of sodium acetate [88]. DMF and dimethyl sulphoxide (DMSO) are solvents which have a reaction-promoting effect but are not very volatile and are not so easy to remove as the volatile acetonitrile or methylene chloride. DMF is also used for derivatization of clomipramine and its N-desmethyl metabolite. The serum extract, in 1.5 ml heptane, is treated with 20 μl of DMF, 20 μl of pyridine and 100 μl PFPA for an hour at 60 °C. Excess reagents are destroyed with 2 ml of alkaline buffer, the organic layer is separated and blown dry with nitrogen, and the residue is dissolved in 20 μl of pyridine for analysis [89].

Steroids are generally converted to the HFB derivatives, which are more volatile and chromatographically more favourable than the parent compounds. These O-HFB derivatives can be analysed with great sensitivity using GC with ECD. In some instances dehydration has been observed as a side reaction, so the milder the acylation conditions the better. Steroids are usually acylated in solvents. The dry steroid (up to 1 mg) is treated with 1:4 HFBA in acetone (250 μl) for one hour at room temperature, the mixture is blown dry with nitrogen and the derivatives are taken up in benzene for GC analysis [90]. In these conditions 17-hydroxyls and the enol form of 3-ketosteroids are acylated. The acetone concentration used is optimal. In a method for determination of testosterone in blood, the steroid extract is dried down and treated with HFBA (20 μl), n-hexane (1 ml) and THF (40 μl) at 50 °C for 30 minutes [91]. A similar recipe [92] involves a reaction time that is probably unnecessarily long. These conditions, using non-polar solvents, leave the 3-keto groups intact and unacylated, so that they can be derivatized to, e.g., methoximes if required.

Vapour phase acylation procedures, as described earlier with acetyl chloride [33], have also been successfully used with the similarly volatile and reactive perfluoroacyl anhydrides on a variety of compounds including several steroids. The 17-hydroxyls reacted very rapidly, but the enol groups of the 3-keto compounds took variably longer times to react completely, and so it may be preferable to convert them to methoximes.

Trimethylamine has been used as a basic catalyst for the derivatization of phenols. The phenols (up to 1 mg) in benzene (500 μl) were mixed with 0.1 M trimethylamine in benzene (100 μl) and acylated with HFBA (10 μl) at room temperature for 10 minutes. Excess reagents were removed by a 30 second wash with pH 6 phosphate buffer, the phases were separated by centrifugation, and the benzene solution was taken for gas chromatography. As with other O-perfluoroacyl derivatives, these compounds could be analysed with excellent sensitivity using the ECD [92].

Vitamers of the B_6 group (pyridoxine, pyridoxamine and pyridoxic acid lactone) could be acylated in THF. Pyridoxal was first converted to the methoxime, and the vitamers (up to 0.5 mg) were then dissolved in THF/TFAA 2:1 (300 μl) for 10 minutes at room temperature [93]. For ECD it would be better to make the HFB derivatives, and excess reagent would have to be removed in a stream of nitrogen before taking the residue up in a suitable solvent for GC analysis.

A lot of interest in catecholamine metabolism has centred on their metabolic end-products 3-methoxy-4-hydroxyphenylethylene glycol (MHPG) and 3-methoxy-4-hydroxyphenylethanol (MHPE). These compounds have been analysed most often as their TFA or PFP derivatives, with electron capture detection for sensitivity. Many different recipes have been published, all variants of the anhydride in ethyl acetate procedure, using different temperatures and reaction times. If 30 minutes at room temperature are sufficient [94], one does not need 40 minutes at 65 °C [95]. Pyridine is not recommended because it gives rise to interference, just as with acetic anhydride (see above). At the end of the reaction the mixture is evaporated to dryness with nitrogen, and the residue is taken up in ethyl acetate. These derivatives are susceptible to hydrolysis, and up to 2% of acetic anhydride may be included in the final solution to protect the derivatives against moisture.

The other end-products of catecholamine metabolism, the phenolic acids, have to be esterified as well as perfluoroacetylated to make them accessible to GC analysis with electron capture detection, and esterification is usually done first [96, 97] although it can be done second [99].

4.5.4 Simultaneous acylation and esterification

The best solution is to do both derivatizations simultaneously, based on the finding by Smith and Tsai that TFAA is a good esterification catalyst [99]. This procedure is applicable not only to phenolic acids [100–102] but also to various other acids, such as pyrrolidone-carboxylic acids [103], γ-glutamyl amino acids [104] and to amino acids and peptides, e.g. glycine [104] and glutamine and glutamic acid [105] and enkephalins [106, 107] in brain; all these derivatizations used hexafluoro-isopropanol together with either PFPA or TFAA. The compounds to be derivatized (up to 1 mg) are treated with a mixture of the alcohol in PFPA (100–250 μl) at 60–75 °C for 15–40 minutes. After evaporation to dryness in a stream of nitrogen, acylation may be completed by repeating the process but without any alcohol, after which the solution is again blown down and the residue is dissolved in ethyl acetate for GC analysis with ECD. In the original procedure pentafluoropropanol was used, because of its electron-capturing properties, but other alcohols may be substituted, the conditions being adapted to give optimum results. A selection of simultaneous esterification and acylation derivatizations is also given in the previous chapter.

4.5.5 Acylation of amides

Primary amines can carry two perfluoroacyl groups if the derivatization is appropriately done, although the second group is generally more sensitive to hydrolysis than the first. It follows that amides, as monoacyl derivatives, can also be derivatized further, although this may not apply if the acyl group already present is bulky. For bisperfluoroacylation, the amine (up to 0.5 mg) is dissolved in benzene (200 μl) and 0.3 M trimethylamine in benzene (0.2 μl) is added. Acylation is with TFAA or HFBA (25 μl) for 30 minutes at room temperature [108]. As with the method for phenols [92], a quick wash with pH 6 phosphate buffer is used to get rid of excess reagents, which seems incompatible with the sensitivity of the second acyl group to hydrolysis, but the authors claim that it is safe. Amides have been trifluoroacetylated in carbon disulphide. The amide (e.g. phenacetin, up to 0.4 mg) is dissolved in carbon disulphide (500 μl) and treated with TFAA (20 μl) for 10 minutes at room temperature. With a FID the reaction mixture may be injected directly, but for use with the ECD the solvent and anhydride have to be evaporated off with nitrogen as usual [109].

N,N'-Ethylenebis-stearamide and -oleamide were derivatized with 4% TFAA in chloroform for 90 minutes at 115 °C [110].

4.5.6 Hydrolysis of sulphate esters and glucuronides with simultaneous perfluoroacylation

As with acetolysis, the perfluoroacyl anhydrides too can be used for simultaneous hydrolysis and acylation. Thus, ether glucuronides, after methylation with methanolic diazomethane on the milligram scale, can be acylated by treatment of the dried residue with, e.g., PFPA (100 μl) in ethyl acetate (20 μl) at 65 °C for one hour. The mixture is evaporated to dryness with nitrogen and the residue is dissolved in ethyl acetate for analysis by GC [111].

Perfluoroacyl anhydrides can also be used to replace the sulphate ester group with a perfluoroacyl group. An amount of the sulphate (about 20 μg) is treated with a 1:1 mixture of ethyl acetate and anhydride (200 μl) for 30 minutes at room temperature (aromatic sulphates) or 70 °C (non-aromatic sulphates). After evaporation to dryness in a stream of nitrogen, the residue is dissolved in ethyl acetate (50 μl) for analysis by GC or by GC–MS [112]. The authors investigated the reaction mechanism and found it to be acid-catalysed, and established that pyridine inhibited it, but the yield and course of the reaction depended considerably on the exact structure of the compounds studied.

4.5.7 Commentary

These practical examples show that there is an underlying similarity among all these perfluoroacylating reactions involving anhydrides, and this should make derivatization with anhydrides straightforward: above all, they recommend themselves for the ease with which excess reagents can be removed by evaporation. All that needs to be added is that, as always, time spent on optimizing the conditions (reagent concentration, solvent and catalyst, temperature and time) is usually a good investment, because in this way yields may be maximized and reaction times minimized.

4.6 Alternative perfluoroacylation procedures

4.6.1 The use of perfluoroacylimidazoles

As mentioned earlier, acylated imidazoles have the advantage of reacting in a non-acidic environment, and are therefore useful for acylating acid-sensitive compounds such as those where dehydration might be caused by the use of anhydrides. Imidazole reagents can acylate alcoholic and primary and secondary amino groups. The reagents are sensitive to moisture and readily hydrolyse to imidazole and an acid, and this makes it easy to get rid of excess reagent after the reaction is complete, provided the products are stable enough to allow a brief wash with an aqueous solution. Both acidic and basic wash solutions have been used, depending on the sensitivity of the derivatives to those conditions.

Indoleamines and indole alcohols, typically acid-sensitive compound, may be smoothly acylated with HFB-imidazole. The indolic compounds (1–2 mg) are dissolved in HFB-imidazole (100–200 μl) and heated in a reaction vial with a PTFE-lined screw-cap at 80 °C for 2–3 hours. After cooling, the products are extracted into hexane (3 × 5 ml). The hexane extracts are combined and chilled to −14 °C to precipitate residual reagent. The hexane layer is decanted, the precipitate is washed with hexane and the combined hexane solutions are concentrated to 500 μl for analysis by GC [113, 114]. Alternatively, excess reagent may be decomposed with water (1 ml) in the presence of toluene (2 ml), the aqueous layer being extracted with three further 2 ml portions of toluene. These extracts are washed once more with water, separated and filtered through filter paper before concentration for analysis [115, 116]. A novel way of separating the layers is to freeze the two layers in a solid CO_2/acetone bath and to withdraw the upper layer from the ice with a Pasteur pipette [117]. Phenolic acids, after esterification, may be acylated in a similar procedure, but here with 10% PFP-imidazole in ethyl acetate for 10 minutes at 70 °C [118].

Metoclopramide in plasma has been smoothly acylated with HFB-imidazole for gas chromatography with electron capture detection. The extract of the drug (from 1 ml of plasma) is heated with 20 μl of HFB-imidazole for 90 minutes at 75 °C and then cooled. Bicarbonate–carbonate buffer of pH 10 (2 ml) is added, the derivative is extracted into 1 ml of hexane and the extract is evaporated to dryness. The residue is dissolved in 50 μl of hexane immediately prior to analysis [119]. A similar method was used for clebropride, although on a smaller scale, and using only 65 °C for 15 minutes [120].

An interesting finding was that one can use HFB-imidazole to derivatize a mixture of an alcohol and an acid—here ethylhexanol and ethylhexanoic acid. The alcohol was smoothly acetylated; the acid displaced the HFB group and could be analysed as ethylhexanoyl-imidazole [121].

4.6.2 Combined acylation and silylation reactions

A combined acylation and silylation can be done in a non-acidic environment using HFB- or TFA-imidazole and trimethylsilylimidazole in a 'single pot' reaction. The sample, e.g. a phenolic amine (1 mg), is dissolved in acetonitrile (100 μl) and TMS-imidazole (200 μl) is added, followed by heating at 60 °C for 3 hours. After cooling, HFB- or TFA-imidazole (100 μl) is added, with a further similar heating step. The reaction mixture may be extracted into hexane as before [122]. In an interesting variant of this procedure, the action of HFB-imidazole in the presence of catalytic amounts of heptafluorobutyric acid can actually displace TMS groups. The compound e.g. a steroid (0.5 mg), is first trimethylsilylated with TMS-imidazole (20 μl) for an hour at 100 °C. After cooling, HFB-imidazole (20 μl) and heptafluorobutyric acid (2 μl) are added and the mixture is heated at 50 °C. Benzene (200 μl) is added and the mixture is washed with ice-cold 0.6 M sodium bicarbonate (200 μl) and afterwards with a similar volume of water. The benzene solution is dried over anhydrous sodium sulphate before analysis [123].

4.6.3 N-Methylbis(trifluoroacetamide) (MBTFA)

The reagent MBTFA was used first by Dönike for introducing the TFA group under mild conditions [124].

$$\begin{matrix} CF_3CO \\ & \diagdown \\ & N-CH_3 \\ & \diagup \\ CF_3CO \end{matrix} \qquad (14)$$

Primary and secondary amines react readily, hydroxyls rather less so. The compound to be analysed (1–2 mg) is dissolved in MBTFA (500 μl) and left at room temperature for 30 minutes. Higher temperatures may be necessary for the less reactive classes of compounds: this needs to be established in practice. Compounds that do not readily dissolve may be dissolved in a 1:4 mixture of MBTFA with acetonitrile, pyridine, dimethyl sulphoxide, THF etc. Excess of MBTFA does not have to be removed; it elutes early in GC analysis. A combined procedure can be done in which the compound to be analysed, such as a phenolic amine, is silylated with bis(trimethylsilyl)trifluoroacetamide (BSTFA, 50 μl) at room temperature overnight (or for shorter periods at higher temperatures), followed by acylation with MBTFA (5–50 μl) at 80 °C for 5 minutes. Any TMS-amino groups get acylated in these conditions [124–126]. Acylation with MBTFA can also be done on-column, which is even quicker [127].

4.6.4 Pentafluorobenzoyl (PFBzO) derivatives

These derivatives are stable, easy to make and chromatograph well, giving very sensitive analyses for amines and phenols with the ECD. The reagent PFBzO-Cl must be stored dry in a refrigerator to minimize hydrolysis and halogen exchange reactions. Pentafluorobenzoic acid is formed as a byproduct, and if its peak on the chromatogram interferes with one of the product peaks it can be removed by a rapid wash of the reaction solution with 1 M NaOH. The easiest way of separating the phases is by freezing in a solid CO_2/acetone bath as mentioned earlier.

The amine extract (up to 1 mg) and an equimolar amount of trimethylamine are dissolved in ethyl acetate (1 ml) and a 50% molar excess of PFBzO-Cl is added dropwise, with shaking after each drop. After 2 hours at room temperature, the reaction solution is evaporated to dryness *in vacuo* and the residue is dissolved in ethyl acetate for analysis by GC [128]. Amines generally do not require a vast excess of PFBzO-Cl; too much reagent will lead to a large peak of pentafluorobenzoic acid which may be a nuisance to have to remove. Thus, β-phenylethylamine (0.1 mg), or its hydrochloride, is treated with 5% PFBzO-Cl in ether (100 µl) for 5 minutes at room temperature. The solution is evaporated to dryness in a vacuum desiccator over flake NaOH and the residue is taken up in ethyl acetate for GC analysis with electron capture detection [13]. A similar procedure for phenylethylamine was described by Baker, Rao and Coutts, but their reagent was pentafluorobenzenesulphonyl chloride [129]. A sensitive GC–MS procedure for analysing platelet activating factor (O-alkyl-2-acetyl-glycerophosphocholine) extracted from blood used 3 millimoles of PFBzO-Cl in 150 µl of acetonitrile at 350 °C for 60 minutes. After evaporation to dryness, the product was dissolved in 3 ml of heptane and washed twice with 1 ml of 5% aqueous citric acid, the heptane layer was again dried down, and the residue was taken up in 100 µl of heptane for analysis [130]. A 2% solution of PFBzO-Cl was also sufficient to derivatize rimantadine and its hydroxylated metabolites extracted from plasma and urine. A small amount of triethanolamine was used as basic catalyst and chloride acceptor, and this enabled the reaction to be completed in 20 minutes at room temperature [131].

A base-catalysed procedure for thiodiglycol also used only a small amount of PFBzO-Cl: 10 µl of PFBzO-Cl and 50µl of pyridine, and the reaction only took 5 minutes at room temperature [132]. Phenols are derivatized in a procedure involving conversion to the sodium salt first: the phenol (25 mg) is dissolved in ether (10 ml) and an excess of solid sodium hydride is added in small portions until no more effervescence occurs. PFBzO-Cl 1:9 in ether (about 300–500 µl, 10% molar excess) is added and the mixture is stirred at room temperature for 4 hours. The solution if filtered and evaporated to dryness, and the residue is dissolved in a suitable solvent for analysis by TLC, or by GC with either a FID or after dilution, with an ECD [133]. Alcohols too require more drastic derivatization: Crabtree *et al.* recommend 50 °C for an hour, using 0.48 ml of PFBzO anhydride in the presence of 20 µl of pyridine [134].

4.6.5 Derivatization of amides to PFBzO derivatives

The derivatization of amides has been mentioned. Although amides form either TMS or perfluoroacyl derivatives, PFBzO is probably the derivative of choice; it is stable [108] and readily prepared in good yield, and the products chromatograph well. The amide (0.2 mg) in hexane (500 µl) is mixed with 1.4 M trimethylamine in hexane (200 µl) and PFBzO-Cl (25 µl) is added. If the amide does not completely dissolve, acetone (100 µl) may be added. After 2 hours at room temperature excess reagents are removed *in vacuo*, and the residue is dissolved in ethyl acetate for analysis by GC with electron capture detection [108].

PFBzO-imidazole (PFBzO-Im) may also be used: the manufacturers recommend that the sample of the compound to be derivatized (0.1–2 mg) is taken up in benzene (200–500 µl) and PFBzO-Im (5–10 µl) is added. The mixture is heated at 60 °C for 15–30 minutes. Sterically hindered compounds may, however, take 2–6 hours. The mixture is evaporated to dryness *in vacuo* and the residue is extracted with hexane. The hexane extract may be injected for GC with ECD or concentrated first if necessary. For less sensitive analysis with an FID instrument the reaction mixture may be injected directly.

4.7 Miscellaneous acylations

4.7.1 General

The acylation reagents presented so far are those most commonly used for the chromatographic analysis of compounds with amino, hydroxyl and thiol groups. They have been selected because they give derivatives with useful properties such as ease of preparation, stability, good chromatographic characteristics and sensitivity of detection, e.g., with the ECD. A host of other possible acyl derivatives has been or could be made; many are well known from earlier applications, for example in peptide chemistry. These include phthaloyl, benzyloxycarbonyl, t-butyloxycarbonyl, succinyl and chloroacetyl derivatives. Broadly speaking, whatever else they may be useful for, they have not caught on for chromatographic applications and will therefore not be considered. Other acylating reagents have been used in chromatography for rendering the parent compounds either 'visible', by producing coloured or UV-absorbing derivatives such as the 'dabsyl' derivatives described for the analysis of amino acids (see Chapter 8), or fluorescent, by producing fluorescent derivatives such as the 'dansyl' derivatives described in Chapter 9. Details of their preparation will be found in those chapters. A few acyl derivatives that *have* been used in chromatography are included here for completeness.

4.7.2 Pivaloyl derivatives

Pivaloyl derivatives are stable and easy to made, and the reagents, either pivalic anhydride or pivaloyl chloride, are also stable. The pivaloyl group, with its screen of methyl groups analogous to the trimethylsilyl group, yields derivatives in which the polarity of amino groups is effectively masked, and which therefore show good chromatographic properties.

The base, or amino acid mixtures or their hydrochlorides (ca. 1 mg), is dissolved in a small amount of methanol or THF (10–20 µl) and is heated with pivalic anhydride (200 µl) and triethylamine (10–15) at 70–110 °C for 30 minutes. Alternatively, a mixture of pivalic anhydride, triethylamine and methanol (20:1:1, 250 µl) is used. The reaction mixture, if cloudy, is centrifuged and the supernatant is evaporated to dryness in a stream of nitrogen or *in vacuo*. The product may be taken up in a suitable solvent (hexane, benzene, chloroform etc.) for analysis [135–137].

$$R-NH_2 + \begin{matrix}(CH_3)_3C-CO \\ (CH_3)_3C-CO\end{matrix} \Big\rangle O \longrightarrow$$

$$R-NH-CO-C(CH_3)_3 + (CH_3)_3C-COOH \quad (15)$$

If pivaloyl chloride is used, the amine or its hydrochloride (up to 5 mg) is suspended in chloroform (200 µl) and 'light' magnesium carbonate (50 mg) is added. To the stirred suspension is added a 110% molar excess of pivaloyl chloride via a microlitre syringe, and the reaction is allowed to proceed at room temperature for 30 minutes. After centrifugation, the supernatant is evaporated to dryness in a stream of nitrogen or *in vacuo*, and the residue is dissolved in a suitable solvent for analysis by TLC or GC [138].

$$R-NH_2 + (CH_3)_3C-CO-Cl \longrightarrow$$
$$R-NH-CO-C(CH_3)_3 + HCl \quad (16)$$

4.7.3 Trichloroacetylation

The trichloroacetyl derivatives of amines are among the most electron-capturing derivatives [139]. The amine or amine extract (up to 1 mg) is treated with trichloroacetyl cloride (50 µl) in hexane (1 ml) at room temperature for 20 minutes. The hexane solution is washed with half its volume of 1 M NaOH and centrifuged. The supernatant may be concentrated or injected directly [140]. A similar procedure has been developed for amantadine, which is initially extracted from plasma or urine (1 ml). The extract is dissolved in a suitable solution containing an toluence standard, and is treated with 10 µl of freshly made up 2% trichloroacetyl chloride in toluene for 30 minutes at 70 °C. Excess reagent is removed by shaking with 1 ml of 1 M NaOH for 5 minutes, and the organic layer is dried and analysed directly [141].

4.7.4 Benzoyl and benzenesulphonyl derivatives

Benzoyl and benzenesulphonyl derivatives have not been widely applied for chromatographic analyses, but because they can be detected in the ultraviolet they have found some use in HPLC. They are also used to make derivatives from volatile compounds as a means avoiding evaporative losses. They are readily made under Schotten–Baumann conditions. Thus, an amine solution (3 ml) is made alkaline with an equal volume of 7.5 M NaOH, and 50 µl of benzoyl choride is vigorously shaken with this solution until consumed. The product is extracted twice with 2 ml of diisopropyl ether, and the ether extracts are blown dry with nitrogen before being taken up in the ether for analysis [142]. A similar procedure was used for amines by Terashi *et al.* but using benzenesulphonyl chloride [143].

Pyridine has also been used as the basic catalyst and chloride acceptor for benzoylation reactions. In one procedure, amines, volatile alcohols or thiols isolated by benzene extraction, are benzoylated with pyridine (1 ml) and benzoyl chloride (0.5 ml) by shaking intermittently at room temperature for several hours. The pyridine phase is extracted with 2 M HCl and the excess benzoyl chloride is hydrolysed with water for 12 hours. After shaking with 2 M sodium carbonate to remove benzoic acid, the benzene solution is dried and concentrated for analysis [144]. Aminoglycoside antibiotics are derivatized to the benzoyl derivatives in a similar reaction using 90 µl of pyridine and 10 µl of benzoyl chloride at 80 °C for 30 minutes. The pyridine is evaporated in a stream of nitrogen and excess benzoyl chloride is converted to methyl benzoate with methanol, again at 80°C, for 10 minutes. The product is cleaned up for analysis by a rather involved solvent extraction procedure [145].

4.7.5 Preparation of 3,5-dinitrobenzoates

Alcohols or amines, or their concentrated extracts (up to 10 mg), are dissolved in benzene (100 µl) and heated on a boiling water bath with a solution of 3,5-dinitrobenzoyl chloride and pyridine in benzene (1:2:8, 300 µl) for 30 minutes. After cooling, the solution is washed twice with a double volume of 2 M KOH, water, 5 M HCl and water again. The benzene solution is dried over sodium sulphate and spotted for TLC [146].

In an alternative method for GC, the 3,5-dinitrobenzoyl chloride is used at 10% in benzene without pyridine. For GC with a FID the washings described for the previous method may be omitted, but for ECD the excess reagent must first be removed [147].

4.7.6 Derivatization with chloroformates

Chloroformates have been widely used to derivatize amines because the derivatives are easily made, even in buffered aqueous solution, and have useful chromatographic properties, especially volatility. The derivatization reactions were studied by Ahnfeldt and Hartvig, who found the trichloroethyl chloroformate an order of magnitude more reactive than alkyl chloroformates. The reactions went faster with the secondary amine studied [148]. Methyl chloroformate was used to derivatize adrenaline and noradrenaline in plasma (1 ml) buffered at pH 7.4. Methyl chloroformate (20 µl) was added and the mixture was stirred for 30 seconds. After 5 minutes the product was extracted into methylene chloride, the extract was evaporated to dryness and the residue was silylated for analysis by gas chromatography [149]. A very similar procedure was developed by Japanese workers around the same time, using ethyl chloroformate to derivatize both amino and hydroxyl groups [150].

Tertiary amines may be chromatographed without derivatization, but their polar nature tends to be troublesome (streaking in TLC, tailing in GC). Treatment with a chloroformate can displace the smallest one of the groups attached to the nitrogen, particularly if it is a methyl group, to form a carbamate, as can be seen from reaction (17).

$$R_1R_2N-CH_3 + C_6F_5-CH_2-O-CO-Cl \longrightarrow R_1R_2N-CO-O-CH_2-C_6F_5 \quad (17)$$

The amine (up to 1 mg) is dissolved in heptane (200 µl) and treated with pentafluorobenzyl chloroformate (pentafluorobenzyloxycarbonyl chloride, 50 µl) and powdered anhydrous sodium carbonate (ca. 10 mg) in a screw-capped reaction vial a PTFE linear. The reaction mixture is heated at 100 °C for 30 minutes, cooled and shaken with 1 M NaOH (1 ml). For GC with a FID the upper layer may be used directly. For GC with an ECD the supernatant from the reaction is concentrated in vacuo, and the residue is dissolved in heptane (2 ml), washed with 1 M NaOH and water (2 × 1 ml) and dried over anhydrous sodium sulphate [151, 152]. A similar procedure used trichloroethyl chloroformate [153].

5. Conclusions

Acylation is one of the most important derivatization procedures because so many compounds of interest have amino and/or hydroxyl groups. The polar nature of the amino group is most effectively masked and thus rendered accessible to chromatographic analysis by acylation, which has therefore remained one of the favoured approaches. In the period since the previous version of this chapter there has been a great extension of the most popular methods, essentially those described above, to fresh applications, but less development of new acylation methods, and this seems likely to be the pattern for some time to come.

6. References

[1] D. P. N. Satchell, Rev. Chem. Soc., **17**, 160 (1963).
[2] C. F. Poole, Lab. Pract., **25**, 309 (1976).
[3] C. F. Poole, J. Chromatogr., **118**, 280 (1976).
[4] M. Pailer and W. J. Hübsch, Monatsh. Chem., **97**, 1541 (1966).
[5] J. P. Thenot and E. C. Horning, Anal. Lett., **5**, 519 (1972).
[6] M. Butler and A. Darbre, J. Chromatogr., **101**, 51 (1974).
[7] R. F. Adams, F. L. Vandemark and G. J. Schmidt, J. Chromatogr. Sci., **15**, 63 (1977).
[8] I. Horman and F. J. Hesford, Biomed. Mass Spectrom., **1**, 115 (1974).
[9] G. Ertinghausen, C. W. Gehrke and W. A. Aue, Sep. Sci., **2**, 681 (1967).
[10] W. Vogt, K. Jacob, A. B. Ohnesorge and G. Schwertfeger, J. Chromatogr., **199**, 191 (1980).
[11] R. M. Smith, A. A. Ghani, D. G. Haverty, G. S. Bament, A. Y. Chamsi and A. G. Fogg, J. Chromatogr., **455**, 349 (1988).
[12] J. E. Arnold and H. M. Fales, J. Gas Chromatogr., **3**, 131 (1965).
[13] K. Blau, I. M. Claxton, G. Ismahan and M. Sandler, J. Chromatogr., **163**, 135 (1979).
[14] D. G. Saunders and L. E. Vanatta, Anal. Chem., **46**, 1319 (1974).
[15] C. D. Pfaffenberger, J. Safranek, M. G. Horning and E. C. Horning, Anal. Biochem., **63**, 501 (1975).

[16] I. Fleming and J. B. Mason, *J. Chem. Soc.*, 2509 (1969).
[17] J. S. Sawardeker, J. H. Sloneker and A. Jeanes, *Anal. Chem.*, **39**, 121 (1967).
[18] L. F. Prescott, *J. Pharm. Pharmacol.*, **28**, 807 (1971).
[19] G. Österlöf and A. Nyhem, *J. Chromatogr.*, **183**, 487 (1980).
[20] H. Maurer and K. Pfleger, *J. Chromatogr.*, **428**, 43 (1988); **430**, 31 (1988).
[21] J. R. Coulter and C. S. Hann, *J. Chromatogr.*, **36**, 42 (1968).
[22] R. F. McGregor, G. M. Brittin and M. S. Sharon, *Clin. Chim. Acta.*, **48**, 65 (1973).
[23] R. F. Adams, *J. Chromatogr.*, **95**, 189 (1974).
[24] R. Wachowiak and K. A. Connors, *Anal. Chem.*, **51**, 27 (1979).
[25] M. Hagopian, R. I. Dorfman and M. Gut, *Anal. Biochem.*, **2**, 387 (1961).
[26] R. Laverty and D. F. Sharma, *Br. J. Pharmacol.*, **24**, 538 (1965).
[27] E. Röder and J. Merzhäuser, *Anal. Chem.*, **34**, 272 (1974).
[28] R. T. Coutts, E. E. Hargesheimer and F. M. Pasutto, *J. Chromatogr.*, **195**, 105 (1980).
[29] M. Balikova and J. Kohlicek, *J. Chromatogr.*, **497**, 159 (1989).
[30] S. Hara and Y. Matsushima, *J. Biochem. (Tokyo)*, **71**, 907 (1972).
[31] J. B. Knaak, J. M. Eldridge and L. J. Sullivan, *J. Agric. Food Chem.*, **15**, 605 (1967).
[32] G. D. Paulson, R. G. Zaylskie and M. M. Dockter, *Anal. Chem.*, **45**, 21 (1973).
[33] D. P. Schwartz, *Anal. Biochem.*, **71**, 24 (1976).
[34] T. S. Stewart and C. E. Ballou, *Biochemistry*, **7**, 1855 (1968).
[35] G. D. Paulson and C. E. Portnoy, *J. Agric. Food Chem.*, **18**, 180 (1970).
[36] D. P. Rose and P. A. Toseland, *Clin. Chim. Acta*, **17**, 235 (1967).
[37] D. P. Schwartz and C. Allen, *J. Chromatogr.*, **208**, 55 (1981).
[38] K. Blau, *Chem. Ind. (London)*, 33 (1963).
[39] L. Farkas, J. Morgós, P. Sallay and I. Rusznák, *J. Chromatogr.*, **168**, 212 (1979).
[40] M. Bergmann and F. Stern, *Ber. Dtsch. Chem. Ges.*, **63B**, 437 (1930).
[41] G. Quadbeck, *Angew. Chem.*, **68**, 361 (1956).
[42] P. S. S. Dawson, *Biochim. Biophys. Acta*, **111**, 51 (1965).
[43] R. W. H. Edwards, H. L. J. Makin and T. M. Barratt, *J. Endocrinol.*, **30**, 181 (1964).
[44] H. L. J. Makin, *J. Endocrinol.*, **47**, 55 (1970).
[45] D. J. H. Trafford and H. L. J. Makin, *Clin. Chim. Acta*, **40**, 421 (1972).
[46] K. Heyns and H.-F. Grützmacher, *Z. Naturforsch.*, **16B**, 293 (1961).
[47] J. O. Thomas, *Tetrahedron Lett.*, 335 (1967).
[48] G. Losse, A. Losse and J. Stöck, *Z. Naturforsch.*, **17B**, 785 (1962).
[49] E. J. Bourne, C. E. M. Tatlow and J. C. Tatlow, *J. Chem. Soc.*, 1367 (1950).
[50] F. Weygand and R. Geiger, *Chem. Ber.*, **64**, 136 (1956).
[51] F. Weygand and E. Csendes, *Angew. Chem.*, **64**, 136 (1952).
[52] D. D. Clarke, S. Wilk and S. E. Gitlow, *J. Gas Chromatogr.*, **4**, 310 (1966).
[53] G. Skarping, L. Renman and B. E. F. Smith, *J. Chromatogr.*, **267**, 315 (1983).
[54] H.-B. Lee, *J. Chromatogr.*, **457**, 267 (1988).
[55] A. Darbre and K. Blau, *J. Chromatogr.*, **17**, 31 (1963).
[56] A. Darbre and A. Islam, *Biochem. J.*, **106**, 923 (1968).
[57] R. T. Coutts, R. Dawe, G. R. Jones, S.-F. Liu and K. K. Midha, *J. Chromatogr.*, **190**, 53 (1980).
[58] V. Rovei, M. Sanjuan and P. D. Hrdina, *J. Chromatogr.*, **182**, 349 (1980).
[59] Y. Okano, T. Kadota, J. Nagata, A. Matsuda, S. Iijima, T. Takahama and T. Miyata, *J. Chromatogr.*, **310**, 251 (1984).
[60] A. C. Tas, J. Odink, M. C. Ten Noever De Brauw, J. Schrijver and R. G. J. Jonk, *J. Chromatogr.*, **310**, 243 (1984).
[61] D. Roach and C. W. Gehrke, *J. Chromatogr.*, **44**, 269 (1969).
[62] F. E. Kaiser, C. W. Gehrke, R. W. Zumwalt and K. C. Kuo, *J. Chromatogr.*, **94**, 113 (1974).
[63] C. W. Moss, M. A. Lambert and F. J. Diaz, *J. Chromatogr.*, **60**, 134 (1971).
[64] J. F. March. *Anal. Biochem.*, **69**, 420 (1975).
[65] G. Gamerith, *J. Chromatogr.*, **256**, 267 (1983).
[66] G. Gamerith, *J. Chromatogr.*, **256**, 326 (1983).
[67] W. A. Koenig, L. C. Smith, P. F. Crain and J. A. M. McCloskey, *Biochemistry*, **10**, 3968 (1971).
[68] I. L. Martin and G. B. Ansell, *Biochem. Pharmacol.*, **22**, 521 (1973).
[69] H. C. Curtius, M. Wolfensberger, U. Redweik, W. Leimbacher, R. A. Maibach and W. Isler, *J. Chromatogr.*, **112**, 523 (1975).
[70] J. Segura, F. Artigas, E. Martinez and E. Gelpi, *Biomed. Mass Spectrom.* **3**, 91 (1976).
[71] H. C. Curtius, H. Farmer and F. Rey, *J. Chromatogr.*, **199**, 171 (1980).
[72] T. M. Trainer, P. Vouros, P. Lampen, J. L. Neumayer, R. Baldessarini and N. S. Kula, *J. Chromatogr.*, **457**, 257 (1988).
[73] M. Ervik, K. Kylberg-Hansse and P.-O. Lagerström, *J. Chromatogr.*, **182**, 341 (1980).
[74] A. Marzo and E. Treffer, *J. Chromatogr.*, **345**, 390 (1985).
[75] B. Sjöquist and E. Magnuson, *J. Chromatogr.*, **183**, 17 (1980).
[76] P. Decker and H. Schweer, *J. Chromatogr.*, **236**, 369 (1982).
[77] E. Bailey and E. J. Barron, *J. Chromatogr.*, **183**, 25 (1980).
[78] F. T. Delbecke, M. Debackere, N. Desmet and F. Maertens, *J. Chromatogr.*, **426**, 194 (1988).
[79] T. Walle and H. Ehrsson, *Acta Pharm. Suec.*, **7**, 389 (1970).
[80] D. A. Garteiz and T. Walle, *J. Pharm. Sci.*, **61**, 1728 (1972).
[81] A. F. Cockerill, D. N. B. Malle, D. J. Osborne and D. M. Price, *J. Chromatogr.*, **114**, 151 (1975).
[82] C. E. Kientz and A. Verweij, *J. Chromatogr.*, **355**, 229 (1986).
[83] C. E. Kientz and A. Verweij, *J. Chromatogr.*, **355**, 253 (1986).
[84] Z. Tamura and T. Imanari, *Chem. Pharm. Bull.*, **15**, 246 (1967).

[85] W. A. König, H. Bauer, W. Voelter and E. Bayer, *Chem. Ber.*, **106**, 1905 (1973).
[86] M. Vilkas, I.-J. Hui, G. Boussac and M.-C. Bonnard, *Tetrahedron Lett.*, (14) 1441 (1966).
[87] T. Ueno, N. Kurihara and M. Nakajima, *Agric. Biol. Chem.*, **31**, 1189 (1967).
[88] A. Sioufi, F. Pommier and J. P. Dubois, *J. Chromatogr.*, **428**, 71 (1988).
[89] L. A. Dehennin and R. Scholler, *Steroids*, **13**, 739 (1969).
[90] W. P. Collins, J. M. Sisterson, E. N. Koullapis, M. D. Mansfield and I. F. Somerville, *J. Chromatogr.*, **37**, 33 (1968).
[91] L. A. Dehennin, A. Reifstock and R. Scholler, *J. Chromatogr. Sci.*, **10**, 224 (1972).
[92] H. Ehrsson, T. Walle and H. Brötell, *Acta Pharm. Suec.*, **8**, 319 (1971).
[93] T. Imanari and Z. Tamura, *Chem. Pharm. Bull.*, **15**, 896 (1967).
[94] C. Braestrup, *J. Neurochem.*, **20**, 519 (1973).
[95] L. Fellows, P. Riederer and M. Sandler, *Clin. Chim. Acta*, **59**, 255 (1971).
[96] E. Änggard and G. Sedvall, *Anal. Chem.*, **41**, 1250 (1969).
[97] B. L. Goodwin, C. R.J. Ruthven and M. Sandler, *Clin. Chim. Acta*, **62**, 439 (1975).
[98] S. W. Dziedzic, L. M. Bertani, D. D. Clarke and S. E. Gitlow, *Anal. Biochem.*, **47**, 592 (1972).
[99] R. V. Smith and S. L. Tsai, *J. Chromatogr.*, **61**, 29 (1971).
[100] C.-G. Fri, F.-A. Wiesel and G. Sedvall, *Life Sci.*, **14**, 2469 (1974).
[101] E. Watson and S. Wilk, *Anal. Biochem.*, **59**, 441 (1974).
[102] J. B. Brooks, C. C. Alley and J. A. Liddle, *Anal. Chem.*, **46**, 1930 (1974).
[103] S. Wilk and M. Orlowski, *FEBS Lett.*, **33**, 157 (1973).
[104] S. Wilk and M. Orlowski, *Anal. Biochem.*, **69**, 100 (1975).
[105] A. Lapin and M. Karobath, *J. Chromatogr.*, **193**, 95 (1980).
[106] M. Wolfensberger, U. Redweik and H. C. Curtius, *J. Chromatogr.*, **172**, 471 (1979).
[107] E. Peralta, H.-Y. Tang and E. Costa, *J. Chromatogr.*, **190**, 43 (1980).
[108] H. Ehrsson and H. Brötell, *Acta Pharm. Suec.*, **8**, 591 (1971).
[109] H. Ehrsson and B. Mellström, *Acta Pharm. Suec.*, **9**, 107 (1972).
[110] P. A. Metz, F. L. Morse and T. W. Theyson, *L. Chromatogr.*, **479**, 107 (1989).
[111] H. Ehrsson, T. Walle and S. Wikström, *J. Chromatogr.*, **101**, 206 (1974).
[112] S. Murray and T. A. Baillie, *Biomed. Mass Spectrom.*, **6**, 82, (1979).
[113] J. Vessman, A. M. Moss, M. G. Horning and E. C. Horning, *Anal. Lett.*, **2**, 81 (1969).
[114] G. Zweig and J. Sherma (Editors), *Handbook of Chromatography*, Vol. 2, CRC Press, Cleveland (1972), pp. 220–221.
[115] F. Benington, S. T. Christian and R. D. Morin, *J. Chromatogr.*, **106**, 435 (1975).
[116] S. T. Christian, F. Benington, R. D. Morin and L. Corbett, *Biochem. Med.*, **14**, 191 (1975).

[117] P. H. Degen, J. R. Do Amaral and J. D. Barchas, *Anal. Biochem.*, **45**, 634 (1972).
[118] F. Karoum, J. C. Gillin, R. J. Wyatt and E. Costa, *Biomed. Mass Spectrom.*, **2**, 183 (1975).
[119] L. M. Ross-Lee, M. J. Eadie, F. Bochner, W. D. Hooper and J. H. Tyrer, *J. Chromatogr.*, **183**, 175 (1980).
[120] P. R. Robinson, M. D. Jones and J. Maddock, *J. Chromatogr.*, **432**, 153 (1988).
[121] T. Gorski, T. J. Goehl, C. W. Jameson, B. Collins, J. Bursey and R. Moseman, *J. Chromatogr.*, **509**, 383 (1990).
[122] M. G. Horning, A. M. Moss, E. A. Boucher and E. C. Horning, *Anal. Lett.*, **1**, 311 (1968).
[123] H. Miyazaki, M. Ishibashi, C. Mori and N. Ikekawa, *Anal. Chem.*, **45**, 1164 (1973).
[124] M. Dönike, *J. Chromatogr.*, **78**, 273 (1973).
[125] M. Dönike, *J. Chromatogr.*, **103**, 91 (1975).
[126] G. Schwedt and H. H. Bussemas, *J. Chromatogr.*, **106**, 440 (1975).
[127] A. S. Christopherson, E. Hovland and K. E. Rasmussen, *J. Chromatogr.*, **234**, 107 (1982).
[128] G. R. Wilkinson, *Anal. Lett.*, **3**, 289 (1970).
[129] G. B. Baker, T. S. Rao and R. T. Coutts, *J. Chromatogr.*, **381**, 211 (1986).
[130] K. Yamada, O. Asano, T. Yoshimura and K. Katayama, *J. Chromatogr.*, **433**., 243 (1988).
[131] F. A. Rubio, N. Chroma and E. K. Fukuda, *J. Chromatogr.*, **497**, 147 (1989).
[132] R. M. Black and R. W. Read, *J. Chromatogr.*, **449**, 261 (1988).
[133] N. K. McCallum and R. J. Armstrong, *J. Chromatogr.*, **78**, 303 (1973).
[134] D. V. Crabtree, A. J. Adler and G. J. Handelman, *J. Chromatogr.*, **466**, 251 (1989).
[135] J. E. Stouffer, *J. Chromatogr. Sci.*, **7**, 124 (1969).
[136] E. M. Volpert, N. Kundu and J. B. Dawidzik, *J. Chromatogr.*, **50**, 507 (1970).
[137] N. N. Nihei, M. C. Gershengorn, T. Mitsuma, L. R. Stringham, A. Cordy, B. Kuchmy and C. S. Hollander, *Anal. Biochem.*, **43**, 433 (1971).
[138] J. C. Cavadore, G. Nota, G. Prota and A. Previero. *Anal. Biochem.*, **60**, 608 (1974).
[139] E. Änggard and A. Hankey, *Acta Chem. Scand.*, **23**, 3110 (1969).
[140] J. S. Noonan, P. W. Murdick and R. S. Ray, *J. Pharmacol. Exp. Ther.*, **168**, 205 (1969).
[141] A. Sioufi and F. Pommier, *J. Chromatogr.*, **183**, 33 (1980).
[142] G. A. R. Decroix, J. G. Gobert and R. De Deurwaerder, *Anal. Biochem.*, **25**, 523 (1968).
[143] A. Terashi, Y. Hanada, A. Kido and R. Shinohara, *J. Chromatogr.*, **503**, 369 (1990).
[144] L. Gasco and R. Barrera, *Anal. Chim. Acta*, **61**, 253 (1972).
[145] T. Harada, M. Iwamori, Y. Nagai and Y. Nomura, *J. Chromatogr.*, **337**, 187 (1985).
[146] I. M. Hais and K. Macek, *Paper Chromatography*, Academic Press, New York (1963) pp. 832–833.
[147] W. G. Galetto, R. E. Kempner and A. D. Webb, *Anal. Chem.*, **38**, 34 (1966).

[148] N. O. Ahnfelt and P. Hartvig, *Acta Pharm. Suec.*, **17**, 307 (1980).
[149] O. Gyllenhaal, L. Johansson and J. Vessmann, *J. Chromatogr.*, **190**, 347 (1980).
[150] S. Yamamoto, K. Kakuno, S. Okahara, H. Kataoka and M. Makita, *J. Chromatogr.*, **194**, 399 (1980).
[151] P. Hartvig and J. Vessman, *Anal. Lett.*, **157**, 223 (1974).
[152] P. Hartvig and J. Vessman, *Acta Pharm. Suec.*, **11**, 115 (1974).
[153] P. Hartvig, K.-E. Karlsson, L. Johansson and C. Lindberg, *J. Chromatogr.*, **121**, 235 (1976).

4

Advances in Silylation

R. P. Evershed
Department of Biochemistry, University of Liverpool, P.O. Box 147, Liverpool

1. INTRODUCTION ... 52
2. GENERAL FEATURES OF SILYLATION REACTIONS ... 52
   2.1 Mechanistic considerations ... 52
   2.2 Practical considerations ... 53
3. GAS CHROMATOGRAPHY OF SILYL DERIVATIVES ... 54
4. PREPARATION OF SILYL DERIVATIVES AND APPLICATIONS TO THE GAS CHROMATOGRAPHY OF ORGANIC COMPOUNDS ... 55
   4.1 Trimethylsilyl (TMS) derivatives ... 55
      4.1.1 N,O-Bis(trimethylsilyl)trifluoroacetamide (BSTFA) ... 57
      4.1.2 N,O-Bis(trimethylsilyl)acetamide (BSA) ... 58
      4.1.3 N-Methyl-N-trimethylsilyltrifluoroacetamide (MSTFA) ... 58
      4.1.4 N-Trimethylsilyldiethylamine (TMSDEA) ... 59
      4.1.5 Trimethylsilylimidazole (TMSIM) ... 59
      4.1.6 Trimethylchlorosilane (TMCS) ... 59
      4.1.7 Hexamethyldisilazane (HMDS) ... 59
      4.1.8 Mixtures of silyl donors ... 60
   4.2 Applications of trimethylsilylation to the analysis of organic compounds ... 60
      4.2.1 Short-chain organic acids and related compounds ... 60
      4.2.2 Prostaglandins and other eicosanoids ... 61
      4.2.3 Steroids, bile acids and related compounds ... 63
      4.2.4 Fatty acids and related acyl lipids ... 68
      4.2.5 Carbohydrates and polyols ... 70
      4.2.6 Phenols ... 72
      4.2.7 Nitrogen-containing compounds ... 73
         4.2.7.1 Nucleic acid constituents ... 73
         4.2.7.2 Amines ... 74
         4.2.7.3 Amino acids ... 74
         4.2.7.4 Porphyrins ... 74
   4.3 Silyl derivatives other than TMS ... 75
      4.3.1 t-Butyldimethylsilyl (TBDMS) derivatives ... 75
      4.3.2 t-Butyldimethylchlorosilane (TBDMCS) ... 75
      4.3.3 N-t-Butyldimethylsilylimidazole (TBDMSIM) ... 76
      4.3.4 N-t-Butyldimethylsilyl-N-methyltrifluoroacetamide (MTBSTFA) ... 76
   4.4 Applications of t-butyldimethylsilylation to the GC analysis of organic compounds ... 76
      4.4.1 Short-chain organic acids and related compounds ... 76
      4.4.2 Prostaglandins and other eicosanoids ... 78
      4.4.3 Steroids, bile acids and related compounds ... 79
      4.4.4 Fatty acids and related acyl lipids ... 81
      4.4.5 Nitrogen-containing compounds ... 82
         4.4.5.1 Nucleic acid constituents ... 82
         4.4.5.2 Amino acids ... 83
         4.4.5.3 Alkylporphyrins ... 85
   4.5 Other alkylsilyl and aryldimethylsilyl derivatives ... 85
      4.5.1 Isopropyldimethylsilyl (DMIPS) ... 88
      4.5.2 Allyldimethylsilyl (ADMS) ... 88
      4.5.3 Cyclic silyl derivatives ... 90
      4.5.4 Alkoxyalkyl (or aryl)silyl derivatives ... 92

Handbook of Derivatives for Chromatography
Edited by K. Blau and J. M. Halket © 1993 John Wiley & Sons Ltd

4.6 Halogenated silyl derivatives ... 93
    4.6.1 Pentafluorophenyldimethylsilyl (flophemesyl) ethers ... 93
    4.6.2 Halocarbondimethylsilyl ethers ... 95
4.7 Nitrogen-containing silyl derivatives ... 96
    4.7.1 Cyanoethyldimethylsilyl (CEDMS) ... 96
    4.7.2 Picolinyldimethylsilyl (PICSI) ethers ... 97
5. CONCLUSIONS ... 99
6. REFERENCES ... 100

1. Introduction

Silylation is the most versatile technique currently available for enhancing GC performance by blocking protic sites, thereby reducing dipole–dipole interactions and increasing volatility. The introduction of a silyl group(s) can also serve to enhance mass spectrometric properties, by producing either (i) more favourable diagnostic fragmentation patterns of use in structure investigations, or (ii) characteristic ions of use in trace analyses employing selected ion monitoring and related techniques. In the fifteen years or so that have elapsed since Colin Poole's chapter [1] in the first edition of this volume, considerable advances have occurred in all the fields of analytical chemistry associated with the use of silylation of organic compounds. Notable developments include the routine use of high resolution fused-silica capillary columns coated with immobilized stationary phases, and the wider availability of GC–MS instruments, particularly the relatively low cost benchtop quadrupole and ion trap system [2]. Chapter 14 provides some background on the MS detector and its applications. The GC–MS content of the present chapter reflects the increasing importance of this combined technique.

It is the intention of this contribution to build upon the chapter on this subject presented in the earlier volume by providing an account of developments that have occurred since the mid to late 1970's. This will involve placing a greater emphasis upon the preparation and use of silyl derivatives other than trimethylsilyl (TMS), which preoccupied the greater part of the earlier edition. TMS derivatives will not be neglected as their convenience of preparation from readily available derivatizing agents, and their generally desirable GC and MS properties, ensure that they will continue to find wide application. This contribution is not intended to be a comprehensive review of the silylation literature. The aim is to provide a reference work for use by the practising analyst. A certain amount of background information, included in the earlier edition, is repeated for the sake of completeness. It would be inappropriate to proceed without drawing the reader's attention to Daniel Knapp's excellent *Handbook of Analytical Derivatization Ractions* [3] published in 1979, which contains an impressive compound by compound listing of derivatization procedures, including numerous silylation methods. Alan Pierce's book *Silylation of Organic Compounds* [4], published in the same year, also contains much useful information and numerous references relating to chromatographic analysis. Other works containing descriptions of silylation include articles describing derivatives for the analysis of pharmaceuticals by GC [5], a general description of commonly used silylation reagents [6] and a review of trialkylsilyl derivatives other than TMS, covering the literature up to 1979 [7].

2. General features of silylation reactions

2.1 Mechanistic considerations

The usefulness of silylation derives from the ease of reaction with many of the common protic functional groups present in naturally occurring and synthetic organic compounds. Silyl derivatives are formed by the displacement of the active proton in OH, NH and SH groups [8] (see Figure 1). The general reaction (1) for the formation of such trialkylsilyl derivatives is:

$$R_3Si-X + R'-H \longrightarrow R_3Si-R' + HX \quad (1)$$

The most common situation is that in which R = methyl, from which TMS derivatives result. The range of other possible silyl derivatives is discussed in detail below. The reaction is viewed as a nucleophilic attack by the more electronegative heteroatom upon the silicon atom of the silyl donor, producing a bimolecular transition state. The leaving group (X) must possess low basicity, the ability to stabilize a negative charge in the transition state (2), and little or no tendency for $\pi(p-d)$ back–bonding between itself and the silicon atom.

$$H-Y + \underset{/}{\overset{\backslash}{-}}Si-X \longrightarrow \begin{bmatrix} \overset{\delta+}{Y} \cdots \overset{\backslash}{\underset{/}{Si}} \cdots \overset{\delta-}{X} \\ | \\ H \end{bmatrix} \longrightarrow Y-\underset{\backslash}{\overset{/}{Si}}- + HX \quad (2)$$

The ideal leaving group (X) must be such that it is readily lost from the transition state during reaction but possesses sufficient chemical stability in combination with the alkylsilyl group to allow long term storage of the derivatizing agent for use as required. As the

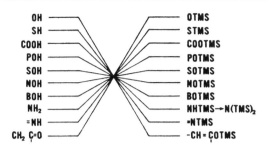

Figure 1. Functional groups which form TMS derivatives (reproduced from reference 8 with permission)

formation of the transition state is reversible, the derivatization will only proceed to completion if the basicity of the leaving group X exceeds that of Y. A close correlation has been observed between the pK_b of Y and the silyl-donor strength of N-silyl derivatives [9, 10]. The ease of derivatization of various functional groups for a given silylating agent follows the order

alcohol > phenol > carboxylic acid > amine > amide

Within this sequence reactivity towards a particular silylating reagent will also be influenced by steric hindrance; hence the ease of reactivity for alcohols follows the order:

primary > secondary > tertiary

and for amines:

primary > secondary

2.2 Practical considerations

It should be assumed that all silylation reagents and derivatives are sensitive to the hydrolytic effects of moisture. Consequently, reactions should always be carried out in sealed reaction vials, ideally with Teflon-lined caps. Many of the commercially available glass reaction vials are suitable, provided they can withstand the heating required to achieve complete silylation. Disposable vials are preferred for trace analysis work in order to avoid problems of cross-contamination. TMS derivatives are more susceptible to hydrolysis than those derivatives containing more sterically crowded alkyl substituents of the silicon atom. The direct trimethylsilylation of hydroxy compounds present in aqueous solutions has been performed by employing a large excess of derivatizing agent [11]. The greater chemical stability of the sterically crowded derivatives is of value in enhancing sample recoveries from the work-up following derivatization. For example, the hydrolytic stability of such derivatives enables water washing steps to be used in sample work-up [12, 13]. In addition, derivatives containing sterically crowded silyl groups are less susceptible to decomposition during adsorption chromatography and reversed phase HPLC prior to GC and GC–MS analysis. Examples exist of the use of a phenyldi-t-butylsilyl derivative to enhance HPLC behaviour [14, 15].

Solvent may be required, both in the reaction medium for derivatization and subsequently for sample dilution prior to analysis. As the silylation of active hydrogens occurs readily, solvents containing such moieties (see Figure 1) must be avoided. In many instances the silylation reagent itself can be an adequate solvent; however, this is not always the case. Pyridine is especially suitable for use as a solvent in view of its catalytic properties (see below). Other commonly used solvents include dimethylformamide (DMF), dimethyl sulphoxide (DMSO), tetrahydrofuran (THF) and acetonitrile. When considering the addition of solvent to the derivatization mixture it should be borne in mind that solubilization of the analyte(s) prior to derivatization is not absolutely essential, as this may occur as silylation proceeds. Where substantial dilution is required following derivatization, the use of valuable derivatizing agent as solvent would be wasteful. The normal practice is to dilute the sample with an inert solvent, e.g. hexane. Heating is often used to achieve effective derivatization. When considering heating, due consideration must be given to the thermal stability of the analytes and silylation reagents. The suitability of reaction vials also warrants careful consideration where high temperatures are employed. Reinforced reaction vials suitable for carrying out reactions at temperatures in excess of 100 °C are commercially available. Indications of temperatures required for specific derivatizations are given in the applications sections below.

Less conventional approaches to silylation include the use of vapour phase derivatization methods. For example, short-chain alcohols (methanol to propanol) present in air collected in stainless steel canisters (5.45 l) were silylated in the vapour phase prior to GC analysis [16]. TMSIM and BSA were the most effective of a range of commercial derivatizing agents that were tested in this procedure. Short-chain carboxylic acids, trapped from air in cartridges containing glass beads coated with sodium hydroxide, were silylated *in situ* with TMSIM (60 °C, 20 min) prior to GC analysis [17]. The so-called 'on-column' derivatization technique offers convenience in some instances. For example, morphine has been determined quantitatively by packed column GC as its bis-TMS derivative, prepared by co-injection (flash heater injection port temperature = 275 °C) of the sample solutions with neat TMSIM [18]. Likewise, enol-TMS ethers of carbonyl compounds (ketones and aldehydes) [19] and O-TMS ethers of alkoxy alcohols [20] have been prepared by injecting BSTFA containing 1% TMCS immediately (4–5 s) after introduction of the sample. A novel double injection derivatization technique (3) has been applied to the capillary GC and GC–MS analysis of phenolalkylamines [21]. The phenolalkylamines are first

reactions between underivatized substrates and excess silyl donor in hot injection ports must be carefully considered in quantitative analyses, particularly in reactions involving sterically crowded alkylsilyl groups [24]. Where problems of this nature are encountered, on-column or programmed temperature vaporizing injectors may provide a practical solution.

O-trimetylsilylated in the splitless injection port (temperature set to 275 °C) by co-injection of the sample with N-methyl-N-trimethylsilyltrifluoroacetamide (MSTFA), then N-acylated 'on-column' (oven temperature at 130 °C) with N-methylbis(trifluoroacetamide) introduced in a second injection.

MSTFA was also the derivatizing agent of choice for the direct derivatization of flutroline in plasma extracts on the tip of a moving needle injector prior to GC–MS selected ion monitoring (SIM) assay [22]. TMS derivatives of formaldehyde condensates, formed in a gas phase reaction with BSTFA, were used for the separation and quantification of methylal (dimethoxymethane), methylene glycol and a series of poly(oxymethylene) glycol monomethyl ethers by capillary GC–MS using ammonia chemical ionization [23]. The possibility of accelerated

3. Gas chromatography of silyl derivatives

Method development guidelines for chemical derivatization for gas chromatography have recently been summarized by McMahon [25]. Ideally, silyl derivatives should form in quantitative yield under mild conditions, such that only a single peak is seen on the chromatogram for each analyte. The use of a large excess of derivatizing agent and solvent (where necessary) can help to minimize problems of interference by moisture or other sample impurities. Earlier problems with the impurities present in derivatizing agents masking sample components are now largely overcome through improvements in the production of reagents and by the use of capillary GC columns and temperature programming. However, the analyst must still be aware of this problem when investigating relatively volatile substances eluting at low temperatures. Peaks in the chromatogram arising from the derivatizing agent are readily assigned by analysis of a reagent 'blank'.

The increasing tendency towards the use of silyl derivatives other than TMS introduces the problem of enhanced molecular weight. Some earlier workers drew attention to the fact that these increases in molecular weight may be prohibitive in the case of higher molecular weight polyfunctional compounds; the temperatures required to effect elution of these derivatives in a reasonable analysis time (if at all) were said to be beyond the maximum operating temperature of the stationary phases available at that time. In view of the cross-linked high temperature stable stationary phases that are now widely available, a reassessment of many high molecular weight silyl derivatives that were previously thought intractable may be warranted. The widespread use of high resolving power capillary columns allows the efficiency of derivatization reactions to be more accurately assessed than was possible with packed columns. Moreover, the use of such columns reduces the need for having available a wide range of columns containing different stationary phases, as was usually the case with packed columns. The apolar dimethyl polysiloxanes (e.g. SE-30, OV-101, DB-1, BP-1 and HP-1) are the most commonly used stationary phases for the GC analysis of silyl derivatives, as they possess the necessary inertness, chemical and thermal stabilities, and resolving power. Other more polar polysiloxane stationary phases that can be used where extra selectivity is required include SE-54, DB-5 (94% methyl 5% phenyl 1% vinyl) and OV-17 (50% methyl 50% phenyl). Where still greater polarity is required, then dimethyl polysiloxane stationary phases containing a proportion of cyanopropyl groups can be used, such as OV-1701 (88% methyl 6% phenyl 6% cyanopropyl). Polar stationary phases containing active protons, such as the polyethylene glycols (e.g. Carbowax 20 M) and free fatty acid phases (FFAP), *must be avoided*, as these will react readily with silylation reagents and derivatives. The recent trend is towards the routine use of cross-linked (also referred to as 'immobilized' or 'bonded') stationary phases because of their enhanced chemical and thermal stability compared with their non-bonded counterparts.

The problem of fouling of the flame ionization detector by SiO_2 deposits formed as a by-product of the combustion of silylation reagents must be considered. Removal of the excess derivatizing agent prior to GC analysis serves to minimize this problem. However, when reagent is not removed, and a significant volume of silylation reagent is injected repeatedly, contamination of the detector is inevitable. Raising the detector temperature will not usually eliminate this problem. Dismantling and thorough cleaning of the detector will be required to restore optimum detector performance. A number of authors have indicated the advantages of using N,O-bis(trimethylsilyl)trifluoroacetamide (BSTFA) for GC rather than other silylation reagents. Firstly, the by-products of derivatization with BSTFA are sufficiently volatile to interfere rarely with analyte peaks in chromatograms. Secondly, the presence of fluorine atoms in the reagent reputedly causes less fouling of the detector by SiO_2 compared with other reagents. In practice, however, fouling of the detector will occur even when employing BSTFA for silylation. This problem can be conveniently eliminated by evaporating the excess reagent under a stream of blown nitrogen, followed by sample dissolution with an appropriate solvent prior to GC analysis. N-Methyl-N-trimethylsilylheptafluorobutyramide, and the corresponding pentafluoropropionamide and pentafluorobenzamide, were originally developed as reagents with an increased fluorine content compared to BSTFA, with the intention of reducing SiO_2 fouling of the FID [26, 27]; however, these reagents have yet to find widespread application [28].

4. Preparation of silyl derivatives and applications to the gas chromatography of organic compounds

Table 1 lists the common names, abbreviations and structures of silyl derivatives used in chromatography. There has been a substantial expansion in the range of silyl reagents that have been considered for use in GC analysis since the late 1970's [1,7]. Many of the derivatives have not, however, found routine application, and TMS derivatives have maintained their popularity. The following section describes the general properties of the most commonly used reagents for trimethylsilylation. Subsequent sections will then deal with the preparation and applications of silyl derivatives other than TMS.

4.1 Trimethylsilyl (TMS) derivatives

For gas chromatography and mass spectrometry, the addition of the TMS group(s) to polar compounds confers thermal and chemical stability in addition to enhanced volatility. Although dimethylsilyl ether derivatives are more volatile, they lack the chemical stability required for routine application [1]. The properties of the most commonly used reagents for trimethylsilylation are given below.

Table 1 Structure of silyl derivatives

Derivative	Silyl groups $\begin{array}{c} R^1 \\ | \\ R^2-Si \\ | \\ R^3 \end{array}$	Abbreviation
Dialkylsilyl		
Dimethylsilyl	$R^1 = R^2 = CH_3; R^3 = H$	DMS
Trialkylsilyl		
Trimethylsilyl	$R^1 = R^2 = R^3 = CH_3$	TMS
Triethylsilyl	$R^1 = R^2 = R^3 = C_2H_5$	TES
Tri-n-propylsilyl	$R^1 = R^2 = R^3 = C_3H_7$	TnPS
Tri-n-butylsilyl	$R^1 = R^2 = R^3 = C_4H_9$	TnBS
Tri-n-hexylsilyl	$R^1 = R^2 = R^3 = C_6H_{13}$	TnHS
Alkyldimethylsilyl		
Ethyldimethylsilyl	$R^1 = C_2H_5; R^2 = R^3 = CH_2$	EDMS
n-Propyldimethylsilyl	$R^1 = C_3H_7; R^2 = R^3 = CH_3$	n-PDMS
Isopropyldimethylsilyl	$R^1 = (CH_3)_2CH; R^2 = R^3 = CH_3$	IPDMS
t-Butyldimethylsilyl	$R^1 = (CH_3)_3C; R^2 = R^3 = CH_3$	TBDMS
Allyldimethylsilyl	$R^1 = CH_2=CHCH_2; R^2 = R^3 = CH_3$	ADMS
N,N-diethylaminodimethylsilyl	$R^1 = N(C_2H_5)_2; R^2 = R^3 = CH_3$	DADS
Aryldimethylsilyl		
Phenyldimethylsilyl	$R^1 = C_6H_5; R^2 = R^3 = CH_3$	PhDMS
Benzyldimethylsilyl	$R^1 = C_6H_5CH_2; R^2 = R^3 = CH_3$	BZDMS
t-Butyldiphenylsilyl	$R^1 = (CH_3)_3C; R^2 = R^3 = C_6H_5$	TBDPS
Alkoxydimethylsilyl		
Methoxydimethylsilyl	$R^1 = CH_3O; R^2 = R^3 = CH_3$	
Ethoxydimethylsilyl	$R^1 = C_2H_5O; R^2 = R^3 = CH_3$	
Propoxydimethylsilyl	$R^1 = C_3H_7O; R^2 = R^3 = CH_3$	
Butoxydimethylsilyl	$R^1 = C_4H_9O; R^2 = R^3 = CH_3$	
Pentoxydimethylsilyl	$R^1 = C_5H_{11}O; R^2 = R^3 = CH_3$	
2-Methylbutoxidimethylsilyl	$R^1 = CH_3CH_2CH(CH_3)CH_2O; R^2 = R^3 = CH_3$	
2,2,2-Trifluoroethoxydimethylsilyl	$R^1 = CF_3CH_2O; R^2 = R^3 = CH_3$	
2-(N,N-diethylamino)ethoxydimethylsilyl	$R^1 = (CH_3)_2NC_2H_4O; R^2 = R^3 = CH_3$	
Alkoxyethylmethylsilyl		
Methoxyethylmethylsilyl	$R^1 = CH_3O; R^2 = C_2H_5; R^3 = CH_3$	
Ethoxyethylmethylsilyl	$R^1 = C_2H_5O; R^2 = C_2H_5; R^3 = CH_3$	
Propoxyethylmethylsilyl	$R^1 = C_3H_7O; R^2 = C_2H_5; R^3 = CH_3$	
Alkoxydiethylsilyl		
Methoxydiethylsilyl	$R^1 = CH_3O; R^2 = R^3 = C_2H_5$	
t-Butylmethoxyphenylsilyl		
t-Butylmethoxyphenylsilyl	$R^1 = (CH_3)_3C; R^2 = CH_3O; R^3 = C_6H_5$	TBMPS
Picolinyldimethylsilyl		
Picolinyldimethylsilyl	$R^1 = 3\text{-}CH_2O\text{-}C_5H_3N; R^2 = R^3 = CH_3$	PICSI
Cyclotetramethylenealkylsilyl		
Cyclotetramethyleneisopropylsilyl	R^1 to $R^2 = (CH_2)_4; R^3 = (CH_3)_2C$	TMIPS
Cyclotetramethylene-t-butylsilyl	R^1 to $R^2 = (CH_2)_4; R^3 = (CH_3)_3C$	TMTBS

Table 1 (continued)

Derivative	Silyl groups $\begin{array}{c} R^1 \\ | \\ R^2-Si \\ | \\ R^3 \end{array}$	Abbreviation
Dialkylmonomethylsilyl		
Di-isopropylmethylsilyl	$R^1 = CH; R^2 = R^3 = (CH_3)_2CH$	MDIPS
Halocarbondimethylsilyl		
3,3,3-Trifluoropropyldimethylsilyl	$R^1 = CF_3(CH_2)_2; R^2 = R^3 = CH_3$	
3,3,4,4,5,5,5-heptafluoropentyldimethylsilyl	$R^1 = CF_3(CF_2)_2(CH_2)_2; R^2 = R^3 = CH_3$	
Nonafluorohexyldimethylsilyl	$R^1 = C_4F_9(CH_2)_2; R^2 = R^3 = CH_3$	DMNFHS
Pentafluorophenyldimethylsilyl	$R^1 = C_6F_5; R^2 = R^3 = CH_3$	Flophemesyl
2-Pentafluorophenylethyldimethylsilyl	$R^1 = C_6F_5(CH_2)_2; R^2 = R^3 = CH_3$	
Chloromethyldimethylsilyl	$R^1 = ClCH_2; R^2 = R^3 = CH_3$	CMDMS
Bromomethyldimethylsilyl	$R^1 = BrCH_2; R^2 = R^3 = CH_3$	BMDMS
Iodomethyldimethylsilyl	$R^1 = ICH_2; R^2 = R^3 = CH_3$	IMDMS
Halocarbonmethylsilyl		
Di(chloromethyl)methylsilyl	$R^1 = R^2 = ClCH_2; R^3 = CH_3$	DCMMS
Pentafluorophenyl-t-butylmethylsilyl	$R^1 = C_6F_5; R^2 = (CH_3)_3C; R^3 = CH_3$	
Pentafluorophenylchloromethylsilyl	$R^1 = C_6F_5; R^2 = ClCH_2; R^3 = CH_3$	
Cyanoethyldialkylsilyl		
2-Cyanoethyldimethylsilyl	$R^1 = CN(CH_2)_2; R^2 = R^3 = CH_3$	CEDMS
Cyclic silyl derivatives		
Di-t-butylsilylene	[cyclic structure: R-O-Si(C(CH$_3$)$_3$)$_2$-O-R]	DTBS
Diethylsilylene*	[cyclic structure: R-O-Si(C$_2$H$_5$)$_2$-O-R]	DES
*Forms diethylhydrogensilyl derivative with an isolated hydroxyl group	$R^1 = R^2 = C_2H_5; R^3 = H$	DEHS
Dimethylsilylene	[cyclic structure: R-O-Si(CH$_3$)$_2$-O-R]	

4.1.1 N,O-Bis(trimethylsilyl)trifluoroacetamide (BSTFA)

[Structure (4): CF$_3$–C(=N–Si(CH$_3$)$_3$)–O–Si(CH$_3$)$_3$]

BSTFA is the most widely used reagent for trimethylsilylation. The reagent was first prepared by Stalling et al. [29], and is now available from a large number of proprietary sources, both as the pure compound and mixed with catalysts, commonly TMCS. Other catalysts that have been used with BSTFA include trifluoroacetic acid [30, 31], hydrogen chloride [32], potassium acetate [33, 34], piperidine [35, 36], O-methylhydroxylamine hydrochloride [37] and pyridine [31, 38]. Pyridine is especially useful as a solvent on account of its ability to act as an HCl acceptor in silylation reactions involving organochlorosilanes [39]. However, there are some instances in which silylation reactions have been found to be slower in pyridine than in other solvents.

BSTFA is very versatile, reacting with all the common protic sites present in organic materials. However, when used alone BSTFA may not derivatize some amides, secondary amines and hindered hydroxyl groups. The addition of TMCS (1% v/v is usually sufficient) as catalyst will generally ensure effective derivatization of these latter functional groups. The donor strength of BSTFA is comparable with that of N,O-bis(trimethylsilyl)acetamide (BSA; see below for further description of this reagent) and it is generally safe to substitute BSTFA/TMCS mixtures for BSA/TMCS mixtures in derivatization procedures. As BSTFA is a liquid at room temperature, derivatizations are usually carried out without solvent. The best results are obtained when the reaction products are soluble in the reaction medium. When problems of solubility are encountered the use of a solvent should be considered (see above), although the dissolution of analytes prior to silylation is not always essential, as this can occur as derivatizations proceed. A notable advantage of BSTFA is the volatility of the by-products of the derivatization, trimethylsilyltrifluoroacetamide and trifluoroacetamide, which usually elute early in analyses, often with the GC solvent front. The use of BSTFA in combination with other silylation reagents will be discussed below in Section 4.1.8.

4.1.2 N,O-Bis(trimethylsilyl)acetamide (BSA)

$$\text{CH}_3-\overset{\overset{\displaystyle O-Si\diagup}{|}}{\underset{\underset{\displaystyle -Si\diagup}{|}}{C}}=N \tag{5}$$

BSA was first reported by Birkofer et al. in 1963 [40]. It is a more potent TMS donor than HMDS or TMSDEA (see below) and is one of the most commonly used silylating reagents. The reactivity of BSA is similar to that of BSTFA and MSTFA (see below), readily silylating non-sterically hindered alcohols, carboxylic acids, amino acids, amides, amines and enols. Although not as volatile as the by-products of silylation using BSTFA, those of BSA (N-trimethylsilylacetamide and acetamide) are sufficiently volatile not to interfere in the majority of GC analyses.

When derivatization does not proceed smoothly, the use of a suitable solvent can help to produce efficient silylation. As with BSTFA, the addition of TMCS (usually 1–20%) as catalyst helps to enhance the effectiveness of silylation. Although silylation reactions involving BSA are normally carried out under anhydrous conditions, it has been found that the presence of 1% of water can substantially increase the reaction rate [41, 42]. This catalytic activity was thought not to be due directly to the water, but to the trimethylsilanol formed by hydrolysis of the BSA [41]. Other catalysts that have been used with BSA include oxalic acid [43], trifluoroacetic acid [30], hydrochloric acid [44], potassium acetate [33] and trimethylbromosilane [45]. The use of BSA together with other silylating reagents is discussed below in Section 4.1.8.

4.1.3 N-Methyl-N-trimethylsilyltrifluoroacetamide (MSTFA)

$$CF_3-\overset{\overset{\displaystyle O}{\|}}{C}-\underset{\underset{\displaystyle CH_3}{|}}{N}-Si\diagup \tag{6}$$

MSTFA has become one of the most important silylating reagents. Introduced by Donike [46], MSTFA has similar silyl donor power to BSA and BSTFA and can be used to silylate all protic functional groups. MSTFA and its by-products (largely N-methyltrifluoroacetamide) are more volatile than BSA and BSTFA, hence this reagent is of value in the GC analysis of compounds that would otherwise be obscured in chromatograms. The enhanced silyl donor power of MSTFA derives at least in part from its relatively high polarity, which means it can be used to silylate the hydrochloride salts of amines or amino acids directly [46]. Trifluoroacetic acid has been used as solvent with MSTFA for the silylation of highly polar compounds [47]. Like BSA and BSTFA, the silylation power of MSTFA can be enhanced by the use of catalysts, most commonly TMCS [46, 48, 49]. Potassium acetate, trimethylbromosilane (TMBS) and trimethyliodosilane (TMIS) have been used to catalyze the formation of TMS-enol ethers in ketosteroids [49, 50], while TMSIM catalyses MSTFA in the trimethysilylation of indolyl-NH-containing compounds [51, 52]. Further applications involving the use of MSTFA are presented below.

4.1.4 N-Trimethylsilyldiethylamine (TMSDEA)

$$\diagup\!\!\!\!-\mathrm{Si}-\mathrm{N}\diagdown\!\!\!\!- \quad (7)$$

TMSDEA is a moderate strength silyl donor, intermediate between HMDS/TMCS (weaker silyl donor than TMSDEA) and BSTFA, BSA and MSTFA (stronger donors). The strongly basic nature of TMSDEA, combined with its high volatility and that of its by-products, makes it particularly useful for the silylation of low molecular weight carboxylic acids and amino acids [53–58]. TMSDEA will react with a wide range of compound types. The use of acidic catalysts such as TMCS [55, 58], trifluoroacetic acid [55] or silica–alumina [55, 56] can serve to enhance the silylation power of TMSDEA. A mixture comprising TMSDEA/BSTFA/TMCS/pyridine (30:99:1:1000 v/v/v/v) has been found simultaneously to trimethylsilylate acidic, neutral and basic metabolites of tyrosine and tryptophan [58]. N-trimethylsilyldimethylamine (TMSDMA) possesses similar silyl donor strength to TMSDEA. By analogy with TMSDEA, the high volatility of TMSDMA makes it potentially useful for the silylation of low molecular weight compounds, such as amino acids [56].

4.1.5 Trimethylsilylimidazole (TMSIM)

$$\diagup\!\!\!\!-\mathrm{Si}-\mathrm{N}\diagdown\!\!\!\!- \quad (8)$$

TMSIM was first prepared by Birkofer and Ritter [59] in 1965, and is generally considered to be the strongest reagent available for the silylation of hydroxyl groups [1, 3–7]. Significantly, though, unlike BSTFA, BSA and MSTFA, TMSIM does not react with aliphatic amines, although the reagent will silylate less basic amines and amides (see Refs. 1 and 3 and Section 4.2.7). This selectivity can be used to advantage in analytical studies.

The low volatility of TMSIM compared with many other commonly used silylating reagents means that the reagent should not be injected directly into capillary columns. Adsorption chromatography, employing either TLC or mini-columns can be used to purify the silylated analytes from the excess reagent prior to GC and GC–MS analysis. Elution through Sephadex LH-20 is also employed to clean-up samples (see application sections below for further details of column dimensions and eluents). The rapid decomposition of TMSIM in air, to give imidazole and hexamethyldisiloxane, is a further undesirable feature of this reagent that necessitates its careful handling when adding it to samples and during purification steps prior to GC analysis.

The use of TMSIM has been favoured by some workers in situations where analytes contain ketone groups that would readily form enol-TMS ethers. The advantage of using TMSIM derives from its strong silyl donor properties, which permit quantitative reactions to be achieved without need for the addition of acid catalysts that might otherwise encourage enol-TMS ether formation. The imidazole that is a by-product of silylation when using TMSIM is weakly amphoteric and does not readily promote enol-silyl ether formation [60].

4.1.6 Trimethylchlorosilane (TMCS)

$$\diagup\!\!\!\!-\mathrm{Si}-\mathrm{Cl} \quad (9)$$

TMCS was one of the first silylation reagents prepared [61]. Although continuing to find wide use in synthetic chemisty, TMCS is rarely used alone in analytical applications. The common analytical use of TMCS is in mixtures, e.g. HMDS/TMCS/pyridine or BSA/TMSIM/TMCS, which function as very powerful broad-spectrum silylation reagents (see below). The role of TMCS as a silylation catalyst is well recognized, as already mentioned. The facile reaction of TMCS with F^- ions has found application to the determination of fluorides in air, exhaust gases, bone ash and water [62]. The use of TMBS and TMIS as silylation catalysts has been discussed briefly above; their analytical use is discussed below in the appropriate sections.

4.1.7 Hexamethyldisilazane (HMDS)

$$\diagup\!\!\!\!-\mathrm{Si}-\overset{\mathrm{H}}{\mathrm{N}}-\mathrm{Si}-\diagdown\!\!\!\!- \quad (10)$$

Like TMCS, HMDS was one of the earliest used silylating reagents [63]. The analytical uses of HMDS have been extensively reviewed by Pierce [4]. The liquid nature of HMDS at room temperature, combined with its favourable solvating properties for many substrates, generally obviates the need for a solvent. Although not as strong a silyl donor as many other reagents, it has the desirable property of reacting more selectively in some instances. TMCS is often used as a catalyst in analytical work to enhance the silylating power of HMDS.

4.1.8 Mixtures of silyl donors

The catalytic properties of TMCS and its use in conjunction with other silylation reagents has been alluded to above. Other blends of reagents that are used to produce strong broad-spectrum silylating reagents for complex molecules include BSTFA/TMSIM/TMCS, MSTFA/TMCS/TMSIM, BSTFA/TMSDEA/TMCS and BSA/TMSIM/TMCS. The last-named mixture is one of the most potent silylating reagents available. These mixed reagents can be prepared from the pure materials or alternatively purchased, pre-mixed in the required proportions, from a variety of commercial sources. Various trade names are used by manufacturers and care should be taken to ensure the desired reagent formulation is obtained. Applications of these mixed reagents are discussed below in the appropriate sections.

4.2 Applications of trimethylsilylation to the analysis of organic compounds

4.2.1 Short-chain organic acids and related compounds

GC and GC–MS are used extensively to profile organic acids in body tissues and fluids in an effort to detect disease states. Trimethylsilylation is probably the most commonly used derivatization technique for the study of the organic acid fraction of urine, which can comprise: mono- and polycarboxylic acids, mono- and polyhydroxy acids, keto acids, phenols, phenolic acids, and conjugates of organic acids, particularly with glycine [64]. BSTFA and BSA are the most commonly used TMS donors, both with and without solvent, and TMCS as catalyst. GC–MS is especially usefull because it can provide unambiguous compound identifications, often with the aid of computer searching of libraries of mass spectra of TMS derivatives [64, 65]. De Leenheer and co-workers use an approach [66] to organic acid profiling for the diagnosis of inborn errors of metabolism by GC and GC–MS which involves first converting the keto acids to stable oxime derivatives, followed by trimethylsilylation using BSTFA containing 1% TMCS in dry pyridine (4:1 v/v) with the reaction carried out at 60 °C for 30 min. Some investigators suggest that the protection of keto groups may not always be necessary prior to silylation. For example, a procedure for the determination of α-ketoadipic acid in biological media containing lower molecular weight carboxylic acids and carbohydrates recommends the use of silylation with BSTFA alone, unless complex mixtures are involved [67]. A mixture of derivatives was produced by silylation performed in this way. The configuration of the TMS derivatives of 2-oxocarbonic acid isomers produced by derivatization with BSTFA/pyridine (1:1 v/v) was investigated by a combination of GC, GC–MS and nuclear magnetic resonance spectroscopy [68]. The investigation showed that the two peaks seen in the GC analysis of the TMS derivatives of 3-methyl-2-oxovalerianic acid were the result of enolization during the silylation, leading to a mixture of E- and Z-isomers (11 and 12). In contrast, 4-methyl-2-oxovalerianic acid gives only a single product as a result of hydrogen bonding in the enol-ether derivative (13).

A re-investigation by capillary GC and GC–MS as TMS and/or TMS oxime derivatives prepared using BSTFA, of the metabolic pattern of organic acids in maple syrup urine disease provides an indication of the advantages that accrue from the use of capillary rather than packed columns [69], and the use of perdeuterated silylating reagents, such as N,O-bis(perdeuterotrimethylsilyl)acetamide produces TMS-d_9 derivatives which can assist in mass spectral interpretation by indicating the number of silylated positions in unknown compounds through mass shifts of key ions. Applications of this technique include the investigation of metabolites in haemodialysis fluid [70] and the mass spectrometric identification of 2-hydroxydodecanedioic acid and its homologues in the urine of patients undergoing hopantenate therapy during acute episodes [71]. An alternative method of determining the number of silylated positions based on the relative abundances of isotopic clusters determined by mass spectrometry is discussed in Section 4.5. GC–MS is also of use in this area when stable isotope labelled compounds are used as metabolic tracers or as internal standards for quantitative analyses. However, an investigation of branched-chain α-keto acids as their O-TMS quinoxalinol (see 14 below) and N-methylquinoxalone derivatives drew attention to the problems of the high natural abundances of the ^{29}Si (5.1%) and ^{30}Si (3.35%) isotopes which can preclude the use of silyl derivatives in situations where trace enrichments of stable isotopes in metabolites are encountered [72]. Despite this, ketone body kinetics were readily determined using D-(−)-3-hydroxy[4, 4, 4-^{2}H$_3$]butyrate tracer [73]; GC–MS with selected ion monitoring (SIM) being used to measure [^{2}H$_3$] trace enrichment in TMS derivatives of 3-hydroxybutyrate and acetoacetate in small (300 µl) blood samples. A mass spectrometric approach was also used to analyse branched-chain α-keto acids and metabolites in maple syrup urine disease after sodium borodeuteride reduction *in situ* of the once-labelled hydroxy acids, followed by addition of a more extensively labelled analogue as internal standard. Extracts were silylated prior to quantification by GC–MS–SIM [74]. A similar approach has also been demonstrated for the quantitative determination of lactic, pyruvic, 3-hydroxybutyric and acetoacetic acids as TMS derivatives in a single analysis [75]. In a study of dicarboxylic

4.2.2 Prostaglandins and other eicosanoids

Trimethylsilylation continues to be the most widely used method of blocking free hydroxyl groups in this class of compounds. Other derivatives that have been explored are discussed below in the appropriate sections. Methods for the preparation of TMS derivatives have been extensively investigated and documented [3]. The methyl esters–TMS ethers (or methyloxime–methyl esters–TMS ethers for compounds containing a keto group) continue be prepared for GC–MS analyses employing electron impact ionization (EI). Some recent examples of the preparation, use and properties of this latter derivative appear in the literature [35, 79–98]. An alternative approach that is employed to improve the sensitivity and selectivity in GC–MS analysis of prostanoids uses pentafluorobenzyl esters [99–117] rather than methyl esters. Trimethylsilylation is then used in conjunction with the pentafluorobenzylation to block the remaining free hydroxyl groups.

Derivatization of prostanoids from biological fluids [102]
Dry residues from solid-phase extraction columns

$$\text{(14)}$$

acid metabolism in man, metabolites of orally administered dicarboxylic acids (containing 6, 8, 10, 12, 14 and 16 carbon atoms) excreted in urine were measured as TMS derivatives by GC–MS–SIM using [^{2}H$_4$]adipic acid as internal standard [76]. In plants the pool size of glycollic acid during photorespiration was determined by stable isotope dilution mass spectrometry following trimethylsilylation [77]. The authors chose to use packed column rather than capillary GC, as the broader peaks allowed increased sampling of spectra from the GC peak and avoided intensity distortion. Langenbeck *et al.* [78] used to O-TMS quinoxalinol derivatives (14) to investigate aliphatic α-keto acids in urine. Derivatives were prepared by incubating (1 h, 70 °C) α-keto acids in 0.5 ml of urine (diluted with 0.5 ml of 4 N hydrochloric acid) with 1 ml of a 1% solution of o-phenylenediamine in 2 N hydrochloric acid. After neutralization, extraction and drying, silylation was achieved by treatment (30 min, 70 °C) of the residue with 50 µl each of BSTFA and pyridine. The α-keto acids derivatized in this way were analysed by GC and GC–MS. SIM of the ions at *m/z* 217, 232 and 245 was used to enhance the detection of the biologically interesting aliphatic α-keto acids.

were treated with 100 µl of methoxyamine hydrochloride in dry pyridine (5 mg/ml). The mixture was allowed to stand overnight at ambient temperature and the pyridine was evaporated under nitrogen. To the residue were added 30 µl of acetonitrile, 10 µl of 35% pentafluorobenzyl bromide in acetonitrile and 10 µl of N,N-diisopropylethylamine, and the reaction mixture was heated at 40 °C for 15 min. The reagents were evaporated under nitrogen, the residue was dissolved in 400 µl of dichloromethane, and the solution was applied to a short column of Sephadex LH-20 pre-swollen in dichloromethane. The derivatized prostanoids were eluted with 3 ml of dichloromethane and the solvent was evaporated under nitrogen. Prostanoid hydroxyl groups were converted to silyl ethers by adding 50 µl of BSTFA and allowing to stand overnight at ambient temperature. The silylating reagent was evaporated under nitrogen and the residue was dissolved in a high boiling hydrocarbon solvent (n-dodecane or n-tetradecane) for GC–MS analysis.

BSTFA is the preferred reagent for the trimethylsilylation of prostanoids. Although neat BSTFA was used in the above protocol some workers favour the use of an equal volume of solvent, such as pyridine [106, 108, 110, 111] or acetonitrile [117] at room temperature. Slightly elevated temperatures are employed to accelerate the silylation; for example, hydroxyeicosatetraenoic acids (HETEs) and thromboxanes were converted to TMS derivatives by heating with BSTFA/pyridine (2:1 v/v) for 20 min at 60 °C [110], while in another study reaction of thromboxane B_2 (TXB_2) was carried out at 40 °C for 1 h [106]. Lawson et al. [111] report the quantitative conversion of the PFB esters of TXB_2 metabolites to their TMS ethers in 15 min at room temperature using BSTFA/pyridine (1:1 v/v). Exactly analogous conditions are used in the preparation of

series prostaglandins is a major disadvantage of using piperidine as catalyst, as it may lead to misinterpretation of the composition of biological extracts. Hence the recommended approach is to protect keto groups by reaction with O-methylhydroxylamine hydrochloride prior to trimethylsilylation for routine analyses. TMS-enol ether formation and the use of oximes to protect carbonyl groups will be discussed further below in relation to the silylation of steroids (Section 4.2.3). Direct solid-phase isolation of prostaglandin E_2 from plasma using XAD-2 resin, followed by in situ oximation substantially reduces the sample preparation of this analyte for analysis of MO–PFB–TMS derivatives by negative-ion chemical ionization (NICI) GC–MS [122]. Further

(15)

methyl esters–TMS ethers and methoxime–methyl esters–TMS ethers except that extracts are treated initially with ethereal diazomethane to produce the methyl esters (see Ref. 3 and Chapter 2).

Uobe et al. [35] re-investigated the methods described earlier by Nicosia and Galli [119] and Rosello and co-workers [120, 121], who used mixtures of TMSIM/piperidine and BSTFA/piperidine respectively in an effort to obtain single trimethylsilylation products for compounds containing keto groups, thus avoiding preparation of the methoximes. However, it was found that reaction of TMSIM/piperidine with the methyl ester of prostaglandin E_2 (PGE_2) yielded a mixture of products, including TMS-9-enol-PGE_2-Me, TMS-PGB_2-Me and TMs-11-piperidyl-PGA_2-Me (15).

Other commonly used silylation mixtures, e.g. HMDS/TMCS/pyridine, BSA/pyridine, BSA/acetonitrile/pyridine, BSTFA/pyridine and TMSIM/pyridine, also yield mixtures of reaction products [120, 121]. The production of B series prostaglandins through enolization of A

attempts to improve sample preparations have involved the use of 0.3 g columns of a special silica gel (Silicar CC-4; Mallinckrodt, St. Louis, MO, USA) to separate various arachidonic acid metabolites [123]. Recovered HETEs and prostaglandins were analysed by GC–MS–SIM after conversion to MO–methyl ester–TMS or MO–PFB–TMS derivatives. Catalytic hydrogenation of unsaturated hydroxy fatty acids was always performed prior to derivatization.

The mass spectral properties of the various derivatives of eicosanoids have been widely investigated with a view to structure elucidation of new metabolites and the use of GC–MS for their detection and quantification in biological tissues and fluids. The EI spectra of the TMS derivatives are complex, although they contain abundant structure information, they exhibit relatively low abundance high mass fragment ions. In contrast the NICI spectra of the PFB derivatives are dominated by abundant [M − PFB]$^-$ anions, which allow low picogram detection limits to be reached in GC–MS–SIM.

The presence of the TMS groups has little influence on fragmentation in the NICI mode [91, 100, 124]. The favourable GC–MS behaviour of the PFB–TMS and MO–PFB–TMS derivatives has made this the method of choice for the validation of other less specific assay techniques such as radioimmunoassay (RIA) [112, 117]. The GC–MS–MS behaviour of prostanoid ME–TMS, MO–ME–TMS [96–98, 125], PFB–TMS and MO–PFB–TMS [116, 125] derivatives has been investigated with a view to improving the selectivity of GC–MS analyses. Indications are that this approach is capable of achieving high sensitivities and selectivities in the analysis of prostanoids.

Selectivity and sensitivity in trace analyses have also been enhanced by use of immunoaffinity columns to provide highly purified extracts for subsequent trimethylsilylation and GC–MS analysis [126–128]. Compounds eluted from such columns can be converted directly to the pentafluorobenzyl–TMS or methoxime–PBF–TMS (for compounds containing keto groups) derivatives prior to GC–MS–SIM employing negative-ion chemical ionization with ammonia as reagent gas. Mass chromatograms obtained by SIM at nominal mass resolution (quadrupole MS) for the $[M - 181]^-$ anion for MO–PBF–TMS–6-keto $PGF_{1\alpha}$ (m/z 614) and a $[^2H_4]$ labelled internal standard were essentially interference free, with detection limits in the low picogram range being readily attained [128]. An immunoaffinity extraction of thromboxane B_2 has also been described, although in this latter instance GC–MS analysis was performed after preparation of the MO–PFB–TBDMS derivative [129, 130] (the use of TBDMS derivatives is discussed in detail in Section 4.4.2).

4.2.3 Steroids, bile acids and related compounds

Despite the investigations that have been performed on the use of other silyl derivatives, trimethylsilylation continues to be employed routinely for the analysis of steroids and related compounds. The hydroxyl groups occurring in steroids differ markedly in their rate of silylation, due to their nature (primary, secondary or tertiary) and differing steric environments. The trimethylsilylation of commonly occurring sterols and related triterpenoids containing an unhindered 3β-hydroxyl group is readily achieved by treating pure compounds or biological extracts with BSTFA containing 1% TMCS at 60 °C in 30 min with or without added pyridine (1:1 to 1:10 v/v). BSA can also be used either alone or together with HMDS and TMCS (10:10:5 v/v/v) by heating at 60 °C for 30–60 min [131]. The TMS derivatives of sterols display excellent chromatographic behaviour on capillary columns coated with apolar stationary phases, and produce EI mass spectra containing abundant structure information concerning the nature of the sterol nucleus and side chain (see Ref. 132 and references therein). Examples of the silylation of other steroids using BSTFA include the determination of oestradiol [133] and catechol oestrogens [134]. Tetrahydroaldosterone (11β,18-epoxy-3α,18,21-trihydroxy-5β-pregrene-20-one) and aldosterone (11β,21-dihydroxy-3,20-dioxopregn-4-en-18-al) are also effectively silylated by reaction with BSA/pyridine/TMBS (4:5:1) overnight at 40 °C [135]. Cortisol has also been determined in human plasma by stable isotope dilution GC–MS following treatment of the dimethoxime cortisol derivative (16) with BSA (50 μl) for 2h at 100 °C [136, 137].

(16)

The wider application of TMS derivatives derives from the possibility of achieving complete silylation of more hindered hydroxyl groups in steroids, particularly in polyhydroxylated compounds. TMSIM is generally the preferred reagent for the silylation of more hindered hydroxyl groups. For example, in ecdysteroids (polyhydroxylated arthropod moulting hormones), the ease of silylation of the hydroxyl groups varies in the positional order 2, 3, 22, 25 > 20 > > 14 [60]. Compounds containing a 14α-hydroxyl group can only be silylated under forcing conditions, e.g. 20-hydroxyecdysone was fully silylated in 15 h at 100 °C [138, 139] using neat TMSIM (17). The degree of silylation was confirmed by GC–MS employing ammonia chemical ionization; the hexa-TMS-20-hydroxyecdysone yielding a pseudomolecular ion $[M + H]^+$ (m/z 913; 35%; base peak m/z 171). GC–MS analysis also confirmed the lack of enol-ether formation when fresh reagent was used and care was taken to exclude air and moisture during sample handling [139, 140]. The EI spectra of the TMS ethers of ecdysteroids have been discussed [60, 141]. The presence of characteristic ions in the EI spectra of ecdysteroids at m/z 561 (20-hydroxy compounds) and 567 (compounds lacking a hydroxyl group at C-20) has been used to advantage in GC–MS analyses employing SIM [138, 139, 141, 142]. The addition of 1% of TMCS to TMSIM was found to catalyse silylation of the 14α-hydroxyl group, reducing the reaction time to 4 h at 100 °C [60]. Significantly though, this catalytic effect was only evident in the silylation of pure compounds. The addition of larger quantities of TMCS was found to cause formation of the enol-TMS ether. Solid potassium acetate has also been reported to increase the rate of reaction of TMSIM, enabling the 14α-hydroxyl group to be silylated in 2 or 3 h at room temperature [60, 143].

Where other functionalities are present in steroidal compounds, trimethylsilylation is frequently used in conjunction with other derivatization methods. Apart from hydroxyl groups, ketone groups are the other most commonly occurring functionality in steroids. When keto groups are present, conversion to methoxime derivatives is employed as a standard procedure [144] to avoid the formation of artefacts during trimethylsilylation via enolisation of the carbonyl function (see also Chapter 6 and Ref. 1, 3 and 145–154). Interestingly, a very significant catalytic effect has been demonstrated for the reaction of methandienone (17β-hydroxy-17-methylandrosta-1,4-dien-3-one) with various silylating reagents in the presence of methoxyamine hydrochloride [155]. The most rapid (< 10 minutes) quantitative silylation was obtained with TMSIM in the presence of methoxyamine hydrochloride and pyridine at 80 °C.

Problems can be encountered due to steric hindrance in the formation of methoximes. For example, oximation of the C-20 oxo group in corticosteroids has been found to be inhibited by the presence of a 16β-methyl group [151]. In this latter study, subsequent silylation of dexamethasone using TMSIM (80 °C, 2 h) yielded predominantly the bis-MO–tris-TMS derivatives, and also the mono-MO–tris-TMS and bis-MO–tris-TMS derivatives of betamethasone in a 3:1 ratio. In the absence of the 16-methyl group, e.g. for prednisolone, derivatization posed no such problem [152]. GC–MS using NICI was the method of choice for analysis of the MO–TMS derivatives.

Although recognized as potentially disadvantageous because of the possibility of producing mixtures of derivatives, efforts have been made to optimize the reaction conditions for the quantitative conversion of keto steroids to their TMS-enol ethers [35, 50, 155, 156]. The preparation of TMS-enol ethers provides an alternative to oximes for protecting ketone groups for chromatography. Interestingly, the GC resolution of the TMS-enol–TMS ether derivatives of four stereoisomeric 3-hydroxy-19-norandrostan-17-ones formed by treatment with MSTFA/TMSIM (100:1) at 60 °C for 30 min was inferior to that achieved for the corresponding TMS and MO–TMS derivatives [147]. Perdeuterated TMS-enol–TMS-ethers were prepared by treating the steroid with BSA-

(17)

d_9/pyridine (1:1 v/v) at 75 °C for 1 h in the presence of anhydrous sodium acetate [147]. TMS-enol–TMS ethers were found to be of use for the quantitative packed column GC–MS SIM determination of 19-norandrosterone in urine [157]. Silylation of dexamethasone(9α-fluoro-11β, 17α, 21-trihydroxy-16α-methylpregna-1,4-diene-3,20-dione) using BSTFA in the presence of sodium acetate yielded a pure tetra-TMS derivative, with no apparent need for prior preparation of the methoxime [158]. Silylation in this latter instance produced the TMS-enol ether of the 20-one moiety (18), leaving the 3-one group unreacted. Quantification was achieved by packed column GC–MS–SIM using [$^{13}C_6, ^2H_3$]dexamethasone as internal standard.

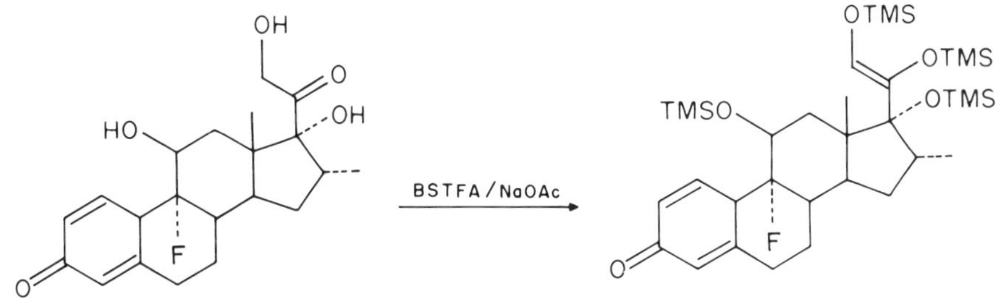

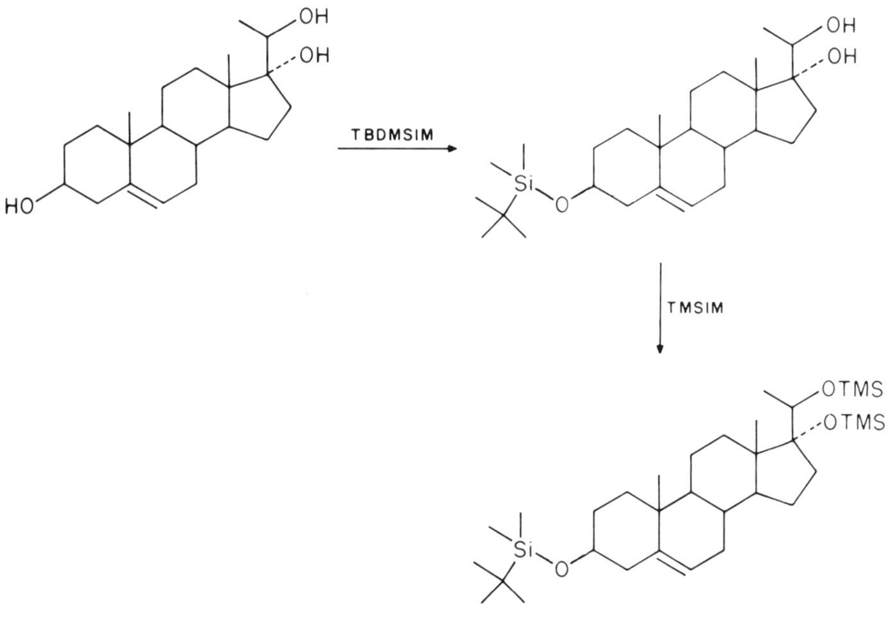

7α, 12α/β, 15β and 17α(tertiary) positions does not occur owing to steric hindrance, and subsequent trimethylsilylation using TMSIM, yielding mixed TBDMS–TMS ethers (19), is required before GC–MS analysis. The enol-TBDMS ethers are reported to have better GC–MS properties than the O-methyloximes, which are prone to decomposition when small amounts of steroid are analysed [159, 160].

(18)

The aromatization of the A ring of norethynodrel (a 3-keto-5,10-ene-nor-19-methyl steroid; 20) during trimethylsilylation is one of the most unusual reactions associated with the enolization of keto steroids [161]. Related to this is the investigation by Abdel-Baky *et al.* [162] of the effect of BSTFA and TMSIM on a variety of steroidal ketones, ketols, enones, and enone epoxides which are putative intermediates in catechol oestrogen biosynthesis. GC–MS data, supported by ^{1}H NMR, led the authors to conclude that aromatic derivatives of

(19)

A procedure for the efficient preparation of enol-TBDMS ethers for the GC–MS analysis of steroids and bile acids is described in Section 4.4.3 [12]. Significantly, tert-butyldimethylsilylation of hydroxyl groups in the 1β,

(20)

3-keto-4,5-epoxides of nor-19-methyl steroids will form under conditions routinely employed in silylations. The aromatization was found to be more strongly favoured by the presence of the 4,5-epoxide function than the corresponding 4,5-olefin. The yield of aromatic silylated products was increased by the use of more basic silylating reagents and/or higher reaction temperatures, with the type of aromatic silylated product(s) being influenced by the basicity of the silylating reagent and the presence of additional functionalization of ring A. The lack of correlation between GC–MS and NMR data led the authors to conclude that some chemical transformations must be induced thermally in the injection port of the gas chromatograph (see Section 2.2 for further discussions of silylation reactions occurring in the GC injection port)

The formation of cyclic methaneboronates of steroids containing vicinal diol groups can be carried out prior to trimethylsilylation of remote silyl groups [131]. The selective formation of cyclic boronates avoids problems of steric crowding in the silylation of adjacent hydroxyl groups and produces derivatives with shorter retention times and abundant $M^{+\cdot}$ ions.

Preparation of methylboronate esters–trimethylsilyl ethers [131] Methaneboronic acid (1 molar proportion) in dry pyridine was added to the steroid diol (100 µg) and the mixture was kept at 60 °C for 30 min. Silylation was carried out after removal of the solvent by treating the methylboronate with BSTFA (5 µl) and heating at 60 °C for 2 min or with BSTFA in DMF (20 µl, 1:3 v/v) for 60 °C for 5 min. The solution was evaporated to dryness and the residue was dissolved in ethyl acetate for GC and GC–MS analysis.

The use of molar equivalents of methylboronic acid and steroid diol affords only the cyclic ester. Acyclic boronate esters will be formed at remote hydroxyl groups if excess reagent is used. While these latter derivatives display undesirable GC properties, the acyclic boronate groups are readily displaced by trimethylsilylation; the cyclic methaneboronate will be unaffected under the mild silylation conditions described above. n-Butyl boronate–methoxime–TMS derivatization has been employed in the quantitative determination of fourteen adrenal-cortical steroids in blood by capillary GC–MS [149]. Silylation was achieved by reaction of the steroid boronate–methoxime derivatives with BSTFA/TMSIM/ TMCS (3:3:1 v/v/v) for 10 min at 100 °C. Minor oxygenated serum sterols have been silylated by treatment with a large excess of a mixture of BSA/HMDS/TMCS (40:40:15 v/v/v) prior to GC and GC–MS analysis [163]. Compounds bearing adjacent hydroxyl groups, e.g. 5-cholest-5-ene-3β,4α-diol, were analysed as both bis-TMS and cyclic methaneboronate derivatives. A combination of TMS ethers and boronate ester derivatives of vitamin D metabolites have been prepared for their GC and GC–MS analysis (21). Mass spectrometry was used in the initial structure elucidation of 24,25-dihydroxy-vitamin-D$_3$ [164] and the GC–MS properties of this and related compounds as their TMS ethers have been described [165, 166]. The EI mass spectrum of the 3-TMS ether–24,25-methylboronate and n-butylboronate derivatives of 24R,25-dihydroxycholecalciferol (cyclized isomer) displayed enhanced high mass ion abundance compared with that obtained from the corresponding tris-TMS derivatives [167], (see also Chapter 14). Silylation was achieved with a mixture of BSTFA/TMSIM/ TMCS (3:3:2 v/v/v) in 30 min at room temperature. The TMS-ether n-butylboronate combination has been used in the stable isotope dilution GC–MS–SIM assay of vitamin D metabolites in plasma [168, 169]. As these derivatives were found to be somewhat unstable, GC–MS analyses were performed immediately after their preparation. Indications were that a cyclic methylboronate did not form across the hydroxyls at the 25 and 26 carbons, or that subsequent removal occurred on reaction with BSTFA [170]. 3-TMS ether derivatives of vitamins D$_2$ and D$_3$ form readily by reaction with BSTFA at room temperature, although more rigorous conditions were required to silylate sterically hindered C-25 hydroxyls. Derivatives of the 25-hydroxylated metabolites can be prepared either by reaction with TMSIM at 50 °C or with BSTFA containing 20% of TMCS at 60 °C. The latter procedure is less time consuming but gives somewhat increased non-specific interferences in GC–MS analyses.

TMS derivatives are used routinely in the GC and

GC–MS analysis of bile acids [171]. Trimethylsilylation is usually preceded by esterification of the carboxyl groups, most commonly to the methyl ester derivative, e.g. by use of diazomethane (see also Chapter 2). Treatment of methyl esters with a mixture of HMDS/TMCS/pyridine (2:1:3 v/v/v) for 30 min at 60 °C [172] is still widely used to achieve silylation of the commonly occurring bile acids. This method provides silylation of all hydroxyl groups, albeit at varying rates. For specialist applications, alternative methods of esterification are used (see Chapter 2). For example, isobutyl ester TMS ethers have been used in simultaneous analyses of neutral sterols and bile acids in rat liver epithelial cell lines [173]. The sterols and bile acids were fully resolved on an OV-1 coated capillary column following silylation with MSTFA/TMBS/pyridine (80:10:10 v/v/v) at 110 °C for 3.5 to 4 h. The relatively forcing conditions for silylation were required to overcome the steric hindrance due to the isobutyl moiety. Mahara and co-workers [174, 175] preferred to use TMSIM in acetonitrile at 38 °C for 60 min to silylate bile acid methyl esters, including those unusual compounds bearing 1β-hydroxyl groups. As in the case of other steroids and vitamin D metabolites, 'mixed' cyclic methaneboronate TMS ether derivatives of bile acid methyl esters can be prepared. The isomeric types of cis-6,7-diols in the methyl 3α,6,7-trihydroxycholanoate group all yield reasonably stable cyclic alkaneboronates [176]. The selectivity of cyclic boronate formation for vicinal diols (or other sufficiently proximal hydroxyls) is of potential value in the analysis and characterization of new bile acids formed by metabolic hydroxylations. The conversion of oxo functions to methyloximes, followed by trimethylsilylation of hydroxyl groups, is used in the analysis of ketonic bile acids [177]. Although bile acid glucuronide [178] and glucoside [179] metabolites can be analysed by GC–MS as intact conjugates following trimethylsilylation using HMDS/TMCS/pyridine, their mass spectra lack fragmentations indicating the position of the glycosidic linkage.

4.2.4 Fatty acids and related acyl lipids

Carboxylic acid groups and unhindered hydroxyl groups are easily silylated by most of the common derivatizing agents, such as BSTFA, BSA or TMSIM. High temperature gas chromatography has long been used to profile total lipid extracts [180, 181]. Myher and Kuksis [182] described a convenient approach to the analysis of plasma total lipid extracts. Samples are treated with phospholipase C to dephosphorylate phospholipids, then silylated using a mixture of HMDS/TMCS/pyridine (5:2:15 v/v/v) for 30 min at room temperature. Mild conditions are preferred for the silylation of acyl glycerols as some reagents can cause isomerization. The use of on-column injection and capillary GC columns affords good recoveries and carbon number resolution of all the major lipid classes, including fatty acids (as TMS esters), sterols (as TMS ethers), monoacylglycerols (as bis-TMS ethers), fatty acid amides of sphingosines (as TMS ethers), diacylglycerols (as TMS ethers) and ceramides (as TMS ethers [181, 182]. Evershed and co-workers have used trimethylsilylation of total lipids followed by high temperature GC and GC–MS to demonstrate the survival of highly preserved intact acyl lipids, sterols and long-chain alkyl compounds in archaeological materials (Figure 2) [183, 184]. In this latter work trimethylsilylation was achieved by treating total lipid extracts with excess BSTFA/TMCS (99:1 v/v) for 10 min at 70 °C. The derivatized extracts were diluted with hexane and injected directly into the GC. Although total lipid extracts of fat samples can also be silylated using TMSIM/pyridine (15 min, room temperature), the requirement to remove the derivatizing agents and by-products prior to GC analysis is disadvantageous [185]. Further developments of this methodology have centred on the use of capillary columns coated with polar stationary phases to resolve different molecular species according to their degree of unsaturation, as well as by carbon number [185–188].

Although intact phospholipids are generally too thermally unstable to be submitted to GC, ether-linked lysophosphatidic acid and a range of analogues have been analysed following trimethylsilylation without prior hydrolysis [189]. Phospholipids were dissolved in dry benzene and derivatized by heating with BSTFA/TMCS (98:2 v/v) for 30 min at 60 °C. High resolution mass spectrometry confirmed that 1-O-hexadecyl-2-lyso-sn-glycero-3-phosphate formed the tris-TMS derivative as shown below (33) under the above reaction conditions. The TMS derivatives of 1-O-hexadecyl(16:0)-

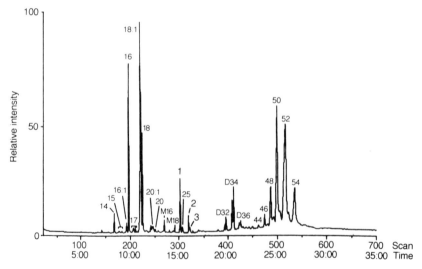

Figure 2. GC–MS total ion chromatogram of the total lipid extract of a potsherd recovered from the fill of a medieval ditch. Peak identities: 14–18, 20 and 25 are saturated fatty acids (as TMS ester derivatives) containing 14–18, 20 and 25 acyl carbon atoms respectively; 16:1, 18:1 and 20:1 are mono-unsaturated fatty acids (TMS esters) containing 16, 18 and 20 acyl carbon atoms, respectively; D32, D34 and D36 are diacylglycerols (TMS ethers) containing 32, 34 and 36 acyl carbon atoms, respectively; 44, 46, 48, 50, 52 and 54 are triacylglycerols containing 44, 46, 48, 50, 52 and 54 acyl carbon atoms, respectively; 1 is nonacosane, 2 is nonacosan-15-one, and 3 is nonacosan-15-ol (TMS ether). Analyses were performed using a 12 m × 0.22 mm i.d. BP-1 coated (immobilized dimethyl polysiloxane, 0.1 μm film thickness) fused-silica capillary. After on-column injection and a 2-minute delay the GC oven temperature was programmed from 50 to 350 °C at 10 °C/min. (Reproduced from reference [183] with permission).

and 1-O-octadecyl(18:0)-2-lyso-sn-glycero-3-phosphate eluted from a short (1 m × 1 mm i.d.) column packed with 1.5% OV-1 on Chromosorb Q (80–100 mesh) in ca. 1.9 and 2.9 min respectively, at an oven temperature of 290 °C (He carrier) without apparently suffering thermal degradation. The EI and CI mass spectra that were produced were of use for structure determination.

$$\begin{array}{l} \text{—O—}\sim\sim\sim\sim\sim\sim \\ \text{—OTMS} \\ \text{—O—}\overset{O}{\underset{OTMS}{\overset{\|}{P}}}\text{—OTMS} \end{array} \quad (22)$$

Long-chain α-mycolic acids isolated from various *Mycobacterium* spp. have also been determined intact by GC after methylation and trimethylsilylation with BSTFA/pyridine (2:1 v/v) for 30 min at 80 °C [190]. The mono-TMS ether derivatives (23) of methyl α-mycolates containing up to 86 carbon atoms eluted after over 2 h from a short (0.3–0.4 m × 3 mm i.d.) packed (1% OV-101 on Gas-Chrom Q) column at a GC oven temperature of 320–340 °C. These compounds are probably the longest chain natural lipids yet determined by GC–MS as TMS derivatives. The mass spectra of the TMS ether derivatives of the methyl mycolates display highly diagnostic fragmentations, particularly in relation to the nature and points of substitution of alkyl chains and hydroxyl groups.

$$CH_3C_nH_{2n}-\underset{TMSO}{CH}-\underset{\underset{CH_3}{\overset{|}{(CH_2)_{21}}}}{CH}-CO_2CH_3 \quad (23)$$

As already indicated in some of the above examples, the mass spectra of the TMS derivatives of many acyl lipids and related long-chain alkyl compounds can provide much useful structural information. For example, the TMS derivatives of mono- and diacylglycerols provide important information concerning the nature of the acyl moieties substituted on to the glycerol backbone, thus allowing positional isomers to be differentiated [191–193, 200]. The position of substitution of hydroxyl groups in long-chain compounds is generally relatively easy to determine [193, 200]. For example, lipid peroxidation products (hydroperoxides) can be determined by GC and GC–MS after reduction and trimethylsilylation (24) [194–196].

(24)

One of the standard methods of determining the exact double bond position in unsaturated fatty acids and other long-chain alkenyl compounds involves oxidation of the olefinic bonds with osmium tetroxide followed by trimethylsilylation [193, 197–200]. The original double bond position is deduced from the m/z values for fragment ions arising from cleavage between the carbon atoms that were originally connected by the double bond. As space does not permit full discussion of the mass spectral behaviour of these and other long chain compounds, the reader is directed to Refs. 191–193 and 200, and references contained therein.

4.2.5 Carbohydrates and polyols

Carbohydrates and polyols are relatively easy to trimethylsilylate, with most of the commonly used silyl donors being readily able to achieve full silylation. The major shortcoming in the use of silylation for this class of compounds arises from the multiple reaction products produced from natural sugars due to anomer formation and pyranose–furanose interconversion. Reducing sugars such as glucose exist in solution as an equilibrium mixture of forms known as anomers. Interconversion between the anomers occurs via the open chain form, with mutarotation resulting from the ready opening and closing of the hemiacetal ring. Each single sugar usually gives rise to five tautomeric forms: two pyranose, two furanose and one open chain form [201]. The anomers have different physical properties and are generally separated by GC (Figure 3). In order to minimize this interconversion, mild and rapid derivatization conditions are recommended. Equilibrium mixtures of anomers are generally to be expected in reducing sugars isolated from natural materials. The capillary GC behaviour of tautomeric equilibrium mixtures of sugars including aldopentoses [202, 203], aldohexoses [204] and ketohexoses [205] as their TMS derivatives has been thoroughly investigated. The tautomer equilibrium mixtures were identified by a combination of GC–MS and NMR data. Ring size can be clearly assigned from MS data [206], e.g. TMS derivatives of furanoses are characterized by very intense fragment ions at m/z 217, and pyranoses by the value of the ratio of m/z 204/217, which is always greater than 1 expect for arabinoses and riboses (between 0.7 and 1). The GC properties of the TMS derivatives of 17 disaccharides were prepared by treatment with HMDS after achieving mutarotation equilibrium in pyridine containing 0.2 M 2-hydroxypyridine for 15 h at 40 °C [207]. This study showed that the analysis of a complex mixture of disaccharides on a high resolution capillary column is relatively straightforward. The increased complexity of the chromatogram due to anomeric peaks is not a problem, provided the mutarotation equilibrium of each compound is taken into account (e.g. Figure 3).

Mixtures of HMDS/TMCS in pyridine are commonly used to silylate sugars [208–211]. Although pure sugars (glucose, mannose and xylose) were readily silylated using a mixture of HMDS/TMCS/pyridine (2:1:10 v/v/v), giving single product peaks by GC [211], small amounts of the anomeric forms were observed when the proportion of TMCS was doubled. A disadvantage of using HMDS/TMCS is the ammonium chloride precipitate, which can potentially contaminate the GC column if the reaction mixture is injected directly into the chromatograph [212]. This problem can be overcome either by extracting the derivatives into hexane [213] or by use of an alternative derivatizing agent. BSTFA, usually in combination with TMCS (10:1 v/v), is also widely used for the silylation of carbohydrates and polyols prior to GC and GC–MS analysis [208, 207, 214–221]. A notable advantage of using BSTFA derives from the possibility of injecting the reaction mixture directly for GC without detrimental effects to the anlytical column.

The preparation of alditols, by treatment of monosaccharides with sodium borohydride, avoids problems of anomer formation. Alditols have been analyzed as TMS derivatives rather than the acetate derivatives that are more commonly prepared [222]. The facile formation of TMS alditols with Tri-Sil 'Z' (Pierce Chemical Co.;

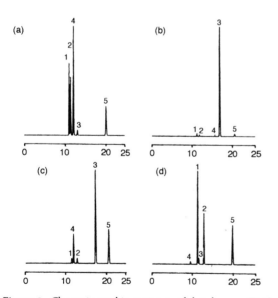

Figure 3. Chromatographic patterns of ketohexose TMS ethers on OV-17 at 150 °C. (a) Fructose; (b) sorbose; (c) tagatose; (d) psicose. Peaks: 1 = α-furanose; 2 = β-furanose; 3 = α-pyranose; 4 = β-pyranose; 5 = acyclic form. (Reproduced from reference 205 with permission).

TMSIM/pyridine) in a few minutes at room temperature may be advantageous, in view of the longer reaction times or higher temperatures required for acetylation. Use of the Tri-Sil 'Z' reagent was of use in quantitative analyses owing to its ability to remove traces of moisture from the reaction mixture [222]. Preparation of novel silylaldonitrile derivatives (25) has been described for the determination of aldoses by GC–MS [220]. Reaction of sugars with hydroxylamine-O-sulphonic acid yielded aldonitriles, which were silylated with BSTFA/pyridine (1:1 v/v). The silylaldonitrile derivatives of different sugars are readily separated by capillary GC, and yield mass spectra which contain ions suitable for use in structure investigations and quantitative studies. A review of the mass spectrometric behaviour of the TMS (and other) derivatives of sugars can be found in Ref. 223 and references given in Chapter 14.

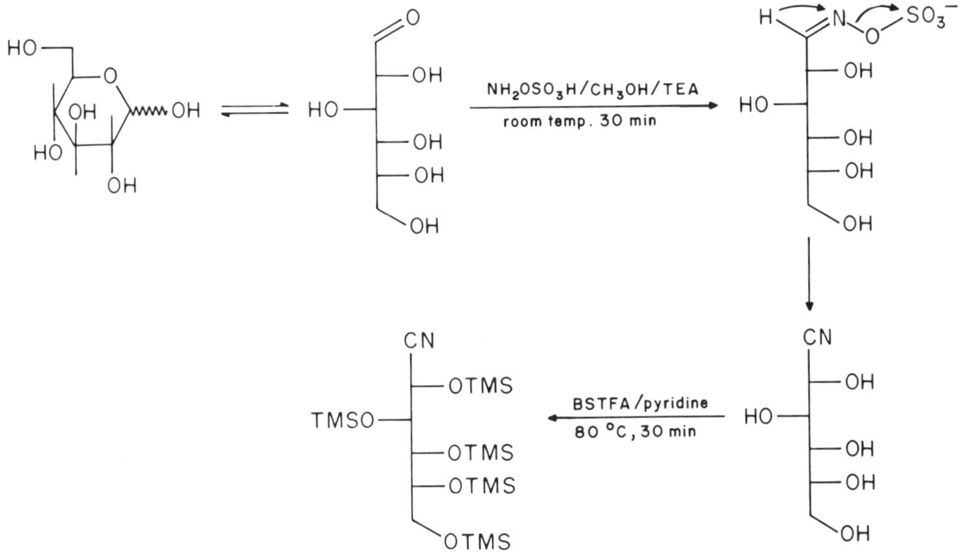

Among other recent analytical applications is a simple GC procedure for the simultaneous analysis of sugars and organic acids in extracts of vegetables and fruit using TMS oxime and TMS derivatives [224]. Trimethylsilylation was achieved after methyloxime preparation by treating pure compounds or extracts with HMDS/trifluoroacetic anhydride (9:1 v/v) for 1h at 100 °C. Among natural products, the stable anthrone O-glucoside was fully silylated by treatment with BSA/acetonitrile (1:1 v/v) for 1h at 90 °C [225] for its determination in skin-care cosmetics. Interestingly, defunctionalization of plant glucosinolates occurs to give pertrimethylsilyl desulphoglucosinolates (26) directly by their treatment with BSTFA/TMCS (10:11 v/v) for 15 min at 120 °C [215].

Also of interest in the phytochemical field are the favourable results obtain from the GC analysis of cyclitols and polyols as both their TMS and TFA derivatives [226] silylation being achieved with BSA/TMCS (19:1) with heating at 75 °C for 20 min. Interfering monosaccharides can be converted to oximes or dithioacetals prior to silylation (see Chapters 6 and 7).

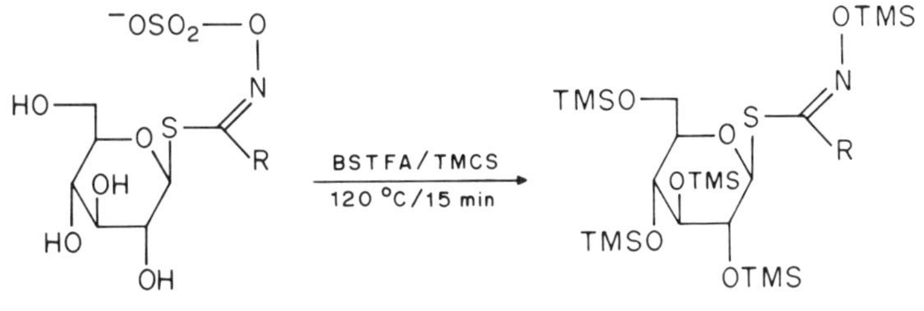

(25)

(26)

4.2.6 Phenols

Unhindered phenolic hydroxyl groups are readily silylated with most of the commonly used reagents, such as BSTFA or BSA. Hindered phenols require more vigorous conditions for silylation. Hoffman and co-workers [227, 228] found that trifluoroacetic acid was more effective than TMCS in catalysing the silylation of hindered phenols. 2-t-Butyl phenol was silylated in 20 min at room temperature using BSTFA/trifluoroacetic acid (20:1 v/v). A novel silylation reagent, N-trimethylsilyl-2-oxazolidinone [229] (SOA), has been found to be useful for the GC analysis of phenolic alcohols, particularly polyhydroxylactones of the leucodrin [230] and conocarpin [231] type. The most hindered hydroxyl function in this series occurs in the case of reflexin [232], where the tertiary hydroxyl group is severely hindered back and front. Full conversion to the pentakis-TMS derivative of reflexin was obtained with SOA in 2h at 72 °C with 0.1 M TsOH in either DMF or pyridine solution. When pyridine/HCl was used, more than 2h at 102 °C was required to obtain full silylation [233].

The wide occurrence of phenolics in nature, drug preparations and environmental samples means that trimethylsilylation has found wide application to their characterization, detection and quantification. For example, in an interlaboratory comparison of the GC–MS–SIM measurement of 11-nor-Δ^9-tetrahydrocannabinol-9-carboxylic acid in urine, the TMS-ether–TMS-ester was prepared by reaction with excess BSA at 50 °C for 15 min [234]. SIM was also used in the determination of residues of anabolic (oestrogenic) drugs in meat. Trimethylsilylation of the phenolic hydroxyl groups was achieved by reaction with BSA/TMCS (9:1 v/v) for 1h at 70 °C [235]. The possibility of using a co-injection flash heater derivatization for the silylation of oestrogenic compounds serves to emphasize the relative ease of silylating phenolic compounds. Diethylstilboestrol, dienoestrol, hexoestrol, zeranol, taleranol, zearalanone, zearalenone, zearalenol, oestradiol and oestriol were silylated by co-injecting a mixture of sample and BSTFA, or BSTFA containing 2% TMSIM, into the heated (260 °C) injection port of the GC unit operating in the splitless mode [236]. Derivatization performed thus was found to be both rapid and reliable in this instance. Further discussion of vapour phase derivatization techniques is given in Section 2.2.

In the field of natural product chemistry, the phenols and monoterpenes commonly found in volatile oils can be silylated using Tri-sil^{TM} or BSA prior to GC analysis. All compounds were silylated readily, although BSA produced a complex solvent peak which overlapped with TMS derivatives eluting at short retention times [237]. Long-chain substituted benzene-1,2-diol (catechol) lipids from the sap of the Burmese lac tree (Melanorrhooea usitata) have been analysed by capillary GC as their bis-O-TMS derivatives prepared by heating with BSTFA for 20 min at 70 °C [238]. More complex mixtures of phenolics, comprising chalcones, dihydrochalcones, flavones and flavanones, can be readily silylated by treatment with BSTFA containing 1% of TMCS [239]. The TMS derivatives of flavanone aglycones have been found to exhibit double peaks in GC analyses owing to isomerization between flavanones and their corresponding chalcones (27) [240]. This isomerization can occur during derivatization and GC injection. Minimal isomerization was observed when performing derivatizations with HMDS/TMCS (2:1 v/v) at room temperature for 30 min, then using cold on-column injection in GC and GC–MS analyses.

The prominent $[M-15]^+$ ions generated by the TMS derivatives were found to be of particular value in resolving the complex mixtures of phenolics occurring in plant exudates by SIM [239]. The mass spectra of the TMS derivatives of ginger constituents have been discussed in detail [241]. Both phenolic and non-phenolic hydroxyl groups were derivatized by heating samples with a mixture of BSTFA/TMCS/acetonitrile (2:1:2:v/v/v) for 10 min at 60 °C. Some problems with the interpretation of the mass spectra were resolved by derivatization with [^{2}H$_{18}$]BSA/acetonitrile for 10 min at 60 °C. The mass spectra of the TMS derivatives of 2-methoxyphenols show abundant $[M-30]^+$ ions due to the consecutive loss of two methyl radicals [242]. This distinctive *ortho* effect is valuable in the identification of isomeric phenolic compounds. Novel steryl ferulates have been characterized from seeds by GC–MS following trimethylsilylation with BSTFA cont. 1%

(27)

TMCS/pyridine (1:1 v/v) at ambient temperature for 15 min [243]. The mild reaction conditions were employed to minimize isomerization of the allylic double bond in the ferulate moiety.

4.2.7 Nitrogen-containing compounds

4.2.7.1 Nucleic acid constituents

Nucleic acid components, i.e. purine and pyrimidine bases, nucleosides and nucleotides, are amenable to analysis by GC and GC–MS as their TMS derivatives. The TMS derivatives are readily prepared [244] and possess good GC properties [245, 246] and well understood mass spectrometric fragmentation behaviour [247–250]. For the analysis of DNA and RNA hydrolysates, BSTFA is preferred to TMSIM or BSA for trimethylsilylation [244, 251–253]. Biological samples are generally fractionated by chromatography or solvent partitioning prior to silylation. A 1000 molar excess of BSTFA is recommended to overcome the effects of potentially interfering salts and hydrolysis by-products [254]. Somewhat different silylation conditions are required for different nucleic acid components. Acetonitrile was the most suitable of a wide range of solvents that were tested [253].

Preparation of the TMS derivatives of nucleic acid bases, nucleosides and nucleotides [251]   After thoroughly drying the sample overnight in a vacuum desiccator in the presence of P_2O_5, transfer a 50 μg portion to a suitable reaction vial, and add BSTFA containing 1% TMCS (40 μl) and pyridine (10 μl). Heat the tightly capped reaction vial at 100 °C for 1h. After cooling to room temperature the reaction mixture can be analysed directly by GC or GC–MS. Smaller scale derivatizations can be performed using melting point capillary tubes in place of the screw-capped reaction vials.

TMS derivatives of nucleic acid components are amenable to analysis on either packed or capillary columns [252, 253]. However, the latter are preferred on account of their greater resolving power. Problems appear to be encountered during the analysis of complex nucleotides, such as those containing polar modifications in the carbohydrate or aglycone, which decompose during chromatography [244]. Nucleoside monophosphates are not amenable to GC analysis as their TMS derivatives. Some undesired side reactions have also been encountered during trimethylsilylation reactions, such as the incorporation of oxygen at the C-8 position of 7-methylguanosine [255] and the dehydration of reduced pyrimidine nucleosides to their aromatic analogues [256]. Applications of TMS derivatives in this field include the characterization of a wide range of novel natural nucleotides in RNA and DNA (Ref. 257, and references therein), and the GC–MS analysis of free radical-induced products of pyrimidines and purines in DNA (Ref. 246 and references therein). Structures 28 to 30 are the TMS derivatives of dimeric products resulting from γ-irradiation of thymine, identified by GC–MS. Applications employing TMS derivatives other than TMS for the analysis of nucleic acid constituents are discussed below in the appropriate sections. Owing to the special problems that exist in the chromatographic analysis of cytidine and some of its analogues, the preparation of mixed derivatives has been proposed for GC–MS analysis [258, 259].

4.2.7.2 Amines

Silyl derivatives are less commonly prepared for the analysis of amino compounds than acyl derivatives. Indeed, whereas N-acyl derivatives tend to be more stable than O-acyl derivatives, the reverse situation exists for TMS derivatives of these moieties [3]. The reactivity of N-TMS groups is reflected in the use of N-TMS compounds as trimethylsilylating reagents (see above). The greater stability of N-TBDMS derivatives compared with the corresponding N-TMS derivatives is discussed below in Section 4.4.5. Powerful silyl donors are recommended, as amino groups are amongst the most difficult protic sites to trimethylsilylate. An additional problem in the silylation of amines arises from the possibility of replacing both protons in primary amines, resulting in the formation of mono- and bis-TMS products.

Primary and secondary amines can be trimethylsilylated with either BSTFA or BSA in the presence of catalytic amounts of TMCS. For example, histamine and its metabolite 4-imidazoleacetic acid are fully trimethylsilylated with BSA containing 4% TMCS, heating at 60 °C for 30 min [260]. Full silylation of the primary and secondary amine functions in histamine was confirmed by mass spectrometry. The N^1-function in indoles, e.g. tryptamine, N,N-diethyltryptamine, serotonin and tryptophan, can be trimethylsilylated using MSTFA/acetonitrile/TMSIM (95:95:10 v/v/v, 60 °C, 5 min) [51]. Heating (100 °C, 15 min) tryptamines with BSTFA containing 1% TMCS also selectively introduced a TMS group at the indole nitrogen [261], while serotonin yielded the bis-TMS derivative. Likewise, melatonin was trimethylsilylated selectively at the indole nitrogen by reaction with a mixture of BSA/TMCS/TMSIM (3:2:3 v/v/v) at room temperature for 15 min [262]. O-Acetyl or O-trifluoroacetyl derivatives of the primary amine function in indolealkylamines such as serotonin and tryptophan are conveniently prepared prior to trimethylsilylation of the phenolic oxygen or indolic nitrogen groups [52, 263].

4.2.7.3 Amino acids

Gehrke and co-workers have thoroughly investigated the conditions required for the study of the 20 major protein amino acids [263–266]. The best results were obtained by heating amino acids in sealed tubes with BSTFA/acetonitrile (1:1) for 2.5 h at 150 °C. The choice of solvent was found to have considerable influence on the degree of substitution of the nitrogen and, hence, on the number of peaks appearing in chromatograms. With the exception of glycine, arginine and glutamic acid, the N-TMS–TMS-ester derivatives of the other 17 amino acids were formed in 15 min at 150 °C. Conditions for the preparation and properties of the corresponding TBDMS derivatives of amino acids are given below (Section 4.4.5.2). N-TMS–butyl ester derivatives provide an alternative to the above [267]. Conversion to the butyl ester is achieved by heating with 3 M HCl in n-butanol for 15 min at 150 °C. After removal of excess reagent, silylation is performed by reaction with excess BSTFA/acetonitrile (1:1) at 150 °C for 90 min. Although 22 protein amino acids were derivatized under these conditions, the arginine derivative was found to decompose during GC. The short retention times of the alanine and glycine derivatives make BSTFA the reagent of choice in order to avoid problems of co-elution with reagent by-products. Trimethylsilylation of C_4–C_{11} ω-amino acid hydrochlorides using BSA alone (80 °C for 1–2 h, or 90 °C for 0.5–1 h) yields the tris-TMS derivatives (31). The free ω-amino acids yield only the bis-TMS derivatives with BSA, whereas a mixture of both the bis and tris derivatives is observed when acetonitrile is used as solvent [268]. The presence of the hydrochloride was presumed to catalyse the reaction. Attempts to trimethylsilylate iminodicarboxylic acids using BSA or BSTFA yielded mixtures of bis- and tris-TMS derivatives depending on the reaction conditions employed [269]. Again, silylation was greatly enhanced by the presence of HCl (i.e. when iminodicarboxylic acids were derivatised as their hydrochlorides). The formation of bis- and tris-TMS derivatives was presumed to be due to steric hindrance of the N-substituent group. Mixed derivatives were also observed during the trimethylsilylation of a variety of imino derivatives of alanine, owing to steric hindrance of the N-substituent group [270]. In order to avoid the problem of forming a mixture of derivatives, the possibility for achieving simultaneous N-trifluoroacetylation and O-trimethylsilylation of some iminodicarboxylic acids has been demonstrated by use of a mixture of N-trifluormethyl-α,α,α-trifluoroacetamide and BSTFA as derivatizing agent [271].

$$HCl \cdot NH_2(CH_2)_n CO_2 H \xrightarrow[80\,°C, 1-2\,hr]{BSA} (TMS)_2 N(CH_2)_n CO_2 TMS \quad (31)$$

4.2.7.4 Porphyrins

Novel TMS derivatives have been investigated for the GC and GC–MS analysis of porphyrins. The approach is based on the bis(trimethylsiloxy)silicon(IV) derivatives originally investigated by Boylan and co-workers for the analysis of the alkylporphyrins that occur widely in sedimentary materials [272, 273]. The derivatization proceeds with the reaction of free base porphyrins with hexachlorodisilane in toluene to form Si(IV) porphyrins [273–275]. The resulting dichloro Si(V) porphyrins are then converted to the corresponding dihydroxy compounds by hydrolysis with methanolic sodium hydroxide. After extraction into dichloromethane and evaporation to dryness trimethylsilylation is performed by reaction with BSTFA/pyridine (1:1 v/v) overnight at 70 °C. The derivatives so formed are amenable to GC and GC–MS analysis on packed columns [273] and capillary columns coated with apolar stationary phases [276]. The bis(trimethylsiloxy)Si(IV) derivatives of the alkylporphyrins that were investigated exhibited Kovats retention indices in the range 3200–3500 [275, 277]. In a search for new metalloporphyrin derivatives that might permit facile GC and GC–MS analysis of porphyrins, the method was applied to the preparation of other alkylporphyrin derivatives, such as trimethylsiloxy derivatives of Al(III), Ga(III) and Sn(IV) [275–278]. The bis(trimethylsiloxy)-Sn(IV) porphyrin derivatives, although readily prepared, were too unstable to survive GC analysis. Porphyrins bearing two polar carboxylate substituents (as methyl ester derivatives) were also amenable to analysis as their bis(trimethylsiloxy)Si(IV) derivatives [279]. Compounds that were analysed included deuteroporphyrin-IX dimethyl ester, mesoporphyrin-IX dimethyl ester and rhodoporphyrin-XV. Although these derivatives have not been extensively applied to the analysis of porphyrins from natural sources, the corresponding bis(t-butyldimethylsiloxy)Si(IV) derivatives have been effectively applied to the analysis of sedimentary alkylporphyrins (see Section 4.4.5.3 for further discussions).

4.3 Silyl derivatives other than TMS

The search for alternative silyl derivatives to TMS was an active area of investigation in the late 1970's and 1980's. Table 1 provides an indication of the range of silyl derivatives that has been tested. It should be stressed that few of these derivatives have been widely applied, and still fewer are in commercial production. The basic aim of using silyl derivatives other than TMS remains that of enhancing the thermal stability and increasing the inertness of the analyte relative to the chromatographic system. However, a number of other advantages can accrue from choosing derivatives other than TMS [7], namely: (i) enhanced hydrolytic stability; (ii) improved separations in chromatographic analyses; (iii) improved sensitivity and selectivity when used with selective detectors, e.g. element specific detectors or GC–MS selected ion monitoring, and (iv) mass spectra containing ions of greater diagnostic potential for use in structure investigations.

The general principles underlying the preparation and use of trimethylsilyl derivatives also apply to other silyl derivatives. However, changing the nature of the substituents on the silicon atom will affect the chemical and physical properties of the resulting silyl derivative. The major inpetus to the development of new silyl derivatives arose largely from the needs of synthetic chemistry for derivatives that were more stable to hydrolysis than the TMS ethers. The pioneering work of Corey and co-workers [280, 281] revealed the potential of isopropyldimethylsilyl (IPDMS) and t-butyldimethylsilyl (TBDMS) compounds for silylation. During their synthetic studies of prostaglandins, they showed the TBDMS and IPDMS derivatives to be respectively about 10^4 and about 10^2 to 10^3 times more stable to hydrolysis than the corresponding TMS ethers. The stability of the TBDMS ethers was demonstrated by their resistance to aqueous and alcoholic base hydrolysis under the conditions for acetate saponification, chromium trioxide–pyridine oxidation, mild chemical reduction (Zn–CH_3OH) and phosphorylation [282–285]. The stability of TBDMS-enol ethers was used to advantage in the isolation of ketone enolates from aqueous solution [286].

The steric effects of silyl groups of differing bulk and geometry have a major influence on the rate and extent of silylation reactions. Increasing the bulk of substituents on the silicon atom impedes the access of the reagent to functional groups. However, once formed, silyl derivatives containing bulky substituents will tend to resist chemical degradation by impeding the access of reagents to the reaction centre; e.g. the increased bulk of a TBDMS group compared with a TMS group is largely responsible for the increased hydrolytic stability of the former, as discussed above.

4.3.1 t-Butyldimethylsilyl (TBDMS) derivatives

Some of the properties of TBDMS derivatives have already been alluded to above in the general discussion of the use of silyl derivatives other than TMS in synthetic studies. The early promise of TBDMS derivatives is now being fulfilled through their routine application in the analytical field.

4.3.2 t-Butyldimethylchlorosilane (TBDMCS)

$$\text{—Si—Cl} \qquad (32)$$

TBDMS derivatives were originally prepared by reaction of one equivalent of tert-butyldimethylchlorosilane (TBDMCS) and two equivalents of imidazole in dimethylformamide added in excess to the analyte [282]. A number of problems exist in using this reagent routinely in analytical applications, including (i) its ineffectiveness in silylating thiols, primary and secondary amines and slightly hindered hydroxyl groups; (ii) the use of imidazole as an acid scavenger can interfere with the quantitative silylation of carboxylic acids, and (iii) possible reduction in yield owing to the need for extraction of the derivative from derivatizing reagents. A further shortcoming arises from the poor solubility of some derivatives, e.g. those of certain steroids, in dimethylformamide used as solvent. Precipitates that form can be redissolved by addition of another solvent, e.g. dichloromethane, after derivatization. An alternative strategy that has been used successfully involves simply replacing the dimethylformamide with pyridine [281, 287].

4.3.3 N-t-Butyldimethylsilylimidazole (TBDMSIM)

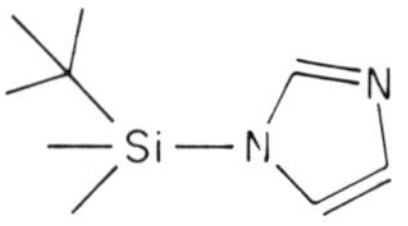

(33)

Tert-Butyldimethylsilylimidazole (TBDMSIM) is presumed to be an intermediate in the above reactions. However, when used alone, TBDMSIM, unlike TMSIM, is a weak silylating agent. Possible advantages of using TBDMSIM include the lack of HCl released by chlorosilane-based reagents, and the selective silylation of hydroxyl groups in the presence of amine functionalities [287]. The addition of TBDMCS (1% [279] and 5–50% [287]) to TBDMSIM greatly increases its reactivity.

4.3.4 N-t-Butyldimethylsilyl-N-methyltrifluoroacetamide (MTBSTFA)

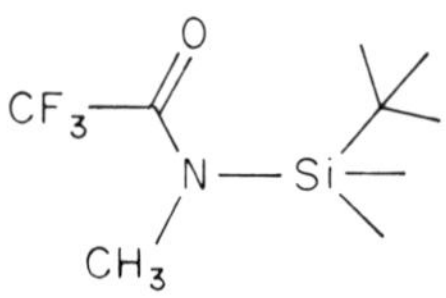

(34)

The development of N-t-butyldimethylsilyl-N-methyltrifluoroacetamide (MTBSTFA) [288], its analytical application [50] and its subsequent commercial availability proved to be a major advance in the routine application of tert-butyldimethylsilylation in analytical chemistry.

The advantages of using MTBSTFA rather than TBDMSIM and TBDMCS include: (i) enhanced reactivity, including its ability to silylate carboxyls, hydroxyls, thiols and primary and secondary amines; (ii) short reaction times, often at room temperature, and (iii) simplified work-up owing to derivatization by-products being neutral and largely volatile, so providing the possibility for direct injection of the reaction mixture for GC analysis.

MTBSTFA is a liquid at room temperature, and so can often be used for derivatizations without solvent. Solvents frequently used include DMF, acetonitrile or pyridine. The catalytic action of small amounts of TBDMCS (ca. 1% v/v) serves to enhance the silylation power of MTBSTFA [289–292]. The addition of 1% v/v tert-butyldimethyliodosilane to MTBSTFA has been found to catalyse the formation of the TBDMS enol-ether of keto steroids, such as testosterone [50]. The selective reaction of secondary amines [288, 289] and the β-hydroxyl groups in secondary amines [290] with MTBSTFA, resulting from steric hindrance of the first introduced TBDMS group towards subsequent silylations, can be used to advantage in analytical work. A mixture of MTBSTFA/TBDMSIM/TBDMCS in acetonitrile (50:5:0.5:100 v/v/v/v) has been used to silylate hydroxy fatty acids [293, 294]. The GC–MS analysis of dihydroxyeicosatetraenoic acid from diseased skin was achieved after micro-scale hydrogenation, followed by silylation for 12 h at room temperature, or 1 h at 45 °C, using the last-mentioned reagent mixture [294].

4.4 Applications of t-butyldimethylsilylation to the GC analysis of organic compounds

4.4.1 Short-chain organic acids and related compounds

Both MTBSTFA [295–299] and TBDMCS/imidazole/DMF [300] reagents have been used for silylation in this field. The best method of preparing TBDMS derivatives is to use MTBSTFA, in view of its ease of reaction and convenience, since there is no need for sample purification prior to GC or GC–MS analysis. The method of derivatization set out below is equally well applied to volatile and non-volatile carboxylic acids.

Solid-phase extraction and TBDMS derivatization of carboxylic acids [298, 299] To 1 ml of aqueous sample or biological sample (e.g. saliva or urine) was added 100 µg/ml of internal standard (*trans*-cinnamic acid dissolved in methanol) and the solution was made basic by saturation with solid sodium bicarbonate.

After extraction with diethyl ether, the ether phase was discarded and the aqueous phase was acidified with concentrated sulphuric acid (0.1 ml) and saturated with sodium chloride (400 mg). The aqueous phase was then loaded onto a Chromosorb P column under nitrogen pressure. The packing was wetted up to 80%. The organic acids were eluted from the column with diethyl ether. The first 1.5 ml of ether eluate was collected in a Reacti-Vial containing 20 µl of triethylamine (TFA) and evaporated to ca. 50 µl under a stream of nitrogen. For derivatization of TEA salts of carboxylic acids, 20 µl of MTBSTFA and 60 µl of isooctane were added to the vial, which was then capped and heated at 60 °C. Under these conditions the hindered hydroxyl and carboxylic groups of tartaric acid and γ-resorcylic acid, as their TEA salts, were quantitatively silylated in 4 h and 8 h respectively. N-Silylation of p-aminobenzoic acid and hippuric acid did not occur, even after prolonged heating with excess of MTBSTFA.

The above procedure overcomes many of the problems associated with the analysis of volatile carboxylic acids by performing derivatizations on the TEA salts and analysing aliquots of the reaction mixture directly. Moreover, the quantitative derivatization of hydroxyl and carboxyl functions makes the method applicable to a wide range of carboxylic acids of physiological importance. Schooley et al. [297] achieved quantitative results for volatile and non-volatile free acids by treatment of ether extracts (1 ml) with MTBSTFA containing 1% TBDMCS (100 µl) at 80 °C for 20 min. Once formed, the TBDMS derivatives of the carboxylic acids investigated were found to be stable for at least six months when stored at 4 °C. The TBDMS derivatives of valproic acid (2-propylpentanoic acid) and twelve of its metabolites, prepared using MTBSTFA, were preferred to the corresponding TMS derivatives for capillary GC–MS on account of improved sensitivity for unsaturated metabolites and shorter analysis times [301]. The bis-TBDMS esters of dicarboxylic acids were prepared by reaction with 100 µl of acetonitrile and 10 µl MTBSTFA at room temperature (22 °C). Figure 4 shows the separation of a range of biologically important short and long chain carboxylic acids and phenolic acids as their TBDMS derivatives.

In common with the mass spectral behaviour of the TBDMS derivatives for other classes of organic com-

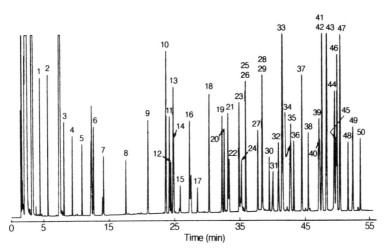

Figure 4. Chromatogram of a mixture of carboxylic acids as the t-butyldimethylsilyl derivatives. GC conditions: DB-1 fused-silica capillary column (30 m × 0.32 mm i.d., 0.25 µm), initially at 60 °C for 2 min, then programmed to 280 °C at 4 °C/min; 0.8 µl sample, injected with split ratio of 15:1; both injector and detector temperatures at 300 °C; nitrogen as the carrier gas at 0.9 ml/min. Peaks: 1 = formic, 2 = acetic, 3 = propionic, 4 = isobutyric, 5 = butyric, 6 = isovaleric, 7 = valeric, 8 = caproic, 9 = enanthic, 10 = benzoic, 11 = caprylic, 12 = lactic, 13 = phenylacetic, 14 = glycollic, 15 = oxalic, 16 = pelargonic, 17 = malonic, 18 = capric, 19 = succinic, 20 = methylsuccinic, 21 = undecanoic, 22 = fumaric, 23 = 5-phenylvaleric, 24 = p-aminobenzoic, 25 = lauric, 26 = mandelic, 27 = adipic, 28 = 3-methyladipic, 29 = tridecanoic, 30 = phenyllactic, 31 = hippuric, 32 = myristic, 33 = p-hydroxybenzoic, 34 = malic, 35 = suberic, 36 = pentadecanoic, 37 = vanillic, 38 = palmitic, 39 = syringic, 40 = tartaric, 41 = margaric, 42 = α-resorcylic, 43 = p-hydroxymandelic, 44 = γ-resorcylic, 45 = stearic, 46 = homogentisic, 47 = protocatechuic, 48 = nonadecanoic, 49 = citric, 50 = arachidic acid. (Reproduced from Ref. 299 with permission).

pound the TBDMS ethers of short-chain organic acids display a prominent $[M-57]^+$ ion which can be used to derive the molecular weight of the acid. Alkyldimethylsilyl (RMe_2Si) derivatives have the general property that the abundance of the $[M-R]^+$ ion is a function of the stability of the leaving radical $R^{\cdot}$. The abundance of $[M-R]^+$ increases in the order TMS < EDMS < PDMS < IPDMS < TBDMS. Comparison of the mass spectra of the TMS and TBDMS ethers of short-chain organic acids shows that the $[M-R]^+$ ion is on average 6.4 times more abundant for the TBDMS esters compared with the TMS esters. This property serves of enhance the sensitivity in GC–MS analyses employing SIM [300]. Similar trends in fragmentation are seen for the bis-TBDMS esters of dicarboxylic acids [296]. By analogy with the O-TMS quinoxalinol derivatives discussed above in Section 4.2.1, the O-TBDMS quinoxalinol derivatives have been prepared for 2-keto acids [480]. The EI spectra of these derivatives display pronounced $[M-57]^+$ ions, which were used for the high sensitivity detection of aliphatic branched 2-keto acids in rat muscle isolates [480]. The high proportion of the ion current carried by the aforementioned ion obviated the need to use CI, which was necessary in analyses using the corresponding TMS derivatives.

4.4.2 Prostaglandins and other eicosanoids

Interest in the biosynthesis, metabolism and physiological roles of prostaglandins and related compounds has prompted extensive investigation of sensitive and selective analytical procedures. An early report in this area [302] described the preparation of TBDMS ethers of prostaglandin methyl esters and methyl ester–oximes by treating samples with 50 μl of TBDMCS (2 M in DMF) and 50 μl of imidazole (2 M in DMF) at 100 °C for 1 h). After cooling the, reaction mixture was applied to the top of a 0.5×3 cm column of Sephadex LH-20 swollen with heptane/ethyl acetate (3:1), and the derivatives were eluted with 4 ml of the same solvent. After evaporation of the solvent, samples were analysed by GC–MS.

An early success of using TBDMS derivatives for prostaglandin analysis was the identification of 19-hydroxyprostaglandin $F_{2\alpha}$. Attempts to confirm the identity of this compound from the EI mass spectrum of the TMS derivative left doubts concerning the number of hydroxyl groups. However, the EI spectrum of the TBDMS derivative exhibited a base peak at m/z 785 which confirmed the presence of the additional hydroxyl group unambiguously [303]. TBDMS derivatives have found wide application in the analysis of prostaglandins, thromboxanes, leukotrienes and related hydroxy fatty acids [304–321]. As with the TMS derivatives, TBDMS derivatives are usually prepared following methylation or pentafluorobenzylation of the carboxylic acid moiety and conversion of keto functions to methyloximes. While some researchers have adopted variants of the original procedure [302] (see also above) for the preparation of TBDMS ethers, others have used TBDMSIM containing 1% of TBDMCS dissolved in DMF [315]. The TBDMS-ester TBDMS-ether of leukotriene B_4 (LTB_4) was readily prepared by heating for 1h at 100 °C; the corresponding 20-COOH and 20-OH metabolites required reaction for 2h at 100 °C for complete silylation. The TBDMS esters offer no significant advantage owing to their relatively poor hydrolytic stability compared with the commonly prepared methyl esters. Moreover, the EI mass spectra of the TBDMS ester of LTB_4 exhibited decreased abundance of high mass ions compared with the methyl ester. The pentafluorobenzyl ester TXB_2-methoxine was converted to its tris-TBDMS derivative (35) by reaction with TBDMSIM (20 μl) in toluene (30μl) together with 500 μg of methoxyamine hydrochloride at 80°C for 2h [316].

More recent reports have described the use of MTBSTFA for silylation in the analysis of prostaglandins, leukotrienes and hydroxy fatty acids [314, 315, 317,

319]. One of these reports described the quantitative determination of urinary prostanoids by capillary GC–high resolution MS [317]. Prostaglandins PGE$_2$ and 6-keto-PGF$_{1\alpha}$ were analysed as their methoxime–TBDMS ether–TBDMS ester derivatives. The derivatives were formed by an unusual derivatization procedure in which the pentafluorobenzyl ester prepared initially was converted to the TBDMS ester by exchange, involving treatment with MTBSTFA (40 µl) in acetonitrile (50 µl) and pyridine (10 µl) followed by incubation at 40 °C for 1.5 h. The derivatives were dissolved in hexane and washed with water prior to GC–MS analysis. The $[M-57]^+$ ions present in the EI spectra were found to carry a significant proportion of the ion current. High resolution SIM ($m/\Delta m = 5000$ and 10 000) of this ion eliminated interferences encountered in analyses performed at low resolution (e.g. $m/\Delta m = 1000$), with detection limits for PGE$_1$ in urine being in the 50–100 pg/ml range. GC–MS with high resolution SIM is considered the method of choice for the microanalysis of prostanoids in biological tissues and fluids [322]. Impressive detection limits can also be achieved by using highly selective sample preparation procedures. For example, an immobilized antibody affinity (immunoaffinity) column has been employed for the one-step purification of TXB$_2$ from urine prior to derivatization. Low picogram detection limits were attained by capillary GC–MS–SIM employing negative ion chemical ionization analysis of column fractions following preparation of the methoxime–pentafluorobenzyl–tris-TBDMS derivatives [316].

4.4.3 Steroids, bile acids and related compounds

Steroids were one of the first classes of compound to be analysed by GC and GC–MS as TBDMS ether derivatives [302]. The high yields that were obtained in this early study, combined with their hydrolytic stability and favourable mass spectrometric properties, make TBDMS derivatives especially useful for analytical applications. A wide variety of reaction conditions has been used for the formation of TBDMS derivatives of steroids. The most commonly used reagent is TBDMSIM, either formed in situ or added as the pure compound. Relatively mild reaction conditions (room temperature, 1 h) were employed when using a TBDMCS/imidazole mixture in dimethylformamide to derivatize 15 monohydroxy and 23 monohydroxymonoketonic steroids of the androstane, pregnane and cholestane series [323]. T-Butyldimethylsilylation of testosterone, testosterone oxime and testosterone methyloxime can be performed using TBDMCS/imidazole/DMF (1:1:16 w/w) at 20 °C overnight [324] but more severe reaction conditions are required to prepare TBDMS ethers of the methyl oximes of cortisol metabolites with TBDMCS/imidazole (80 °C, 1.5 h [325].

As problems are encountered in the trimethylsilylation of 11β or 17α hydroxyls in steroids (see above; Section 4.2.3), it is to be expected that the use of a more sterically crowded trialkylsilyl group will exacerbate problems of achieving full silylation. Although the tris-TMS ether of 5β-pregnane-3α,17α,20α-triol could be formed by reaction with TMSIM by heating overnight at 80 °C, no detectable silylation of the 17α-hydroxyl was observed after heating for 90 h at 80 °C with various TBDMS reagents [287]. Problems can also be encountered in forming TBDMS derivatives in polyhydroxy compounds with adjacent hydroxyl groups. The first silyl groups to become attached may hinder the next incoming silyl group, resulting in slow or incomplete derivatization.

As discussed earlier (see also Chapter 6), oxime derivatives are often prepared for polyfunctional compounds containing ketone groups in order to prevent the formation of mixed enol-silyl ethers. An alternative approach is to use a catalyst to enhance the yield of enol-silyl ethers (see also above, Section 4.2.3, for TMS ethers). The use of potassium acetate in toluene as catalyst is reported to produce a quantitative yield of TBDMS ethers and >96% yield of TBDMS-enol ethers of α,β-unsaturated keto steroids [326]. An alternative to this approach, set out below, uses sodium formate to catalyse enol-silyl ether formation with fewer by-products being produced [12, 13].

Conversion of steroids and bile acids to enol–TBDMS ether derivatives A solution of sodium formation in water (1 mg in 100 µl) was dried under a stream of nitrogen in a 1 ml reaction vial with a Teflon-lined screw-cap. The vial was then heated at 270 °C for 30 min. After cooling, the steroid (ca. 10 µg) was added to the vial in 100 µl of methanol. The solvent was removed under a stream of nitrogen and 100 µl of heptane and 20 µl of TBDMSIM were added. The vial was purged with nitrogen, sealed and heated for 4 h at 100 °C. After heating, the excess reagent was destroyed by adding 20 µl of 2-propanol and heating of the closed vial for 10 min at 100 °C. After cooling the excess reagents were removed according to one of the following procedures.

(i) Solvent extraction Add 0.5 ml of water to the above reaction mixture, then extract three times with 0.5 ml of hexane. Concentrate the hexane extract under a stream of nitrogen and analyse by GC or GC–MS.

(ii) Elution through Lipidex gel Dissolve the reaction

mixture in 1 ml of hexane and apply to a column of Lipidex 1000 (10 × 9 mm i.d.) held in a glass tube between two pieces of Teflon covered by 10 μm stainless steel gauze, with one end piece to allow attachment of a syringe (5 ml) and the other a tip for elution of the sample. Wash the reaction vial with two further volumes of hexane and elute each time through the column using the syringe. Combine the hexane washings, concentrate and analyse by GC and GC–MS. The recovery of the bis-TBDMS derivative of 3β-hydroxy[4-^{14}C]androst-5-en-17-one from such a column was 95% [12].

Gaskell and co-workers have taken advantage of the favourable properties of the TBDMS derivatives to develop highly sensitive and selective GC–MS techniques for the detection and quantification of steroids in biological matrices. Reference 327 contains an overview of the various sample preparation and mass spectrometric strategies developed by this group. The EI mass spectra of the TBDMS ethers of steroids contain abundant high mass ions, e.g. $[M-57]^+$, which are of use in SIM (see also Chapter 14). In one study [324], plasma testosterone was analysed as its TBDMS ether, methyloxime–TBDMS ether or TBDMS oxime–TBDMS ether derivative, using high resolution SIM and metastable peak monitoring techniques. In high resolution SIM, the mass spectrometer is set to detect ions of a selected exact mass, and hence of specified elemental composition, rather than all ions of a particular nominal mass [329]. The abundant high mass ions in the spectra of steroid TBDMS derivatives help to enhance detection limits. Metastable peak monitoring involves the detection of fragmentations occurring in the first field-free region of a double-focusing magnetic sector mass spectrometer and achieves high selectivity as detection depends on the m/z value of both the parent and the daughter ions [330]. TBDMS derivatives of steroids yield favourable metastable decompositions, e.g. $[M]^{+\bullet} \rightarrow [M-C_4H_9]^+$. Figure 5 displays the daughter ion spectrum for the bis-TBDMS derivative of oestradiol showing the successive metastable losses of tert-butyl and dimethylsiloxy radicals from the molecular ion. Gaskell and co-workers were able to combine the advantages of both these techniques by using a tandem double-focusing/quadrupole instrument [331]. Figure 6 summarizes the results of GC–MS analyses of the bis-TBDMS ether of oestradiol in blood plasma using the various techniques discussed

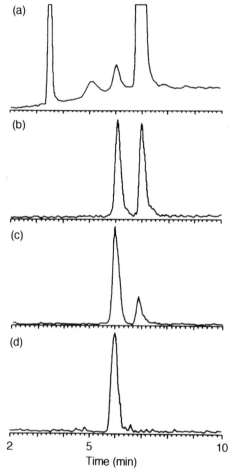

Figure 6. GC–MS analyses of a plasma extract for oestradiol-17β, as the bis-TBDMS ether. (a) Low resolution (1000) selected ion monitoring of m/z 500; (b) selected reaction monitoring of m/z 500 → 443; parent ion resolution 1000; (c) as (b), parent ion resolution 2500; (d) as (b), parent ion resolution 5000. (Reproduced from Ref. 331 with permission).

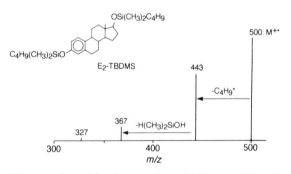

Figure 5. Scan of daughter ions formed following collisional activation of m/z 500 from oestradiol-17β (E_2) bis-t-butyldimethyl-silyl ether. The analysis was performed on a hybrid EBQQ instrument (VG 7070 EQ). Parent ions were selected using the double focusing instrument (EB); collisional activation (air; 35 eV) was effected in the first quadrupole and the spectrum shown resulted from a scan of the second quadrupole. (Reproduced from Ref. 327 with permission).

above to improve analytical specificity. At a parent ion resolution of 5000 and with selected reaction monitoring of the $[M-C_4H_9]^+$ daughter ion, the detection limit for the bis-TBDMS derivative of oestradiol was 10 pg, with the chosen analyte being the only component detected.

4.4.4 Fatty acids and related acyl lipids

The preparation of TBDMS derivatives of fatty acids and acyl lipids, e.g. mono- and diacylglycerols, is straightforward and provides a viable alternative to TMS derivatives for GC analysis. The TBDMCS/imidazole mixture described by Corey and Venkateswarlu [282] continues to be used for the derivatization of fatty acids and other acyl lipids [332-339]. The hydrolytic stability of the TBDMS derivatives allows their purification by washing with water prior to GC or GC-MS analysis. The relatively high chemical stability of TBDMS ethers also reduces losses during adsorption chromatography. Kuksis and co-workers used TLC and $AgNO_3$ TLC to purify the TBDMS ethers of diacylglycerols [333] prior to GC and GC-MS analyses.

Preparation of TBDMS derivatives of fatty acids and diacylglycerols Dissolve fatty acids or diacylglycerols (microgram quantities) in N,N-dimethylformamide (500 µl) and add the silylation mixture (500 µl; prepared by dissolving 1mM TBDMCS and 2.5 mM imidazole in 1 ml of DMF). The reaction mixture is heated in a capped vial at 80 °C for 90 min. After cooling, the reaction mixture is diluted with 5 ml of water and extracted three times with petroleum ether. The extracts are combined and evaporated to dryness, and the residue is dissolved in hexane for GC-MS analysis.

The TBDMS derivatives of fatty acids and acyl lipids possess excellent GC properties. Carbon number resolution is readily attained on capillary columns coated with apolar stationary phases. Resolution of molecular species according to their degree of unsaturation has been achieved on capillary columns coated with a high temperature-stable polarizable stationary phase (Figure 7) [188]. Unlike those of TMS esters, the EI mass spectra of TBDMS esters of fatty acids do not generally exhibit a molecular ion, the highest mass ion usually being $[M-15]^+$. In common with many other TBDMS derivatives, the most abundant high mass ion corresponds to loss of the tert-butyl moiety from TBDMS group [335]. The TBDMS ethers of diacylglycerols display abundant $[M-57]^+$ ions, which can be used to deduce the molecular weight and degree of unsaturation. The 1,3-isomers are identified by the $[M-\text{acyloxymethylene}]^+$ fragments, which are

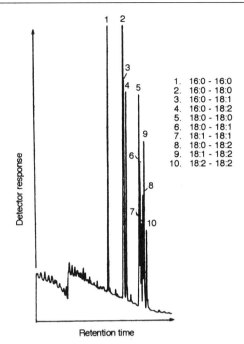

Figure 7. GC analysis of a synthetic mixture of diacylglycerols as their TBDMS ethers on a 25 m × 0.25 mm i.d. capillary column, coated with RSL-300 (methyl 50% phenylsiloxane) polarizable stationary phase. Tentative peak identification as given in the figure. Temperature programming was from 40–290 °C, ballistic; 290 °C, isothermal for 0.5 min; 290–330 °C, 10 °C/min; 330–360 °C, 1 °C/min. Dioleoylglycerol retained 21.23 min. Carrier gas, hydrogen at 1 bar head pressure. Manual on-column injection with fused-silica needle at 40 °C. (Reproduced from Ref. 188 with permission).

absent in sn-1,2- and sn-2,3-diacylglycerols. The abundance ratio of the ion due to losses of acyloxy radicals, $[M-RCO_2]^+$, from positions 1 (or 3) and position 2 indicates the proportions of the reverse isomers [333].

Platelet-activating factor (PAF) is a biologically active phospholipid acting as a mediator in allergic and inflammatory processes. Its TBDMS derivative has been successfully used in the highly sensitive determination of PAF in complex biological matrices using GC-MS-SIM [339-346]. Earlier reports involve the analysis of PAF following hydrolysis with phospholipase C to give the 2-acetyl-1-O-alkylglycerol, which was then converted to its TBDMS derivative by treatment with TBDMCS/imidazole reagent at 100 °C for 12 min. An alternative approach allows the 2-acetyl-1-O-alkylglycerol to stand overnight in order to effect migration of the acetyl group from the 2- to the 3-position [345]. Formation of the TBDMS derivative was achieved by reaction with 100 µl of neat TBDMSIM for 15 min at 110 °C.

Analysis of the 3-acetyl isomer was preferred on account of the formation of a single derivative and its more favourable EI properties, which are of value in structure investigations [345]. GC–MS analysis of the TBDMS derivative using EI with SIM is significantly less sensitive than analysis of the pentafluorobenzoyl derivative in the negative-ion mode [347].

Woollard [335] compared a variety of reagents and reaction conditions for the preparation of TBDMS derivatives of fatty acids, and preferred TBDMS pyrrolidide for the quantitative analysis of arachidonic acid in human skin. However, MTBSTFA can be used very conveniently for the analysis of a wide variety of carboxylated compounds, as demonstrated by Mawhinney et al. in their silylation of a wide range of mono-, di- and tricarboxylic acids [348].

Preparation of TBDMS esters of carboxylic acids [348]
To the carboxylic acids (microgram quantities) contained in a Reacti-vial equipped with a small PTFE-coated stir bar and a PTFE-faced silicone septum were added 10 µl of DMF and 250 µl of MTBSTFA at room temperature with constant stirring. The ammonium salts of mono-carboxylic and dicarboxylic acids were derivatized in the same manner with the exception that 40 µl of DMF were added together with 250 µl of MTBSTFA containing 1% TBDMCS and the samples heated to 60 °C with constant stirring until dissolution of the sample was achieved (< 15 min). The reaction mixture was analysed directly by GC and GC–MS.

Using the above protocol, silylation of twenty saturated monocarboxylated acids, sixteen dicarboxylates (both alkanedioic and alkenedioic), and the tricarboxylates citrate and aconitate was complete upon addition of MTBSTFA at room temperature. The derivatization was also effective when acetonitrile, DMSO, ethyl acetate or chloroform was used as solvent rather than DMF. Derivatization also proceeded to completion in the absence of organic solvent. As can be seen from Figure 4, the TBDMS derivatives of various carboxylic acids produce single product peaks which displayed excellent GC behaviour. The EI mass spectra of all the compounds are dominated by the $[M-57]^+$ ion that is commonly observed in the mass spectra of the TBDMS derivatives of other classes of compound.

4.4.5 Nitrogen-containing compounds

4.4.5.1 Nucleic acid constituents

GC–MS provides a very useful means of analysing nucleic acid constituents. Reviews of the analytical methodologies for the study of nucleic acid constituents, including sample preparation and silylation procedures, are to be found in references 250 and 257. Although trimethylsilylation has found wide application is this field (see above), TBDMS derivatives have also been shown to be of value in structure elucidation and trace analysis employing GC–MS. Westmore and co-workers [24, 287, 349], investigated the properties of various alkylsilyl derivatives including TBDMS for the GC analysis of nucleosides. A range of reaction conditions was used to prepare TBDMS derivatives, including reaction with a mixture of TBDMCS (1 M) and imidazole (2 M) in DMF or pyridine. Quilliam and Westmore [287] drew attention to the importance of steric crowding, which can inhibit full silylation for polyhydroxy compounds with adjacent hydroxyl groups. For example, uridine, a ribonucleoside containing a 2′,3′-*cis*-diol system, reacting for 10 min at room temperature (22 °C) with either of the above reagents, yielded 92 mol% of the bis-*O*-TBDMS derivative and 8 mol% of the 2′,3′,5′-tris-*O*-TBDMS derivative. The bis-*O*-TBDMS derivative comprised almost entirely the 2′,5′- and 3′,5′-bis-*O*-TBDMS-uridine derivatives. Quantitative silylation was achieved by silylating at 80 °C for 1 h (36). Favourable GC behaviour (no detectable decomposition) was observed for the fully silylated derivatives of the deoxynucleosides thymidine, 2′-deoxyuridine and 2′-deoxyadenosine, and for the ribonucleosides uridine, 5-methyl-uridine and adenosine [24, 287, 350]. The derivatives of deoxy- or ribonucleosides derived from cytosine or guanine bases were found not to be suitable for GC (packed column). As with the TBDMS derivatives of other classes of compound, the EI mass spectra of nucleosides yield mass spectra containing much more prominent high mass ions, notably the $[M-57]^+$ ion corresponding to loss of a t-butyl radical, than the corresponding TMS derivatives. Ions reflecting stereochemical control of fragmentation can be used in distinguishing between isomers [349].

TBDMS derivatives of nucleotide constituents have been prepared for investigation using GC and GC–MS in the areas of DNA base damage and cross-link analysis [351, 352], stable isotope tracers [353], quantitative analyses by stable isotope dilution MS [354, 355] and the analysis of a chemotherapeutic/radiosensitizer substance in cellular DNA [356]. Latterly, MTBSTFA has been preferred for the GC–MS analysis of nucleic acid constituents. Lewis and Yudkoff [353] found that optimum yields of the bis-TBDMS derivative of adenine were obtained by heating a mixture of MTBSTFA and acetonitrile for 1 h at 150 °C using either a 1:1 or a 1:3 ratio of solvent to reagent. The presence of 1% of TBDMCS enhanced the yields by 10–20%. Reducing the temperature or duration of heating resulted in a mixture of mono- and bis-TBDMS derivatives. This derivatization technique has been used successfully to determine the enrichment of 6-^{15}NH$_2$ of adenine nucleotides in rat hepatocytes perfused with [^{15}N]alanine, and in the development of a reference method for the RIA of adenosine [357]. The following procedure was used to hydrolyse DNA and prepare TBDMS derivatives for the determination of 5-methylcytosine content [355]. The molar percentage of 5-methylcytosine is accurately determined from the known thymine content of DNA by stable isotope dilution MS.

DNA hydrolysis and TBDMS derivatization Salt-free DNA and internal standards (^{15}N$_4$-5-([^{2}-H$_3$]methyl)-cytosine and [*methyl*-^{2}H$_3$]thymine prepared as described in reference 355 and references therein) are transferred to 5 cm × 3 mm o.d. cleaned and silanized Pyrex tubes flamed-sealed at one end, and dried *in vacuo*. To each tube is then added 10 μl of 98% formic acid, and the tube is flame-sealed, then spun briefly at 12 000 rpm to disperse any trapped air and heated at 180 °C for 1 h. After hydrolysis, the formic acid is removed *in vacuo*.

Immediately prior to derivatization, the dried DNA hydrolysate is treated (in the original hydrolysis tube) with 10 μl of dry pyridine and dried *in vacuo* to remove any trace of water by azeotrope formation. The derivatizing reagent (10 μl of a 1:1 solution of dry pyridine and MTBSTFA containing 1% TBDMCS) is added to the dried sample and the tube is flame-sealed and heated for 20 min at 120 °C. After heating, the solutions are analysed directly by GC–MS SIM of the $[M-57]^+$ ions using a bonded dimethylpolysiloxane or 5% phenyl 95% dimethylpolysiloxane coated capillary column. The bis-TBDMS derivatives of thymine and 5-methylcytosine elute in the Kovats retention index range 1900–2000.

4.4.5.2 Amino acids

A very convenient means of amino acid analysis by GC has been made possible through the development of the MTBSTFA reagent [357–365]. The attraction of using MTBSTFA lies in its ability to achieve direct silylation of amino acids to yield a single product for each of the commonly occurring amino acids.

Peptide hydrolysis and TBDMS derivatization of amino acids [362] Bovine insulin β-chain (100 μg) was weighed into a 1-ml derivatization vial and 500 μl of 6 M hydrochloric acid were added. The closed vial was heated at 110 °C for 24 h. After cooling to room temperature, the solvent was evaporated under a stream of nitrogen and the residue was dried over phosphorus pentoxide *in vacuo*. The dry residue was derivatized directly by the addition of MTBSTFA (50 μl), acetonitrile (50 μl) and ethanethiol (5 μl), then sonicated for 30 min at room temperature followed by heating at 150 °C for 2.5 h. After cooling, aliquots of the reaction mixture were used directly for capillary GC–MS analysis employing selected ion monitoring. Detection limits were at the picomole level. This derivatization method is readily applied to the analysis of amino acids isolated from serum by elution through a column of Dowex 50W-X8 [358].

This derivatization method is reproducible and affords good yields of amino acids as their N,O(S)-TBDMS derivatives. Except for the more basic amino acids, derivatization occurs at room temperature. The higher temperature used in the above procedure ensures complete derivatization of lysine, arginine and histidine. Arginine is the most difficult amino acid to derivatize; however, a single product is obtained under the reaction conditions described above. Derivatives of 27 amino acids (and glutamine and asparagine) were simultaneously chromatographed and well separated in a single analytical run on a 25 m × 0.2 mm glass capillary column coated with OV-1 [356]. The separation of a synthetic mixture of amino acids as their TBDMS derivatives is shown in Figure 8.

MacKenzie *et al.* [365] investigated the effects of temperature, solvent and reagents on the formation of

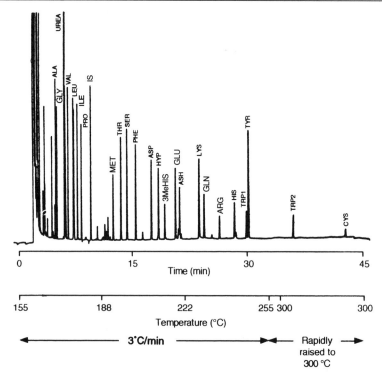

Figure 8. Gas chromatographic separation of t-butyldimethylsilyl derivatives of amino acid standards on a 30-m fused-silica SE-30 capillary column. Standard concentration: 100 μM each. IS = cycloeucine, internal standard. TRP1 and TRP2 are two peaks of tryptophan. (Reproduced from Ref. 363 with permission).

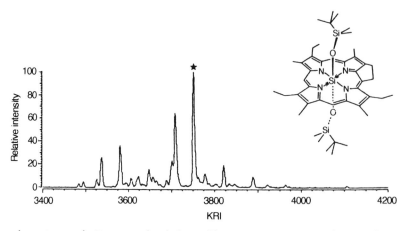

Figure 9. Total ion chromatogram for Boscan crude oil obtained by computer summation of ions in the mass range 550–850. Analyses were performed on a 25 m × 0.3 mm i.d. OV-1 coated (0.17 μm) flexible silica capillary (Hewlett-Packard) using helium carrier gas at a linear flow rate of 50 cm/s. Following on-column injection of ca. 2.5 μg of porpyrins as their (TBDMSO)$_2$Si(IV) derivatives (e.g. see inset, desoxophylloerythroacetioporphyrin derivative, major component of starred peak), the GC oven temperature was programmed ballistically from ambient to 150 °C, then from 150–290 °C at 3 °C/min. The retention time scale has been converted to Kováts' retention indices by computer interpolation from co-chromatographed n-alkanes.

TBDMS derivatives of amino acids. Although quantitative derivatization was achieved with MTBSTFA containing 1% of TBDMCS in DMF by heating at 75 °C for 30 min, two peaks were obtained for arginine. The size of the TBDMS group prevents multiple silylation of the nitrogen atoms. The EI mass spectra are sufficiently characteristic to make the TBDMS derivatives very useful for GC–MS analysis, with detection limits at the picomole level [362, 364]. Methane CI mass spectra have also been found to be of use in confirming the nature of TBDMS derivatives [365].

4.4.5.3 Alkylporphyrins

The preparation of bis(trimethylsilyl)Si(IV) derivatives for the GC–MS analysis of porphyrins has already been discussed in detail in Section 4.2.7. For the routine analysis [366–372] of alkylporphyrins extracted from geological materials, the corresponding bis-TBDMS derivatives are preferred. The bis-TBDMS derivatives (Figure 9) are prepared by treating dihydroxysilicon(IV) porphyrins (prepared from the free base alkylporphyrins as described in Section 4.2.4) with either TBDMCS–imidazole or MTBSTFA. In the case of the latter, two drops of MTBSTFA were added to the dihydroxysilicon(IV) porphyrin (0.1 mg) in dry pyridine (0.3 ml) and the mixture was heated for 10h at 60 °C. In some instances the porphyrin derivatives were vacuum sublimed before GC–MS analysis to remove volatile impurities. The bis-TBDMS derivatives were preferred to the corresponding TMS derivatives on account of their enhanced chemical stability and desirable mass spectrometric properties. Although bis-TBDMS derivatives elute ca. 380 Kóvats' retention index (KRI) units higher than the corresponding bis-TMS derivatives, the difference in temprature requirements is trivial. The naturally occurring alkylporphyrins elute predominantly in the KRI range 3400–4000 (Figure 9). The EI mass spectra of both the TBDMS and the TMS derivatives yielded abundant $M^{+\bullet}$ ions. This bis-TBDMS derivatives yielded an abundant $[M-R_3SiO]^+$ ion (M − 131) that was of use in deconvoluting the complex mixtures of porphyrins that occur in geological materials using mass chromatography [367].

4.5 Other alkylsilyl and aryldimethylsilyl derivatives

The development and applications of isopropyldimethylsilyl, allyldimethylsilyl, diethylsiliconide and diethylsilyl derivatives are discussed in detail in Sections 4.5.1, 4.5.2 and 4.5.3 below. Amongst the other alkyldimethylsilyl (DADS) derivatives that have been considered, dimethylethylsilyl (DMES) and dimethyl-n-propylsilyl (DMNPS) ethers have been the most thoroughly investigated. Both the alkyldimethylsilylimidazoles and the N,O-bis(alkyldimethylsilyl)trifluoroacetamides have been used as silyl donors [373]. As discussed above in Section 4.3, increasing the bulk of the alkyl substituents in the silyl group slows the rate of silylation. Consequently, the persilylation of 5β-pregnane-3α,17α,20α-triol with dimethylethylsilylimidazole (DMESIM) and dimethyl-n-propylsilylimidazole (DMNPSIM) requires ca. 10h at 100 °C for complete reaction, compared with ca. 2–3h for TMSIM. When the reaction temperature was raised to 150 °C, silylation with DMESIM and DMNPSIM was complete within 3h. In common with TMSIM, the axial and equatorial hydroxyl groups at C-3, C-7, C-11, C-12, C-16, C-17, C-20 and C-21 in the steroid skeleton were also silylated with DMESIM and DMNPSIM. As a general guideline the reactivity of alkyldimethylsilyl reagents for steroid hydroxyl groups follows the order:

TMS > ADMS > EDMS > DMNPS ≫ TBDMS

GC/MS was used to compare the silylation products of 44 hydroxylated steroids, including androgens, oestrogens, pregnanes, bile acids, and C_{27} sterols and stanols. The EI mass spectra of cholesterol TMS, DMES and DMNPS ether derivatives are shown in Figure 10 [373]. The trends in fragmentation are as would be expected, with shifts in the molecular ions from m/z 458 to 472 or 486. In addition, the appearance of A-ring fragment ions, containing the C-2 and C-3 carbons, at m/z 129, 143 and 157 further indicate the incorporation of DMES and DMNPS groups into cholesterol. The greater relative abundance of the $[M-29]^+$ and $[M-43]^+$ fragment ions compared with $M^{+\bullet}$ for the DMES and DMNPS derivatives of cholesterol follows the trend observed for the corresponding TBDMS, DMIPS and ADMS derivatives.

Comparison of the incremental methylene unit values for the capillary GC separation of TMS, DMES, DMNPS and DMIPS ether derivatives of a range of prostaglandin methyl esters, prostaglandin methyloxime–methyl esters and TXB_2 methyloxime–methyl ester showed that the latter three derivatives produce enhanced separations compared with the commonly used TMS derivatives [322, 374, 375]. While the TMS derivatives were prepared using BSTFA, imidazole donors were used for the preparation of DMES, DMNPS and DMIPS ethers; silylation being achieved by heating samples with excess reagent for 1h at 60 °C. Although the EI mass spectra of the DMES and DMNPS ethers of eicosanoids produced somewhat greater abundance high mass ions compared with the corresponding TMS derivatives, the intense

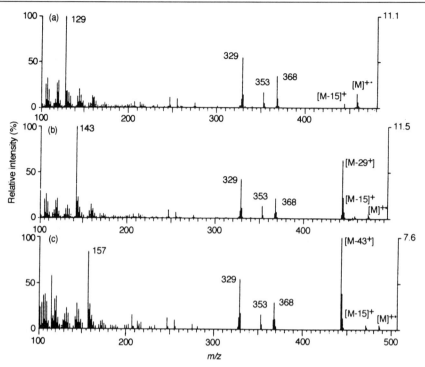

Figure 10. Mass spectra of cholesterol TMS (a), DMES (b) and DMPS (c) ether derivatives. (Reproduced from reference 373 with permission).

$[M-43]^+$ ion produced by the DMIPS ether made this the preferred derivative for microdeterminations on tissues and fluids [322, 374, 375] (see Section 4.5.1 below for further discussions of the use of DMIPS derivatives). The DMES, DMNPS, DMIPS and TMS ether derivatives of the pentafluorobenzoyl (PFB) esters of $PGF_{2\alpha}$ and $PGF_{1\alpha'}$ and the methoxime–PFB esters of PGE_1, PGE_2, 6-keto-$PGF_{1\alpha}$ and TXB_2 were also investigated for selective GC analysis using ECD [376]. The DMNPS ethers gave the best separations when analyses were carried out on an OV-101 coated capillary, and the derivative was stable for at least one week during storage in hexane solution (10 ng ml^{-1}) at room temperature. Application of the technique to the quantitative analysis of PGs in the urine of spontaneously hypertensive rats showed PGE_2 and $PGF_{2\alpha}$ to be present at concentrations of 92 ± 30 and 29 ± 7 ng ml^{-1} respectively, without serious interference from other extractable components.

Recent applications of dimethylalkylsilylation include the use of DMNPS ethers for the analysis of catecholamines [377]. The DMNPS derivatives showed enhanced stability during adsorption chromatography (silica gel) compared with the corresponding TMS derivatives. In GC–MS analyses employing ammonia CI with SIM of the $[M+NH_4]^+$ ion (base peak), the detection limit for dopamine was 2 pg ($S/N=2:1$). Dimethylethylsilylation has also been applied to the analysis of bile acids [378–383]. A range of analytical approaches has been investigated. DMESIM appears to be the reagent of choice for silylation in this area. Derivatization conditions range from treatment of samples with the neat reagent at room temperature for 30 min in the case of bile acid methyl esters [382] to the use of a mixture of DMESIM/hexane/pyridine (50:50:1 v/v/v) with heating to 60 °C for 1h in the case of their corresponding PFB esters [380, 381]. Superior separations were achieved for the methyl ester-DMES ether derivatives compared with the corresponding TMS ethers on an SE-30 coated capillary column [378]. Latterly the ethyl- [379, 382] or PFB-esters [380 381]. DMES ethers appear to have been preferred for the GC–MS–SIM analysis of bile acids in tissues and fluids. Analyses involving the latter derivative use negative ion chemical ionization with SIM of the characteristic $[M-181]^-$ anion which is observed when using isobutane as reagent gas. Where keto-groups are present in bile acids, these are usually converted to methyloximes prior to treatment with DMESIM [382]. Capillary GC can be applied to the analysis of glycine-conjugated bile acids as their methyl ester-DMES ethers without prior hydrolysis [383].

Harvey [384] has studied the GC–MS behaviour of fourteen substituted silyl derivatives of alcohols, steroids and cannabinoids. In addition to derivatives already discussed in this section and elsewhere in this chapter, he examined two aryldimethylsilyl derivatives, namely phenyldimethylsilyl and benzyldimethylsilyl ether, and four tri-n-alkylsilyl derivatives: the triethylsilyl, tri-n-propylsilyl, tri-n-butylsilyl and tri-n-hexylsilyl ethers. The majority of the derivatives that were studied showed good stability on GC analysis, with the shift in retention time reflecting the nature of the substituents on the silicon atom. The trialkylsilyl derivatives produced mass spectra that were similar to those of the alkyldimethylsilyl derivatives. The spectra of the alcohol and steroid derivatives contained extremely abundant $[M-alkyl]^+$ ions, which in many instances carried a greater proportion of the total ion current than the $[M-57]^+$ ions of the corresponding TBDMS derivatives. Interestingly, the TES derivative, which produces the same molecular weight increment shift as TBDMS, gave a more abundant $[M-alkyl]^+$ ion, i.e. $[M-29]^+$. The mass spectra of the trialkylsilyl derivatives also contain ions due to the loss of C_nH_{2n} from one of the alkyl chains [385]. The phenyl-

use in discriminating between phenolic and alcoholic hydroxyl groups [386]. The technique is analogous to that described for the conversion of oestradiol bis-TMS into a mixed TMS and TMS-d_9 by GC by means of the 'sandwich injection' technique [41]. In this application the trimethylsilylated steroid was injected into a packed GC column (injector temperature = 270 °C; oven maintained at 210–250 °C) with either DMESIM or DMNPSIM (37). Phenolic steroids, e.g. oestradiol, are converted into a DMES (or DMNPS) ether, or mixed TMS and DMES (or DMNPS) ether derivative with over 95% recovery. The selective exchange was presumed to depend on the difference in lability between the ethereal TMS linkages to the phenolic and alcoholic hydroxyl groups.

The careful use of mixtures of closely related silyl derivatives can assist in resolving problems arising in GC–MS analyses where spectra lack diagnostic fragmentations or where co-elutions mask compounds of interest [387]. For example, the number of trimethylsilyl groups in the TMS derivative of butyl-Δ^1-tetrahydrocannabinolic acid was determined by comparison of the mass spectra of the TMS and TMS-d_9 derivatives [387]. As part of the same investigation, an interfering peak

(37)

dimethylsilyl derivatives showed somewhat different behaviour to that discussed above for alkyldimethylsilyl derivatives, in eliminating a methyl rather than a phenyl radical from the $M^{+\bullet}$ ion. In common with the ADMS and flophemesyl derivatives discussed below in Sections 4.5.2 and 4.6.1 respectively, the benzyldimethylsilyl derivatives were found to stabilize the $M^{+\bullet}$ ions.

A novel application of dimethylalkylsilyl ethers in conjunction with TMS ethers has been described for

in the GC–MS analysis was selectively shifted to longer retention time by preparation of triethylsilyl and tri-n-propylsilyl derivatives, in order to reveal compounds of interest [388].

A novel computer algorithm has been employed to determine the number of silyl groups in a molecule [389]. This approach provides an interesting alternative to using mixed derivatives or deuterated silyl derivatives. The number of silyl groups present is calculated on the basis of isotope clusters of the derivative. The program is applicable to low resolution mass spectra of derivatives containing up to 10 silicon atoms. The method is limited to those compounds containing up to 20 carbon atoms lacking Cl, Br and S, and when the relative abundance of the molecular ion and/or $[M-15]^+$ ion is greater than or equal to about 30%.

4.5.1 Isopropyldimethylsilyl (DMIPS)

Promising results have been obtained from the analysis of DMIPS ethers of a range of biologically important compounds, including steroids [390], bile acids [391, 392], prostaglandins [374, 376, 393–402, 405] and thromboxanes [374, 376, 393–399, 403–405]. DMIPS ethers possess enhanced hydrolytic stability compared with TMS ethers and are more volatile on GC than TBDMS derivatives, owing to their lower molecular weight. Although DMIPS derivatives have been most commonly prepared using DMIPS-imidazole as the silyl donor, they may also be prepared from N,O-bis(DMIPS)trifluoroacetamide [390]. Derivatization of a range of steroids (including sterols, stanols, androgens, oestrogens and pregnanes) was performed by reaction with a 1:1 (v/v) solution of pyridine and IP-BSTFA at room temperature [390]. Steroids with unhindered hydroxyl groups were silylated smoothly in 1–2h at room temperature without formation of by-products. Steroids containing more hindered hydroxyl groups, e.g. 20α- and 20β-, reacted much more slowly. Heating (50 °C) the reaction mixture significantly reduced the reaction time. IP-BSTFA did not react with carbonyl groups in the steroids, and the rate of reaction of this reagent was slower than that of BSTFA, E-BSTFA and NP-BSTFA [390]. The DMIPSIM reagent has been extensively used by Miyazaki and co-workers. A derivatization procedure for the analysis of prostanoids and thromboxanes is set out below.

Preparation of DMIPS ethers of prostaglandins and thromboxanes Methyl esters of eicosanoids are prepared by treating extracts with 2 ml of freshly prepared diazomethane in methanol for 30 min at room temperature. After evaporation of the solvent and dissolution in 3 ml of n-hexane/ethyl acetate (3:1 v/v), the methyl esters are purified by stepwise elution through a silica gel column (50 mm × 8 mm i.d., 70–230 mesh, E. Merck, Darmstadt, Germany) using 20 ml each of solutions of n-hexane/ethyl acetate 3:1 and 2:1 v/v, 10 ml of 1:1 v/v n-hexane ethyl acetate mixture, and finally 20 ml of ethyl acetate/methanol (99:1 v/v). LTB_4 and TXB_2 elute in the 2:1 v/v and 1:1 v/v fractions respectively, while PGE_2, $PGF_{2\alpha}$ and 6-keto-$PGF_{1\alpha}$ emerge in the later ethyl acetate/methanol

fraction. The dried eluates are then treated with methoxyamine hydrochloride (100 µl of saturated solution in dry pyridine, 60 °C, 1h) to form methyloximes of keto-groups. Then the dried residue from above is treated with DMIPSIM (50 µl) for 1h at room temperature. To remove excess silylating reagents the reaction mixture is passed through a column of Sephadex LH-20 (30 mm × 8 mm i.d.), eluting with 2.5 ml of a mixture of chloroform/hexane/methanol (10:10:1 v/v/v). After solvent evaporation, the residue is dissolved in 100 µl of hexane containing 1% of pyridine (v/v). The LTB_4 methyl ester-DMIPS ether derivative should be further purified by silica gel column chromatography using n-hexane/diethyl ether (4:1 v/v) as the eluting solvent. Finally, the samples are analysed by GC–MS–SIM.

The DMIPS derivatives displayed favourable GC behaviour on both packed and capillary columns. The *syn* and *anti* isomers of the DMIPS derivatives of 6-keto-$PGF_{1\alpha}$ methyloxime are resolved as separate peaks on capillary columns. The mass spectra of the DMIPS ethers of steroids [390] and of prostaglandins and thromboxanes [391] show similar patterns of fragmentation to the corresponding TBDMS derivatives. The EI mass spectra are characterized by the appearance of an $M^{+\bullet}$ ion. Lower mass ions, $[M-15]^+$ and $[M-43]^+$, correspond to the elimination of methyl and isopropyl groups respectively from $M^{+\bullet}$. Sensitive and selective detection of the DMIPS derivatives of eicosanoids extracted from various biological tissues and fluids, e.g. brain [398], gall bladder wall [403], plasma [406] and urine [375, 404], has been achieved by use of high resolution SIM of the $[M-43]^+$ ion in order to eliminate interferences from endogenous contaminants that give ions of the same nominal mass as the analytes.

4.5.2 Allyldimethylsilyl (ADMS)

Several groups have considered the use of allyldimethylsilyl (ADMS) derivatives primarily with a view to improving detection limits in GC–MS analyses employing SIM. The ADMS derivatives of steroids, prostanoids and cannabinoids have been most widely studied. The earlier studies used allyldimethylchlorosilane (ADMCS) with [407, 408] or without [409] imidazole as catalyst to prepare ADMS ethers.

Preparation of ADMS ethers of steroids [407] Steroid (0.5 mg) was allowed to react with 100 µl of ADMCS, 1 M in DMF and 100 µl of imidazole (2 M in DMF). For steroids containing unhindered hydroxy groups, reactions were carried out at room temperature for 1h. Only partial reaction was observed for $3\alpha,11\beta$-

dihydroxy-5β-androstan-17-one after 5h at room temperature, but full silylation was achieved by heating for 1h at 75 °C. 5-pregnene-3β,17α,20α-triol was derivatized in 4h at room temperature or 1h at 75 °C. In all cases the reaction products were purified by elution through a column of Sephadex LH-20 as described above for the TBDMS ethers (see Section 4.4.2).

In one of the initial studies [408] of the mass spectral behaviour of ADMS derivatives of steroids, side reactions including allyldimethylsiloxane formation and intramolecular cyclization were found to occur during the derivatization (38). These side reactions were thought to arise at least in part from the high temperature and long reaction times (e.g. 100 °C for several hours) required for derivatization [408]. Both the above by-products appear to derive from the tendency of the allyl group to be readily displaced through nucleophilic attack, driven by resonance stabilization of the allyl anion (39) [410].

N,O-bis(allyldimethylsilyl)trifluoroacetamide (BASTFA). Prostanoids and steroids were then derivatized by reaction with a 1:1 mixture of pyridine/BASTFA for 60 min at room temperature (free carboxyl groups having been converted to methyl esters by treatment with ethereal diazomethane prior to silylation). Although BASTFA has the advantage of being conveniently removed before analysis by evaporation under a stream of nitrogen, its use is limited to the derivatization of unhindered hydroxyl groups. Poor yields of the ADMS ether of corticosterone (which contains a hindered 11β-hydroxyl group) were obtained even when derivatization was performed in the presence of catalytic amounts of ADMCS [412].

The mass spectra of the ADMS derivatives of steroids and prostanoids exhibit an abundant ion produced by the elimination of the allyl moiety, $[M-41]^+$. The ion currents carried by this latter ion were slightly less than those for the corresponding $[M-57]^+$ ion for TBDMS derivatives. However, the spectra contain more abundant ions characteristic of compound structure. ADMS derivatives possess shorter GC retention times and lower

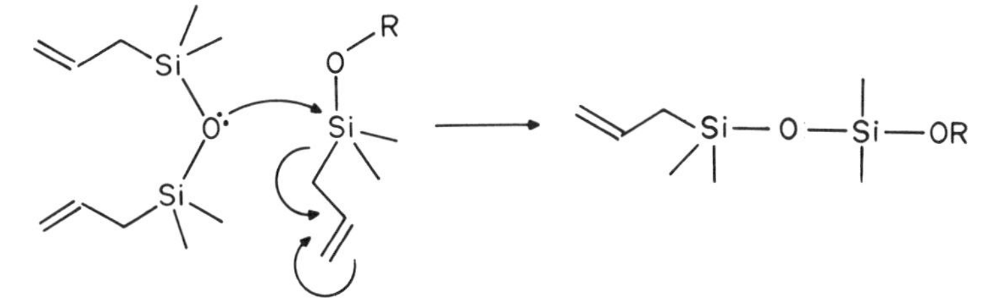

(38)

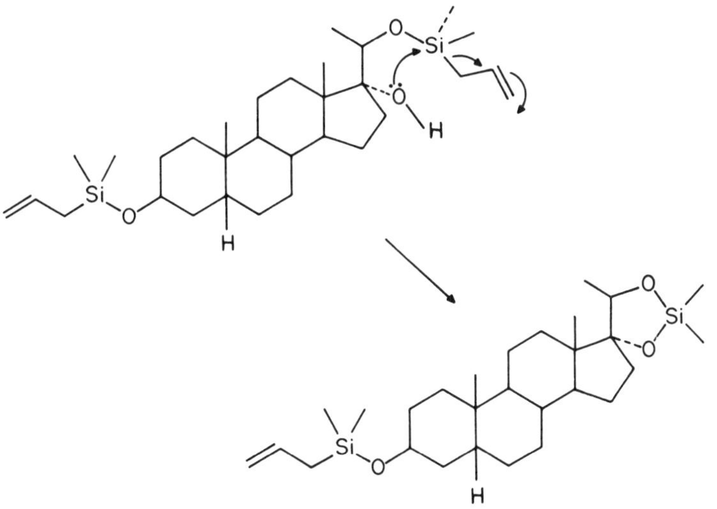

(39)

In an effort to overcome these problems, Steffenrud and Borgeat [411] prepared a new derivatizing agent

molecular weights than the corresponding TBDMS ethers. Picogram detection limits were achieved by SIM of the

[M − 41]⁺ ion for LTB₄, PGE₂, PGF₂α and pregnanediol [412]. The mass spectra of the ADMS derivatives of cannabinoids did not produce such an abundant [M − 41]⁺ ion and so were deemed to offer no special advantage over the TMS derivatives for analyses employing SIM [409]. The use of catalytic hydrogenation to reduce double bonds prior to silylation has been found greatly to enhance the GC sensitivities and relative abundances of the high mass ions in the EI mass spectra of ADMS and TBDMS derivatives of leukotrienes and monohydroxyeicosatetraenoic acids (HETEs) [413, 414]. The favourable GC–MS behaviour of these compounds permits low picogram detection limits in SIM analyses.

4.5.3 Cyclic silyl derivatives

The selective derivatization of bifunctional compounds has been exploited very effectively for the GC and GC–MS analysis of cyclic boronates (see Chapter 7 of this volume). Somewhat analogous cyclic alkylsilylene derivatives can also be prepared. In the late 1960s, dimethylsilylene derivatives of corticosteroids were successfully prepared by Kelly and shown to be amenable to GC analysis [415, 416]. Although dimethylsilylene derivatives of a wide variety of bifunctional compounds can be formed [417–420], their analytical application is limited by their tendency to polymerize and hydrolyse [420]. Greater stability is achieved by introduction of a di-t-butylsilylene (DTBS) group, which reportedly avoids the problems of hydrolysis and polymerization [421]. However, Brooks and co-workers have found that even the DTBS derivatives of acidic and phenolic substrates are relatively unstable in solution [422]. In spite of this they showed the DTBS reagent to be of use for the GC and GC–MS analysis of α- and β-hydroxy acids, alkylsalicylic acids, 1,2-diols, 1,3-diols and catechols [422–424]. DTBS derivatives were prepared by Brooks and Cole [422] as set out below; this is a modification of the method of Trost and Caldwell [421] and Corey and Hopkins [425].

Preparation of di-t-butylsilylene (DTBS) derivatives
The diol, hydroxy acid or amino acid (100 μg) was dissolved in acetonitrile (30 μl). N-Methylmorpholine (20 μl), 1-hydroxybenzotriazole (9 μg) (dried *in vacuo* at 40 °C prior to preparing the stock solution: 3 mg dissolved in 1 ml of acetonitrile; and di-t-butylchlorosilane (3.5 μl) were added sequentially, and the mixture was heated in a Reacti-Vial at 80 °C for 1 h (phenolic substrates and anthranilic acid) or 15 h (other substrates). The resulting solution was diluted to 100 μl with ethyl acetate and used for GC and GC–MS analysis. For the derivatization of sterically

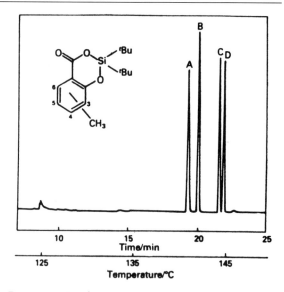

Figure 11. Gas chromatographic separation (25 m, OV-1) of the four isomeric methylsalicylic acid di-tert-butylsilylene derivatives: A, 6-Me; B, 3-Me; C, 5-Me; and D, 4-Me. (Reproduced from Ref. 373 with permission).

hindered 9-methyl-9,10-dihydrophenanthrene-cis-9,10-diol, di-t-butylsilyl di(trifluoromethanesulphonate) was substituted for di-t-butylchlorosilane in the above procedure. Saturated sodium hydrogen carbonate solution (1 ml) was added to the resultant solution and the DTBS derivative was extracted with diethyl ether (3 × 0.5 ml); the extract was dried over anhydrous sodium sulphate and evaporated to dryness. The derivative was redissolved in ethyl acetate (100 μl) for GC and GC–MS analysis.

Preparation of these derivatives allowed useful separation of some isomeric compounds by GC, e.g. the four possible isomers of methylsalicyclic acid DTBS derivatives were resolved to baseline on a capillary column coated with an apolar stationary phase (OV-1; Figure 11). The EI spectra of the DTBS derivatives generally showed $M^{+\bullet}$ ions which were of greater abundance in the case of the aromatic compounds that were tested.

Ishibashi and co-workers prepared a new silylating reagent, N,O-bis(diethylhydrogensilyl)trifluoroacetamide (DEHS–BSTFA), in an effort to overcome the problems of instability encountered in dimethylsilylene derivative [426]. This reagent has the advantage of yielding cyclic diethylsiliconide (DES) derivatives in the case of suitably adjacent hydroxyl groups or diethylhydrogensilyl (DEHS) ethers in the case of isolated hydroxyl groups.

Preparation of diethylsiliconide (DES) or diethylhydrogensilyl (DEHS) derivatives of steroids   Add DEHS–BSTFA

(100 μl) to a solution of steroid (0.1–0.2 mg) in pyridine (100 μl) and allow to stand for 1h at room temperature. Use the solution directly for GC or GC–MS.

A wide range of steroids has been derivatized using this method [426]. Silylations appear to proceed smoothly, with the DEHS–BSTFA reagent exhibiting similar reactivity to BSTFA. Single well-shaped GC peaks were obtained for mono-, di- and trihydroxysteroids. For example, dehydroepiandrosterone afforded a DEHS ether derivative, whereas 5β-pregnane-3α,17α,20α-triol yielded the 3-DEHS ether–17,20-cyclic diethylsilyl ether (40). 16-Epioestriol also yielded a DEHS–DES derivative, whereas oestriol formed a tris-DEHS derivative, thus indicating selectivity of the 1,2-cis-diol group of the steroid D-ring in forming the cyclic DES derivative. The DES–DEHS derivative of 5β-pregnane-3α,17α,20α-triol was stable for at least 7 days at room temperature. The EI mass spectra of DES–DEHS derivatives of steroids generally display abundant $M^{+\bullet}$ ions and few fragment ions.

The DEHS–DES derivatives of pregnanes with a hydroxylated 17β side chain [427], 5β-pregnane-17,20,21-triols [428], hydrocortisone [429], cortisol and 6β-hydroxycortisol [430] have also been prepared. Hydrocortisone was converted to the 3,20-bis-methyloxime derivative prior to reaction with DEHS–BSTFA, which yielded the 11-DEHS-17,20-DES derivative [429]. The relative ease of diethylhydrogensilylation of the 11β-hydroxyl group in hydrocortisone and 5β-pregnane-3α,11β,17α,20α-tetrol [430], compared with trimethylsilylation has been argued to result from the smaller bulk of the DEHS group which was more conducive to the production of the S_N2 transition state [429]. Where alkyloximes, e.g. methyloxime, were used to block ketone groups in steroids, doublets of peaks were seen in GC and GC–MS analyses owing to separation of the syn and anti geometric isomers.

Various prostanoids have also been analyzed using the DEHS–BSTFA reagent [431–434]. For example, derivatisation of PGF$_{2\alpha}$ and 15-keto-PGE$_{2\alpha}$ was performed by treating the methyl esters, or the methyl ester–alkyloxime in the case of the latter, with DEHS–BSTFA for 30 min at room temperature to yield the DEHS–DES methyl ester and DES methyl ester–alkyloxime derivatives respectively. Pure reaction products were obtained for the DEHS–DES methyl esters (41), as judged by the appearance of a single well-shabed peak on capillary GC chromatograms. The DES methyl ester–alkyloximes gave doublets of peaks because of separation of the syn- and anti-isomers. The favourable GC resolution attained for the DEHS–DES derivatives of ketonic PG methyl ester–alkyloximes may be used to advantage in deconvoluting the poorly resolved GC peaks produced by methyloxime derivatives [435, 436].

A recent study focused on the suitability of the DEHS–DES derivatives for the GC and GC–MS analysis of stereoisomeric bile acids [437]. Bile acids were converted to either methyl or pentafluorobenzyl (PFB) esters before treatment with DEHS–BSTFA in pyridine for 60 min at room temperature. Under these reaction conditions bile acids having hydroxyl group(s) at C-3, C-4, C-6, C-7 and/or C-12 yielded a single reaction product on GC analysis. Bile acids possessing isolated hydroxyl groups and diaxial *trans*-diol groups were converted to DEHS ethers, whereas those having vicinal diol groups (except for the diaxial group) were converted into cyclic DES derivatives. The differences in molecular weights of the DES and DEHS derivatives provide opportunities for distinguishing between isomeric bile acids. The favourable isobutane NICI properties of the PFB−DEHS−DES derivative offered considerable scope for the trace analysis of unusual bile acids containing a vicinal glycol group. In common with other derivatives [372, 381], the NICI spectra of the PFB−DEHS−DES derivative are characterized by the presence of carboxylate anions, [M − PFB]$^-$, as base peaks.

4.5.4 Alkoxyalkyl(or aryl)silyl derivatives

The preparation of alkoxydialkylsilyl derivatives raises possibilities for varying the nature of the alcohol (and hence the alkoxy group in the final derivative) in order to induce shifts in GC elution orders for compounds containing different numbers of protic sites. The possibility of varying the nature of the alcohol could also be used to modify the selectivity of detection of a chosen analyte. Two methods of preparing alkoxydimethylsilyl derivatives of steroids and cannabinoids were considered by Harvey [438]. The preferred approach proceeds as follows.

General method for the preparation of alkoxydimethylsilyl ethers The steroid or cannabinoid (20 μg) was dissolved in pyridine bis(diethylamino)dimethylsilane (10 μl) was added, then the mixture was heated at 60 °C for 5 min. The alcohol (R^{1}OH, 1.1 equivalents) was then added and the mixture was heated for a further 5 min at 60 °C. Finally, the excess alcohol was silylated by addition of BSTFA (5 μl) and the mixture was examined directly by packed column GC−MS.

Alkoxydialkylsilyl derivatives were formed containing the following alkyl groups (R^1): CH$_3$−, C$_2$H$_5$−, C$_3$H$_7$−, C$_4$H$_9$−, C$_5$H$_{11}$−, iso-C$_3$H$_7$−, 1-CH$_3$-C$_3$H$_6$−, 2-CH$_3$-C$_4$H$_8$−, CF$_3$CH$_2$−, C$_2$H$_3$− and (CH$_3$)$_2$NC$_2$H$_4$−. Silyl alkyl groups containing CH$_3$− and C$_2$H$_5$− were also prepared. Once formed, alkoxydialkylsilyl derivatives were found to be stable to hydrolysis, presumably as a result of electron donation from the two bound oxygen atoms into the empty silicon *d*-orbitals. It is important, though, to exclude all traces of moisture from the reaction medium, otherwise derivatives are produced containing polymeric dialkylsiloxane chains.

As alluded to above, the preparation of alkoxydialkylsilyl derivatives raises the possibility of varying the nature of the alcohol to modify the selectivity of detection of a chosen analyte. For example, *N,N*-dimethylaminoethanol was used to prepare a nitrogen-containing derivative which could be used with the nitrogen-specific detector, and halogenated alcohols, e.g. 2,2,2-trifluoroethanol, could be used to introduce electrophoretic groups to improve the sensitivity and selectivity to ECD. The mass spectra of these derivatives are very similar to those of the corresponding TMS derivatives are very similar to those of the corresponding TMS derivatives, and the use of [^{2}H$_4$]methanol to produce [^{2}H$_3$]methoxydimethylsilyl ethers can be of value in the interpretation of mass spectra. Harvey [439] also prepared dimethoxymethylsilyl ethers of steroids in an effort to provide hydrolytically stable derivatives displaying favourable mass spectrometric properties.

Preparation of dimethoxymethylsilyl ethers of steroids The derivatives were prepared by heating acetonitrile (200 μl), pyridine (10 μl), diethylamine (50 μl) and dimethoxymethylchlorosilane (50 μl) for 10 min at 60 °C. After cooling, a 10 μl portion of this mixture was added to the steroid (10 μg) and heating was continued for a further 10 min at 60 °C. Alternatively, the latter reaction was performed in the presence of 1 mg of imidazole.

The dimethoxymethyldimethylaminosilane reagent prepared as described above was found to be much less reactive than other alkyldiethylaminosilyl reagents prepared by analogous methods. This was reflected in the relative difficulty encountered in silylating steroids containing axial hydroxyl groups compared with those containing equatorial hydroxyl groups. However, once formed, the dimethoxymethylsilyl derivatives were much more stable than the corresponding TMS derivatives. The mass spectra of the dimethoxymethylsilyl derivative were generally more informative than those of the corresponding TBDMS ethers. While molecular ions were generally present in the EI spectra of the dimethoxymethylsilyl derivatives, they were of lower abundance than those appearing in the spectra of the corresponding TMS ethers. [M − (CH$_3$O)$_2$SiOH]$^+$ ions were present in high abundance in most of the compounds examined. Alkoxyalkylsilyl derivatives have yet to find widespread application in analytical work.

Of relevance to this class of derivatives is t-butylmethoxyphenylbromosilane (TBMPSBr; 42), which was originally developed as a versatile and selective protecting group for alcohols that was stable to hydrolysis [440]. Among the few known analytical applications is its use in the GC−MS analysis of the arachidonic acid metabolite 12,20-dihydroxyeicosatetraenoic acid (12,20-diHETE) as the methyl ester 12-TMS, 20-TBMPS ether [441]. A more recent application was aimed at

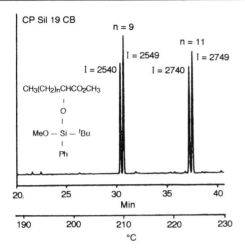

Figure 12. Gas chromatographic separation of the TBMPS derivatives of (±)-methyl 2-hydroxylaurate (n = 9) and (±)-methyl 2-hydroxymyristate (n = 11). Column, CP Sil 19 CB fused-silica capillary (25 m × 0.32 mm i.d; film thickness 0.18 μm); column temperature, programmed from 80 °C (2 min) to 160 °C (1 min) at 30 °C/min, and then at 2 °C/min to 240 °C; helium flow rate 3 ml/min. ^{t}Bu = t-butyl; Me = methyl; Ph = phenyl. (Reproduced from reference 442 with permission).

determining its suitability for converting chiral alcohols into diastereomeric ethers for resolution on achiral columns by GC [442]. Results were presented for the resolution of diastereomers of methyl 2-hydroxylaurate and methyl 2-hydroxymyristate (Figure 12) and of *trans,trans*-α-decalol and *cis,cis*-α-decalol. The unequal proportions of the diastereomers that were formed indicated stereoselectivity in the rate of etherification. This study complements earlier GC separations of diastereomeric silyl ethers [443], and emphasizes the need for the development of optically stable silylation reagents [444–446] before practical applications can be achieved.

$$\text{(42)}$$

4.6 Halogenated silyl derivatives

4.6.1 Pentafluorophenyldimethylsilyl (flophemesyl) ethers

The pentafluorophenyldimethylsilyl derivatives, termed flophemesyl, were developed by Morgan and Poole in the mid-1970's [447–449]. This derivative was found to be particularly suitable for the analysis of sterols by GC with electron capture detection [447]. The sequence of silyl donor power of reagents in pyridine solution for the derivatization of sterols has been established as [7, 450] flophemesylamine > flophemesyl chloride > flophemesyldiethyl amine > flophemesyldisilazane ≫ flophemesylimidazole.

Flophemesylamine (43) was found to be a selective reagent which reacts only with primary and secondary hydroxyl groups in steroids in the presence of unprotected ketone groups.

$$\text{(43)}$$

Preparation of flophemesyl derivatives of steroids [448]
To form sterol flophemesyl ethers add flophemesylamine (20 μl) to the sterol (1–5 mg) in pyridine (20 μl, dried over barium oxide) at room temperature in a Reacti-Vial (Pierce and Warriner) and allow to stand for 15 min at room temperature. Analyse directly by GC using flame ionization detection. For electron capture detection remove the solvent and re-dissolve in hexane.

Flophemesyl derivatives display excellent GC properties with ECD limits in the nanogram range for steroids containing one flophemesyl group, and in picograms for those containing two or more flophemesyl groups. The high sensitivity to ECD detection is presumed to result from the captured electron becoming buried in the π-orbitals of the phenyl ring, which is further stabilized by (p–d) π-bonding with the low energy d-orbitals of silicon. The application of flophemesyl derivatives to the detection of low molecular weight aliphatic alcohols has also been demonstrated [451]. Optimum ECD response to flophemesyl derivatives was found to occur at 350 °C, such that aliphatic alcohols exhibited detection limits in the low femtogram range [451]. This preferred high temperature operation also minimizes contamination of the detector during the analysis of biological extracts. However, it should be noted that repeated injection of silylating reagent into the ECD will eventually lead to drastically reduced response.

Much less thoroughly investigated, but showing considerable promise, is the use of flophemesyl derivatives to improve detection limits in GC analyses with photoionization detection (PID) [452, 453]. Simple aliphatic alcohols dissoved in acetonitrile, pyridine or triethylamine were found to react very rapidly with flophemesyl chloride at room temperature. Detection limits were lowered by 2–3 orders of magnitude compared with the starting alcohol, with calibration plots showing linearity over 5–7 orders of magnitude [452]. The method has been applied to the analysis of steroids in biological fluids; detection limits and precision of capillary GC–PID determinations were similer to those obtained by GC–ECD [453].

The favourable MS properties of flophemesyl derivatives can be used to advantage in GC–MS analyses. The flophemesyl group has a strong influence on fragmentation, producing diagnostic ions which carry a greater percentage of the ion current than the corresponding TMS ethers [449]. For example, the EI mass spectra of steroid flophemesyl ethers contain prominent $M^{+\cdot}$ ions and with a greater percentage of the total ion current associated with the steroid hydrocarbon fragments as compared with the TMS derivative. Characteristic silicon-containing ions corresponding to $[(CH_3)_2Si]^+$ and $[(CH_3)_2SiF]^+$ appear at m/z 58 and 77 respectively. The electron capturing properties of flophemesyl derivatives have been taken advantage of in their application to the capillary GC-MS analysis of steroids employing NICI [454]. Derivatization was carried out using flophemesylamine in dry toluene at 60°C for 15 min. The ammonia NICI spectra of equilin, oestrone and 1-methyl-6-dehydroestrone (used as internal standard) were dominated by intense molecular anions. GC–MS–SIM analyses using these molecular anions allowed measurement of oestrone and equilin in plasma at a concentration of 1 ng ml^{-1}. The flophemesyl derivatives of a series of fatty acid methyl ester chlorohydrins have been prepared and found to be of value for their trace detection by GC–ECD and their characteristic fragmentation patterns in GC–MS analysis [455]. Flophemesyl derivatives have also been found useful for the highly sensitive quantitative analysis of chlorohydrins isolated from polymer systems, using capillary GC with ECD [456]. It appears that flophemesylamine reacts quantitatively with N-nitrosodiethanolamine in toluene upon heating at 60°C for 15 min (44) [458]. The derivative is sufficiently thermally stable to allow the detection of low picogram amounts of nitroso derivative by GC using ECD.

Poole and co-workers investigated the properties of three further reagents structurally related to the flophemesyl reagents [457, 458], namely pentafluorophenylisopropylmethylchlorosilane (ISP-flophemesyl chloride), t-butylpentafluorophenylmethylchlorosilane (t-buflophemesyl chloride) and chloromethylpentafluorophenylmethylchorosilane (CM-flophemesyl chloride). Although the last-named reagent was prepared with a view to its use as a selective reagent for the determining of bifunctional compounds, it was found to polymerize rather than act as a cyclizing reagent for γ-hydroxyamines. The ISP-flophemesyl and t-buflophemesyl derivatives were found to be considerably more stable towards hydrolyses than the flophemesyl and CM-flophemesyl derivatives. The greater bulk of these reagents means that they react less readily with hindered functional groups. The t-buflophemesyl derivative has been applied to the trace analysis of alcohols [457]. Although t-buflophemesyl chloride, in the presence of diethylamine as catalyst, reacted quantitatively at room temperature with unhindered secondary hydroxyl groups, heating at 60 °C for 3 h was required to silylate the tertiary 17β-hydroxyl group in 17α-methyl-17β-hydroxy-5α-androstan-3-one. No reaction occurred between t-buflophemesyl chloride and the 11β-OH group in 11β-hydroxy-4-androstene-3,17-dione dimethoxime or other more hindered hydroxyl groups under the latter reaction conditions.

4.6.2 Halocarbondimethylsilyl ethers

Halomethyldimethylsilyl ether derivatives were introduced in the 1960's because of the possibility for their selective determination at trace levels in GC analyses using the electron capture detector [459, 460]. Under favourable circumstances, ECD limits for halocarbondimethylsilyl ethers can be 3–4 orders of magnitude better than those attainable with FID. The order of response of the ECD to halogens is $I > Br > Cl > F$.

Although the high volatility of fluorinated hydrocarbons is of potential advantage in GC work, they capture electrons poorly. However, the presence of a number of fluorine atoms can produce derivatives that exhibit favourable detection limits. The flophemesyl derivatives discussed above are excellent examples of the use of multiple fluorination to enhance detectability. Fluoroalkylsilanes with fluorine atoms alpha or beta to the silicon are thermally unstable at GC elution temperatures above 200 °C because of fluorine migration.

In their development work on the flophemesyl derivatives, Morgan and Poole [447] also assessed the suitability of trifluoropropylsilyl and heptafluoropentylsilyl derivatives for electron capture detection in GC. Although these derivatives were thermally stable, owing to the presence of the fluorine atom in the γ-position relative to the silicon, they possessed poor ECD responses compared with the flophemesyl derivatives. Rembold and co-workers [461–463] have used the nonafluorohexyldimethylsilyl (NFHDMS) derivatives for the detection and quantification of juvenile hormones (JH's) in insects. The method, summarized in the scheme below (45), relies on opening the epoxide ring by heating the JH fraction with a mixture of trifluoroacetic acid in methanol. The resulting 10-hydroxy-11-methoxy-JH was then converted to the 10-NFHDMS derivative using a mixture of NFHDMCS and pyridine. Detection limits for JHs I, II and III using GC–MS with SIM were ca. 0.01 pmol g^{-1} of fresh insect tissue.

The properties and applications of the halomethyldimethylsilyl ethers introduced by Eaborn et al. [459, 460] have been discussed in detail by Poole and Zlatkis [7]. Significantly, the chloromethyldimethylsilyl (CMDMS) derivatives of steroids possess comparable sensitivity to ECD and FID, while the bromomethyldimethylsilyl (BMDMS) and iodomethyldimethylsilyl (IMDMS) derivatives exhibit subnanogram and low picogram ECD limits, respectively. More recent applications of halomethyldimethylsilyl derivatives include the analysis of the thymine content of DNA [464] following hydrolysis and derivatization at room temperature for 5 min using 1,3-bis(chloromethyl)-1,1,3,3-tetramethyldisilazane. After derivatization, two-dimensional ECD GC allowed the routine determination of low nanogram amounts of thymine in DNA hydrolysates even in the presence of large amounts of RNA. The method can also be used to measure RNA based on the uracil content. Cyclic silyl derivatives result from the treatment of β-hydroxyamines with a mixture of 1,3-bis(chloromethyl)-1,1,3,3-tetramethyldisilazane and chloromethyldimethylchlorosilane (CMDMCS) (2:1 v/v) at 60 °C for 5–15 min (46) [465]. The reaction was quantitative except for quaternary amines and phenacylamine.

Derivatized tyrosine had to be analysed immediately after derivatization to avoid unwanted side reactions. An analogous reagent combination was used in the determination of the therapeutic pyrimidine antimetabolite florafur (N-2'-tetrahydrofurfuryl)-5-fluorouracil and its biotransformation product 5-fluorouracil [466]. Derivatization was achieved on standing overnight at room temperature followed by GC analysis employing ECD. Likewise, chloromethyldimethylsilyl ethers of alcohols, steroids and cannabinoids were prepared by heating for 10 min at 60 °C [384]. Pregnanolone metabolites have been profiled from porcine testicular preparations by GC and GC–MS as chloromethyldimethylsilyl, TMS and TBDMS ethers [467]. A recent application serves to demonstrate the high specificity and sensitivity that can be achieved using halomethyldimethylsilyl ethers in combination for GC with ECD in the analysis of steroids in complex biological matrices [468]. Dehydroepiandrosterone was extracted from human plasma and purified on a Celite column, and the iodomethyldimethylsilyl (IDMS) ether derivative was prepared by the method of Symes and Thomas [469] and Thomas [470], which involves the initial formation of the bromomethyldimethylsilyl derivative. This is then converted to the IDMS ether by halide ion exchange with sodium iodide. GC–MS analysis showed that only the IDMS derivative was formed. The use of capillary GC combined with ECD constitutes an assay that is a potential reference method for the determination of unconjugated dehydroepiandrosterone in biological samples [468].

Preparation of the IDMS ether derivatives of steroids
The silylating reagent was prepared from hexane (2 ml), bromomethyldimethylsilyl chloride (200 µl) and diethylamine (100 µl). Portions of the reagent (100 µl) were added to authentic steroids (nanogram amounts) or plasma extracts and allowed to stand at room temperature for 45 min. After removal of the excess reagent under a stream of nitrogen, 100 µl of a saturated sodium iodide solution in acetone was added and the mixture was heated at 37 °C for 30 min. The reaction mixture was blown to dryness and the residue was dissolved in 0.5 ml of a mixture of hexane/dichloromethane/ethanol/acetone (55:44: 0.5 v/v/v/v). The mixture was applied to a column of alumina (3 g, 200 mm × 6 mm) which was eluted with 7 ml of the same solvent; the derivatives emerged in the second 3.5 ml of eluent. After evaporation of the solvent, the steroid derivatives were dissolved in hexane (100 µl) for analysis by GC and GC–MS.

Some care is required in the preparation of halomethyldimethylsilyl ethers, as their instability gives rise to undesired reaction products. For example, the technique of preparing the IDMS derivatives *in situ* by halide ion exchange overcomes the problems of storage of the unstable IDMS reagents. A commonly used alternative approach to the preparation of halomethyldimethylsilyl derivatives involves the reaction of analytes with a mixture of 1,3-bis(halomethyl)-1,1,3,3-tetramethyldisilazane and the halomethyldimethylchlorosilane, or the use of excess halomethyldimethylchlorosilane and diethylamine in solution in hexane [459, 460, 470]. The use of mild reaction conditions helps to avoid loss of the halogen atom or halomethyl group (as shown below; 47) [465, 471, 472], which can result in greatly diminished electron capturing properties [465, 471] or unnecessarily enhanced molecular weights if dimethylsiloxanes are produced [472].

$$H-R + ClCH_2-\underset{|}{\overset{|}{Si}}-Cl \longrightarrow$$

$$R-\underset{|}{\overset{|}{Si}}-Cl + CH_3Cl \quad (47)$$

4.7 Nitrogen-containing silyl derivatives

4.7.1 *Cyanoethyldimethylsilyl (CEDMS)*

A novel derivatizing agent, 2-cyanoethyldimethyl(diethyl)aminosilane (CEDMSDEA), has been developed by Bertrand and co-workers [473–475] in order to take advantage of the high sensitivity and selectivity offered by the element specific nitrogen–phosphorus GC detector (NPD). Another approach explored by Bertrand and co-workers made use of the silylating reagent N,N-dimethylaminodimethylsilane [476, 477] to form nitrogen-containing derivatives (see also Section 4.5.4 above). However, the latter derivatizing reagent was found to have limited application because of the chemical reactivity of the derivatives so formed, i.e. due the susceptibility of the Si–N bond to hydrolysis.

In contrast, stable cyanoethyldimethylsilyl (CEDMS) derivatives have been prepared for compounds containing a range of functional groups, by adding CEDMSDEA (80 fold excess; prepared according to the procedure given in Ref. 473) to a solution of the compound of interest dissolved in ethyl acetate (*ca.* 1 mM). Quantitative reaction of homologous C_8–C_{18} fatty acids was achieved within a few minutes at room temperature. The use of higher temperatures did not affect the reaction yield, i.e. no secondary reactions or thermal degradation of the derivatives occurred. Quantitative reaction was also achieved with 2-chlorophenol and pentachlorophenol within a few minutes at room temperature, indicating that the acidity of the hydroxyl group has no noticeable effect on the reaction rate. The silylation proceeded much more slowly for the secondary hydroxyl group in cholesterol, with quantitative reaction being achieved only in 17 h at room temperature. At 75 °C, quantitative silylation of cholesterol was achieved in *ca.* 2 h. Similar detection limits (*ca.* 10^{-13} g of N (S/N = 3)) were recorded for all the compounds tested. The analytical performance of the CEDMS derivatives of chlorophenols [474] and phenoxy acids [475] is adequate for

(48)

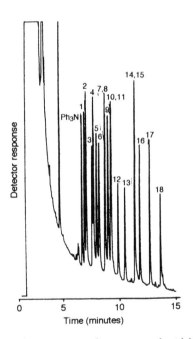

Figure 13. Chromatogram of a mixture of acid herbicides reacted with CEDMSDEA and analysed by NPD. Peak identities are: 1 = 2-(4-chloro-2-methylphenoxy)propionic acid; 2 = 2-methoxy-3,6-dichlorobenzoic acid; 3 = 4-chloro-2-methylphenoxyacetic acid; 4 = 2,3,6-trichlorobenzoic acid; 5 = 2-(2,4-dichlorophenoxy)propionic acid; 6 = (2,6-dichlorophenoxy)acetic acid; 7 = (2,4-dichlorophenoxy)acetic acid; 8 = (2,5-dichlorophenoxy)acetic acid; 9 = naphthaleneacetic acid; 10 = (2,3-dichlorophenoxy) acetic acid; 11 = (3,4-dichlorophenoxy)acetic acid; 12 = (4,6-dichloro-2-methylphenoxy)acetic acid; 13 = 2-(3,4,5-trichloro-phenoxy)propionic acid; 14 = 3-amino-2,5-dichlorobenzoic acid; 15 = (2,4,5-trichlorophenoxy)acetic acid; 16 = (4-chloro-2-methylphenoxy)butyric acid; 17 = (4-chloro-2-oxobenzothiazolin-3-yl)acetic acid; 18 = (2,4-dichlorophenoxyl)butyric acid. (Reproduced from Ref. 475 with permission).

their analysis. The preparation (48) and analysis of a range of acid herbicides as their CEDMS derivatives the GC with NPD is shown below (Figure 13). Pretreatment of glassware with CEDMSDEA reagent was found to be essential to ensure reproducible results [475]. The EI mass spectra of the CEDMS esters of fatty acids exhibit only weak $M^{+\cdot}$ ions ($< 5\%$ relative abundance; cf. TBDMS derivatives). Much more characteristic, however, are ions at m/z 75, 129, $[M - 54]^+$ and $[M - 15]^+$. The latter two ions correspond to the loss of C_2H_4CN and CH_3 radicals, respectively, and are characteristic of a given fatty acid. The mass spectrum of cholesterol showed the $M^{+\cdot}$ ion (m/z 497) as base peak and an ion of only slightly lower abundance at m/z 368 due the loss of CEDMSOH ($[M - 129]^+$). The high abundance, high mass ions observed in the mass spectra of CEDMS derivatives were deemed to be potentially useful for structure investigation or trace analyses employing SIM [473, 475].

4.7.2 *Picolinyldimethylsilyl (PICSI) ethers*

Picolinyldimethylsilyl (PICSI) ethers were developed by Harvey [478] in an effort to provide an improved derivative for the structure investigation of long-chain alcohols. PICSI ethers (Figure 14) were considered in view of their potential to combine the EI fragmentation-directing properties of nicotinate esters [478, 479] with the ease of preparation of silyl ethers. Two methods were used for preparing the PICSI derivatives.

Method (i) The alcohol (10 μg) was dissolved in dry pyridine (10 μl) and bis(diethylamino)dimethylsilane (2 μl) was added. The reaction mixture was then heated for 10 min at 60 °C and cooled and 3-pyridylmethanol (3 μl) was added. The mixture was

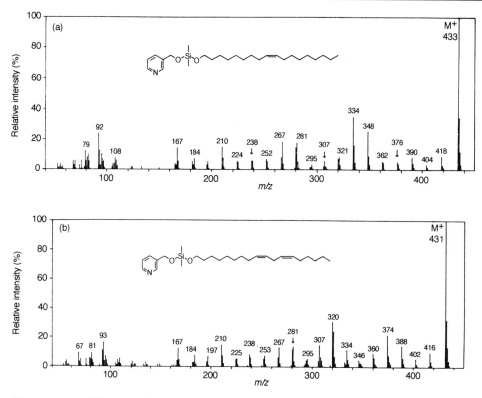

Figure 14. EI mass spectra of the PICSI derivatives of (a) Z-9-octadecenol and (b) Z,Z-9,12-octadecadienol. (Reproduced from Ref. 478 with permission).

heated for a further 10 min at 60 °C, then 1 μl portions were examined by GC–MS.

Method (ii) The alcohol (10 μl) was dissolved in dry pyridine (10 μl) and diethylaminodimethylsilyl-3-pyridylmethanol (5 μl; prepared according to the procedure described by Harvey [478] was added. The mixture was heated at 60 °C for 10 min before analysing portions of the reaction mixture directly by GC–MS.

PICSI derivatives were successfully formed for both saturated and unsaturated long-chain alcohols. Method (i) is analogous to that used in the preparation of alkoxydialkylsilyl derivatives of steroids discussed in Section 4.5.4. Pure derivatives were formed provided that care was taken to exclude water from the reaction medium. The EI (25 eV) mass spectra (e.g. Figure 14) were characterized by very abundant $M^{+\bullet}$ ions (base peak) and a series of radical-induced cleavage ions, the masses and abundances of which gave information on

(49)

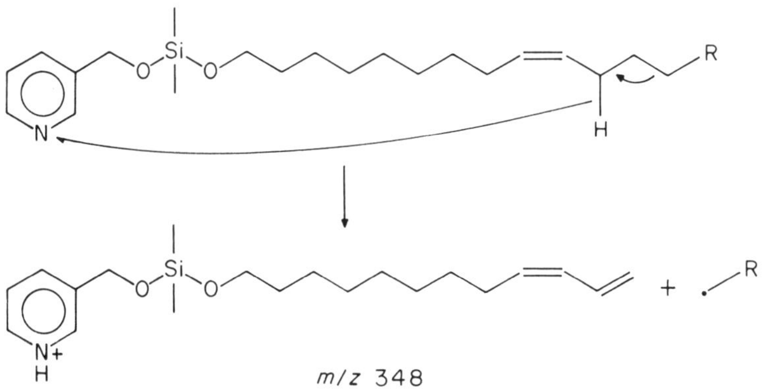

$$\text{(50)}$$

m/z 348

the presence and position of double bonds in mono- and polyunsaturated long-chain alcohols. The most prominent and useful feature in indicating the position of the double bond from the mass spectra shown in Figure 14(a) is the presence of abundant ions at m/z 334 and 348. These ions are formed by abstraction of two allylic hydrogens atoms (49 and 50). In the case of polyunsaturated alcohols, e.g. Z,Z-9,12-octadecadienol (linoleyl alcohol), pairs of ions relating to the position of the Δ^9 and Δ^{12}-double bonds appear at m/z 267/281 and 307/321 respectively [Figure 14(b)]. Even mass ions formed as shown below are not as prominent, with only m/z 374 and 388 being of diagnostic value. Although apparently showing promise, these derivatives have yet to be applied to the analysis of mixtures of long-chain alcohols from biological materials.

5. Conclusions

The late 1970s and early 1980s saw a period of intense activity in the development and testing of new silyl derivatives. As a result of these efforts, there has been a substantial increase in the range of silyl derivatives available to the analytical chemist since the publication of the original edition of this volume. Research activity and the development of further silyl derivatives appears to have slowed. Surprisingly few of the new derivatives have been adopted for routine use, and trimethylsilylation remains the preferred approach for the vast majority of analytical applications.

The range of applications involving the use of the more sterically crowded TBDMS derivatives has increased significantly. The popularity of TBDMS derivatives stems largely from their enhanced chemical stability and favourable mass spectrometric behaviour. The former property can serve to ease sample handling during work-up procedures, while the high proportion of the ion current generally carried by high mass ions in the mass spectra of TBDMS derivatives, e.g. $[M-57]^+$, can help to increase precision and reliability of detection in trace analyses. The steric bulk of the TBDMS moiety can impede the silylation of some crowded protic sites. Other slightly less bulky alkyldimethylsilyl derivatives, such as the DMIPS derivatives, would appear to offer a compromise between the TMS and TBDMS derivatives in terms of chemical stability and favourable mass spectrometric properties. However, like many of the other silyl derivatives that have been developed, the DMIPS derivative has not been widely tested and applied. This and other similar derivatives are worthy of further investigation.

The wider availability of mass spectrometers has resulted in a substantial increase in the range of applications using GC–MS for the sensitive and selective detection of silyl derivatives. The favourable MS–MS behaviour of some silyl derivatives warrants further investigation for use in trace analysis. The increasing availability of MS–MS instrmentation will produce an expansion in the range of applications in the near future. Substantial research effort in the 1970s and 1980s was directed towards the development of derivatives for use with selective detectors other than FID, e.g. ECD. However, the use of silylation in conjunction with other element-specific detectors, e.g. the great potential of the atomic emission detector to monitor specifically very low levels of silicon-containing compounds, or NPD to detect nitrogen-containing silyl derivatives, to enhance selectivity has seen few applications; the increased use of mass spectrometry has probably contributed to this trend. Although many applications will continue to use silyl derivatives for GC–MS there will inevitably be a drift away from the use of derivatization as LC–MS instrumentation becomes more widely available. From the purely analytical standpoint, derivatization should be avoided wherever possible. However, it should be borne in mind that the preparation of derivatives and subsequent sample work-up steps can serve to enhance selectivity. The relatively poor absolute detection limits generally achieved by LC–MS means that GC–MS will continue to be the method of choice for many applications particularly trace analyses. The possibility of using silylation in conjunction with LC–MS or SFC–MS to enhance detection limits, chromatographic behaviour or the structural information content of mass spectra is also worthy of consideration.

6. References

[1] C. F. Poole, in *Handbook of Derivatives for Chromatography*, K. Blau and G. S. King (Editors), Heyden & Sons Ltd., London (1977), p. 152.
[2] R. P. Evershed, in *Specialist Periodical Reports: Mass Spectrometry*, Vol. 10, M.E. Rose (Editor), The Royal Society of Chemistry, London, 1989, p. 181.
[3] D. R. Knapp, *Handbook of Analytical Derivatization Reactions*, John Wiley & Sons, New York, 1979.
[4] A. E. Pierce, *Silylation of Organic Compounds*, Pierce Chemicals, Rockford (1979).
[5] J. D. Nicholson, *Analyst*, **103**, 193 (1978).
[6] R. C. Denney, *Spec. Chem.*, **6**, 6 (1983).
[7] C. F. Poole and A. Zlatkis, *J. Chromatogr. Sci.*, **17**, 115 (1979).
[8] C. F. Poole and W.-F. Sye, S. Singhawangcha, F. Hsu, A. Zlatkis, A. Arfwidsson and J. Vessman, *J. Chromatogr.*, **199**, 123 (1980).
[9] R. Piekos, K. Osmialowski, K. Kobylczyk and J. Grzybowski, *J. Chromatogr.*, **116**, 315 (1976).
[10] R. Piekos, J. Teodorczyk, J. Grzybowski, K. Kobylczyk and K. Osmialowski, *J. Chromatogr.*, **117**, 431 (1976).
[11] D. Valdez, *J. Chromatogr. Sci.*, **23**, 128 (1985).
[12] S. H. G. Andersson and J. Sjövall, *J. Chromatogr.*, **289**, 195 (1984).
[13] S. H. G. Andersson and J. Sjövall, *J. Steroid Biochem.*, **23**, 469 (1985).
[14] C. A. White, S. W. Vass, J. F. Kennedy and D. G. Large, *J. Chromatogr.*, **264**, 99 (1983).
[15] M. A. Quilliam and J. M. Yaraskavitch, *J. Liq. Chromatogr.*, **8**, 449 (1985).
[16] M. Osman, H. H. Hill, Jr., M. Holdren and H. Westberg, *J. Chromatogr.*, **186**, 273 (1979).
[17] T. Aoyama and T. Yashiro, *J. Chromatogr.*, **265**, 57 (1983).
[18] K. E. Rasmussen, *J. Chromatogr.*, **120**, 491 (1976).
[19] A. E. Yatsenko, A. I. Mikaya, L. S. Glebov and V. G. Zaikin, *Bull. Acad. Sci. USSR Div. Chem. Sci. (Engl. Transl.)*, **35** (3, Part 2), 666 (1986).
[20] A. I. Mikaya, A. V. Antonova, V. G. Zaikin, N. S. Prostakov, V. Yu. Rumyantsev, E. V. Slivinskii and S. M. Loktev, *J. Gen. Chem. USSR (Engl. Transl.)*, **54** (3 Part 2), 578 (1984).
[21] A. S. Christophersen, E. Hovland and K. E. Rasmussen, *J. Chromatogr.*, **234**, 107 (1982).

[22] F. C. Faulkner, H. G. Fouda and F. G. Mullins, *Biomed. Mass Spectrom.*, **11**, 482 (1984).
[23] A. Gold, D. F. Utterback and D. S. Millington, *Anal. Chem.*, **56**, 2879 (1984).
[24] M. A. Quilliam, K. K. Ogilvie and J. B. Westmore, *J. Chromatogr.*, **105**, 297 (1975).
[25] D. H. McMahon, *J. Chromatogr. Sci.*, **23**, 426 (1985).
[26] A. Radmacher, German Patent 2,262,842 (30 May 1974).
[27] M. Donike, U.S. Patent 3,954, 651 (4 May 1976).
[28] W. Thies, *Fette Seifen Anstrichm.*, **78**, 231 (1976).
[29] D. L. Stalling, C. W. Gehrke and R. W. Zumwalt, *Biochem. Biophys. Res. Commun.*, **31**, 616 (1968).
[30] N. E. Hoffman and K. A. Peteranetz, *Anal. Lett.*, **5**, 589 (1972).
[31] V. Fell and C. R. Lee, *J. Chromatogr.*, **121**, 41 (1976).
[32] K. Kawashiro, S. Mori and H. Yoshida, *Bull. Chem. Soc. Jpn.*, **57**, 2871 (1984).
[33] E. M. Chambaz, G. Defaye and C. Madani, *Anal. Chem.*, **45**, 1090 (1973).
[34] S. Z. Nicosia, G. Galli, A. Fiecchi and A. Ros, *J. Steroid Biochem.*, **4**, 417 (1973).
[35] K. Uobe, R. Takeda, M. Wato, T. Nishikawa, S. Yamaguchi, T. Koshimura, Y. Kawaguchi and M. Tsutsui, *J. Chromatogr.*, **214**, 177 (1981).
[36] J. Roselló and E. Gelpi, *J. Chromatogr. Sci.*, **16**, 177 (1978).
[37] A. B. Benko and V. Mann, *Anal. Lett.*, **13**, 735 (1980).
[38] T. R. Kemp, R. A. Andersen and T. H. Vaughn, *J. Chromatogr.*, **241**, 325 (1982).
[39] M. Makita and W. W. Wells, *Anal. Biochem.*, **4**, 523 (1963).
[40] L. Birkofer, A. Ritter and W. Giebler, *Angew. Chem.*, **75**, 93 (1963).
[41] E. M. Chambaz and E. C. Horning, *Anal. Biochem.*, **30**, 7 (1969).
[42] M. G. Horning, A. M. Moss and E. C. Horning, *Biochem. Biophys. Acta*, **148**, 597 (1967).
[43] V. A. Joliffe, C. W. Coggins, Jr. and W. W. Jones, *J. Chromatogr.*, **179**, 333 (1979).
[44] J. Marik, A. Capek and J. Kralicek, *J. Chromatogr.*, **128**, 1 (1976).
[45] S. J. Gaskell, A. G. Smith and C. J. W. Brooks, *Biomed. Mass Spectrom.*, **2**, 148 (1975).
[46] M. Donike, *J. Chromatogr.*, **42**, 103 (1969).
[47] M. Donike, *J. Chromatogr.*, **85**, 1 (1973).
[48] M. Donike, *J. Chromatogr.*, **85**, 9 (1973).
[49] H. Gleispach, *J. Chromatogr.*, **91**, 407 (1974).
[50] M. Donike and J. Zimmermann, *J. Chromatogr.*, **202**, 483 (1980).
[51] M. Donike, *Chromatographia*, **9**, 440 (1976).
[52] M. Donike, R. Goln and L. Jaenicke, *J. Chromatogr.*, **134**, 385 (1977).
[53] K. Rühlmann, *J. Prakt. Chem.*, **9**, 315 (1959).
[54] K. Rühlmann, *Chem. Ber.*, **94**, 1876 (1961).
[55] P. S. Mason and E. D. Smith, *J. Gas Chromatogr.*, **4**, 398 (1966).
[56] E. D. Smith and K. L. Shewbart, *J. Chromatogr. Sci.*, **7**, 704 (1969).

[57] J. Hils, V. Hagen, H. Ludwig and K. Rühlmann, *Chem. Ber.*, **99**, 776 (1966).
[58] P. W. Albro and L. Fishbein, *J. Chromatogr.*, **55**, 297 (1971).
[59] L. Birkofer and A. Ritter, *Angew. Chem.*, **77**, 414 (1965).
[60] E. D. Morgan and C. F. Poole, *J. Chromatogr.*, **116**, 333 (1976).
[61] R. O. Sauer, *J. Am. Chem. Soc.*, **66**, 1707 (1944).
[62] G. Hanika, *Z. Ges. Hyg.*, **32**, 151 (1986).
[63] R. C. Osthoff and S. W. Kantor, *Inorg. Synth.*, **5**, 55 (1957).
[64] E. Jellum, *J. Chromatogr.*, **143**, 427 (1977).
[65] S.-F. Yeh, K.-J. Hsiao, S.-H. Hung and K.-U. Chang, *J. Chin. Chem. Soc.*, **33**, 251 (1986).
[66] A. P. De Leenheer, M. F. Lefevre and L. M. R. Thienpont, *J. Pharm. Biomed. Anal.*, **4**, 735 (1986).
[67] A. A. Bornstein, *J. Chromatogr. Sci.*, **18**, 183 (1980).
[68] H. Binder and A. A. Ashy, *J. Chromatogr. Sci.*, **22**, 536 (1984).
[69] C. Jakobs, E. Solem, J. Ek, K. Halvorsen and E. Jellum, *J. Chromatogr.*, **143**, 31 (1977).
[70] T. L. Masimore, H. Veening, W.J.A. Vandenheuvel and D. A. Dayton, *J. Chromatogr.*, **143**, 247 (1977).
[71] M. Matsumoto, T. Kuhara, Y. Inoue, T. Shinka, I. Matsumoto and M. Kajita, *Biomed. Environ. Mass Spectrom.*, **19**, 171 (1990).
[72] A. A. Fernandes, S. C. Kalhan, F. G. Njoroge and G. S. Matousek, *Biomed. Environ. Mass Spectrom.*, **13**, 569 (1986).
[73] P. F. Bougneres, E. O. Balasse, P. Ferré and D. M. Bier, *J. Lipid Res.*, **27**, 215 (1986).
[74] O. A. Mamer, N. S. Laschic and C. R. Scriver, *Biomed. Environ. Mass Spectrom.*, **13**, 553 (1986).
[75] O. A. Mamer, *Biomed. Environ. Mass Spectrom.*, **15**, 57 (1988).
[76] J. S. Svendsen, L. K. Sydnes and J. E. Whist, *Spectrosc. Int. J.*, **3**, 380 (1984).
[77] P. Jolivet, P. Gans and C. Triantaphylides, *Anal. Biochem.*, **147**, 86 (1985).
[78] U. Langenbeck, A. Hoinowski, K. Mantel and H.-U. Möhring, *J. Chromatogr. Biomed. Appl.*, **143**, 39 (1977).
[79] L. J. Roberts, *Methods Enzymol.*, **86**, 559 (1982).
[80] W. C. Hubbard, *Methods Enzymol.*, **86**, 571 (1982).
[81] P. Falardeau and A. R. Brash, *Methods Enzymol.*, **86**, 585 (1982).
[82] M. L. Ogletree, K. Schlesinger, M. Nettleman and W. C. Hubbard, *Methods Enzymol.*, **86**, 607 (1982).
[83] J. Maclouf and M. Rigaud, *Methods Enzymol.*, **86**, 612 (1982).
[84] E. H. Oliw, *FEBS Lett.*, **172**, 279 (1984).
[85] J. M. F. Douse, *J. Chromatogr.*, **348**, 111 (1985).
[86] J. Y. Westcott, K. L. Clay and R. C. Murphy, *Biomed. Mass Spectrom.*, **12**, 714 (1985).
[87] C. J. Hartzell and N. H. Anderson, *Biomed. Mass Spectrom.*, **12**, 303 (1985).
[88] S. Fischer and P. C. Weber, *Biomed. Mass Spectrom.*, **12**, 470 (1985).
[89] H. Gleispach, R. Moser and H. J. Leis, *J. Chromatogr.*, **342**, 245 (1985).
[90] H. Chable-Rabinovitch, J. P. Leblanc, F. Mulliez, A. Outifrakh, P. Chebroux, J. Durand and M. Rigaud, *Prostaglandins Leukotrienes Med.*, **23**, 161 (1986).
[91] R. C. Murphy and T. W. Harper, in *Mass Spectrometry in Biomedical Research*, S. J. Gaskell (Editor), John Wiley & Sons Ltd., Chichester (1986), p. 11.
[92] H. Schweer, H. Seyberth and R. Schubert, *Biomed. Environ. Mass Spectrom.*, **13**, 611 (1986).
[93] C. Moussard, D. Alber and J. C. Henry, *Prostaglandins*, **34**, 79 (1987).
[94] A. Ferreti, V. P. Flanagan and V. B. Reeves, *Anal. Biochem.*, **167**, 174 (1987).
[95] R. W. Walenga, S. Boone and M. J. Staurt, *Prostaglandins*, **34**, 733 (1987).
[96] H. Schweer, H. W. Seyberth and C. O. Meese, *Biomed. Environ. Mass Spectrom.*, **15**, 129 (1988).
[97] H. Schweer, H. W. Seyberth, C. O. Meese and O. Fürst, *Biomed. Environ. Mass Spectrom.*, **15**, 139 (1988).
[98] R. J. Strife and J. R. Simms, *Anal. Chem.*, **60**, 1800 (1988).
[99] B. H. Min, J. Pao, W. A. Garland, J. A. de Silva and M. Parsonnet, *J. Chromatogr.*, **183**, 411 (1980).
[100] S. E. Barrow, K. A. Waddell, M. Ennis, C. T. Dollery and I. A. Blair, *J. Chromatogr.*, **239**, 71 (1982).
[101] K. A. Waddell, J. Wellby and I. A. Blair, *Biomed. Mass Spectrom.*, **10**, 83 (1983).
[102] K. A. Waddell, S. E. Barrow, C. Robinson, M. A. Orchard, C. T. Dollery and I. A. Blair, *Biomed. Mass Spectrom.*, **11**, 68 (1984).
[103] C. R. Pace-Asciak and S. Micallef, *J. Chromatogr.*, **310**, 233 (1984).
[104] H. Gleispach, R. Moser, B. Mayer, H. Esterbauer, U. Skriletz, L. Zimmerman and H. J. Leis, *J. Chromatogr.*, **344**, 11 (1985).
[105] D. B. Cosulich, N. A. Perkinson and V. K. Batra, *J. Pharm. Sci.*, **74**, 76 (1985).
[106] C. O. Meese, C. Fischer, P. Thalheimer and O. Fürst, *Biomed. Mass Spectrom.*, **12**, 554 (1985).
[107] J. A. Lawson, A. R. Brash, J. Doran and G. A. Fitzgerald, *Anal. Biochem.*, **150**, 463 (1985).
[108] C. Fisher and C. O. Meese, *Biomed. Mass Spectrom.*, **12**, 399 (1985).
[109] M. Balazy and R. C. Murphy, *Anal. Chem.*, **58**, 1098 (1986).
[110] H. J. Leis, E. Malle, R. Moser, J. Nimpf, G. M. Kostner, H. Esterbauer and H. Gleispach, *Biomed. Environ. Mass Spectrom.*, **13**, 483 (1986).
[111] J. A. Lawson, C. Patrono, G. Cibattoni and G. A. Fitzgerald, *Anal. Biochem.*, **155**, 198 (1986).
[112] J. Y. Westcott, S. Chang, M. Balazy, D. O. Stene, P. Pradelles, J. Maclouf, N. F. Voelkel and R. C. Murphy, *Prostaglandins*, **32**, 857 (1986).
[113] A. Martineau and P. Falardeau, *J. Chromatogr.*, **417**, 1 (1987).
[114] H. Schweer, C. O. Meese, O. Fürst, P. G. Kule and H. W. Seyberth, *Anal. Biochem.*, **164**, 156 (1987).
[115] N. Shindo, T. Saito and K. Murayama, *Biomed. Environ. Mass Spectrom.*, **15**, 25 (1988).
[116] H. Schweer, H. W. Seyberth, C. O. Meese and O. Fürst, *Biomed. Environ. Mass Spectrom.*, **15**, 143 (1988).

[117] W. R. Mathews, G. L. Bundy, M. A. Wynalda, D. M. Guido, W. P. Schneider and F. A. Fitzpatrick, *Anal. Chem.*, **60**, 349 (1988).
[118] R. C. Murphy, *Prostaglandins*, **28**, 597 (1984).
[119] S. Nicosia and G. Galli, *Anal. Biochem.*, **21**, 192 (1974).
[120] J. Rosello, J. M. Tusell and E. Gelpi, *J. Chromatogr.*, **130**, 65 (1977).
[121] J. Rosello, C. Sunol, J. M. Tusell and E. Gelpi, *Biomed. Mass Spectrom.*, **4**, 237 (1977).
[122] J. M. Rosenfeld, Y. Moharir and R. Hill, *Anal. Chem.*, **63**, 1536 (1991).
[123] B. Mayer, R. Moser, H.-J. Leis and H. Gleispach, *J. Chromatogr.*, **378**, 430 (1986).
[124] C. R. Pace-Asciak, in *Chromatography of Lipids in Biomedical Research and Clinical Diagnosis*, A. Kuksis (Editor), *J. Chromatogr. Library*, Vol. 37, Elsevier, Amsterdam (1987), pp. 107–127.
[125] J.C. Fröhlich, M. Sawada, G. Bochmann, and O. Oetz, in *Advances in Prostaglandin, Thromboxane and Leukotriene Research*, Vol. 16, U. Zor, Z. Naor and F. Kochen (Editor), Raven, New York (1987), p 363–372.
[126] W. Krause, U. Jacobs, P. E. Schulze, B. Nieuweboer and M. Humpel, *Prostaglandins, Leukotrienes Med.*, **17**, 167 (1985).
[127] C. Chiabrando, A. Benigni, A. Piccinelli, C. Carminati, E. Cozzi, G. Remuzzi and R. Fanelli, *Anal. Biochem.*, **163**, 255 (1987).
[128] J. J. Vrbanac, T. D. Eller and D. R. Knapp, *J. Chromatogr.*, **425**, 1 (1988).
[129] J. J. Vrbanac, T. D. Eller, H. L. Hubbard, R. H. Baker, P. V. Halushka, I. A. Blair and D. R. Knapp, *34th Annual Conference on Mass Spectrometry and Allied Topics*, Cincinnati, OH, 1986, p. 110.
[130] H. L. Hubbard, T. D. Eller, D. E. Mais, P. V. Halushka, R. H. Baker, I. A. Blair, J. J. Vrbanac and D. R. Knapp, *Prostaglandins*, **33**, 149 (1987).
[131] C. J. W. Brooks, W. J. Cole, H. B. McIntyre and A. G. Smith, *Lipids*, **15**, 745 (1980).
[132] L. J. Goad, in *Methods in Plant Biochemistry*, Vol. 7, B. V. Charlwood and D. V. Barnthorpe (Editors), Academic Press, New York (1991), p. 369.
[133] G. C. Thorne, S. J. Gaskell and P. A. Payne, *Biomed. Mass Spectrom.*, **11**, 415 (1984).
[134] D. J. Porubek and S. D. Nelson, *Biomed. Environ. Mass Spectrom.*, **15**, 157 (1988).
[135] B. J. Koopman, IJ. G. Lokerse, H. Verwelj, G. T. Nagel, J. C. Van Der Molen, N. M. Drayer and B. G. Wolthers, *J. Chromatogr.*, **378**, 283 (1986).
[136] N. Hirota, T. Furuta and Y. Kasuya, *J. Chromatogr.*, **425**, 237 (1988).
[137] Y. Kasuya, T. Furuta and N. Hirota, *Biomed. Environ. Mass Spectrom.*, **16**, 309 (1988).
[138] R. P. Evershed, M. C. Prescott, L. J. Goad and H. H. Rees, *Biochem. Soc. Trans.*, **15**, 175 (1987).
[139] R. P. Evershed, J. G. Mercer and H. H. Rees, *J. Chromatogr.*, **390**, 357 (1987).
[140] C. R. Bielby, A. R. Gande, E. D. Morgan and I. D. Wilson, *J. Chromatogr.*, **194**, 43 (1980).
[141] R. Lafont, G. Somme-Martin, B. Mauchamp, B. F. Maume and J.-P. Delbeque, in *Progress in Ecdysone Research*, J. A. Hoffman (Editor), Elsevier, Amsterdam (1980), p. 45.
[142] A. H. W. Mendis, M. E. Rose, H. H. Rees and T. W. Goodwin, *Mol. Biochem. Parasitol.*, **9**, 209 (1983).
[143] S. G. Webster, *J. Chromatogr.*, **333**, 186 (1985).
[144] C. J. W. Brooks, *Philos. Trans. R. Soc. London, Ser. A*, **A293**, 53 (1979).
[145] S. J. Gaskell and L. Siekman, *Clin. Chem.*, **32** (3), 536 (1986).
[146] D. De Clerck, H. Diederiek and A. De Loof, *Insect Biochem.*, **14**, 199 (1984).
[147] R. Massé, C. Laliberté and L. Tremblay, *J. Chromatogr.*, **339**, 11 (1985).
[148] C. H. L. Shackleton, *J. Chromatogr.*, **379**, 91 (1986).
[149] K. Ichimura, H. Yamanaka, K. Chiba, T. Shinozuka, Y. Shiki, K. Saito, S. Kusano, S. Ameniya, K. Oyama, Y. Nozaki and K. Kato, *J. Chromatogr.*, **374**, 5 (1986).
[150] N. Blau, M. Zachmann, B. Kempken, W. Staudenmann, M. Möhr and H.-C. Curtius, *Biomed. Environ. Mass Spectrom.*, **14**, 633 (1987).
[151] E. Houghton, M. C. Dumasia and P. Teale, *Analyst*, **113**, 1179 (1988).
[152] E. Houghton, P. Teale, M. C. Dumasia and J. Welby, *Biomed. Mass Spectrom.*, **9**, 459 (1982).
[153] N. Hirota, T. Furuta and Y. Kasuya, *J. Chromatogr.*, **425**, 237 (1988).
[154] N. J. Fairs, R. P. Evershed, P. T. Quinlan and L. J. Goad, *Gen. Compar. Endocrinol.*, **4**, 199 (1989).
[155] E. M. Chambaz, G. Maume, B. Maume and E. C. Horning, *Anal. Lett.*, **1**, 749 (1968).
[156] P. Pfeifer and G. Spiteller, *J. Chromatogr.*, **223**, 21 (1981).
[157] R. Massé, C. Laliberté, L. Tremblay and R. Dugal, *Biomed. Mass Spectrom.*, **12**, 115 (1985).
[158] K. Minagawa, Y. Kasuya, S. Baba, G. Knapp and J. P. Skelly, *J. Chromatogr.*, **343**, 231 (1985).
[159] M. Axelson, *J. Steroid Biochem.*, **8**, 693 (1977).
[160] J. Sjövall, in *Advances in Steroid Analysis '84*, S. Görög (Editor), Elsevier, Amsterdam (1985), p. 359.
[161] R. M. Thompson and E. C. Horning, *Steroid Lipid Res.*, **4**, 135 (1973).
[162] S. Abdel-Baky, P. W. Le Quesne, E. Schwarz and P. Vouros, *Biomed. Mass Spectrom.*, **12**, 679 (1986).
[163] C. J. W. Brooks, J. MacLachan, W. J. Cole and T. D. V. Lawrie, in *Advances in Steroid Analysis '84*, S. Görög (Editor), Elsevier, Amsterdam (1985), pp. 349–358.
[164] H. F. Holick, H. K. Schnoes, H. F. DeLuca, T. Suda and R. J. Cousins, *Biochemistry*, **10**, 2794 (1971).
[165] J. M. Halket and B. P. Lisboa, in *Recent Developments in Mass Spectrometry in Biochemistry and Medicine*, Vol. 1, A. Frigerio (Editor), Plenum, New York (1978), p. 457.
[166] J. M. Halket and B. P. Lisboa, *Acta Endocrinol., Suppl.*, **215**, 120 (1978).
[167] J. M. Halket, I. Ganschow and B. P. Lisboa, *J. Chromatogr.*, **192**, 434 (1980).
[168] R. D. Coldwell, D. J. H. Trafford, H. L. J. Makin, M. J. Varley and D. N. Kirk, *J. Chromatogr.*, **338**, 289 (1985).
[169] R. D. Coldwell, D. J. H. Trafford, M. J. Varley, H. L.

J. Makin and D. N. Kirk, *Biomed. Environ. Mass Spectrom.*, **16**, 81 (1988).
[170] W. C. Kossa, in *Chemical Derivatization in Analytical Chemistry*, Vol. 1, *Chromatography*, R. W. Frei and J. F. Lawrence (Editors), Plenum Press, New York (1981), p. 99.
[171] K. D. R. Setchell and A. M. Lawson, in *Mass Spectrometry*, A. M. Lawson (Editor), Walter de Gruyter, Berlin (1989), pp. 54–125.
[172] M. Makita and W. W. Wells, *Anal. Biochem.*, **5**, 523 (1963).
[173] C. Tsconas, P. Padieu, G. Maume, M. Chessebeuf, N. Hussein and N. Pitoizet, *Anal. Biochem.*, **157**, 300 (1986).
[174] M. Tohma, R. Mahara, H. Takeshita, T. Kurosawa and S. Ikegawa, *Chem. Pharm. Bull.*, **34**, 2890 (1986).
[175] R. Mahara, H. Takeshita, T. Kurosawa, S. Ikegawa and M. Tohma, *Anal. Sci.*, **3**, 449 (1987).
[176] C. J. W. Brooks, G. M. Barrett and W. J. Cole, *J. Chromatogr.*, **289**, 231 (1984).
[177] R. Tandon, M. Axelson and J. Sjövall, *J. Chromatogr.*, **302**, 1 (1984).
[178] B. Almé and J. Sjövall, *J. Steroid Biochem.*, **13**, 907 (1982).
[179] H.-U. Marschall, B. Egestad, H. Matern, S. Matern and J. Sjövall, *FEBS Lett.*, **213**, 411 (1987).
[180] A. Kuksis, *J. Chromatogr.*, **143**, 3 (1977).
[181] A. Kuksis and J. J. Myher, in *Chromatography of Lipids in Biomedical Research and Clinical Diagnosis*, A. Kuksis (Editor), *J. Chromatogr. Library*, Vol. 37, Elsevier, Amsterdam (1987), pp. 1–47.
[182] J. J. Myher and A. Kuksis, *J. Biochem. Biophys. Methods*, **10**, 13 (1984).
[183] R. P. Evershed, C. Heron and L. J. Goad, *Analyst*, **115**, 1339 (1990).
[184] R. P. Evershed, C. Heron and L. J. Goad, *Antiquity*, **65**, 540 (1991).
[185] E. Geeraert, in *Chromatography of Lipids in Biomedical Research and Clinical Diagnosis*, A. Kuksis (Editor), *J. Chromatogr. Library*, Vol. 37, Elsevier, Amsterdam (1987), pp. 48–75.
[186] J. J. Myher and A. Kuksis, *Can. J. Biochem.*, **60**, 638 (1982).
[187] J. J. Myher, A. Kuksis and L.-Y. Yang, *Biochem. Cell Biol.*, **68**, 336 (1990).
[188] A. Kuksis, J. J. Myher and P. Sandra, *J. Chromatogr.*, **500**, 427 (1990).
[189] A. Tokumura, Y. Yoshioka, H. Tsukatani and Y. Handa, *Biomed. Environ. Mass Spectrom.*, **13**, 175 (1986).
[190] T. Baba, K. Kaneda, E. Kusunose, M. Kusunose and I. Yano, *Lipids*, **23**, 1132 (1988).
[191] G. R. Waller (Editor), *Biochemical Applications of Mass Spectrometry*, John Wiley & Sons, New York (1972).
[192] G. R. Waller and O. C. Dermer (Editors), *Biochemical Applications of Mass Spectrometry*, First Supplementary Volume, John Wiley & Sons, New York (1980).
[193] W. W. Christie, *Gas Chromatography of Lipids: A Practical Guide*, The Oily Press, Ayr (1989).
[194] F. J. G. M. Van Kujik, D. W. Thomas, R. J. Stephens and E. A. Dratz, *J. Free Radicals Biol. Med.*, **1**, 215 (1985).
[195] F. J. G. M. Van Kujik, D. W. Thomas, R. J. Stephens and E. A. Dratz, *J. Free Radicals Biol. Med.*, **1**, 387 (1985).
[196] D. W. Thomas, F. J. G. M. Van Kujik, E. A. Dratz and R. J. Stephens, *Anal. Biochem.*, **198**, 104 (1991).
[197] P. Capella and C. M. Zorzut, *Anal. Chem.*, **40**, 1458 (1968).
[198] V. Dommes, F. Wirtz-Peitz and W.-H. Kunau, *J. Chromatogr. Sci.*, **14**, 360 (1976).
[199] J. Myerson, W. F. Haddon and E. L. Soderstrom, *Tetrahedron Lett.*, **23**, 2757 (1982).
[200] R. P. Evershed, in *Lipids: A Practical Approach*, R. J. Hamilton and S. Hamilton (Editors), Oxford University Press, Oxford (1992) pp. 263–308.
[201] S. J. Angyal, *Adv. Carbohydr. Chem. Biochem.*, **42**, 15 (1984).
[202] M. Páez, I. Martinez-Castro, J. Sanz, A. Olano, A. Garcia-Raso and F. Saura-Calixto, *Chromatographia*, **23**, 43 (1987).
[203] A. Garcia-Raso, I. Martinez-Castro, M. Páez, J. Sanz, J. Garcia-Raso and F. Saura-Calixto, *J. Chromatogr.*, **398**, 9 (1987).
[204] I. Martinez-Castro, M. Paez. J. Sanz and A. Garcia-Raso, *J. Chromatogr.*, **462**, 49 (1989).
[205] A. Garcia-Raso, M. Fernandez-Diaz, M. Páez, J. Sanz and J. Martinez-Castro, *J. Chromatogr.*, **471**, 205 (1989).
[206] J. Longren and S. Svensson, *Adv. Carbohydr. Chem. Biochem.*, **29**, 42 (1974).
[207] Z. L. Nikolov and P. J. Reilly, *J. Chromatogr.*, **254**, 157 (1983).
[208] J. Jurenitsch, B. Kopp, I. Gabler-Kolacsek and W. Kubelka, *J. Chromatogr.*, **210**, 337 (1981).
[209] R. Novina, *Chromatographia*, **15**, 704 (1982).
[210] R. Novina, *Chromatographia*, **17**, 441 (1983).
[211] F. Mahmud and E. Catterall, *Pak. J. Sci. Ind. Res.*, **29**, 72 (1986).
[212] M. F. Laker, *J. Chromatogr.*, **163**, 9 (1979).
[213] J. De Neef, *Clin. Chim. Acta*, **26**, 485 (1969).
[214] W. Maciejewicz, M. Daniewski and Z. Mielniczuk, *Chem. Anal. (Warsaw)*, **29**, 421 (1984).
[215] B. W. Christensen, A. Kjaer, J. O. Madsen, C. E. Olsen, O. Olsen and H. Sorensen, *Tetrahedron*, **38**, 353 (1982).
[216] R. H. Dunstan, W. Greenaway and F. R. Whatley, *Proc. R. Soc. London, Ser. B*, **220**, 423 (1984).
[217] J. Turk, B. A. Wolf and M. L. McDaniel, *Biomed. Environ. Mass Spectrom.*, **13**, 237 (1986).
[218] I. Martinez-Castro, M. M. Calvo and A. Olano, *Chromatographia*, **23**, 132 (1987).
[219] G. M. O'Hanlon, O. W. Howarth and H. H. Rees, *Biochem. J.*, **248**, 305 (1987).
[220] F. M. Rubino, *J. Chromatogr.*, **473**, 125 (1989).
[221] D. Küry and U. Keller, *J. Chromatogr.*, **572**, 302 (1991).
[222] A. G. W. Bradbury, D. J. Halliday and D. G. Medcalf, *J. Chromatogr.*, **213**, 146 (1981).
[223] J. P. Kamerling and J. F. G. Vliegenthart, in *Mass Spectrometry*, A. M. Lawson (Editor), Walter de Gruyter, Berlin (1989), p. 177.
[224] M. Morvai, I. M. Perl and D. Knausz, *J. Chromatogr.*, **552**, 337 (1991).
[225] H. Nakamura and T. Okuyama, *J. Chromatogr.*, **509**, 377 (1990).

[226] P. Englmaier, *Fresenius' Z. Anal. Chem.*, **324**, 338 (1986).
[227] N. E. Hoffman and K. A. Peteranetz, *Anal. Lett.*, **5**, 589 (1972).
[228] N. E. Hoffman and R. Y. Yang, *Anal. Lett.*, **5**, 7 (1972).
[229] J. M. Aizpurua and C. Palomo, *Bull. Soc. Chim. Fr. II*, 265 (1982).
[230] G. W. Perold and K. G. R. Pachler, *J. Chem. Soc. C*, 1918 (1966).
[231] P. E. J. Kruger and G. W. Perold, *J. Chem. Soc.*, 2127 (1970).
[232] G. W. Perold, A. J. Hodgkinson and A. S. Howard, *J. Chem. Soc., Perkin Trans. 1*, 2450 (1972).
[233] G. W. Perold, *J. Chromatogr.*, **291**, 365 (1984).
[234] J. M. Rosenfeld, Y. Moharir and S. D. Sandler, *Anal. Chem.*, **61**, 925 (1989).
[235] H.-J. Stan and B. Abraham, *J. Chromatogr.*, **195**, 231 (1980).
[236] T. R. Covey, D. Silvestre, M. K. Hoffman and J. D. Henion, *Biomed. Environ. Mass Spectrom.*, **15**, 45 (1988).
[237] T. J. Betts, K. A. Allan and C. A. Donovan, *J. Chromatogr.*, **291**, 361 (1984).
[238] A. Jefferson and S. Wangchareontrakul, *J. Chromatogr.*, **367**, 145 (1986).
[239] W. Greenaway and F. R. Whatley, *J. Chromatogr.*, **519**, 145 (1990).
[240] C. S. Creaser, M. R. Koupai-Abyazani and G. R. Stephenson, *J. Chromatogr.*, **586**, 323 (1991).
[241] D. J. Harvey, *Biomed. Mass Spectrom.*, **8**, 546 (1981).
[242] D. Krauss, H. G. Mainx, B. Tauscher and P. Bischof, *Org. Mass Spectrom.*, **20**, 614 (1985).
[243] R. P. Evershed, N. Spooner, M. C. Prescott and L. J. Goad, *J. Chromatogr.*, **440**, 23 (1988).
[244] K. H. Schram, *Methods Enzymol.*, **193**, 791 (1990).
[245] S. E. Hattox and J. A. McCloskey, *Anal. Chem.*, **46**, 1378 (1974).
[246] M. Dizardaroglu, *Methods Enzymol.*, **193**, 842 (1990).
[247] E. White, P. M. Krueger and J. A. McCloskey, *J. Org. Chem.*, **37**, 430 (1972).
[248] H. Pang, K. H. Schram, D. L. Smith, S. P. Gupta, L. B. Townsend and J. A. McCloskey, *J. Org. Chem.*, **47**, 3923 (1982).
[249] A. M. Lawson, R. N. Stillwell, M. M. Tacker, K. Tsuboyana and J. A. McCloskey, *J. Am. Chem. Soc.*, **93**, 1014 (1971).
[250] J. A. McCloskey, *Methods Enzymol.*, **193**, 825 (1990).
[251] K. H. Schram and J. A. McCloskey, in *GLC and HPLC Determination of Therapeutic Agents*, K. Tsuji (Editor), Dekker, New York (1979), p. 1149.
[252] C. W. Gehrke and A. B. Patel, *J. Chromatogr.*, **123**, 335 (1976).
[253] C. W. Gehrke and A. B. Patel, *J. Chromatogr.*, **130**, 103 (1977).
[254] A. B. Patel and C. W. Gehrke, *J. Chromatogr.*, **130**, 115 (1977).
[255] D. L. von Minden, R. N. Stillwell, W. A. Koenig, K. J. Lyman and J. A. McCloskey, *Anal. Biochem.*, **50**, 110 (1972).
[256] J. A. Kelly, M. M. Abbasi and J. A. Beisler, *Anal. Biochem.*, **103**, 203 (1980).
[257] J. A. McCloskey, in *Mass Spectrometry in Biomedical Research*, S. J. Gaskell (Editor), John Wiley & Sons, Chichester (1986), p. 75.
[258] U. I. Krahmer, J. G. Liehr, K. J. Lyman, E. A. Orr, R. N. Stillwell and J. A. McCloskey, *Anal. Biochem.*, **82**, 217 (1977).
[259] K. H. Schram, T. Taniguchi and J. A. McCloskey, *J. Chromatogr.*, **155**, 355 (1978).
[260] N. Mahy and E. Gelpi, *J. Chromatogr.*, **130**, 237 (1977).
[261] N. Narasimhachari and K. Leiner, *J. Chromatogr. Sci.*, **15**, 181 (1977).
[262] W. Wilson, W. Sneddon, R. E. Silman, I. Smith and P. Mullen, *Anal. Biochem.*, **81**, 283 (1977).
[263] J. Eyem and L. Bergstedt, *Adv. Mass Spectrom. Biochem. Med.*, **1**, 497 (1976).
[264] C. W. Gehrke, H. Nakamoto and R. W. Zumwalt, *J. Chromatogr.*, **45**, 24 (1969).
[265] C. W. Gehrke and K. Leimer, *J. Chromatogr.*, **53**, 201 (1970).
[266] C. W. Gehrke and K. Leimer, *J. Chromatogr.*, **57**, 219 (1971).
[267] J. P. Hardy and S. L. Kerrin, *Anal. Chem.*, **44**, 1497 (1972).
[268] J. Marik, A. Capek and J. Králicek, *Chromatogr.*, **128**, 1 (1976).
[269] K. Kawashiro, S. Morimoto and H. Yoshida, *Bull. Chem. Soc. Jpn.*, **57**, 2871 (1984).
[270] K. Kawashiro, S. Morimoto and H. Yoshida, *Bull. Chem. Soc. Jpn.*, **58**, 1903 (1985).
[271] K. Kawashiro, S. Morimoto and H. Yoshida, *Bull. Chem. Soc. Jpn.*, **58**, 2727 (1985).
[272] D. B. Boylan and M. Calvin, *J. Am. Chem. Soc.*, **89**, 5472 (1967).
[273] D. B. Boylan, Y. I. A. Alturki and G. Eglinton, in *Advances in Organic Geochemistry 1968*, P. A. Schenck and I. Havenaar (Editors), Pergamon Press, New York (1969), p. 227.
[274] Y. I. A. Alturki, Ph.D. Thesis, University of Bristol, 1972.
[275] P. J. Marriott, J. P. Gill and G. Eglinton, *J. Chromatogr.*, **249**, 291 (1982).
[276] R. A. Alexander, G. Eglinton, J. P. Gill and J. K. Volkman, *J. High Resolut. Chromatogr. Chromatogr. Commun.*, **3**, 521 (1980).
[277] P. J. Marriott, J. P. Gill, R. P. Evershed, G. Eglinton and J. R. Maxwell, *Chromatographia*, **16**, 304 (1982).
[278] R. P. Evershed, G. A. Wolff, G. J. Shaw and G. Eglinton, *Org. Mass Spectrom.*, **20**, 445 (1985).
[279] P. J. Marriott and G. Eglinton, *J. Chromatogr.*, **249**, 311 (1982).
[280] E. J. Corey and R. K. Varma, *J. Am. Chem. Soc.*, **93**, 7319 (1971).
[281] E. J. Corey and T. Ravindranathan, *J. Am. Chem. Soc.*, **94**, 4013 (1972).
[282] E. J. Corey and A. Venkateswarlu, *J. Am. Chem. Soc.*, **94**, 6190 (1972).
[283] K. K. Ogilvie, *Can. J. Chem.*, **51**, 3799 (1973).
[284] H. Hosoda, D. K. Fukushima and J. Fishman, *J. Org. Chem.*, **38**, 4209 (1973).
[285] K. K. Ogilvie, S. L. Beaucage and D. W. Entwistle, *Tetrahedron Lett.*, **16**, 1255 (1976).

[286] G. Stork and P. Hurolick, *J. Am. Chem. Soc.*, **90**, 4462 (1968).
[287] M. A. Quilliam and J. B. Westmore, *Anal. Chem.*, **50**, 59 (1978).
[288] J. L. Kutschinski, Ph.D. Thesis, University of California (1975); *Diss. Abstr. Int.*, B35, 4838 (1975).
[289] T. P. Mawhinney and M. A. Madson, *J. Org. Chem.*, **47**, 3336 (1982).
[290] K. D. Ballard, D. R. Knapp, J. E. Oatis and T. Walle, *J. Chromatogr.*, **277**, 333 (1983).
[291] S. Lewis and M. Yudkoff, *Anal. Biochem.*, **145**, 354 (1987).
[292] S. L. MacKenzie, D. Tenaschuk and G. Fortier, *J. Chromatogr.*, **387**, 241 (1987).
[293] Z. L. Bandi and G. A. S. Ansari, *J. Chromatogr.*, **363**, 402 (1986).
[294] A. I. Mallet, R. M. Barr and J. A. Newton, *J. Chromatogr.*, **378**, 194 (1986).
[295] J. M. Miles, W. F. Schwenk, K. L. McClean and M. W. Haymond, *Anal. Biochem.*, **141**, 110 (1984).
[296] P. D. Marcell, S. P. Stabler, E. R. Podell and R. H. Allen, *Anal. Biochem.*, **150**, 58 (1985).
[297] D. L. Schooley, F. M. Kubiak and J. V. Evans, *J. Chromatogr. Sci.*, **23**, 385 (1985).
[298] K. R. Kim and M. K. Hahn, *Yakhak Hoeji*, **32**, 6 (1988).
[299] K. R. Kim, M. K. Hahn, A. Zlatkis, E. C. Horning and B. S. Middleditch, *J. Chromatogr.*, **468**, 289 (1989).
[300] A. P. J. M. de Jong, J. Elema and B. J. T. van der Berg, *Biomed. Mass Spectrom.*, **7**, 359 (1980).
[301] F. Abbott, J. Kassam, A. Acheampong, S. Ferguson, S. Panesar, R. Burton, K. Farrell and J. Orr, *J. Chromatogr.*, **375**, 285 (1986).
[302] R. W. Kelly and P. L. Taylor, *Anal. Chem.*, **48**, 465 (1976).
[303] P. L. Taylor and R. W. Kelly, *FEBS Lett.*, **57**, 22 (1975).
[304] A. G. Smith, W. A. Harland and C. J. W. Brooks, *J. Chromatogr.*, **142**, 533 (1977).
[305] A. C. Bazan and D. R. Knapp, *J. Chromatogr.*, **236**, 201 (1982).
[306] C. R. Pace-Asciak and N. S. Edwards, *Methods Enzymol.*, **86**, 552 (1982).
[307] R. C. Murphy and K. L. Clay, *Methods Enzymol.*, **86**, 547 (1982).
[308] A. R. Brash, *Methods Enzymol.*, **86**, 579 (1982).
[309] R. L. Maas, D. F. Taber and L. J. Roberts, *Methods Enzymol.*, **86**, 592 (1982).
[310] C. R. Pace-Asciak and N. S. Edwards, *Methods Enzymol.*, **86**, 604 (1982).
[311] J. Mai, B. German and J. E. Kinsella, *J. Chromatogr.*, **254**, 91 (1983).
[312] S. Steffenrud and P. Borgeat, *Prostaglandins*, **28**, 593 (1984).
[313] O. Vesterqvist and K. Green, *Prostaglandins*, **28**, 139 (1984).
[314] S. Steffenrud, P. Borgeat, H. Salari, M. J. Evans and M. J. Bertrand, *J. Chromatogr.*, **416**, 219 (1987).
[315] S. Steffenrud, P. Borgeat, M. J. Evans and M. J. Bertrand, *J. Chromatogr.*, **423**, 1 (1987).
[316] H. L. Hubbard, T. D. Eller, D. E. Mais, P. V. Halushka, R. H. Baker, I. A. Blair, J. J. Vrbanac and D. R. Knapp, *Prostaglandins*, **33**, 149 (1987).
[317] D. A. Herold, J. Savory, M. Kinter, R. Ross and M. R. Wills, *Anal. Chim. Acta*, **197**, 149 (1987).
[318] H. J. Leis, E. Malle, B. Mayer, G. M. Kostner, H. Esterbauer and H. Gleispach, *Anal. Biochem.*, **162**, 337 (1987).
[319] S. Steffenrud, P. Borgeat, M. J. Evans and M. J. Bertrand, *Biomed. Environ. Mass Spectrom.*, **14**, 313 (1987).
[320] H. J. Leis, E. Hohenester, H. Gleispach, E. Malle and B. Mayer, *Biomed. Environ. Mass Spectrom.*, **14**, 617 (1987).
[321] H. Hughes, J. Nowlin, S. J. Gaskell and V. C. Parr, *Biomed. Environ. Mass Spectrom.*, **16**, 409 (1988).
[322] M. Ishibashi, K. Yamashita, K. Watanabe and H. Miyazaki, in *Mass Spectrometry in Biomedical Research*, S. J. Gaskell (Editor), John Wiley & Sons, Chichester (1986), pp. 423–441.
[323] B. P. Lisboa and I. Ganschow, in *Chromatography and Mass Spectrometry in Biomedical Sciences*, 2, A. Frigerio (Editor), Elsevier, Amsterdam (1983), p. 291.
[324] E. M. H. Finley and S. J. Gaskell, *Clin. Chem.*, **27**, 1165 (1981).
[325] T. E. Chapman, G. P. B. Kraan, N. M. Drayer, G. T. Nagel and B. G. Wolthers, *Biomed. Environ. Mass Spectrom.*, **14**, 73 (1987).
[326] I. A. Blair and G. Phillipou, *J. Chromatogr. Sci.*, **16**, 201 (1978).
[327] S. J. Gaskell, V. J. Gould and H. M. Leith, in *Mass Spectrometry in Biomedical Research*, (S. J. Gaskell (Editor), John Wiley & Sons, Chichester (1986), p. 347.
[328] V. J. Gould, A. O. Turkes and S. J. Gaskell, *J. Steroid Biochem.*, **24**, 563 (1986).
[329] D. S. Millington, *J. Steroid Biochem.*, **6**, 239 (1975).
[330] S. J. Gaskell and D. S. Millington, *Biomed. Mass Spectrom.*, **5**, 557 (1978).
[331] S. J. Gaskell, C. J. Porter and B. N. Green, *Biomed. Mass Spectrom.*, **12**, 139 (1985).
[332] G. Phillipou, D. Bigham and R. F. Seamark, *Lipids*, **10**, 714 (1975).
[333] J. J. Myher, A. Kuksis, L. Marai and S. K. F. Yeung, *Anal. Chem.*, **50**, 557 (1978).
[334] K. Satouchi and K. Saito, *Biomed. Mass Spectrom.*, **6**, 87 (1979).
[335] P. M. Woollard, *Biomed. Mass Spectrom.*, **10**, 143 (1983).
[336] H. Parsons, E. M. Emken, L. Marai and A. Kuksis, *Lipids*, **21**, 247 (1986).
[337] F. S. Abbott et al., *J. Chromatogr.*, **375**, 285 (1986).
[338] A. Kuksis and J. J. Myher, *J. Chromatogr.*, **379**, 57 (1986).
[339] K. Satouchi and K. Saito, *Biomed. Mass Spectrom.*, **6**, 396 (1979).
[340] K. Satouchi, M. Oda, K. Yasunaga and K. Saito, *J. Biochem. (Tokyo)*, **94**, 2067 (1983).
[341] K. Satouchi, M. Oda, K. Saito and D. J. Hanahan, *Arch. Biochem. Biophys.*, **234**, 318 (1984).
[342] K. Satouchi, M. Oda, K. Yasunaga and K. Saito, *Biochem. Biophys. Res. Commun.*, **128**, 1409 (1985).
[343] M. Oda, K. Satouchi, K. Yasunaga and K. Saito, *J. Immunol.*, **134**, 1090 (1985).
[344] K. Yasuda, K. Satouchi, R. Nakayama and K. Saito, *Biomed. Environ. Mass Spectrom.*, **16**, 137 (1988).

[345] S. Thurl, J. Offermanns, B. Muller-Werner and G. Sawatzki, *J. Chromatogr.*, **568**, 291 (1981).
[346] K. Saito, R. Nakayama, K. Yasuda, K. Satouchi and J. Sugatani, in *Biological Mass Spectrometry*, A. L. Burlingame and J. A. McCloskey (Editors), Elsevier, Amsterdam (1990), pp. 527–547.
[347] C. S. Ramesha and W. C. Pickett, *Biomed. Environ. Mass Spectrom.*, **13**, 107 (1986).
[348] T. P. Mawhinney, R. S. P. Robinett, A. Atalay and M. A. Madson, *J. Chromatogr.*, **361**, 117 (1986).
[349] M. A. Quilliam, K. K. Ogilvie and J. B. Westmore, *Biomed. Mass Spectrom.*, **1**, 78 (1974).
[350] M. A. Quilliam, Ph.D. Thesis, University of Manitoba, Canada (1977).
[351] M. Dizdarouglu, *J. Chromatogr.*, **295**, 103 (1984).
[352] M. Dizdaroglu, *Anal. Biochem.*, **144**, 593 (1985).
[353] S. Lewis and M. Yudkoff, *Anal. Biochem.*, **145**, 351 (1985).
[354] K. D. Ballard, T. D. Eller, J. G. Webb, W. H. Newman, D. R. Knapp and R. G. Knapp, *Biomed. Environ. Mass Spectrom.*, **13**, 667 (1986).
[355] P. F. Crain, *Methods Enzymol.*, **193**, 857 (1991).
[356] J. Maybaum, M. G. Kott, N. J. Johnson, W. D. Ensminger and P. L. Stetson, *Anal. Biochem.*, **161**, 164 (1987).
[357] W. F. Schwenk, P. J. Berg. B. Beaufrere, J. M. Miles and M. W. Haymond, *Anal. Biochem.*, **141**, 101 (1984).
[358] S. L. MacKenzie and D. Tenaschuk, *J. Chromatogr.*, **322**, 228 (1985).
[359] C. J. Biermann, C. M. Kinoshita, J. A. Marlett and R. D. Steele, *J. Chromatogr.*, **357**, 330 (1986).
[360] T. P. Mawhinney, R. S. R. Robinett, A. Atalay and M. A. Madson, *J. Chromatogr.*, **358**, 231 (1986).
[361] G. Fortier, D. Tenaschuk and S. L. Mackenzie, *J. Chromatogr.*, **361**, 253 (1986).
[362] H. J. Chaves das Neves and A. M. P. Vasconcelos, *J. Chromatogr.*, **392**, 249 (1987).
[363] R. J. Earley, J. R. Thompson, G. W. Sedgwick, J. M. Kelly and R. J. Christophson, *J. Chromatogr.*, **416**, 15 (1987).
[364] H. J. Chaves das Neves, A. M. P. Vasconcelos, J. Rueff Tavares and P. Nogueira Ramos, *J. High Resolut. Chromatogr. Chromatogr. Commun.*, **11**, 12 (1988).
[365] S. L. MacKenzie, D. Tenaschuk and G. Fortier, *J. Chromatogr.*, **387**, 241 (1987).
[366] G. Eglinton, R. P. Evershed and J. P. Gill, *Org. Geochem.*, **6**, 157 (1984).
[367] P. J. Marriott, J. P. Gill, R. P. Evershed, C. S. Hein and G. Eglinton, *J. Chromatogr.*, **301**, 107 (1984).
[368] J. P. Gill, R. P. Evershed, M. I. Chicarelli, G. A. Wolff, J. R. Maxwell and G. Eglinton, *J. Chromatogr.*, **350**, 37 (1985).
[369] G. Eglinton, R. P. Evershed, J. P. Gill and C. S. Hein, *Anal. Proc.*, **22**, 263 (1985).
[370] C. S. Hein, J. P. Gill, R. P. Evershed and G. Eglinton, *Anal. Chem.*, **57**, 1872 (1985).
[371] J. P. Gill, R. P. Evershed and G. Eglinton, *J. Chromatogr.*, **369**, 281 (1986).
[372] S. Kaur, J. P. Gill, R. P. Evershed, G. Eglinton and J. R. Maxwell, *J. Chromatogr.*, **473**, 135 (1989).
[373] H. Miyazaki, M. Ishibashi, M. Itoh and T. Nambara, *Biomed. Mass Spectrom.*, **4**, 23 (1977).
[374] H. Miyazaki, M. Ishibashi, K. Yamashita and M. Katori, *J. Chromatogr.*, **153**, 83 (1978).
[375] K. Watanabe, K. Yamashita, M. Ishibashi, Y. Hayashi, S. Yamamoto and H. Miyazaki, *J. Chromatogr.*, **468**, 383 (1989).
[376] H. Miyazaki, M. Ishibashi, K. Yamashita, I. Ohguchi. H. Saitoh, H. Kurono, M. Shimono and M. Katori, *J. Chromatogr.*, **239**, 595 (1982).
[377] H. Miyazaki, M. Ishibashi, K. Yamashita and M. Yakushiji, *Chem. Pharm. Bull.*, **29**, 796 (1981).
[378] A. Fukunaga, Y. Hatta, M. Ishibashi and H. Miyazaki, *J. Chromatogr.*, **190**, 339 (1980).
[379] M. Aso, K. Miyazaki, J. Yangisawa and F. Nakayama, *J. Biochem. (Tokyo)*, **101**, 1429 (1987).
[380] J. Goto, K. Watanabe, H. Miura, T. Nambara and T. Iida, *J. Chromatogr.*, **388**, 379 (1987).
[381] J. Goto, H. Miurta, M. Inada, T. Nambara, T. Nagakura, and H. Suzuki, *J. Chromatogr.*, **452**, 119 (1988).
[382] T. Eguchi, H. Miyazaki and F. Nakayama, *J. Chromatogr.*, **525**, 25 (1990).
[383] T. Iida, T. Tamaru, F. C. Chang, J. Goto and T. Nambara, *J. Chromatogr.*, **558**, 451 (1991).
[384] D. J. Harvey, *J. Chromatogr.*, **147**, 291 (1978).
[385] D. J. Harvey, *Org. Mass. Spectrom.*, **12**, 473 (1977).
[386] H. Miyazaki, M. Ishibashi, M. Itoh, K. Yamashita and T. Nambara, *J. Chromatogr.*, **133**, 311 (1977).
[387] D. J. Harvey, *J. Pharm. Pharmacol.*, **28**, 280 (1976).
[388] D. J. Harvey and W. D. M. Paton, *J. Chromatogr.*, **109**, 73 (1975).
[389] R. J. Anderegg, A. Brajter-Toth and J. P. Toth, *Anal. Chem.*, **56**, 1351 (1984).
[390] H. Miyazaki, M. Ishibashi and Y. Yamashita, *Biomed. Mass Spectrom.*, **6**, 57 (1979).
[391] A. Fukunaga, Y. Hatta, M. Ishibashi and H. Miyazaki, *J. Chromatogr.*, **190**, 339 (1980).
[392] J. Yanagisawa, M. Itoh, M. Ishibashi, H. Miyazaki and F. Nakayama, *Anal. Biochem.*, **104**, 75 (1980).
[393] H. Miyazaki, M. Ishibashi, K. Yamashita, Y. Nishikawa and M. Katori, *Biomed. Mass Spectrom.*, **8**, 521 (1981).
[394] Y. Harada, K. Tanaka, Y. Uchida, A. Ueno, S. Ohishi, M. Ishibashi and H. Miyazaki, *Prostaglandins*, **23**, 881 (1982).
[395] Y. Harada, K. Tanaka, K. Yamashita, M. Ishibashi, H. Miyazaki and M. Katori, *Prostaglandins*, **26**, 79 (1983).
[396] H. Miyazaki, M. Ishibashi, H. Takayama, K. Yamashita, I. Suwa and M. Katori, *J. Chromatogr.*, **289**, 249 (1984).
[397] K. Yamashita, M. Ishibashi, H. Miyazaki, S. Narumiya, T. Ogorochi and O. Hayaishi, *J. Pharmacobio-Dyn.*, **7**, s-12 (1984).
[398] T. Ogorochi, S. Narumiya, S. Mizuno, N. Yamashita, K. Miyazaki and O. Hayaishi, *J. Neurochem.*, **43**, 71 (1984).
[399] H. Miyazaki, K. Yamashita and M. Ishibashi, in *Advances in Prostaglandin, Thromboxane and Leukotriene Research*, Vol. 15, (O. Hayaishi and S. Yamamoto (Editors)), Raven Press, New York (1985), p. 45.
[400] K. Yamashita, K. Watanabe, Y. Hashimoto, M. Ishibashi, H. Miyazaki, T. Kubodera and T. Takeda, *J. Pharmacobio-Dyn.*, **9**, s-70 (1986).
[401] K. Yamashita, K. Watanabe, M. Ishibashi, H. Miyazaki,

K. Yokota, K. Horie and S. Yamamoto, *J. Chromatogr.*, **399**, 223 (1987).
[402] K. Yamashita, K. Wanatabe, M. Ishibashi, M. Katori and H. Miyazaki, *J. Chromatogr.*, **424**, 1 (1988).
[403] M. Ishibashi, K. Watanabe, N. Harima and S. Krolik, *Biomed. Environ. Mass Spectrom.*, **17**, 133 (1988).
[404] M. Ishibashi, Y. Ohyama, N. Harima and M. Mizugaki, *Biomed. Environ. Mass Spectrom.*, **18**, 787 (1990).
[405] T. Sakurai, H. Ichimiya, H. Miyazaki and F. Nakayama, *J. Chromatogr.*, **571**, 1 (1991).
[406] T. Hirai, M. Fugisaki, M. Igarashi, T. Asada, T. Shiba, S. Takeuchi, R. Yamaguchi, T. Shirai and M. Kayama, *Thrombosis Res.*, **41**, 637 (1986).
[407] G. Phillipou, *J. Chromatogr.*, **129**, 384 (1976).
[408] I. A. Blair and G. Phillipou, *J. Chromatogr. Sci.*, **15**, 478 (1977).
[409] D. J. Harvey, *Biomed. Mass Spectrom.*, **4**, 265 (1977).
[410] C. Eaborn, *Organosilicon Compounds*, Butterworth, London (1960), pp. 140–143.
[411] S. Steffenrud and P. Borgeat, *Prostaglandins*, **28**, 593 (1984).
[412] S. Steffenrud, P. Borgeat, M. J. Evans and M. J. Bertrand, *Biomed. Environ. Mass Spectrom.*, **13**, 657 (1986).
[413] S. Steffenrud, P. Borgeat, M. J. Evans and M. J. Bertrand, *Biomed. Environ. Mass Spectrom.*, **14**, 313 (1987).
[414] S. Steffenrud, P. Borgeat, M. Salari, M. J. Evans and M. J. Bertrand, *J. Chromatogr.*, **416**, 219 (1987).
[415] R. W. Kelly, *J. Chromatogr.*, **43**, 229 (1969).
[416] R. W. Kelly, *Steroids*, **13**, 507 (1969).
[417] M. Wieber and M. Schmidt, *Chem. Ber.*, **96**, 1561 (1963).
[418] R. H. Cragg and R. D. Lane, *J. Organomet. Chem.*, **212**, 301 (1981).
[419] Y. Kita, H. Yasuda, Y. Sugiyama, F. Fukata, J. Haruta and Y. Tamura, *Tetrahedron Lett.*, **24**, 1273 (1983).
[420] J.-C. Pommier, R. Calas and J. Valade, *Bull. Soc. Chim. Fr.*, 1475 (1968).
[421] B. M. Trost and C. G. Caldwell, *Tetrahedron Lett.*, **22**, 4999 (1981).
[422] C. J. W. Brooks and W. J. Cole, *Analyst*, **110**, 587 (1985).
[423] C. J. W. Brooks, W. J. Cole and G. M. Barrett, *J. Chromatogr.*, **315**, 119 (1984).
[424] D. G. Watson, D. S. Rycroft, I. M. Freer and C. J. W. Brooks, *Phytochemistry*, **24**, 2195 (1985).
[425] E. J. Corey and P. B. Hopkins, *Tetrahedron Lett.*, **27**, 4871 (1982).
[426] H. Miyazaki, M. Ishibashi, M. Itoh and K. Yamashita, *Biomed. Mass Spectrom.*, **11**, 377 (1984).
[427] M. Ishibashi, M. Itoh, K. Yamashita, H. Miyazaki and H. Nakata, *Chem. Pharm. Bull.*, **34**, 3298 (1986).
[428] H. Nakata, M. Ishibashi, M. Itoh and H. Miyazaki, *Org. Mass Spectrom.*, **22**, 23 (1987).
[429] M. Ishibashi, T. Irie and H. Miyazaki, *J. Chromatogr.*, **399**, 197 (1987).
[430] M. Ishibashi, H. Takayama, Y. Nakagawa and N. Harima, *Chem. Pharm. Bull.*, **36**, 845 (1988).
[431] M. Ishibashi, K. Watanabe, H. Miyazaki and S. Krolik, *Chem. Pharm. Bull.*, **34**, 3510 (1986).
[432] M. Ishibashi, K. Watanabe, H. Miyazaki and S. Krolik, *Yakugakuzasshi*, **106**, 118 (1986).
[433] M. Ishibashi, K. Watanabe, H. Miyazaki and S. Krolik, *J. Chromatogr.*, **391**, 183 (1987).
[434] K. Watanabe, M. Ishibashi, N. Harima and S. Krolik, *Chem. Pharm. Bull.*, **37**, 140 (1989).
[435] P. G. Devaux, M. G. Horning, R. M. Hill and E. C. Horning, *Anal. Biochem.*, **41**, 70 (1971).
[436] T. A. Baillie, C. J. W. Brooks and E. C. Horning, *Anal. Lett.*, **5**, 351 (1972).
[437] J. Goto, Y. Teraya, T. Nambara and T. Iida, *J. Chromatogr.*, **585**, 281 (1991).
[438] D. J. Harvey, *Biomed. Mass Spectrom.*, **7**, 278 (1980).
[439] D. J. Harvey, *J. Chromatogr.*, **196**, 156 (1980).
[440] Y. Guindon, R. Fortin, C. Yoakim and J. W. Gillard, *Tetrahedron Lett.*, **25**, 4717 (1984).
[441] A. I. Mallet, R. M. Barr and J. A. Newton, *J. Chromatogr.*, **378**, 194 (1986).
[442] C. J. W. Brooks, W. J. Cole and R. A. Anderson, *J. Chromatogr.*, **514**, 305 (1990).
[443] B. Feibush and L. Spialter, *J. Chem. Soc. B*, 115 (1971).
[444] R. J. P. Corriu, C. Guérin and J. J. E. Moreau, *Top. Stereochem.*, **15**, 43 (1984).
[445] R. J. P. Corriu, C. Guérin and J. J. E. Moreau, in *The Chemistry of Organic Silicon Compounds*, Vol. 1, S. Patai and Z. Rappoport (Editors), John Wiley & Sons Ltd, Chichester (1989), p. 305.
[446] R. J. P. Corriu and G. Royo, *J. Organomet. Chem.*, **14**, 291 (1968).
[447] E. D. Morgan and C. F. Poole, *J. Chromatogr.*, **89**, 225 (1974).
[448] E. D. Morgan and C. F. Poole, *J. Chromatogr.*, **104**, 351 (1975).
[449] C. F. Poole and E. D. Morgan, *Org. Mass Spectrom.*, **10**, 537 (1975).
[450] C. F. Poole, A. Zlatkis, W.-F. Sye, S. Singhawangcha and E. D. Morgan, *Lipids*, **15**, 734 (1980).
[451] P. M. Burkinshaw, E. D. Morgan and C. F. Poole, *J. Chromatogr.*, **132**, 548 (1977).
[452] I. S. Krull, M. Swartz and J. N. Driscoll, *Anal. Lett.*, **17**, 2369 (1984).
[453] I. S. Krull, M. Swartz and J. N. Driscoll, *Anal. Lett.*, **18**, 2619 (1985).
[454] P. R. Robinson, M. D. Jones and J. Maddock, *J. High Resolut. Chromatogr. Chromatogr. Commun.*, **10**, 6 (1987).
[455] J. Gilber and J. R. Startin, *J. Chromatogr.*, **189**, 86 (1980).
[456] J. Gilbert and J. R. Startin, *Eur. Polym. J.*, **16**, 73 (1980).
[457] C. F. Poole, S. Singhawangcha, L.-E. Chen Hu, W.-F. Sye, R. Brazell and A. Zlatkis, *J. Chromatogr.*, **187**, 331 (1980).
[458] C. F. Poole, W.-F. Sye, S. Singhawangcha, F. Hsu, A. Zlatkis, A. Arfwidsson and J. Vessman, *J. Chromatogr.*, **199**, 123 (1980).
[459] C. Eaborn, D. R. M. Walton and G. S. Thomas, *Chem. Ind. (London)*, 827 (1967).
[460] C. Eaborn, C. A. Holder, D. R. M. Walton and B. S. Thomas, *J. Chem. Soc. C*, 2502 (1969).
[461] H. Rembold and B. Lackner, *J. Chromatogr.*, **323**, 355 (1985).
[462] H. Rembold, *Chimia*, **39**, 348 (1985).

[463] M. Bownes and H. Rembold, *Eur. J. Biochem.*, **164**, 709 (1987).
[464] J. Stadler, *Anal. Biochem.*, **86**, 477 (1978).
[465] C.-G. Hammer, *Biomed. Mass Spectrom.*, **5**, 25 (1978).
[466] C. Pantarotto, R. Fanelli, S. Filippeschi, T. Facchinetti, F. Spreafico and M. Salmona, *Anal. Biochem.*, **97**, 232 (1979).
[467] T. K. Kwan, N. F. Taylor and D. B. Gower, *J. Chromatogr.*, **301**, 189 (1984).
[468] L. Chabraoui, B. Mathian, M. C. Patricot and A. Revol, *J. Chromatogr.*, **567**, 299 (1991).
[469] E. K. Symes and B. S. Thomas, *J. Chromatogr.*, **116**, 163 (1976).
[470] B. S. Thomas, *J. Chromatogr.*, **56**, 37 (1971).
[471] L. M. Cummins, in *Recent Advances in Gas Chromatography*, I. I. Domsky and J. A. Perry (Editors), Marcel Dekker, New York (1971), p. 313.
[472] J. B. Brooks, J. A. Liddle and C. C. Alley, *Anal. Chem.*, **47**, 1960 (1975).
[473] M. J. Bertrand, S. Stefanidis and B. Sarrasin, *J. Chromatogr.*, **351**, 47 (1986).
[474] M. J. Bertrand, S. Stefanidis, A. Donais and B. Sarrasin, *J. Chromatogr.*, **354**, 331 (1986).
[475] M. J. Bertrand, A. W. Ahmed, B. Sarrasin and V. N. Mallet, *Anal. Chem.*, **59**, 1302 (1987).
[476] M. J. Bertrand, R. Massé and R. Dugal, in *Mass Spectrometry and Combined Techniques in Medicine, Clinical Chemistry and Clinical Biochemistry*, M. Eggstein and H. M. Liebich (Editors), Tübingen, Germany (1977), p. 256.
[477] R. Massé, R. Dugal and M. J. Bertrand, in *Recent Developments in Chromatography and Electrophoresis*, 10, A. Frigerio and M. McCamish (Editors), Elsevier, Amsterdam (1980), p. 169.
[478] D. J. Harvey, *Biomed. Environ. Mass Spectrom.*, **14**, 103 (1987).
[479] D. J. Harvey, *Spectrosc. Int. J.*, **8**, 211 (1990).
[480] U. Langenbeck, H. Luthe and G. Schaper, *Biomed. Mass Spectrom.*, **12**, 507 (1985).

5

Alkylation

Pavol Kováč

National Institute of Diabetes and Digestive and Kidney Diseases, National Institutes of Health, Bethesda, MD 20892, USA

---

1. INTRODUCTION — 109
2. ALKYLATION PROCEDURES — 110
   - 2.1 Alkylation with alkyl halides and silver oxide — 111
   - 2.2 Alkylation with alkyl halides and silver oxide in N,N-dimethylformamide — 111
   - 2.3 Alkylation with alkyl halides and barium oxide in N,N-dimethylformamide — 114
   - 2.4 Alkylation with alkyl halides and sodium hydride in N,N-dimethylformamide — 115
   - 2.5 Alkylations with alkyl halides in dimethyl sulfoxide — 115
     - 2.5.1 Alkylation with alkyl halides and methylsulfinyl anion in dimethyl sulfoxide (Hakomori methylation) — 115
     - 2.5.2 Alkylation with alkyl halides and a solid base in dimethyl sulfoxide — 116
   - 2.6 Alkylation with alkyl halides and sodium hydride in ether-type solvents — 116
   - 2.7 Alkylation with alkyl halides and sodium hydride in N,N-dimethylacetamide — 117
   - 2.8 Alkylation with diazoalkanes — 117
   - 2.9 Lewis acid-catalyzed alkylation with diazoalkanes — 118
   - 2.10 Alkylation with N,N-dimethylformamide dialkyl acetals — 119
   - 2.11 Alkylation with 3-alkyl-1-p-tolyltriazenes — 120
   - 2.12 Alkylation with alkyl fluorosulfonates — 120
   - 2.13 Alkylation with trimethylanilinium hydroxide — 120
   - 2.14 Alkylation with trialkyloxonium fluoroborates — 121
   - 2.15 Alkylation with alcohols in the presence of acid catalysts — 122
     - 2.15.1 Formation of benzyl ethers from substances related to catechols — 122
     - 2.15.2 Condensation of simple alcohols with free sugars (Fischer's method of glycosidation) — 122
   - 2.16 Preparation of pentafluorobenzyl derivatives — 124
     - 2.16.1 Formation of carbamate derivatives using pentafluorobenzyl chloroformate — 124
     - 2.16.2 Formation of carbamate derivatives using pentafluorobenzyl-hydroxylamine hydrochloride — 125
     - 2.16.3 Formation of pentafluorobenzyl derivatives using pentafluorobenzyl bromide — 125
   - 2.17 Alkylation by means of reductive condensation of amines with aldehydes — 126
   - 2.18 Alkylation with alkyl trichloroacetimidates — 126
3. CONCLUSIONS — 126
4. REFERENCES — 127

1. Introduction

In this chapter the term 'alkylation' refers to the small scale reaction of organic substances that contain a reactive hydrogen, e.g. R–COOH, R–OH, R–SH, R–NH–R', R–NH$_2$, R–CONH$_2$, R–CONH–R' and R–CO–CH$_2$–CO–R', with a derivatizing agent. Replacement of such a hydrogen with an alkyl group is important in chromatographic analysis because of the decreased polarity of the derivative as compared with the parent substance, facilitating analysis by chromatographic tech-

Handbook of Derivatives for Chromatography
Edited by K. Blau and J. M. Halket © 1993 John Wiley & Sons Ltd

niques. The decrease in polarity and intermolecular association is particularly important for analysis by gas chromatography and mass spectrometry, as it permits their application to the analysis of compounds not amenable to these techniques because of low volatility. Another reason for derivatization prior to the analysis by chromatography is that mixtures of closely related compounds that show poor separation before derivatization may often be resolved by use of a proper derivative. The right choice of a particular type of derivative may be of special importance when dealing with heat-sensitive substances, in which case the purpose of the derivatization is to 'protect'. Consequently, in these situations we would be performing 'protective alkylation'.

No chapter on alkylation methodology can hope to cover all facets of the topic, let alone make reference to all papers which bear on the subject. Here, attention has been focused on the most commonly used types of derivatives and procedures which, owing to their reliability, have found wide application. Since esterification with alcohols in the presence of acid catalysts is dealt with in Chapter 2, the method is considered here only with respect to simple glycosidation of sugars and the formation of benzyl ethers from substances related to catechol. Silyl ether derivatives are dealt with in Chapter 4.

One of the most important areas of chromatography where alkylation has been applied concerns carbohydrates. In that area, methyl derivatives, particularly, have been of the utmost value in structural determination as a means of 'labeling' free hydroxyl groups. In this context, the value of the trideuteromethyl groups resulting from trideuteromethylation in the elucidation of structurally significant patterns of fragmentation of various partially methylated cabohydrate derivatives upon impact with electrons cannot be overemphasized. Methods which have been used for alkylation of carbohydrates can often be applied successfully in other areas, such as glycolipid, lipid, peptide, glycoprotein and protein chemistry. As substances submitted to the derivatization procedure often contain more than one type of reactive functional group, it is important to consider the specificity of the reagent and the possible side reactions which may occur.

2. Alkylation procedures

Haworth originally devised the procedure of methylation in aqueous solution with dimethyl sulfate and sodium hydroxide [1]; similarly, the method of benzylation [2] with benzyl chloride and powdered alkali hydroxide (with or without an additional solvent) and the old methylation procedure of Purdie and Irvine [3] are slow reactions: several treatments with the alkylating agent are often required with substrates bearing more than one active site. These methods are, however, still of utility where more powerful procedures cannot be applied because of substrate instability or side reactions, and can equally well be carried out with other reactive halides to introduce other alkyl groups. Alkylations in the presence of an aqueous solution of the base are, however, only seldom used now except, perhaps, in conjunction with the cheap alkylating reagents dialkyl sulfates. This is mainly because water would be expected to consume the alkylation reagent at a higher rate than the substrate to be alkylated. When alkylations of ionized hydroxyl functions are carried out in an inert organic solvent, the drier the solvent, the more efficient the alkylation. Purdie's technique was considerably improved by Kuhn et al. [4], who used N,N-dimethylformamide as a polar but aprotic solvent. Still more powerful alkylation techniques have been devised: when a substrate containing a replaceable hydrogen atom is treated with sodium hydride in a suitable solvent, molecular hydrogen is evolved slowly, and the resultant sodium salt reacts rapidly with alkyl halides (R'−X). Probably the most remarkable advance in protective alkylation was achieved by Hakomori [5], who developed this approach and used the strongly basic methylsulfinyl carbanion in dimethyl sulfoxide for the methylation of glycolipids and polysaccharides. Hydroxyl functions are ionized under these conditions and react readily with an alkyl halide. Rapid alkylations of other functional groups by this method probably proceed analogously. The general belief that the methylsulfinylmethanide ($[CH_3SOCH_2]^-$) generated from sodium hydride and dimethyl sulfoxide is the effective base in the Hakomori methylation has been questioned. The conclusions drawn from an extensive study [6] on the permethylation of carbohydrates in dipolar aprotic solvents suggested a new, more efficacious method for the permethylation of sugars (procedure 2.5.2).

Diazoalkanes are well established as alkylation agents. They preferentially alkylate moderately acidic functional groups. It would be beyong the scope of this chapter to discuss all the reactions of diazoalkanes with the whole spectrum of organic substances. The range of possible reactions is extensive; the possibility of side reactions during alkylation of new compounds with diazoalkanes should therefore always be considered, and the data obtained should be interpreted with care. An important extension to the use of diazoalkanes was the finding that in the presence of Lewis acids, such as boron trifluoride etherate [7], hydrogen tetrafluoroborate [8], aluminum chloride [7], stannous chloride [9] and several other subtances [10], aliphatic alcohols can also be extensively alkylated with these substances. The procedure has been widely used for the derivatization of hydroxyl group(s)-containing organic compounds.

The widespread use of gas chromatography and mass spectrometry necessitated fast, reliable and simple new derivatization techniques. Apart from silylation, most procedures for the formation of ethers from alcohols do not fulfil these requirements, but several useful reagents, most of them commercially available, have been introduced for rapid alkylation of the more acidic functions.

Many of the procedures given in the following sections are generally applicable. Comments are made on all the methods described (see also Table 1), and it is hoped that with a little chemical intuition readers will be able to choose the reaction conditions best suited to their particular application.

2.1 Alkylation with alkyl halides and silver oxide

R—COOH ⟶ R—COOR'
R—OH ⟶ R—OR'
R—SH ⟶ R—SR'

Silver oxide is added to a solution of the substrate in an excess of alkyl halide. The mixture is shaken in the dark until the reaction is complete, and the progress of the reaction is monitored by, e.g., TLC. Fresh portions of silver oxide are added as necessary at intervals of 2–3 h. The reaction mixture is filtered, the solids are washed with a suitable non-hydroxylic solvent and the combined filtrate is concentrated. If necessary, the procedure is repeated using this partially alkylated product.

Note: Alkyl iodides and bromides are far more reactive than chlorides and are the reagents of choice. When the substance is insoluble in the alkyl halide used, a small amount of a suitable additional solvent may be used but, clearly, hydroxylic solvents must only be used as a last resort. If hydroxylic solvents have to be used, the extra amount of alkylating agent that will be consumed in alkylating the solvent has to be allowed for.

The procedure will convert any non-hindered carboxylic function (or its salt) to the corresponding alkyl ester in minutes, and phenolic or thiol groups will also be alkylated rapidly. Alcoholic groups are alkylated more slowly. The process may be accelerated by stirring the reaction under reflux in the presence of a desiccant. The addition of dimethyl sulfide to the reaction mixture for methylation of hydroxylic compounds has been shown [11] to affect the rate of O-alkylation markedly, and to yield permethylated products [12] in cases where the standard Purdie procedure gives only partially methylated products. Evidence has been presented [11] against the trimethylsulfonium iodide being the alkylating agent in this case, and it has been suggested that the enhancement of the rate of O-alkylation in the presence of dimethyl sulfide is due to a modification of silver oxide, e.g. by complex formation, in such a manner as to convert it to a more efficient base. This is supported by the fact that degradation by β-elimination of uronic acid derivatives, similar to a strong base-induced analogous reaction, was observed to have taken place when substances of this class were methylated with Purdie's reagent in the presence of dimethyl sulfide [13, 14]. The conditions of procedure 2.1 are probably the mildest available for the alkylation of aliphatic hydroxyl groups. When the substrate contains O-acetyl groups these survive but may migrate; N-acetyl groups are also stable to the process so that, when N-acetamidodeoxy sugars or related substances are alkylated in this way, O-alkylated N-acetamido derivatives are produced, the N-acetamido group remaining intact [15]. The procedure is not recommended for the alkylation of free sugars because oxidative degradation by silver oxide may occur.

2.2 Alkylation with alkyl halides and silver oxide in N,N-dimethylformamide

—CONH— ⟶ —CON—R'

For further reactions see Section 2.1.

Silver oxide and the alkyl halide (1:20, w/v) are added to a solution of the substrate in N,N-dimethylformamide, and the suspension is stirred at room temperature until the reaction is complete. The mixture is filtered, the solids are washed with N,N-dimethylformamide, and dichloromethane (4 volumes) is added to the filtrate.

Table 1 A summary of derivatizing agents and procedures for application to the alkylation of specific types of compounds

Reagent	Procedure No.	Recommended for	Not recommended for	Comments
Alkyl halide, silver oxide (no solvent)	2.1	C(1)-protected esterified uronic acids; substances which tend to undergo base-promoted degradation	Free sugars and other easily oxidizable substances	O-Acyl groups may migrate
Alkyl halide, silver oxide in DMF	2.2	General: certain oligopeptides, amino acids, free sugars [21], esterified uronic acids [24]	Peptides containing glutamic acid or tryptophan residues	O-Acyl groups migrate; modification of sulfur-containing residues in peptides may occur [19, 20]
Alkyl halide, barium oxide and/or barium hydroxide in DMF	2.3	General	Substances sensitive to base-catalyzed degradation	O-Acyl groups are replaced with O-alkyl groups
Alkyl halide, barium oxide in DMF	2.3	General		O-Acyl migration is promoted and faster than during procedures involving Ag_2O
Alkyl halide, sodium hydride in DMF	2.4	General	Esterified uronic acids and other substances are likely to be modified under strongly basic conditions	C-Alkylation instead of O-alkylation may occur [27], O-acyl groups are replaced with O-alkyl groups
Alkyl halide, sodium hydride in DMSO	2.5.1	General, substances insoluble in common organic solvents, polysaccharides, lipopolysaccharides, peptides, sterically hindered functions, amides, amino acids	Esters, and as stated for procedure 2.4; peptides containing sulfur-amino acids, histidine and arginine [32]	O-Acyl groups are replaced with O-alkyl groups. For N-isopropyl amino acids, isopropyl ester formation, (see Chapter 2)
Alkyl halide, solid base in DMSO	2.5.2	As for procedure 2.5.1	As for procedure 2.5.1	Compared to procedure 2.5.1, this procedure is simpler to perform, since it bypasses the preparation of methylsulfinyl anion
Alkyl halide, sodium hydride in ether-type solvents	2.6	General; preferred over procedures 2.4 and 2.5 when substrates are soluble in ether-type solvents	As for procedure 2.4	For the derivatization of highly reactive functions, sodium hydride may be replaced with potassium carbonate (see procedure 2.16.3 and Ref. 138); 1,2-dimethoxyethane is the preferred solvent
Alkyl halide, sodium hydride in DMF	2.7	Peptides [44]	As for procedure 2.4	
Diazoalkanes	2.8	Selective alkylation of acidic functions in the presence of aliphatic hydroxyls; esterification		The reaction should be 'alkyl homogeneous' (see **Note**, procedure 2.8); de-O-acetylation

Table 1 (continued)

Reagent	Procedure No.	Recommended for	Not recommended for	Comments
		of acidic polysaccharides [46, 47]		may occur in the presence of lower alcohols [56]
Diazoalkanes in the presence of Lewis acid	2.9	Less reactive functions; alkylation in the presence of base-sensitive substituents	Extremely acid-labile substances	O-Acyl migration does not generally occur; however see [179] polymethylene is the usual by-product of methylation
DMF–dialkyl acetals	2.10	Sterically hindered carboxylic acids, aldehydes, phenols, amines		cis-Diols will be simultaneously acetalized [25]; acetals may undergo trans-acetalization
3-Alkyl-p-tolyltriazenes	2.11	Certain carboxylic acids and phenols		
Alkyl fluorosulfonates	2.12	General		Ether cleavage and non-specific alkylation may occur
Trialkylanilinium hydroxide	2.13	N-Alkylation of barbiturates [84], other sedatives, phenolic alkaloids and related substances	Convenient flash heater (or on-column) methylation for GLC analysis; the injector temp. should be set at 250–300 °C	
Trialkyloxonium fluoroborates	2.14	Exocyclic O-alkylation in mesoionic ring systems [92]	N-Acyl derivatives, when these groups are to be preserved	N-Deacylation occurs [93, 94]
Alcohols in the presence of an acid catalyst	2.15.1	Catechol and related substances		Two electron donating functions in the aromatic ring are essential for the alkylation to occur; possible risk of racemization of optically active substances
Alcohols in the presence of an acid catalyst	2.15.2	Free sugars		Losses due to degradations commencing at the reducing end of the sugars do not occur when these are first converted to glycosides; acid-labile substituents are alcoholyzed and those at the reducing end are replaced with the alkyl group to give glycosides; oligosaccharides and polymeric, glycosidically linked substances are alcoholyzed to variable extents giving lower glycosides

(continued)

Table 1 (continued)

Reagent	Procedure No.	Recommended for	Not recommended for	Comments
Pentafluorobenzyl-chloroformate	2.16.1	Derivatization of tertiary amines for ECD–GLC analysis and mass spectrometry		Convenient introduction of fluorine when perfluoro acid chlorides or anhydrides cannot be used
O-(1,2,3,4,5,6)-Pentafluorobenzyl-hydroxylamine hydrochloride	2.16.2	Derivatization of steroids [26] and other ketones for ECD–GLC analysis and mass spectrometry		Convenient fluorine-introducing alkylation for ketones
Pentafluorobenzyl bromide	2.16.3	Preparation of pentafluorobenzyl ethers and esters for ECD–GLC and MS analysis; N-alkylation of barbiturates [138] and sulfonamides [126]		
Aldehydes, Pd/C catalyst	2.17	Amino acids		N-Alkyl formation should be combined with esterification before GC and/or MS; aromatic nitroacids are reduced and N-alkylated in one operation
Alkyl trichloroacetimidates	2.18	Alkylations, under neutral conditions, of substances bearing acid- or alkali-labile substituents [136, 137]		Solubility of the substance to be alkylated in a non-polar solvent is of advantage

The organic phase is washed with 5% KCN solution to remove silver salts which may have dissolved, then with water, and is dried over anhydrous sodium sulfate to give, after concentration, the peralkylated product. Any residual N,N-dimethylformamide present does not normally interfere with the chromatographic analysis.

Note: Complete substitution of some less reactive substrates may need several days' reaction; repeated treatment with the reagent is commonly required. The procedure has been applied [16] to the permethylation of N-acyl oligopeptides, and was used in a number of cases for amino acid sequence determination. The presence of any existing N-methylamino acid residues in the peptides could be detected by mass spectrometry of the permethylated product when the above procedure was carried out with trideuteromethyl iodide [17, 18]. However, partial chain cleavage of certain peptides and modification of sulfur-containing residues have been observed [19, 20] during this alkylation process and, consequently, the method is no longer employed for the permethylation of peptides. On the other hand, it has been established [21] that the methylation of unprotected aldoses, ketoses and uronic acids proceeds smoothly because the substitution of the hemiacetal hydroxyl group occurs before the oxidative degradation, due to the presence of silver oxide, can take place. Consequently, the methylation of free sugars may be carried out reliably by this method. The same procedure has been applied successfully [22] to the methylation of N-protected amino acids: the corresponding N-protected N-methylamino acid methyl esters are obtained in excellent yield.

2.3 Alkylation with alkyl halides and barium oxide in N,N-dimethylformamide

For the chemistry see Section 2.2.

An excess of alkyl halide is added to a solution of the substrate in N,N-dimethylformamide (1:1 to 1:20 w/v) followed by finely divided barium oxide, and the mixture is stirred vigorously in a flask equipped with an efficient condenser (not necessary for high boiling halides, in which case the flask is closed with a drying tube, or for milligram scale reactions, where the reaction is carried out in a screw-capped vial). The temperature of the reaction mixture rises slowly, depending on the scale of the reaction, and the reaction is checked for completion one hour after the mixture has cooled to room temperature. For isolation, the mixture is diluted with a suitable water-immiscible solvent, the suspension is filtered, the solids are washed, and the combined filtrates are washed with water (for water-soluble substances it may be necessary to backwash the aqueous washings). The desired product is obtained by concentration of the organic phase.

Note: Compared with procedures 2.1 and 2.2, alkylation is faster under these more strongly basic conditions. When working under scrupulously dry conditions, the O-acyl groups present in the substrate survive but O-acyl migration is promoted. In the modified version [23] of the procedure, barium oxide is replaced with barium hydroxide or barium oxide–barium hydroxide (1:1) as the base, here O-acyl groups are completely replaced by O-alkyl groups. Variable results have been reported as to the completeness of N-alkylation using this procedure, and prolonged or repeated treatment with the reagent may be necessary.

2.4 Alkylation with alkyl halides and sodium hydride in N,N-dimethylformamide

$$R-CONH_2 \longrightarrow R-CON(R')_2$$

For further reactions see Section 2.5.

Powdered sodium hydride (≈ 3 equiv per H to be replaced) is added to a solution of the substrate in N,N-dimethylformamide, and the suspension is swirled for 15 min, with the exclusion of atmospheric moisture and carbon dioxide (KOH drying tube). The mixture is then cooled in ice, and the halide (≈ 2 equiv per H to be replaced) is introduced dropwise. The reaction mixture is stirred, again with the exclusion of moisture and CO_2, for 1–2h (the content of the reaction vessel thickens sometimes, so that stirring becomes impossible, but later a clear solution is obtained), after which time the reaction is normally complete. Methanol is then added cautiously to destroy the excess of the alkylating reagents and, when effervescence ceases, the mixture is partitioned between dichloromethane and water. The product is isolated by concentration of the organic phase.

Note: We have used powdered sodium hydride without difficulty and preferred it over dispersions in mineral oil, particularly when working on the milligram scale. When the reagent is obtained from commercial sources as a dispersion in oil, the oil must be removed before use by washing the dispersion with dry ether under nitrogen on a sintered-glass funnel. When dimethyl sulfoxide is used instead of N,N-dimethylformamide, procedure 2.5 should be followed.

This procedure is a more powerful method of alkylation than those described earlier. Sterically hindered groups will, as a rule, be quantitatively alkylated. However, the strongly basic reaction conditions of the procedure should be taken into account whenever sodium hydride is used as the base. Substances known to undergo base-catalysed transformations will do so under these conditions. Thus, esterified uronic acid derivatives treated with sodium hydride will be more or less converted to 4,5-unsaturated-4-deoxyhexuronates [24]. As these unsaturated sugars are not [25] the final products of the base-promoted β-elimination degradation of esterified hexuronic acids, one cannot expect the formation of a single product when such compounds are alkylated by this procedure. The same holds for the very efficient alkylation procedure described next (Section 2.5). Under the conditions of procedure 2.4, degradation by β-elimination of uronic acid derivatives was reported [26] not to occur to such an extent as when dimethyl sulfoxide was used as the solvent (Section 2.5), but it is still safer to alkylate this class of substance using the procedure described in Section 2.2 (see Ref. 24). Some substances may undergo unwanted C-alkylation instead of, or in addition to, O-alkylation, as for example when derivatives of L-ascorbic acid were treated with an alkyl halide in the form of their sodium salt [27]. C-Methylation as well as other unwanted side reactions are more likely to be encountered when working with sub-milligram quantities of the substrate, as was the case for instance with peptides [28], because in such cases the use of too large an excess of the base is difficult to avoid.

2.5 Alkylations with alkyl halides in dimethyl sulfoxide

2.5.1 Alkylation with alkyl halides and methylsulfinyl anion in dimethyl sulfoxide (Hakomori methylation)

$$R-CO-NH_2 \longrightarrow R-CO-N(R')_2$$
$$R-CO-NH-CH_3 \xrightarrow{CD_3I} R-CO-N(CH_3)(CD_3)$$

The base is prepared by heating sodium hydride in dry dimethyl sulfoxide at 50 °C in a flask protected from atmospheric moisture and CO_2 with a KOH drying tube. The solution becomes green, and evolution of hydrogen gas ceases after approximately 1 h, after which time a solution of the substrate in dry dimethyl sulfoxide is added. The mixture is stirred for 20 min at 50 °C and, after cooling to room temperature, the halide is introduced drop by drop with stirring, which is continued for 1 h. (The reaction is exothermic and, therefore, for alkylations with low boiling halides the reaction flask is held in an ice bath during the addition of the halide.) When the reaction is complete, often indicated by the formation of a clear solution, water is added cautiously and the solution is extracted with dichloromethane. The dichloromethane solution is backwashed with water, and the organic phase is dried with anhydrous sodium sulfate and concentrated. Polymeric material is most conveniently isolated by dialysis and freeze-drying.

CAUTION: Although no difficulties have been experienced in our laboratory, care must be taken when heating sodium hydride in dry dimethyl sulfoxide, as there have been reports of violent explosions when the anion was prepared on a large scale [28, 29]. The amounts of the base and of the halide should be in more than 50% excess over the number of equivalents of replaceable hydrogen atoms.

Note: The important difference between the alkylation procedures involving sodium hydride in polar aprotic solvents and those involving alkyl halides and metal salts is that O-acyl groups are completely replaced by O-alkyl groups [30]. N-Acyl groups survive and N-alkyl-N-acylamido derivatives are produced. This alkylation can be performed on an extremely small amount of material and, thus, the procedure provides a rapid means for methylation linkage analysis in the polysaccharide [30], lipopolysaccharide [31] and peptide [20, 32] fields.

Although it has been claimed [33] in the earlier application of this procedure to acidic carbohydrates that elimination reactions do not occur under the conditions used, convincing evidence has been presented [24, 26, 34] showing that base-sensitive substances will undergo side reactions during treatment with this reagent, particularly on repeated [35] alkylation, i.e. when complete substitution is not achieved in one step. Consequently, substrates where base-promoted unwanted transformations are likely to occur should not be alkylated in this way, or substances which on alkylation yield such structures should not be realkylated using this procedure. Another complication which may be encountered when other functional groups of esters are alkylated in the presence of the methylsulfinyl anion is the formation of methylsulfinyl ketones [36]. These difficulties may be largely overcome by saponification of the esters before treatment with the base. The ester formed will be exposed to the strong base for only a short time, which will minimize unwanted side reactions.

2.5.2 Alkylation with alkyl halides and a solid base in dimethyl sulfoxide

Finely powdered NaOH (20 mg; or the equivalent amount of KOH or NaH), followed by methyl iodide (0.1 mL) is added to a solution of the sample (4–5 mg) in dimethyl sulfoxide (0.3–0.5 mL). The mixture is stirred at room temperature in a closed vial for 6 min (7 min for KOH and 15 min for NaOH). After addition of water (1 mL), the mixture is extracted with dichloromethane, which is then backwashed with water. The organic phase, containing the per-O-methylated material, is dried and analyzed as required.

Note: The advantages of the method involving a solid base rather than the methylsulfinyl anion lie in the experimental simplicity and in a virtually theoretical yield of the fully alkylated product. Also noteworthy is the absence of non-carbohydrate peaks from gas chromatograms of the products [6]. The method has been successfully applied to the alkylation of non-reducing carbohydrates [37–39], phenols [40], amides, acids and simple alcohols [41, 42].

2.6 Alkylation with alkyl halides and sodium hydride in ether-type solvents

For typical reactions see Section 2.5.1.

Sodium hydride (3 equiv per H to be replaced) is added to a solution of the substrate (substrate solvent ratio ≈ 1:20, w/v) contained in a small flask, which is then protected from atmospheric moisture and CO_2 with a KOH drying tube. The flask is shaken occasionally and, when effervescence ceases, the halide (2 equiv per H to be replaced) is introduced with stirring, which is continued for 15–60 min. When the reaction is complete, the excess of the etherification reagents is destroyed by the addition of methanol, and the mixture is ready for chromatographic analysis. Alternatively, the mixture is diluted with water and the organic solvents are removed with a rotary evaporator, whereupon the solid product often separates. It is filtered off and dissolved in dichloromethane, and the dichloromethane solution is washed in a separatory funnel with water until neutral. The product is isolated by concentration of the organic phase (dried over anhydrous calcium chloride or sodium sulfate). For water-soluble substances, the aqueous phase obtained after evaporation of organic solvents is extracted with dichloromethane and treated as above.

Note: This is probably the simplest of the very efficient methods of alkylation, possessing the advantage over procedures involving N,N-dimethylformamide or dimethyl sulfoxide in that the isolation of substances can be performed on a very small scale without the polar aprotic solvents interfering. Sterically hindered functional groups will be quantitatively converted to the fully substituted products, although in this case, or when alkylation is carried out with less reactive halides, it may be necessary to prolong the reaction time or to apply gentle heating. Thus, for instance, tritylated carbohydrate derivatives were successfully converted [43] to per O-methyl and per-O-benzyl derivatives in excellent yield. The procedure is a matter of choice when the substrate is soluble in any suitable anhydrous ether-type solvent. We prefer 1,2-dimethoxyethane over other solvents in this class. It is a very good solvent for many types of compounds, some so polar one would not expect them to be appreciably soluble in it. More importantly, some alkylations which were sluggish using ether as the solvent proceeded smoothly in 1,2-dimethoxyethane. When the commercial product is colourless, dry solvent suitable for a small scale experiment can be obtained by keeping a few milliliters of 1,2-dimethoxyethane over powdered sodium hydride until effervescence ceases, with occasional swirling. When addition of a fresh portion of the drying agent fails to cause effervescence, indicating that all moisture, alcohols and peroxides have been removed, sodium hydride is allowed to settle, and the dry solvent is withdrawn, with the aid of a dry pipette, into a dry vial to perform the desired alkylation. Discoloration of the solvent following the addition of sodium hydride indicates the presence of a large amount of impurities, and the dry solvent should be distilled from sodium hydride. For possible side reactions see the notes to procedures 2.4 and 2.5.1.

2.7 Alkylation with alkyl halides and sodium hydride in N,N-dimethylacetamide

For typical reactions see Section 2.5.1.

Sodium hydride is added to N,N-dimethylacetamide and the solution is heated under a KOH drying tube at 120–125 °C until gas evolution ceases. The solution is cooled to room temperature and added to a solution of the substrate in N,N-dimethylacetamide, followed immediately by the halide. The mixture, protected from contact with air, is stirred for 1h. For the substrate: solvent:reagent ratios and work-up see Section 2.4.

Note: The procedure is closely related to the one using N,N-dimethylformamide as the solvent. Methylation of a peptide under these conditions was reported [44] to be cleaner than when methylsulfinyl carbanion was used as the base (procedure 2.5.1), although this may be due to the difference in the reaction time or the actual amount of the base.

2.8 Alkylation with diazoalkanes

$$R-COOH \xrightarrow{CH_2N_2} R-COOMe$$

$$Ph-OH \xrightarrow{CH_2N_2} Ph-OMe$$

A solution of diazoalkane is added slowly to a solution of the substrate until a faint yellow color persists and the evolution of nitrogen gas ceases. Concentration of the solution affords virtually pure alkylated product.

CAUTION: As diazomethane and related substances are both toxic and explosive, all work with them should be carried out behind a safety shield in an efficient hood. For further details concerning the preparation and handling of diazoalkanes see Chapter 2.

Notes: Except in the etherification of aliphatic alcohols in the presence of a Lewis acid, in which case dichloromethane is most frequently used as the solvent (Section 2.9), ethereal and ether–alcoholic solutions are the most commonly used forms of diazoalkanes in alkylation reactions. The substrate may be dissolved in ether (preferred), alcohol, alcohol–water [45] or dimethyl sulfoxide, a solvent which has been successfully applied for acidic polysaccharides [46, 47].

Methanol is reported [48, 49] to catalyze the methylation of certain hydroxy compounds which are stable towards ethereal diazomethane. It has been suggested that diazomethane reacts with lower alcohols to give a new powerful alkylating agent, in order to explain the observed [48] n-propylation with ethereal diazomethane in the presence of n-propanol. Therefore, when diazoalkylation is carried out in the presence of alcohols, the reaction mixture should be 'alkyl-homogeneous'.

Diazoalkanes alkylate acidic and enolic groups rapidly and other groups with replaceable hydrogen slowly. Carboxylic and sulfonic acids, as well as phenols and enols, will be almost instantaneously converted to the corresponding alkyl derivatives when treated with the reagent. In this way N-alkyl (methyl and ethyl) derivatives of several barbiturates possessing acidic NH groups have been prepared [50, 51] by treatment with diazomethane and diazoethane, respectively. Catalysts are normally necessary with aliphatic alcohols and other substances bearing low reactivity hydrogen (Section 2.9), so that many carboxylic acids and phenols can be selectively alkylated by means of plain diazoalkanes in alcoholic solution. Occasionally, however, as in the case of tartaric acids and their esters, the hydroxyl groups were smoothly methylated [52, 53] by diazomethane in ether. Further, aliphatic alcohols were also reported to be smoothly methylated by diazomethane when hexane or heptane replaced ether as the solvent [54].

Diazoalkanes are among the most versatile reagents available to the organic chemist. Their high reactivity towards a number of types of organic substance has to be taken into account when they are used for alkylation, because in the presence of other groups which combine with this class of substance the resulting derivative may be far from the one expected. One most unusual example may be cited [55]: when adenosine was treated with diazoethane in aqueous 1,2-dimethoxyethane, mixed methylation and ethylation occurred as a result of solvent participation in the diazoalkylation reaction.

When O-acylated substances are alkylated with diazoalkanes in the presence of lower alcohols, the product should not be left for long in the solution containing the excess of the reagent, since de-O-acylation may occur [56].

In view of the commercial availability of deuterodiazomethane precursor, diazoalkylation is a powerful means for the specific labeling of various classes of organic substances.

2.9 Lewis-acid-catalyzed alkylation with diazoalkanes

$$R\text{---}OH \xrightarrow[BF_3 \cdot Et_2O]{CH_2N_2} R\text{---}OMe$$

The substrate is dissolved in dichloromethane (1:10 to 1:50, w/v, depending upon the solubility at low temperature), and the solution is cooled to $-10\,°C$. The solution is magnetically stirred while boron trifluoride etherate (1–3 drops, or its solution in dichloromethane) is added, followed by a solution of the diazoalkane in dichloromethane, until a faint yellow color persists for at least 5 min. The solution is stirred at below $0\,°C$, protected from atmospheric moisture, and further portions of the reagents are periodically added until a suitable test shows satisfactory conversion of the starting material into the alkylated product. The solution is filtered to remove any polymeric material formed as a by-product, and the filtrate is washed successively with a solution of sodium bicarbonate and water, dried over anhydrous sodium sulfate, and concentrated.

Note: Alkylations of partially O-acylated substances, such as carbohydrates, with an alkyl halide and metal salts (Sections 2.1–2.3) frequently give rise to acyl group migration. On the other hand, the more powerful alkylations cause partial or complete replacement of the acyl groups with the alkyl groups. However, the use of diazomethane–boron trifluoride etherate reagent to methylate O-acylated sugars bearing free hydroxyl groups has been repeatedly shown [57–65] to give the corresponding methyl ethers without the migration of base-labile substituents, and it appears that this is the only safe methylation agent when such groups are present in the substrate (see also procedure 2.12).

Apart from complications due to the side reactions already mentioned, the Lewis acid may cause unwanted acid-catalyzed transformations. When the substrate is likely to show acid lability, to avoid a high local concentration of the catalyst, the addition of a dilute solution of the catalyst in dichloromethane (1:100 to 1:1000, v/v), rather than the neat catalyst, is recommended. Furthermore, when the stereochemistry of the substrate is favorable, addition of diazomethane instead of, or in addition to, simple alkylation may occur. In this way, when derivatives of methyl α-D-galacturonate having the hydroxyl group in the 4-position unsubstituted were methylated under the conditions of procedure 2.9, in addition to the formation of the C(4)-O-methyl ether (major product), addition of diazomethane to the methoxycarbonyl group followed by cyclization occurred, finally giving bicyclic tetrahydrofurone dimethyl acetal derivatives as by-products [66].

Variable amounts of the catalyst with respect to the amount of the substrate, solvent and alkylating agent have been used and the reaction, when carried out with carbohydrates bearing more than one free hydroxyl group, as rarely been complete (70–90% conversion of the starting material to the fully substituted product is normally achieved). Although any Lewis acid would probably catalyze diazoalkylations of groups which, without a promoter, are resistant towards this type of reagent, boron trifluoride etherate is the one most frequently used. A successful use of $SnCl_2 \cdot 2H_2O$ (10^{-3} mol g^{-1}) in the benzylation of nucleosides with phenyldiazomethane in 1,2-dimethoxyethane has been reported [67]. The same catalyst was effective in cases where boron trifluoride etherate failed to catalyze the methylation of nucleosides with diazomethane because of the presence of the basic heterocycle, which may have effectively complexed the boron trifluoride reagent. Dichloromethane has been used as the solvent in most Lewis acid-catalyzed diazoalkylations, and the reaction has been conducted at temperatures ranging from -20 to $0\,°C$. Large excess of the catalyst and/or temperatures above $0\,°C$ should be avoided, as polymerization of diazoalkanes, i.e. the loss of the alkylating agent, will then be more pronounced. For side reactions see **Note**, procedure 2.8.

2.10 Alkylation with N,N-dimethylformamide dialkyl acetals

A concentrated solution of the substrate in a suitable solvent is treated with an excess of the reagent (1:4 to 1:6, w/v), and the mixture, protected from contact with air, is heated at $50-60\,°C$ for about 15 min. The excess of the reagent is removed under reduced pressure and a solution of the residue in a water-immiscible solvent is washed with water, dried over anhydrous sodium sulfate, and concentrated.

Note: *N,N*-Dimethylformamide dimethyl acetals are moisture-sensitive. They are hydrolyzed to give *N,N*-dimethylformamide and the corresponding alcohol. Therefore the derivatization should be carried out under scrupulously dry conditions. The reaction is, as a rule, complete immediately upon dissolution of the substrate, and therefore heating is necessary only with samples which do not dissolve at room temperature. The final simple purification should not be omitted, as extraneous peaks on gas chromatograms have been reported to appear when the reaction mixtures were directly injected [68]. Pyridine, benzene, alcohols, halogenated hydrocarbons, N,N-dimethylformamide, acetonitrile and tetrahydrofuran have been used as the solvent.

Carboxylic acids [69], phenols [69], and thiols [70] quickly react to give the corresponding alkyl derivatives. Thus, simple amino acids [71] are full derivatized, N,N-dimethylamino acid alkyl esters being produced. Free amines and substances containing the –CO–NH– group give [72], respectively, N,N-dialkylamino and N-alkyl derivatives. Although hydroxyl groups on hydroxyl substituted amino acids were reported [71] not to react under the conditions of derivatization with N,N-dimethylformamide dialkyl acetals, O-alkylation in other systems has been observed [72], as was the simultaneous acetalization of cis-diol groups [72].

Sterically hindered acids, e.g. trimethylbenzoic acid, react almost as rapidly with this type of reagent as unhindered acids, e.g. benzoic acid, and similarly hindered functions can be expected to react in the same way. This type of reagent will condense with active methylene compounds, and is capable of exchange reactions to generate new acetals. This fact has to be taken into account when substrates amenable to such reactions are being derivatized. A wide variety of N,N-dimethylformamide dialkyl acetals is now commercially available.

2.11 Alkylation with 3-alkyl-1-p-tolyltriazenes

$$R-COOH \xrightarrow{R'-NH-N=N-C_6H_4-CH_3} R-COOR'$$

$$Ph-OH \xrightarrow{R'-NH-N=N-C_6H_4-CH_3} Ph-OR'$$

The solution of the substrate (preferably in an ether-type solvent) is treated with a solution of the triazene (1.1 equiv per H to be replaced, or a larger excess for groups assumed to be less reactive) and the reaction mixture is stirred for 1h while nitrogen is evolved. When the conversion is complete, as often indicated by ceased effervescence, excess of the reagent is destroyed by careful addition of 10% hydrochloric acid. The mixture is quantitatively transferred into a separating funnel and washed successively with 10% hydrochloric acid, then with a dilute solution of sodium bicarbonate, and finally with water. The organic phase is dried over a suitable desiccant and concentrated.

Note: Gentle heating accelerates the reaction. For isolation of pure substances it may be necessary to decolorize the final solution with a little charcoal or silicic acid.

Carboxylic acid esters can be conveniently prepared in this manner using many commercially available 3-alkyl-1-p-tolyltriazenes. Alkylation of phenols, thiols and other less reactive functional groups requires more vigorous reaction conditions [73].

CAUTION: The compounds of this class have been reported to exhibit carcinogenic activity and should be handled accordingly.

2.12 Alkylation with alkyl fluorosulfonates

$$R-OH \xrightarrow{CF_3SO_3Et} R-OEt$$

$$R_2-NH \xrightarrow{CF_3SO_3Et} EtR_2\overset{+}{-}NHCF_3SO_3^-$$

$$\text{pyridine} \xrightarrow{CF_3SO_3Me} \text{N-methylpyridinium } CF_3SO_3^-$$

A mixture of the substrate and an alkyl fluorosulfonate (1.1 equiv per H to be replaced) in a suitable inert solvent, protected from atmospheric moisture, is kept at room temperature for 5–16h; or the mixture is heated in a sealed tube at 100 °C for 2–8h. Alternatively, a solution of the substance to be alkylated, the trifluoroalkylsulfonate (3 equiv per OH to be alkylated) and a catalytic amount of mercuric cyanide in dry dichloromethane is heated under reflux for 24h.

Note: Alkyl fluorosulfonates are extremely powerful alkylating agents. Their high reactivity may therefore result in non-specific alkylation. Among the possible side reactions, alkylation of aromatic substances and O-alkylation by ether cleavage may be cited [74]. However, methylation of a partially acetylated substrate was reported [75] to have occurred in excellent yield without migration of the acetyl groups. Numerous data for this type of alkylation are available in the literature [74–84].

2.13 Alkylation with trimethylanilinium hydroxide (TMAH)

$$\text{\textbackslash}N-H \xrightarrow{TMAH} \text{\textbackslash}N-Me$$

will methylate substances bearing replaceable protons attached to nitrogen. The reagent has been accepted as the preferred N-alkylation reagent for barbiturates and related substances. The approaches to the analysis of barbiturates have been evaluated [87]. It has been found by chemical ionization studies [88] that in the GC analysis of phenobarbital two additional products (B, C) were formed besides the expected N,N-dimethyl derivative (A). Although the amount of the second phenobarbital decomposition product (C) is insignificant, the amount of B is large and has to be taken into account in the quantitative analysis. The extent of formation of B is affected by the concentration of the derivatizing agent and the time the drug is in contact with trimethylanilinium hydroxide prior to GC injection. Following a detailed study, a procedure has been developed [88] under the conditions of which the amount of B formed is reproducible and can be used for more precise quantitation of phenobarbital. Another application is the detection of xanthine bases [85] and phenolic alkaloids [85].

The substance is dissolved in a 0.2 M alcoholic solution of trimethylanilinium hydroxide (1:5 w/v; for the effect of concentration of the derivatizing agent upon the quantitation of barbiturates see **Note** and Ref. 22) and, after 2 min, the solution is injected into the gas chromatography for on-column reaction and GC analysis.

Note: Upto 1000-fold molar excess of the reagent has been used [85, 86]. Trimethylanilinium hydroxide

2.14 Alkylation with trialkyloxonium fluoroborates

A solution of the substrate in an inert solvent (such as dichloromethane) is treated with an equimolar amount of trialkyloxonium fluoroborate, at a suitable temperature and with the exclusion of moisture. After concentration, the product often crystallizes in the form of a fluoroborate salt, which can be decomposed with a solution of sodium bicarbonate. Alternatively, when they do not affect other functions present, alkoxides in alcohols can be used. For the conversion of liquid substances, solid trialkyloxonium fluoroborate is added, and the mixture is stirred until the reagent dissolves. Warming to a suitable temperature may be necessary at this step.

Note: Triethyloxonium fluoroborate and its methyl analogue can readily be prepared in the laboratory according to Meerwein [89, 90]. Alkylation using this method is limited to the lower members of the series because of the difficulty of preparation of the higher trialkyloxonium fluoroborates. This class of reagent alkylates [91] alcohols, phenols and carboxylic acids, to give the corresponding O-alkyl derivatives, and ethers, sulfides, nitriles, ketones, esters, and amides are alkylated on oxygen, nitrogen or sulfur to give -onium fluoroborates. O-Alkylation on an exocyclic oxygen occurs readily with this reagent, even in mesoionic ring systems [92] where, most of the time, alkylation cannot be achieved with alkyl halides. Temperatures ranging from -80 to $+100\,^\circ\mathrm{C}$ have been applied in the alkylations with this type of reagent. Triethyloxonium fluoroborate has proved to be useful not only for alkylations but also for mild N-deacylation, a reaction which otherwise requires drastic conditions. Thus, de-N-benzoylation [93] and de-N-acetylation [94] have been achieved using this reagent under conditions which left ester, acetal and glycosidic linkages unaffected. Hence, N-deacetylation can be selectively carried out with O-acetyl groups present, which provides an important means for N-acyl group interchange. Acylamino acids and peptides were esterified with an excess of triethyloxonium fluoroborate [95].

2.15 Alkylation with alcohols in the presence of acid catalysts

2.15.1 Formation of benzyl ethers from substances related to catechols

$R^1 = CH_2NH_2, CH_2OH; R^2 = H, CH_3; R_3^3 = H$

When R^1 is COOH, the ethyl ester is simultaneously formed.

A solution of the substrate in alcoholic hydrochloric acid (1–3 M, substrate reagent ratio 1:20 to 1:100, w/v) is heated in a sealed vial at 90–100 °C for 2h. For complete dehydration, several successive portions of dry alcohol are evaporated under reduced pressure from the solution of the product.

Note: Substrates related to catechol, when treated with acidified alcohols, undergo O-alkylation at the benzylic hydroxyl to give the corresponding benzyl alkyl ethers. Several substances of this class have been successfully analyzed by GC–MS as O-ethyl ethers following this alkylation [96]. The reaction, which occurs with [97] methanol, ethanol, n-butanol, and n- and 2-propanol, forms derivatives which protect the sensitive benzyl hydroxyl group of this class of substance before further derivative formation. Losses due to polymerization, enamine formation and other degradations can be avoided in this way. The ease with which deutero analogues are prepared when the reaction is carried out with deutero-alcohols is useful for specific ion monitoring GC–MS assay. The reaction may proceed via a carbonium ion intermediate; if this is so, racemization of optically active compounds could occur during this process.

2.15.2 Condensation of simple alcohols with free sugars (Fischer's method of glycosidation)

A solution of the sugar in the chosen alcohol containing the acid catalyst (for substrate/alcohol/catalyst ratios see **Note** below) is heated at the boiling point for a suitable period of time and, after removal of the acid, the solution is concentrated.

Note: Treatment of higher sugars with anhydrous alcohols containing an acid catalyst (unlike that of the simplest members of the sugar series, glycolaldehyde and glyceraldehyde, which in this way give dialkyl acetals) results in the formation of mixed cyclic acetals termed glycosides. This represents the simplest but least selective method for the preparation of glycosides, and it is a convenient means for the protection of the potential aldehyde or keto function of free sugars. The method is applicable to lower aliphatic alcohols but not to phenols. Hydrogen chloride (0.01–5%, w/v) is by far the most commonly used acid catalyst for this reaction, although other acids, including Lewis-type acids [98] and cation exchange resins in their H^+ form [99], have also been used. When substances sensitive to mineral acids, such as ketoses, are to be converted to glycosides, losses due to decomposition may be minimized by catalyzing the reaction with acetic acid [100]. This has the additional advantage that acetic acid can be readily removed from the reaction mixture by evaporation under reduced pressure. For the isolation of the products, the mineral acid is neutralized, most conveniently with an anion exchange resin, and the solutions are concentrated under reduced pressure. Although occasionally [98–101] products were isolated without the removal of the acid catalyst, this step should not be omitted when the products are to be directly chromatographed or analysed by mass spectrometry.

Variable concentrations of the sugars in the acidified alcohol (0.5–15%), different temperatures (3–100 °C) and various reaction times (1 h–5 days) have been employed. Although the first parameter appears to have little effect on the outcome of the reaction, at least within the range indicated, the latter two significantly affect the rate at which the equilibrium between the α- and β-forms of pyranosides and furanosides is established. A mixture of anomers is always obtained if the structure of the sugar molecule is appropriate. Bishop, with his co-workers [102–105], thoroughly studied the glycosidation of sugars and did complete GC analyses of equilibrium mixtures of glycosides formed from common sugars. It follows from his conclusions that pentoses tend to form furanosides more extensively than do hexoses, and that substitution of free hydroxyl groups with methyl groups makes this tendency even more pronounced. Although exceptional cases exist, pyranosides are generally the principal constituents of the glycosidation mixtures when the reaction is allowed to go to completion. The course of the reaction is such that the decrease in the free sugar concentration is accompanied by a rapid but transient build-up of furanosides, which then slowly isomerize to the pyranosides. Thus, when Fischer's glycosidation reactions are prevented from reaching their equilibrium positions, thermodynamically less stable products may be isolated. It follows that, for the preparation of glycoside mixtures rich in furanosides, mild reaction conditions should be used (concentration of the acid below 0.1% [106] and 40–60 °C or 0.5–1% and ambient or sub-ambient temperature [107, 108]).

Insignificant amounts of dimethyl acetals of higher sugars are formed, although their presence in the reaction

Table 2 Sources of GLC and TLC data of some carbohydrate alkyl derivatives[a]

	References
GLC data of	
O-acetyl-O-methyl-D-glucononitriles	139
methylated sugar derivatives, a review	140
permethylated alditols and aldonic acids	141
partially methylated alditols as TMS ethers	142
partially ethylated and methylated alditol acetates	143
methyl O-methyl-β-D-arabinofuranosides and methyl O-methyl-β-D-arabinopyranosides	144
methyl tri-O-acetyl-O-methyl-α-D-mannosides	145
O-methylmannoses	146
2-deoxy-O-methyl-2-methylamino-D-glucoses and D-galactoses	147
methyl (methyl O-methyl-α-D-glucopyranosid)uronates	148
methyl (methyl 4-deoxy-O-methyl-β-L-*threo*-hex-4-enopyranosid)uronates	149
TLC data of	
carbohydrate O-methyl derivatives	144, 150
various carbohydrate derivatives, qualitative and quantitative, a review	151, 152
methyl-O-methyl-D-arabinosides	153
substituted methyl-α-D-glucopyranosides	154–157
O-methyl-D-glucoses	155
O-methyl-D-galactoses	158
mono-O-methyl-D-fructoses	159
substituted methyl β-D-maltosides	160
aryl tetra-O-acetyl-D-glycopyranosides	161
halogenated aryl β-D-glucosides and thioglucosides	162
O-methyl-D-glucosamine	163
O-acetyl-O-methyl alditol acetates	164
branched-chain sugars and their methyl glycosides	165
alditol acetates from partially methylated and ethylated sugars, per-O-methyl alditols of simple sugars and oligosaccharides, acetylated aldononitriles derived from methylated sugars, methyl glycosides of methylated sugars, acetates of partially methylated sugars	118

[a]For further data see Refs. 116 and 117.

Table 3. Examples of protective alkylation in GLC of various classes of organic substances

Class of substance	Reference
Steroids	125, 166
Amino acids	71
Catecholamines	96
Sulfonamides	126, 167
Barbiturates	50, 87, 88, 138, 168
Phenols	169–174
Organic acids	175, 176
GLC, TLC and HPLC of various derivatives of simple sugars and oligosaccharides	118
GLC and MS of carbohydrates	177
Modern guidelines for methylation analysis of carbohydrates	178

mixtures of Fischer's glycosidation has been demonstrated [109–111]. This is particularly so at equilibrium, and it appears that the true acetals are not the primary products from which glycosides are subsequently formed.

Glycosidation of uronic acids has not been studied in such detail as that of neutral sugars, but sufficient information has accumulated [111–115] to indicate that the reaction follows the same general pattern as in the case of aldoses. The carboxylic function is esterified before the glycosides are formed, and furanoside formation is perhaps more pronounced.

Examination of the products of Fischer's glycosidation by GC–MS reveals the significant fact that a single product is not formed when the anomeric center of sugars is derivatized in this way. GC is a powerful means of separating the various forms of sugar glycosides. Numerous data concerning the chromatographic properties of these substances are summarized in the excellent reviews by Dutton [116, 117] and, more recently, in the monograph by Churms [118]. For further references see also Tables 2 and 3. Anomeric isomers are normally not distinguished by mass spectrometry, but pyranoid and furanoid forms follow different fragmentation pattern so that their presence can be clearly demonstrated [12, 119–121].

2.16 Preparation of pentafluorobenzyl derivatives

2.16.1 Formation of carbamate derivatives using pentafluorobenzyl chloroformate

$$(R)_3N \xrightarrow{C_6F_5CH_2OCOCl} C_6F_5CH_2OCO-N(R)_2$$

The solution of a tertiary amine in n-heptane (1:200, w/v) is treated with the chloroformate (50 μL per mg of the substrate) and a pinch of anhydrous sodium carbonate. The stoppered tube is then heated for 1h at 100 °C and, after cooling, a little 1 M NaOH solution is added. After shaking, the n-heptane phase contains the derivatized product.

2.16.2 Formation of benzyloxime derivatives using pentafluorobenzyl hydroxylamine hydrochloride

The ketone is dissolved in a stock solution of the hydrochloride in pyridine, and the stoppered tube is swirled at 65 °C for 30 min. The pyridine is removed by a stream of nitrogen and a little cyclohexane is added. After addition of a little water and shaking, the phases are allowed to separate. The top cyclohexane layer is either chromatographed immediately or transferred with the aid of a pipette to another tube, to be dried over a few grains of anhydrous sodium sulfate. These conditions are discussed fully in Chapter 6.

Note: This class of derivatives is commonly used for electron-capture gas chromatographic analysis. The above two reagents are recommended for derivatizing substances which cannot be conveniently derivatized using perfluoro acid chlorides or anhydrides.

Pentafluorobenzyl chloroformate has been used for the direct GC determination of tertiary amines [122, 123]. The method seems to be an improvement over the original [124] time-consuming procedure for the derivatization of these substances.

The above-described oxime formation has been successfully used in steroid chemistry [125] in the nanogram range. For the formation of oximes, it is convenient to have handy a stock solution of the hydrochloride in pyridine containing 50 mg mL^{-1} of the reagent. For derivatization, 1 mg of the sample is dissolved in 0.2 mL of the stock solution. Suitable dilutions of the stock solution with pyridine are made for the reactions at microgram and nanogram levels.

2.16.3 Formation of pentafluorobenzyl derivatives using pentafluorobenzyl bromide

$$R\text{—OH} \xrightarrow{C_6F_5CH_2Br} R\text{—OCH}_2C_6F_5$$

$$R\text{—COOH} \xrightarrow{C_6F_5CH_2Br} R\text{—COOCH}_2C_6F_5$$

$$C_6H_5\text{—SO}_2\text{—NH}_2 \xrightarrow{C_6F_5CH_2Br} C_6H_5\text{—SO}_2\text{—N(CH}_2C_6F_5)_2$$

Pentafluorobenzyl bromide (≈ 1.1 equiv per H to be replaced) is added to the solution of the substrate (≈ 50 μg) in a suitable solvent (0.5 mL), followed by anhydrous potassium carbonate (slight excess based on the amount of the bromide), and the mixture is heated in a water bath near the boiling point for 1h. A small amount of iso-octane is added and the mixture is concentrated to ≈ 1–2 mL. When the solids settle, the remaining solution is transferred onto a column of 1 g of silica gel poured from a slurry in hexane. The solids are mixed with 1 mL of hexane and the washings are added to the column followed by 5–10 mL 5% benzene in hexane to remove the excess of pentafluorobenzyl bromide reagent. The derivatized substance, held on the column during these operations, is then eluted with a benzene/hexane mixture of appropriate polarity.

Note: The method involving pentafluorobenzyl bromide is a convenient procedure for making esters and ethers, and has accordingly been used in trace analysis. Sulfonamides have also been derivatized with the reagent and determined as N-pentafluorobenzyl derivatives [126]. The mildly basic conditions involving potassium carbonate as the base may not be sufficient for the conversion of sterically hindered groups, in which case stronger basic conditions (procedure 2.6) are recommended. The convenient clean-up procedure [127], which removes the excess of the reagent and other interfering substances, is an improvement on the original directions, where this was achieved by distillation [128, 129]. The proper ratio of the benzene/hexane mixture for the elution of the products from the silica gel micro-column should first be established by TLC.

CAUTION: Pentafluorobenzyl bromide is a strong lachrymator and should be handled with appropriate precautions; for routine work, a stock solution of the reagent in acetone (1:50, w/v) may be prepared [127].

An alternative reagent for the purposes just described is hexafluorobenzene. Under the conditions of procedure 2.6, it readily converted free hydroxyl groups into pentafluorophenyl ethers [130].

It goes without saying that, as with any work at the nanogram level, any glassware including the pipettes and the syringes used in the preparation of derivatives must be most throughly cleaned.

2.17 Alkylation by means of reductive condensation of amines with aldehydes

$$NH_2-CH_2-CO-NH- \xrightarrow{R'-CH=O, H_2}$$

$$(R'-CH_2)_2N-CH_2-CO-NH-$$

[structure: 3-nitrobenzoic acid] $\xrightarrow{CH_2O, H_2}$ [structure: 3-(dimethylamino)benzoic acid]

A mixture of the substrate and 10% palladium-on-charcoal catalyst (1:1, w/w) in ethanol or aqueous ethanol is stirred in the presence of the requisite aldehyde (30–400% excess) under hydrogen, at room temperature and at ordinary pressure, until either reduction ceases or slightly more than the theoretical amount of hydrogen is consumed (3–16h). The mixture is then heated to boiling and filtered, the catalyst is washed with hot ethanol, and he combined filtrates are concentrated to dryness to give the dialkylamino compound with some residual aldehyde.

Note: The value of this procedure lies mainly in the fact that when formaldehyde is used for the condensation N,N-dimethylamino acids can be conveniently prepared in this way [131], whereas methylation of amino acids with methyl iodide in the presence of alkali leads to betaines, occasionally accompanied by monoalkylation. In the case of methylation, commercial 40% formaldehyde in water serves as both the reagent and the solvent, and its amount should be at least twice the theoretical amount. Combined with suitable esterification [132, 133], volatile derivatives of many amino acids suitable for anlysis by GC are produced.

With higher straight-chain aldehydes, glycine and alanine furnish the corresponding N,N-dialkylamino acids, but other amino acids, when the reaction is carried out at ambient or sub-ambient temperature, undergo mono-alkylation owing to steric hindrance. When dialkylation is the aim this can often be achieved by conducting the condensation at a higher temperature for prolonged reaction time. With aldehydes branched in the α-position, N-monoalkyl derivatives are produced [134]. The method has been successfully extended to the methylation of oligopeptides which undergo alkylation only at the end amino acid unit, leading to the identification of the N-terminal amino acid in polypeptides [135].

2.18 Alkylation with alkyl trichloroacetimidates

$$Cl_3C-\underset{OR}{\overset{NH}{\underset{\|}{C}}} + R^1-OH \longrightarrow R^1-OR$$

$R = CH_3$, $CH_2C_6H_5$, or $CH_2CH=CH_2$

A catalytic amount of trifluoromethanesulfonic acid is added to a stirred solution of the substrate to be alkylated in carbon tetrachloride (or dichloromethane), followed by the addition of a solution of the derivatizing reagent alkyl trichloroacetimidate (2 equiv per OH to be alkylated) in cyclohexane or in a mixture of cyclohexane dichloromethane. The mixture is kept at room temperature overnight. After neutralization with triethylamine, the mixture is concentrated and chromatographed. To avoid the appearance during chromatography of peaks due to the decomposition and/or excess of the reagent, or when the product is to be isolated, the neutralized mixture is partitioned between dichloromethane and aqueous sodium hydrogen carbonate solution, the organic phase is processed conventionally and the residue, obtained after concentration, is chromatographed.

Note: The best results have been obtained when the reactions have been conducted in non-polar solvents. The suitability of these alkylations depends, therefore, on the solubility of the substance to be alkylated under these conditions. The method has been successfully applied to compounds bearing acid- or alkali-labile functional groups [136, 137].

3. Conclusions

The various alkylation procedures described here not only have a protective function, but also lead to derivatives which are useful for analysis by chromatographic methods. In the case of methylation reactions this is of practical value, particularly in GC and GC–MS, because of the small increase in molecular weight and the consequent volatility of the resulting methylated derivatives. This gains in significance when one is dealing with polyfunctional compounds, which is why it is of particular importance in the carbohydrate field.

It is for reasons such as these that alkylation reactions continue to be applied in all branches of chromatography.

4. References

[1] W. N. Haworth, *J. Chem. Soc.*, **107**, 8 (1915).
[2] G. Zemplén, Z. Csürös and S. Angyal, *Ber. Dtsch. Chem. Ges.*, **70**, 1848 (1937).
[3] T. Purdie and J. C. Irvine, *J. Chem. Soc.*, **83**, 1021 (1903).
[4] R. Kuhn, H. Trischmann and I. Löw, *Angew. Chem.*, **67**, 32 (1955).
[5] S. Hakomori, *J. Biochem. (Tokyo)*, **55**, 205 (1964).
[6] I. Ciucanu and F. Kerek, *Carbohydr. Res.*, **13**, 209 (1984).
[7] E. Miner and W. Rundel, *Angew. Chem.*, **70**, 105 (1958).
[8] M. Neeman, M. C. Caserio, J. D. Roberts and W. S. Johnson, *Tetrahedron*, **6**, 36 (1959).
[9] M. J. Robins and R. S. Naik, *Biochim. Biophys. Acta*, **246**, 341 (1971).
[10] M. J. Robins, A. S. K. Lee and F. Norris, *Carbohydr. Res.*, **41**, 304 (1975).
[11] B. Bannister and P. Kováč, *VIIth International Symposium on Carbohydrate Chemistry*, Bratislava, August 1974, Abstracts, Organic Chemistry Section, p. 8.
[12] V. Kováčik and P. Kováč, *Chem. Zvesti*, **27**, 662 (1973).
[13] P. Kováč, *Carbohydr. Res.*, **22**, 464 (1972).
[14] P. Kováč, J. Hirsch, R. Palovčík, I. Tvaroška and S. Bystrický, *Collect. Czech. Chem. Commun.*, **41**, 3119 (1976).
[15] R. Jeanloz, *Adv. Carbohydr. Chem.*, **13**, 189 (1958).
[16] B. C. Das, S. D. Gero and E. Lederer, *Biochem. Biophys. Res. Commun.*, **29**, 211 (1967).
[17] B. C. Das, S. D. Gero and E. Lederer, *Nature (London)*, **217**, 547 (1968).
[18] D. W. Thomas, E. Lederer, M. Bodzanszky, J. Izdebski and I. Muramatsu, *Nature (London)*, **220**, 580 (1968).
[19] D. W. Thomas, B. C. Das, S. D. Gero and E. Lederer, *Biochem. Biophys. Res. Commun.*, **32**, 519 (1968).
[20] D. W. Thomas, *Biochem. Biophys. Res. Commun.*, **32**, 483 (1968).
[21] H. G. Walker, Jr., M. Gee and R. M. McCready, *J. Org. Chem.*, **27**, 2100 (1962).
[22] R. K. Olsen, *J. Org. Chem.*, **35**, 1912 (1970).
[23] R. Kuhn, H. H. Baer and A. Seeliger, *Justus Liebigs Ann. Chem.*, **611**, 236 (1958).
[24] G. O. Aspinall and P. E. Barron, *Can. J. Chem.*, **50**, 2203 (1973).
[25] J. Hirsch, P. Kováč and V. Kováčik, *J. Carbohydr. Nucleosides, Nucleotides*, **1**, 431 (1974).
[26] Z. Tamura and T. Imanari, *Chem. Pharm. Bull.*, **12**, 1386 (1964).
[27] K. J. A. Jackson and J. K. N. Jones, *Can. J. Chem.*, **43**, 450 (1965).
[28] F. A. French, *Chem. Eng. News*, (April 11) 48 (1966).
[29] G. L. Olson, *Chem. Eng. News*, (June 13) 7 (1966).
[30] H. Bjorndal, C. G. Hellerqvist, B. Lindberg and S. Svensson, *Angew. Chem., Int. Ed. Engl.*, **9**, 610 (1970).
[31] H. Nikaido, *Eur. J. Biochem.*, **15**, 57 (1970).
[32] H. R. Morris, D. H. Williams and R. P. Ambler, *Biochem. J.*, **125**, 189 (1971).
[33] D. M. W. Anderson and G. M. Cree, *Carbohydr. Res.*, **2**, 162 (1966).
[34] R. Toman, Š. Karácsonyi and M. Kubačková, *Carbohydr. Res.*, **43**, 111 (1975).
[35] D. M. W. Anderson, I. C. M. Dea, P. A. Maggs and A. C. Munro, *Carbohydr. Res.*, **5**, 489 (1967).
[36] E. J. Corey and M. Chaykovsky, *J. Am. Chem. Soc.*, **87**, 1345 (1965).
[37] R. N. Shah, J. Baptista, G. R. Perdomo, J. P. Carver and J. J. Krepinsky, *J. Carbohydr. Chem.*, **6**, 645 (1987).
[38] P. Kováč and L. Lerner, *Carbohydr. Res.*, **184**, 87 (1988).
[39] P. Fügedi and P. Nánási, *J. Carbohydr. Nucleosides, Nucleotides*, **8**, 547 (1981).
[40] R. J. Gillis, *Tetrahedron Lett.*, 3503 (1976).
[41] R. A. W. Johnstone and M. E. Rose, *Tetrahedron*, **35**, 2169 (1979).
[42] D. R. Benedict, T. A. Bianchi and L. A. Cate, *Synthesis*, 428 (1979).
[43] P. Kováč, *Carbohydr. Res.*, **184**, 87 (1973).
[44] K. I. Agarwal, G. Kenner and R. C. Sheppard, *J. Am. Chem. Soc.*, **91**, 3096 (1969).
[45] R. Kuhn and H. A. Baer, *Chem. Ber.*, **86**, 724 (1953).
[46] V. Zitko and C. T. Bishop, *Can. J. Chem.*, **44**, 1275 (1966).
[47] B. A. Dmitriev, L. V. Backinowsky, Yu. A. Knirel, V. L. Lvov and N. K. Kochetkov, *Izv. Akad. Nauk SSSR, Ser. Khim.*, 2235 (1974).
[48] A. Schönberg and A. Mustafa, *J. Chem. Soc.*, 746 (1946).
[49] C. M. Williams and C. C. Sweeley, in *Biomedical Applications of Gas Chromatography*, Vol. 1, H. A. Szymanski (Editor), Plenum Press, New York (1964), p. 231.
[50] M. G. Horning, K. Letratanangkoon, J. Nowlin, W. G. Stillwell, R. N. Stillwell, T. E. Zion, P. Kellaway and R. M. Hill, *J. Chromatogr. Sci.*, **12**, 630 (1974).
[51] J. G. Cook, C. Riley, R. F. Nunn and D. E. Budger, *J. Chromatogr.*, **6**, 182 (1961).
[52] G. Hesse, F. Exner and H. Hertel, *Justus Liebigs Ann. Chem.*, **609**, 60 (1957).
[53] O. T. Schmidt and H. Kratt, *Ber. Dtsch. Chem. Ges.*, **74**, 33 (1941).
[54] H. Meerwein, T. Bersin and W. Burneleit, *Ber. Dtsch. Chem. Ges*, **62**, 1006 (1929); cf. H. Meerwein and G. Hinz, *Justus Liebigs Ann. Chem.*, **484**, 1 (1930).
[55] L. M. Pike, M. K. A. Khan and F. Rottman, *J. Org. Chem.*, **39**, 3674 (1974).
[56] H. Bredereck, R. Sieber and I. Kamphenkel, *Chem. Ber.*, **89**, 1169 (1956).
[57] I. O. Mastronardi, S. M. Flematti, J. O. Deferrari and E. G. Gros, *Carbohydr. Res.*, **3**, 177 (1966).
[58] J. O. Deferrari, E. G. Gros and I. O. Mastronardi, *Carbohydr. Res.*, **4**, 432 (1967).
[59] G. J. F. Chittenden, *Carbohydr. Res.*, **31**, 127 (1973).

[60] E. J. Bourne, I. R. McKinley and H. Weigel, *Carbohydr. Res.*, **25**, 516 (1972).
[61] A. Lipták, *Acta Chim. Acad. Sci. Hung.*, **66**, 315 (1970).
[62] P. Kováč, *Carbohydr. Res.*, **20**, 418 (1971).
[63] P. Kováč and Ž. Longauerová, *Chem. Zvesti*, **26**, 71 (1972).
[64] J. Hirsch, P. Kováč and V. Kováčik, *Chem. Zvesti*, **28**, 833 (1974).
[65] P. Kováč and R. Palovčík, *Carbohydr. Res.*, **36**, 379 (1974).
[66] P. Kováč, R. Palovčík, V. Kováčik and J. Hirsch, *Carbohydr. Res.*, **44**, 205 (1975).
[67] L. F. Christensen and A. D. Broom, *J. Org. Chem.*, **37**, 3398 (1972).
[68] *Chromatography Lipids*, a publicity series available from Supelco Corp., Bellefonte, PA, USA, Vol. VII, No. 4, p. 2 (1973).
[69] H. Vorbrüggen, *Angew. Chem., Int. Ed. Engl.*, **2**, 211 (1963).
[70] A. Holý, *Tetrahedron Lett.*, **7**, 585 (1972).
[71] J. P. Thenot and E. C. Horning, *Anal. Lett.*, **5**, 519 (1972).
[72] J. Žemlička, *Collect. Czech. Chem. Commun.*, **28**, 1060 (1963).
[73] V. Ya. Pochinok and A. P. Limarenko, *Ukr. Khim. Zh.*, **21**, 628 (1955); *Chem. Abstr.*, **50**, 11270e (1956).
[74] T. Gramstad and R. N. Haszeldine, *J. Chem. Soc.*, 4069 (1957).
[75] J. M. Berry and L. D. Hall, *Carbohydr. Res.*, **47**, 307 (1976).
[76] M. G. Ahmed, R. W. Adler, G. H. James, M. L. Sinnot and M. C. Whiting, *Chem. Commun.*, 1533 (1968).
[77] M. G. Ahmed and R. W. Adler, *J. Chem. Soc., D*, 1389 (1969).
[78] R. F. Borch, *Chem. Commun.*, 442 (1968).
[79] T. Kametani, K. Takahashi and K. Ogasawara, *Synthesis*, 473 (1972).
[80] R. L. Hansen, *J. Org. Chem.*, **30**, 4322 (1965).
[81] J. Burdon and V. C. R. McLoughlin, *Tetrahedron*, **21**, 1 (1965).
[82] L. D. Hall and D. C. Miller, *Carbohydr. Res.*, **47**, 299 (1976).
[83] R. U. Lemieux and T. Kondo, *Carbohydr. Res.*, **35**, C4 (1974).
[84] J. Arnarp, L. Kenne, B. Lindberg and J. Lönngren, *Carbohydr. Res.*, **44**, C5 (1975).
[85] E. Brochmann-Hansen and T. O. Oke, *J. Pharm. Sci.*, **58**, 370 (1969).
[86] J. MacGee, *Clin. Chem.*, **17**, 357 (1971).
[87] G. Kananen, R. Osiewicz and I. Sunshine, *J. Chromatogr. Sci.*, **10**, 283 (1972).
[88] R. Osiewicz, V. Aggarwal, R. M. Young and I. Sunshine, *J. Chromatogr.*, **88**, 157 (1974).
[89] H. Meerwein, *Org. Synth.*, **46**, 113 (1966).
[90] H. Meerwein, *Org. Synth.*, **46**, 120 (1966).
[91] L. F. Fieser and M. Fieser, *Reagents for Organic Syntheses*, Vol. 1 (see also Vol. II and Vol. III), John Wiley & Sons, Inc., Chichester (1967).
[92] K. T. Potts, E. Houghton and S. Husain, *J. Chem. Soc., D*, 1025 (1970).
[93] H. Muxfelt and W. Rogalski, *J. Am. Chem. Soc.*, **87**, 933 (1965).
[94] S. Hanessian, *Tetrahedron Lett.*, 1549 (1967).
[95] O. Yonemitsu, T. Hamada and Y. Kanaoka, *Tetrahedron Lett.*, 1819 (1969).
[96] N. Narasimhachari, *J. Chromatogr.*, **90**, 163 (1974).
[97] W. Weg and G. S. King, personal communication.
[98] W. Rhoads and P. G. Gross, *Z. Naturforsch., Teil B*, **28**, 647 (1973).
[99] J. E. Cadotte, F. Smith and D. Spriesterbach, *J. Am. Chem. Soc.*, **74**, 1501 (1952).
[100] L. Stankovič, K. Linek and M. Fedoroňko, *Carbohydr. Res.*, **35**, 242 (1973).
[101] M. N. Oldham and H. Honeyman, *J. Chem. Soc.*, 986 (1946).
[102] C. T. Bishop and F. P. Cooper, *Can. J. Chem.*, **40**, 224 (1962).
[103] C. T. Bishop and F. P. Cooper, *Can. J. Chem.*, **41**, 2743 (1963).
[104] V. Smirnyagin, C. T. Bishop and F. P. Cooper, *Can. J. Chem.*, **43**, 3109 (1965).
[105] V. Smirnyagin and C. T. Bishop, *Can. J. Chem.*, **46**, 3085 (1968).
[106] I. Augestad and E. Berner, *Acta Chem. Scand.*, **8**, 251 (1954).
[107] S. Baker and W. W. Haworth, *J. Chem. Soc.*, 365 (1925).
[108] E. E. Percival and R. Zobrist, *J. Chem. Soc.*, 4307 (1952).
[109] B. D. Heard and R. Barker, *J. Org. Chem.*, **33**, 740 (1968).
[110] R. J. Ferrer and L. R. Hatton, *Carbohydr. Res.*, **6**, 75 (1968).
[111] K. Larsson and G. Pettersen, *Carbohydr. Res.*, **34**, 323 (1974).
[112] H. W. H. Schmidt and H. Neukom, *Helv. Chim. Acta*, **47**, 865 (1965).
[113] H. W. H. Schmidt and H. Neukom, *Helv. Chim. Acta*, **49**, 510 (1966).
[114] L. N. Owen, S. Peat and W. J. G. Jones, *J. Chem. Soc.*, 339 (1941).
[115] E. F. Jansen and R. Jang, *J. Am. Chem. Soc.*, **68**, 1475 (1946).
[116] G. G. S. Dutton, *Adv. Carbohydr. Chem. Biochem.*, **28**, 11 (1973).
[117] G. G. S. Dutton, *Adv. Carbohydr. Chem. Biochem.*, **30**, 9 (1974).
[118] *CRC Handbook of Chromatography*, Vol. 1, S. C. Churms (Editor), CRC Press Boca Raton (1982).
[119] N. K. Kochetkov and O. S. Chizhov, *Methods Carbohydr. Chem.*, **6**, 540 (1972).
[120] K. Heyns, K. R. Sperling and H. F. Grützmacher, *Carbohydr. Res.*, **9**, 79 (1969).
[121] V. Kováčik and P. Kováč, *Carbohydr. Res.*, **24**, 23 (1972).
[122] P. Hartvig, J. Vessman and C. Svahn, *Anal. Lett.*, **7**, 223 (1974).
[123] P. Hartvig and J. Vessman, *Acta Pharm. Suec.*, **11**, 115 (1974).
[124] J. Vessman, P. Hartvig and M. Molander, *Anal. Lett.*, **6**, 699 (1973).
[125] K. T. Koshy, D. G. Kaiser and A. K. VanDer Slik, *J. Chromatogr. Sci.*, **13**, 97 (1975).
[126] O. Gyllenhaal and J. Ehrsson, *J. Chromatogr.*, **107**, 327 (1975).
[127] L. G. Johnson, *J. Assoc. Off. Anal. Chem.*, **56**, 1503 (1973).
[128] F. K. Kawahara, *Anal. Chem.*, **40**, 1009 (1968).
[129] F. K. Kawahara, *Anal. Chem.*, **40**, 2073 (1968).

[130] A. H. Haines and K. C. Symes, *J. Chem. Soc., Perkin Trans. 1*, 153, (1973).
[131] R. E. Bowman and H. H. Stroud, *J. Chem. Soc.*, 1342 (1950).
[132] K. Blau and A. Darbre, *Biochem. J.*, **88**, 8P (1963).
[133] E. K. Doms, *J. Chromatogr.*, **105**, 79 (1975).
[134] R. E. Bowman, *J. Chem. Soc.*, 1346 (1950).
[135] R. E. Bowman, *J. Chem. Soc.*, 1349 (1950).
[136] T. Iversen and D. R. Bundle, *J. Chem. Soc., Chem. Commun.*, 1240 (1981).
[137] H. P. Wessel, T. Iversen and D. R. Bundle, *J. Chem. Soc., Perkin Trans.*, **122**, 47 (1985).
[138] W. Dünges, H. Heinmann and K. J. Netter, in *Gas Chromatography*, S. G. Perry (Editor), Applied Science, London (1972).
[139] D. Anderle and P. Kováč, *Chem. Zvesti*, **30**, 335 (1976).
[140] H. G. Jones, *Methods Carbohydr. Chem.*, **6**, 25 (1972).
[141] J. N. C. Whyte, *J. Chromatogr.*, **87**, 163 (1973).
[142] B. H. Freeman, A. M. Stephen and P. Van Der Bijl, *J. Chromatogr.*, **73**, 29 (1972).
[143] D. P. Sweet, P. Albersheim and R. H. Shapiro, *Carbohydr. Res.*, **40**, 199 (1975).
[144] P. A. Mied and Y. C. Lee, *Anal. Biochem.*, **49**, 534 (1972).
[145] B. Fournet, Y. Leroy and J. Montreuil, *J. Chromatogr.*, **92**, 185 (1974).
[146] B. Fournet and J. Montreuil, *J. Chromatogr.*, **75**, 29 (1973).
[147] P. A. J. Gorin and R. J. Magus, *Can. J. Chem.*, **49**, 2583 (1971).
[148] D. Anderle and P. Kováč, *J. Chromatogr.*, **91**, 463 (1974).
[149] D. Anderle, P. Kováč and J. Hirsch, *J. Chromatogr.*, **105**, 206 (1975).
[150] M. Gee, *Anal. Chem.*, **35**, 350 (1963).
[151] R. E. Wing and J. N. B. Miller, *Methods Carbohydr. Chem.*, **6**, 42 (1972).
[152] H. Scherz, G. Stehlik, E. Bancher and K. Kaindl, *Chromatogr. Rev.*, **10**, 1 (1968).
[153] S. C. Williams and J. K. N. Jones, *Can. J. Chem.*, **45**, 275 (1972).
[154] H. B. Sinclair, *J. Chromatogr.*, **64**, 117 (1972).
[155] G. W. Hay, B. A. Lewis and F. Smith, *J. Chromatogr.*, **11**, 479 (1963).
[156] A. Lipták and I. N. Jodál, *Acta Chim. Acad. Sci., Hung.*, **69**, 103 (1971).
[157] P. J. Brennan, *J. Chromatogr.*, **59**, 231 (1971).
[158] M. L. Wolfrom, D. L. Palin and R. M. de Lederkremer, *J. Chromatogr.*, **17**, 488 (1965).
[159] V. Prey, H. Berbalk and M. Kausz, *Mitrochim. Acta*, 449 (1962).
[160] R. T. Sleeter and H. B. Sinclair, *J. Chromatogr.*, **49**, 543 (1970).
[161] T. D. Audichya, *J. Chromatogr.*, **57**, 255 (1971).
[162] J. L. Garraway and S. E. Cook, *J. Chromatogr.*, **46**, 134 (1970).
[163] Y. C. Lee and I. Scocca, *Anal. Biochem.*, 3924 (1971).
[164] Y. M. Choy, G. G. S. Dutton, K. B. Gibney, S. Kabir and I. N. C. Whyte, *J. Chromatogr.*, **72**, 13 (1972).
[165] R. I. Ferrier, W. G. Overend, G. A. Rafferty, H. M. Wall and N. R. Williams, *J. Chem. Soc. C*, 1092 (1968).
[166] R. B. Clayton, *Biochemistry*, **1**, 357 (1962).
[167] M. Ervik and K. Gustav, *Anal. Chem.*, **47**, 39 (1975).
[168] J. Pecci and T. I. Giovanniello, *J. Chromatogr.*, **109**, 163 (1975).
[169] W. Dünges, *Anal. Chem.*, **45**, 963 (1973).
[170] W. Dünges, *Chromatographia*, **6**, 196 (1973).
[171] C. Landault and G. Guiochon, *Anal. Chem.*, **39**, 713 (1967).
[172] A. C. Bhattacharyya, A. Bhattacharjee, O. K. Guha and A. N. Basu, *Anal. Chem.*, **40**, 1873 (1968).
[173] L. V. Semenchenko and V. T. Kaplin, *Zavod. Lab.*, **29**, 801 (1967).
[174] H. O. Henkel, *J. Chromatogr.*, **20**, 596 (1965).
[175] P. A. Bond and M. Cagnasso, *J. Chromatogr.*, **109**, 389 (1975).
[176] C. B. Johnson and E. Wong, *J. Chromatogr.*, **109**, 403 (1975).
[177] C. J. Biemann and G. D. McGinnis (Editors), *Analysis of Carbohydrates by GLC and MS*, CRC Press, Boca Raton (1989).
[178] M. F. Chaplin and J. F. Kennedy (Editors), *Carbohydrate Analysis, a Practical Approach*, IRL Press, Oxford (1986).
[179] S. Manna and B. H. McAnalley, *Carbohydr. Res.*, **222**, 261 (1991).

6

Derivative Formation by Ketone–Base Condensation

Roberta H. Brandenberger and **Hans Brandenberger**
Branson Research, Lindenhofrain 8, CH-8708 Männedorf, Switzerland

1. ANALYSIS OF CARBONYL COMPOUNDS AS SCHIFF BASES 131
2. ANALYSIS OF PRIMARY AMINES AS SCHIFF BASES 133
3. ANALYSIS OF PRIMARY AMINES AS MUSTARD OILS 134
4. SELECTED PROCEDURES 136
   4.1 Schiff bases from carbonyl compounds 136
   4.2 Schiff bases from amines 137
   4.3 Isothiocyanates from primary amines 138
5. REFERENCES 139

The Schiff base or enamine reaction has been employed to modify aldehydes and ketones by condensation with a primary amine (reaction 1) and to condense primary amines with a carbonyl compound (reaction 2). A similar reaction between primary amines and carbon disulfide yields isothiocyanates or mustard oils (reaction 3). These three possibilities for derivative formation will be discussed in this chapter.

$$\begin{array}{c} R' \\ R'' \end{array}\!\!>\!\!C=O + H_2N-R \longrightarrow \begin{array}{c} R' \\ R'' \end{array}\!\!>\!\!C=N-R + H_2O \quad (1)$$

$$R-NH_2 + O=\!\!\underset{\underset{R''}{|}}{\overset{\overset{R'}{|}}{C}} \longrightarrow R-N=\!\!\underset{\underset{R''}{|}}{\overset{\overset{R'}{|}}{C}} + H_2O \quad (2)$$

$$R-NH_2 + S=C=S \longrightarrow R-N=C=S + H_2S \quad (3)$$

Handbook of Derivatives for Chromatography
Edited by K. Blau and J. M. Halket © 1993 John Wiley & Sons Ltd

Of course, enamine formation by routes (1) and (2) involves the same reaction scheme. However, the purposes and requirements for its application to the isolation, identification and quantitative determination of carbonyl compounds or amines differ. Therefore, (1) and (2) will be treated separately.

1. Analysis of carbonyl compounds as Schiff bases

There is a large number of classical reagents for preparing derivatives from aldehydes and ketones. Among the most common are hydroxylamine (formation of oximes), phenylhydrazine, o- and p-nitrophenylhydrazine and 2,4-dinitrophenylhydrazine (formation of the corresponding hydrazones), as well as semicarbazide, 4-phenylsemicarbazide and thiosemicarbazide (yielding the respective semicarbazones). Well known procedures for synthesizing these derivatives can be found in many textbooks and analytical monographs [1–5].

Before the age of chromatography, the characterization of unknown compounds necessitated the preparation of crystalline derivatives with definite melting points. For this purpose, hydroxylamine and phenylhydrazine are not always ideal reagents, particularly for low molecular weight carbonyl compounds. These derivatives possess rather low melting points or may even be oils at room temperature. The oximes are often very soluble and therefore difficult to purify, and mixtures of stereoisomers may also be formed. With phenylhydrazine, variations in the reaction procedure can lead to different derivatives: α-diketones yield mono- and dihydrazones and osazones; β-diketones form pyrazoles; α-hydroxy-aldehydes and -ketones are oxidized in the presence of excess phenylhydrazine and condense with a second molecule of reagent to give phenylosazones, useful in the study of sugars.

The characterization of carbonyl derivatives by melting point is more easily effected with the different nitrophenylhydrazones or with the semicarbazones. Extensive melting point lists are also available for these derivatives [5].

Thiosemicarbazones are of special interest, as they form insoluble compounds with monovalent metals such as silver, copper and mercury, which may be an aid to their separation from mixtures and to their quantitative determination. The colour of the nitrated phenylhydrazine derivatives can be characteristic of the type of carbonyl group. p-Nitrophenylhydrazones of aliphatic compounds are yellow or brown, those of aromatic compounds red. 2,4-Dinitrophenylhydrazones of saturated aliphatic aldehydes and ketones are yellow or orange, those of α,β-unsaturated carbonyls are red, and those of simple aromatic compounds are yellow, orange or red. They are usually formed in excellent yields.

With the development of chromatographic separation and characterization methods, the classical carbonyl reagents have not been abandoned. However, the reasons for preparing these derivatives have changed to a large degree, and that, of course, influences the choice of reagent. Today, the main purposes of derivative formation are:

(1) to isolate the carbonyl compounds from complex mixtures or from dilute solutions;
(2) to eliminate aldehydes and ketones from a mixture undergoing chromatographic investigation, i.e. by using a pre-column reaction;
(3) to convert carbonyls to compounds visible on paper, thin-layer plates or columns, or to compounds which can easily be revealed by spraying with a colour- or fluorescence-producing substance (see Chapters 8 and 9);
(4) to change the volatility and/or increase the differences between chemically related compounds for easier chromatographic separation;
(5) to improve the sensitivity of the chromatographic detection procedure, i.e. by formation of compounds with high electron affinity;
(6) to convert the aldehydes and ketones to derivatives which can more readily be identified by combined gas chromatography–mass spectrometry.

A few examples will serve to illustrate these points.

(1) The carbonyls from tobacco smoke [6, 7], urine [8] and aroma concentrates [9], the dicarbonyls from oxidized whole milk powder [10] and the keto acids from blood and urine [11] have been collected as 2,4-dinitrophenylhydrazones and removed by filtration or by extraction before chromatography. By mixing the 2,4-dinitrophenylhydrazones with α-ketoglutaric acid and heating to 240–260 °C, the aldehydes and ketones can be freed (flash exchange), swept into a column and separated by GC [12, 13]. Volatile carbonyls from air or water were trapped in a solution of 2,4-dinitrophenylhydrazine, extracted and analyzed by HPLC [14–16]. Steroids were isolated from blood plasma, tissues and urine as O-methyloxime– and O-benzyloxime–trimethylsilyl derivatives in good yields, prior to their identification by GC–MS and determination by mass specific detection [17–19].

(2) Semicarbazide has been used to remove carbonyls from mixtures in a pre-column reaction. The reagent was deposited on diatomaceous earth in a short plug of the GC injection block. Upon injection of the extract, the plug retained most or all of the carbonyl compounds [20].

(3) The coloured p-nitro- and 2,4-dinitrophenylhydrazones are ideal derivatives for the separation and characterization of carbonyls by paper, thin-layer and column chromatography. The oximes can easily be revealed on the thin-layer plates by spraying with solutions of copper(II) chloride or copper(II) acetate (alcoholic) or iron(III) chloride [21, 22]. For the visualization of carbonyl compounds by means of fluorogenic labeling by Schiff base formation with reagents such as dansylhydrazine, we refer the reader to Chapter 9.

(4) Good results with GC of keto acids have been obtained by changing the volatility and polarity of the compounds by means of conversion to silylated O-oximes including methyloximes, ethyloximes and pentafluorobenzyloximes [23–25].

(5) If a separation of derivatives is to be effected by GC using a flame ionization or other general purpose detector, the non-substituted oximes or phenylhydrazones might be preferred over the nitro-substituted derivatives on account of their better volatility and lower retention properties. On the other hand, the detection limits can be improved by a factor of about 500 if the reagent phenylhydrazine is replaced by 2,4-dinitrophenylhydrazine and the FID by an electron capture detector [26]. Halogen-substituted phenylhydrazones can give even lower detection limits. Estrone in blood plasma has been determined with ECD in pg amounts after conversion to its pentafluorophenylhydrazone 3-methyl ether [27]. The pentafluorobenzyloximes are equally good derivatives for electron capture GC. The pentafluorobenzyloxime trimethylsilyl esters from keto acids proved to be more suitable for GC than

the corresponding ethoxime trimethylsilyl derivatives because of their longer retention times. They eluted in a chromatographic region with less interference from by-products of the extracts. Furthermore, the formation of double peaks due to the presence of *syn*- and *anti*-isomers was less pronounced [25].

(6) Hydrazones and other Schiff bases from carbonyl compounds have been analyzed and characterized by combination GC–MS. In a recent paper [28], ng amounts of carbonyl compounds in different materials (vapors from decomposing polymers, engine exhaust, liquid soaps) were characterized and measured as 2,4-dinitrophenylhydrazones by reversed phase HPLC with simultaneous detection by UV monitoring at 365 nm and by positive- and negative-ion MS (chemical ionization with methane as reagent). The negative-ion information is especially interesting. The 2,4-dinitrophenylhydrazones give an intense anion at m/z 182 which characterizes the dinitrophenylhydrazine residue. The ketone derivatives yield strong negative molecular ions with relative intensities between 60 and 100%, but aldehydes give only weak molecular anions. Since negative-ion mass spectrometry by electron attachment has – for compounds with high electron affinity such as dinitro-derivatives – even better sensitivity that classical electron capture detection, this paper gives the basis for a highly sensitive trace ·detection method with very good specificity for characterizing carbonyls.

2. Analysis of primary amines as Schiff bases

The number of reagents that has been used for the identification of primary amines is possibly larger than for any other class of compounds [5]. But most of these classical derivatives lie beyond the scope of this chapter. They are covered in other parts of the book: acylation in Chapter 3, silylation in Chapter 4 and alkylation in Chapter 5. The different nitrophenyl derivatives are discussed in Chapter 8, and Chapter 9 presents the options for preparing fluorescent derivatives. The only derivatization fitting into the framework of this chapter is the condensation with carbonyls resulting in the formation of Schiff bases or, in the case of aromatic amines, anils.

In the past, Schiff base condensations of primary amines have been carried out mainly with benzaldehyde and *p*-nitrobenzaldehyde, less often with 2,4-dinitrobenzaldehyde. Derivative formation occurs rapidly on warming for 10–30 min without or with a solvent (usually alcohol or acetic acid) [5]. Benzylimines from aliphatic amines are stable oils which can be distilled without decomposition under atmospheric pressure. Aromatic benzylimines as well as aliphatic and aromatic *p*-nitro- and 2,4-dinitrobenzylimines are crystalline solids. The nitro-compounds are yellow.

These Schiff bases are obtainable in good yields. They are easy to purify by recrystallization and possess sharp melting points. However, with the development of gas chromatographic techniques, reasons for derivative formation other than characterization by melting point have predominated. At present the main objectives are:

(1) the isolation of primary amines from complex mixtures or dilute solutions as a first purification or as a pre-concentration step;
(2) their conversion to coloured or to fluorescent compounds for easier visibility, be it before · or after a separation by paper, thin-layer or column chromatography;
(3) their transformation to less polar and/or more stable derivatives better suited for chromatographic separations;
(4) their conversion to derivatives which can be detected by more sensitive procedures, i.e. as strongly absorbing or fluorescent compounds for HPLC, as compounds with a high intensity IR absorption band for GC–IR combination, or as compounds with high electron affinity for electron capture detection in GC;
(5) their transformation to derivatives better suited for identification by mass spectrometry and quantification by mass specific detection (see Chapter 12 and 14).

Again, we would like to illustrate these points with a few examples. We will see that, quite unlike the situation existing for the analysis of carbonyl compounds, the classical reagents have been largely replaced.

(1) Acetone, butanone, cyclobutanone and similar ketones have been used to prepare Schiff bases from long-chain amines, aromatic amines, diamines, amino-phenols and catecholamines in biological media [29–34], and from hydrazines in air [35], prior to separation by GC. When electron capture detection was used, pentafluorobenzaldehyde was the chosen reagent. This procedure was applied to the analysis of traces of amines in water samples [36] and biological extracts [37,38], as well as hydrazine in tobacco smoke, technical maleic hydrazide and pyrolysis products [39].

(2) The yellow *p*-nitro- and 2,4-dinitrobenzylimines are good derivatives for making primary amines visible in paper and thin-layer chromatography. It is somewhat surprising that this possibility has not been used more often, but this might have to do with the insolubility of these derivatives in most cold solvents.

(3) Chromatographic separation and quantification of free primary amines is hampered by their polarity, which can cause peak tailing and which favors irreversible adsorption [40,41]. The formation of non-polar and more stable derivatives has therefore been the subject of many investigations. The enamines fulfill these requirements to a large extent. Even quite complex compounds such as pentafluorobenzaldehyde derivatives of catecholamines and related substances of biological importance, their hydroxyl groups blocked by silylation, are formed quantitatively, and can be separated by GC without decomposition [42]. Very dilute solutions of the corresponding derivatives of dopamine and norepinephrine could be kept at room temperature for two weeks. Chromatographic checks failed to indicates signs of decomposition, except for the Schiff base of 3,4-dimethoxyphenylethylamine [43].

(4) Attention has been centered on the halogenated derivatives of primary amines. This development started with the introduction of electron capture detection systems. With pentafluoro-derivatives for example, detection by electron capture is over 1000 times more sensitive than by flame ionization. If hydroxyl groups are present, they are usually trimethylsilylated, which augments the electron affinity of the product. Picogram amounts of catecholamines have been separated on methylsilicone phases and detected by electron capture after derivatization with pentafluorobenzaldehyde and silylation [37,43]. Such methods can be a great aid in the ultra-trace analysis of primary amines in biological specimens.

(5) The mass spectra of primary amines are often not ideal for their characterization, especially if these compounds are present only in low concentrations. Enamine formation can also be of help in this respect. The mass spectra of Schiff bases show the molecular ion and some fragments of analytical importance, permitting an easy identification of the initial amine. In analogy to the work with electron capture GC, derivatization with pentafluorobenzaldehyde has also been used in this context [36]. However, this seems only justified for negative-ion MS and not for the usual positive-ion work. In this case, the additional electron attracting substituent may only introduce unnecessary analytical complications.

3. Analysis of primary amines as mustard oils

The mustard oil reaction is a well known test for the presence of primary amines. By the action of carbon disulfide under mild conditions (room temperature), the amines are converted to dithiocarbamic acids (reaction 4), which decompose on heating with a heavy metal salt to the so-called mustard oils or isothiocyanates, easily detectable by a very intense characteristic odor, and metal sulfide [44–46]. Secondary amines are also converted to dithiocarbamic acids by carbon disulfide (reaction 5), but these derivatives do not yield isothiocyanates; tertiary amines do not react at all.

$$R-NH_2 + CS_2 \longrightarrow R-NH-CS-SH \xrightarrow[\text{heat}]{M^+} R-N=C=S \quad (4)$$

$$\begin{array}{c}R'\\R''\end{array}\!\!\!>\!\!NH + CS_2 \longrightarrow \begin{array}{c}R'\\R''\end{array}\!\!\!>\!\!N-CS-SH \quad (5)$$

The conversion of the weakly UV-absorbing primary and secondary amines to the more strongly absorbing dithiocarbamic acids has been used to reveal the free amines on thin-layer plates [47]. The plates were sprayed with dilute ammonia to free the bases from their salts before exposure to carbon disulfide.

In 1967, we found that β-aryl-substituted alkylamines and carbon disulfide react at room temperature with the direct formation of the corresponding isothiocyanates [48, 49]. There is reason to believe that this reaction proceeds according to a ketone–base condensation (reaction 6).

$$R-NH_2 + CS_2 \longrightarrow R-N=C=S + H_2S \quad (6)$$

β-Phenylisopropyl isothiocyanate, resulting from such a direct condensation of amphetamine (β-phenylisopropylamine) with carbon disulfide, has been isolated *in vitro* and characterized by UV, IR and positive- and negative-ion MS [48, 50]. The reaction has successfully been extended to β-phenylethyl- and β-phenylisopropylamines having various substituents [49, 51–55], to some of the corresponding indolyl derivatives [56, 57] and to some β-phenoxyalkylamines [50, 58]. Most of these compounds have biological significance. Some are drugs of abuse (amphetamine, mescaline, STP), a few are pharmaceuticals (mexiletine), and several are biogenic amines (catecholamines such as norephedrine, dopamine, norepinephrine serotonin and phenylethylamine, a putrefaction product often found in analyses of biological fluids). Since no systematic investigation of the range of the ketone–base condensation of primary amines to form mustard oils has been published so far, the number of compounds which can be characterized by this simple approach may be considerably larger than reported up to now.

In the field of chromatography, the mustard oils of primary amines have substantial analytical value.

(1) The isothiocyanates are much less polar than the corresponding amines. Their chromatographic separation is therefore easier; less tailing and less irreversible adsorption occur, and lower detection limits can be obtained.

(2) Primary amines of relatively low molecular weight, such as amphetamine, are quite volatile; losses occur easily during the concentration of extracts by evaporation. The vapor pressure of the corresponding isothiocyanates is considerably lower, and solutions of these derivatives can thus be evaporated to dryness without loss.

(3) The high stability of the isothiocyanates, in the dry state and in solution, has already been pointed out.

(4) Quite often, both the free amine and the corresponding isothiocyanate can be separated by chromatography on the same support. The peak shift technique can therefore be used as an additional means of identification, as has been described for amphetamine [48]. This approach will also permit differentiation between primary and secondary amines with similar retention behavior, such as (in GC) amphetamine and methylamphetamine.

(5) Some very sensitive detection methods have been developed for the identification of mustard oils in thin-layer chromatography, as discussed below.

(6) For observing the sulfur-containing isothiocyanates in GC, specific detectors can be employed. The use of an electron capture system has been described [48]. A sulfur-specific photometric detector might even be more useful in order to improve sensitivity and selectivity. A nitrogen-selective detection system has been used recently for the same purpose [59].

(7) The importance of mass spectrometry, especially in combination with GC, for the identification and trace analysis of primary amines as isothiocyanates will be discussed in more detail below. The identification power of positive-ion MS for the Schiff bases is already quite high, but it can be improved by incorporating negative-ion mass spectrometry into the analytical scheme. GC with mass specific detection of positive and/or negative ions is certainly the most sensitive approach for the trace detection of this class of compounds [50].

(8) Finally, we have also to mention the excellent potential of infrared analysis for the detection of isothiocyanates. This functional group is one of the strongest IR absorbers. Its strong band around 2100 cm^{-1} has a molar absorption coefficient of over 5×10^4, that is, two orders of magnitude higher than the molar absorption coefficient of the primary amino group (a weak absorber, especially in the gas phase). This permits the detection of primary amines as mustard oils by combined GC–FTIR spectrometry. With the simple light pipe technique, detection limits in the lower ng range can be obtained [60]; even lower limits are certainly possible with the more sophisticated matrix isolation GC–FTIR approach.

Many of the biologically important β-phenylalkylamines and also indolylalkylamines contain alcoholic and/or phenolic hydroxyls. For a separation by GC, these polar groups must be derivatized. This can be done by silylation. We use a one-step method [49]. Both reagents, carbon disulfide and the silylating agent (usually trimethylsilylimidazole), are added to a solution of the amine in a polar solvent such as dimethylformamide or pyridine. A two-step procedure has also been described [56, 57]. The free bases are dissolved in ethyl acetate and treated with carbon disulfide, the solution is evaporated to dryness, and mustard oil is dissolved in pyridine and silylated at 100 °C. If the reaction time is sufficiently long, this procedure also silylates the indole nitrogen. Trifluoroacetylation has also been used to block the hydroxyls of the mustard oils [60].

The use of the mustard oil reaction prior to separation by TLC is mentioned in a few publications [54, 59]. One of these deals with the detection of 3,4-dimethoxyphenylethylamine, the dimethyl ether of dopamine, in urine samples. The compound is converted to the isothiocyanate and separated on silica gel plates. These are then sprayed with a mixture of equal volumes of sulfuric acid and methanol and irradiated with UV light. An intense fluorescence develops and permits the detection of ng amounts of 3,4-dimethoxyphenylethylamine [54]. The derivative can be extracted with methanol for quantitative determination by spectrofluorimetry (excitation at 365 nm, fluorescence at 465 nm). The only other isothiocyanates which yielded a similar fluorescence were those of 3-methoxytyramine and 4-O-methyldopamine. Silica gel thin-layer plates have also been used to separate norephedrine from norpseudoephedrine as isothiocyanates [59].

The mass spectrometric degradation of primary amines by electron impact is often complex. The molecular ion, if visible, is generally of low intensity. A large number of fragment ions is formed, but none are good tracer ions for mass specific detection; their intensity is low and their diagnostic value is usually modest. In the isothiocyanates, the different charge distribution stabilizes the molecular ions and favors the formation of (usually) only two high intensity fragments resulting from the cleavage of the ethyl chain between the α- and β-carbons (Figure 1). These ions can be monitored during a chromatographic run with a prefocused mass spectrometer as detector. Detection limits in the lowest pg range can be obtained, as long as column adsorption does not interfere [49,51,53]. In order to concentrate

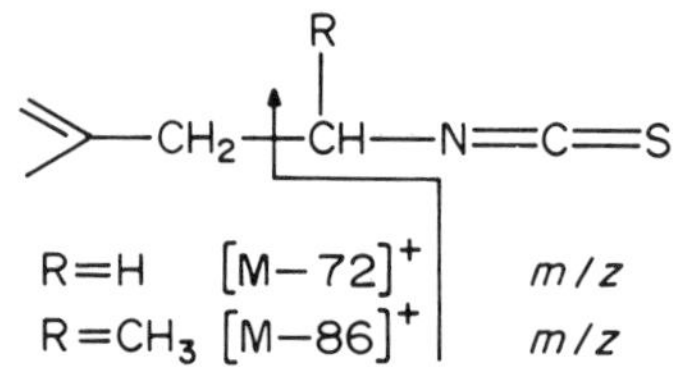

Figure 1. Cleavage of β-phenylalkyl isothiocyanates by positive-ion electron impact MS

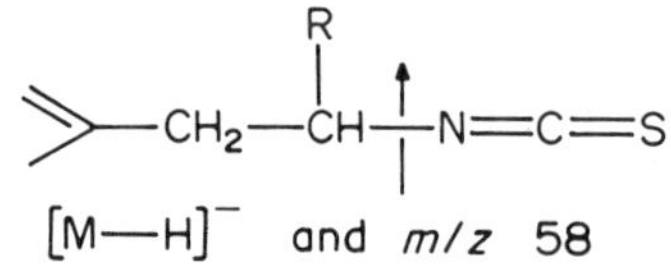

R=H $[M-72]^+$ m/z
R=CH$_3$ $[M-86]^+$ m/z

Figure 2. Cleavage of β-phenylalkyl isothiocyanates by negative-ion low pressure chemical ionization MS

$[M-H]^-$ and m/z 58

the mass spectrometric information in only a few intense ions, we recommend working with a low electron ionization voltage. The tracer ions are (Figure 1):

- for phenylethyl derivatives: M^+, $[M-72]^+$ and m/z 72;
- for phenylisopropyl derivatives: M^+, $[M-86]^+$ and m/z 86.

Recently, we have also included negative-ion MS by chemical ionization into the analytical scheme [50]. This produces two main fragments, the negative ion with m/z 58 resulting from the isothiocyanate function and $[M-1]^-$ as indication of the molecular mass of the mustard oil Figure 2. In mass specific detection, the anion m/z 58 can be used to detect all isothiocyanates, the cations m/z 72 and 86 allow differentiation between phenylethyl and phenylisopropyl isothiocyanates, and M^+ and $[M-1]^-$, as well as $[M-27]^+$ and $[M-86]^+$, are for identifying the individual compounds. Since modern instrumentation permits recording positive and negative ions simultaneously, it has become possible to obtain all the described information in a single run, either by recording from each peak the entire positive- and negative-ion spectra or, for trace analysis in the lower ng and pg range, by monitoring the selected positive and negative ions side by side.

4. Selected procedures

4.1 Schiff bases from carbonyl compounds

4.1.1 2,4-Dinitrophenylhydrazones from carbonyl compounds [26]

The carbonyl compounds (100 µl of each) were shaken with 100 ml of saturated 2,4-dinitrophenylhydrazine (DNPH) in aqueous 2 M HCl. The mixture was allowed to stand overnight at room temperature. The precipitate was removed by filtration, washed with 2 M HCl and H$_2$O and dried over silica gel in a vacuum desiccator. The derivatives were dissolved in ethyl acetate for GC with FID or in benzene for GC with ECD: 2% SE-30 or 12% F-60, was used as stationary phase.

4.1.2 2,4-Dinitrophenylhydrazones from carbonyls in liquid soaps, organic binders and engine exhaust [28]

Standards were prepared by dissolving carbonyls in acetonitrile and adding aliquots to DNPH and 0.01 N perchloric acid in acetonitrile. Liquid soap samples (2–80 mg in 5 ml of H$_2$O) were added to an acetonitrile solution of DNPH containing perchloric acid catalyst and allowed to react for 45 min. A ceramic mat with an organic binder (300 mg) was heated to 500 °C for 10 min in the furnace of a thermal analyzer while it was purged with filtered air. The air was bubbled through an impinger containing 10 ml of DNPH–perchloric acid solution in acetonitrile. Engine exhaust samples were collected by bubbling the exhaust through impingers filled with water. A 100 µl portion of the solution was added to 5–10 ml of DNPH–perchloric acid reagent in acetonitrile. Portions (10 µl) of these solutions were analyzed by HPLC on a column of Spherisorb ODS using gradient elution with acetonitrile water (60:40 to 100:0) and two detection systems in series;

- first, UV monitoring at a wavelength of 365 nm;
- then recording of positive- and negative-ion mass spectra obtained by chemical ionization with methane after introduction into the ion source by a moving belt interface.

4.1.3 N,N-Dimethylhydrazones of steroids [41]

The steroid (0.1–1.0 mg) was dissolved in 0.1–0.2 ml of anhydrous N,N-dimethylhydrazine and about 0.05 ml of glacial acetic acid was added as catalyst, if necessary. After 1–2 h at room temperature, the excess reagent was removed in a stream of N$_2$. The residue was dissolved in tetrahydrofuran and the solution was used directly for GC on a column of 1% SE-30 on Gas-Chrom P.

The position of the carbonyl group and the reaction conditions determined the extent of the condensation reaction. A quantitative reaction in 1–2 h at room temperature occurred for 3-ones. A catalytic amount of acetic acid was necessary for 20-, 16- and 17-ones. The 11-keto group failed to react even after 12 h.

4.1.4 Benzyloximes and p-nitrobenzyloximes of C_1 to C_7 carbonyls [62]

Aliquots of carbonyl compounds (200–500 μg each) in aqueous solution were placed in 15 ml vials with Teflon-lined screw-caps. The volume was adjusted to 10 ml with distilled water and 50 mg of O-benzylhydroxylamine (BHA) or p-nitrobenzylhydroxylamine was added, followed by 10 drops of triethylamine. The sealed vials were heated at 65–70 °C for 1 h and allowed to cool. The solutions were transferred to 60 ml separatory funnels, acidified with 2 ml of 2 N HCl and extracted with three 3 ml portions of diethyl ether. The ether was evaporated to near dryness using N_2 at room temperature and the volume was reconstituted to 1 ml. Aliquots of 0.1–0.3 μl were injected for GC with N-selective thermionic detection.

4.1.5 O-Benzyloximes of steroids [19]

The steroid (1 mg) was dissolved in 0.5 ml of a dry pyridine solution of BHA (20 mg ml^{-1}). The solution was heated at 70 °C overnight in a screw-capped tube fitted with a Teflongasket. The solvent was evaporated under N_2, the residue was dissolved in 0.5 ml of bis-(trimethylsilyl)trifluoroacetamide (BSTFA) and the solution was heated at 100 °C for 3 h. GC: 1% Dexsil 300 or SE-30 on Gas-Chrom P.

Alternatively, the concentration of BHA was increased to 50 mg ml^{-1} in dry pyridine and the O-benzyloxime was silylated with N-trimethylsilylimidazole (TMSI) at 150–160 °C for 3 h. GC-MS showed that the major metabolites of the adrenocortical steroid hormones were converted into O-benzyloxime and trimethylsilyl derivatives.

4.1.6 O-Methyloxime–trimethylsilyl derivatives of steroids from urine [61]

Urine (25 ml) was adjusted to pH 5.2 with acetic acid or NaOH, and 0.5 ml of phosphate buffer, pH 5.2, was added. To 10 ml of this solution were added 0.25 ml of *Helix pomatia* enzyme preparation and 4 drops of $CHCl_3$. The mixture was incubated at 37 °C for 24 h and then extracted twice with 20 ml of EtOAc. The combined EtOAc layers were extracted twice with 20 ml of NaOH (0.1 M) and twice with 20 ml of H_2O. The EtOAc layer was dried over $MgSO_4$. The solvent was removed until 0.25–0.5 ml remained. The residue was transferred quantitatively, using the minimum amount of EtOAc, to a small test the tube with a Teflonstopper and a pear-shaped base. Internal standards and reference compounds were added, and EtOAc was removed under a stream of dry N_2 by heating to 40 °C. Then 200 μl of 10% methoxyamine in pyridine was added to the residue and the mixture was allowed to react for 15 min at 60 °C. The pyridine was removed in a N_2 stream and 100 μl of bis(trimethylsilyl)acetamide/trimethylchlorosilane (BSA/TMCS) (4:1) was added. The mixture was kept for 1 h at 70 °C, with subsequent GC on SE-30 or OV-1.

4.2 Schiff bases from amines

4.2.1 Enamines from amines and acetone or other ketones [31]

The amine hydrochloride (0.5–1 mg) was dissolved in 0.05 ml of dimethylformamide (DMF). Solid K_2CO_3 or $KHCO_3$ (2 mg) was added, followed by 0.5 ml of acetone (or other ketone). The mixture was shaken at room temperature for 3 h. Samples were injected directly into the gas chromatograph; 10% F-60 or 7% F-60 + 1% EGSP-Z on Gas-Chrom P was used.

For enamines from amines containing OH groups, to 0.5–1 mg amine in 0.05 ml DMF was added 0.15 ml of hexamethyldisilazane (HMDS), and the mixture was allowed to stand for 30 min at room temperature.

A ketone–HMDS mixture was prepared by adding 1 ml of HMDS to 10 ml of acetone (or other ketone), heating to boiling and cooling. A 0.4 ml portion of the ketone–HMDS mixture was added to the DMF solution and the reaction mixtures was allowed to stand for 12 h. Precipitates were separated by centrifugation. The reaction products were stable for several days at −5 °C. Analyses were effected by GC or GC–MS.

4.2.2 Pentafluorobenzaldehyde derivatives from water-soluble primary amines [36]

Aqueous samples (0.5 ml) were mixed with 1 ml of acetonitrile containing 40 μg of pentafluorobenzaldehyde (PFB) and heated for 30 min at 85 °C. The reaction mixtures was cooled and, after addition of 0.5 ml of water, extracted with 10 ml of hexane. The organic layer was concentrated to 1 ml and analyzed by GC–MS or by GC with multiple-ion monitoring. Detection limits of 10 ppb could be obtained. For lower limits, 5 ml of aqueous sample and 10 ml of acetonitrile containing 0.5 mg of PFB were used.

4.2.3 Pentafluorobenzaldehyde–trimethylsilyl derivatives of catecholamines [43]

The amine and PFB were heated in acetonitrile solution for 1 h at 60 °C to form a Schiff base. A further 1 h at

60 °C after the addition of BSA converted all OH groups to O-TMS groups. The reaction mixture was injected directly for GC with FID or diluted with hexane for GC with ECD on 5% SE-30 or 5% OV-17.

Epinephrine was converted to two isomeric substituted tetrahydroisoquinolines, which interfered with the norepinephrine derivative peak on SE-30. OV-17 separated the norepinephrine and normetanephrine derivatives from each other and from the epinephrine derivative. The 3,4-dimethoxyphenethylamine PFB derivative showed signs of decomposition during separation on SE-30.

4.2.4 Pentafluorobenzaldehyde–trimethylsilyl derivatives of catecholamines [37]

Microgram level

A sample of catecholamines in redistilled DMF (1 µg per 50 µl) was added to 100 µl of an acetonitrile solution of repurified PFB (100 µg ml^{-1}). The mixture was heated at 85 °C for 5 min. BSA (2 µl) was then added. Complete reaction was obtained in 15 min at room temperature. Derivatives were extracted with 850 µl of hexane and injected into the gas chromatograph for analysis on 1% SE-30 or 1% OV-17 on Gas-Chrom P with ECD and mass specific detection.

Nanogram level

Samples of 1–10 ng of catecholamines were introduced into silanized glass tubes, and 10 µl of an acetonitrile solution of repurified PFB (4 µg ml^{-1}) was added. The tubes were heated at 85 °C for 15 min. After the addition of 1 µl of BSA and 9 µl of hexane, each tube was sealed again and heated at 60 °C for 5 min. The hexane phase was injected into the chromatograph.

4.2.5 Determination of hydrazines in air by Schiff base formation with acetone [35]

Samples of air (2 l) were pumped through a cooled micro-impinger containing 2 ml of acetone with 2 µl of glacial acetic acid. A 3 µl portion of the impinger solution was analyzed by GC with a thermionic nitrogen detector. MS was used for identification of the hydrazone derivatives.

4.3 Isothiocyanates from primary amines

4.3.1 Isothiocyanates of biologically active amines [49]

Amphetamine was extracted with ether from alkaline aqueous solutions (body fluids). If interfering compounds were present, the solution was first steam distilled, then the distillate was made alkaline and extracted. The ether extract was mixed with an equal volume of CS_2, allowed to stand for 1 h and evaporated to a small volume.

Mescaline (3,4,5-trimethoxyphenylethylamine) and STP or DOM (2,5-dimethoxy-4-methyl amphetamine) were extracted with ether from alkaline aqueous solution. The ether extracts were mixed with CS_2 and allowed to stand for several hours or heated at 50 °C for 2 h, then evaporated to a small volume.

The reactions with dopamine, norephedrine and noradrenaline were carried out in DMF. The solutions were mixed equal volumes of CS_2 and TMSI, then warmed under erflux for 3 h and evaporated to small volume. Pyridine, formamide and tetrahydrofuran were also satisfactory solvents for converting dopamine.

The reaction mixtures were analyzed by GC with FID or ECD [48], by GC–MS and GC with mass-specific detection [49], by GC with combined positive- and negative-ion MS or mass specific detection [50] and by GC–FTIR spectrometry [60].

4.3.2 Isothiocyanates of biologically active amines [56,57]

A solution containing 1 mg of free base in 5 ml of EtOAc was shaken with 0.5 ml of CS_2 for 30 min, the mixture was evaporated to dryness under reduced pressure and the residue was dissolved in EtOAc (1 ml). Aliquots (1 µl) were used for GC on 1% SE-30 or 2.5% OV-225 on Gas-Chrom Q.

For phenolic and indoleamines, a 100 µg amount of the isothiocyanate derivative was reacted with BSTFA/TMCS (99:1) at 90 °C for 15 min. For 2- or 7-methyl substituted tryptamines, it was necessary to extend the reaction time to 1 h at 100 °C to ensure complete silylation of the indole nitrogen, presumably because of steric hindrance from the alkyl substituent.

Perdeutero-TMS derivatives were prepared by reacting the isothiocyanate derivatives with a mixture of d_{18}-BSA and d_9-TMS (10:1).

Trifluoroacetyl derivatives were prepared by treating the isothiocyanate compounds in EtOAc with a few drops of trifluoroacetic anhydride and allowing the solution to stand at room temperature for 30 min.

5. References

[1] A. J. Vogel, revised by B. S. Furniss, A. J. Hannaford, P. W. G. Smith and A. R. Tatchell, *Vogel's Textbook of Practical Organic Chemistry*, 5th edition, Longman, London; John Wiley & Sons Inc., New York (1989).
[2] N. D. Cheronin, J. B. Entrikin and E. W. Hodnett, *Semimicro Qualitative Organic Analysis: The Systematic Identification of Organic Compounds*. 3rd edition, Krieger, Melbourne (1983).
[3] R. L. Shriner, R. C. Fuson, D. Y. Curtin and T. C. Morrill, *The Systematic Identification of Organic Compounds, A Laboratory Manual*, 6th Edition, John Wiley & Sons Inc., New York (1980).
[4] D. J. Pasto and C. Johnson, *Organic Structure Determination*, Prentice Hall, Inglewood, UK (1969).
[5] Staff of Hopkin and Williams, *Organic Reagents for Organic Analysis*, 2nd Edition, Hopkin Williams, Chadwell Heath, Essex (1950).
[6] D. A. Buyske, I. H. Owen, P. Wilder and M. E. Hobbs, *Anal. Chem.*, **28**, 910 (1956).
[7] D. L. Manning, M. P. Maskarinec, R. A. Jenkins and A. H. Marshall, *J. Assoc. Off. Anal. Chem.*, **66**(1), 8 (1983).
[8] M. Yancey, R. Stuart, D. Wiesler and M. Novotny, *J. Chromatogr.*, **382**, 47 (1986).
[9] L. Gasco, R. Barrera and F. de la Cruz, *J. Chromatogr. Sci.*, **7**, 228 (1969).
[10] E. A. Corbin, *Anal. Chem.*, **34**, 1244 (1962).
[11] D. Cavallini, N. Frontali and G. Toschi, *Nature (London)*, **163**, 568; **164**, 792 (1949).
[12] J. W. Ralls, *Anal. Chem.*, **32**, 332 (1960).
[13] R. I. Stephens and A. P. Teszler, *Anal. Chem.*, **32**, 1047 (1960).
[14] D. Grosjean and K. Fung, *Anal. Chem.*, **54**, 1221 (1982).
[15] Z. Meng and R. L. Tanner, *Energy Res. Abstr.*, **9**(5), Abstr. No. 29197 (1984); *Chem. Abstr.*, **102**, 83566 (1985).
[16] F. VanHoof, A. Wittvox, E. VanDuggenhout and J. Janssens, *Anal. Chim. Acta*, **169**, 419 (1985).
[17] W. L. Gardiner and E. C. Horning, *Biochim. Biophys. Acta*, **115**, 524 (1966).
[18] E. C. Horning, M. G. Horning, N. Ikekawa, E. M. Chambaz, P. I. Juakonmaki and C. J. W. Brooks, *J. Gas Chromatogr.*, **5**, 283 (1967).
[19] P. G. Devaux, M. G. Horning, R. M. Hill and E. C. Horning, *Anal. Biochem.*, **41**, 70 (1971).
[20] D. A. Cronin, *J. Chromatogr.*, **64**, 25 (1972).
[21] M. Hranisavljevic-Jakovljevic, I. Pejkovic-Tadic and A. Stojiljkovic, *J. Chromatogr.*, **12**, 70 (1963).
[22] S. Nesic, Z. Nikic, I. Pejkovic-Tadic and M. Hranisavljevic-Jakovljevic, *J. Chromatogr.*, **76**, 185 (1973).
[23] H. J. Sternowsky, J. Roboz, F. Hutterer and G. Gaull, *Clin. Chim. Acta*, **47**, 371 (1973).
[24] E. C. Horning and M. G. Horning, *J. Chromatogr. Sci.*, **9**, 129 (1971).
[25] G. Hoffmann and L. Sweetman, *J. Chromatogr.*, **421**, 336 (1987).
[26] H. Kallio, R. R. Linko and J. Kartaranta, *J. Chromatogr.*, **65**, 355 (1972).
[27] J. Attal. S. M. Hendeles and K. B. Eik-Nes, *Anal. Biochem.*, **20**, 394 (1967).
[28] K. L. Olson and S. J. Swarin, *J. Chromatogr.*, **333**, 337 (1985).
[29] E. Brockmann-Hanssen and A. Baerheim Svendsen, *J. Pharm. Sci.*, **51**, 938 (1962).
[30] C. J. W. Brooks and E. C. Horning, *Anal. Chem.*, **36**, 1540 (1964).
[31] P. Capella and E. C. Horning, *Anal. Chem.*, **38**, 316 (1966).
[32] E. C. Horning, M. G. Horning, W. J. A. VandenHeuvel, K. L. Knox, B. Holmstedt and C. J. W. Brooks, *Anal. Chem.*, **36**, 1546 (1964).
[33] R. Toyoda, T. Nakagawa and T. Uno, *Bunseki Kagaku*, **22**, 914 (1973); *Chem. Abstr.*, **80**, 59136d (1974).
[34] W. J. A. VandenHeuvel, W. L. Gardiner and E. C. Horning, *Anal. Chem.*, **36**, 1550 (1964).
[35] J. R. Holtzclaw, S. L. Rose, J. R. Wyatt, D. P. Rounbehler and D. H. Fine, *Anal. Chem.*, **56**, 2952 (1984).
[36] M. J. Avery and G. A. Junk, *Anal. Chem.*, **57**, 790 (1985).
[37] J. C. Lhuguenot and B. F. Maume, *J. Chromatogr. Sci.*, **12**, 411 (1974).
[38] B. F. Maume, P. Bournot, J. C. Lhuguenot, C. Baron, F. Barbier, G. Maume, M. Prost and P. Padieu, *Anal. Chem.*, **45**, 1073 (1973).
[39] Y. Liu, I. Schmeltz and D. Hoffmann, *Anal. Chem.*, **46**, 885 (1974).
[40] L. D. Metcalfe, *J. Chromatogr. Sci.*, **13**, 516 (1975).
[41] W. J. A. VandenHeuvel and E. C. Horning, *Biochim. Biophys. Acta*, **74**, 560 (1963).
[42] J. C. Lhuguenot and B. F. Maume, in *Mass Spectrometry in Biochemistry and Medicine*, A. Frigerio and N. Castagnoli (Editors), Raven Press, New York (1974), p. 111.
[43] A. C. Moffat and E. C. Horning, *Biochim. Biophys. Acta*, **222**, 248 (1970).
[44] A. W. Hofmann, *Ber. Dtsch. Chem. Ges.*, **1**, 25, 169 (1868).
[45] L. F. Fieser and M. Fieser, *Organic Chemistry*, D. C. Heath, Boston (1944), p. 615.
[46] P. Karrer, *Lehrbuch der Organischen Chemie*, Vol. 13, Georg Thieme, Stuttgart (1959), pp. 148, 491.
[47] H. M. Stevens and P. D. Evans, *Acta Pharmacol. Toxicol.*, **32**, 525 (1973).
[48] H. Brandenberger and E. Hellbach, *Helv. Chim. Acta*, **50**, 958 (1967).
[49] H. Brandenberger and D. Schnyder, *Fresenus Z. Anal. Chem.*, **261**, 297 (1972).
[50] M. Yamada and H. Brandenberger, in *TIAFT 23, Proceedings of the International TIAFT Meeting and the Second World Congress on New Compounds in Biological and Chemical Warfare*, B. Heyndrickx (Editor), State University of Ghent (1986), p. 165.
[51] H. Brandenberger, in *Proceedings of the International Symposium on Gas Chromatography and Mass Spectrometry*, A. Frigerio (Editor), Tamburini Editore, Milan (1972), pp. 39–66.
[52] H. Brandenberger, *Pharm. Acta Helv.*, **45**, 394 (1970).

[53] H. Brandenberger, in *Clinical Biochemistry II*, A. C. Curtius and Roth, (Editors), Walter De Gruyter, New York (1974), p. 1465.

[54] N. Narasimhachari, J. Plaut and K. Leiner, *J. Chromatogr.*, **64**, 341 (1972).

[55] N. Narasimhachari and P. Vouros, *J. Chromatogr.*, **70**, 135 (1972).

[56] N. Narasimhachari and P. Vouros, *Anal. Biochem.*, **45**, 154 (1972).

[57] N. Narasimhachari and P. Vouros, *Biomed. Mass Spectrom.*, **1**, 367 (1974).

[58] H. Brandenberger, in *TIAFT 24, Proceedings of the 24th International Meeting*, Banff, G. R. Jones and P. P. Singer (Editors), University of Alberta Printing Services, Edmonton (1988), pp. 330–339.

[59] P. J. VanderMerwe, in *TIAFT 26, Abstracts of the 26th International Meeting*, University of Glasgow, Scotland (1989), p. 43

[60] H. Brandenberger and P. Wittenwiler, in *TIAFT 23, Proceedings of the International TIAFT Meeting and the Second World Congress on New Compounds in Biological and Chemical Warfare*, B. Heyndrickx (Editor), State University of Ghent (1986), pp. 231–239.

[61] P. Sandra, M. Versele and E. Vanluchene, *Chromatographia*, **8**, 499 (1975).

[62] D. F. Magin, *J. Chromatogr.*, **178**, 219 (1979).

7

Formation of Cyclic Derivatives

Karl Blau
Department of Chemical Pathology, Bernhard Baron Memorial Research Laboratories, Queen Charlotte's & Chelsea Hospital, London

and

André Darbre
Formerly, Department of Biochemistry, King's College, London

1. INTRODUCTION	142	
2. ACETALS AND KETALS	142	
2.1 Methylenedioxy derivative of an anhydro-sugar	143	
2.2 Isopropylidenes of glycerol	143	
2.3 Isopropylidenes of glyceryl ethers	143	
2.4 Isopropylidenes of diols	143	
2.5 Isopropylidenes of hydroxy acids	143	
2.6 Isopropylidenes of sugars	143	
2.7 Formation of butylidene alditols	144	
2.8 Benzylidene derivatives of sugars	144	
2.9 Benzylidene derivatives of alditols	144	
2.10 Benzylidene derivative of sucrose	144	
2.11 Sugar cyclopentylidene and cyclohexylidene derivatives	144	
2.12 Sugar cyclohexylidene–ethylene acetal derivatives	144	
2.13 Bismethylenedioxy derivative of a steroid	144	
2.14 Isopropylidenes of oestriols	145	
2.15 Isopropylidene of 3β-mercapto-5α-cholestan-5β-ol	145	
3. CYCLIC SILICONIDES	145	
3.1 Siliconides of steroids using DMCS	145	
3.2 Siliconides of steroids with DMDAS	146	
3.3 Cyclic di-t-butylsilyl derivatives	146	
4. CYCLIC BORONATES	146	
4.1 Cyclic boronate derivatives of steroids	146	
4.2 Butaneboronates of carbohydrates	146	
4.3 Alkylboronates of polyols or polyol methyl ethers	147	
4.4 Boronates of diols	147	
4.5 Boronates of 3-methoxy-4-hydroxyphenylethylene glycol	147	
4.6 Boronates of hydroxyamines and diamines	147	
4.7 Boronates of catecholamines	147	
4.8 Methylboronates of sphingosines and ceramides	147	
4.9 Butylboronates of prostaglandins	147	
4.10 Halogen-substituted boronate derivatives and the electron capture detector	147	
5. QUINOXALINOLS FROM α-KETO ACIDS	147	
5.1 Derivatization of α-keto acids to quinoxalinols	148	
6. CYCLIZATION REACTIONS OF GUANIDINO GROUPS	148	
6.1 δ-Pyrimidinylornithine from arginine and malondialdehyde	148	
6.2 Reactions of guanidines with hexafluoroacetylacetone	148	
7. OTHER MISCELLANEOUS HETEROCYCLIC DERIVATIVES	149	
7.1 Epoxides of monounsaturated and diene fatty acids	149	
7.2 Episulphide or dithiocarbonate of a steroid	149	
7.3 Carbonates of diols	149	
7.4 Carbonate derivatives of sugars	150	
7.5 Cyclization of amino alcohols to oxazolidin-2-ones	150	
7.6 Cyclization of hydroxy acids to dioxolanones	150	
8. CYCLIC DERIVATIVES OF AMINO ACIDS	150	
9. SEQUENCING OF PEPTIDES AND POLYPEPTIDES	151	
10. CONCLUSIONS	153	
11. REFERENCES	153	

1. Introduction

The preparation of derivatives for chromatography usually involves the reaction of each individual reactive group with a specific derivatizing group. Multiple reactive groupings may be derivatized simultaneously with the same group, as with the silylation or acylation of sugars, to give a common example, but the 'one reactive group–one derivatizing group' approach is the most widely used, as shown by the many examples in the various chapters of the handbook. However, when a compound is polyfunctional it is also possible to use reagents which react simultaneously with adjacent or proximal reactive groupings to bridge them into a cyclic derivative. There may be several reasons for doing this: it may produce a chromatographically suitable derivative in a single step, a cyclic structure may stabilize a sensitive molecule, derivatization to a cyclic derivative may avoid too great an increase in molecular weight and the consequent decrease in volatility, and a ring structure may make a derivative more detectable, e.g. by UV/VIS or fluorescence spectrometry or by mass spectrometry, where ring systems often yield stable and highly diagnostic fragment ions such as the tropylium ion of mass 91 characteristic of many aromatic compounds.

Cyclization may also occur as a result of an intramolecular rearrangement, sometimes unintentionally, as in the cyclization of vitamin D_2 in the heated injection port of a gas chromatograph [1], which was evident from the appearance of unexpected extra peaks. Condensation or polymerization reactions may also lead to cyclic derivatives, for example the formation of diketopiperazines from dipeptide esters or the production of paraldehyde from three molecules of formaldehyde. However, this chapter is mainly concerned with the deliberate formation of a cyclic derivative, most often by a bridging type of reagent ('bidentate ligand'), and the emphasis will be on analysis and quantitation by chromatographic methods. Earlier work was well reviewed by Poole and Zlatkis [2].

The requirements for a satisfactory derivatization to a cyclic derivative are set out in Table 1. Apart from the first three, the rest are common to derivatization procedures for chromatography.

2. Acetals and ketals

Cyclic acetals and ketals are normally prepared from diols, triols and polyhydroxy compounds by reaction with an aldehyde or ketone, and a wide variety have been used (see Table 2).

The reaction of a diol with neighbouring hydroxyls is shown in Figure 1.

Table 1

Requirements for conversion to cyclic derivatives

At least one pair of reactive groupings
Spatial separation appropriate for ring formation
Stable configuration of resulting ring
High and reproducible yield of cyclic derivative
Cyclization to occur readily (mild conditions)
Formation of a single, unique, derivative
Rapid and convenient derivatization reaction
Derivative to have good chromatographic properties

Table 2. Aldehydes and ketones used for acetal and ketal formation

Aldehydes	Ketones
Formaldehyde	Acetone
Acetaldehyde	2-Butanone
n-Propionaldehyde	Cyclopentanone
n-Butyraldehyde	Cyclohexanone
Glyoxal	Cyclohexane-1,2-dione
Benzaldehyde	Hexafluoroacetone
2-Furaldehyde	Trifluoromethylacetone
p-Anisaldehyde	Bis(chlorodifluoro)acetone
p-Tolualdehyde	
3,7-Dimethyl-2,6-octadienal	
Cinnamaldehyde	

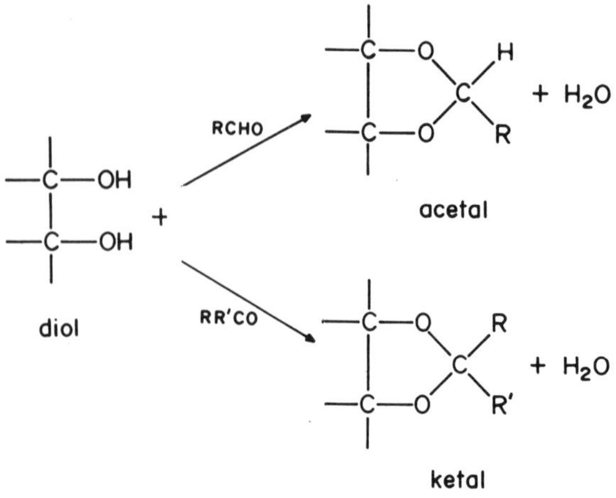

Figure 1. Examples of acetal and ketal formation.

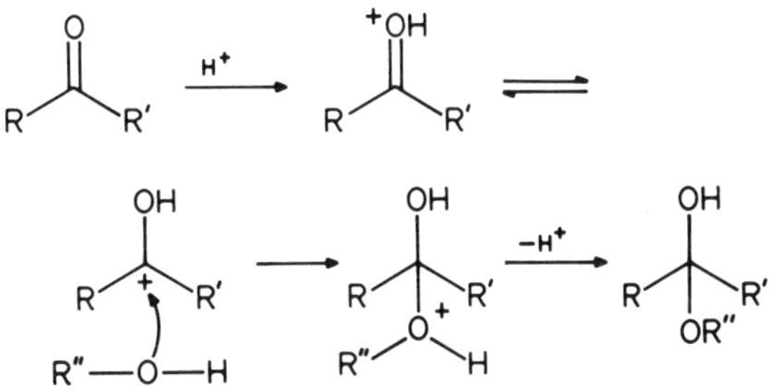

Figure 2. Acid catalyst and hemiacetal formation.

In the carbohydrate field, isopropylidene derivatives, often called acetonides, have been extensively used, formed by the reaction mechanism shown in Figure 2 and subsequent ring formation.

Acetals are formed in good yield, usually under acid conditions, and water formed during the reaction may be removed by azeotropic distillation or by a dehydrating agent such as anhydrous sodium sulphate [3]. They are stable to many reagents but are sensitive to hydrolysis. Methylene acetals of aldoses are relatively more stable; further derivatization of other reactive groups in the molecule is therefore possible.

2.1 Methylenedioxy derivative of an anhydro-sugar

The sugar (up to 35 mg) is dissolved in 200 µl of 37% aqueous formaldehyde and 200 µl of concentrated HCl. The mixture is kept in a desiccator over concentrated sulphuric acid for 72 h and the solid residue is extracted with ethyl acetate [4].

2.2 Isopropylidenes of glycerol

Dry glycerol (10 mg) is treated with 60 µl of 0.3 M HCl in dry acetone and 4 mg of powdered anhydrous sodium sulphate, and the mixture is shaken at intervals over a

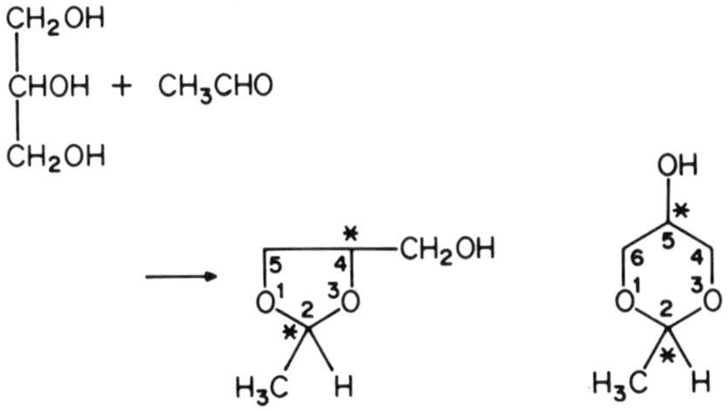

Figure 3. Condensation of glycerol and acetaldehyde.

12 h period at room temperature, then neutralized with lead carbonate and centrifuged. The supernatant may be analysed directly, or the product may be isolated by fractional distillation from silver oxide in a micro-still, or by preparative gas chromatography [7].

Acetal and ketal formation has been widely used in carbohydrate chemistry [8–13], not only for structural and analytical purposes, but also for chromatography [11] and mass spectrometry [14–16]. The formation of ring systems is largely dependent on the availability of cis-1,2- and cis-1,3-diol groupings, and the formation of bi- and tricyclic ring systems is possible where more than one such diol grouping is present.

2.3 Isopropylidenes of glyceryl ethers

1-O-Octadecylglycerol (40 mg) is suspended in 1 ml of acetone, and 5 µl of 12 M perchloric acid is added. After 20 min at room temperature, water is added dropwise until the solution starts to go cloudy. The mixture is extracted with four volumes of ether, and the ether is separated, washed with water until free of acid, and evaporated to dryness $in\ vacuo$. The residue is dissolved in hexane, and may be purified on a small silicic acid column before gas chromatographic analysis [17].

2.4 Isopropylidenes of diols

The diols (produced for example by oxidation of a double bond, but any vicinal diols will work equally well), on a scale of a few milligrams, are dissolved in 1 ml of acetone in a reaction vial, and 50 mg of anhydrous copper sulphate is added. The vial is heated at 50 °C for two hours with occasional shaking, and the supernatant solution can be analysed directly by GC or GC–MS [18, 19].

2.5 Isopropylidenes of hydroxy acids

The hydroxy acid or acids (ca. 25 mg) in 250 µl of 0.3 M HCl in dimethoxypropane and 30 µl of methanol, are kept in the dark at room temperature for 24 h. The solution is then passed over a small column of 0.3 g of basic, washed, activated alumina and eluted with hexane. The eluate is evaporated to dryness and the products are taken up in a suitable solvent and analysed by GC [20].

2.6 Isopropylidenes of sugars

The sugar (up to 100 mg) is suspended in 5 ml of acetone containing 50 µl of concentrated sulphuric acid and shaken until dissolved. After 20 h at room temperature, the acid is neutralized with solid sodium carbonate. The supernatant may be analysed directly or evaporated to dryness, the derivative being dissolved in a suitable solvent [4]. Alternatively, a small aliquot of a sugar or mixture of sugars isolated from some biological matrix is evaporated to dryness and shaken with 1.5 ml of acetone containing 10 µl of concentrated sulphuric acid for 2 h at room temperature, then the solution is neutralized with solid sodium bicarbonate. The supernatant solution may be analysed directly [15]. There are variations on this theme that can be tried out, and the results and the time course can be monitored by thin-layer chromatography; charring after a sulphuric acid spray is the easiest way of detection. In some cases boric acid (10% molar excess) may be included in the reaction mixture in order to 'complex out' some of the hydroxyls.

2.7 Formation of butylidene alditols

The alditol (up to 100 mg) in 47% aqueous HBr (50 µl) is shaken with butyraldehyde (100 µl) for 2.5 h at room temperature. The reaction mixture is neutralized with NaOH and evaporated to dryness *in vacuo*. The derivatives are extracted with hexane and dried over $CaCl_2$. This solution can be used for TLC (butanone saturated with water, or benzene/methanol 9:1). For gas chromatography any underivatized hydroxyls are first silylated [21].

2.8 Benzylidene derivatives of sugars

α-Methylmannoside (20 mg) is added to 100 µl of freshly distilled benzaldehyde and heated at 150 °C under a blanket of CO_2 for 2.5 h. Excess benzaldehyde is evaporated *in vacuo* and the residue is poured into 200 µl of ethanol, whereupon the derivative crystallizes out. Two new asymmetric centres are introduced. The product may be analysed by TLC or GC [22].

2.9 Benzylidene derivatives of alditols

The sugar (up to 100 mg) in concentrated HCl (250 µl) is stirred or shaken with benzaldehyde (250 µl) at room temperature for 15 h. The mixture is diluted with chloroform (500 µl) and the solution is washed to neutrality with successive equal volumes of 0.6 M sodium bicarbonate. The chloroform solution is dried and evaporated to dryness. The residue may be dissolved in a suitable solvent for TLC, with detection by spraying with 0.1 M potassium permanganate in 1 M H_2SO_4 and heating [4].

2.10 Benzylidene derivative of sucrose

Sucrose (25 mg) in dry pyridine (500 µl) is treated with benzylidene bromide (28 µl) at 85 °C for an hour; then a further 10 µl of benzylidene bromide is added and the reaction mixture is heated at 95 °C for a further 30 min. The product is acetylated with 50 µl of acetic anhydride at room temperature for 5 h. The solution is poured into ice–water and extracted with dichloromethane. After water washing, the lower layer is dried over anhydrous sodium sulphate; it may be used for analysis by TLC or further purified on a silica gel column [23].

2.11 Sugar cyclopentylidene and cyclohexylidene derivatives

These are made with cyclopentanone or cyclohexanone and an acidic condensation catalyst [24]. The sugar (30 mg) and, e.g., cyclohexanone (350 µl) are treated with 5 µl of concentrated sulphuric acid at room temperature for 24 h. The acid is neutralized with sodium bicarbonate, the solution is filtered and the excess cyclohexanone is evaporated under reduced pressure. The residue may be recrystallized from ethanol [25].

2.12 Sugar cyclohexylidene–ethylene acetal derivatives

The sugar (100 mg) is suspended in toluene (20 µl), cyclohexanone ethylene acetal (700 µl) and a drop of concentrated sulphuric acid, and the mixture is refluxed for 6 h, then neutralized with solid sodium bicarbonate, filtered and evaporated to dryness under reduced pressure. Analysis is by TLC using toluene/ethanol (9:1) as solvent [26].

2.13 Bismethylenedioxy derivative of a steroid

In the steroid field the same kinds of derivatization reactions are possible. Formaldehyde can react to form cyclic derivatives when there is a dihydroxyacetone side chain (Figure 4).

Cortisone (1 mg) is dissolved in 500 µl of chloroform and 130 µl of 12 M HCl is added, followed by 130 µl of 37% formaldehyde. The reaction mixture is agitated at 5 °C for 48 h. The solution is neutralized with 5 M NaOH. The chloroform layer is dried over anhydrous sodium sulphate, and may be analysed directly by gas chromatography [27].

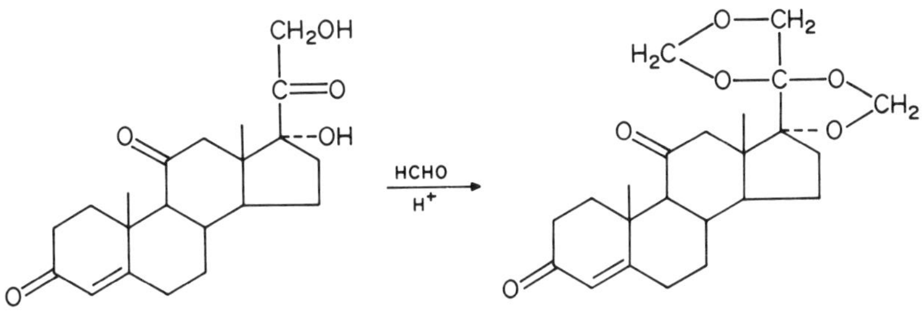

Figure 4. Cyclic derivatives of corticosteroids.

2.14 Isopropylidenes of oestriols

The formation of isopropylidene derivatives (acetonides) is also possible, and those of 3,16α-oestradiol and cis (but not trans) epimeric oestriols have been successfully separated [28, 29]. Under normal GLC conditions 17-hydroxycorticosteroids undergo pyrolysis to C-19 ketosteroids. This can be avoided if a thermally stable C-20-21 side chain can be formed by cyclization (see also siliconides and boronates, below) under acid conditions.

The steroid (100 μg) is dissolved in 1 ml of dry acetone and 0.5 ml of acetone previously saturated with dry HCl gas at 0 °C. After 15 min at room temperature, the mixture is neutralized with 0.6 M sodium bicarbonate, then filtered and evaporated to dryness in vacuo. The derivative is extracted from the residue with diethyl ether for analysis by TLC or GC [29].

2.15 Isopropylidene of 3β-mercapto-5α-cholestan-5β-ol

The mercaptosterol (40 mg) in acetone (3 ml) is refluxed with 6 mg of p-toluenesulphonic acid for 4 h. The product is isolated on a small Florisil column by hexane and hexane benzene (9:1) elution. It may be either characterized or analysed by GC [30].

In the lipid field, too, isopropylidene derivatives can be made from monoglycerides [31], glyceryl ethers [17] and hydroxy acids [32, 33] for analysis by GC. The position of double bonds can be established by osmium tetroxide oxidation to glycols with the subsequent cyclization to isopropylidene derivatives [34].

3. Cyclic siliconides

Cyclic siliconides are analogous to acetonides in structure, and can most readily be made with dichlorodimethylsilane (DMCS) in pyridine [35, 36]. Cyclic siliconide derivatives of steroids can similarly be made by reacting the dihydroxyacetone side chain of, e.g., cortisone, cortisol or betamethasone with dimethyldiacetoxysilane (DMDAS) and triethylamine [37] as shown in Figure 5. Other reagents for preparing cyclic siliconides include tetramethoxysilane and triethoxysilane [38].

3.1 Siliconides of steroids using DMCS

The steroid (100 μg) in pyridine (10 μl) is mixed with benzene (20 μl) and DMCS (910 μl) in a reaction vial and left for two hours at 40 °C. The solution may be directly injected for GC or GC–MS [39].

Another, similar, procedure is for steroids with cis-diol groupings, and uses a mixture of DMCS, pyridine and benzene in the ratio of 1:1:8 [35], or even more simply DMCS 1:1 in pyridine [36], for 1 h at room temperature. The proportions of reagents to steroid are not critical, although an excess of DMCS is essential. The reaction mixure may be directly analysed by GC.

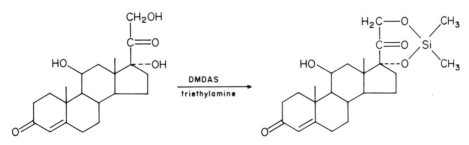

Figure 5. Siliconide formation of corticosteroids.

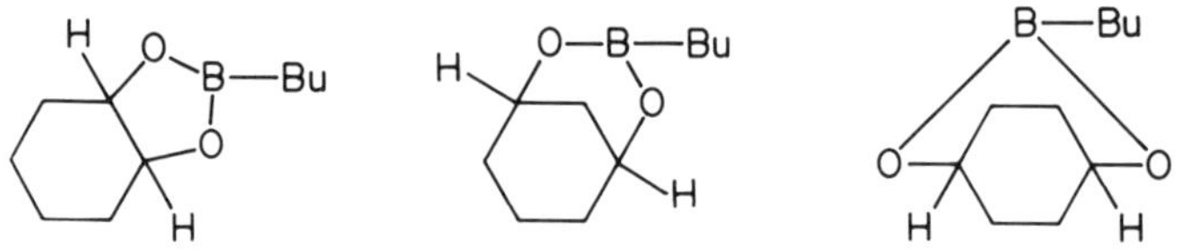

Figure 6.

Alternatively, the steroid (10 µg) is dissolved in 1,1,3,3-tetramethyldisilazane (10 µl) and DMCS (10 µl of a 20% solution in benzene) and incubated for 12 h in a sealed tube at 40 °C. Again, samples may be injected directly, or the reaction mixture may be evaporated to dryness, the derivative being dissolved in carbon disulphide for analysis [39].

3.2 Siliconides of steroids with DMDAS

A few mg of the steroids in a reaction vial are treated with 20 µl of a hexane solution containing 2% DMDAS and 2% triethylamine for 2 h at 40 °C, and the products are analysed directly [37]. A similar procedure, on the micro scale, may be used for salicylic and anthranilic acids and other *ortho*-hydroxy aromatic acids [40].

3.3 Cyclic di-t-butylsilyl derivatives

Diols and o-hydroxy acids can also be converted to the cyclic di-t-butylsilyl derivatives, which have certain advantages, particularly for GC–MS. The diols or hydroxy acids (100 mg) are dissolved in a mixture of acetonitrile (30 µl) and N-methylmorpholine (20 µl), or alternatively just N-methylmorpholine alone (50 µl), and treated with 3.5 µl of di-t-butyldichlorosilane and 1 µl of a 0.3% solution, in acetonitrile, of the silyl transfer catalyst 1-hydroxybenzotriazole for 15 h at 80 °C [41].

4. Cyclic boronates

Boric acid has long been known to form complexes with sugars [42, 43], and more recently substituted boronic acids (benzeneboronic, butylboronic, cyclohexylboronic, etc.) have been found to be valuable bidentate ligands for use in chromatography. The reactions occur readily and quickly under mild conditions to give cyclic boronates with five-, six- or seven-membered heterocyclic rings (Figure 6).

As expected, the six-membered rings are the most stable, followed by the five- and seven-membered rings [44]. Alkaneboronic acids are stable in air but sensitive to moisture. This may be useful after chromatographic separation, when the original parent compound may be recovered by solvolysis [45–50]. The methylboronates show only a modest mass increment and yield the most volatile cyclic boronate derivatives, although the t-butylboronates are not much less volatile [49, 50]. The methylboronate derivatives may in certain cases be more volatile than the corresponding silyl derivatives, and an advantage with the boronates of some compounds is that they yield single chromatographic peaks as against multiple peaks from the TMS derivatives [51]. Cyclic butylboronate derivatives have been widely used in gas chromatography of carbohydrates, but in some cases underivatized hydroxyls need to be silylated [52, 53], and this applies also in the prostaglandin field [54–57]. Phenylboronates of steroids [50, 58, 59], carbohydrates [45, 60–62], nucleotides and nucleosides [48, 63] and various amino compounds [64, 65] have also been prepared. 4-Iodobutaneboronic acid has been proposed as a selective reagent for the trace determination of bifunctional compounds [66], and halogen-containing benzeneboronic acids have been used as derivatives for sensitive analyses using gas chromatography with the electron capture detector [67, 68].

4.1 Cyclic boronate derivatives of steroids

The steroid or mixture of steroids (10 µmol) and an appropriate boronic acid (11 µmol of alkyl- or benzeneboronic acid) are dissolved in 1 ml of ethyl acetate and left at room temperature for 5 min. In obstinate cases a greater excess of the boronic acid may be used. The solution may be analysed directly by TLC, GC or GC–MS. If necessary, the solution may be evaporated to dryness for silylation in 0.1 ml of dry pyridine using hexamethyldisilazane (100 µl) with a trace of trimethylchlorosilane for 5 min at room temperature. Unconsumed boronic acids present no problems in GC or GC–MS, but elute as the trimeric anhydrides (boroxines) [50].

4.2 Butaneboronates of carbohydrates

The sugar or mixture of sugars (about 1 mg) is dissolved in 1 ml of pyridine containing 5 mg of butaneboronic acid. If dissolution is sluggish, the mixture may be warmed: reaction is essentially complete once the sugar has dissolved, and the solution may be injected directly for GC or GC–MS analysis [69].

4.3 Alkylboronates of polyols or polyol methyl ethers

The polyols or polyol methyl ethers (up to 500 µg) in 1 ml of a 10% solution of the appropriate boronic acid (butaneboronic, benzeneboronic, etc.) in pyridine are heated for 5 min at 100 °C, and the products may be analysed directly [70, 71]. A similar procedure, but with acetone as solvent, has been described for dianhydrogalactitol [72].

4.4 Boronates of diols

Equimolar amounts of the diol and butylboronic or benzeneboronic acid are mixed in acetone solution and left at room temperature for 10 minutes. The solution may be analysed directly by GC [58].

4.5 Boronates of 3-methoxy-4-hydroxyphenylethylene glycol

The glycol (5–40 µg), extracted from a biological matrix, e.g. urine, was taken to dryness and dissolved in 300 µl of 2,2-dimethoxypropane, which acts both as solvent and water scavenger. Either methylboronic or butaneboronic acid (up to 300 µl) is added, together with an internal standard such as phenanthrene, and the reaction is left to proceed at room temperature for 15 min. The underivatized phenolic group must be silylated to avoid tailing in gas chromatographic analyses, but this may be achieved at the last moment by injecting the reaction mixture together with BSTFA [73, 74].

4.6 Boronates of hydroxyamines and diamines

The aminoalcohol or amine is dissolved in anhydrous dimethylformamide, acetone or pyridine and treated with a 10% molar excess of the butane- or benzeneboronic acid at room temperature for 15 min— or longer, depending on the nature of the compound to be analysed [75].

4.7 Boronates of catecholamines

The sample (1 mg) is dissolved in pyridine or dimethylformamide, 1.5–2 molar equivalents of the selected boronic acid (methyl, n-butyl, cyclohexyl or phenyl) are added and the mixture is left to react overnight at 22 °C.

The solution may be analysed directly or evaporated to dryness, the residue being taken up in ether [76–78].

4.8 Methylboronates of sphingosines and ceramides

The lipid is dissolved in pyridine and a 10% excess of methaneboronic acid is added. The mixture is left at room temperature for 10 min. The nature of the starting material governs whether or not further derivatization (acylation, silylation or esterification) is necessary. If it is, then the solution is first evaporated to dryness in a stream of nitrogen; otherwise the sample may be analysed directly by GC or GC–MS [51, 79].

4.9 Butylboronates of prostaglandins

In this procedure the prostaglandins are first converted to the methoximes with methoxyamine, and taken to dryness in a stream of nitrogen. n-Butylboronic acid (30 µl) in dimethoxypropane (75 µl) is added to the methoxime derivatives, and the mixture is heated for 2 min at 60 °C. The solution is again evaporated to dryness, and the residue is silylated immediately (20 µl of Tri-Sil Z, 5 min at 60 °C) before analysis by GC or GC–MS [54, 55].

4.10 Halogen-substituted boronate derivatives and the electron capture detector

Although p-bromobenzeneboronates [80] and p-iodobenzeneboronates [66] were investigated earlier, more recently the 2,4-dichlorobenzeneboronates [68, 81] and 3,5-bis(trifluoromethyl)benzeneboronates [81] have been preferred as electron capturing derivatives for the sensitive analysis of drugs such as guaiphenesin or alprenolol. Typically, the residue from an extract of 2 ml of plasma is treated with dichlorobenzeneboronic acid in acetonitrile (1 ml of a solution containing 20 mg l^{-1}) and heated for 5 min at 50 °C. The reaction mixture is evaporated to dryness and the residue is dissolved in 200 µl of hexane for analysis by gas chromatography with electron capture detection.

5. Quinoxalinols from α-keto acids

Compounds with keto groups often give more than one derivative on silylation, acylation or esterification because of the possibility of geometrical isomerism, and yields may be unsatisfactory. For α-keto acids the forma-

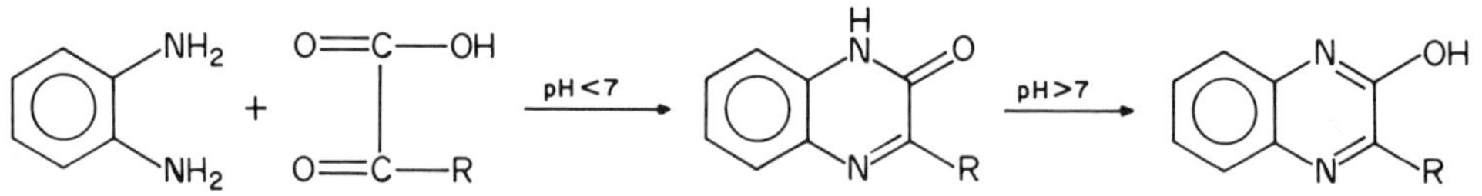

Figure 7. Mechanism of reaction of α-keto α acids and o-phenylenediamine.

tion of cyclic quinoxalinols is one of the best ways of getting round these problems [82–84]. The keto acids react with o-phenylenediamine to give the hydroxyquinoxalinols in good yield [85, 86] (Figure 7).

The hydroxyl groups were silylated to derivatives giving single peaks with excellent chromatographic properties. With oxaloacetic acid, decarboxylation to the pyruvic acid derivative occurred [86]. Glyoxylic acid is not derivatized to the quinoxalinol, but may instead be condensed with a diamine, e.g. N,N-diphenylethylenediamine, to yield a cyclic diphenylimidazolidine (Figure 8), and this may be silylated for gas chromatographic analysis [87].

5.1 Derivatization of α-keto acids to quinoxalinols

Langenbeck [86] stabilized urinary keto acids with dithionite, which has become the general practice, and for derivatization followed the procedure of Nielsen [88] and of Mowbray and Ottaway [89]. Urine is treated with $Na_2S_2O_4$ (4 mg ml^{-1}). To 2 ml of this is added the internal standard, α-ketovaleric acid (20 mM, 50 µl). The resulting spiked urine (500 µl) is heated with 4 M HCl (500 µl) and 1 ml of a 1% solution of o-phenylenediamine in 2 M HCl in a sealed reaction vial for an hour at 70 °C. Exactly 1.5 g of ammonium sulphate is added, and the solution is extracted twice with 5 ml of chloroform. The combined chloroform extract is dried for 2 h over anhydrous sodium sulphate, filtered and evaporated to dryness. The dry residue is shaken for exactly 1 min with 1 ml of ether and the ethereal solution is analysed [90]. This procedure has stood the test of time [91], but it has been suggested that oxygen should be excluded; in addition to also adding mercaptoethanol to the urine, Livesey and Edwards suggest a blanket of nitrogen gas for the reaction [92].

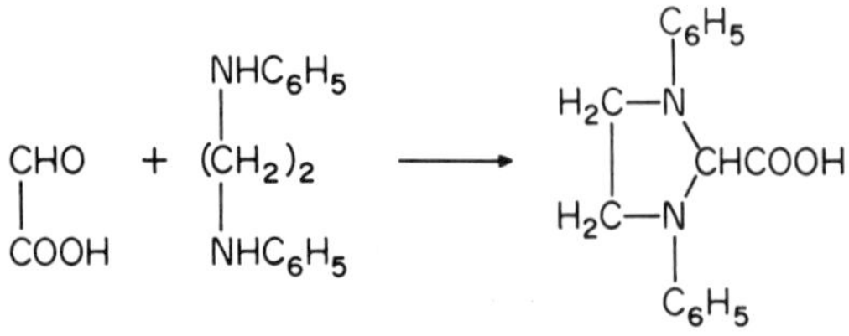

Figure 8.

The quinoxalinols may be analysed by TLC or HPLC, and are UV-absorbing and fluorescent [89]. For gas chromatogrphy silylation is necessary [86, 90]. Electron capturing derivatives have been made using p-chloro-, 4,5-dichloro- and 3,4,5,6-tetrafluoro-1,2-phenylenediamines [84].

6. Cyclization reactions of guanidino groups

Guanidino compounds can be readily cyclized by reaction with malondialdehyde, either as such or in the form of its diethyl acetal, 1,1,3,3-tetraethoxypropane, giving substituted pyrimidines, which may have improved chromatographic properties: increased detectability (UV absorption after TLC) and potentially also increased volatility for analysis by GC [93, 94].

6.1 δ-Pyrimidinylornithine from arginine and malondialdehyde

A solution of arginine (1 mmol) and 1,1,3,3-tetraethoxypropane (10% molar excess) in 8 ml of 12 M HCl is allowed to stand for 3 h at room temperature, then evaporated to dryness on a rotary evaporator. The solution may be analysed by TLC at this stage, or further purified on a column of Dowex 50-X8 by elution with a pyridine/acetic acid gradient. Elution is monitored by UV absorption at 315 nm. Gas chromatographic analysis may be carried out after further derivatization of the ornithine side chain [94].

6.2 Reaction of guanidines with hexafluoroacetylacetone (Figure 9)

The guanidino-compound, either a drug such as 3-hydroxyguanfacine, debrisoquine or guanadrel [95–98] or a naturally occurring guanidine such as δ-guanidinovaleric acid [99], is extracted from urine or plasma (2 ml), and the extract is evaporated to dryness. To the residue is added a 10% solution of hexafluoroacetylacetone in methanol (200 µl) and the mixture is heated for an hour or two at 100 °C. The solution is evaporated to dryness, and the product may be analysed by TLC, or further

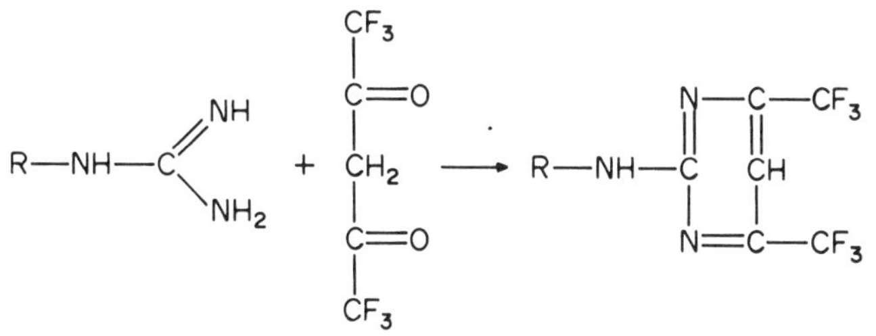

Figure 9. Cyclization of a guanidine to a pyrimidine.

derivatized for GC or GC–MS by esterification to the methyl [95] or to the N^δ-trifluoroacetylated butyl ester [99].

7. Other miscellaneous heterocyclic derivatives

Ethylene oxide or oxirane is the simplest α-epoxide (for reviews see Refs. 100–103). Epoxides or oxirane derivatives may generally be formed by treatment of ethylenic compounds or cis-diols with peracids [104–106]. Episulphide analogues can also be formed [107]. Epoxides can be analysed by TLC. GC or MS.

7.1 Epoxides of monounsaturated and diene fatty acids

The monounsaturated fatty acid ester (3 µl) is mixed with peracetic acid (150 µl) and allowed to stand for 2–3 h at room temperature. For dienes, twice the amount of peracetic acid and at least twice the reaction time are needed. Samples may be analysed directly or neutralized with soldium bicarbonate and extracted with hexane for subsequent GC analysis [108, 109].

7.2 Episulphide or dithiocarbonate of a steroid

3β-Hydroxycholestan-2α-ethylxanthate (10 mg) is dissolved in ethanol (2.5 ml) and an excess of sodium borohydride (1 mg) is added. After 48 h at room temperature the mixture is diluted with water and cholestane episulphide is extracted with hexane [107]. This reaction may, however, also give the thiocarbonate (Figure 10).

7.3 Carbonates of diols

Cyclic carbonates may be made from a glycol and phosgene (Figures 11 and 12).

The diol (1 mmol) and dry pyridine (150 µl) were dissolved in alcohol-free chloroform (1 ml), and phosgene was gently bubbled through the solution for 1 h at 0 °C and then for a further 2–3 h at room temperature. The organic solution was extracted with an equal volume of water several times until neutral, then dried over anhydrous calcium sulphate [110, 111].

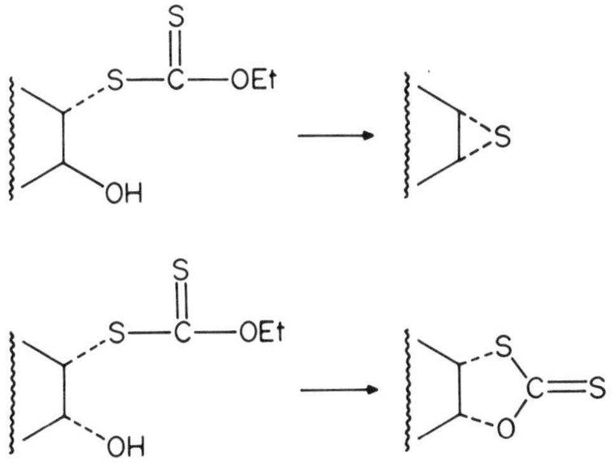

Figure 11.

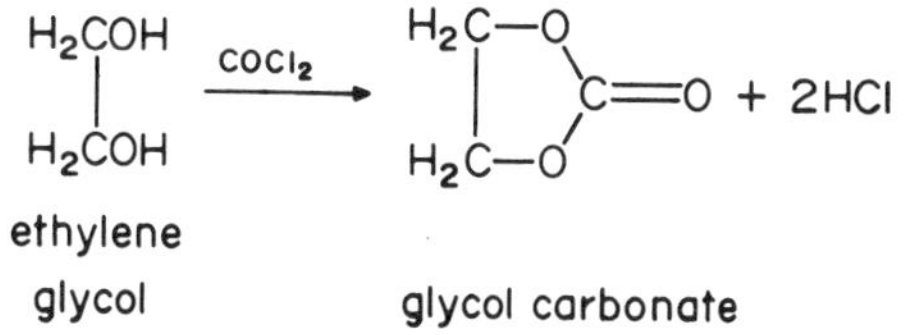

Figure 10.

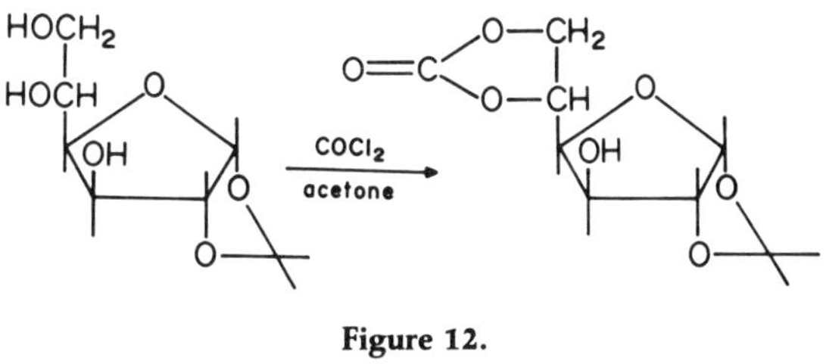

Figure 12.

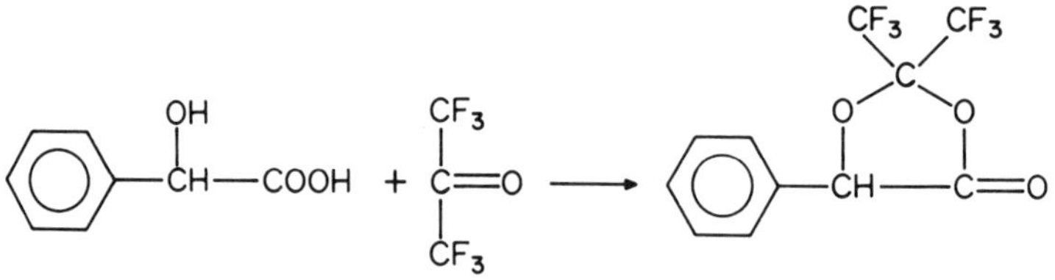

Figure 13. Cyclization of mandelic acid to a 1,3-dioxolanone.

7.4 Carbonate derivatives of sugars

The carbonate and thiocarbonate derivatives used in the carbohydrate field have been reviewed [112]. The cyclic carbonates are sensitive to alkali and stable to acid hydrolysis, and in these respects are complementary to the cyclic isopropylidene acetals. Cyclic carbonates may be made from sugars by the action of chloroformate esters. The original phosgene method is also applicable, either directly or in a solvent. The solvent usually used is toluene, or a reaction-promoting solvent such as pyridine or quinoline may be used.

The furanose ring seems preferred, so that for example D-mannopyranose yields the D-manno furanose-2,3-carbonate [113]. Cyclization between *trans*-hydroxyl groups is not usual.

6-O-Trityl-D-mannose (10 mg), pyridine (70 μl) and a solution of phosgene (2.4 mg) in toluene (20 μl) are vigorously shaken together for 30 min in an ice bath at 0 °C. The vial is taken out of the ice bath and, after 30 min at room temperature, the suspension in the vial is centrifuged and the supernatant is evaporated to a syrup. This is dissolved in chloroform and purified on a small silicic acid column, with chloroform elution. The clean-up may be monitored by gas chromatography, the appropriate fractions being combined and evaporated to a small volume for analysis [113].

7.5 Cyclization of amino alcohols to oxazolidin-2-ones

Phosgene can also used to form the cyclic oxazolidinones from amino alcohols and hydroxy acids.

The sample (0.5–1 mg) in methylene chloride (200 μl) is treated with 20% phosgene in toluene (50 μl) for 1 h at room temperature. The solution is evaporated to dryness and the residue is dissolved in methylene chloride (200 μl) for analysis [114]. Similar procedures have been described using reactions mixtures buffered at alkaline pH values, and varying the conditions to optimize the reactions [115, 116].

7.6 Cyclization of hydroxy acids to dioxolanones

This reaction is carried out as above, but using a solution of the hydroxy acid (0.5–1 mg) in dioxan (200 μl) instead of methylene chloride [114]. Reaction with bis-(chlorodifluoro)acetone also converts α-hydroxy to halogen-containing dioxolanones, which can be sensitively analysed by GC with ECD.

As an example, mandelic acid (50 μg) in acetonitrile (30 μl) and pyridine (5 μl) is treated in a reaction vial with hexafluoroacetone (10 μl) for 15 min at room temperature. The solution is evaporated to dryness in a stream of nitrogen at 80 °C [117] (Figure 13).

8. Cyclic derivatives of amino acids

Amino acids may be converted to various cyclic derivatives for chromatography, but with one exception (phenylthiohydantoin derivatives, described in the next section) these approaches were not successful and are only briefly outlined here for the sake of completeness. Apart from this, they concern only those interested in amino acid chemistry, who are directed to the original references. The reasons for making such derivatives were to maximize the chemical differences among the amino acids and to minimize increases in molecular weight on derivatization that might overshadow the chemical differences among the group; the impetus for these derivatization schemes thus came from the quest for the best way to analyse amino acids. However, for the most part they are still analysed, without derivatization, by ion exchange chromatography. More recently, HPLC of various highly coloured or fluorescent amino acid derivatives has gained ground (see Chapters 8 and 9).

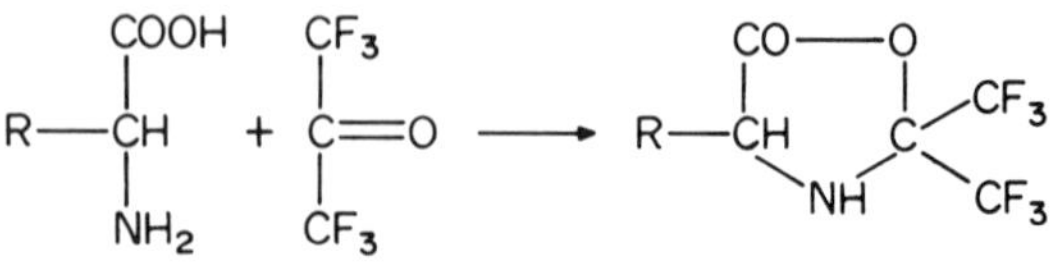

Figure 14. Cyclization of amino acids to oxazolidinones.

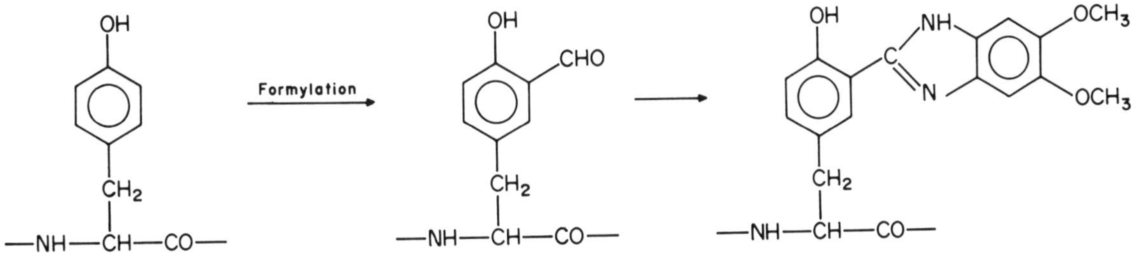

Figure 15. Labelling of tyrosine residues as dimethyloxybenzimidazoles.

Gas chromatography of amino acids has not really caught on, although for certain applications this option remains a viable alternative [118].

Just as with hydroxy acids, hexafluoroacetone or dichlorotetrafluoroacetone converts amino acids to substituted oxazolidin-5-ones [119–122]. The mass spectra of these derivatives have been described [123].

Trifluorooxazolidinones are also produced by reacting amino acids with trifluoroacetic anhydride [124–126]. Cyclization was also achieved with 1,3-dicyclohexylcarbodiimide (DCCI) [127]. The mass spectra of these compounds are also available [128].

Finally, an interesting way of specifically labelling tyrosine residues involves creating a dimethyoxybenzimidazole ring *ortho* to the phenolic hydroxyl [129] (Figure 15).

9. Sequencing of peptides and polypeptides

The only remaining significance of cyclic amino acid derivatives now lies in peptide sequencing and the chromatography of the individual amino acid residues cleaved off as cyclic amino acid derivatives, which are identified on the basis of their retention parameters. It follows that the same procedures that are applied to the identification of cyclic derivatives, following the cleavage of individual amino acids from peptides in sequence studies, can also be applied to the separation of mixtures of amino acids following conversion to the same cyclic derivatives.

Nowadays most peptide sequence work is done on

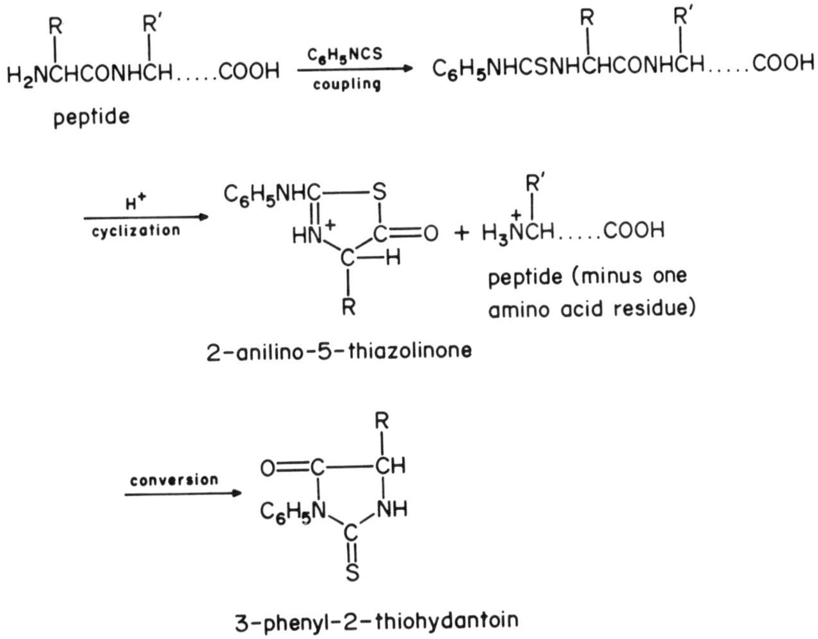

Figure 16. Sequential degradation of a protein from the N-terminus by the Edman method.

commercial machines called 'sequenators' or 'peptide sequencers', and the whole package comes complete with standards of the appropriate amino acid cyclic derivatives, so their preparation is not required. For details of the latest chemical procedures it is best to consult the manufacturers' literature. A brief account is all that is appropriate here; a general survey of protein and peptide sequencing is available in recent monographs [130, 131].

The standard and most widely used procedure is the Edman degradation [132–134], in which the N-terminal amino acid is treated with phenyl isothiocyanate and converted to a phenylthiourea. The residue is then split off under relatively mild conditions and cyclized, ultimately yielding the corresponding phenylthiohydantoin (PTH), which is the cyclic derivative identified by its chromatographic retention properties; the whole scheme is shown in Figure 16. The procedure has been worked out for operation linked to an ion exchange resin matrix, and is usually mechanized for consistency, speed and convenience. The chemistry has undergone continuous development: peptide sequencers now achieve high yields at each step with only minimal side reactions, so that long peptide sequences may be reliably determined. There seems, little point, therefore, in attempting peptide sequencing in any other way.

As already indicated, procedures based on conversion to PTH derivatives have also been published for the analysis of amino acids by HPLC [135–140].

Stepwise degradation from the C-terminal end of peptide chains has also been studied, based on the Schlack and Kumpf reaction [141, 142], leading to the cleavage of the last amino residue as the cyclic thiohydantoin (Figure 17).

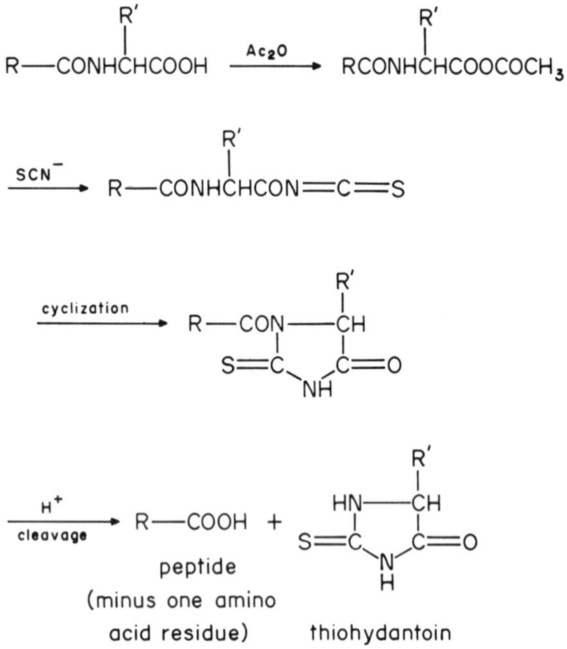

$$R-CONHCHCOOH \xrightarrow{Ac_2O} RCONHCHCOOCOCH_3$$ (R')

$$\xrightarrow{SCN^-} R-CONHCHCON=C=S$$ (R')

$$\xrightarrow{cyclization} R-CON-CH-C=O, S=C-NH$$ (R')

$$\xrightarrow{H^+ \text{ cleavage}} R-COOH + \text{thiohydantoin}$$

peptide (minus one amino acid residue)

Figure 17. Thiohydantoin method for the sequential degradation of a protein from the carboxyl terminus.

Again, solid-state procedures have been worked out [143–145] and the products have been identified by chromatographic methods [146–150], but this approach has not been mechanized, nor has it received anything like the degree of development of the N-terminal approach. The same is true of stepwise degradation from the C-terminal end of a peptide by the iminohydantoin [151] procedure (Figure 18).

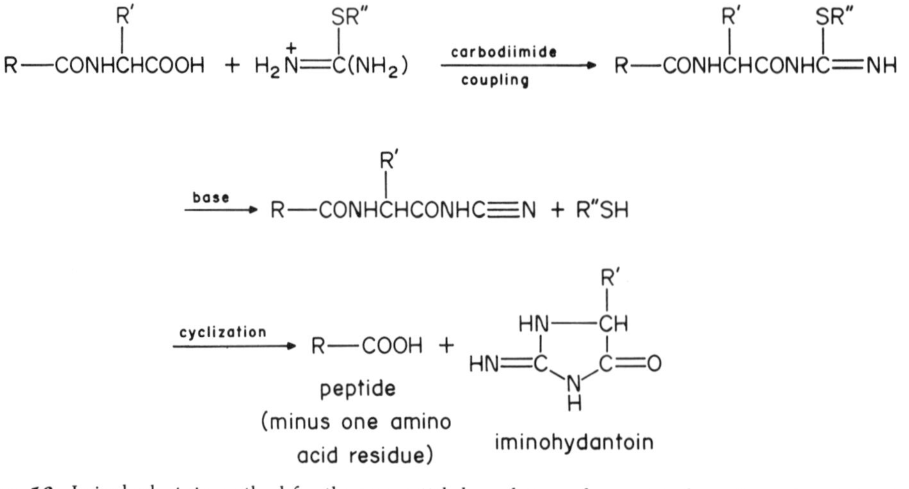

$$R-CONHCHCOOH + H_2\overset{+}{N}=C(NH_2) \xrightarrow[\text{coupling}]{\text{carbodiimide}} R-CONHCHCONHC=NH$$

$$\xrightarrow{\text{base}} R-CONHCHCONHC\equiv N + R''SH$$

$$\xrightarrow{\text{cyclization}} R-COOH + \text{iminohydantoin}$$

peptide (minus one amino acid residue)

Figure 18. Iminohydantoin method for the sequential degradation of a protein from the carboxyl terminus.

Degradation from the C-terminal of the polypeptide chain has been superseded by the success of the N-terminal approach, where the Edman degradation and its recent evolution holds the field as having stood the test of time. However, there may still be circumstances that make C-terminal sequence studies desirable, and it may then be useful to be able to build on what has already been done in this area.

10. Conclusions

With such a variety of procedures, it may be hard to decide which to select for a given application, assuming that one has a compound that looks as if it might yield a useful cyclic derivative. Butaneboronic acid recommends itself, because butaneboronates are formed quickly and easily under relatively mild conditions (Section 4). Similarly, derivatization to cyclic siliconides might also bear looking at (Section 3). In the case of α-keto acids, formation of the quinoxalinols, with subsequent silylation where necessary, is one of the best derivatization schemes (Section 5). For guanidines, the method of choice is reaction with hexafluoroacetylacetone (method 6.2). Finally, the chapter overall contains a wide (though not exhaustive) selection of other cyclization approaches, and where such an approach looks promising for a particular application, it is worth trying.

11. References

[1] H. Ziffer, W. J. A. VandenHeuvel, E. O. A. Haahti and E. C. Horning, *J. Am. Chem. Soc.*, **82**, 6411 (1960).
[2] C. F. Poole and A. Zlatkis, *J. Chromatogr.*, **184**, 99 (1980).
[3] A. N. De Belder, *Adv. Carbohyd. Chem.*, **20**, 219 (1965).
[4] J. Kuszmann, P. Sohar and G. Horvath, *Carbohydr. Res.* **50**, 45 (1976).
[5] C. T. Bishop, F. P. Cooper and R. K. Murray, *Can. J. Chem.*, **41**, 2245 (1963).
[6] G. Aksnes, P. Albriktsen and P. Juuvik, *Acta Chem. Scand.*, **19**, 920 (1965).
[7] E. Fischer and E. Pfähler, *Ber. Dtsch. Chem. Ges.*, **53**, 1606 (1920).
[8] S. A. Barker and E. J. Bourne, *Adv. Carbohydr. Chem.*, **7**, 138 (1952).
[9] S. J. Angyal and L. Anderson, *Adv. Carbohyd. Chem.*, **14**, 136 (1959).
[10] R. J. Ferrier, *Adv. Carbohydr. Chem.*, **20**, 67 (1965).
[11] J. W. Berry, in *Advances in Chromatography*, Vol. 2, J. C. Giddings and R. A. Keller (Editors), Marcel Dekker, New York (1966), p. 271.
[12] R. F. Brady, *Adv. Carbohydr. Chem.*, **26**, 197 (1971).
[13] J. D. Wander and D. Horton, *Adv. Carbohydr. Chem.*, **32**, 15 (1976).
[14] K. Biemann, H. K. Schnoes and J. A. McCloskey, *Chem. Ind. (London)*, 448 (1963).
[15] S. Morgenlie, *Carbohydr. Res.*, **41**, 285 (1975).
[16] D. C. DeJongh and K. Biemann, *J. Am. Chem. Soc.*, **86**, 67 (1964).
[17] D. J. Hanahan, J. Ekholm and C. M. Jackson, *Biochemsitry*, **2**, 630 (1963).
[18] J. A. McCloskey and M. J. McClelland, *J. Am. Chem. Soc.*, **87**, 5090 (1965).
[19] V. Dommes, F. Wirtz-Peitz and W.-H. Kuhau, *J. Chromatogr. Sci.*, **14**, 360 (1976).
[20] J. L. O'Donnell, L. E. Gast, J. C. Cowan, W. J. De Jarlais and G. E. McManis *J. Am. Oil Chem. Soc.*, **44**, 652 (1974).
[21] T. G. Bonner, D. Lewis and L. Yüceer, *Carbohydr. Res.*, **49**, 119 (1976).
[22] G. J. Robertson, *J. Chem. Soc.*, 330 (1934).
[23] R. Khan, *Carbohydr. Res.*, **32**, 375 (1974).
[24] V. M. Micovic and S. Stojilkovic, *Tetrahedron*, **14**, 186 (1958).
[25] K. Heyns and J. Lenz, *Chem. Ber.*, **94**, 348 (1961).
[26] H. Paulsen, H. Salzburg and H. Redlich, *Chem. Ber.*, **109**, 3598 (1976).
[27] M. A. Kirschner and H. M. Fales, *Anal. Chem.*, **34**, 1548 (1962).
[28] M. N. Huffman and M. H. Lott, *J. Biol. Chem.*, **215**, 627 (1955).
[29] J. C. Touchstone, M. Breckwoldt and T. Murawec, *J. Chromatogr.*, **59**, 121 (1971).
[30] T. Komeno, K. Tori and K. Takeda, *Tetrahedron*, **21**, 1635 (1965).
[31] A. G. McInnes, N. H. Tattrie and M. Kates, *J. Am. Oil Chem. Soc.*, **37**, 7 (1960).
[32] S. G. Batrakov, A. N. Ushakov and V. L. Sadovskaya, *Bioorg. Khim.*, **2**, 1095 (1976).
[33] O. D. Hensens, P. W. Khong and K. G. Lewis, *Aust. J. Chem.*, **29**, 1549 (1976).
[34] H. B. S. Conacher, *J. Chromatogr. Sci.*, **14**, 405 (1976).
[35] R. W. Kelly, *Tetrahedron Lett.*, 967 (1969).
[36] J. K. Findlay, L. Sieckmann and H. Breuer, *Biochem. J.*, **137**, 263 (1974).
[37] R. W. Kelly, *J. Chromatogr.*, **43**, 229 (1969).
[38] R. C. Mehrotra and R. P. Narain, *Indian J. Chem.*, **5**, 444 (1967).
[39] R. W. Kelly, *Steroids*, **13**, 507 (1969).
[40] M. Wieber and M. Schmidt, *Chem. Ber.*, **96**, 1561 (1963).
[41] C. J. W. Brooks, W. J. Cole and G. M. Barrett, *J. Chromatogr.*, **315**, 119 (1984).
[42] J. Böseken, *Adv. Carbohydr. Chem.*, **4**, 189 (1949).
[43] A. B. Foster, *Adv. Carbohydr. Chem.*, **12**, 81 (1957).
[44] R. A. Bowise and O. C. Musgrave, *J. Chem. Soc.*, 3945 (1963).
[45] R. J. Ferrier, *J. Chem. Soc.*, 2325 (1961).
[46] R. J. Ferrier, D. Prasad, A. Rodowski and I. Sangster, *J. Chem. Soc.*, 3330 (1964).
[47] R. J. Ferrier and D. Prasad, *J. Chem. Soc.*, 7429 (1965).
[48] A. M. Yurkevich, I. I. Kolodkina, L. S. Varshavskaya, V. I. Bolodulina-Shvetz, I. P. Rudakova and N. A. Preobrazhenski, *Tetrahedron*, **25**, 477 (1969).

[49] C. J. W. Brooks and D. J. Harvey, *Biochem. J.*, **114**, 15P (1969).
[50] C. J. W. Brooks and D. J. Harvey, *J. Chromatogr.*, **54**, 193 (1971).
[51] S. J. Gaskell, C. G. Edmonds and C. J. W. Brooks, *Anal. Lett.*, **9**, 325 (1976).
[52] P. J. Wood and I. R. Siddiqui, *Carbohydr. Res.*, **19**, 283 (1971).
[53] P. J. Wood, I. R. Siddiqui and J. Weisz, *Carbohydr. Res.*, **42**, 1 (1975).
[54] C. Pace-Asciak and L. S. Wolfe, *J. Chromatogr.*, **56**, 129 (1971).
[55] R. W. Kelly, *Anal. Chem.*, **42**, 2097 (1973).
[56] R. W. Kelly and P. L. Taylor, *Anal. Chem.*, **48**, 465 (1976).
[57] C. Pace-Asciak, *J. Am. Chem. Soc.*, **98**, 2348 (1976).
[58] C. J. W. Brooks and J. Watson, *J. Chem. Soc., Chem. Commun.*, 952 (1967).
[59] C. J. W. Brooks and J. Watson, in the *Gas Chromatography 1968*, C. L. A. Harborne and R. Stock (Editors), Institute of Petroleum, London (1969), p. 129.
[60] M. L. Wolfrom and J. Solms, *J. Org. Chem.*, **21**, 815 (1956).
[61] R. J. Ferrier, D. Prasad and A. Rudowski, *J. Chem. Soc.*, 858 (1965).
[62] J. M. Sugihara and C. M. Bowman, *J. Am. Chem. Soc.*, **80**, 2443 (1958).
[63] J. J. Dolhun and J. L. Wiebers, *J. Am. Chem. Soc.*, **91**, 7755 (1969).
[64] R. Hemming and D. G. Johnson, *J. Chem. Soc.*, 466 (1964).
[65] M. Pailer and W. Fenzl, *Monatsh. Chem.*, **92**, 1294 (161).
[66] C. F. Poole, S. Singhawangcha and A. Zlatkis, *Analyst*, **104**, 82 (1979).
[67] C. F. Poole, L. Johansson and J. Vessman, *J. Chromatogr.*, **194**, 365 (1980).
[68] S. Singhawangcha, C. F. Poole and A. Zlatkis, *J. Chromatogr.*, **183**, 433 (1980).
[69] F. Eisenberg, *Carbohydr. Res.*, **19**, 135 (1971).
[70] M. P. Rabinowitz, J. I. Bodin and P. Reisberg, *J. Pharm. Sci.*, **63**, 1601 (1974).
[71] D. L. Sondack, *J. Pharm. Sci.*, **64**, 128 (1975).
[72] T. Kimura, L. A. Sternson and T. Higuchi, *Clin. Chem.*, **22**, 1639 (1976).
[73] P. Biondi and M. Cagnasso, *Anal. Lett.*, **9**, 507 (1976).
[74] M. Cagnasso and P. A. Biondi, *Anal. Biochem.*, **71**, 597 (1976).
[75] C. J. W. Brooks and I. Maclean, *J. Chromatogr. Sci.*, **9**, 18 (1971).
[76] G. M. Anthony, C. J. W. Brooks and B. S. Middleditch, *J. Chromatogr. Sci.*, **7**, 623 (1969).
[77] G. M. Anthony, C. J. W. Brooks, I. Maclean and I. Sangster, *J. Pharm. Pharmacol.*, **22**, 205 (1970).
[78] C. J. W. Brooks, G. M. Anthony and B. S. Middleditch, *Org. Mass Spectrom.*, **2**, 1023 (1969).
[79] S. J. Gaskell, C. G. Edmonds and C. J. W. Brooks, *J. Chromatogr.*, **126**, 5912 (1976).
[80] H. G. Kuivila, A. H. Keough and E. J. Soboczenski, *J. Org. Chem.*, **19**, 780 (1954).

[81] C. F. Poole, L. Johansson and J. Vessman, *J. Chromatogr.*, **194**, 365 (1980).
[82] N. E. Hoffman and T. A. Killinger, *Clin. Chem.*, **14**, 807 (1968); *Anal. Chem.*, **41**, 162 (1969).
[83] N. E. Hoffman, K. M. Gooding, K. M. Sheadan and C. A. Tylenda, *Res. Commun. Chem. Pathol. Pharmacol.*, **2**, 87 (1971).
[84] A. Frigerio, P. Martelli, K. M. Baker and P. A. Biondi, *J. Chromatogr.*, **81**, 139 (1973).
[85] D. C. Morrison, *J. Am. Chem. Soc.*, **76**, 4483 (1954).
[86] U. Langenbeck, H.-U. Möhring and K.-P. Dieckmann, *J. Chromatogr.*, **115**, 65 (1975).
[87] R. A. Chalmers and R. W. E. Watts, *Analyst*, **97**, 951 (1972).
[88] K. H. Nielsen, *J. Chromatogr.*, **10**, 463 (1963).
[89] J. Mowbray and J. H. Ottaway, *Biochem. J.*, **120**, 171 (1970).
[90] U. Langenbeck, A. Mench-Hoinowski, K.-P. Dieckmann, H.-U. Möhring and M. Petersen, *J. Chromatogr.*, **145**, 185 (1978).
[91] P. A. David and M. Novotny, *J. Chromatogr.*, **452**, 623 (1988).
[92] G. Livesey and W. T. E. Edwards, *J. Chromatogr.*, **337**, 98 (1985).
[93] D. J. Brown, *The Pyrimidines*, Interscience, New York (1962), p. 32.
[94] T. P. King, *Biochemistry*, **5**, 3454 (1966).
[95] M. Guevret, C. Julien-Larose, J. R. Kiechel and D. Lavène, *J. Chromatogr.*, **233**, 181 (1982).
[96] S. L. Malcolm and T. R. Marten, *Anal. Chem.*, **48**, 807 (1976).
[97] P. Erdmansky and T. J. Coehl, *Anal. Chem.*, **47**, 750 (1975).
[98] D. G. Kaiser, G. J. Vangressen, J. A. Shah and D. J. Weber, *J. Chromatogr.*, **435**, 135 (1988).
[99] I. Yokoi, Y. Watanabe, A. Edaki and A. Mori, *Life Sci.*, **41**, 1305 (1987).
[100] N. R. Williams, *Adv. Carbohydr. Chem.*, **25**, 109 (1970).
[101] J. Defaye, *Adv. Carbohydr. Chem.*, **25**, 181 (1970).
[102] S. Soltzberg, *Adv. Carbohydr. Chem.*, **25**, 229 (1970).
[103] R. E. Parker and N. S. Isaacs, *Chem. Rev.*, **59**, 737 (1959).
[104] D. Swern, *Org. React.*, **7**, 378 (1953).
[105] D. Swern and G. B. Dickel, *J. Am. Chem. Soc.*, **76**, 1957 (1954).
[106] M. Nakajima, I. Tomida, N. Kurihara and S. Takei, *Chem. Ber.*, **92**, 173 (1959).
[107] D. A. Lightner and C. Djerassi, *Tetrahedron*, **21**, 583 (1965).
[108] E. A. Emken, *Lipids*, **6**, 686 (1971).
[109] E. A. Emken, *Lipids*, **7**, 459 (1972).
[110] P. Brown and C. Djerassi, *Tetrahedron*, **24**, 2949 (1968).
[111] P. C. Lauterburg, J. G. Pritchard and R. L. Vollmer, *J. Chem. Soc.*, 5307 (1963).
[112] L. Hough, J. E. Priddle and R. S. Theobald, *Adv. Carbohydr. Chem.*, **15**, 91 (1960).
[113] A. S. Perlin, *Can. J. Chem.*, **42**, 1365 (1964).
[114] W. A. König, E. Steinbach and K. Ernst, *J. Chromatogr.*, **301**, 129 (1984).
[115] O. Gyllenhaal, W. A. König and J. Vessman, *J. Chromatogr.*, **350**, 328 (1985).

[116] O. Gyllenhaal and J. Vessman, *J. Chromatogr.*, **395**, 445 (1987).
[117] J. M. Midgley, R. Andrew, D. G. Watson, N. MacDonald, J. L. Ried and D. A. Williams, *J. Chromatogr.*, **527**, 259 (1990).
[118] K. Blau, in *Amino Acid Analysis*, J. M. Rattenbury (Editor), Ellis Horwood, Chichester (1981), pp. 48–65.
[119] F. Weygand, K. Burger and K. Engelhardt, *Chem. Ber.*, **99**, 1461 (1966).
[120] P. Hušek, *J. Chromatogr.*, **91**, 475 (1974).
[121] P. Hušek, *J. Chromatogr.*, **91**, 483 (1974).
[122] P. Hušek and V. Felt, *J. Chromatogr.*, **305**, 442 (1984).
[123] R. Liardon, U. Ott-Kuhn and P. Hušek, *Biomed. Mass Spectrom.*, **6**, 381 (1979).
[124] F. Weygand, *Freseniis Z. Anal. Chem.*, **205**, 406 (1964).
[125] F. Weygand and U. Glöckler, *Chem. Ber.*, **89**, 653 (1956).
[126] O. Grahl-Nielsen and B. Móvik, *Biochem. Med.*, **12**, 143 (1975).
[127] O. Grahl-Nielsen and E. Solheim, *Anal. Chem.*, **47**, 333 (1975).
[128] V. Ferrito, R. Borg, J. Eagles and G. R. Fenwick, *Biomed. Mass Spectrom.*, **6**, 499 (1979).
[129] J. Ishida, M. Kai and Y. Ohkura, *J. Chromatogr.*, **356**, 171 (1986).
[130] J. B. C. Findlay and M. J. Geisow, *Protein Sequencing: A Practical Approach*, IRL Press, Oxford (1989).
[131] G. Allen. *Sequencing of Proteins and Peptides: Laboratory Techniques in Biochemistry and Molecular Biology*, 2nd edition, Elsevier, Amsterdam (1989).
[132] P. Edman, *Acta Chem. Scand.*, **4**, 283 (1950).
[133] P. Edman, *Acta Chem. Scand.*, **10**, 761 (1956).
[134] P. Edman and G. Begg, *Eur. J. Biochem.*, **1**, 80 (1967).
[135] N. D. Johnson, M. W. Hunkapiller and L. E. Hood, *Anal. Biochem.*, **100**, 335 (1979).
[136] J. Simmons and D. H. Schlesinger, *Anal. Biochem.*, **104**, 254 (1980).
[137] J. Fohlman, L. Rask and P. A. Peterson, *Anal. Biochem.*, **106**, 22 (1980).
[138] S. M. Rose and B. D. Schwartz, *Anal. Biochem.*, **107**, 206 (1980).
[139] W. G. Kruggel and R. V. Lewis, *J. Chromatogr.*, **342**, 376 (1985).
[140] R. F. Ebert, *Anal. Biochem.*, **154**, 431 (1986).
[141] P. Schlack and W. Kumpf, *Hoppe Seyler's Z. Physiol. Chem.*, **154**, 125 (1926).
[142] S. G. Waley and J. Watson, *J. Chem. Soc.*, 2394 (1951).
[143] A. Darbre and M. Rangarajan in *Solid-Phase Methods in Protein Sequence Analysis, Proc. Int. Conf., 1st*, R. A. Laursen (Editor), p. 131, Pierce Chemical Co., Rockford (1975), p. 131.
[144] M. Rangarajan and A. Darbre, *Biochem. J.*, **157**, 307 (1976).
[145] M. Williams and B. Kassell, *FEBS Lett.*, **54**, 353 (1975).
[146] S. Yamashita, *Biochim. Biophys. Acta*, **229**, 301 (1971).
[147] S. Yamashita, and N. Ishikawa, in *Chemistry and Biology of Peptides, Proc. Am. Pept. Symp., 3rd*, J. Meienhofer (Editor), Ann Arbor Science, Ann Arbor (1972), p. 701.
[148] M. Rangarajan and A. Darbre, *Biohem. J.*, **147**, 435 (1975).
[149] M. Rangarajan, R. E. Ardrey and A. Darbre, *J. Chromatogr.*, **87**, 499 (1973).
[150] T. Suzuki, S. Matsui and K. Tuzimura, *Agric. Biol. Chem.*, **36**, 1061 (1972).
[151] G. E. Tarr, in *Solid-Phase Methods in Protein Sequence Analysis, Proc. Int. Conf., 1st*, R. A. Laursen (Editor), Pierce Chemical Co., Rockford (1975), p. 139.

8

Coloured and UV-absorbing Derivatives

Famei Li
Department of Clinical Biochemistry, King's College School of Medicine and Dentistry, London

and

C. K. Lim
MRC Toxicology Unit, Medical Research Council Laboratories, Woodmansterne Road, Carshalton, Surrey

1. REAGENTS FOR AMINES AND AMINO ACIDS   158
  1.1 Acyl chlorides   158
    1.1.1 *p*-Methoxybenzoyl chloride   159
    1.1.2 *m*-Toluoyl chloride   159
    1.1.3 *p*-Nitrobenzoyl chloride   159
    1.1.4 Benzoyl chloride   159
  1.2 Arylsulphonyl chlorides   160
    1.2.1 Toluenesulphonyl chloride (TSCl)   160
    1.2.2 Benzenesulphonyl chloride (BSCl)   160
    1.2.3 Dimethylaminoazobenzenesulphonyl chloride (DABSCl)   160
  1.3 Nitrobenzenes   161
    1.3.1 1-Fluoro-2,4-dinitrobenzene (FDNB)   161
    1.3.2 Trinitrobenzenesulphonic acid (TNBS)   162
    1.3.3 4-Fluoro-3-nitrobenzotrifluoride (FNBT)   162
  1.4 Isocyanates and isothiocyanates   162
    1.4.1 Phenyl isocyanate (PIC) and naphthyl isocyanate (NIC)   162
    1.4.2 Phenyl isothiocyanate (PITC) and naphthyl isothiocyanate (NITC)   162
    1.4.3 4-*N,N'*-Dimethylaminoazobenzene-4'-isothiocyanate and *p*-phenylbenzoyl isothiocyanate   163
  1.5 Miscellaneous   163
    1.5.1 *p*-Nitrobenzyl bromide (*p*-NBBr)   163
    1.5.2 Dansyl chloride (Dns-Cl)   164
    1.5.3 *o*-Phthaldialdehyde (OPA)   164
    1.5.4 Ninhydrin   164

2. REAGENTS FOR CARBOXYLIC ACIDS   164
  2.1 Phenacyl bromide, naphthacyl bromide and their analogues   164
  2.2 *N*-Methylphthalimide derivatives   166
  2.3 *O-p*-Nitrobenzyl-*N,N'*-diisopropylisourea (*p*-NBDI)   167
  2.4 Derivatization with aromatic amines after conversion into acyl chlorides   167
  2.5 Miscellaneous   167
    2.5.1 2-Nitrophenylhydrazine   167
    2.5.2 *o*-Phenylenediamine   167
    2.5.3 Pyridinium dichromate   167
    2.5.4 2-Methylquinoxanol derivative   167
    2.5.5 Imidazole   168

3. REAGENTS FOR HYDROXY COMPOUNDS   168
  3.1 Acyl chlorides   168
    3.1.1 Benzoyl chloride   168
    3.1.2 *p*-Methoxybenzoyl chloride   169
    3.1.3 *p*-Nitrobenzoyl chloride   169
    3.1.4 3,5-Dinitrobenzoyl chloride   169
    3.1.5 Anthracene-9-carbonyl chloride   169
  3.2 Phenyl isocyanate (PIC)   169
  3.3 Silyl chlorides   170

4. REAGENTS FOR CARBONYL COMPOUNDS   170
  4.1 2,4-Dinitrophenylhydrazine (2,4-DNPH)   170
  4.2 *p*-Nitrobenzylhydroxylamine   171
  4.3 1-Phenyl-3-methyl-5-pyrazolone (PMP)   171

5. CONCLUSIONS   172

6. REFERENCES   172

Coloured and UV-absorbing derivatives are prepared for chromatography to improve the detectivity of compounds which do not possess a chromophore or fluorophore.

Such derivatives have been used in thin-layer chromatography, paper chromatography and high-performance liquid chromatography (HPLC). In the present chapter, the emphasis is on those applied to HPLC. The subject has been extensively reviewed [1–8].

There is a wide choice of reagents for the introduction of an increased absorptivity into a target compound, and the reaction can be performed either pre- or post-column. Pre-column derivatization to some extent governs the choice of the HPLC system, and is therefore not restricted by the requirement of compatibility with the mobile phase. The reagent itself can also be coloured or UV-absorbing, provided that it can be removed or separated by HPLC.

In post-column derivatization (see also Chapter 15), the HPLC mobile phase must be miscible and compatible with the reaction mixture in the reactor so that it dose not cause precipitation or interfere with the derivatization reaction. The reagent itself must have significant differences in spectral properties from the products. Post-column derivatization reactions are usually rapid, and fluorescent rather than UV-absorbing labels are preferred.

The labels used in UV–visible (UV–VIS) derivatization are usually highly conjugated aromatic compounds which react with the analytes to give derivatives with high absorptivity, thus allowing highly sensitive detection.

1. Reagents for amines and amino acids

Amino groups exist in many biologically important compounds, including such endogenous compounds as amino acids and polyamines. There are also many pharmaceutical amines, for example amikacin, gentamicin, amphetamine, methylphenidate and phenmetrazine. Direct HPLC analysis with UV–VIS detection of these amines is limited by the weak or non-existent UV–VIS absorption of these compounds. Only by converting amines to strongly absorbing derivatives can they be determined by HPLC with high sensitivity when present in biological and clinical samples at low concentrations.

Primary and secondary amino groups are usually derivatized as aromatic derivatives by nucleophilic substitution reactions. The common labelling reagents are acyl chlorides, arylsulphonyl chlorides, isothiocyanates and nitrobenzenes. Tertiary amines are rather difficult to detect and their determination involves quaternization.

1.1 Acyl chlorides

1.1.1 p-Methoxybenzoyl chloride

Acyl chlorides react readily with amino groups to form amides. The reactions (Figure 1) are carried out in an alkaline medium (Schotten–Baumann conditions). The use of amide derivatives for the HPLC analysis of amines has been widely studied. Clark and Wells [9] described several acyl chlorides such as p-nitrobenzoyl, p-methoxybenzoyl, p-methylbenzoyl and p-chlorobenzoyl chlorides. The spectrophotometric properties of their benzamides were investigated. The methoxybenzamides showed the most intense absorption at 252 nm, and therefore most fixed-wavelength (254 nm) UV detectors can conveniently be used for detection. The reaction of p-methoxybenzoyl chloride with some amines and the molar absorptivities of the derivatives are shown in Table 1.

Both the amines and the acyl chlorides were dissolved in tetrahydrofuran and then reacted in potassium carbo-

Table 1 The reaction of p-methoxybenzoyl chloride with amines and the spectrometric properties of the amides

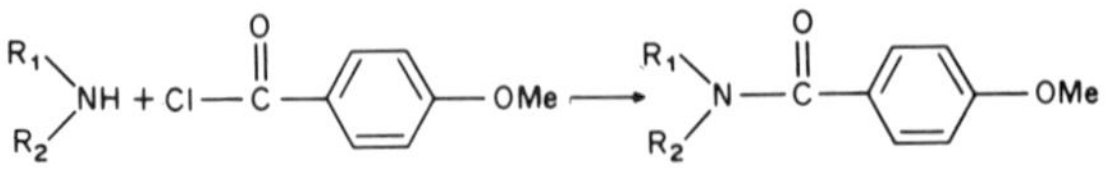

R_1	R_2	$\varepsilon_{252\,nm}$
$C_6H_5CHCH_3$	H	17500
n-C_3H_7	H	15100
$C_6H_5CH_2$	H	18000
$C_6H_5CH_2$	CH_3	11200 (242 nm)
$C_6H_5CH_2CH_2$	H	18300

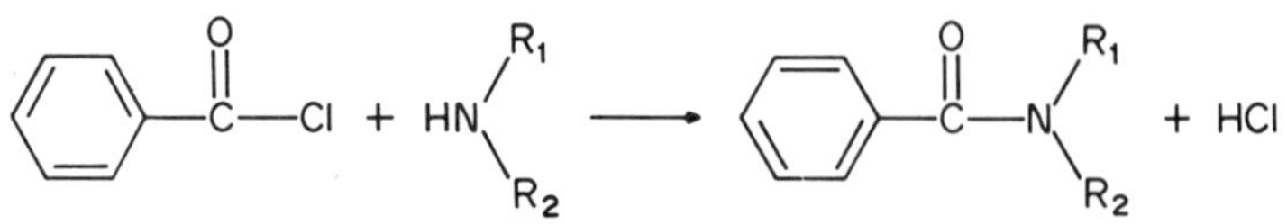

Figure 1. The formation of amide derivatives by reaction of primary or secondary amines with benzoyl chloride.

nate at a pH about 8. For primary amines the reaction was carried out at 50 °C for 3 h, and for secondary amines the reaction mixture was refluxed for 12 h. The amides formed were extracted with chloroform and subjected to reversed phase HPLC (RP-HPLC) analysis. The detection limit for amphetamine was 3 ng and the recovery was 90.8 ± 4.4%.

1.1.2 m-Toluoyl chloride

Polyfunctional amines can be analysed by HPLC as amides [10]. Several acyl chlorides have been investigated for measuring piperazine. These are benzoyl, m-nitrobenzoyl, m-toluoyl, p-t-butylbenzoyl, phenoxyacetyl and p-anisoyl chloride. Among them only m-toluoyl chloride gave a reaction that was free of by-products. It also provided derivatives that were soluble in water-immiscible solvents and were easily separated. The reaction was carried out in pyridine at room temperature and completed in 5 min. Neither water nor ammonia interfered to any extent. After clean-up, the amides were extracted with dichloromethane and subjected to RP-HPLC (C_{18} column with acetonitrile/water 40:60 v/v, as eluent and UV detector set at 254 nm). The derivatives were found to be stable for at least 6 months. Eleven polyfunctional amines, including 2,5- or 2,6-dimethylmorpholine, 1,2-diaminoethane, monoethanolamine and piperazine were studied. Because the polyfunctional amines are labelled at each active hydrogen (Figure 2), the absorbances of their derivatives are high and so are the sensitivities of detection. The detection limit was about 5 ng and the recovery exceeded 94%.

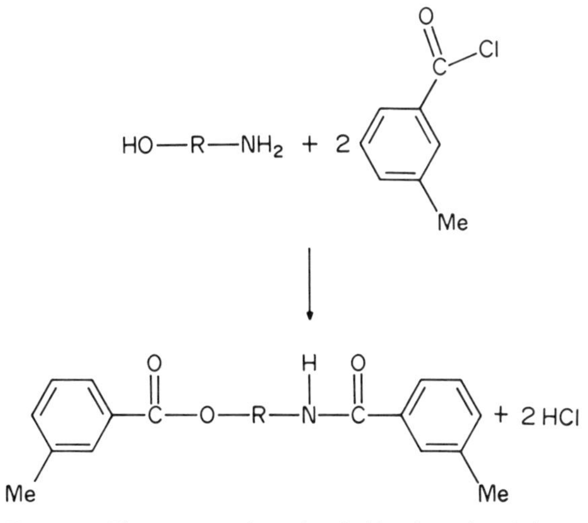

1.1.3 p-Nitrobenzoyl chloride

Metformin is a biguanide compound with high polarity and poor chromatographic properties. Reaction with p-nitrobenzoyl chloride resulted in the triazine derivative (Figure 3), which has lower polarity and is suitable for HPLC separation and detection at 280 nm [11]. Metformin in acetonitrile and 20% sodium hydroxide was treated with p-nitrobenzoyl chloride. The reaction was continued for 6 h at room temperature, with a yield of 94%. The derivative, in acetonitrile, was directly injected for RP-HPLC. Metformin in urine samples can be determined by using a similar procedure.

p-Nitrobenzoyl chloride has also been used as a derivatizing agent for ketamine (a parenteral anaesthetic) and two of its derivatives [12]. The formation of p-nitrobenzamides greatly enhances detectability at 254 nm after separation on a C_{18} column.

1.1.4 Benzoyl chloride

Benzoyl chloride is the preferred derivatizing agent for polyamines, as it reacts with most of the naturally occurring diamines and polyamines. Redmond and Tseng [13] described a procedure involving the addition of benzoyl chloride to the sample suspended in an alkaline solvent. Furniss et al. [14] recommended that following the reaction the mixture should be washed with alcohol to remove excess benzoyl chloride. Benzoylated poly-

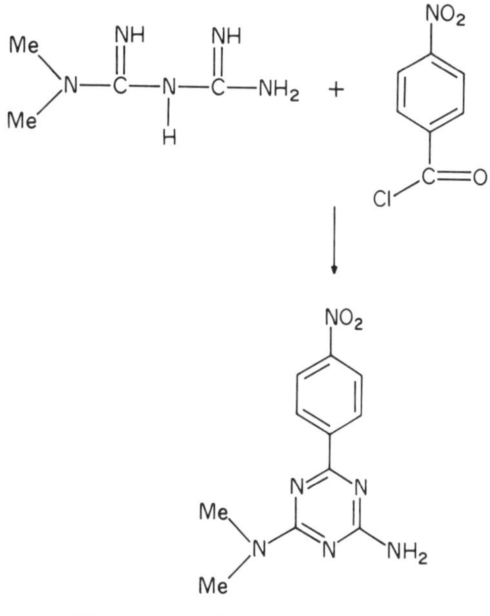

Figure 2. The reaction of m-toluoyl chloride with polyfunctional amines.

Figure 3. The reaction of metformin with p-nitrobenzoyl chloride.

amines could be stored for up to three weeks at $-20\,°C$. The derivatives of polyamines (putrescine, cadaverine, spermidine and spermine) were separated by RP-HPLC using methanol/water (60:40 v/v) as the mobile phase. The detection limit at 254 nm was 100 pmol of polyamine [15]. This method has been applied to determine various polyamines in vertebrate and invertebrate tissues [15], rat brain tissue [16], cell cultures [17] and tumour tissues [18]. The assay procedures have been optimized and evaluated [18], and an example is given below.

A pellet of cultured cells was obtained by centrifugation and thoroughly mixed with 1 ml of 0.3 M perchloric acid (PCA) and the internal standard (1,6-hexanediamine). Tumour tissues were homogenized in 5–10 volumes of cold 0.3 M PCA in the presence of internal standard and centrifuged. In glass tubes, 2 ml of 2 M NaOH and 10 µl of benzoyl chloride were added to 100–400 µl of the sample supernatant. After incubation for 30 min at room temperature the derivatives were extracted with 2 ml of chloroform. The solvent was evaporated and the residue was dissolved in 100–500 µl of methanol.

Two C_{18} columns and a guard column were used for separation. The mobile phase was methanol/water (65:35 v/v). The benzoylated putrescine, 1,6-hexanediamine, spermidine and spermine were separated in 10 min. Detection was at the absorption maximum of 229 nm (instead of 254 nm) with a detection limit of 1 pmol. The linear range was up to 100 nmol of polyamine and the recoveries for putrescine, spermidine and spermine were $104.7 \pm 3.4\%$, $97.3 \pm 3.7\%$ and $88.8 \pm 5.5\%$ respectively. Considerably lower spermidine and putrescine concentrations were found in tumour tissues obtained from animals treated with DL-α-difluoromethylornithine for seven days compared with untreated controls.

1.2 Arylsulphonyl chlorides

1.2.1 Toluenesulphonyl chloride (TSCl)

Similarly to acyl chlorides, arylsulphonyl chlorides react with primary and secondary amines. A number of arylsulphonyl chlorides have also been investigated by Clark and Wells [9]. The reaction and the conditions of derivatization used were similar to those for acyl chlorides. The spectrophotometric properties of some typical benzenesulphonamides are given in Table 2.

The determination of polyamines based on the formation of tosylated derivatives was reported by Sugiura et al. [19]. The tosylation resulted not only in an increase in the absorptivity but also in the improvement of HPLC separation. Toluenesulphonyl chloride was dissolved in acetone and 0.5 M $NaHCO_3$ was added. The reaction was carried out at 70 °C for 1 h. The reaction

Table 2 Structures and spectrophotometric properties of various benzenesulphonamides

R_1	R_2	λ_{max}	ε_{max}	$\varepsilon_{254\,nm}$
$C_6H_5CHCH_3$	NO_2	268.5	10800	8500
$n\text{-}C_3H_7$	NO_2	266.5	10100	8430
$C_6H_5CHCH_3$	OCH_3	241.0	15300	4660
$n\text{-}C_3H_7$	OCH_3	239.0	16300	3330
$C_6H_5CHCH_3$	Cl	233.5	12600	1630
$n\text{-}C_3H_7$	Cl	231.5	14700	1060

mixture was washed with n-hexane to remove TSCl, which interfered with the peaks of some of the products. After extraction with chloroform the derivatives were isocratically separated by RP-HPLC in 20 min with UV detection at 254 nm.

1.2.2 Benzenesulphonyl chloride (BSCl)

Gentamicin, a polyfunctional amino compound, was determined by using RP-HPLC after labelling with benzenesulphonyl chloride [20]. The derivatization reaction was complete in 10 min at 75 °C. The HPLC separation was carried out on a C_{18} column with acetonitrile/methylene dichloride/water/methanol as the mobile phase. Separation of gentamicin isomers was not achieved.

1.2.3 Dimethylaminoazobenzenesulphonyl chloride (DABSCl)

Dimethylaminoazobenzenesulphonyl chloride is one of the derivatization agents for amino acids [21]. The reaction is shown in Figure 4. The resulting sulphonamides show an absorbance maximum at 420–450 nm, which allows detection at long wavelength and excludes interference from endogenous substances in clinical samples to a large extent. In the procedure developed by Lammens and Verzele [21], the derivatization was performed at 50 °C and at pH 8.9 for a short time, and gave a quantitative yield. Reversed-phase HPLC with isocratic elution and detection at 464 nm was employed; 17 amino acids were separated, but no separation between histidine and arginine was achieved. The method has been used for amino acid assay in urine [22] and in brain tissue [23, 24]. For determination of amino acids in tissues, the tissue was suspended in dilute perchloric acid, and the supernatant after centrifugation was used for derivatiz-

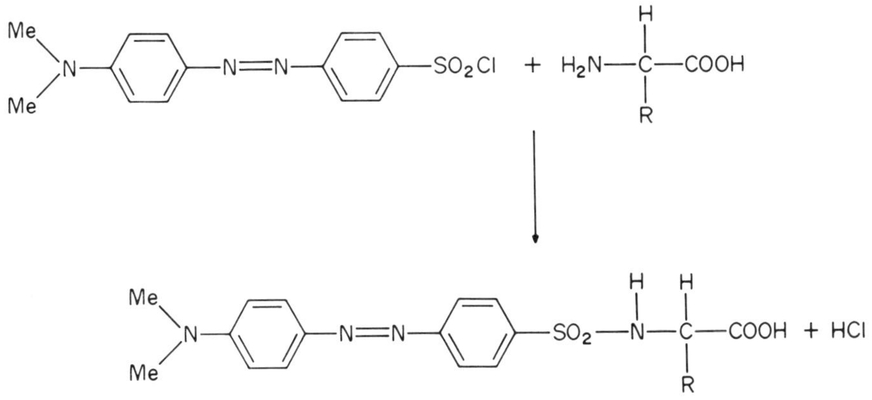

Figure 4. The reaction of dimethylaminoazobenzenesulphonyl chloride with amines.

ation at 70 °C for 10 min. Some bis-DABS-amino acids (lysine, histidine and tyrosine) were formed [24]. The derivatives were separated on an ODS column by gradient elution (acetate buffer/acetonitrile).

For urinary amino acids [22], 0.5 ml of urine was mixed with 0.5 ml of 0.5 M sodium bicarbonate and 2 ml of DABSCl (2 g l^{-1} in acetone). The reaction mixture was maintained at pH 8.5–9.0 and 25–26 °C for 30 min. The filtrates were directly applied to a C_{18} column for HPLC separation. Interfering amino compounds present in the urine also form derivatives with DABSCl, but may be removed by extraction with chloroform. Ammonia reacts readily with DABSCl, resulting in a derivative coeluted with DABS-methionine. Furthermore, ammonia may consume so much DABSCl that it seriously interferes with the derivatization of amino acids. Therefore the ammonia should be removed by lyophilizing the samples at pH 8.9 before derivatization. This method allowed the detection of urinary amino acid down to concentrations of about 16 mg l^{-1}.

1.3 Nitrobenzenes

1.3.1 1-Fluoro-2,4-dinitrobenzene (FDNB)

Nitrobenzenes are most commonly used for the derivatization of amino compounds. The 4-nitrobenzoyl group has been shown to impart high UV absorptivity to amines. 1-Fluoro-2,4-dinitrobenzene has been used as a label for aminoglycosides (Figure 5) such as neomycin [25, 26], fortimicin A [27], amikacin [28, 29], tobramycin [30–33], gentamicin and sissomicin [31–33].

The derivatization of fortimicin A [27] with FDNB was carried out in 1% potassium phosphate buffer, pH 9.0. The reaction was complete in 15 min at 85 °C. The derivative was solubilized by addition of acetonitrile and the pH was adjusted to 6.5. The derivative was

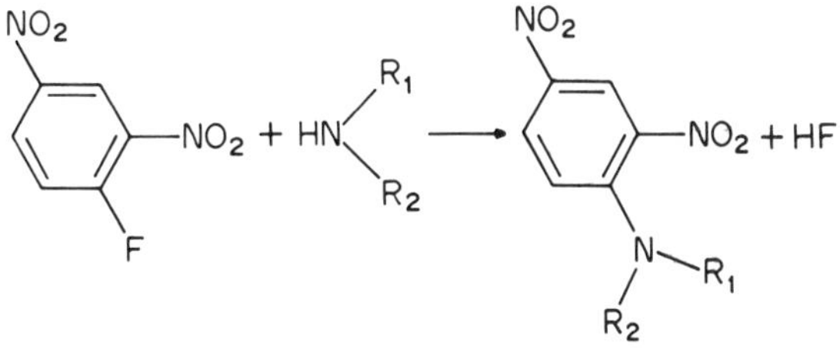

Figure 5. The reaction of 1-fluoro-2,4-dinitrobenzene with amines.

stable for more than 20 h at room temperature. The derivatized fortimicin A could be separated from other isomers in 15 min on an ODS column with 0.02 M K_2HPO_4, pH 9.0, and acetonitrile as eluent.

Tobramycin (Tb) was derivatized with FDNB at pH 8 and 70 °C for 10 min. The conversion into Tb(DNB)$_5$ was 90% complete with a minimum of side reactions. The derivative was separated on reversed-phase column (C_{18}) and detected at 254 nm [30]. The kinetics of derivatization with FDNB and the hydrolysis of the derivatives have been thoroughly investigated. To prevent the derivatization of hydroxyl groups and to ensure the formation of totally N-derivatized product, accurately standardized reaction conditions are essential.

A typical example of FDNB derivatization is the determination of amikacin in human serum [28, 29]. In a pre-column derivatization method, the aminoglycoside was extracted from serum by using a cation exchange column [29]. The eluate was treated with FDNB in dimethyl sulphoxide (DMSO) at pH 10. The derivative was then separated on an RP-C_{18} column with water/acetonitrile as mobile phase and monitored at 365 nm. The recovery of amikacin was only 72%, and kanamycin was used as an internal standard. The sensitivity was 1 mg l^{-1} and 200 μl of serum was needed.

The derivatization of amino functions with FDNB was also applied to amino acid analysis. The amino acid of interest, N^6,N^6,N^6-trimethyllysine, in urine [34] was isolated by ion exchange chromatography prior to derivatization. The reaction was carried out in alkaline solution at 37 °C for 1h. Reversed-phase ion-pair HPLC with heptanesulphonate as the ion-pairing agent was used for the separation, and detection was at 405 nm. The linear range was 10–150 nmol ml^{-1}.

1.3.2 Trinitrobenzenesulphonic acid (TNBS)

2,4,6-Trinitrobenzenesulphonic acid has similarly been used for labelling aminoglycosides in biological fluids, e.g. amikacin [35] and tobramycin [36] in serum. In order to obtain a single N-trinitrophenyl derivative (Figure 6) in quantitative yield, one requires a large excess of TNBS dissolved in acetonitrile, a temperature above 70 °C at pH 9.5–10.0 and a reaction time of not less than 30 min. The derivatives were isolated with a Bond-Elut C_{18} cartridge and separated on a C_{18} column at 50 °C. The procedure provided almost quantitative recovery [36]. An advantage of TNBS is its selectivity, because it reacts with hydroxyl groups only with difficulty [37]. TNBS has also been used for amino acid labelling in alkaline medium [38]. The analysis was performed on a reversed-phase or an ion exchange column; detection was at 254 nm, with a detection limit of 25 pmol.

1.3.3 4-Fluoro-3-nitrobenzotrifluoride (FNBT)

4-Fluoro-3-nitrobenzotrifluoride [39] was used for labelling diamines and polyamines such as putrescine, spermidine and spermine to give the nitrotrifluoromethylphenyl derivatives (Figure 7). The labelled amines were separated by RP-HPLC at a constant temperature of 40 °C, and detection was at 242 or 410 nm. FNBT does not react with secondary amines or molecules containing polar groups, such as amino acids [39].

1.4 Isocyanates and isothiocyanates

1.4.1 Phenyl isocyanate (PIC) and naphthyl isocyanate (NIC)

Phenyl isocyanate [40] reacts with aliphatic and aromatic primary and secondary amines to yield N,N'-disubstituted ureas (Figure 8). The reaction is quantitative and completed in a few minutes. A single, very stable derivative is obtained even with primary amines. Water and alcohols also react with PIC but the amino group appears to be the most reactive [40]. The derivatization was carried out in N,N-dimethylformamide (DMF), and an

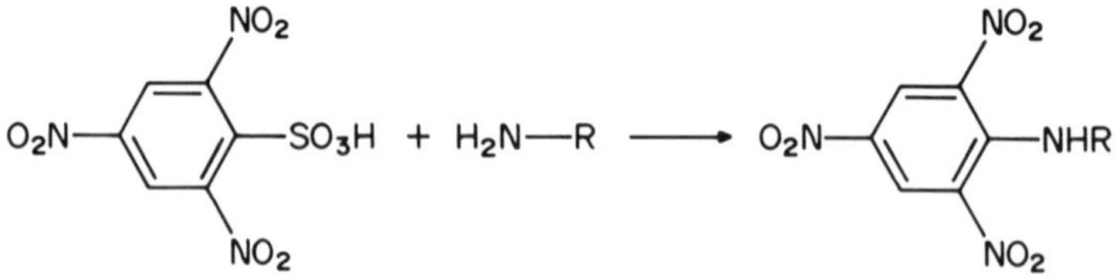

Figure 6. The formation of N-trinitrophenyl derivatives by reaction of trinitrobenzenesulphonic acid with amines.

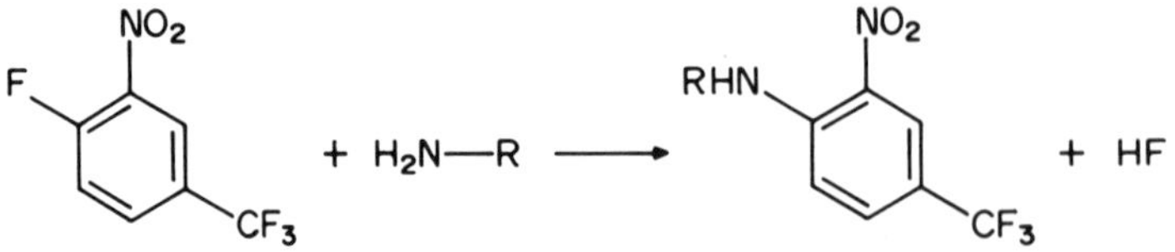

Figure 7. The reaction of 4-fluoro-3-nitrobenzotrifluoride with amines.

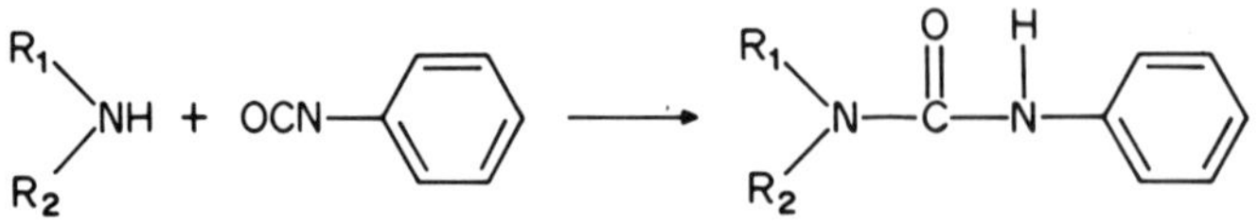

Figure 8. The reaction of phenylisocyanate with amines.

aliphatic alcohol was added to destroy the excess of reagent. The separation was by RP HPLC with detection at 240–260 nm. The detection limit was 1–10 ng. Naphthyl isocyanate (NIC) was used for the derivatization of amino acids, allowing detection by either fluorescence or UV absorption [41].

1.4.2 Phenyl isothiocyanate (PITC) and naphthyl isothiocyanate (NITC)

Isothiocyanates are widely used for derivatizing primary amines. They possess higher selectivity and are less reactive towards water, hydroxyl and carboxylic groups compared with isocyanates [42–44]. Labelling with phenyl isothiocyanate (PITC) or naphthyl isothiocyanate (NITC) has been used for the determination of C_3–C_{10} alkylamines in ethanol [45].

PITC reacts with amino acids to form phenylthiohydantoins (PTH) [44]. Peptides and proteins can similarly react with PITC, and the method is widely used in the Edman method for the sequencing of proteins and peptides. The reaction involves both the amino and the carboxylic groups in the molecule as follows: (i) coupling of PITC to the terminal NH_2 group; (ii) cleavage of the peptide bond in acidic medium; (iii) conversion of the labile thiazolinone derivative into the more stable PTH derivative. The derivatized amino acids can be detected at 254 or 269 nm. The earlier HPLC separations of PTH-amino acids were by adsorption chromatography [46–49]. Since the pioneering work of Zimmerman et al. [50], who separated 20 common PTH-amino acids in 20 min, reversed phase HPLC has become one of the most commonly applied methods for the identification of PTH-amino acids. Most methods [49–55] required long analysis times and/or relatively high temperatures. However, a binary system achieved the complete separation of common PTH-amino acids in 14 min at relatively low temperature (35 °C) with good reproducibility [56]. Lottspeich [57] also described an isocratic system for the separation of PTH-amino acids in 15 min which was capable of resolving the difficult Gln/Ser pair.

Derivatization with PITC may alternatively be used to detect primary and secondary amino acids as the phenylthiocarbamoyl (PTC) derivatives (Figure 9) [58–62]. The PITC derivatization reagent, made up fresh each day, consists of ethanol/triethylamine/water/PITC (7:1:1:1 by vol) and is stored at −20 °C under nitrogen to prevent degradation. The derivatization procedure is identical for all dried samples (standard mixtures, hydrolysates, serum and urine extracts). PITC reagent (100 μl) is mixed throughly with the dried sample and allowed to react for 5 min at room temperature. The mixture is evaporated to dryness under vacuum and the

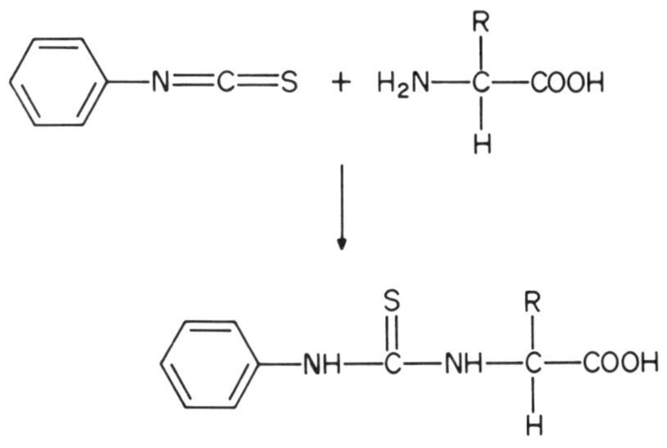

Figure 9. The formation of phenylthiocarbamoyl derivatives of amino acids.

residue is dissolved in 500 μl of 0.01 M sodium acetate buffer, pH 6.4, vortex-mixed with 200 μl of dichloromethane and centrifuged. The aqueous phase is then used for HPLC analysis. There are numerous methods for the separation of PTC-amino acids using mainly C_{18} [61, 62] or a commercially available application-specific column [60]. The PTC-amino acids are detected at 254 nm.

1.4.3 4-N,N'-Dimethylaminoazobenzene-4'-isothiocyanate and p-phenylbenzoyl isothiocyanate

The derivatization of 6-amino-2-methyl-2-heptanol (heptaminol) with 4-N,N'-dimethylaminoazobenzene-4'-isothiocyanate was used for the HPLC determination of this amino compound in pharmaceutical preparations [63, 64]. Labelling with p-phenylbenzoyl isothiocyanate [65] may also be a good technique for the HPLC determination of amines.

1.5 Miscellaneous

1.5.1 p-Nitrobenzyl bromide (p-NBBr)

Tertiary aliphatic amines are rather difficult to detect, but they can be detected at 254 nm after quaternization with p-NBBr [66]. Pilocarpine and isopilocarpine were derivatized in a sealed ampoule for 24h at 40 °C, then separated by ion-pair HPLC. Tertiary amines can also be determined by post-chromatographic derivatization. Kudoh et al. described a method [67] involving derivatization with 1% citric acid in acetic anhydride at 120 °C, separation on a chemically bonded silica gel Nucleosil $5N(CH_3)_2$, column, and detection at 550 nm.

1.5.2 Dansyl chloride (Dns-Cl)

The derivatives of amino acids with dansyl chloride are usually detected fluorimetrically. However, the fluorescence is quenched by phosphate ions. In RP-HPLC with phosphate buffer/acetonitrile as the mobile phase, detection was by UV absorption at 250 nm [68].

1.5.3 o-Phthaldialdehyde (OPA)

The reaction between o-phthaldialdehyde and primary amines was employed by Radjai and Hatch to detect amino acid derivatives by UV absorption at 340 nm [69]. The detection limit was 0.5 nmol.

1.5.4 Ninhydrin

Automatic amino acid analysers using ninhydrin for post-column reaction are widely used. LePage and Rocha [70] separated amino acids by using ion-pair chromatography with dodecylsulphonate as the ion pairing agent, and derivatized them with ninhydrin or 1,2,3-perinaphthindantrione (Peri) at 140 or 100 °C respectively. The reaction with Peri was not influenced by the low pH of the eluent, which is necessary for separation of the amino acids, but the absorptivities of ninhydrin derivatives were much higher. Luo and Siena [71] applied the reaction with ninhydrin for determining methylated amino acids in urine after separation on an ion exchange column.

2. Reagents for carboxylic acids

Carboxylic acids are a large group of naturally occurring compounds which include fatty acids, prostaglandins, bile acids and other organic acids, all of which are important in biological systems. Most of these compounds do not have strong UV or visible absorption. Commonly, carboxylic acids are converted to their UV–VIS-absorbing esters by reaction with aromatic halides. Phenacyl bromide (PBr), naphthacyl bromide (NBr) and their analogues are most often used as labels. Carboxylic acids can also form amides with aromatic amines after being first converted to the corresponding acyl chloride.

2.1 Phenacyl bromide, naphthacyl bromide and their analogues

The molar absorption coefficients [72] of some phenacyl and naphthacyl bromides at λ_{max} or the detection wavelength are given in Table 3. To improve the sensitivity and selectivity of the alkylation reaction of carboxylic acids, catalysts such as crown ethers [73, 74] and tertiary amines [75] were introduced. The crown ether complexes the cation of a neutralized acid, enabling dissolution of carboxylate anion in an aprotic solvent. The activated anion displaces the α-bromine in phenacyl bromide (PBr) by a nucleophilic substitution reaction (Figure 10). The carboxylic acids are neutralized with potassium hydroxide prior to derivatization [73, 74]; alternatively, potassium carbonate [76] or sodium hydrogen carbonate [77] may be used. Tertiary amines catalyse the reaction, and N,N'-diisopropylethylamine [75] and triethylamine [78] are often used. Another catalyst in carboxylic acid alkylation is potassium fluoride [79], with which pre-neutralization of the acids is not required. The derivatization can be performed in any aprotic solvent, such as benzene [73], acetonitrile [73, 80–82] or DMF [75, 83, 84]. The HPLC analysis of the derivatives can be performed in either normal phase [80, 82] or reversed phase mode [73, 78]. Generally the detection limits are in the nanogram region. The ease of preparation, using a range of solvents, and the quantitative yields make this derivatiz-

Table 3 Absorption characteristics of some phenacyl reagents

Reagent	Molar absorption coefficient	Wavelength of detection/λ_{max}
α-Bromo-2-acetonaphthone	37000	254 nm
α-p-Dibromo-acetophenone	28000	260 nm
α-Bromoacetophenone	35000	243 nm
α-Bromo-p-nitroacetophenone	14590	263 nm
	20000	266 nm

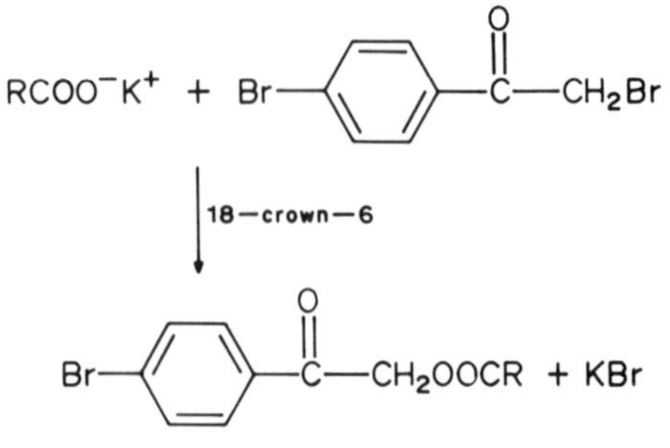

Figure 10. The reaction of phenacyl bromide with carboxylic acids catalysed by a crown ether.

ation technique useful in a wide variety of situations involving fatty acids, organic acids and other biologically important compounds containing acidic functional groups.

Fatty acids can be derivatized with p-bromophenacyl [73, 85, 86], phenacyl [87–89], naphthacyl [90, 91] or p-nitrophenacyl [74] bromide. To prepare p-bromophenacyl esters [73], fatty acids (0.001–0.5 mM) were dissolved in methanol or water and neutralized to a phenolphthalein end-point by methanolic KOH. The solvent was removed by either a rotary evaporator or lyophilization. A 3–10-fold molar excess of alkylating agent, p-bromophenacyl bromide/18-crown-6 (20:1) in acetonitrile, was then added. The mixture was stirred continuously in a sealed Reacti-Vial at 80 °C for 15 min. After cooling, the solution containing the derivatives was directly subjected to RP-HPLC with UV detection at 260 nm. The mobile phase was usually acetonitrile/water [87, 92, 93], methanol/water [73, 88, 94, 95] or acetonitrile/methanol/water [86, 89]. The separation of phenacyl derivatives of palmitoleic ($C_{16:1}$) and arachidonic ($C_{20:4}$) acids was not achieved with the acetonitrile/water system [87, 89], and elution with methanol/water could not resolve linolenic ($C_{18:3}$) and myristic ($C_{14:0}$) acids [88, 89]. A ternary mobile phase [89] containing a mixture of acetonitrile, methanol and water seemed to be best for the separation of phenacyl derivatives of fatty acids.

Hanis et al. [89] developed a method capable of separating 22 biologically relevant fatty acids and suitable for determining the fatty acid composition of rat adipose tissue and blood vessel walls. Fat was extracted from adipose tissue and saponified with 25% (w/v) potassium hydroxide in 96% ethanol. For saponification of 50 mg of fat, 1 ml of KOH solution was used. The saponification was carried out in a tightly closed tube in a boiling water bath for 1 h. After cooling and acidification to pH 2 with 3 M HCl, free fatty acids were extracted twice with n-hexane/diethyl ether (1:1). Phenacyl bromide (25 μl of a 10 mg ml^{-1} solution in acetone) and ethylamine (25 μl of a 10 mg ml^{-1} solution in acetone) were added to 10–100 μg of dried fatty acid from the saponification reaction. The mixture was heated in a boiling water bath for 5 min. The excess of phenacyl bromide was reacted with acetic acid (40 μl of a 2 mg ml^{-1} solution in acetone). The solvent was evaporated under nitrogen at 40 °C. The products were then dissolved in methanol for HPLC analysis. The products were stable for up to 8 weeks when stored at 4 °C. This derivatization procedure provided high speed and reproducible and quantitative yields.

The separation was performed on a Separon SGX C_{18} bonded spherical silica (5 μm) column coupled with an octadecylsilica guard column by gradient elution with water/methanol/acetonitrile as the solvent mixture. Baseline separation of the phenacyl esters of 22 saturated, monounsaturated and polyunsaturated fatty acids ($C_{6:0}$–$C_{22:6}$) was complete within 80 min. This method was more efficient in the resolution of cis- and trans-isomers of fatty acids than the GC method [89]. Detection was at 242 nm, with detection limits for short- and long-chain fatty acids of 0.8 and 12 ng respectively. Linearity was observed up to 100 ng and the recoveries were 92–106%.

Derivatization methods with phenacyl bromides or naphthacyl bromides have been developed for both low and high molecular weight acids. The carboxylic acids of low molecular weight [96] in an aqueous sample were extracted with diethyl ether at pH < 2. Trace amounts of water containing high concentrations of HCl and salts were eliminated by deep-freezing at −20 °C for 30 min. This step was found to be important, as the derivatizing agent p-bromophenacyl bromide decomposes in the presence of chloride ion.

Biotin [97], its biogenetic precursor dethiobiotin and its sulphoxide and sulphone metabolites have been derivatized with p-bromophenacyl bromide or 4-bromomethylmethoxycoumarin with dibenzo-18-crown-6 as the catalyst. The reaction products were separated on a C_{18} column. Both UV (254 nm) and fluorescence detection were used, with detection limits of 10 and 1 ng ml^{-1} respectively. The derivatization has been applied to the analysis of valproic acid in serum and plasma [76, 77, 98, 99]. The acid was extracted from biological matrices with organic solvents. Alric et al. [76] developed a procedure involving solvent demixing in which the extraction of valproic acid was achieved by saturation with potassium chloride following addition of acetonitrile and HCl to the plasma. The derivatization was performed in the acetonitrile phase after separation from the aqueous phase. The recovery was over 80%. The analysis was carried out on a C_{18} column with acetonitrile/water as the mobile phase.

A method for determining valproic acid in patients' plasma has been described [98]. The plasma from patients receiving valproate therapy was separated within 2 h of blood collection. The internal standard cyclohexanecarboxylic acid and 2 ml of pentane were added to 0.25 ml of plasma. After separation, the aqueous solution was acidified with 0.25 ml of 1 M sulphuric acid, and the valproic acid was extracted with 4 ml of pentane and derivatized at 50–55 °C for 1 h with 20 μl of phenacyl bromide and 20 μl of triethylamine. After evaporation of the solvent, the residue was dissolved in 250 μl of the mobile phase for HPLC separation on a Lichrosorb RP-18 column with methanol/water (3:1 v/v) or acetonitrile/water (2:1 v/v) as eluent. The derivatized valproate was well separated from the internal standard and the reagent in 6 min. Detection was at 246 nm. Other commonly prescribed drugs did not interfere with the determination.

Some biologically active eicosanoids present at picogram levels in blood and other tissues were determined as their p-(9-anthroyloxy)phenacyl esters [78]. Prostaglandins were also derivatized with p-bromophenacyl bromide in acetonitrile in the presence of N,N-diisopropylethylamine at room temperature [80, 82]. The use of high temperature and alkali salts must be avoided to prevent dehydration or isomerization. Normal phase chromatography was used to separate prostaglandins and their epimers [80, 82]. A silica adsorbent was employed as the stationary phase and a ternary solvent (methylene chloride, 1,3-butanediol or teramyl alcohol, and water) was used as the eluent. This procedure was suitable for the determination of prostaglandin E_1 and the impurities in bulk drug and pharmaceutical formulations. Morozowich and Douglas [100] used two coupled silica columns with a ternary eluent of hexane, methylene chloride and methanol to separate 10 F-series isomers and 8 E-series isomers. The UV detector was set at 254 nm. RP-HPLC on a C_{18} column with 50% acetonitrile in water as eluent separated the p-bromophenacyl esters of prostaglandins A_2, B_2, D_2, E_2 and F_2 and of 15-methylprostaglandin B_2 in 50 min [101].

Bile acids can also be converted into phenacyl [81] and p-bromophenacyl [102] esters. A bile acid sample with 50% excess of 0.15 M triethylamine in acetonitrile was warmed briefly and 0.1 M phenacyl bromide (50% molar excess) in acetonitrile was added. The reaction mixture was maintained at 80–90 °C for 45–60 min. The separation of phenacyl esters of lithocholic, deoxycholic, chenodeoxycholic, ursodeoxycholic, hyodeoxycholic and cholic acids was carried out on an ODS column. The excess derivatizing agents and interfering biological compounds from bile samples were first eluted with n-heptane/dioxane (90:10 v/v), and the separation of derivatives of bile acids was accomplished with n-heptane/dioxane/isopropanol (70:25:5, by vol). The detection limits at 254 nm for the derivatives were 5–10 ng [81].

Alkylation can also be carried out on other acidic compounds, such as barbiturates. 2-Naphthacyl derivatives of barbiturates [103] were formed in acetone at 30 °C in 30 min, using caesium carbonate as catalyst. The derivatives of six barbiturates were separated in 10 min by RP-HPLC. Down to 1 ng of N,N-dinaphthacyl phenobarbitore could be detected at 249 nm, providing the potential for the determination of barbiturates in small plasma or serum samples at concentrations well below the therapeutic range.

Some new phenacyl labels [104] for carboxylic acids have been synthesized recently, and 1-(4-hydroxyphenyl)-2-bromoethanone (4-HBE) was found to be a useful UV derivatizing agent. 4-HBE reacted with aliphatic and aromatic acids. A 4:2:1 molar ratio of 4-HBE, triethylamine and organic acid was used. The reaction was carried out at 80 °C for 1 h with 76% conversion. HPLC separation was carried out on an Ultrasphere RP-18 column with methanol/water (60:40 v/v) as eluent. Detection was at 289 nm, which avoided most of the interference found at lower wavelengths. The detection limit was 1 ng for the derivative of benzoic acid.

2.2 N-Methylphthalimide derivatives

N-Chloromethylphthalimide (ClMPI), N-chloromethyl-4-nitrophthalimide (ClMNPI) and N-chloromethylisatin (ClMIS) react quantitatively with fatty acids, dicarboxylic acids and barbiturates (Figure 11) [105]. The reactivity of these labels is due to the high mobility of the chlorine atom. This reaction is similar to those with phenacyl bromide. For a complete reaction it is necessary to convert the acids to alkali metal or ammonium salts. Triethylamine or a crown ether is used as catalyst. Aprotic solvents such as acetonitrile, methanol and diethyl ether are suitable reaction media. The reaction is complete within 30–40 min at 60 °C. The disadvantage of these labels is reactivity to alcohols and primary and secondary amines, and as a result the selectivity is limited. HPLC separation of phthalimidomethyl esters was performed on a reversed phase column (C_8) with acetonitrile/water in various proportions as the mobile phase. Detection was at 254 nm.

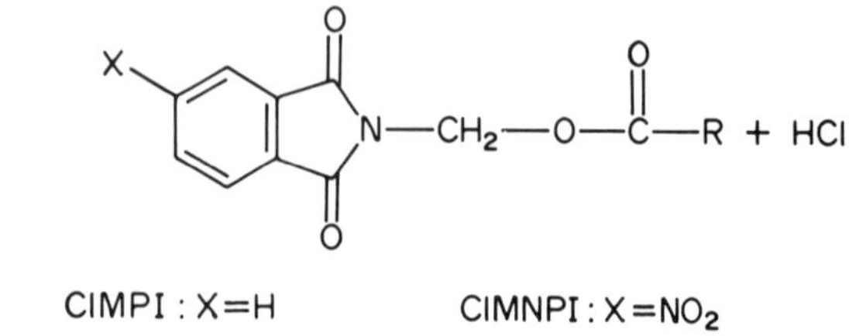

Figure 11. The reaction of N-chloromethylphthalimide with carboxylic acids.

ClMPI : X=H   ClMNPI : X=NO$_2$

2.3 O-p-Nitrobenzyl-N,N'-diisopropylisourea (p-NBDI)

p-NBDI is another alkylation agent. The p-nitrobenzyl esters of bile acids were prepared by reaction with p-NBDI in methylene chloride in a sealed vial [106]. The derivative was redissolved in chloroform/methanol (1:1 v/v) and chromatographed on a Partisil-10 column with 2% isopropanol in isooctane as eluent. This technique was applied successfully to the analysis of faecal bile and acid metabolites.

2.4 Derivatization with aromatic amines after conversion into acyl chlorides

Carboxylic acids can be converted to the corresponding acyl chlorides, which react with an amine containing chromophore. A few methods have been developed for conversion of acids to acyl chlorides. Hoffman and Liao [107] used a mixture of triphenylphosphine and carbon tetrachloride or polystyryldiphenylphosphine to prepare acyl chlorides of fatty acids. The reaction was carried out at 80 °C for 5 min. The intermediate acyl chlorides were then converted to amides with p-nitroaniline or p-methoxyaniline. Thionyl chloride [108] and oxalyl chloride [109] were also applied to prepare acyl chloride. The conversion with oxalyl chloride was quantitative in 30 min at 70 °C. The commonly used amines were 1-naphthylamine [109] and p-chloroaniline [110]. The final derivatives were separated by RP-HPLC with either gradient [108] or isocratic elution [109, 110].

Free fatty acids in human serum were derivatized with 1-naphthylamine after being converted to acyl chlorides [109]. Serum (0.5 ml) was mixed with 0.1 ml of methanol containing an internal standard and 1.4 ml of 1/15 M phosphate buffer and poured into a column packed with 1 g of Extrelut. The adsorbed fatty acid was recovered by elution with 10 ml of chloroform. After removal of solvent, the residue was dissolved in 0.6 ml of benzene. A solution of oxalyl chloride in benzene (2%) was added to the fatty acids and the mixture was allowed to react at 70 °C for 30 min. The solvent was removed at reduced pressure. Then naphthylamine solution (0.1 ml) and triethylamine solution (0.01 ml) were added. The reaction was carried out at 30 °C for 15 min, and 2 μl of the mixture was injected onto a μBondapak C_{18} column at 40 °C. The mobile phase was methanol/water (81:19 v/v) and detection was at 280 nm. Each fatty acid was quantitatively converted into its acid chloride and the overall recovery of naphthyl amides by this method was 94–106%. The main free fatty acids in human serum (14:0, 16:0, 16:1, 18:0, 18:1, 18:2) and the internal standard (17:0) were separated in 30 min.

2.5 Miscellaneous

2.5.1 2-Nitrophenylhydrazine

Derivatization of acids with 2-nitrophenylhydrazine has been described in a number of publications [111–115]. N,N'-Dicyclohexylcarbodiimide (DCCI) [111] or 1-ethyl-3-(dimethylaminopropyl)carbodiimide (EDCI) [114, 115] was used for the activation of the carboxylic acid function. Normally the derivatives were detected at 550 nm. Miwa and Yamamoto [115] applied this method to the determination of urinary dicarboxylic acids, which reacted with 2-nitrophenylhydrazine to form monohydrazides. The acids were extracted from acidified urine with ethyl acetate and dried. After derivatization, the monohydrazides of the dicarboxylic acids were first separated from the hydrazides of monocarboxylic acids and other interfering substances by a two-step extraction. The recoveries varied from 27.6–103.1%, depending on the polarity and the water solubility of the acids. The separation of monohydrazides of 11 straight- and branched-chain dicarboxylic acids was performed on an ODS column at 40 °C by isocratic elution with acetonitrile-phosphate buffer containing counter-ions such as tetramethylammonium as mobile phase. Detection was at 400 nm.

2.5.2 o-Phenylenediamine

A method [116] was developed for determining formate as benzimidazole after reaction with o-phenylenediamine at 130 °C for 2 h in 1 M perchloric acid. The benzimidazole was extracted from the reaction mixture with ethyl acetate/ethanol (9:1, v/v). The HPLC system was a C_{18} column, isocratic elution with a mixture of phosphate buffer (pH 2.1) and acetonitrile, and detection at 267 nm. This procedure was employed to determine the formate concentration in human urine and rat liver and urine, and to measure the activity of formaldehyde dehydrogenase [116].

2.5.3 Pyridinium dichromate

Pyridinium dichromate [117] was used for the oxidation of prostaglandins from seminal fluid to the corresponding 15-oxoprostaglandin derivatives, which were detected at 230 nm. The prostaglandins were extracted with a solid-phase C_{18} column prior to derivatization, and the derivatives were separated by RP-HPLC.

2.5.4 2-Methylquinoxanol derivative

D-Lactic acid in biological samples was converted into 2-methylquinoxanol by a series of reactions [118]. The derivative was determined by HPLC with either fluorescence or UV (334 nm) detection. By using this method, D-lactic acid in plasma, human urine and rat liver was determined.

2.5.5 Imidazole

A post-chromatographic derivatization method has been developed for penicillins, and was applied to the analysis of amoxicillin [119]. The compound was extracted from plasma and urine and separated by RP-HPLC with phosplate buffer (pH 8.0) methanol (92:8, v/v) as the mobile phase. The derivatization was accomplished with an aqueous solution of imidazole (33%) and mercuric chloride (0.11%), pH 7.2, containing Brij 35 (0.12%). The derivative was detected at 310 nm.

3. Reagents for hydroxy compounds

Hydroxy compounds include alcohols, carbohydrates, steroids and phenols. These compounds, especially carbohydrates and steroids, are important in biological systems. Mono- to tetrasaccharides of neutral sugars are of interest to the food industry. Many oligosaccharides are important components of body fluids, such as milk and urine, and of cell surface and secreted glycoconjugates. Various diseases such as mannosidosis, diabetes and lysosomal storage diseases, are related to sugar secretion. Steroid hormones are produced in man by the adrenals, gonads and placenta. The analysis of steroids in whole blood, plasma, urine or tissue is of diagnostic and clinical importance. Steroids containing an α,β-unsaturated ketone moiety can be detected at λ_{max} between 230 and 270 nm. Unfortunately, a number of important steroids, including those in physiological fluids, are present with a saturated ring structure and the carbonyl reduced to a hydroxyl group. Derivatization prior to HPLC analysis is necessary for these compounds. Hydroxyl groups are readily converted to esters by reaction with acyl chlorides, and can also be derivatized with methylsilyl chlorides or phenylisocyanate.

3.1 Acyl chlorides

Among the derivatization methods for hydroxyl groups, esterification with benzoyl chloride and its analogues to introduces a strongly UV-absorbing chromophore is the most widely used. A number of aromatic acyl chlorides, such as benzoyl [120–123], p-chlorobenzoyl [106], p-methoxybenzoyl [124, 125], p-nitrobenzoyl [120, 126, 127], and 3,5-dinitrobenzoyl chloride [128, 129], have been described as lebels for hydroxyl-containing compounds. The reaction is shown in Figure 12.

3.1.1 Benzoyl chloride

Fitzpatrick and Siggia developed a procedure for the HPLC analysis of benzoate derivatives of seven steroids [120]. The steroids (0.5–50 mg) were dissolved in 4 ml of pyridine, and a threefold molar excess of benzoyl chloride was added. The benzoylation reaction was accomplished by keeping the mixture at 80 °C for 15 min. After extraction, the derivatives were separated on Corasil C_{18} and Permaphase ODS columns and monitored at 230 nm. The analysis of urine after enzymatic hydrolysis was unsuccessful because of a large amount of interfering hydroxylated material in the sample. Benzoate derivatives were also used for labelling carbohydrates, alcohols and glycols [121]. The carbohydrates were benzoylated in pyridine. The anomers of sugars were equilibrated by heating the solution at 62 °C for 1 h, and benzoyl chloride was added to the solution cooled to room temperature. The derivatization was carried out at 62 °C for 45 min. Methanol was added to decompose the excess benzoyl chloride. The anomers were separated by normal phase HPLC on a Corasil II column with diethyl ether/hexane mobile phase. Separation was achieved by increasing the diethyl ether content from 0 to 99% in 110 min. An RP-HPLC method was developed for separating the benzoate esters of sapogenins extracted from plant material [123]. The derivatives were chromatographed on a C_{18} column, eluted with acetonitrile/water and detected at 235 nm. White et al. [130] used a modified procedure for the analysis of carbohydrates, amino sugars and glycosides. The derivatives were extracted with dichloromethane and analysed by normal phase HPLC. Resolution of the various anomers was achieved in 25 min. However, the derivatization time was very long (19 h). Either 230 nm (maximum absorbance) or 254 nm was used for monitoring, with detection limits of 2 and 20 ng respectively.

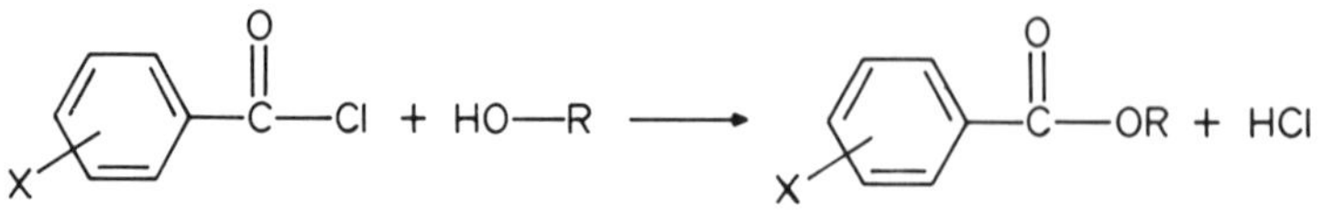

Figure 12. The reaction of benzoyl chlorides with hydroxy compounds.

3.1.2 p-Methoxybenzoyl chloride

p-Methoxybenzoyl chloride was used as a derivatization agent for two pharmaceutical products, hexachlorophene and pentaerythritol. The derivatization was carried out in sodium hydroxide medium. Both derivatized products were separated by normal phase HPLC and detected at 254 nm. Hexachlorophene formed a di-p-methoxybenzoate derivative detectable in amounts less than 40 ng [124]. Pentaerythitol formed a tetra-p-methoxybenzoate derivative. This method was applied to determine the pentaerythritol concentration in plasma [125]. Proteins in plasma (5 ml) were precipitated with 10% trichloroacetic acid. After centrifugation at 2000 g for 15 min, the supernatant was mixed with 2 ml of 10% NaOH and 0.7 ml of p-methoxybenzoyl chloride. The reaction mixture was maintained at 40 °C for 60 min, then mixed with 4 ml of mobile phase and directly injected onto a silica column with n-heptane/chloroform (3:2 v/v) as mobile phase. The detection limit was as low as 12 ng.

3.1.3 p-Nitrobenzoyl chloride

p-Nitrobenzoyl chloride is another benzoylation reagent for hydroxyl functional groups. Mono-, di- and trisaccharides were labelled with p-nitrobenzoyl chloride in pyridine [126]. The derivatization required 75 min. The derivatives showed maximum absorbance at 260 nm. The relatively apolar derivatives permitted rapid (10–15 min) isocratic separation on polar adsorbents (silica or alumina). n-Hexane/ethyl acetate (3:1 v/v) plus 5% dioxane was used as the mobile phase. Quantitation was performed by using the equation $C_{sugar} = (A_{sugar}/A_{standard})$ $(C_{standard}/f)$, where C is the concentration and f is a factor related to the molecular weight and the number of hydroxyl groups in the molecules of analyte and internal standard. The quantitative results showed relative standard. deviations of 0.6–4.5%. Nachtmann et al. [127] applied p-nitrobenzoyl chloride to the analysis of digitalis glycosides and their aglycones. The labelling reagent solution in pyridine was freshly prepared every day. A large excess of p-nitrobenzoyl chloride was necessary, and the surplus was removed after the reaction by adding water. The analytes reacted rapidly, and the reaction was completed in 10 min at room temperature. At high temperature partial decomposition of derivatives occurred. The products were extracted into chloroform at pH 8 and the solvent was evaporated under reduced pressure. The purified derivatives were than subjected to HPLC with a silica gel column and an eluent of n-hexane/methylene chloride/acetonitrile (10:3:3 by vol). This procedure was suitable for trace analysis of digitalis glycosides. The derivatization of steroids with p-nitrobenzoyl chloride has also been investigated [120].

3.1.4 3,5-Dinitrobenzoyl chloride

3,5-Dinitrobenzoyl chloride is similarly able to label hydroxy compounds. A method [129] has been developed for derivatizing cardiac glycosides, which are drugs for the treatment of congestive heart failure. The glycosides (e.g. digitoxin, digoxin and β-methyldigoxin) and their metabolites (disdigitoxoside, monodigitoxoside, its glycone and its 3-epimer) were labelled at room temperature in pyridine over 2 h. Separation was by reversed phase HPLC on an ODS column with acetonitrile/methanol/water (3:1:1 by vol) as eluent. Detection at 230 nm give linear response for 1.3–35 pmol of derivative. The detection limits for digoxin and digoxigenin were 0.6 and 0.2 ng respectively. Accurate results were obtained for blood and urine samples with gitoxin as internal standard. 3.5-Dinitrobenzoyl chloride was also used for determining diethylene glycols in various polyethylene glycols. Ginseng saponins can be derivatized with 3,5-dinitrobenzoyl chloride as well as benzoyl, p-nitrobenzoyl or naphthoyl chloride [131].

3.1.5 Anthracene-9-carbonyl chloride

Recently, Bayliss et al. [132] described the synthesis of anthracene-9-carbonyl chloride and its aplications as a label for fluorescence and UV absorbance detection of hydroxy compounds. The preparation and properties of esters of short-chain alcohols, diols, trichothecene mycotoxins and sterols were investigated. Anthracene-9-carbonyl chloride was prepared from commercial anthracene-9-carboxylic acid. Derivatization was carried out in acetonitrile free from water or active hydrogen compounds. The reaction rate was dependent on the structure of the alcohol. The derivatization of diethylene glycol was complete at ambient temperature within 10 min (0.25 M reagent) or 30 min (0.1 M reagent) without a catalyst, but required 1 h for cholesterol, testosterone and the trichothecene T-2 toxin. For sterically hindered alcohols such as t-butanol and 17α-methyltestosterone, more than 10 h, or refluxing for 1 h, was needed to complete the reaction. The derivatives had absorption maxima at 250 nm. Both normal and reversed phase HPLC were applied to the separation of the derivatives.

3.2 Phenyl isocyanate (PIC)

Phenyl isocyanate reacts with hydroxy compounds similarly to amines, giving an alkylpheylurethane (Figure 13). This derivatization method can be applied to alcohols and sugars.

Bjorkqvist and Toivonen developed a procedure for derivatizing 2-ethylhexanol and other aliphatic alcohols

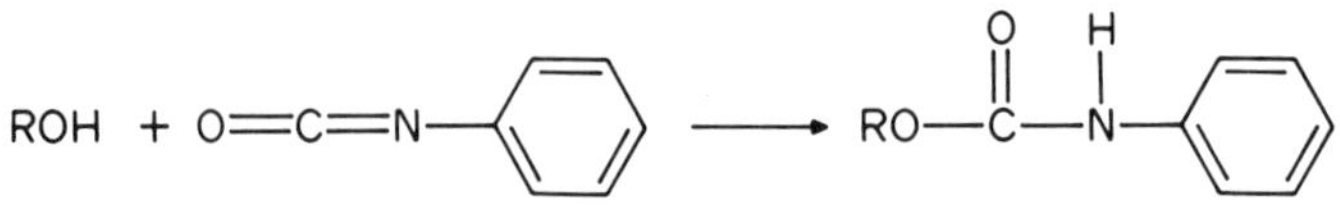

Figure 13. The reaction of phenylisocyanate with hydroxy compounds.

[133]. The alcohol (0.5–10 mg) was dissolved in 1 ml of dimethylformamide, and 0.5 ml of PIC was added. The reaction was completed by vigorously shaking the mixture and allowing it to stand for 15 min. No catalysts were needed. The excess reagent was destroyed by adding 0.5 ml of methanol. The derivatives were separated on a C_{18} column at 50 °C and detected at 230 nm. A gradient of acetonitrile in water eluted the derivatives of alcohols from propanol to heptanol in 20 min.

A sensitive method based on the fact that polyols react with PIC forming strongly UV-absorbing derivatives was applied to the determination of sorbitol and galactitol at the nanogram level in various biological samples [134]. The lenses, sciatic nerves and red cells from normal and diabetic rats, as well as human skin fibroblasts, were homogenized in 2 ml of cold water. An internal standard was added to each sample, and protein was eliminated by adding ethanol to 70% of the final volume. The supernatant after centrifugation was collected and lyophilized. Pyridine (250 μl) and 500 μl of phenylisocyanate were added to samples of lenses and sciatic nerves, but only 100 μl of pyridine and 200 μl of PIC were added to samples of red cells and fibroblasts. The reaction was carried out at 55 °C for 1 h. The excess reagent, which could react with water in the HPLC eluent, was eliminated by adding 250 μl of methanol. RP-HPLC with a pre-column and a UV detector set at 240 nm was used for analysis of the derivatives. A mixture of acetonitrile and 0.01 M K_2HPO_4, pH 7.0, was used as the mobile phase, with a stepwise gradient for separation of sorbitol from red cells and isocratic elution (60:40 v/v) for other samples. Major interfering peaks from biological samples eluted before galactitol and sorbitol. The recoveries of both galactitol and sorbitol in sciatic nerves were quantitative; in lenses, however, they were only around 40%. The detection limit was as little as 3 ng.

3.3 Silyl chlorides

Phenyldimethylsilyl chloride has been introduced as a reagent for labelling alditols and mono- and di-saccharides [135]. The derivatization was carried out in N,N-dimethylformamide. Imidazole solution (200 μl, 0.33 g ml^{-1}) was added to a solution of the carbohydrate (<10 mg). The mixture was then heated at 100 °C for 1 h. After cooling the solution on ice, the label was added and the reaction mixture was maintained at ambient temperature for over 6 h or at 100 °C for 1 h. Normal phase HPLC with a mobile phase of ethyl acetate/hexane was applied. The molar absorptivities of derivatives at λ_{max} (240–280 nm) were below 1400. To improve the sensitivity, the derivatives could be extracted with hexane and concentrated by evaporation.

4. Reagents for carbonyl compounds

4.1 2,4-Dinitrophenylhydrazine (2,4-DNPH)

Aldehydes, ketones, ketosteroids and sugars contain the carbonyl function group. In earlier studies on the HPLC separation of these compounds, 2,4-dinitrophenylhydrazine was widely used as a derivatization reagent to enhance their detectivity. 2,4-DNPH reacts with carbonyl groups to form 2,4-dinitrophenylhydrazones (Figure 14).

Carey and Persinger [128] described the derivatization of a number of simple aliphatic carbonyl compounds such as formaldehyde and acetaldehyde with 2,4-DNPH. The derivatives were analyzed on a Corasil II column and monitored at 254 nm. Two HPLC systems were reported for separating the 2,4-DNPH derivatives of thirteen carbonyl compounds ranging from acetaldehyde and acetone to benzaldehyde and salicylaldehyde [136]. One system used a column packed with 1% tris-(2-

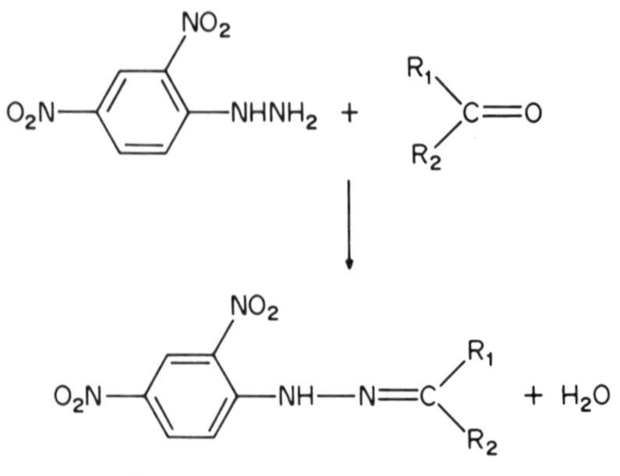

Figure 14. The reaction of 2,4-dinitrophenylhydrazine with carbonyl compounds.

cyanoethoxy)propane (TCEP) on Zipax and a mobile phase of hexane. The hexane was saturated with TCEP and a pre-column with 30% TCEP stationary phase was used to prevent stripping of TCEP from the analytical column. A permaphase ETH column was used at 40 °C with a gradient of hexane and chloroform. The detection limit was 5 ng and the relative precision was 2.5%. This method was applied to the analysis of carbonyl compounds in car exhaust gases. Acetaldehyde in blood has also been analysed as the 2,4-DNPH derivative [137].

Steroidal ketones have been separated as their 2,4-DNPH derivatives [138, 139]. The neutral 17-keto steroids are of clinical interest in certain disease states: the urinary 17-keto steroid level in women is indicative of adrenal activity, and that in men is a composite measure of adrenal and testicular function. Samples of urine and blood hydrolysates can be derivatized with 2,4-DNPH and analyzed by HPLC. In the work of Fitzpatrick et al. [138], the labelled 11-hydroxy-17-keto steroids were chromatographed on a C_{18} column and eluted with ethanol/water, which was able to resolve 11β-hydroxyandrosterone from 11β-hydroxyaetiocholanolone. The derivatives of four epimeric isomers of androsterone and dehydroepiandrosterone were separated from each other by using 1.5% β,β'-oxydipropionitrile on a Zipax column and an isooctane mobile phase. This system had the disadvantage of stripping the stationary phase, so that saturation of the isooctane with β,β'-oxydipropionitrile and a pre-column were necessary. The urine sample was solvolysed and hydrolysed enzymatically, and was then extracted with ethyl acetate. After evaporation of the solvent, the residue reacted with 2,4-DNPH at a pH above 8.0. Urine could also be hydrolysed, the liberated free steroids simultaneously being extracted into benzene under reflux conditions. The steroid conjugates in urine and blood plasma were extracted and analysed by the same technique.

Bile acids are usually derivatized in the same way as carboxylic acids. However, in conjugated bile acids the carboxylic functions are not available for derivatization. Reid and Baker [140] developed a method for labelling these compounds using the hydroxyl group on the steroid nucleus. The hydroxyl group at C-3α or C-7α was oxidized to a keto function with hydroxysteroid dehydrogenase and the formed keto bile salts were isolated by means of a Sep-pak C_{18} cartridge. The oxidation of the hydroxyl group was achieved by incubating the bile salt solution with 3α- or 7α-hydroxysteroid dehydrogenase and oxidized nicotinamide adenine dinucleotide at 37 °C for 1 h in pyrophosphate buffer (pH 9.5). 2,4-Dinitrophenylhydrazine (400 μl; 1 mg ml^{-1} in ethanol) and two drops of concentrated HCl were added to the dried keto bile acids. Derivatization was accomplished at 60 °C in 10 min. A disadvantage of this method is the formation of isomers during the derivatization. The HPLC analysis was performed on a C_{18} column with a 7 min linear gradient of methanol/ 10 mM, KH_2PO_4 (from 40:60 to 90:10 at a flow rate of 2 ml min^{-1}). Most separations were completed in 12 min, and the detection limit was 20 pmol at 340 nm. Naturally occurring keto bile salts can be derivatized directly with 2,4-DNPH.

4.2 p-Nitrobenzylhydroxylamine

p-Nitrobenzylhydroxylamine reacts with keto functional groups (Figure 15) similarly to 2,4-DNPH. Prostaglandins extracted from biological samples were determined via the p-nitrobenzyloximes [141]. The carboxyl group of the prostaglandins was esterified with excess diazomethane prior to the derivatization. After evaporation of excess reagent, the keto function of the esters was labelled with a 10-fold molar excess of p-nitrobenzylhydroxylamine in pyridine. The reaction was complete in 2 h at 40 °C or overnight at room temperature. RP-HPLC was performed with a C_{18} column, acetonitrile/ water eluent and detection at 254 nm. The background absorbance was low, giving a detection limit of 25 ng, but the reproducibility of the reaction was limited.

4.3 1-Phenyl-3-methyl-5-pyrazolone (PMP)

Honda et al. [142] reported that 1-phenyl-3-methyl-5-pyrazolone reacted with reducing carbohydrates almost quantitatively to yield strongly UV-absorbing derivatives. This reaction differs from other derivatization reactions via the hydroxy group of carbohydrates, such as benzoylation or, p-bromobenzoylation, in that it gives no stereoisomers. This method is especially useful for the analysis of component monosaccharides of gly-

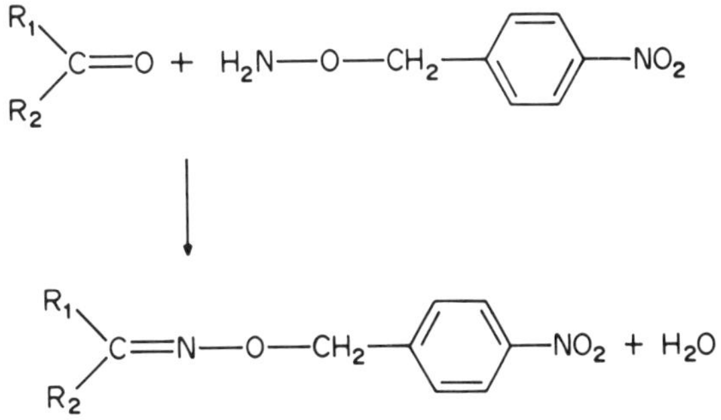

Figure 15. The reaction of p-nitrobenzylhydroxylamine with keto compounds.

coproteins, and is also suitable for the separation of reducing oligosaccharides or maltodextrins.

Glycoproteins were hydrolysed with 2 M trifluoroacetic acid for 6 h in bioling water bath to yield aldoses, or with 4 M HCl for amino sugars. To a sample of aldoses or reducing oligosaccharides (10–500 pmol each) were added a 0.5 M methanolic solution (50 μl) of PMP and 0.3 M sodium hydroxide (50 μl). The mixture was kept for 30 min at 70 °C. After cooling and neutralization, the solution was evaporated to dryness, and the derivatives were extracted into aqueous solution with water/chloroform.

The separation and retention behaviour of these derivatives was investigated by using various stationary phases and solvents. For the best separation of selected monosaccharides, a Capcell-Pak C_{18} column and a mixture of 0.1 M phosphate buffer (pH 5) and acetonitrile (4:1 v/v) were the best choice. The use of 18% acetonitrile in 0.1 M phosphate buffer (pH 7.0) shortened the analysis time but gave slightly poorer resolution. Detection was at 245 nm, the absorption maximum, with a detection limit of 1 pmol. This procedure provided good reproducibility with a relative standard deviation of less than 2.3%.

5. Conclusions

The introduction of a wide variety of UV–VIS-absorbing aromatic systems has been applied to all the common reactive grouping found in chemical compounds to improve their detectability and, at the same time, if possible, thier chromatographic properties. The greater part of this effort has been devoted to HPLC separation methods, because such predominantly aromatic derivatives are particularly compatible with HPLC. Some notable successes of this approach have been described here, and it is of course especially successful where the introduced aromatic system has a high E_{max} or a high absorbance at the wavelength of detection.

Although other, more sensitive modes of detection, such as fluorescence, may be strong competitors, the availability of spectrophotometers in every laboratory almost guarantees that UV/VIS colorimetric methods will be the automatic first choice for solving analytical problems involving derivatization for chromatography. This is clearly shown by the wealth of applications described, and is likely to continue in the future.

6. References

[1] R. E. Majors, H. G. Barth and C. H. Lochmuller, Anal. Chem., **54**, 323R (1982).
[2] K. Imai, H. Maiyano and T. Toyo'oka, Analyst, **109**, 1365 (1984).
[3] R. E. Majors, H. G. Barth and C. H. Lochmuller, Anal. Chem., **56**, 300R (1984).
[4] H. Lingman, W. J. M. Underberg, A. Takadata and A. Hulshoff, J. Liq. Chromatogr., **8**, 789 (1985).
[5] D. Perrett and S. R. Rudge, J. Pharm. Biomed. Anal., **3**, 3 (1985).
[6] I. S. Krull, C. M. Selavka, C. Duda and W. Jacobs, J. Liq. Chromatogr., **8**, 2845 (1985).
[7] K. Imai, Adv. Chromatogr., **27**, 215 (1987).
[8] J. A. P. Meulendijk and W. J. M. Underberg, Detection Oriented Derivatization Techniques in Liquid Chromatography, Marcel Dekker, New York (1990).
[9] C. R. Clark and M. M. Wells, J. Chromatogr. Sci., **16**, 333 (1978).
[10] S. L. Wellons and M. A. Carey, J. Chromatogr., **154**, 219 (1978).
[11] M. S. F. Ross, J. Chromatogr., **133**, 408 (1977).
[12] L. L. Needham and M. M. Kochhar, J. Chromatogr., **144**, 220 (1975).
[13] J. W. Redmond and A. Tseng, J. Chromatogr., **170**, 479 (1979).
[14] B. S. Furniss, A. J. Hannaford, V. Rogers, P. W. G. Smith and A. R. Tatchell, Vogel's Textbook of Practical Organic Chemistry, Longman, London (1978).
[15] S. Asotra, P. V. Mladenov and R. Burke, J. Chromatogr., **408**, 227 (1987).
[16] R. Porta, M. Camardells, V. Gentile and A. Desantis, J. Neurochem., **42**, 321 (1984).
[17] J. R. Clarks and A. S. Tyms, Med. Lab. Sci., **43**, 258 (1986).
[18] C. F. Verkoelen, J. C. Romijn and F. H. Schroeder, J. Chromatogr., **426**, 41 (1988).
[19] T. T. Sugiura, S. Hayashi, S. Kawai and T. Ohno, J. Chromatogr., **110**, 385 (1975).
[20] N. E. Larsen, K. Marinelli and A. M. Heilesen, J. Chromatogr., **221**, 182 (1980).
[21] J. Lammens and M. Verzele, Chromatographia., **11**, 376 (1978).
[22] J.-K. Lin and C.-H. Wang, Clin. Chem., **26**, 579 (1980).
[23] J.-Y. Chang, P. Martin, R. Bernasconi and D. G. Braun, FEBS Lett., **132**, 117 (1981).
[24] J-Y. Chang, R. Knecht and D. G. Braun, Methods Enzymol., **91**, 41 (1983).
[25] K. Tsuji, J. F. Goetz, W. Vanmeter and K. A. Gusciora, J. Chromatogr., **175**, 141 (1979).
[26] P. Helboe and S. Kryger, J. Chromatogr., **235**, 215 (1982).
[27] L. Elrod Jr., L. B. White and C. W. Wong, J. Chromatogr., **208**, 357 (1981).
[28] L. T. Wong, A. R. Beaubien and A. P. Pakuts, J. Chromatogr., **231**, 145 (1982).
[29] D. M. Barends, J. S. Blauw, H. M. Smith and A. Hulshoff, J. Chromatogr., **276**, 385 (1983).
[30] D. M. Barends, J. S. Blauw, C. W. Mijnsbergen, C. J. L. R. Govers and A. Hulshoff, J. Chromatogr., **322**, 321 (1985).
[31] D. M. Barends, C. L. Zwaan and A. Hulshoff, J. Chromatogr., **182**, 201 (1980).

[32] D. M. Barends, C. L. Zwaan and A. Hulshoff, *J. Chromatogr.*, **222**, 316 (1981).
[33] D. M. Barends, C. L. Zwaan and A. Hulshoff, *J. Chromatogr.*, **225**, 417 (1981).
[34] C. L. Hoppel, D. E. Weir, A. P. Gibbons, S. T. Ingalls, A. T. Britain and F. M. Brown, *J. Chromatogr.*, **272**, 43 (1983).
[35] P. M. Kabra, P. K. Bhatnagar and M. A. Nelson, *J. Chromatogr.*, **307**, 224 (1984).
[36] P. M. Kabra, P. K. Bhatnagar, M. A. Nelson, J. H. Wall and L. J. Marton, *Clin. Chem.*, **29**, 672 (1983).
[37] D. J. Edwards, *Handbook of Derivatives for Chromatography*, K. Blau and G. S. King (Editors), Heyden, London (1977), p. 394.
[38] W. L. Caudill and R. M. Wightman, *Anal. Chim. Acta*, **141**, 269 (1978).
[39] B. P. Spragg and A. D. Hutchings, *J. Chromatogr.*, **258**, 289 (1983).
[40] B. Bjorkqvist, *J. Chromatogr.*, **204**, 109 (1981).
[41] A. Neidle, M. Banay-Schwartz, S. Sacks and D. S. Dunlop, *Anal. biochem.*, **180**, 291 (1986).
[42] N. Seiler and L. Demisch, *Handbook of Derivatives for Chromatography*, K. Blau and G. S. King (Editors), Heyden, London (1977), p. 370.
[43] J. F. Lawrence, *Chemical Derivatization in Analytical Chemistry, Vol. 2, Separation and Continuous Flow Techniques*, R. W. Frei and J. F. Lawrence (Editors), Plenum Press, London (1982), p. 220.
[44] J. F. Lawrence and R. W. Frei, *Chemical Derivatization in Liquid Chromatography*, Elsevier, Amsterdam (1976), p. 113.
[45] J. Iskierko, E. Soezewinsky and B. Kanadys-Sobieraj, *Chem. Anal.*, Warsaw, **25**, 955 (1980).
[46] P. A. Graffeo, A. Haag and B. L. Karger, *Anal. Lett.*, **6**, 505 (1973).
[47] E. W. Matthews, P. G. H. Byfield and I. MacIntyre, *J. Chromatogr.*, **110**, 369 (1975).
[48] C. Bollet and M. Caude, *J. Chromatogr.*, **121**, 323 (1976).
[49] J. A. Rodkey and C. D. Bennett, *Biochem. Biophys. Res. Commun.*, **72**, 1407 (1976).
[50] C. L. Zimmerman, E. Appella and J. J. Pesano, *Anal. Biochem.*, **77**, 569 (1977).
[51] M. Abrahamsson, K. Groningsson and S. Castensson, *J. Chromatogr.*, **154**, 313 (1978).
[52] P. W. Moser and E. E. Rickli, *J. Chromatogr.*, **176**, 451 (1979).
[53] N. D. Johnsson, M. W. Hunkapillar and L. E. Hood, *Anal. Biochem.*, **100**, 335 (1979).
[54] G. E. Tarr, *Anal. Biochem.*, **111**, 27 (1981).
[55] D. Hawke, Y. Pau-Miau and J. E. Shively, *Anal. Biochem.*, **120**, 302 (1982).
[56] P. Pucci, G. Sannia and G. J. Marino, *J. Chromatogr.*, **270**, 371 (1983).
[57] F. Lottspeich, *J. Chromatogr.*, **326**, 321 (1985).
[58] R. L. Heinrikson and S. C. Meredith, *Anal. Biochem.*, **136**, 65 (1984).
[59] J. F. Davey and R. S. Ersser, *J. Chromatogr.*, **528**, 9 (1990).
[60] S. A. Cohen and D. J. Strydom, *Anal. Biochem.*, **174**, 1 (1988).
[61] L. E. Lavi, J. S. Holcenberg, D. E. Cole and J. Jolivot, *J. Chromatogr.*, **377**, 155 (1986).
[62] P. Furst, L. Pollack, T. A. Graser, H. Godel and P. Stehle, *J. Chromatogr.*, **449**, 557 (1990).
[63] D. Tocksteinova, J. Churacek, J. Slosar and L. Skalik, *Mikrochim. Acta*. **I**, 507 (1978).
[64] M. Bauer, L. Mailhe and L. Nguyen, *J. Chromatogr.*, **292**, 468 (1984).
[65] Y.-Z. Chen, *Mikrochim. Acta*, **I**, 343 (1980).
[66] A. K. Mitra, C. L. Baustian and T. J. Mikkelson, *J. Pharm. Sci.*, **69**, 257 (1980).
[67] M. Kudoh, I. Mutoh and S. Fudano, *J. Chromatogr.*, **261**, 293 (1983).
[68] J. M. Wilkinson, *J. Chromatogr. Sci.*, **16**, 547 (1978).
[69] M. K. Radjai and R. T. Hatch, *J. Chromatogr.*, **196**. 319 (1980).
[70] J. N. LePage and E. M. Rocha, *Anal. Chem.*, **55**, 1360 (1980).
[71] M. F. Lou and M. Siena, *Biochem. Med.*, **25**, 309 (1981).
[72] B. Tracey, in *HPLC of Small Molecules, a Practical Approach*, C. K. Lim (Editors), IRL Press, Oxford (1986), p. 72.
[73] H. D. Durst, M. Milano, E. J. Kikta, Jr., S. A. Conelly and E. Grushka, *Anal. Chem.*, **47**, 1797 (1975).
[74] E. Grushka, H. D. Durst and E. J. Kikta, Jr., *J. Chromatogr.*, **112**, 673 (1975).
[75] R. F. Borch, *Anal. Chem.*, **47**, 2437 (1975).
[76] R. Alric, M. Cociglio, J. P. Blayac and R. Puech, *J. Chromatogr.*, **224**, 289 (1981).
[77] W. F. Kline, D. P. Enagonio, D. J. Reeder and W. E. May, *J. Liq. Chromatogr.*, **5**, 1697 (1982).
[78] W. D. Watkins and M. B. Peterson, *Anal. Biochem.*, **125**, 30 (1982).
[79] L. Nagels, C. Debeuf and E. Esmans, *J. Chromatogr.*, **190**, 411 (1980).
[80] P. H. Zoutendam, P. B. Bowman, J. L. Rumph and T. M. Ryan, *J. Chromatogr.*, **283**, 281 (1984).
[81] F. Stellaard, D. L. Hachey and P. D. Klein, *Anal. Biochem.*, **87**, 359 (1978).
[82] P. H. Zoutendam, P. B. Bowman, T. M. Ryan and J. L. Rumph, *J. Chromatogr.*, **283** 273 (1984).
[83] H. C. Jordi, *J. Liq. Chromatogr.*, **1**, 215 (1978).
[84] N. E. Bussell and R. A. Miller, *J. Liq. Chromatogr.*, **2**, 697 (1979).
[85] P. T. S. Pei, W. C. Kossa, S. Ramachandran and R. S. Henley, *Lipids*, **11**, 814 (1976).
[86] J. Halgunset, E. W. Lund and A. Sunde, *J. Chromatogr.*, **237**, 496 (1982).
[87] R. Wood and T. Lee, *J. Chromatogr.*, **254**, 237 (1983).
[88] K. Korte, K. R. Chien and M. L. Casey, *J. Chromatogr.*, **375**, 225 (1986).
[89] T. Hanis, M. Smrz, R. Klir, K. Macek, J. Klima, J. Base and Z. Deyl, *J. Chromatogr.*, **452**, 443 (1988).
[90] M. J. Cooper and M. W. Anders, *Anal. Chem.*, **46**, 1849 (1974).
[91] W. Distler, *J. Chromatogr.*, **192**, 240 (1980).
[92] J. Ryan and T. W. Honeyman, *J. Chromatogr.*, **312**, 461 (1984).
[93] H. Miwa, M. Yamamoto, T. Nishida, K. Nunoi and M. Kikuchi, *J. Chromatogr.*, **416**, 237 (1987).

[94] V. P. Agrawall, R. Lessire and P. K. Stumpf, *Arch. Biochem. Biophys.*, **230**, 580 (1984).

[95] M. Yamaguchi, R. Matsunaga, S. Hara, M. Nakamura and Y. Ohkura, *J. Chromatogr.*, **357**, 27 (1986).

[96] R. L. Patience and J. D. Thomas, *J. Chromatogr.*, **234**, 225 (1982).

[97] P.-L. Desbene, S. Coustal and F. Frappier, *Anal. Biochem.*, **128**, 359 (1983).

[98] R. N. Gupta, P. M. Keane and M. L. Gupta, *Clin. Chem.*, **25**, 1984 (1979).

[99] J. P. Moody and S. M. Allan, *Clin. Chim. Acta*, **127**, 263 (1983).

[100] W. Morozowich and S. L. Douglas, *Prostaglandins*, **10**, 19 (1975).

[101] F. A. Fitzpatrick, *Anal. Chem.*, **48**, 499 (1976).

[102] G. Mingrone, A. V. Greco and S. Passi, *J. Chromatogr.*, **183**, 277 (1980).

[103] A. Hulshoff, H. Roseboom and J. Renema, *J. Chromatogr.*, **186**, 535 (1979).

[104] R. K. Munns, J. E. Roybal, W. Shimoda and J. A. Hurlbut, *J. Chromatogr.*, **422**, 209 (1988).

[105] W. Linder and W. Santi, *J. Chromatogr.*, **176**, 55 (1979).

[106] B. Shaikh, N. J. Pontzer, J. E. Molina and M. I. Kelsey, *Anal. Biochem.*, **85**, 47 (1978).

[107] N. E. Hoffman and J. C. Liao, *Anal. Chem.*, **48**, 1104 (1976).

[108] M. Ikeda, K. Shimada and T. Sakaguchi, *Chem. Pharm. Bull.*, **30**, 2258 (1982).

[109] M. Ikeda, K. Shimada and T. Sakaguchi, *J. Chromatogr.*, **272**, 251 (1983).

[110] J. M. Bissinger, K. T. Rullo, J. T. Stoklosa, C. M. Shearer and N. J. DeAngelis, *J. Chromatogr.*, **268**, 102 (1983).

[111] J. W. Munson and R. Bilous, *J. Pharm. Sci.*, **66**, 1403 (1977).

[112] T. Momose, *Chem. Pharm. Bull.*, **26**, 1627 (1980).

[113] H. Miwa and M. Yamamoto, *J. Chromatogr.*, **351**, 275 (1986).

[114] R. Horikawa and T. Tanimura, *Anal. Lett.*, **15**, 1629 (1982).

[115] H. Miwa and M. Yamamoto, *Anal. Biochem.*, **170**, 301 (1988).

[116] S. Ohmori, I. Sumii, Y. Toyonaga, K. Nakata and M. Kawase, *J. Chromatogr.*, **426**, 15 (1988).

[117] J. Doehl and T. Greibrokk, *J. Chromatogr.*, **529**, 21 (1990).

[118] S. Ohmori and T. Iwamoto, *J. Chromatogr.*, **431**, 239 (1988).

[119] J. Carlqvist and D. Westerlund, *J. Chromatogr.*, **164**, 373 (1979).

[120] F. A. Fitzpatrick and S. Siggia, *Anal. Chem.*, **45**, 2310 (1973).

[121] J. Lehrfeld, *J. Chromatogr.*, **120**, 141 (1976).

[122] P. F. Daniel, D. F. Defeudis, I. T. Lott and R. H. McCluer, *Carbohydr. Res.*, **97**, 161 (1981).

[123] W. Higgis, *J. Chromatogr.*, **121**, 329 (1976).

[124] P. J. Porcaro and P. Shubiak, *Anal. Chem.*, **44**, 1865 (1972).

[125] L. D. Bighley, D. E. Wurster, D. Cruden-Loeb and R. V. Smith, *J. Chromatogr.*, **110**, 375 (1975).

[126] F. Nachtmann and K. W. Budna, *J. Chromatogr.*, **136**, 278 (1977).

[127] F. Nachtmann, H. Spitzy and R. W. Frei, *Anal. Chem.*, **48**, 1576 (1976).

[128] M. A. Carey and A. F. Persinger, *J. Chromatogr. Sci.*, **10**, 537 (1972).

[129] Y. Fujii, R. Oguri, A. Misuhashi and M. Yamazaki, *J. Chromatogr. Sci.*, **21**, 495 (1983).

[130] C. A. White, J. F. Kennedy and B. T. Golding, *Carbohydr. Res.*, **76**, 1 (1989).

[131] H. Besso, Y. Saruwatari, K. Futamura, T. Kunihiro, T. Fuwa and O. Tanaka, *Planta Med.*, **37**, 226 (1979).

[132] M. A. J. Bayliss, R. B. Homer and M. J. Shepherd, *J. Chromatogr.*, **445**, 393 (1988).

[133] B. Bjorkqvist and H. Tovivonen, *J. Chromatogr.*, **153**, 265 (1978).

[134] J.-M. Dethy, B. Callaert-Deveen, M. Janssens and A. Lenaers, *Anal. Biochem.*, **143**, 119 (1984).

[135] C. A. White, S. W. Vass, J. F. Kennedy and D. G. Large, *J. Chromatogr.*, **264**, 99 (1983).

[136] L. J. Papa and L. P. Turner, *J. Chromatogr. Sci.*, **10**, 747 (1972).

[137] C. Lynch, C. K. Lim, M. Thomas and T. J. Peters, *Clin. Chim. Acta*, **130**, 117 (1983).

[138] F. A. Fitzpatrick, S. Siggia and J. Dingham, *Anal. Chem.*, **44**, 2211 (1972).

[139] R. A. Henry, J. A. Schmidt and J. F. Diekman, *J. Chromatogr. Sci.*, **9**, 513 (1971).

[140] A. D. Reid and P. R. Baker, *J. Chromatogr.*, **260**, 115 (1983).

[141] F. A. Fitzpatrick, M. A. Wynalda and D. G. Kaiser, *Anal. Chem.*, **49**, 1032 (1977).

[142] S. Honda, E. Akao, S. Suzuki, M. Okuda, K. Kakehi and J. Nakamura, *Anal. Biochem.*, **180**, 351 (1989).

9

Fluorescent Derivatives

Nikolaus Seiler
Marion Merrell Dow Research Institute 16, rue d'Ankara 67084 Strasbourg Cédex, France

1. INTRODUCTION	176	
1.1 The choice of fluorigenic reagents	176	
2. SULPHONYL CHLORIDES	177	
2.1 5-Dimethylaminonaphthalene-1-sulphonyl chloride (Dns-Cl)	177	
2.2 5-Di-n-butylaminonaphthalene-1-sulphonyl chloride (Bns-Cl)	183	
2.3 6-N-Methylanilinonaphthalene-2-sulphonyl chloride (Mns-Cl)	184	
2.4 2-p-Chlorosulphophenyl-3-phenylindone (Dis-Cl)	185	
2.5 Other sulphonyl chlorides	186	
3. CARBONYL CHLORIDES	186	
3.1 Chloroformates	186	
3.2 [R,S]-2-(p-Chlorophenyl)-α-methyl-5-benzoxazoleacetyl chloride	187	
4. HALOGENONITROBENZOFURAZANS	187	
4.1 4-Chloro-7-nitro-2,1,3-benzoxadiazole (4-chloro-7-nitrobenzofurazan; NBD-Cl)	187	
4.2 4-Fluoro-7-nitrobenzofurazan (NBD-F)	189	
4.3 Properties of NBD derivatives	189	
4.4 Applications	189	
5. ISOCYANATES AND ISOTHIOCYANATES	190	
5.1 9-Isothiocyanatoacridine	190	
5.2 Fluorescein isothiocyanate	190	
5.3 Other isothiocyanates	191	
6. FLUORESCAMINE	191	
6.1 History	191	
6.2 Reaction and reaction conditions	192	
6.3 Properties of the fluorophore	192	
6.4 Applications	192	
7. 2-METHOXY-2,4-DIPHENYL-3(2H)-FURANONE) (MDPF)	192	
8. SCHIFF BASE-FORMING AND RELATED REAGENTS	192	
8.1 Pyridoxal and pyridoxal 5-phosphate	192	
8.2 2-Fluorenecarboxaldehyde and 1-pyrenecarboxaldehyde	193	
8.3 o-Phthaldialdehyde/alkylthiol (OPA/R-SH) reagents	193	
8.4 2-Acetylbenzaldehyde	195	
8.5 ω-Formyl-o-hydroxyacetophenone and benzopyrone	195	
8.6 Benzoin, a reagent for the determination of compounds containing guanidino groups	196	
9. FURTHER REAGENTS FOR THE LABELLING OF AMINO COMPOUNDS	196	
9.1 N-Succinimidyl-2-naphthoxyacetate	196	
9.2 N-Succinimidyl-1-naphthylcarbamate	196	
9.3 5-(4,6-Dichloro-1,3,5-triazine-2-yl) aminofluorescein (DTAF)	196	
10. REAGENTS FOR CARBONYL COMPOUNDS	196	
10.1 Dansylhydrazine	196	
10.2 4-Hydrazino-7-nitrobenzofurazan (Nbd-H)	197	
10.3 4'-Hydrazino-2-stilbazole (4H2S)	197	
10.4 Semicarbazide	197	
10.5 2-Aminopyridine	198	
10.6 1,2-Diphenylethylenediamine (DPE)	198	
10.7 2-Cyanoacetamide	198	
10.8 Quinoxalinol formation with o-phenylenediamine	198	
10.9 Analogues of o-phenylenediamine	199	
11. CONDENSATION OF ALIPHATIC ALDEHYDES WITH 1,3-DIKETONES AND AMMONIA TO LUTIDINE DERIVATIVES	199	
11.1 Derivative formation with CHD	199	

Handbook of Derivatives for Chromatography
Edited by K. Blau and J. M. Halket © 1993 John Wiley & Sons Ltd

11.2 Properties ... 199
12. SULPHYDRYL REAGENTS ... 199
    12.1 Fluorobenzofurazan-4-sulphonate (SBD-F) and 4-fluoro-7-sulphamoyl-benzofurazan ... 199
    12.2 Dansylaziridine (Dns-A) ... 200
    12.3 N-Substituted maleimides ... 200
    12.4 Bimanes ... 201
13. FLUORESCENCE LABELLING OF ACIDIC FUNCTIONS ... 201
    13.1 4-Bromomethyl-7-methoxycoumarin (Br-Mmc) ... 201
    13.2 Other fluorescent coumarin derivatives ... 202
    13.3 Other fluorophores for acidic functions ... 202
    13.4 Fluorescence labelling of carboxylic acids after activation ... 202
14. FLUORESCENCE LABELLING OF ALCOHOLS ... 203
    14.1 7-Methoxycoumarin-3-carbonyl azide (3-MCCA) and 7-methoxycoumarin-4-carbonyl azide (4-MCCA) ... 203
    14.2 7-(Chlorocarbonylmethoxy)-4-methylcoumarin ... 203
    14.3 Nitrile group-containing reagents ... 203
    14.4 Naphthaleneboronic acid and phenanthreneboronic acid ... 203
15. FLUORESCENCE LABELLING OF ADENINE AND ITS DERIVATIVES BY CHLORO- OR BROMOACETALDEHYDE ... 203
16. CONCLUSIONS ... 204
17. REFERENCES ... 204

1. Introduction

It was realized early on that the fluorescence exhibited by certain substances was a valuable property for their analysis because it was measurable at lower concentrations than optical absorbance, because it was linear with concentration over a wide concentration range, and because its relatively high specificity permitted determination even in the presence of other substances. Numerous analytical methods for the determination of compounds with native fluorescence have therefore been developed. Many methods were also developed where a compound of interest was converted to a fluorescent product by a specially devised chemical reaction to enable it to be determined by the sensitive technique of fluorimetry. Our main concern is with the application of reagents designed to make fluorescent derivatives of compounds of interest, in order to permit their determination or detection by fluorimetric methods allied to chromatographic (or electrophoretic) separation processes.

1.1 The choice of fluorigenic reagents

Sensitivity is mormally one of the most important aspects of fluorescence methods. High fluorescence efficiency of the derivatives is therefore necessary. With all methods in which a reagent reacts with a certain functional group, specificity is normally limited by the efficiency of the separation method. From a practical viewpoint, any reagent for fluorescent derivative formation should fulfil some or all of the following prerequisites.

(1) Rapid quantitative reaction under mild conditions, preferably in water or at least in aqueous media.
(2) Formation of relatively non-polar derivatives in order to permit their isolation and concentration by exraction into organic solvents.
(3) Specificity for a given functional group.
(4) Excess of the reagent shoud be easily separable from its reaction products.
(5) The derivatives should have favourable chromatographic properties.
(6) High fluorescence efficiency; emission at a wavelength long enough to avoid the generally bluish background fluorescence of solvents and adsorbents, but within the high range of sensitivity of the available detectors.
(7) High molar absorption coefficient; no overlapping of excitation and emission bands.
(8) If the reagent is to be used for end-group determination of peptides and proteins, or if conjugates of certain compounds are to be studied, the derivatives must be stable to hydrolytic cleavage.

The history of fluorescence labelling of small molecules for the purpose of identification by separation followed by quantitative determination began with the application of 5-dimethylaminonapthalene-1-sulphonyl chloride (Dns-Cl) as a fluorescent end-group reagent in protein chemistry by Gray and Hartley [1] in 1963. This was the first successful attempt to replace the non-fluorescent napthalene-2-sulphonyl chloride of Fischer and Bergell [2, 3] and Sanger's 2,4-dinitrofluorobenzene [4, 5] by a fluorescent label, and improved the sensitivity of end-group determinations by more than an order of magnitude [6, 7]. Thin-layer chromatography as a separation method, quantitative methods, and applications of this reagent to amine analysis were all introduced by Seiler and co-workers [7–10].

Although the measurement of fluorescence polarization and decay are still powerful tools in the study of the physical chemistry of proteins, the application of fluorescence labelling of proteins has lost some of its importance, since improved instrumentation permits measurement of the native fluorescence of proteins. Most of the fluorescent reagents developed for this purpose reacted with the amino groups of proteins. As these same reagents were applied for microanalytical purposes, initially only a few fluorescent labels were applied for functional groups other than primary or secondary amino groups, but this has recently changed.

The measurement of low concentrations of compounds by fluorimetry is normally the main purpose of fluorescence labelling of compounds which have no native properties that allow sensitive detection. Where there is fluorescence quenching by known or unknown components of the sample solutions or a lack of specificity of the fluorescence measurements, the application of alternative methods for the determination of fluorescent derivatives may be necessary. In such cases fluorescence may be used to monitor the chromatographic separation even though alternative quantitation methods are applied. Radioactivity measurement and mass spectrometry are frequently used. If a reagent is selected for a certain analytical purpose, besides its reactivity and the fluorescence and chromatographic characteristics of its derivatives, its accessibility to radioactive labelling and its suitability for qualitative and quantitative mass spectrometry should therefore also be considered.

2. Sulphonyl chlorides

2.1 5-Dimethylaminonaphthalene-1-sulphonyl chloride (Dns-Cl)

2.1.1 History

Weber [11] introduced Dns-Cl (dansyl chloride) in 1952 as a reagent for the preparation of fluorescent conjugates of proteins, and used it for the study of protein structure by fluorescence polarization measurements. Shore and Pardee [12] showed the value of Dns conjugates for the study of energy transfer from tryptophan-containing proteins. Subsequently, Dns-Cl was used for the preparation of fluorescent antibodies in numerous studies [13], and for the study of active centres of enzymes. The stoichiometric inhibition of α-chymotrypsin by reaction with Dns-Cl suggested a method for the identification of the constituent amino acid in question, namely degradation of the fluorescent enzyme and identification of the fluorescent fragment. From this, Gray and Hartley developed an end-group method with Dns-Cl as reagent [1, 6, 14], analogous to the method of Sanger [4, 5], and used high-voltage paper electrophoresis to separate the dansylated amino acids. Subsequently Seiler and Wiechmann developed two-dimensional TLC systems for this purpose [8], and Boulton and Bush [15] devised paper chromatographic systems. With the introduction of polyamide sheets by Woods and Wang [16], the method became a routine one and found broad application. The dansylation procedure was first applied to the identification of biogenic amines in 1965 [9]. Methods for the quantitative estimation of Dns derivatives were first published in 1966 [10]. Nowadays dansylation is mostly used in conjunction with high performance liquid chromatography (HPLC).

2.1.2 Purification of Dns-Cl

About 200 mg of commercially available Dns-Cl in 2–3 ml of toluene is applied to a column (2 × 20 cm) prepared by filling a glass column with a slurry of silica gel in toluene. Toluene elutes Dns-Cl ahead of its derivatives and decomposition products. The solvent is evaporated in vacuo, giving orange crystals of m.p. 69 °C, which are easily soluble in alcohols, ketones, aromatic hydrocarbons and chloroform, less readily soluble in aliphatic hydrocarbons and very slightly soluble in water. The fluorescence efficiency of Dns-Cl is low.

2.1.3 Reaction of Dns-Cl

As an aromatic sulphonyl chloride, Dns-Cl reacts with primary and secondary amino groups even at slightly alkaline pH; it reacts at higher pH with phenols, imidazoles, and even slowly with alcohols. Derivatives with barbiturates, purines and carbamates are also known. Thiol compounds form the corresponding disulphides [7, 17].

The reaction with amines is formulated in Figure 1. Polyfunctional molecules react completely under the usual conditions: O,N-bis-Dns-tyramine, N,N-bis-Dns-histamine, tri-Dns-spermidine, tetra-Dns-spermine and tri-Dns-norepinephrine are the products when excess of the reagent is used. Amino acids react first with the amino group. The Dns-amino acids tend, however, to react with excess Dns-Cl to form mixed anhydrides, especially at high pH. These tend to break down to CO, Dns-NH$_2$, the aldehyde with one carbon less than the parent amino acid and dimethylaminonapthalene-1-sulphonic acid [18] (Figure 2). γ-Amino acids under the same conditions form the γ-lactams. [17, 19–21].

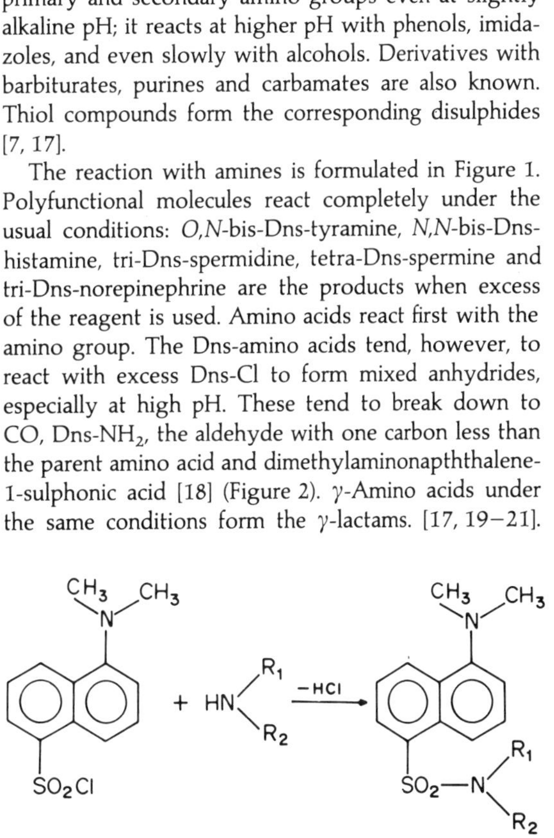

Figure 1. Reaction of Dns-Cl with an amine.

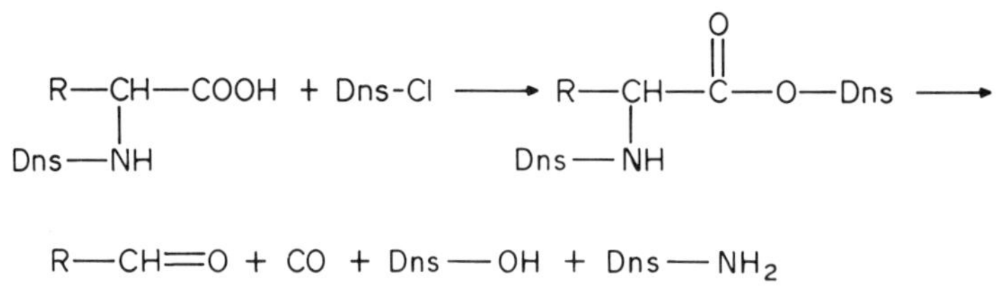

$$\text{R—CH—COOH} + \text{Dns-Cl} \longrightarrow \text{R—CH—}\overset{\displaystyle O}{\overset{\|}{C}}\text{—O—Dns} \longrightarrow$$
$$\underset{\text{Dns—NH}}{} \qquad\qquad\qquad\qquad \underset{\text{Dns—NH}}{}$$

$$\text{R—CH}=\text{O} + \text{CO} + \text{Dns—OH} + \text{Dns—NH}_2$$

Figure 2. Decomposition of α-amino acids with excessive Dns-Cl.

Steric hindrance, as occurs with the branched-chain aliphatic amino acids, favours decomposition. Amino acids with secondary amino groups (proline, hydroxyproline) and β-amino acids form relatively stable derivatives. The breakdown reactions with excess Dns-Cl are a disadvantage in the determination of free amino acids in tissues and body fluids. Lactam formation can be exploited for the assay of γ-aminobutyric acid and its derivatives [19, 21].

With primary amines side reactions are unusual. It is normally possible to obtain stoichiometric amounts of the reaction products. At high pH and elevated temperatures tertiary amines may be attacked to a significant extent, forming the Dns derivative of the secondary amine by elimination of an alkyl or aryl residue [22]. Dns-dimethylamine may be formed by direct electrophilic attack or by a mechanism analogous to that of the Bucherer reaction.

The rate of dansylation increases with increasing pH, but is paralleled by an increased rate of hydrolysis of the Dns-Cl. Optimal conditions are those under which the reactive groups most effectively compete for the limited amount of reagent available. Labelling of most amino acids, amines and phenols is optimal at pH 9.5–10 (at room temperature). At pH values below 8 the unreactive protonated form of the amino groups predominates.

Dns-Cl is only very slightly soluble in water, and so dansylation is mostly done in acetone/water mixture. Other organic solvents, e.g. dioxane, have been used. For the dansylation of barbiturates, ethyl acetate has been suggested.

2.1.4 Procedures for the use of Dns-Cl

2.1.4.1 General procedure for derivative formation with amines [7, 17, 21]

Tissue extracts in 0.2 M perchloric acid (100–500 μl) or similar volumes of other sample solutions are used. Three times the sample volume of a solution of Dns-Cl in acetone or dioxane (10 mg ml^{-1}) is added to each sample. The mixture is saturated with $Na_2CO_3 \cdot 10H_2O$, and the stoppered tubes are stored overnight at room temperature. Shaking of the tubes is advisable, especially during the initial phase of the reaction. By sonicating in an ultrasonic cleaning bath, the reaction can be completed within 2–3 h at room temperature. Since Dns derivatives are light-sensitive, unnecessary exposure to bright light should be avoided.

In order to remove excess reagent, 5 mg of proline dissolved in 20 μl of water is added to each sample if amines or γ-aminobutyric acid are to be determined. After sonication for about 3 min or storage for 30 min, the Dns-amine derivatives are extracted with 3–5 volumes of toluene.

For the estimation of amino acids the excess of Dns-Cl must be minimized, because of the partial breakdown of the Dns-amino acids and the need to separate dimethylaminonaphthalenesulphonic acid from the amino acid derivatives. If this acid is formed in large amounts, it may cause serious difficulties during chromatographic separation. Excess reagent, side products of the dansylation reaction and Dns-amine derivatives are removed, at least partially, if n-heptane is used as solvent, and nearly completely if toluene is used. Toluene extraction may, however, remove traces of the Dns-amino acids as well. The extraction of Dns-NH$_2$ is not complete under these conditions. After neutralization with NaH$_2$PO$_4$, most of the Dns-amino acids can be extracted from the water phase with ethyl acetate. Exceptions are the derivatives of aspartate, arginine, taurine, mono-Dns-lysine and mono-Dns-ornithine. Dns-glutamate also remains in the aqueous layer but N-Dns-pyrrolidonecarboxylic acid, the reaction product of glutamate with excess reagent [23], is extractable. Other non-extractable compounds of importance are O-Dns-choline and O-Dns-bufotenine.

2.1.4.2 Recommended procedure for peptides [6]

A solution of the peptide (0.5–5 nmol) is transferred to a small test tube. After drying *in vacuo*, the peptide is redissolved in 15 μl of 0.2 M NaHCO$_3$ and dried a second time. The peptide is then dissolved in 15 μl of water. The pH at this point should be 8.5–9.0. If it is correct, 15 μl of Dns-Cl (5 mg ml^{-1} of acetone) are added and the reaction is allowed to proceed for 1 h at 37 °C or 2 h at room temperature.

The solutions are dried *in vacuo*, and 100 μl of 6.1 M HCl is added to each tube. The tubes are sealed and hydrolysis is effected for 6–18 h at 105 °C. The HCl is removed *in vacuo* over NaOH pellets. Most of the material in the peptide hydrolysates is a mixture of sodium chloride and Dns-OH with small amounts of Dns-amino acids and Dns-NH$_2$. Two extractions with water-saturated ethyl acetate (100 μl) will remove the neutral Dns-amino acids almost completely, Dns-aspartate and Dns-glutamate by approximately 80%, moderate amounts of mono-Dns-lysine and O-Dns-tyrosine, and only traces of Dns-histidine, Dns-arginine and Dns-cysteic acid.

2.1.4.3 Derivative formation of amino acids

In spite of the aforementioned fragmentation reaction, reproducible results can be obtained by dansylation under carefully controlled conditions.

(a) Aliquots of the dried residue of tissue extracts in ethanol [24] are mixed with the same volume of 0.5 M NaHCO$_3$. To these solutions four volumes of Dns-Cl solution (6 mg ml^{-1} in acetone) are added. Derivative formation is complete within 3–4 h. If the samples are left overnight yields do not decrease [25].

(b) Reaction of amino acids in acetonitrile (33.3%)/water (66.6%) in the presence of lithium carbonate (26.7 μmol ml^{-1}; pH 9.5) and 0.5 mg ml^{-1} Dns-Cl was reported to give high yields within 2 h at room temperature. Alternatively, the samples are heated for 5 min at 95 °C. Aliquots (25 μl) are used directly for separation by HPLC [26].

2.1.4.4 Reaction of barbiturates in ethyl acetate [27]

Blood samples are extracted with acetone/diethyl ether (1:1 v/v). To 60 μl of extract are added 5 μl of ethyl acetate and the solution is dried by addition of some pellets of molecular sieve 4 Å. Acetone and diethyl ether are removed to a residual volume of about 5 μl by careful evaporation. The remaining solution is refluxed for 2 h after the addition of Dns-Cl (15 μg) and K$_2$CO$_3$ (2 mg). The reaction mixture is diluted to 100 μl and aliquots are used for chromatographic separation.

Other imides, but also primary and secondary amines, react with Dns-Cl in a similar way.

2.1.5 Properties of Dns-derivatives

2.1.5.1 General properties

The Dns derivatives are yellow crystalline solids, mostly readily soluble in organic solvents and only slightly soluble in water. The NH-group of the derivatives of ammonia and of primary amines can be deprotonated

Table 1 Stability of Dns-derivatives to hydrolysis. (A) Hydrolysis with 6.1 M HCl at 105 °C for 18 h, according to Gray [6]; (B) hydrolysis with 5 M methanolic KOH for 30 min at 50 °C according to Seiler and Deckardt [28]

Derivative	Percent remaining (A)	Percent remaining (B)	Derivative	Percent remaining (B)
Dns-glycine	82		Dns-dimethylamine	100
Dns-α-alanine	93		Dns-piperidine	100
Dns-β-alanine		49	Dns-ethanolamine	100
Dns-serine	65		bis-Dns-cystamine	60
Dns-threonine		22	bis-Dns-putrescine	100
Dns-valine	100		tri-Dns-spermidine	100
Dns-leucine	100		tetra-Dns-spermine	100
Dns-isoleucine	100	78	Dns-aniline	96
Dns-proline	23		Dns-benzylamine	100
Dns-hydroxyproline		6	Dns-β-phenylethylamine	100
Dns-aspartate		8	Dns-β-tryptamine	100
Dns-tryptophan	0	82	bis-Dns-histamine	0
bis-Dns-lysine		80	Dns-phenol	0
O,N-bis-Dns-tyrosine		0	O,N-bis-Dns-tyramine	100a
Dns-ammonia		100	O,N-bis-Dns-synephrine	100a
Dns-methylamine		100	O,N-bis-Dns-5-hydroxytryptamine	0

aRecovered as N-Dns-derivative.

[10]; the solubility of these compounds in water therefore increases with increasing pH.

Dns-amides are stable to acid hydrolysis, and fairly stable even to hydrolysis with bases. However, transfer of dansyl groups to other bases occurs in mild alkali; this reaction must be borne in mind when hydrolysis is carried out in alkaline media. In Table 1 some data concerning the stability of Dns derivatives to hydrolysis are summarized. Phenolic esters are much less stable than the sulphonamides, therefore it is possible to split off the O-Dns from O,N-bis-Dns-aminophenols with methanolic KOH [28]. This is of analytical interest because the N-Dns-derivatives of tyramine, synephrine etc. have much higher fluorescence efficiencies than the corresponding O,N-bis-Dns-derivatives. The derivatives of imidazoles and esters are unstable, both in acid and alkaline media. Dns-derivatives of imidazoles, can, however, be selectively cleaved with formic acid [29]. This feature of Dns-imidazole derivatives can be exploited in the deter-mination of histamine and methylhistamines [30].

2.1.5.2 Spectral characteristics

The spectral properties of Dns derivatives are dependent on their structures, but are also changed by solvents, adsorption onto active surfaces, and pH. As a rule, fluorescence emission maxima of compounds with primary amino groups are at shorter wavelengths than those with secondary amino groups (Table 2). Imidazoles and the dansylated lactams fluoresce at longer wavelengths than phenol esters. For instance, Dns-γ-aminobutyric acid has its excitation maximum (in methanol) at 354 nm and its fluorescence maximum at 534 nm, whereas the corresponding lactam (Dns-2-oxopyrroline) is maximally activated at 367 nm and emits at 567 nm [21]. Consequently Dns-amine and amino acid derivatives fluoresce greenish to yellow whereas the dansylated phenols, imidazoles and 2-oxopyrrolidines shown an orange fluorescence.

The quantum yields of amine and amino acid Dns derivatives are of the same order of magnitude (Table 2). Phenol and imidazole derivatives, however, have much lower fluorescence efficiencies. In Table 3, the spectral characteristics of the Dns derivatives of an amine, an aminophenol and a phenol are compared. Interestingly, the aminophenol derivative (O,N-bis-Dns-p-tyramine) shows much lower fluorescence efficiency than the analogous amine derivative although it bears two apparently independent fluorophores, and possesses

Table 2 Fluorescence efficiency of some Dns and Bns derivatives[a]; emission maxima in parentheses[b]

Derivative	Solvent			
	Ethyl acetate		Water	
ε-Dns-lysine			0.026	
Dns-proline			0.053	
Dns-glycine			0.065	
Dns-glutamic acid			0.066	
Dns-tryptophan	0.54	(510 nm)	0.068	(578 nm)
bis-Dns-cystine			0.087	
Dns-methionine			0.088	
Dns-valine			0.091	
Dns-methylamine	0.61	(502 nm)		
Dns-dimethylamine	0.53	(510 nm)		
Dns-piperidine	0.59	(508 nm)		
Dns-pyrrolidine	0.59	(508 nm)		
Bns-methylamine	0.72	(500 nm)		
Bns-dimethylamine	0.64	(505 nm)		
Bns-piperidine	0.68	(505 nm)		
Bns-pyrrolidine	0.68	(503 nm)		

[a] Bns = 5-di-n-butylaminonaphthalene-1-sulphonyl
[b] The values of the Dns-amino acids are from Chen [31] and those of the Bns-amine derivatives from Seiler et al [32].

Table 3 Comparison of the spectral characteristics of Dns-β-phenylethylamine. O,N-bis-Dns-p-tyramine and Dns-phenol in methanol as solvent [7, 10]

	Absorption maximum (nm)	ε	Fluorescence excitation (nm)	Fluorescence emission (nm)	Relative fluorescence efficiency
Dns-β-phenylethylamine	336	4530	360	521	100
	252	14100			
O,N-bis-Dns-p-tyramine	344	7880	362	525	58
	254	26800			
Dns-phenol	350	4050	375	540	10
	255	14100			

only one excitation and one fluorescence maximum which lie between the corresponding bands of the Dns-amine and the Dns-phenol (Table 3). Similar observations can be made, e.g., with the Dns-histamine derivatives. Since the two fluorophores are not connected by a system of conjugated double bonds, and the spectral characteristics are not changed by dilution, one might assume an intramolecular interaction of the π-electron systems of the fluorophores.

If in acidic solutions the dimethylamino group of the Dns-derivative is protonated, fluorescence efficiency decreases to a low level [10]. Protonation may be at least partially responsible for the decreased fluorescence efficiency of Dns derivatives adsorbed onto active silica gel, because spraying with bases increases the fluorescence efficiency very considerably, besides stabilizing the Dns derivatives against oxidative degradation (most probably light-induced) [7]. In strongly alkaline media, the amino group of the Dns derivatives of ammonia and of primary amines is ionized, and a hypsochromic shift of the absorption and fluorescence bands is observed [10].

The fluorescence efficiency of Dns derivatives is strongly solvent-dependent: increased fluorescence efficiency and a hypsochromic shift of the emission maximum are observed with decreasing dielectric constant. In Table 4 are summarized data reported by Chen [31] for Dns-tryptophan.

Table 4 Fluorescence efficiency and emission maximum of Dns-D,L-tryptophan in solvents of different dielectric constant, according to Chen [31]

Solvent	Solvent dielectric constant (debye)	Emission maximum (nm)	Fluorescence efficiency
Water	78.5	578	0.068
Glycerol	42.5	553	0.18
Ethylene glycol	37.0	543	0.36
Propylene glycol	32.0	538	0.37
Dimethylformamide	36.5	517	0.59
Ethanol	25.8	529	0.50
n-Butanol	19.2	519	0.50
Acetone	21.5	513	0.35
Ethyl acetate	6.1	510	0.54
Chloroform	5.1	508	0.41
Dioxane	3.0	500	0.70

Oxygen significantly quenches the fluorescence of Dns derivatives, which also decompose upon irradiation with UV light, especially if adsorbed onto active surfaces (silica gel thin-layer plates). In the case of tri-Dns-catecholamines, the photodecomposition products may be of some analytical value [33].

2.1.6 Separation procedures

High-voltage paper electrophoresis was the first method introduced for the identification of dansylated amino acids and for the separation of peptides [1, 6]; subsequently, electrophoretic separations on thin layers were also reported [34, 35]. These methods are, however, unsuitable for the separation of amine derivatives. TLC on silica gel has been widely used for the separation of amino acid, peptide and amine derivatives. A large variety of solvent systems of varying polarity is now available, which allows the separation of even complex mixtures of amino acids [7, 36–38] and amines [7, 17, 39]. Fingerprinting or identification of Dns-peptides can also be carried out on silica gel layers [29, 40, 41]. For the separation of Dns-peptides, electrophoresis in polyacrylamide gel and on cellogel strips was recommended [42, 43]. Alumina [44] kieselguhr [45] and cellulose thin-layers [46] have also been tried.

Dns derivatives can be separated by TLC on polyamide sheets and by partition paper chromatography. Derivatives of amines and amino acids move together, so that they cannot be separated as easily from each other as on active layers. The original solvents of Woods and Wang [16] have been only slightly modified for the separation of the protein amino acids [6] and for the dansylated amino acids and amines from a variety of tissues [47–51]. Polyamide sheets are good for end-group determinations of peptides and proteins because of the high detection sensitivity (small spots, high fluorescence efficiency), the stability of the Dns derivatives and rapid chromatographic development. Reversed phase TLC is another potentially useful method.

An advantage of all surface chromatographic procedures is that autoradiographs can be prepared. Furthermore, these methods are convenient if the specific radioactivity of amino group containing compounds is to be determined [49].

Disregarding early work, where a variety of materials was tried out, including silica gel [52], glass capillaries [53] and other [54–57], the majority of investigators have based their separations of Dns-amino acids on reversed phase HPLC. Most commercial columns have been used and a great variety of elution modes have been tested [25, 26, 51, 58–69]. The gradients used consist of an aqueous phase (water or buffer) and an organic solvent (methanol, acetonitrile, etc.). The numerous published methods differ only in their details. In general, separations were improved with more recent reversed phase columns. In order to improve reproducibility, isocratic separations of Dns-amino acids have been tried [70], and the use of microbore columns to improve sensitivity was suggested [71, 72]. The tendency to replace surface chromatographic methods by reversed phase HPLC is typical not only of amino acid separations but also for all other applications of Dns-Cl as a fluorigenic reagent (see below).

2.1.7 Quantitative estimation

For the quantitative assay of Dns derivatives, four methods are at our disposal: optical absorption, fluorimetry, quantitative mass spectrometry and the application of radioactive Dns-Cl.

Absorbance. The intense absorption bands of Dns derivatives in the ranges 250—255 nm and 335—350 nm [7] (see Table 3) are suitable for quantitative measurements in column effluents, but absorbance measurements are less sensitive than fluorimetric methods.

Fluorimetry. Fluorimetry of chromatographically separated Dns derivatives is the most obvious method for quantitative measurement. Extraction of the derivatives from the adsorbent and *in situ* fluorescence scanning of thin-layer chromatograms are equally suitable, as is the monitoring of fluorescence in column effluents. The principles and applications of procedures for the estimation of Dns derivatives on thin-layer chromatograms were first reported by Seiler and Wiechmann [7, 10, 17], and for paper chromatograms by Boulton [73]. Some modifications for the assay of Dns-amino acids were subsequently published [49, 74—79]. Reproducibility is normally within $\pm 5\%$.

Amino acid derivatives are best extracted from silica gel layers with methanol or with a mixture of methanol/25% NH_3 (95:5) [7, 10, 17] or chloroform/methanol/acetic acid (7:2:2) [77]. Peptides are extractable with acetone/water (1:1) [78], and amine derivatives are extractable with less polar solvents, e.g. ethyl acetate, benzene/acetic acid (99:1), benzene/triethylamine (95:5) [10, 17], or with dioxane if, as well as fluorescence, radioactivity is to be measured by liquid scintillation counting [80]. Chloroform is suitable for the extraction of Dns-amino acids from polyamide sheets [40]. Excitation and emission wavelengths are, if possible, adjusted to those of the individual Dns-derivatives; however, for almost every compound, excitation can generally be achieved using the 365 nm mercury line.

Since Dns derivatives are labile on active layers, spraying with a solution of triethanolamine in 2-propanol (1:4) [10] or impregnation with paraffin oil or silicone [81] is advantageous, as it increases the fluorescence and the stability of the derivatives. For the evaluation of two-dimensional thin-layer chromatograms, an apparatus was devised which allows two-dimensional fluorimetry [74, 75]. Fluorescence intensity is linearly related to the amount of Dns derivative up to at least 10 nmol ml^{-1} or 1 nmol per spot. For monitoring column effluents, optimum excitation and emission wavelengths may change considerably with the composition of the solvent gradient [69]. Usually, excitation between 340 and 395 nm and fluorescence measurement between 500 and 530 nm is suitable for both amino acid and amine derivatives.

Instead of excitation by UV light, Dns derivatives (and other fluorescent compounds) can be excited by chemiluminescence, using the reaction with bis(2,4,6-trichlorophenyl) oxalate and hydroperoxide [72, 82].

Quantitative mass spectrometry. Mass spectrometry is almost indispensable for the identification of small amounts of unknown compounds. Chromatographic criteria alone are insufficient for unambiguous identification. Other physical methods normally require more than nanomole quantities of material.

Since electron impact (EI) mass spectra of Dns derivatives usually include the molecular ions, these derivatives are quite useful for the mass spectrometric identification of unknown compounds, although only a few typical fragments characteristic of the labelled molecule are formed (Figure 3). The main fragment ion (m/z 170 or 171) is that formed by cleavage of the S—C bond of the sulphonyl group [20, 83—87]. Ionization methods other than EI, although useful for the identification of Dns derivatives [88, 89], have not been systematically studied.

The quantitative evaluation of the molecular ion, or of a diagnostic fragment ion, is much more specific than fluorimetry. Although underivatized compounds can be determined with a method called integrated ion current monitoring [90], it is often advantageous to use Dns derivatives [91]. The background at the mass range of the molecular ions of Dns derivatives is normally very low, so that erroneous peak identification is much less likely than with low molecular weight compounds. For detailed descriptions of suitable procedures see Refs. 91 and 92. Picomole amounts of Dns derivatives can be measured with these methods.

Radioactive Dns-Cl. [N-Methyl-^{14}C]Dns-Cl (specific radioactivity 10—30 Ci mol^{-1}) and [G-^{3}H]Dns-Cl (specific radioactivity 3—10 Ci $mmol^{-1}$) are available from several commercial sources. The labelled reagents can be applied just like unlabelled Dns-Cl. Their application allows the substitution of fluorimetry by the more convenient and automated liquid scintillation counting. The fluorescence

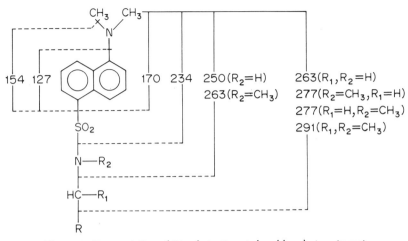

Figure 3. Fragmentation of Dns-derivatives induced by electron impact.

of the derivatives is used only for visualization of the spots on the chromatograms. The adsorbent is scraped off, or the layer is cut out with a razor blade in the case of polyamide sheets, and placed directly into the counting vial. Methanol (0.5–1 ml) is added and, after about 15 min, 10 ml of the liquid scintillator. In comparison with fluorimetry, the sensitivity is about one order of magnitude greater if the reagent with the highest available specific radioactivity is used. Detection sensitivity can be further improved by preparing autoradiographs of the chromatograms.

For quantitative derivative formation Dns-Cl has to be employed in large excess, so that even micro techniques for the preparation of the derivatives are expensive. Users of labelled Dns-Cl are thus frequently tempted to use too little reagent. More importantly, the specificity of the method is not increased, as compared with fluorimetry, by the use of a labelled reagent. Nevertheless, both [^{14}C] and [^{3}H]Dns-Cl have been used for the identification or estimation of amino acids and amines [93, 94].

A considerable improvement of the application of labelled Dns-Cl lies in the use double isotope techniques [95–100]. A known amount of the compound to be determined is added to the tissue sample as its ^{14}C-labelled analogue (with high specific radioactivity) as the internal standard. The compound is now isolated by the usual procedure and the derivative is formed with (G-^{3}H)Dns-Cl. The Dns derivative is purified extensively by the usual chromatographic procedures, and the radioactivity of the two isotopes is measured. The amount of non-radioactive compound present in the sample can be calculated from the ^{3}H/^{14}C radio. Incomplete derivative formation or losses during the isolation procedure do not influence the result as long as the isolated material is homogeneous. The coefficient of variation for the isotope ratio in blanks containing only 4-amino[^{14}C]butyrate was 6.2% [96].

2.1.8 Applications

Because of its wide applicability, sensitive detection and the favourable chromatographic properties of its derivatives, Dns-Cl is still a much used fluorigenic reagent. Only with the advent of fully automated pre-column derivatization have other procedures assumed comparable importance.

Applications of Dns-Cl range from the determination of small amounts of amines, amino acids, phenols and catechols in tissues and body fluids to the quantitative assay of drugs and drug metabolites; from the evaluation of peptide sequences to the characterization of active centres of enzymes. The selected applications in Table 5 may help to identify appropriate procedures for particular analytical problems.

2.2 5-Di-n-butylaminonaphthalene-1-sulphonly chloride (Bns-Cl) [32]

Bns-Cl is an analogue of Dns-Cl. It is somewhat more stable in storage, but its reaction with amines is slightly slower. The fluorescence excitation maxima of its derivatives are shifted a little towards longer wavelengths and emission maxima towards shorter wavelenghts, and the fluorescence efficiency is about 10% higher compared with Dns-derivatives. Its uses are the same as those of Dns-Cl [165]. A suitable solvent system has been reported for the complete separation on polyamide sheets of the Bns derivatives of all the amino acids normally found in

Table 5 Applications of Dns-Cl

Application	Remarks	Reference
Separation and identification of amino acids	In protein hydrolysates, tissues, and body fluids	25, 51, 62, 63 66, 69, 70, 104–7
	Selective method for secondary amino acids	108
	Methods for 4-aminobutyric acid	109–113
	Separation of optical isomers	114–117
	Determination of N-termini of peptides	6, 58, 59, 61, 118–20
	Determination of C-termini	121, 122
	Determination of formyl and acetyl groups of proteins and peptides	123
Determination and identification of biogenic amines and their metabolism	Catecholamines	124–127
	Vanillylmandelic acid	128
	Aliphatic and araliphatic amines	17, 129–132
	Polyamines	81, 133–7
	Histamine, N-methylhistamine	30
	β-Phenylethylamine	138
	Tryptamine	139
	Tyramine	140
	Serotonin, bufotenin	141
	Hexosamines	142
Determination of aminophospholipids		143, 144
Determination of phenolic compounds	Oestrogens	145
	Cannabinoids	146
	Morphine	147
Determination/identification of drugs		27, 148–164

proteins [166]. Bns-Cl was synthesized to overcome difficulties in the estimation of piperidine in tissues [167], and to improve quantitative mass spectrometry by increasing the molecular weight of the reagent and decreasing its polarity. Bns derivatives can be separated on thin-layer chromatograms with low-polarity systems [166–167]. More importantly they are extracted from reaction mixtures and adsorbents with less polar solvents than the corresponding Dns derivatives. This means that all amino acid Bns derivatives can be extracted from reaction mixtures with ethyl acetate, and that material extracted for quantitative mass spectrometry contains fewer impurities. In addition, Bns derivatives are eluted from reversed phase columns at higher methanol concentration, and the increased fluorescence efficiency gives about threefold better detection sensitivity.

The greatest advantage of the Bns derivatives is their characteristic fragmentation pattern on electron impact (Figure 4). Bns derivatives are preferentially cleaved in the butyl side chain, forming a base peak at $[M-43]^+$ which retains the complete information of the derivatized molecule. This is in contrast with the fragment formation of Dns derivatives (Figure 3). Besides the $[M-43]^+$ ion, the molecular ion $(m^{+\bullet})$ is observed with about the same abundance as for Dns derivatives, so that derivatization with Bns-Cl not only allows the determination of smaller amounts but facilitates the identification of compounds in mixtures by means of the characteristic ions M^+ and $[M-43]^+$, the observed intensities of which are in a fixed radio. In contrast with Dns-Cl, Bns-Cl can also be used for the determination of the biologically important amines methylamine and dimethylamine [32]. It has also been advantageously used for the determination of dopamine and some of its derivatives by EI and field desorption mass spectrometry [168].

2.3 6-N-Methylanilinonaphthalene-2-sulphonyl chloride (Mns-Cl)

Mns-Cl (Figure 5) was introduced by Cory *et al.* [169] for N-terminal protein labelling. Its characteristics as a reagent are quite similar to those of Dns-Cl and Bns-Cl, as are the procedures for derivative formation and application. Mns derivatives have considerably higher molar absorption coefficients than the corresponding Dns

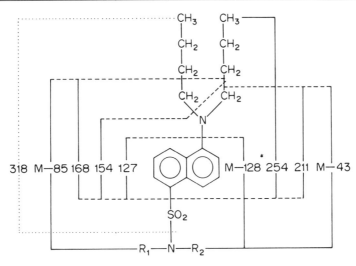

Figure 4. Fragmentation of Bns-derivatives induced by electron impact.

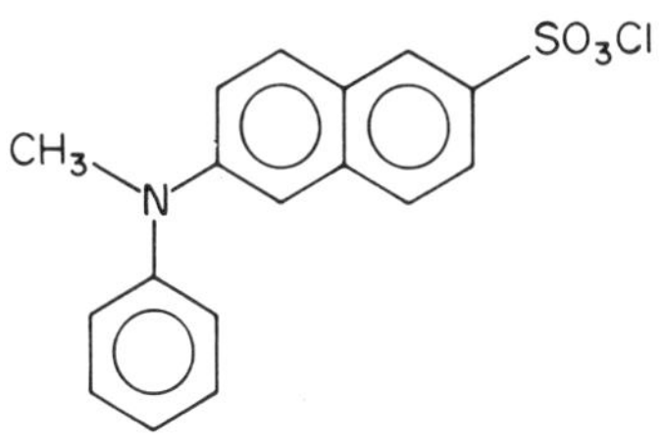

Figure 5. 6-N-Methylanilinonaphthalene-2-sulphonylchloride (Mns-Cl).

derivatives. For Mns-NH$_2$, $\varepsilon_{255} = 4.2 \times 10^4$ and $\varepsilon_{321} = 2.3 \times 10^4$, as measured in n-propanol [170]; the corresponding values for Dns-NH$_2$ are $\varepsilon_{252} = 1.3 \times 10^4$ and $\varepsilon_{333} = 0.43 \times 10^4$ in methanol [7]. The maximun of fluorescence emission is at shorter wavelenght than that of the corresponding Dns derivatives, around 450 nm, with an excitation maximum near 321 nm, whereas Dns derivatives fluorescence maximally at wavelengths about 500 nm.

2.4 2-p-Chlorosulphophenyl-3-phenylindone (Dis-Cl)

Direct sulphonation of 2,3-diphenylindone leads to an almost quantitative conversion into 2-p-chlorosulphophenyl-3-phenylindone [171].

Dis-Cl is in some respect similar to Dns-Cl. It reacts with primary and secondary amines, amino acids, imidazoles, phenols and even with alcohols to form the corresponding amides or esters [171]. The sulphonamides are stable to hydrolytic cleavage with 6 M HCl. Dis-Cl is therefore a suitable reagent for the determination of the amino end-groups of peptides and proteins. The procedure [172] is quite similar to the dansylation method of Gray [6].

Dis-Cl forms orange to red derivatives, which can be detected on chromatograms with about the same sensitivity as the dinitrophenyl derivatives. However, Dis derivatives can be rearranged to strongly green fluorescing 1-phenyl-3-p-sulphophenylisobenzofuran derivatives, by reaction with alcoholic KOH or with sodium ethylate solution (Figure 6) [173]. This improves the detection sensitivity, permitting the detection of amino acids in the range 0.1–1 pmol.

2.4.1 Reaction

Dis derivatives are prepared by mixing equal volumes of the amino acid solution in 0.1 M NaHCO$_3$ with the solution of Dis-Cl (1 mg ml^{-1} in acetone). If the sulphonyl chloride is precipitated, acetone is added until a clear soluton is obtained [172]. After being left for 3 h in a closed tube at room temperature, the solution is evaporated to dryness. The residue is dissolved in 1 ml of acetone methanol (1:1). Aliquots of this solution are used for chromatographic separation.

2.4.2 Applications

For end-group determinations, Ivanov and Vladovska-Yukhonovska [173] worked out TLC systems for the separation of the Dis derivatives of the usual amino acids. After development, the chromatograms are dried at 105 °C and then sprayed with a solution of sodium ethylate (50 mg ml^{-1} in 96% ethanol). The fluorescent

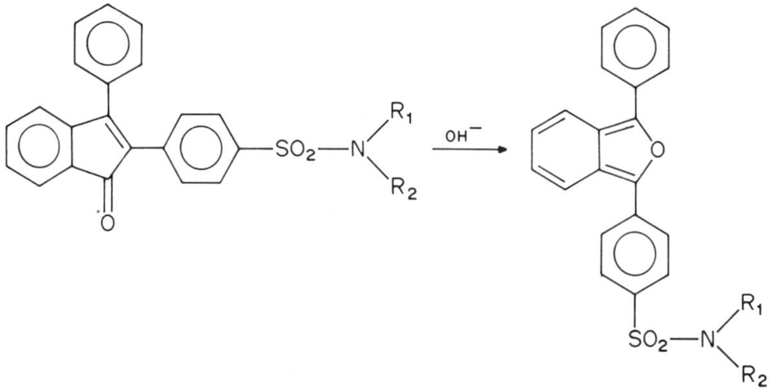

Figure 6. Formation of 1-phenyl-3-*p*-sulphophenylisobenzofuran derivatives from sulphophenyl-3-phenylindone derivatives.

spots are immediately visualized under a UV lamp (365 nm).

Direct (*in situ*) fluorescence quantitation of the spots is not feasible. Durko *et al.* [174] recommend the extraction of the chromatoraphically separated Dis derivatives with acetone, evaporation of the extract and addition of 5 ml of sodium ethylate solution 25 min before fluorescence measurement. For the pyridoxamine derivative, excitation was 410 nm and the emission was measured at 480 nm.

2.5 Other sulphonyl chlorides

1,2-Naphthylenebenzimidazole-6-sulphonyl chloride was prepared by Tockensteinova *et al.* [175] and chromatographic methods (HPLC) and TLC for the separation of its derivatives with aliphatic amines were reported [176].

8-Methoxyquinoline-5-sulphonyl chloride was recommended as a fluorigenic reagent for the determination of amines, amino acids and peptides [177]. These sulphonyl chlorides have no obvious advantages over the related compounds.

3. Carbonyl chlorides

Chlorides of carboxylic acids are considerably more reactive than sulphonyl chlorides. They react not only with primary and secondary amines, but under suitable conditions also with tertiary amines and alcohols. From the standpoint of specificity they have, therefore, disadvantages compared with the fluorigenic sulphonyl chlorides, especially because they are sensitive to hydrolysis. However, owing to their fast reaction, they are suitable for fully automated pre-column derivatization.

3.1 Chloroformates

Chloroformates have been used for derivative formation of amines and alcohols in GLC [178, 179]. Their reaction with primary and secondary amino groups and alcohols is shown in Figure 7. For liquid chromatographic separations chloroformates with a suitable fluorophore have been suggested.

3.1.1 9-Fluorenylmethyl chloroformate (FMOC-Cl)

FMOC-Cl was introduced as a reagent for the protection of amino groups during peptide synthesis [180]. Subsequently, Moye and Boning [181] demonstrated its usefulness for analytical purposes. The compound is commercially available.

3.1.1.1 Derivative formation of amino acids [182]

The reaction is carried out at pH 7.7 (in 0.02 M borate buffer) in order to prevent extensive reaction with phenolic or alcoholic hydroxyls. In this way tyrosine can be measured as a monosubstituted derivative. To 0.5 ml of sample solution is added 0.5 ml of reagent (15 mM FMOC-Cl in acetone). After about 40 s the mixture is extracted with three 2 ml portions of n-pentane. The aqueous solution with the amino acid derivatives is ready for chromatographic separation. If the reaction volumes are reduced to 10–50 μl and the amount of the reagent is minimized, it is not necessary to extract the surplus reagent, even though it is fluorescent. The chromatographic procedure must be suitable to separate the derivatives from the reagent and its hydrolysis product.

Another possibility is off-line solid-phase derivative formation: a solid phase reagent is packed into small

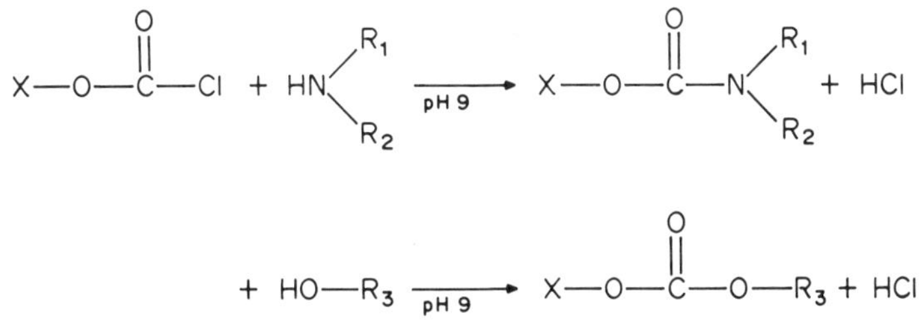

$$X-O-\overset{\overset{O}{\|}}{C}-Cl + HN\overset{R_1}{\underset{R_2}{\diagdown}} \xrightarrow{pH\ 9} X-O-\overset{\overset{O}{\|}}{C}-N\overset{R_1}{\underset{R_2}{\diagdown}} + HCl$$

$$+ HO-R_3 \xrightarrow{pH\ 9} X-O-\overset{\overset{O}{\|}}{C}-O-R_3 + HCl$$

Figure 7. Reaction of chloroformates with an amine and an alcohol. X = fluorophore

cartridges, and the sample solution is passed through. The derivatives can then be eluted and monitored. A procedure for derivative formation of the natural polyamines using new polymeric benzotriazole derivatives loaded with FMOC-Cl has been reported [183].

3.1.1.2 Properties

FMOC derivatives of amino acids are stable at room temperature and in daylight for at least two weeks. The excitation maximum (in an elution buffer of tetrahydrofuran/acetonitrile containing sodium acetate) is at 266 nm and the emission maximum at 310 nm [184]. The quantum yield of amino acid derivatives is in the range 0.3–0.4. The low excitation and emission wavelengths are a disadvantage.

3.1.1.3 Applications

FMOC-Cl has been used for the determination of amino acids [182, 184, 185], natural β-lactams and their intermediates [186] and certain herbicides [187], among others. Perliminary work has been published on a polyamine assay [183].

3.1.2 2-Naphthyl chloroformate (NCF-Cl)

The preparation of NCF-Cl has been described by Raiford and Inman [188]. When tertiary amines are reacted with NCF-Cl in the presence of K_2CO_3 in benzene in a sealed vial at 100 °C for 1 h, the amines undergo dealkylation and fluorescent carbamates are obtained in high yield [189]. These derivatives are suitable for chromatographic separation and determination by fluorimetry (excitation at 275 nm; emission at 335 nm) in the picomole range. Although not yet widely applied, there is little doubt that NCF-Cl can be used similarly to the other fluorescent chloroformates. Its usefulness for the determination of a number of drugs with tertiary amino groups has been demonstrated [189].

3.1.3 2-Dansylethyl chloroformate [190].

2-Dansylethyl chloroformate is used in the same way as FMOC-Cl; it reacts rapidly with hydroxyl groups if the pH of the reaction medium is around 9. The fluorescence characteristics of the derivatives of dansylethyl chloroformate are similar to those of the dansylamides, and therefore the reagent should be preferable to FMOC-Cl and NCF-Cl.

3.2 [R,S]-2-(p-Chlorophenyl)-α-methyl-5-benzoxazoleacetyl chloride

[R, S]-2-(p-Chlorophenyl)-α-methyl-5-benzoxazoleacetyl chloride can be used similarly to chloroformates and sulphonyl chlorides. It reacts with primary and secondary amines and with alcohols. Excitation and emission maxima are at 310 nm and 365 nm, respectively. Solvents for the TLC and HPLC separation the derivatives of a number of amines and drugs have been reported [191].

4. Halogenonitrobenzofurazans

4.1 4-Chloro-7-nitro-2,1,3-benzoxadiazole (4-chloro-7-nitrobenzofurazan; NBD-Cl)

NBD-Cl was noted as a possible fluorescent label during the examination of benzofurazans as potential antileukaemic agents and blockers of thiol groups [192]. NBD-Cl is prepared by nitrating 4-chlorobenzofurazan [193]. It is commercially available. NBD-Cl is a stable, non-fluorescent pale yellow solid, m.p. 97 °C, readily soluble in organic solvents and somewhat more soluble in water than Dns-Cl.

As an aryl halide with an activated halogen, NBD-Cl is closely related to dinitrofluorobenzene. In aqueous solution or in organic solvents it reacts with primary and secondary amino groups, it reacts less readily with phenolic OH and thiol groups under alkaline conditions (Figure 8).

Table 6 Spectral characteristics of NBD derivatives

Compound	Ethyl acetate		50 mM Sodium citrate (with 1 mM EDTA), pH 7.0		Methanol		Reference
	Absorption maximum (nm)	Emission maximum (nm)	Absorption maximum (nm)	Emission maximum (nm)	Absorption maximum (nm)	Colour (on silica gel)	
N-Acetyl-S-NBD-lysteine	464		425	545			202
N,S-bis-NBD-cysteine methyl ester		512					192
S-NBD-Glutathione			420	540			202
Phosphorylase b			430	525			202
N-NBD-Cyclohexylamine			475	545			202
S-NBD-Thioglycollic acid	464	512					192
O-NBD-p-Nitrophenyl ether	464	524					192
O-NBD-N-Acetyltyrosine	464	512					192
N-NBD-Glycine	464	512					192
N-NBD-Methylamine	470 (464)	524 (512)					192, 194
N-NBD-Dimethylamine	464	522 (512)					192, 194
N-NBD-Diethylamine	475	527					194
N-NBD-Pyrrolidine	476	523					194
N-NBD-Piperidine	473	531					194
N-NBD-Benzylamine	464	504				red–orange	194
N-NBD-Aniline	464	524					194
N-NBD-Ephedrine					476	orange	203
N-NBD-Methamphetamine					478	orange	203
N-NBD-Norephedrine					464	yellow	203
N-NBD-Methylphenidate					475	orange	203
N-NBD-Amphetamine					464	yellow	203

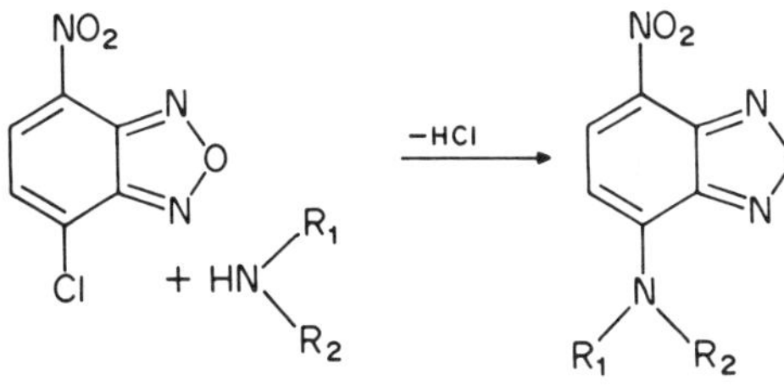

Figure 8. Derivative formation with Nbd-Cl.

For derivative formation with amines the following procedure was recommended [194]: 25–500 μl of the amine solution (1–20 μg of amine) is mixed with 4 volumes of a 0.5 mg ml^{-1} solution of NBD-Cl in methanol, then 50–100 μl of 0.1 M NaHCO$_3$ solution is added. The reaction is completed by heating at 55 °C for 60 min. Yields of NBD-piperidine and NBD-dimethylamine were 95%.

Lawrence and Frei [195] recommend a two-phase system of water/isobutyl methyl ketone, but formation of by-products under these conditions seems to be worse than in methanol [194]. Ethyl acetate and chloroform have also been suggested as reaction media: for the estimation of methamphetamine, a 100 ml portion of the 24-h urine was adjusted to pH 9 and extracted with 50 ml of ethyl acetate. The ethyl acetate phase was concentrated to 10 ml, NBD-Cl (5 mg) was added, together with 1 drop of 2 M NaOH, and the mixture was left at room temperature for 4 h [196]. Amphetamine and morphine were derivatized in the same way using chloroform instead of ethyl acetate [197, 198].

NBD derivatives can be extracted after dilution of the reaction mixtures with water, using ethyl acetate or dichloromethane. Removal of excess reagent can be achieved by chromatography [194]: the reaction mixture is applied to a silica gel column (1.25 cm × 60 cm: particle size 10–40 μm). With cyclohexane ethyl acetate (1:1) the non-fluorescent NBD-Cl is the first compound to be eluted.

It has been pointed out [199] that, in methanol-containing media, NBD-Cl partly solvolyses to form NBD-OCH$_3$, which also reacts with amino functions to yield fluorescent derivatives.

Non-polar 7-nitro-4-benzofurazanyl ethers react faster than NBD-Cl; polar ethers are less reactive, but owing to their increased solubility in water they do not require the presence of an organic solvent.

4.2 4-Fluoro-7-nitrobenzofurazan (NBD-F)

In order to reduce the reaction time and to increase the yield, NBD-F (m.p. 53.5 °C) was introduced [200]. One minute is sufficient for the complete reaction of primary and secondary amino acids in 50% ethanol/0.1 M borate buffer (pH 8.0) at 60 °C in the dark. Hydrochloric acid is added (to pH 1.0) in order to terminate the reaction.

4.3 Properties of NBD-derivatives

NBD-derivatives are stable towards hydrolysis with 6 M HCl for 4 h at 110 °C [201], so that end-groups of peptides and proteins can be determined with NBD-Cl.

In contrast to most other fluorescent labels, the absorption maximum of the NBD derivatives is in the visible region. Unfortunately the excitation and emission bands overlap. In Table 6 some spectral chracteristics of several NBD derivatives have been summarized. The fluorescence excitation and emission maxima of a large number of amino acid derivatives have been published by Fager et al. [201].

The quantum efficiencies of NBD-hydroxyproline range from 0.01 in water (pH 9) to 0.80 in isobutyl methyl ketone [204]. Tryptophan does not yield a fluorescent product [205].

4.4 Applications

Only pre-chromatographic applications of NBD-Cl and NBD-F are considered here.

Although the reagent is non-fluorescent, it nevertheless interferes with fluorescence measurements, so that the excess has to be removed before fluorimetry. Usually chromatographic procedures have been applied for this purpose. TLC, and more recently reversed phase HPLC, have been used for the determination of primary and secondary amines, amino acids, and drugs in the picomole and subpicomole range. Applications are summarized in Table 7.

Table 7 Applications of halogenonitrobenzofurazans

Application	Remarks	Reference
Separation and identification of amino acids	In protein hydolysates	201, 202, 207, 208
	Automated method	206
	Hydroxyproline in collagen	204, 209
Amines		194, 196, 203, 205, 210
	Identification by mass spectrometry	213
Drugs		174, 196, 197, 198, 211, 212

5. Isocyanates and isothiocyanates

Isocyanates and isothiocyanates react with primary amines to give urea and thiourea derivatives respectively. However, isocyanates react quite readily with water and alcohols to give urethanes. For this reason they are generally replaced by the less reactive isothiocyanates. However, α-naphthylisocyanate was recommended for the determination of lipids with a hydroxyl function [214] and may also find application in amino acid analysis.

5.1 9-Isothiocyanatoacridine

9-Isothiocyanatoacridine is prepared from 9-chloroacridine by replacement of the chloro group by the isothiocyanate group [215].

The reaction of 9-isothiocyanatoacridine with primary and secondary aliphatic amines leads to several fluorescent products. The fluorescence of one these products could be related to the amount of amine present [216]. On chromatographic evidence the fluorophore is a cyclization product, 2-alkylamino-1,3-thiazinol[6,5,4-kl] acridine, which is produced by photooxidation from the initially formed thiourea derivative [217].

All amines with $pK_a > 9.3$ were successfully determined with the following procedure, but no compound with $pK_a < 5.3$ could be measured. In practice, 5 ml of a toluene solution of the amine (free base) is mixed with 2 ml of a solution of 9-isothiocyanatoacridine (0.8 mg ml^{-1} in toluene), and stored overnight at room temperature. In the morning the mixture is exposed for a limited period to indirect sunlight. Experiments to maximize the yield of cyclized product by UV irradiation of the reaction mixture in a photochemical reactor were unsuccessful: the required excess of reagent was readily converted into acridone, resulting in excessive background fluorescence. Water interferes with the method, but limited amounts of water can be tolerated, so permitting analysis of aqueous solutions of amines [216]. Quantification is by fluorimetry of the toluene reaction mixture (excitation at 295–310 nm; emission at 500–525 nm). When the reaction mixture was mixed with acidic ethanol (0.5 ml of the toluene solution, 4.5 ml of ethanol and 0.1 ml of conc. HCl), there was a pronounced increase in the fluorescence of the amine reaction solution in comparison with the blank, but at the cost of a decrease in the stability of the derivative.

Linearity of response of *in situ* fluorescence measurements was found over the range 3–25 µg (recovery 98.9%; standard deviation 9.9%).

5.2 Fluorescein isothiocyanate

Fluorescein isothiocyanate was proposed by Maeda *et al.* as a fluorescent end-group reagent [218] analogous to phenylisothiocyanate in the Edman method. A procedure for the determination of free amino acids by formation of fluorescent thiohydantoins with fluorescein isothiocyanate was reported by Kawauchi *et al.* [219].

5.2.1 Preparation

Nitrofluorescein is prepared by the method of Coons and Kaplan [220] from resorcinol and 4-nitrophthalic acid. The two isomers formed are separated by fractional crystallization. The nitro group is reduced as described by McKinney *et al.* [221]. Finally, fluorescein isothiocyanate is obtained from aminofluorescein by treatment with thiophosgene [222].

5.2.2 The use of fluorescein isothiocyanate [219, 223]

The reaction of fluorescein isothiocyanate with an amino acid is depicted in Figure 9. In the first step the fluorescein thiocarbamoyl amino acid is formed by addition of the amino acid to the isothiocyanate group. This reaction product could be used for the assay of the amino acids. However, the thiocarbamoyl amino acids are usually converted into the corresponding thiohydantoins by acidification.

Fluorescein isothiocyanate (0.3 mmol) is dissolved in 5 ml of acetone with a trace of pyridine. This solution (0.05 ml) is added to an amino acid solution (0.5 ml, about 10 µmol) in 0.2 M carbonate/bicarbonate buffer, pH 9, and allowed to react at 25 °C for 4 h. The reaction is stopped by acidifying with glacial acetic acid to pH 4.5, and the precipitate formed is dissolved in about 0.5 ml

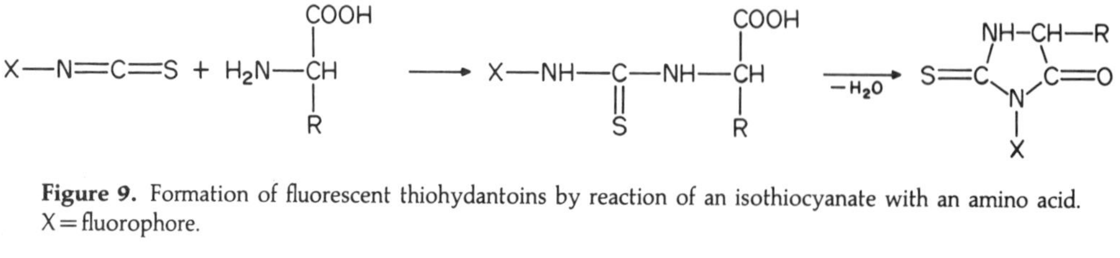

Figure 9. Formation of fluorescent thiohydantoins by reaction of an isothiocyanate with an amino acid. X = fluorophore.

of acetone. To this solution 0.2 ml of 6 M HCl is added to convert the fluorescein thiocarbamoyl derivative to the fluorescein thiohydantoin. The solutions of the fluorescent thiohydantoins are immediately used for chromatographic separation. On silica gel G, unreacted fluorescein isothiocyanate runs behind the solvent front and aminofluorescein, a decomposition product, stays at the origin in the solvent systems devised by Kawauchi et al. [219].

Sulphhydryl groups react with fluorescein isothiocyanate, and phenolic groups (of tyrosine) react to a small extent under the above reaction conditions. Fluorescein thiohydantions in amounts exceeding 1 nmol are visible as yellow spots on TLC plates and show up as intense greenish-yellow fluorescent spots under UV light.

5.3 Other isothiocyanates

4-Dimethylamino-1-naphthylisothiocyanate [224], 4-(benzyloxycarbonylaminomethyl)phenylisothiocyanate [225] and 4-(dimethylaminonaphthalene-1-sulphonylamino)phenylisothiocyanate [226, 227] have been introduced in order to improve the sensitivity of the Edman method. These reagents are used in the same way as the other isothiocyanates, and allow end-group determinations in the picomole·range.

6. Fluorescamine

Fluorescamine (4-phenylspiro[furan-2(3H), 1'-phthalan]-3,3'-dione) is a non-fluorescent compound, with m.p. 154 °C. It is readily soluble in acetone, dioxane, tetrahydrofuran, dimethyl sulphoxide, etc. It is only slightly soluble in water, but it is hydrolysed in aqueous media to non-fluorescent products. A 0.5 M solution in acetone is commercially available.

6.1 History

Udenfriend and co-workers showed that phenylacetaldehyde, which is formed by oxidative decarboxylation of phenylalanine by ninhydrin, reacts with excess of ninhydrin and with a primary amino group to give a fluorescent product [228]. The elucidation of this mechanism and of the structure of the fluorescent reaction product [229] led to the synthesis of fluorescamine [230].

6.2 Reaction and reaction conditions

Compounds with nucleophilic functional groups (primary and secondary amines, alcohols, water, etc.) are capable of reacting with fluorescamine, but only primary amines form fluorescent products. Fluorescamine is, therefore, a specific reagent for compounds with primary amino groups.

The reaction of fluorophore formation is shown in Figure 10. In a reversible, rapid reaction, the primary amine is added to the double bond of the reagent. The addition product is then transformed in a multistep rearrangement to the fluorophore [231]. For the reaction of aliphatic primary amines and peptides a pH of 8–8.5 is adequate, whereas proteins and amino acids are normally reacted at pH 9. At higher pH, hydrolytic

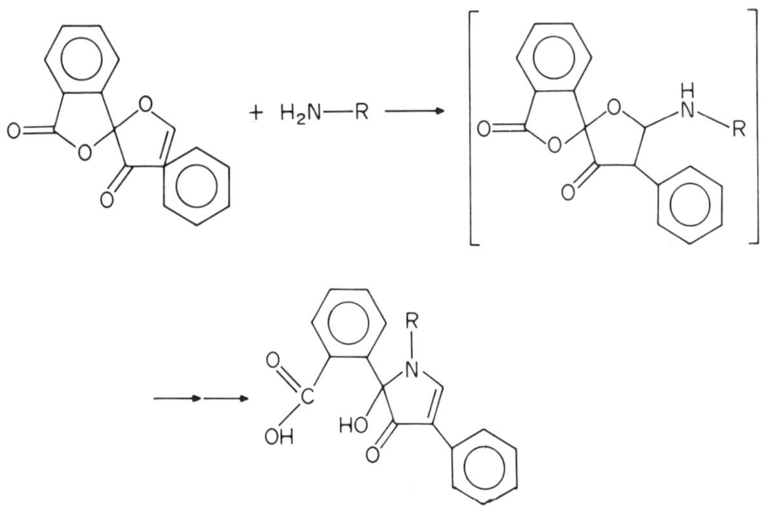

Figure 10. Derivative formation with fluorescamine.

cleavage of the reagent competes with the formation of the fluorophore, so that yields of the fluorophore decrease. Buffer and salt concentrations and organic solvents in the reaction mixture influence fluorophore formation as well, but their influence is less pronounced than that of pH.

The estimation of amines, amino acids, peptides and proteins is carried out by essentially the same procedure [232, 233]. Three volumes of a solution of a primary amine, peptide or protein in 0.05 M phosphate buffer, pH 8–9.5 (or a 0.05 M sodium borate buffer, pH 8–9.5) are mixed under vigorous stirring with 1 volume of fluorescamine solution (0.28 mg ml^{-1} in acetone or 0.56 mg ml^{-1} in dioxane). Fluorescence measurement is carried out 5–30 min after mixing. Yields of the fluorophore are 80–90%. There is a linear relationship between the concentration of the amine and the observed fluorescence up to 25–30 nmol ml^{-1}.

6.3 Properties of the fluorophore

The reaction products of some amines and peptides with fluorescamine have been synthesized as crystalline solids [232]. Solutions of these compounds are stable in neutral and mildly alkaline medium in the dark. In acidic solutions the fluoroescence rapidly deteriorates. Fluroescence efficiency is highest at acidic pH because of the highly fluorescent protonated form of the fluorophore. In freshly prepared solutions, the fluorescence intensity decreases in two distinct steps with increasing pH, reflecting the presence of two dissociable acidic functions in the molecule with pK values of 3.8 and 11.6.

Fluorescence measurements are carried out in the range of greatest stability of the fluorophore (pH 4.5–10.5), corresponding to the molecule with the dissociated carboxyl group. In this range the absorption maximum is at 390 nm and fluorescence emission is at 475 nm. Quantum yields of amine and peptide derivatives in ethanol are normally in the range 0.2–0.34 [232].

6.4 Applications

Fluorescamine has found its main application as a post-column reagent in the estimation of amines, amino acids or peptides separated by column chromatography, especially in atuomated amino acid analysis, where the reagent (0.15 mg ml^{-1} of fluorescamine in acetone) is continuously added to the column effluent buffered with sodium borate (0.16 M; pH 9.6) [234]. It can also be used for the estimation of proteins [233, 235] and their visualization in gel electrophoretograms [236].

Pre-column derivative formation with fluorescamine was successfully applied to the determination of biogenic amines [237–240], polyamines [241–243], 3-methylhistidine [244], aminocaproic acid [245], peptides (oxytocin, vasopressin [246, 247] and some drugs [248–256]. α-Amino acids gave double peaks on reversed phase columns, probably because of intramolecular cyclization [257].

7. 2-Methoxy-2,4-diphenyl-3(2H)-furanone (MDPF)

MDPF is closely related in structure to fluorescamine. Both primary and secondary amines react with MDPF in acetonitrile, but only primary amines form fluorescent pyrrolinones. Secondary amines form non-fluorescent aminodienones. By reacting with ethanolamine, the latter produce fluorescent products. Their stability is adequate for chromatographic separation [258, 259].

8. Schiff base-forming and related reagents

8.1 Pyridoxal and pyridoxal 5-phosphate

Pyridoxal and pyridoxal 5-phosphate are the only naturally occurring compounds which have been used as fluorescent labels. These compounds form Schiff bases with primary amino groups. The reduction of the C=N bond with sodium borohydride (NaBH$_4$) leads to the formation of fluorescent pyridoxyl derivatives [260, 261].

8.1.1 Reaction

The reaction of pyridoxal with a primary amine and the reduction of the Schiff base to the pyridoxyl derivative are shown in Figure 11. As only primary amino groups form Schiff bases, the reaction should be specific for compounds with primary amino groups. However, proline and hydroxyproline react as well and, as noted [261], even secondary amines can be brought to reaction, especially in non-aqueous solvents. A ketoenimine is assumed to be an intermediate of the reaction sequence in this case. Histidine reacts first via the primary amino group to form a Schiff base, which is rearranged to a non-reducible imidazolopyridine derivative.

The following procedure has been suggested. Amino acids and primary amines are dissolved in 0.5 M phosphate buffer, pH 9.3. This solution is mixed with the same volume of a freshly prepared 0.4 M pyridoxal solution in the same buffer (pyridoxal/amino acids = 5:1).

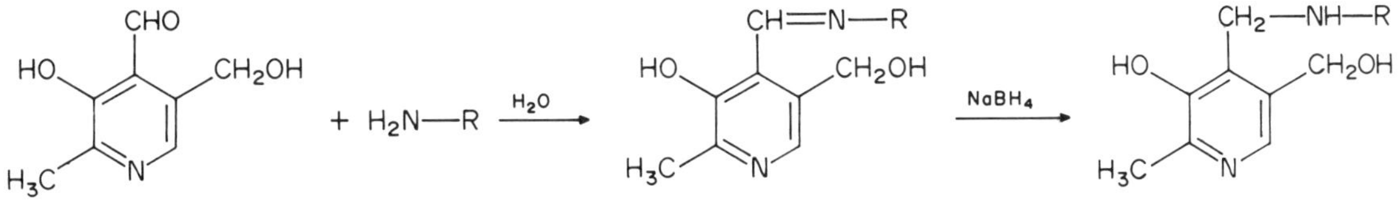

Figure 11. Formation of a Schiff base with pyridoxal and its reduction to a pyridoxyl derivative.

After heating for 30 min at 80 °C, 100 mg of NaBH$_4$ dissolved in 1 ml of 0.1 M NaOH is added. The excess of NaBH$_4$ is decomposed by acidification to pH 1–2 with hydrochloric acid.

The yield of α-amino acid derivatives is > 90%. Lysine and ornithine form dipyridoxyl derivatives. Recovery of the proline and hydroxyproline derivatives is about 25%.

For derivative formation with amines, pyridoxal 5-phosphate is preferred to pyridoxal in order to allow ion echange chromatographic separation. The reaction mixtures are submitted immediately to chromatographic separation.

8.1.2 Properties of pyridoxyl derivatives

Pyridoxyl derivatives are stable at 0 °C for some weeks if protected from light. Their fluorescence decays within minutes if the solutions are irradiated at the wavelength of the absorption maximum.

Pyridoxylamino acids exhibit spectral characteristics similar to those of pyridoxamine: absorption maxima at 255 and 328 nm, and fluorescence emission at 400 nm. Fluorescence efficiency is pH-dependent, with maximum fluorescence at pH 5.28 (except for the histidine derivative, which shows maximum fluorescence at pH 12). Molar absorption coefficients and fluorescence efficiencies are, with few exceptions, the same for all pyridoxyl amino acids. Between 10 and 100 pmol of an amino acid can be determined in the effluent from an amino acid analyser [261].

The pyridoxyl residue increases the retention time, but most amino acid derivatives are eluted from the cation exchange column in the same order as the free amino acids. For continuous fluorescence measurement, one part of the effluent is mixed with 49 parts of sodium citrate buffer, pH 5.28.

One of the advantages of the pyridoxal method is the possibility of using sodium borotritide (NaBT$_4$), which is available with a specific activity of about 116 Ci mmol^{-1}. Thus, radioactively labelled pyridoxyl derivatives can be obtained, allowing the detection of about 0.1 pmol of an amine or amino acid.

8.2 2-Fluorenecarboxaldehyde and 1-pyrenecarboxaldehyde

2-Fluorenecarboxaldehyde and 1-pyrenecarboxaldehyde are commercially available, and form strongly fluorescent Schiff bases with primary amines. These can be separated by TLC, and presumably also by other chromatographic methods, and determined sensitively (excitation at 450 nm; emission at 500 nm). Preliminary experiments demonstrate the potential usefulness of these reagents for the determination of histamine and amphetamine [262].

8.3 o-Phthaldialdehyde/alkylthiol (OPA/R-SH) reagents

o-Phthaldialdehyde (OPA) is a commercially available reagent. It has been used for the assay of histamine [263, 264] and spermidine [265, 266]. A number of other biologically important compounds, among them glutathione, arginine, agmatine and 5-hydroxy- and 5-methoxy-indole derivatives and histidine-containing peptides, form fluorescent derivatives with OPA [267–274].

A much more general application of OPA was introduced when Roth [275] showed that nearly all amino acids can form fluorescent condensation products in the presence of an alkylthiol, such as 2-mercaptoethanol. Continuous reaction of the column effluent with OPA/2-mercaptoethanol or other thiol at pH 10, followed by recording of fluorescence intensity (excitation at 340–345 nm, emission at 455 nm), became a very widely used method for the determination of amino acids [276–298], peptides [280, 291, 294–299], biogenic amines and polyamines [300–317], antibiotics [318–323] carbamates [324] and other primary amino group-containing compounds. OPA/R-SH reagents have to a considerable extent replaced ninhydrin for post-column derivative formation.

With the advent of automated devices for sample preparation [325–327], OPA/R-SH reagents have attracted much interest for pre-chromatographic derivative formation, despite the instability of the reaction products.

8.3.1 Reaction of OPA and thiols with compounds containing primary amino groups

At pH 10 in aqueous media (usually borate buffer) compounds with primary amino groups react with OPA in the presence of a thiol to give 1-alkylthio-2-alkylisoindoles [328–331] (Figure 12). If OPA/R-SH is present in concentrations exceeding that of the amine by 2–3 orders of magnitude, the reaction is usually complete within 1–2 min at room temperature.

2-Mercaptoethanol [275] is most frequently used, but 3-mercapto-1-propanol [332] ethanethiol [329] and 3-mercaptopropionic acid [314] are also in use. Ethanethiol reacts more slowly than 2-mercaptoethanol, but quantitative reaction can be obtained [333].

The use of mercaptopropionic acid or of ethanethiol is imperative if amino acids with primary amino groups are reacted first to form isoindoles, and if then a chloroformate [184] or another fluorigenic reagent which is capable of reacting with hydroxyl groups is used for the labelling of proline and hydroxyproline.

The following procedure [334] is typical for the manual derivatization of amino acids and primary amines.

The sample solution (50 μl) is mixed with 4 volumes of reagent. At a timed 2 min interval, a 20 μl aliquot is injected into the HPLC system.

Extraction with chloroform or ethyl acetate (100 μl) removes excessive reagent and may be advantageous.

OPA reagent [335]

OPA (27 mg) is dissolved in 0.5 ml of ethanol, and 5 ml of 0.1 M sodium tetraborate (or 0.4 M boric acid adjusted to pH 9.5 with NaOH) are added, followed by 20 μl of mercaptoethanol (or the equivalent amount of another thiol). The reagent is kept overnight before use, and 10 μl of mercaptoethanol are added each week to maintain maximum yield.

If an automated system for sample preparation is used, it is possible to decrease the reaction volumes considerably [184].

8.3.2 Properties of 1-alkylthio-2-alkylisoindoles

1-Alkylthio-2-alkylisoindoles usually have quantum yields of 0.3–0.5 [336], but the derivatives of the individual amino acids show quite great variation in their fluorescence intensity [337]; the isoindole from glycine (and ethanethiol) shows the highest response, those from aspartate and glutamate a low response. The poudcts from cysteine, cystine and lysine are also known to

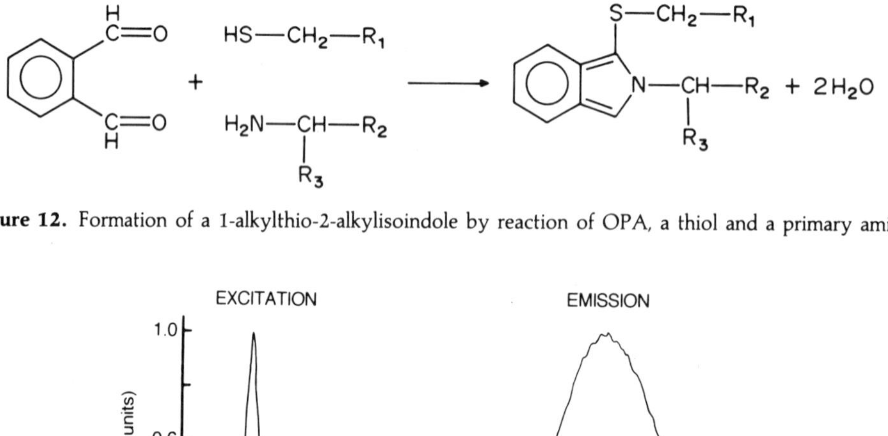

Figure 12. Formation of a 1-alkylthio-2-alkylisoindole by reaction of OPA, a thiol and a primary amine.

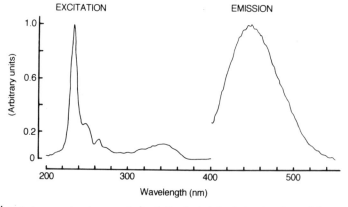

Figure 13. Excitation and emission spectra (uncorrected) of the isoindole derivative formed by reaction of glycine with OPA/2-mercaptoethanol.

fluoresce only weakly [275]. Owing to the quenching of fluorescence by carboxamide groups [336], the method is less sensitive for the determination of peptides than for amino acids, unless detergents (sodium dodecyl sulphate and dimethyl sulphoxide) are present [280, 336]. The amino acid derivatives show constant fluorescence intensity in the pH range 6.0–11.5 [275, 329] and this is the preferred range for post-chromatographic derivatization.

1-Alkylthio-2-alkylisoindoles exhibit several excitation maxima (Figure 13). Usually the maximum at 340 nm is used, but Schuster [184] recommends the much more intense maximum at 230 nm for excitation of fluorescence in order to improve the detection sensitivity. The emission maximum at 455 nm is universally used.

The isoindoles are unstable, and decompose via an intramolecular rearrangement to non-fluorescent 2,3-dihydro-1H-isoindol-1-ones [328, 330]. The degree of instability depends somewhat on the thiol and the amine component [332, 335]. Cooper et al. [338] published a detailed study of the on-column stability of the isoindoles.

8.3.3 Scope and limitations of the OPA/R-SH reaction

Secondary amines do not form isoindoles. Attempts have been made to overcome this difficulty by reaction with chloramine T [279], N-chlorosuccinimide [339] or sodium hypochlorite [276, 315].

Using chiral thiols for derivative formation, isoindoles are formed from amino acids and amino alcohols. BOC-L-cysteine, N-acetyl-L-cysteine, N-acetyl-D-penicillamine [340] and 1-thio-β-D-glucose [341] have been shown to be suitable reagents which allow the separation of most amino acid enantiomers using reversed phase column chromatography. Thus, OPA/N-acetyl-L-cysteine has been used among others for the separation of enantiomers of aspartate [342], baclofen [343], norepinephrine, dopa [344] and lombricine [345].

Advantages of pre-column derivative formation with OPA/R-SH include the relatively short run time for amino acid separations and the sensitivity of the method. Compared with post-column derivative formation, a considerably lower background fluorescence is observed. If properly timed (e.g. by automated sample preparation [184, 346–349]), the method allows the routine determination of pmol, or even fmol, quantities of certain amino acids.

An alternative to fluorescence detection, with comparable sensitivity, is provided by the use of an electrochemical detector [334].

8.3.4 Applications of pre-column derivative formation with OPA/R-SH

The method finds increasing application for the determination of amino acids. Factors affecting the retention of the isoindole derivatives by reversed phase columns have been investigated [335, 337]. Usually columns with C_{18} brushes are used [350]. Numerous gradient elution systems suitable for the separation of amino acid-derived isoindoles have been published; their number is still increasing with the availability of new column types. The various references [184, 337, 339–365] are a representative selection of those available, most of which involve minor variations of the basic method.

In addition to more or less complete amino acid separations, suitable chromatographic systems for the rapid determination of selected amino acids [4-aminobutyric acid (GABA) [336], histidine and 3-methylhistidine [367] or a group of selected amino acids, such as those involved in the urea cycle (arginine, citrulline, ornithine, agmatine) [368], have been reported. Pre-column derivative formation with OPA/R-SH reagents is also suited to the automated establishment of peptide maps [369]. A number of authors have published methods for the sensitive determination of aminoglycoside antibiotics [365, 370–373]. The method has also found application for the determination of histamine and its methylation products [374–376], catecholamines and serotonin [377–379] and polyamines [380].

8.4 2-Acetylbenzaldehyde (OAB)

Based on considerations of the mechanisms involved in the degradation of the isoindoles formed from OPA, a thiol and a primary amine, it was concluded that 1-alkylthio-2,3-dialkylisoindoles are more stable than the usual 1,2-dialkylisoindoles. For their formation, o-ketobenzaldehydes are suitable. The preparation of OAB (and of 2-benzoylbenzaldehyde) was reported, and it was demonstrated that the condensation product of 4-aminobutyrate with OAB/ethanethiol is much more stable than the corresponding derivative obtained with OPA/ethanethiol. In preliminary experiments, 50 fmol of 4-aminobutyric acid could be detected using HPLC as separation method [381]. Thus, it appears that OAB/ethanethiol is a promising replacement for the OPA/R-SH reagent.

8.5 ω-Formyl-o-hydroxyacetophenone and benzopyrone

ω-Formyl-o-hydroxyacetophenone and benzo-γ-pyrone react with primary and secondary aliphatic and aromatic amines to give β-aminovinyl-*o*-hydroxyphenyl ketones. Kostka [382] has utilized this reaction for the derivative formation of amines and their sensitive detection in TLC. The chromatographic characteristics of a number of enamines were studied using ethyl acetate/benzene (1:5), chloroform/xylene (4:1) and acetone/xylene (1:9) and Silica gel G layers. Down to 0.1–1.0 µg of the derivatives can be detected in UV radiation.

8.6 Benzoin, a reagent for the determination of compounds containing guanidino groups

Benzoin (2-hydroxy-2-phenylacetophenone) condenses with guanidino compounds to form 2-substituted amino-4,5-diphenylimidazoles [383]. These imidazoles are suitable for chromatographic separation and determination by fluorescence measurement (excitation at 325 nm, emission at 425 nm).

8.6.1 Derivative formation [384]

The sample (200 µl in water) is mixed with 100 µl of benzoin solution (4 mM in methyl cellosolve), and 100 µl of an aqueous solution containing 0.1 M 2-mercaptoethanol and 0.2 M sodium sulphite and 200 µl of 2.0 M potassium hydroxide. The mixture is heated in a boiling water bath for 5 min, then cooled on ice, and 200 µl of a 1:1 mixture of 2 M HCl + 0.5 M Tris/HCl buffer, pH 9.2 is added. Aliquots (100 µl) are used for chromatographic separations. 2-Mercaptoethanol stabilizes the resultant imidazole derivatives and sodium sulphite suppresses background fluorescence.

8.6.2 Applications

Methods for the separation of a series of guanidino compounds, such as creatine, creatinine arginine, argininosuccinic acid, agmatine etc. [384, 385] and guanidine-containing peptides (angiotensin [386], leupeptin [387]) and their sensitive determination in body fluids have been reported.

9. Further reagents for the labelling of amino compounds

9.1 *N*-Succinimidyl-2-naphthoxyacetate

This compound was originally used for the detection of amino acids on paper chromatograms. More recently, it has been used for the detemination of aminophospholipids [388]. Reaction is achieved in chloroform in the presence of triethylamine at room temperature. With a 3-fold excess of the reagent reaction is complete within 2 h. The labelled derivatives are detected at 342 nm in the column eluent after fluorescence excitation at 228 nm.

9.2 *N*-Succinimidyl-1-naphthylcarbamate

This compound has been synthesized by Nimura *et al.* [389]. In 0.5 M borate buffer, pH 9.5, it reacts with amino acids within 1 min at room temperature to form naphthylcarbamoyl derivatives. The rapid reaction and the removal of excessive reagent by hydrolysis make *N*-succinimidyl-1-naphthylcarbamate suitable for automated pre-column derivative formation [390]. The intense fluorescence (excitation at 290 nm, emission at 370 nm) allows the determination of sub-picomole quantities. A HPLC system suitable for the separation of the commonly occurring amino acids has been worked out [390].

9.3 5-(4,6-Dichloro-1,3,5-triazine-2-yl)-aminofluorescein (DTAF)

DTAF was originally suggested for the cyto- and histochemical detection of proteins [391]. Recently, its use for the labelling of secondary amines has been suggested, and a method for the determination of desipramine in plasma in the fmol range was reported [392, 393].

10. Reagents for carbonyl compounds

Aldehydes and ketones readily react with hydrazine and hydrazines, respectively, to give the corresponding hydrazones, so it seemed reasonable to devise fluorescent hydrazines for the fluorescence labelling of carbonyl compounds by analogy with the known coloured 2,4-dinitrophenylhydrazine.

10.1 Dansyhydrazine (Dns-H) (Figure 14)

Dns-H has been suggested independently by three different groups [17, 394, 395] as a fluorescent reagent for

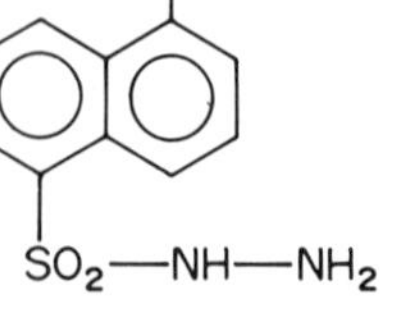

Figure 14. Dansylhydrazine (Dns-H).

carbonyl compounds. It is prepared by addition of a concentrated solution of Dns-Cl to a solution of a large excess of hydrazine monohydrate in acetone. The rate of addition of Dns-Cl is matched to the progress of the reaction, which can be followed by the disappearance of the yellow colour of Dns-Cl. The reaction mixture is diluted with water and the product is extracted with ethyl acetate. Pale yellow crystals, m.p. 126 °C, are obtained on evaporation to dryness, and are soluble in practically all organic solvents but only slightly soluble in water [17] Dns-H is commercially available.

10.1.1 *Derivative formation with steroids* [394]

The dry residue of the steroid extract, in a glass-stoppered centrifuge tube, is treated with ethanolic HCl (0.65 ml l^{-1}) followed by 0.2 ml of the mono-Dns-hydrazine solution (2 mg ml^{-1} in ethanol). The tube is heated for 10 min in a boiling water bath and then cooled to room temperature. A solution of 5 mg ml^{-1} of sodium pyruvate in ethanol (0.2 ml) is mixed with the contents of the tube to react with the excess of mono-Dns-hydrazine. After 15 min at room temperature, 3 ml of 0.5 M aqueous NaOH is added and the steroid derivatives are extracted with 6 ml of diethyl ether. If the amounts are sufficient, and if a suitable method is available to separate the reaction products from the excess of reagent, aliquots of the reaction mixture can be applied to chromatographic separation immediately after completion of the reaction. Kawasaki *et al.* [396] recommend a 0.1% solution of trichloroacetic acid in benzene as reaction medium instead of ethanol/HCl.

10.1.2 *Derivative formation with sugars*

Based on the method of Avigad [397], the following optimized method was recommended [398]. To 100 µl of aqueous sample (or standard) are added 10 µl of trichloroacetic acid (10% in water) and 50 µl of a 5% solution of Dns-H in acetonitrile. The sealed tube is heated at 65 °C for 20 min. The reaction is stopped by immersion of the tube in an ice bath. After 1:1 dilution with water, 10–300 µl of the reaction mixture are used for HPLC. Since the products are of limited stability, chromatographic separation within 2 h is advisable.

10.1.3 *Mode of application*

Hydrazones formed with Dns-H have fluorescence characteristics similar to those of the Dns-amides. Excitation of fluorescence can be achieved either at 360–370 nm or at 240 nm. Fluorescence emission is measured at 525–540 nm. Both TLC and HPLC have been used as separation methods.

Methods for the determination of 3-keto- and 17-ketosteroids and 17-ketosteroid glucuronides and sulphates have been reported [394, 395, 399–401]. Bile acids can be determined after enzymatic oxidation of the 3-OH to a keto group [396].

Mono- and polysaccharides have been determined with great precision (S. D. ± 0.4% at the 0.5 nmol level) [398, 402, 403].

10.2   4-Hydrazino-7-nitrobenzofurazan (Nbd-H)

Substitution of the halogen atom of Nbd-Cl by a hydrazine residue yields a reagent analogous to Dns-H [404] which is also suited to the determination of aldehydes and ketones. The hydrazones are formed in 99% yield.

10.3   4′-Hydrazino-2-stilbazole (4H2S)

4H2S is used by analogy with other hydrazines as a selective reagent for carbonyl functions [405, 406]. Detection limits of α-keto acids are in the pmol range.

10.4   Semicarbazide

Pyridoxal 5′-phosphate (PLP) forms with semicarbazide a fluorescent semicarbazone. A relatively simple and reliable method for the determination of PLP was based on this reaction [407].

10.4.1 *Reaction conditions* [407]

Perchloric acid tissue extracts are adjusted to pH 6–7 with 3 N KOH. Aliquots (2 ml) of each neutralized extract are mixed with 0.5 ml of aqueous 0.2 M semicarbazide·HCl, and heated in a boiling water bath for 5 min. To the cooled solutions is added 0.5 ml of 0.2 M H_3PO_4. The filtered solutions are ready for chromatographic separation.

10.4.2 *Properties*

The semicarbazones of pyridoxal and PLP can be separated by HPLC using a reversed phase column and determined in the column effluent by fluorimetry (excitation maxima at 333 and 380 nm, fluorescence maximum at 446 nm). It is advantageous to use the 380 nm maximum for fluorescence excitation. The detection limit is about 0.2 µg of PLP per ml of plasma or per g of tissue [407].

10.5 2-Aminopyridine

2-Aminopyridine reacts with reducing sugars to form Schiff bases that can be reduced to fluorescent 2-aminopyridyl derivatives [408]. This reaction has been used for the determination of pmol quantities of neutral and amino sugars and of muraminic acid [409].

Glycoconjugates are hydrolysed by heating with trifluoroacetic acid/HCl. Subsequently, the amino sugars are acetylated by reaction with acetic anhydride in sodium bicarbonate. The sugars are then separated on a small Dowex-50 W-2 column (H^+ form). The dried column eluate is reacted with 5 μl of a 2-aminopyridine solution (0.5 g of 2-aminopyridine, 0.4 ml of conc. HCl, 11 ml of water; pH 6.2) by heating for 13–15 min at 100 °C, followed by reduction achieved by adding 2 μl of an aqueous solution of sodium cyanoborohydride (20 mg ml^{-1}) and heating at 90 °C for 8 h. Excess reagent is removed by gel permeation chromatography, and the 2-aminopyridyl derivatives are separated by HPLC. The 2-aminopyridyl derivatives are detected in the column eluate by fluorescence excitation at 320 nm and emission measurement at 400 nm.

10.6 1,2-Diphenylethylenediamine (DPE)

By analogy with the well known ethylenediamine method for the fluorimetric determination of catecholamines [410], the use of DPE was suggested to convert catecholamines into highly fluorescent products [411].

The synthesis of DPE was described by Nohta et al. [411]. After the usual clean-up by alumina adsorption, cation exchange column chromatography or other suitable procedure [412], 150 μl of the sample solution (adjusted to pH 6.0–6.5) are mixed with 10 μl of a 20 mM solution of potassium ferricyanide and 150 μl of acetonitrile. Subsequently, 50 μl of the DPE solution (21.2 mg ml^{-1} in 0.1 M HCl) are added and the reaction mixture is allowed to stand at 37 °C for 40 min. After cooling to room temperature, 100 μl aliquots are injected onto the column. Fluorescence intensity is recorded at 485 nm after excitation at 345 nm. Procedures for the determination of norepinephrine, epinephrine and dopamine in plasma [412] and urine [413] were reported.

The method allows the detection of 2 fmol of the usual catecholamines in a 100 μl injected volume.

10.7 2-Cyanoacetamide

2-Cyanoacetamide reacts with reducing carbohydrates and polyphenols to form fluorescent condensation products [414]. Usually this reaction is used as post-column detector reaction for the determination of sugars [414] and catecholamines [415] but it can also be used for pre-chromatographic derivative formation [416]. The condensation product with D-glucose has an excitation maximum at 331 nm and an emission maximum at 383 nm.

10.8 Quinoxalinol formation with o-phenylenediamine: a method for the determination of α-keto acids

Spikner and Towne [417] observed that the condensation of α-keto acids with o-phenylenediamine yields highly fluorescent quinoxalinols (Figure 15). This reaction attracted much attention during the last decade as a sensitive method for the determination of α-keto acids [418–421]. See also Chapter 7.

10.8.1 Reaction conditions

A preparative method for standard quinoxalinol derivatives was described by Hayashi et al. [422].

For the preparation of analytical samples the following procedure is recommended [419]. The deproteinized biological sample (serum or tissue extract) (0.5 ml) is mixed with 1.5 ml of a 0.13% solution of o-phenylenediamine (1.33 mg ml^{-1} in 3 N HCl; prepared daily), 5 μl of mercaptoethanol is added, and the volume is adjusted to 3 ml with water. The mixture is heated for 30 min in boiling water. In order to terminate the reaction, the tubes are cooled on ice and 0.5 g of anhydrous sodium sulphate is added. The quinoxalinols are extracted with three 3 ml portions of ethyl acetate. The pooled extracts are dried over anhydrous sodium sulphate and evaporated to dryness. The residue is dissolved in 0.2 ml of methanol and the solution is centrifuged and filtered through a 0.45 μm pore size filter. The filtrate is ready for HPLC using a reversed phase column.

In order to improve the chromatographic separation, a pre-purification step for biological samples is recom-

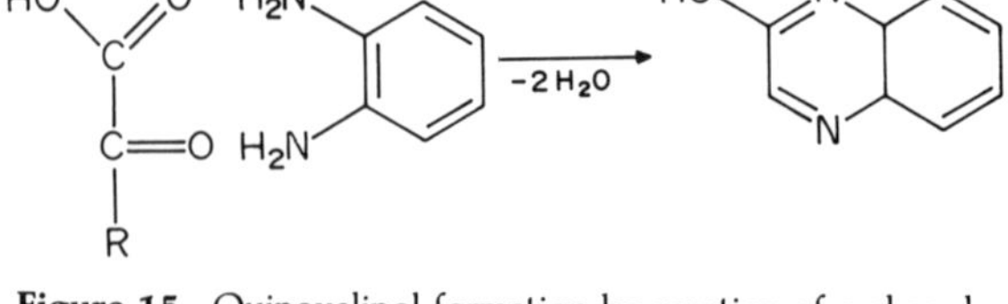

Figure 15. Quinoxalinol formation by reaction of o-phenylenediamine with an α-keto acid.

mended using anion exchange column chromatography [421].

10.8.2 Properties

The quinoxalinol derivatives of α-keto acids are stable and sufficiently volatile to allow their separation by gas chromatography [423–427]. Their fluorescence characteristics depend somewhat on the α-keto acid: excitation maxima of the quinoxalinol derivatives of most keto acids (pyruvic, α-ketobutyric, α-ketoisovaleric, α-ketoisocaproic and α-keto-β-methylvaleric acid) in methanol are at 375–377 nm, and emission maxima at 407–407.5 nm. The derivative of α-ketoglutaric acid has an emission maximum at 412 nm. Only for the derivative of glyoxylic acid do the values differ considerably; its excitation maximum is at 381 nm and its emission maximum is at 390 nm [419].

Separation of these α-keto acid derivatives can be achieved by isocratic elution using methanol/water as eluent [419] or a water/acetonitrile gradient [421]. The intense fluorescence of the quinoxalinols permits the determination of around 10 pmol per sample.

10.9 Analogues of o-phenylenediamine

Wang et al. [428] suggested the use of 1,2-diamino-4,5-methylenedioxybenzene (DMB) as a condensation reagent for the determination of α-keto acids in serum and urine.

The synthesis of the compound was reported [429]; it is now commercially available.

The derivative-forming reaction is very similar to that used for reaction with o-phenylenediamine: samples (0.5 ml) are heated with 0.5 ml of reagent solution at 100 °C for 50 min. Aliquots (5 μl) are immediately injected into the chromatograph.

The reagent solution contains 0.5 mmol of DMB, 0.1 mol of 2-mercaptoethanol and 2.8 mmol of sodium hydrosulphite per 100 ml of 0.4 M hydrochloric acid.

The major advantage of DMB is the superior sensitivity: for most α-keto acids detection limits of 6–29 fmol per 5 μl sample volume are reported [428]. Similarly, 1,2-diamino-4,5-dimethoxybenzene and 1,2-diamino-4,5-ethylenedioxybenzene can be used for the determination of α-keto acids.

Further applications include the sensitive determination of aromatic aldehydes having a phenolic hydroxyl or compounds containing formyl groups, such as Forfenicine [430], or of compounds which allow the quantitative introduction of a formyl group, such as tyrosine-containing peptides [431].

11. Condensation of aliphatic aldehydes with 1,3-diketones and ammonia to form lutidine derivatives

Aldehydes condense with 1,3-diketones [acetylacetone, 1,3-cyclohexanedione (dihydroresorcinol; CHD); 5,5-dimethyl-1,3-cyclohexanedione (dimedone)] and ammonia to form fluorescent lutidine derivatives [432]. These reactions of the diketones have been suggested for precolumn derivatization of aldehydes [433–435].

11.1 Derivative formation with CHD [435]

The reagent comprises an aqueous solution containing 25 g of ammonium acetate, 1 g of CHD and 8 ml of conc. HC in 100 ml. In order to remove contaminants, it is heated for 1 h at 60 °C and, after cooling, passed through a C_{18} Bond-Elut cartridge.

Equal volumes of reagent and sample are heated in sealed tubes for 1 h at 60 °C. After cooling on ice, an aliquot is used for analysis by HPLC.

11.2 Properties

The optimum pH for reaction is around 5 for CHD, whereas formaldehyde reacts with dimedone best at about pH 7. CHD and dimedone derivatives have much higher fluorescence quantum yields than those of acetylacetone derivatives [432].

Suitable systems for separation of the homologous series of aliphatic aldehydes have been worked out for all three lutidines [433, 435]. Excitation of fluorescence at 385 nm with emission measurement at 460 nm is suitable, although the derivatives of various aldehydes have slightly differing excitation and emission maxima. The detection limit is in the fmol range.

12. Sulphhydryl reagents

Among the large number of fluorescent sulphhydryl directed reagents known to protein chemists, some have found application as fluorigenic reagents for the sensitive determination of thiols.

12.1 7-Fluorobenzofurazan-4-sulphonate (SBD-F) and 4-fluoro-7-sulphamoylbenzofurazan

Nbd-Cl (see Section 4.1) reacts rapidly with thiol-containing compounds (N-acetylcysteine, glutathione, proteins) dissolved in 50 mM sodium citrate buffer containing 1 mM EDTA at pH 7 [436]. Since derivative formation of amines requires a pH of 9, Nbd-Cl can be used as a thiol reagent as well.

In order to improve selectivity towards thiols, SBD-F [437] and 4-fluoro-7-sulphamoylbenzofurazan [438] have been introduced.

12.1.1 Derivative formation and separation of thiols [439]

After deproteinization of e.g., fresh blood with 10% trichloroacetic acid containing 1 mM EDTA, 20 µl of deproteinized sample, 80 µl of 1 mM EDTA in water, 0.2 ml of 2.5 M borate buffer (pH 9.5) containing 4 mM EDTA and 0.1 ml of SBD-F (2 mg ml^{-1} in borate buffer) are mixed and heated for 1h. An aliquot can be used immediately for separation by HPLC.

In order to reduce oxidized thiols before derivative formation, tri-n-butylphosphine is used [439].

For the separation of the usual thiols (cysteine, homocysteine, cysteamine, glutathione, N-acetylcysteine), isocratic and gradient elution modes have been suggested using reversed phase columns [439].

Fluorescence excitation is achieved at 385 nm and emission is measured at 515 nm. Detection limits are in the range 0.07–1.4 pmol.

12.2 Dansylaziridine (Dns-A)

Dns-A reacts with strong nucleophiles, such as thiols, with opening of the aziridine ring and formation of stable fluorescent thioethers (Figure 16) [440]. Functions having weaker nucleophilic properties, such as phenols, alcohols and amines, do not react.

Dns-A is prepared by reaction of Dns-Cl with equimolar amounts of ethyleneimine in benzene in the presence of triethylamine [440]. The reagent is commercially available.

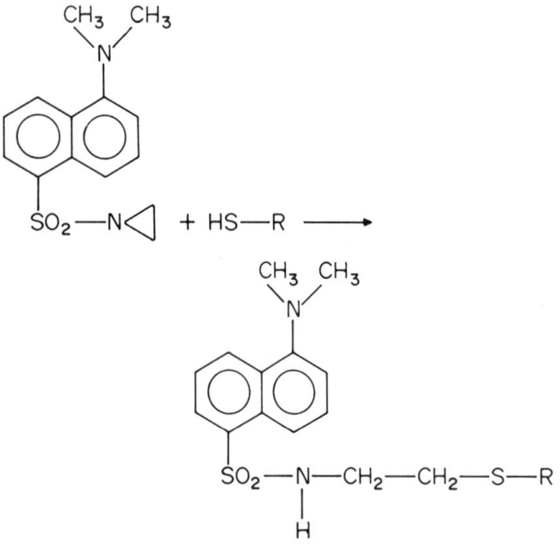

12.2.1 Derivative formation and applications

The reaction is carried out by mixing the sample solution in phosphate buffer, pH 8.2 with the same volume of Dns-A (3 mM in methanol) and heating for 30 min at 60 °C [441]. Aliquots are immediately used for chromatographic separation.

Thiols can be detected at the pmol level with excitation at 338 nm and emission at 540 nm. Amino acids and peptides containing sulphyhdryl groups, as well as other thiols such as penicillamine, can be analysed, using HPLC as separation method [441, 442].

12.3 N-Substituted maleimides

Maleimides react, as shown in Figure 17, by addition of thiols to the 3,4-double bond. By introduction of a fluorigenic substituent on the imide nitrogen, numerous fluorescent maleimides have been prepared in the past, and they usually form derivatives with high quantum yields [443–445]. However, the derivatives are mostly unstable; they react by rearrangement, and thus form more than one fluorescent derivative.

N-(1-Pyrenyl)maleimide and N-(7-dimethylamino-4-methylcoumarinyl)maleimide [446, 447] have been used for the determination of N-acetylcysteine. Cysteine and glutathione have been determined using N-(9-acridinyl)-maleimide [448, 449], and D-penicillamine with N-[(p-2-benzoxazolyl)phenyl]maleimide [450]. More recently, N-[4-(6-dimethylamino-2-benzofuranyl)phenyl]maleimide (DBPM) was synthesized and used for the rapid chromatographic determination of some natural thiols [451, 452].

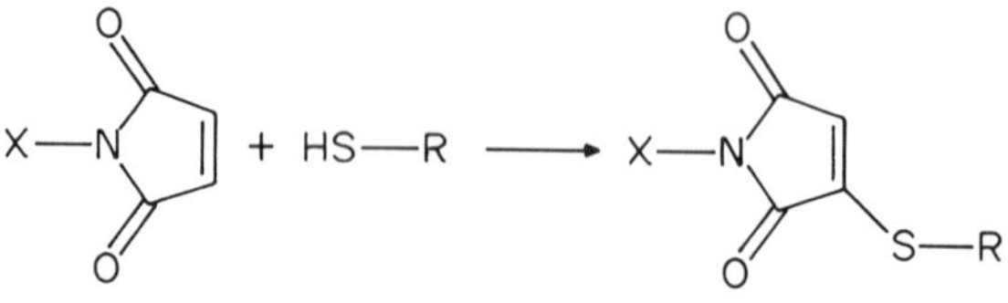

Figure 16. Reaction of dansylaziridine with a thiol.

Figure 17. Reaction of an N-substituted maleimide with a thiol.
X = fluorophore.

12.3.1 Derivative formation and mode of application

Derivative formation with maleimides is performed directly in the biological fluid (plasma, urine, etc.) at a pH of 8–10 using a 5–10-fold excess of the reagent. The mixture is subjected to chromatography after dilution with the mobile phase.

Detection sensitivity by fluorescence measurement is usually in the order of 10–100 pmol ml^{-1}. For DBPM derivatives, a detection limit in the fmol range was reported [452]. More recently, a method for the determination of DBPM derivatives by chemiluminescence was reported [453].

12.4 Bimanes

Monobromotrimethylammoniobimane (3,7-dimethyl-4-bromoethyl-6-trimethylammoniomethyl-1,5-diazabicyclo[3.0.0]octa-3,6-diene-2,8-dione; a commercial product), monobromobimane and dibromobimane have been introduced as reagents for pre-chromatographic derivactive formation of thiols [454–457]. The reaction is shown in Figure 18.

12.4.1 Reaction and application

Since the fluorophores are not stable at akaline pH and are sensitive to light, reaction is carried out as close as possible to neutral pH in the dark. Usually, 1 equiv of thiol is reacted with 2 equiv of the bimane in 10 mM ammonium carbonate buffer pH 8.0 (containing 1 mM EDTA). The reaction is complete within 3 min. Chromatographic systems are available (ion exchange and reversed phase) for separation of the derivatives of a large number of thiols [455, 456], which can be detected in the picomole range (excitation at 375 nm; emission at 480 nm). Quantum yields (in water) are 0.07–0.09 [458].

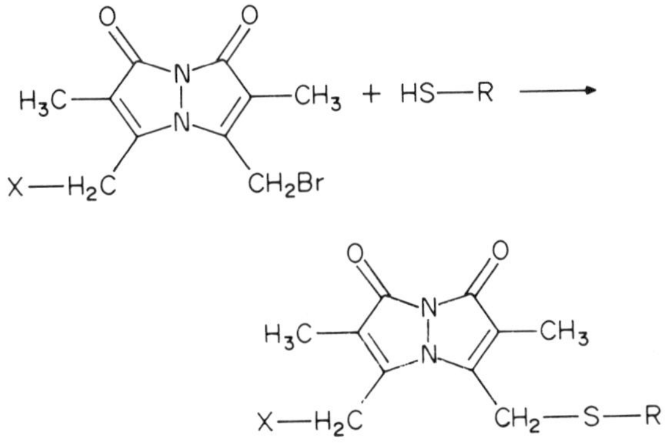

13. Fluorescence labelling of acidic functions

Alkylation of carboxylate anions by 4-bromomethyl-7-methoxycoumarin (Br-Mmc) (Figure 19) has been suggested by Dünges [459, 460] as a method for the fluorescence labelling of carboxylic acids and other compounds with acidic functions. Subsequently, a considerable number of related reagents have been suggested.

Although carboxylic acids can be readily determined by GLC methods, fluorescence labelling with Br-Mmc has nevertheless gained some importance for the sensitive derivatization of fatty acids [461–463] dicarboxylic acids, α-keto acids and other organic acids [460, 464], prostaglandins and thromboxanes [465], glucuronides [466], giberellins [467], imides [461, 468, 469] and phenols [460].

13.1 4-Bromomethyl-7-methoxycoumarin (Br-Mmc)

Derivatives are prepared by refluxing with an excess of Br-Mmc in the presence of a crown ether and anhydrous K_2CO_3 in an aprotic solvent (e.g. acetone or acetonitrile).

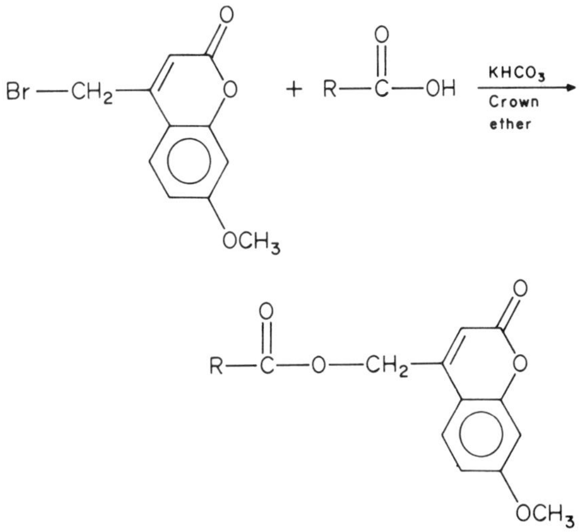

Figure 18. Reaction of bimanes with a thiol.
X = H, monobromobimane.
X = Br, dibromobimane.
X = N–(CH$_3$)$_3$, monobromotrimethylammoniobimane.

Figure 19. Reaction of Br-Mmc with a carboxylic acid.

Suitable instrumentation for derivative formation in microlitre volumes has been described [470]. The crown ether accelerates the alkylation reaction by acting as a phase-transfer catalyst [471]. Instead of K_2CO_3, other bases may be used (KOH, Li_2CO_3). Yields are quantitative. A modified derivatization procedure suitable for automated samples preparation has been reported [472].

13.1.1 Properties of Mmc esters

Br-Mmc and its derivatives are light-sensitive; the derivatives are therefore preferably formed in the dark. Br-Mmc itself shows little fluorescence.

Fluorescence characteristics of Mmc-carboxylic acid esters have been reported by Lloyd [473]. Excitation maxima in methanol and water methanol (9:1) were at 323 and 325 nm, respectively, and the corresponding fluorescence maxima were at 395 and 402 nm (corrected spectra). Quantum yields were 0.08–0.095 (methanol) and 0.054–0.44 (water/methanol, 9:1). TLC separation is possible, but owing to the sensitivity of the Mmc esters to light column chromatographic methods, especially on reversed phases, are much more suitable. Mmc esters were detected in column eluates (excitation at 340 nm; emission at 420 nm) in pmol quantities [461].

13.2 Other fluorescent coumarin derivatives

13.2.1 4-Bromomethyl-6,7-dimethoxycoumarin (Br-Mdmc)

Br-Mdmc was developed to improve the fluorescence quantum yield of the derivatives. In contrast to Mmc esters, which show differences in quantum yield depending on the chain length of the carboxylic acid, all Mdmc esters have approximately the same quantum yield (0.64 in water), and the fluorescence emission is not much affected by pH or salt concentration [474].

Br-Mdmc has also been used for derivative formation from imides, by analogy with the application of Br-Mmc. A method for the sensitive determination of 5-fluorouracil was reported [475].

13.2.2 4-Bromomethyl-7-acetoxycoumarin (Br-Mac)

Br-Mac reacts with carboxylic acids as described for Br-Mmc. In order to enhance the detection sensitivity, the Mac esters are separated by reversed phase column chromatography. The column eluate is mixed with 0.1 M borate buffer, pH 11 and heated in a reaction coil at 50 °C to split off the acetyl group hydrolytically. The resulting 7-hydroxycoumarin derivatives are intensely fluorescent, allowing the determination of fmol amounts of carboxylic acids (excitation at 365 nm; emission at 460 nm) [476].

13.2.3 4-Diazomethyl-7-methoxycoumarin

4-Diazomethyl-7-methoxycoumarin is a rather stable reagent which allows convenient derivative formation: carboxylic acids form esters by heating in acetonitrile. With alcohols, fluorescent ethers are obtained in dichloromethane in the presence of HBF_4 as catalyst [477].

13.2.4 N,N'-Dicyclohexyl-O-(-7-methoxycoumarin-4-yl)methylisourea and N,N'-diisopropyl-O-(7-methoxycoumarin-4-yl)methylisourea

These reagents react with carboxylic acids to form Mmc esters by heating in aprotic solvents (benzene, dioxane) at 80 °C [478, 479]. For the derivative formation of α-keto acids, hydrazone formation with dimethylhydrazine and esterification with one of these reagents was suggested.

13.3 Other fluorophores for acidic functions

The chemical principles applied in the labelling of acidic functions using fluorescent coumarin derivatives were also used in conjunction with other fluorophores. The conditions of the alkylation reactions are therefore the same. Consequently, these reagents suffer equally from a lack of selectivity: usually all acidic functions (carboxyl, phenol, thiol, imide and hydroxyl) can be brought to reaction.

The following fluorescent alkylating agents have been introduced as pre-chromatographic reagents: 9-bromomethylacridine [480], 3-bromomethyl-6,7-dimethoxy-1-methyl-2(1H)-quinoxalinone [481], naphthacyl bromide (2-bromoacetonaphthone) [482, 483], p-(anthroyloxy)-phenacyl bromide (panacyl bromide) [484], 1-bromoacetylpyrene [485], 9-chloromethylanthracene [486], and 9-anthryldiazomethane [487, 488].

A method for the determination of arachidonic acid metabolites with 9-anthryldiazomethane has been reported [489].

13.4 Fluorescence labelling of carboxylic acids after activation

If the carboxylic acid function is activated, fluorescent reagents may be used that are not sufficiently reactive to give high yields with the carboxyl function itself.

The following reagents have been suggested for the activation of carboxylic acids in analytical procedures: oxalyl chloride [490], N-ethyl-N'-(3-dimethylaminopropyl)carbodiimide [491, 492], N,N'-carbonyldiimidazole [493], and 2-bromo-1-methylpyridinium iodide [493].

The activated carboxylic acids react with fluorescent alcohols and amines.

13.4.1 Alcohols suitable for fluorescence labelling of activated carboxylic acids

9-Hydroxymethylanthracene [493], 2-dansylaminoethanol [494] and 4-hydroxymethyl-7-methoxycoumarin (HO-Mmc) [495] have been used for ester formation.

13.4.2 Amines suitable for fluorescence labelling of activated carboxylic acids

1-Naphthylamine [490] and 9-aminophenanthrene [496] have been reacted with acid chlorides in the presence of triethylamine.

1-4-(Dimethylamino-1-naphthyl)ethylamine (DANE) is a chiral amine. Both D- and L-forms may be used for amide formation with activated carboxylic acids [491]. A method for the separation of the enantiomers of naproxen has been reported [492].

9,10-Diaminophenanthrene (DAP) forms 2-substituted phenanthrimidazoles upon direct condensation with a carboxylic acid. The sample (in 10 μl of toluene or chloroform) is mixed with 0.5 ml of a solution 2 mg ml^{-1}) of DAP in methyl polyphosphate and heated in a sealed tube at 85 °C for 3–6 min.

Phenanthrimidazoles have fluorescence excitation maxima at 254–255 nm and overlapping emission maxima at 367 and 382 nm [497]. Their high fluorescence quantum yield (0.68 in methanol) renders them suitable for the sensitive determination of carboxylic acids.

14. Fluorescence labelling of alcohols

14.1 7-Methoxycoumarin-3-carbonyl azide (3-MCCA) and 7-methoxycoumarin-4-carbonyl azide (4-MCCA)

3-MCCA and 4-MCCA were developed for the labelling of hydroxyl functions [498]. They react with primary and secondary alcohols in dichloromethane to form carbamic esters, which are sufficiently stable to be separated by reversed phase column chromatography. For cholesterol, a detection limit of 50 fg per 100 μl sample volume was reported.

14.2 7-(Chlorocarbonylmethoxy)-4-methylcoumarin

7-(Chlorocarbonylmethoxy)-4-methylcoumarin is a reagent related to the above mentioned carbonyl azides [499]. It is prepared from 7-hydroxy-4-methylcoumarin and bromoacetic acid. The resulting carboxylic acid is transformed into the carbonyl chloride by refluxing with thionyl chloride. Derivatives (of prostaglandins and steroids) are formed in carefully dried methylene chloride containing 4-dimethylaminopyridine as catalyst. The esters formed are sufficiently stable for chromatographic separation.

14.3 Nitrile group-containing reagents

4-Dimethylamino-1-naphthoylnitrile [500], 1- and 9-anthroylnitrile [501] and 2-methyl-1,1'-binaphthalene-2'-carbonylnitrile [502] have been suggested as fluorescent labels for primary and secondary alcohols. 2-Methyl-1,1-binaphthalene-2'-carbonylnitrile is a chiral reagents, and is therefore suited for the separation of enantiomers. Quantitative reaction is achieved by heating in the presence of triethylamine, whereby the diastereomeric esters are formed.

Separations of the diastereomers were performed using a silica gel column (excitation at 342 nm, emission at 420 nm) [502].

14.4 Naphthaleneboronic acid and phenanthreneboronic acid

Naphthaleneboronic acid was used to form fluorescent esters with bifunctional alcohols (pinacol; cis-1,2-cyclohexanediol), and phenanthreneboronic acid was reacted with ecdysone and ecdysterone. The boronates are suitable for separation by TLC and HPLC [503].

15. Fluorescence labelling of adenine and its derivatives by chloro- or bromoacetaldehyde

Chloroacetaldehyde was shown to react with adenine, forming 1-N^6-ethenoadenine [504]. Subsequently the reaction was used by a number of investigators for the sensitive determination of a series of adenine derivatives, using ion exchange column chromatography [505] or HPLC [506–509] as separation method.

Derivative formation is improved by using bromoacetaldehyde as reagent [510]. To a mixture of adenine derivatives, about 10 μM in 200 μm citrate buffer pH 3 (or 1 M acetate buffer, pH 3.5–5.5, or 1 M phosphate buffer, pH 6.0–7.0) is added 40 µl of a 1.2 M solution of bromoacetaldehyde in water, and the tightly closed vial is heated at 80 °C for 15 min. The solutions are stored at 4 °C until an aliquot is used for separation by HPLC.

Bromoacetaldehyde may be prepared as described by Schukovskaya et al. [511].

Secrist et al. [512] have studied the fluorescence characteristics of 1-^{6}N-ethenoadenosines in detail. They have a high fluorescence quantum yield: 0.56 in 0.25 M phosphate buffer. Of the three absorption maxima, that at 300 nm is normally used for excitation. Since it is shifted to lower wavelength with decreasing pH, optimization is necessary according to the pH of the eluent used. Wagner et al. [509] recommend excitation of fluorescence at 270–275 nm for an eluent of pH 4.3. Fluorescence emission is measured at 410 nm.

Suitable separation systems have been reported for the sensitive determination of cAMP [510] and a selection of S-adenosylmethionine derivatives [509] in urine and tissues, among other matrices.

16. Conclusions

The number of fluorigenic reagents that has been introduced to analytical chemistry during the last decade is impressive. The reason has been, with little doubt, the successful application of the early fluorigenic reagents in HPLC. This has led many investigators to search for similar but improved methods. For most classes of compounds suitable reagents are now available for prechromatographic labelling, which permits the routine detection and determination of picomole quantities of the fluorescent derivatives.

Admittedly not all reagents mentioned in this chapter have been adequately explored. Some of the new reagents may turn out to have no advantages over known compounds. On the other hand, even old, but yet unexplored, reagents may turn out, on closer investigations, to have characteristics which favour certain applications. For example, rapid derivative formation, preferably at room temperature, is a feature required for automated sample preparation, although not of the same importance in manual methods. Consequently, known reagents have recently been adopted which are less specific than those previously favoured or even, perhaps do not form stable derivatives. Examples are the chloroformates and the isoindoles which are formed by reaction of OPA and a thiol with primary amines.

It is also evident that fluorescence labelling has not remained of equal importance for all classes of biologically important non-fluorescent compounds. For example, the determination of catecholamines was a classical domain of fluorescence methods. With the advent of simple and sensitive electrochemical detectors, fluorimetric methods have lost much of their importance. Similarly, volatile compounds will mainly continue to be determined by GLC, eventually coupled with mass spectrometry, even though suitable fluorigenic labels are available.

The main domain of fluorescence labelling was and still is in amino acid analyses, where the greatest progress has been made by introduction of new reagents and, especially, by improvement of the separation methods. Several fully automated procedures of sample preparation, separation and quantitative evaluation are now at our disposal, routinely allowing the determination of pmol and even fmol quantities of the commoner amino acids. Related methods for the determination of aliphatic amines, thiols, keto acids and sugars have also been improved.

In the sensitive analysis of peptides [513] and in other fields of analytical chemistry, where the complexity of the samples to be analyzed does not allow complete separation, selective reagents are of especial importance. Labelling of thiol groups (see Section 12), reaction of guanidino residues [386, 387] or the formylation of tyrosine are reactions suitable for the selective labelling of groups of peptides. In this area, specific fluorigenic reactions that lead to unknown reaction products may even be acceptable. As an example, N-terminal tyrosine-containing peptides react with hydroxylamine and Co^{2+} to give a fluorescent product [514]. However, the goal must remain the development of reagents forming well defined products which allow the unambiguous identification of the labelled compound by mass spectrometry or an equivalent method. Only by consulting methods independent of chromatographic criteria may one escape erroneous peak identification, a danger which increases with the increased sensitivity of available methods. The combination of sensitive detection with unambiguous identification methods is indispensable whenever new or unidentified products appear in a chromatogram.

Fluorescence labelling as a method for the sensitive determination of chemicals and natural products is now firmly established as a major weapon in the arsenal of analytical chemists. There is still scope for considerable improvements with respect to sensitivity, selectivity and reliability.

17. References

[1] W. R. Gray and B. S. Hartley, *Biochem. J.*, **89**, 59 (1963).
[2] E. Fischer and P. Bergell, *Ber. Dtsch. Chem. Ges.*, **35**, 3779 (1903).

[3] E. Fischer and P. Bergell, *Ber. Dtsch. Chem. Ges.*, **36**, 2592 (1903).
[4] F. Sanger, *Biochem. J.*, **39**, 507 (1945).
[5] F. Sanger and H. Tuppy, *Biochem. J.*, **49**, 463 (1951).
[6] W. R. Gray, *Methods Enzymol.*, **25**, 121 (1972).
[7] N. Seiler, *Methods Biochem. Anal.*, **18**, 259 (1970).
[8] N. Seiler and J. Wiechmann, *Experientia*, **20**, 559 (1964).
[9] N. Seiler and M. Wiechmann, *Experientia*, **21**, 203 (1965).
[10] N. Seiler and M. Wiechmann, *Fresenius' Z. Anal. Chem.*, **220**, 109 (1966).
[11] G. Weber, *Biochem. J.*, **51**, 155 (1952).
[12] V. G. Shore and A. B. Pardee, *Arch. Biochem. Biophys.*, **62**, 355 (1956).
[13] A. H. Coons, *Immunofluorescence, Public Health Reports, U.S.* **75**, 937 (1960).
[14] W. R. Gray, *Methods Enzymol.*, **25**, 333 (1972).
[15] A. A. Boulton and I. E. Bush, *Biochem. J.*, **92**, 11 (1964).
[16] K. R. Woods and K. T. Wang, *Biochim. Biophys. Acta*, **133**, 369 (1966).
[17] N. Seiler and M. Wiechmann, in *Progress in Thin-Layer Chromatography and Related Methods*, A. Niederwieser and G. Pataki (Editors), Vol. 1, Ann Arbor-Humphrey Science (1970), p. 94.
[18] D. J. Neadle and R. J. Pollit, *Biochem. J.*, **97**, 607 (1965).
[19] N. Seiler and M. Wiechmann, *Hoppe-Seyler's Z. Physiol. Chem.*, **349**, 588 (1968).
[20] N. Seiler, H. H. Schneider and K.-D. Sonnenberg, *Fresenius' Z. Anal. Chem.*, **252**, 127 (1970).
[21] N. Seiler, in *Research Methods in Neurochemistry*, N. Marks and R. Rodnight (Editors), Vol. 3, Plenum Press, New York (1975), p. 409.
[22] M. Wiechmann, *Hoppe-Seyler's Z. Physiol. Chem.*, **358**, 967 (1977).
[23] N. Seiler, M. Wiechmann, H. A. Fischer and G. Werner, *Brain Res.*, **28**, 317 (1971).
[24] J. W. T. Dickerson and S. K. Pao, *Biol. Neonate*, **25**, 114 (1975).
[25] V. T. Wiedmeir, S. P. Porterfield and C. E. Hendrich, *J. Chromatogr.*, **231**, 410 (1982).
[26] Y. Tapuhi, D. E. Schmidt, W. Lindner and B. L. Karger, *Anal. Biochem.*, **115**, 123 (1981).
[27] W. Dünges, G. Naundorf and N. Seiler, *J. Chromatogr. Sci.*, **12**, 655 (1974).
[28] N. Seiler and K. Deckardt, *J. Chromatogr.*, **107**, 227 (1975).
[29] Z. Tamura, T. Nakajima, T. Nakayama, J. J. Pisano and S. Udenfriend, *Anal. Biochem.*, **52**, 595 (1973).
[30] A. Yamamotodani, T. Seki, H. Wada and M. Taneda, *J. Chromatogr.*, **144**, 141 (1977).
[31] R. F. Chen, *Arch. Biochem. Biophys.*, **120**, 609 (1967).
[32] N. Seiler, T. Schmidt-Glenewinkel and H. H. Schneider, *J. Chromatogr.*, **84**, 95 (1973).
[33] R. W. Frei, *J. Chromatogr.*, **165**, 75 (1979).
[34] M. S. Arnott and D. N. Ward, *Anal. Biochem.*, **21**, 50 (1967).
[35] B. P. Sloan, *J. Chromatogr.*, **42**, 426 (1969).
[36] A. Niederwieser, *Methods Enzymol.*, **25**, 60 (1971).
[37] R. S. Fager and C. B. Kutina, *J. Chromatogr.*, **76**, 268 (1973).
[38] H. T. Nagashawa, P. S. Fraser and J. A. Elberling, *J. Chromatogr.*, **44**, 300 (1969).
[39] N. Seiler and M. Wiechmann, *J. Chromatogr.*, **28**, 351 (1967).
[40] J. Langner, *Hoppe-Seyler's Z. Physiol. Chem.*, **347**, 275 (1966).
[41] R. S. Atherton and A. R. Thompson, *Biochem. J.*, **111**, 797 (1969).
[42] T. Kato and M. Makoto, *Anal. Biochem.*, **66**, 515 (1975).
[43] M. O. Creighton and J. R. Trevithick, *Anal. Biochem.*, **50**, 255 (1972).
[44] I. S. Forrest, S. D. Rose, L. G. Brookes, B. Halpern, V. A. Bacon and I. A. Silberg, *Agressologie*, **11**, 127 (1970).
[45] K. Igarashi, I. Izumi, K. Hara and S. Hirose, *Chem. Pharm. Bull.*, **22**, 451 (1974).
[46] R. L. Munier and A. M. Drapier, *Chromatographia*, **5**, 306 (1972).
[47] G. L. Moore and R. S. Antonoff, *Anal. Biochem.*, **39**, 260 (1971).
[48] G. Briel, V. Neuhoff and M. Maier, *Hoppe-Seyler's Z. Physiol. Chem.*, **353**, 540 (1972).
[49] J. Airhart, S. Sibiga, H. Sanders and E. A. Khairallah, *Anal. Biochem.*, **53**, 132 (1973).
[50] P. Teichgräber, I. Krautschick, R. Arnold and D. Biesold, *Pharmazie*, **29**, 186 (1974).
[51] M. de L. A. Barcelon, *J. Chromatogr.*, **238**, 175 (1982).
[52] E. Bayer, E. Grom, B. Kaltenegger and R. Uhlmann, *Anal. Chem.*, **48**, 1106 (1976).
[53] G. Nota, G. Marino, V. Buonocore and A. Ballio, *J. Chromatogr.*, **46**, 103 (1970).
[54] N. Seiler and H. H. Schneider, *J. Chromatogr.*, **59**, 367 (1971).
[55] T. Seki and H. Wada, *J. Chromatogr.*, **102**, 251 (1974).
[56] Z. Iskandarani and D. J. Pietrzyk, *Anal. Chem.*, **53**, 489 (1981).
[57] T. Yamabe, N. Takai and H. Nakamura, *J. Chromatogr.*, **104**, 359 (1975).
[58] S. Winer and A. Tishbee, *J. Chromatogr.*, **213**, 501 (1981).
[59] N. Kaneda, M. Sato and K. Yagi, *Anal. Biochem.*, **127**, 49 (1982).
[60] L. N. Mackey and T. A. Beck, *J. Chromatogr.*, **240**, 455 (1982).
[61] C. De Jong, G. J. Hughes, E. Van Wieringen and K. J. Wilson, *J. Chromatogr.*, **241**, 345 (1982).
[62] B. Grego and M. T. W. Hearn, *J. Chromatogr.*, **255**, 67 (1983).
[63] B. Oray, H. S. Lu and R. W. Gracy, *J. Chromatogr.*, **270**, 253 (1983).
[64] R. Bongiovanni and W. Duthon, *J. Liq. Chromatogr.*, **1**, 617 (1978).
[65] R. A. Miller, N. E. Bussell, J. R. Wynkoop, R. Judson, R. Bongiovanni and T. M. Boehm, *J. Chromatogr. Sci.*, **18**, 235 (1981).
[66] H. Kneifel and A. S. Julich, *J. Liq. Chromatogr.*, **6**, 1395 (1983).
[67] L. Zecca and P. Ferrario, *J. Chromatogr.*, **337**, 391 (1985).
[68] F. J. Marquez, A. R. Quesada, F. Sanchez, Jimenez and I. Nunez de Castro, *J. Chromatogr.*, **380**, 275 (1986).

[69] J. T. Työppönen, *J. Chromatogr.*, **413**, 25 (1987).
[70] K.-T. Hsu and B. L. Currie, *J. Chromatogr.*, **166**, 555 (1978).
[71] E. M. Koroleva, V. G. Maltsev and B. G. Belenkii, *J. Chromatogr.*, **242**, 145 (1982).
[72] K. Miyaguchi, K. Honda and K. Imai, *J. Chromatogr.*, **316**, 501 (1984).
[73] A. A. Boulton, *Methods Biochem. Anal.*, **16**, 327 (1969).
[74] V. V. Nesterov, B. G. Belinsky and L. G. Senyutenkova, *Biokhimiya*, **34**, 666 (1969).
[75] V. A. Spivak, V. M. Orlov, V. V. Scherbukhin and J. M. Varshavsky, *Anal. Biochem.*, **35**, 227 (1970).
[76] V. A. Spivak, V. V. Scherbukhin, V. M. Orlov and J. M. Varshavsky, *Anal. Biochem.*, **39**, 271 (1971).
[77] V. A. Spivak, M. I. Levjant, S. P. Katrukha and J. M. Varshavsky, *Anal. Biochem.*, **44**, 503 (1971).
[78] C. Gros and B. Labouesse, *Eur. J. Biochem.*, **7**, 463 (1969).
[79] J. P. Zanetta, G. Vincendon, P. Mandel and G. Gombos, *J. Chromatogr.*, **51**, 441 (1970).
[80] N. Seiler and B. Knödgen, *Hoppe-Seyler's Z. Physiol. Chem.*, **352**, 97 (1971).
[81] N. Seiler and B. Knödgen, *J. Chromatogr.*, **164**, 155 (1979).
[82] S. Kobayashi, J. Sekino, K. Honda and K. Imai, *Anal. Biochem.*, **112**, 99 (1981).
[83] J. Reisch, H. Alfes, N. Jantos and H. Möllmann, *Acta Pharm. Suec.*, **5**, 393 (1968).
[84] C. R. Creveling, K. Kondo and J. W. Daly, *Clin. Chem.*, **14**, 302 (1968).
[85] G. Marino and V. Buonocore, *Biochem. J.*, **110**, 603 (1968).
[86] N. Seiler, H. H. Schneider and K.-D. Sonnenberg, *Anal. Biochem.*, **44**, 451 (1971).
[87] H. Egg, H. Ockenfels, H. Thomas and F. Zilliken, *Z. Naturforsch., Teil B*, **26**, 229 (1971).
[88] N. Seiler, F. N. Bolkenius, B. Knödgen and K. Haegele, *Biochem. Biophys. Acta*, **676**, 1 (1980).
[89] S. Sarhan, F. Dezeure and N. Seiler, *Int. J. Biochem.*, **19**, 1037 (1987).
[90] A. E. Jenkins and J. R. Majer, in *Mass Spectrometry*, Butterworth, London (1968), p. 253.
[91] N. Seiler and B. Knödgen, *Org. Mass Spectrom.*, **7**, 97 (1973).
[92] D. A. Durden, B. A. Davis and A. A. Boulton, *Biomed. Mass Spectrom.*, **1**, 83 (1974).
[93] N. N. Osborne, *Prog. Neurobiol.*, **1**, 301 (1973).
[94] V. Neuhoff, in *Micromethods in Molecular Biology*, V. Neuhoff (Editor), Chapman and Hall, London (1973), p. 87.
[95] A. J. Kennedy and M. J. Voades, *J. Neurochem.*, **23**, 1093 (1974).
[96] S. R. Snodgrass and L. L. Iversen, *Nature (London), New Biol.*, **241**, 154 (1973).
[97] P. M. Beart and S. R. Snodgrass, *J. Neurochem.*, **24**, 821 (1975).
[98] M. H. Joseph and J. Halliday, *Anal. Biochem.*, **64**, 389 (1975).
[99] M. Recasens, J. Zwiller, G. Mack, J. P. Zanetta and P. Mandel, *Anal. Biochem.*, **82**, 8 (1977).
[100] T. J. Paulus and R. H. Davis, *Methods Enzymol.*, **94**, 36 (1983).
[101] K. Crowshaw, S. J. Jessup and P. W. Ramwell, *Biochem. J.*, **103**, 79 (1967).
[102] N. N. Osborne, P. H. Wu and V. Neuhoff, *Brain Res.*, **74**, 175 (1974).
[103] P. J. Roberts, P. Keen and J. F. Mitchell, *J. Neurochem.*, **21**, 199 (1973).
[104] J. M. Wilkinson, *J. Chromatogr. Sci.*, **16**, 547 (1978).
[105] G. J. Schmidt, D. C. Olson and W. Slavin, *J. Liq. Chromatogr.*, **2**, 1031 (1979).
[106] B. Oray, H. S. Lu and R. W. Gracy, *J. Chromatogr.*, **270**, 253 (1983).
[107] P. Martin, C. Polo, M. D. Cabezudo and M. V. Dabric, *J. Liq. Chromatogr.*, **7**, 539 (1984).
[108] H. Tsuchiya, T. Hayashi, N. Tatsumi, T. Fukita and N. Takagi, *J. Chromatogr.*, **339**, 59 (1985).
[109] N. Seiler and B. Wiechmann, *Hoppe-Seyler's Z. Physiol. Chem.*, **350**, 1493 (1969).
[110] N. Seiler and B. Eichentopf, *Biochem. J.*, **152**, 201 (1975).
[111] N. N. Osborne, H. Zimmermann, M. J. Dowdall and N. Seiler, *Brain Res.*, **88**, 115 (1975).
[112] G. E. Griesmann, W.-Y. Chan and O. M. Rennert, *J. Chromatogr.*, **230**, 1493 (1969).
[113] L. Zecca, F. Zambotti, N. Zonta and P. Mantegazza, *J. Chromatogr.*, **233**, 307 (1982).
[114] W. Lindner, J. N. Le Page, G. Davies, D. E. Seitz and B. L. Karger, *J. Chromatogr.*, **185**, 323 (1979).
[115] S. K. Lam and F. K. Chow, *J. Liq. Chromatogr.*, **3**, 1579 (1980).
[116] Y. Tapuhi, N. Miller and B. L. Karger, *J. Chromatogr.*, **205**, 325 (1981).
[117] S. K. Lam and A. Karmen, *J. Chromatogr.*, **239**, 451 (1982).
[118] G. Schmer, *Hoppe-Seyler's Z. Physiol. Chem.*, **348**, 199 (1967).
[119] K. Worowski, *Wiad. Chem.*, **27**, 789 (1973).
[120] L. Casola, G. Di Matteo, G. Di Prisco and F. Cervone, *Anal. Biochem.*, **57**, 38 (1974).
[121] P. Nedkov and N. Genov, *Biochem. Biophys. Acta*, **127**, 544 (1966).
[122] B. Mesrob and V. Holeysovsky, *Collect. Czech. Chem. Commun.*, **32**, 1976 (1967).
[123] G. Schmer and H. Kreil, *Anal. Biochem.*, **29**, 186 (1969).
[124] G. Schwedt and H. H. Bussemas, *Fresenius' Z. Anal. Chem.*, **283**, 23 (1977).
[125] G. Schwedt and H. H. Bussemas, *Fresenius' Z. Anal. Chem.*, **285**, 381 (1977).
[126] R. W. Frei, M. Thomas and I. Frei, *J. Liq. Chromatogr.*, **1**, 443 (1978).
[127] H. Tsuchiya, M. Tatsumi, N. Takagi, T. Koike, H. Yamaguchi and T. Hayashi, *Anal. Biochem.*, **155**, 28 (1986).
[128] K. Yamada, E. Kayame and Y. Aizawa, *J. Chromatogr.*, **223**, 176 (1981).
[129] N. Seiler, *J. Chromatogr.*, **63**, 97 (1971).
[130] P. Lehtonen, *J. Chromatogr.*, **314**, 141 (1984).
[131] A. R. Hayman, D. O. Gray and S. V. Evans, *J. Chromatogr.*, **325**, 462 (1985).
[132] M. L. Henriksen and T. Laijoki, *J. Chromatogr.*, **333**, 220 (1985).

[133] N. Seiler and T. Schmidt-Glenewinkel, *J. Neurochem.*, **24**, 791 (1975).
[134] N. Seiler, B. Knödgen and F. Eisenbeiss, *J. Chromatogr.*, **145**, 29 (1978).
[135] P. Scherer and H. Kneifel, *J. Bacteriol.*, **154**, 1315 (1983).
[136] C. Stefanelli, D. Carati and C. Rossoni, *J. Chromatogr.*, **375**, 49 (1986).
[137] P. M. Kabra, H. K. Lee, W. P. Lubich and L. J. Marton, *J. Chromatogr.*, **380**, 19 (1986).
[138] A. A. Boulton and J. R. Majer, in *Research Methods in Neurochemistry*, N. Marks and R. Rodnight (Editors), Vol. 1, Plenum, New York, (1972), p. 341.
[139] S. R. Snodgrass and A. S. Horn, *J. Neurochem.*, **21**, 687 (1973).
[140] S. Axelsson, A. Björklund and N. Seiler, *Life Sci.*, **13**, 1411 (1973).
[141] N. Seiler and K. Bruder, *J. Chromatogr.*, **106**, 159 (1975).
[142] A. A. Galoyan, B. K. Mesrob and V. Holeysovsky, *J. Chromatogr.*, **24**, 440 (1966).
[143] S. S.-H. Chen, A. Y. Kou and H.-H. Y. Chen, *J. Chromatogr.*, **208**, 339 (1981).
[144] L. F. Congote, *J. Chromatogr.*, **253**, 276 (1982).
[145] L. P. Penzes and G. W. Oertel, *J. Chromatogr.*, **51**, 325 (1970).
[146] S. R. Abbott, A. Abu-Shumays, K. O. Loeffler and I. S. Forrest, *Res. Commun. Chem. Pathol. Pharmacol.*, **10**, 9 (1975).
[147] F. Tagliaro, A. Frigerio, R. Dorizzi, G. Lubli and M. Marigo, *J. Chromatogr.*, **330**, 323 (1985).
[148] B. Schultz and S. H. Hansen, *J. Chromatogr.*, **228**, 279 (1982).
[149] A. T. R. Williams, S. A. Winfield and R. C. Belloli, *J. Chromatogr.*, **240**, 224 (1982).
[150] J. F. Lawrence, C. Renault and R. W. Frei, *J. Chromatogr.*, **121**, 343 (1976).
[151] R. W. Frei, J. F. Lawrence, J. Hope and R. M. Cassidy, *J. Chromatogr. Sci.*, **12**, 40 (1974).
[152] P. J. Meffin, S. R. Harapat and D. C. Harrison, *J. Pharm. Sci.*, **66**, 583 (1977).
[153] R. F. Adams, G. J. Schmidt and F. L. Vandemark, *Clin. Chem.*, **23**, 1226 (1977).
[154] E. Johnson, A. Abu-Shumays and S. R. Abbott, *J. Chromatogr.*, **134**, 107 (1977).
[155] G. Powis and M. M. Ames, *J. Chromatogr.*, **181**, 95 (1980).
[156] K.-S. Hui, M. Hui, K.-P. Cheng and A. Lajtha, *J. Chromatogr.*, **222**, 512 (1981).
[157] A. J. Varghese, *Anal. Biochem.*, **110**, 197 (1981).
[158] J. P. Sommadossi, M. Lemar, J. Necciari, Y. Sumirtapura, J. P. Cano and J. Gaillot, *J. Chromatogr.*, **228**, 205 (1982).
[159] A. A. Kumar, R. J. Kempton, G. M. Anstead, E. M. Price and H. J. Freisheim, *Anal. Biochem.*, **128**, 191 (1983).
[160] R. W. Frei, W. Santi and M. Thomas, *J. Chromatogr.*, **116**, 365 (1976).
[161] S. Courte and N. Bromet, *J. Chromatogr.*, **224**, 162 (1981).
[162] N. A. Farid and S. M. White, *J. Chromatogr.*, **275**, 458 (1983).
[163] A. T. Rhys Williams and S. A. Winfield, *Analyst*, **107**, 1092 (1982).
[164] J. Meyer and P. Portmann, *Pharm. Acta Helv.*, **57**, 12 (1982).
[165] H. Kamimura, H. Sasaki and S. Kawamura, *J. Chromatogr.*, **225**, 115 (1981).
[166] N. Seiler and B. Knödgen, *J. Chromatogr.*, **97**, 286 (1974).
[167] N. Seiler and H. H. Schneider, *Biomed. Mass Spectrom.*, **1**, 387 (1974).
[168] W. D. Lehman, H. D. Beckey and H.-R. Schulten, *Anal. Chem.*, **48**, 1572 (1976).
[169] R. P. Cory, R. R. Becker, R. Rosenbluth and T. Isenberg, *J. Am. Chem. Soc.*, **90**, 1643 (1968).
[170] J. Rosmus and Z. Deyl, *Chromatogr. Rev.*, **13**, 163 (1971).
[171] Ch. P. Ivanov, *Monatsh. Chem.*, **97**, 1499 (1966).
[172] Ch. P. Ivanov and Y. Vladovska-Yukhonovska, *J. Chromatogr.*, **71**, 111 (1972).
[173] Ch. P. Ivanov and Y. Vladovska-Yukhonovska, *Biochim. Biophys. Acta*, **194**, 345 (1969).
[174] I. Durko, Y. Vladovska-Yukhonovska and Ch. P. Ivanov, *Clin. Chim. Acta*, **49**, 407 (1973).
[175] D. Tockensteinova, J. Slosar, J. Urbanek and J. Churacek, *Mikrochim. Acta*, II, 193 (1979).
[176] P. Jandera, H. Pechova, D. Tockensteinova, J. Churacek and J. Kralovsky, *Chromatographia*, **16**, 275 (1982).
[177] A. I. Tochilkin, G. A. Davydova, I. N. Gracheva, I. R. Kovelman and S. I. Kirillova, USSR Pat. Appl. 3,404,186 (1982).
[178] P. Hartvig and C. M. Svahn, *Anal. Chem.*, **48**, 390 (1976).
[179] M. Makita, S. Yamamoto and M. Kono, *J. Chromatogr.*, **120**, 129 (1976).
[180] L. A. Carpino and G. Y. Han, *J. Org. Chem.*, **37**, 3404 (1972).
[181] H. A. Moye and A. J. Boning Jr., *Anal. Lett.*, **12**, 25 (1979).
[182] S. Einarsson, B. Josefsson and S. Lagerkvist, *J. Chromatogr.*, **282**, 609 (1983).
[183] C.-X. Gao, T.-Y. Chou, S. T. Colgan, I. S. Krull, C. Dorschel and B. Bidlingmeyer, *J. Chromatogr. Sci.*, **26**, 449 (1988).
[184] R. Schuster, *J. Chromatogr.*, **431**, 271 (1988).
[185] T. Nasholm, G. Sandberg and A. Ericson, *J. Chromatogr.*, **396**, 225 (1987).
[186] A. J. Shah and M. W. Adlard, *J. Chromatogr.*, **424**, 325 (1988).
[187] C. J. Miles, L. R. Wallace and H. A. Moye, *J. Assoc. Off. Anal. Chem.*, **69**, 458 (1986).
[188] L. C. Raiford and G. O. Inman, *J. Am. Chem. Soc.*, **56**, 1586 (1934).
[189] G. Gübitz, R. Wintersteiger and A. Hartinger, *J. Chromatogr.*, **218**, 51 (1981).
[190] A. Takadate, M. Iwai, H. Fujino, K. Tahara and S. Goya, *Yakugaku Zasshi*, **103**, 962 (1983).
[191] H. Spahn, H. Weber, E. Mutschler and W. Möhrke, *J. Chromatogr.*, **310**, 167 (1984).
[192] P. B. Ghosh and M. W. Whitehouse, *Biochem. J.*, **108**, 155 (1968).

[193] A. J. Boulton, P. B. Ghosh and A. R. Katritzky, *J. Chem. Soc. B*, 1004 (1966).
[194] H. J. Klimisch and L. Stadler, *J. Chromatogr.*, **90**, 141 (1974).
[195] J. F. Lawrence and R. W. Frei, *Anal. Chem.*, **44**, 2046 (1972).
[196] D. Clasing, H. Alfes, H. Möllmann and J. Reisch, *Z. Klin. Chem. Klin. Biochem.*, **7**, 648 (1969).
[197] J. Montforte, R. J. Bath and I. Sunshine, *Clin. Chem.*, **18**, 1329 (1972).
[198] F. Van Hoof and A. Heyndrick, *Anal. Chem.*, **46**, 286 (1974).
[199] L. Johnson, S. Lagerkvist, P. Lindroth, M. Ahnhoff and K. Martinsson, *Anal. Chem.*, **54**, 939 (1982).
[200] K. Imai and Y. Watanabe, *Anal. Chim. Acta*, **130**, 377 (1981).
[201] R. S. Fager, C. B. Kutina and E. W. Abrahamson, *Anal. Biochem.*, **53**, 290 (1973).
[202] Y. Watanabe and K. Imai, *J. Chromatogr.*, **309**, 279 (1984).
[203] J. Reisch, H. J. Kommert, H. Alfes and H. Möllmann, *Fresenius' Z. Anal. Chem.*, **247**, 56 (1969).
[204] M. Ahnoff, I. Grundevik, A. Arfwidsson, J. Fonselius and B. A. Persson, *Anal. Chem.*, **53**, 485 (1981).
[205] T. Toyo'oka, I. Watanabe and K. Imai, *Anal. Chim. Acta*, **149**, 305 (1983).
[206] I. Watanabe and K. Imai, *J. Chromatogr.*, **239**, 273 (1982).
[207] J. H. Wolfram, J. I. Feinberg, R. C. Doerr and W. Fiddler, *J. Chromatogr.*, **132**, 37 (1977).
[208] G. J. Krol, J. M. Banovsky, C. A. Mannan, R. E. Pickering and B. T. Kho, *J. Chromatogr.*, **163**, 383 (1979).
[209] H. Kotaniguchi, M. Kawakatsu, T. Toyo'oka and K. Imai, *J. Chromatogr.*, **420**, 141 (1987).
[210] H. J. Klimisch and D. Ambrosius, *J. Chromatogr.*, **121**, 93 (1976).
[211] E. Besenfelder, *J. High Resolut. Chromatogr. Chromatogr. Commun.*, **4**, 237 (1981).
[212] M. Roth, *Clin. Chim. Acta*, **83**, 273 (1978).
[213] J. Reisch, H. Alfes, H. J. Kommert, N. Jantos, H. Möllmann and E. Clasing, *Pharmazie*, **25**, 331 (1970).
[214] J. Krüger, H. Rabe, G. Reichmann and B. Rüstow, *J. Chromatogr.*, **307**, 387 (1984).
[215] P. Kristan, *Chem. Zvesti*, **15**, 641 (1961).
[216] J. E. Sinsheimer, D. D. Hong, J. T. Stewart, M. L. Fink and J. H. Burckhalter, *J. Pharm. Sci.*, **60**, 141 (1971).
[217] A. DeLeenheer, J. E. Sinsheimer and J. H. Burckhalter, *J. Pharm. Sci.*, **62**, 1370 (1973).
[218] H. Maeda, N. Ishida, H. Kawauchi and K. Tuzimura, *J. Biochem.*, **65**, 777 (1969).
[219] H. Kawauchi, H. Tuzimura, H. Maeda and N. Ishida, *J. Biochem.*, **66**, 783 (1969).
[220] A. H. Coons and M. H. Kaplan, *J. Exp. Med.*, **91**, 1 (1950).
[221] R. M. McKinney, J. T. Spillane and G. W. Pierce, *J. Org. Chem.*, **27**, 3986 (1962).
[222] J. L. Riggs, R. J. Seiwald and J. H. Burckhalter, *Am. J. Pathol.*, **34**, 1081 (1958).
[223] K. Muramato, H. Kamiya and H. Kawauchi, *Anal. Biochem.*, **141**, 446 (1984).
[224] H. Ichikawa, T. Tanimura, T. Nakajima and Z. Tamura, *Chem. Pharm. Bull.*, **18**, 1493 (1970).
[225] J. J. L'Italien and S. B. H. Kent, *J. Chromatogr.*, **283**, 149 (1984).
[226] S. W. Jin, G.-X. Chen, Z. Palacz and B. Wittmann-Liebold, *FEBS Lett.*, **198**, 150 (1986).
[227] H. Hirano and B. Wittmann-Liebold, *Hoppe-Seylers Z. Biol. Chem.*, **367**, 1259 (1986).
[228] K. Samejima, W. Dairman and S. Udenfriend, *Anal. Biochem.*, **42**, 222 (1971).
[229] M. Weigele, J. F. Blount, J. P. Tengi, R. C. Czajkowski and W. Leimgruber, *J. Am. Chem. Soc.*, **94**, 4052 (1972).
[230] M. Weigele, S. L. DeBernardo, J. P. Tengi and W. Leimgruber, *J. Am. Chem. Soc.*, **94**, 5927 (1972).
[231] S. Stein, P. Böhlen and S. Udenfriend, *Arch. Biochem. Biophys.*, **163**, 400 (1974).
[232] S. De Bernardo, M. Weigele, V. Toome, K. Manhart, W. Leimgruber, P. Böhlen, S. Stein and S. Udenfriend, *Arch. Biochem. Biophys.*, **163**, 390 (1974).
[233] P. Böhlen, S. Stein, W. Dairman and S. Udenfriend, *Arch. Biochem. Biophys.*, **155**, 213 (1973).
[234] S. Stein, P. Böhlen, J. Stone, W. Dairman and S. Udenfriend, *Arch. Biochem. Biophys.*, **155**, 202 (1973).
[235] C. Schwabe, *Anal. Biochem.*, **53**, 484 (1973).
[236] W. L. Ragland, J. L. Pace and D. L. Kemper, *J. Int. Res. Commun.*, **1**, 7 (1973).
[237] K. Imai, *J. Chromatogr.*, **105**, 135 (1975).
[238] G. Schwedt, *J. Chromatogr.*, **118**, 429 (1976).
[239] S. Kamata, K. Imura, A. Okada, Y. Kawashima, A. Yamatodani, T. Watanabe and H. Wada, *J. Chromatogr.*, **231**, 291 (1982).
[240] T. Yonekura, S. Kamata, M. Wasa, A. Okada, A. Yamatodani, T. Watanabe and H. Wada, *J. Chromatogr.*, **427**, 320 (1988).
[241] K. Samejima, *J. Chromatogr.*, **96**, 250 (1974).
[242] K. Samejima, M. Kawase, S. Sakamoto, M. Okada and Y. Endo, *Anal. Biochem.*, **76**, 392 (1976).
[243] M. Kai, T. Ogata, K. Haraguchi and Y. Ohkura, *J. Chromatogr.*, **163**, 151 (1979).
[244] S. J. Wassner, J. L. Schlitzer and J. B. Li, *Anal. Biochem.*, **104**, 284 (1980).
[245] N. A. Farid, *J. Pharm. Sci.*, **68**, 249 (1979).
[246] K. A. Gruber, S. Stein, L. Brink, A. Radhakrishnan and S. Udenfriend, *Proc. Natl. Acad. Sci. USA*, **73**, 1314 (1976).
[247] G. Szokan, *J. Liq. Chromatogr.*, **5**, 1493 (1983).
[248] G. J. DeJong, *J. Chromatogr.*, **183**, 203 (1980).
[249] A. J. Sedman and J. Gal, *J. Chromatogr.*, **232**, 315 (1982).
[250] S. E. Walker and P. E. J. Coates, *J. Chromatogr.*, **223**, 131 (1981).
[251] Y. C. Sumirtapura, C. Aubert, Ph. Coassolo and J. P. Cano, *J. Chromatogr.*, **232**, 111 (1982).
[252] T. K. Narayanan and L. M. Greenbaum, *J. Chromatogr.*, **306**, 109 (1984).
[253] N. F. Swynnerton, E. P. McGovern, D. J. Mangold, J. A. Nino, E. M. Gause and L. Fleckenstein, *J. Liq. Chromatogr.*, **6**, 1523 (1983).
[254] F. Veronese, R. Lazzarini, O. Schiavon and A. Bettero, *Farmaco, Ed. Prat.*, **37**, 390 (1982).

[255] K. Groeningsson and M. Widahl-Naesman, *J. Chromatogr.*, **291**, 185 (1984).
[256] M. Abrahamsson, *J. Pharm. Biomed. Anal.*, **4**, 399 (1986).
[257] W. McHugh, R. A. Sandman, W. G. Haney, S. P. Sood and D. P. Wittmer, *J. Chromatogr.*, **124**, 376 (1976).
[258] H. Nakamura and Z. Tamura, *Anal. Chem.*, **52**, 2087 (1980).
[259] H. Nakamura, K. Takagi, Z. Tamura, R. Yoda and Y. Yamamoto, *Anal. Chem.*, **56**, 919 (1984).
[260] N. Lustenberger, H. W. Lange and K. Hempel, *Angew. Chem.*, **84**, 255 (1972).
[261] H. W. Lange, N. Lustenberger and K. Hempel, *Fresenius Z. Anal. Chem.*, **261**, 337 (1972).
[262] T. K. Hwang, J. N. Miller, D. T. Burns and J. W. Bridges, *Anal. Chim. Acta*, **99**, 305 (1978).
[263] P. A. Shore, A. Burkhalter and V. H. Cohen, *J. Pharmacol. Exp. Ther.*, **127**, 182 (1959).
[264] A. H. Anton and D. F. Sayre, *J. Pharmacol. Exp. Ther.*, **166**, 285 (1969).
[265] L. T. Kremzner, *Anal. Biochem.*, **15**, 270 (1966).
[266] R. Hakanson and A. L. Rönnberg, *Anal. Biochem.*, **54**, 353 (1973).
[267] R. P. Maickel and F. P. Miller, *Anal. Chem.*, **38**, 1937 (1966).
[268] L. Juhlin and W. B. Shelley, *J. Histochem. Cytochem.*, **14**, 525 (1966).
[269] B. Ehinger and R. Thunberg, *Exp. Cell. Res.*, **47**, 116 (1967).
[270] R. Hakanson, L. Juhlin, C. Owman and B. Sporrong, *J. Histochem. Cytochem.*, **18**, 93 (1970).
[271] D. Aures, R. Fleming and R. Hakanson, *J. Chromatogr.*, **33**, 480 (1968).
[272] D. Turner and S. L. Wightman, *J. Chromatogr.*, **32**, 315 (1968).
[273] W. B. Shelley and L. Juhlin, *J. Chromatogr.*, **22**, 130 (1966).
[274] L. Edvinsson, R. Hakanson, A. L. Rönnberg and F. Sundler, *J. Chromatogr.*, **67**, 81 (1972).
[275] M. Roth, *Anal. Chem.*, **43**, 880 (1971).
[276] M. Roth and A. Hampai, *J. Chromatogr.*, **83**, 353 (1973).
[277] P. E. Hare, *Methods Enzymol.*, **47**, 3 (1977).
[278] P. Böhlen, *Methods Enzymol.*, **91**, 17 (1983).
[279] D. G. Drescher and K. S. Lee, *Anal. Biochem.*, **84**, 559 (1979).
[280] J. R. Benson and P. E. Hare, *Proc. Natl. Acad. Sci. USA*, **72**, 619 (1975).
[281] C. J. Little, J. A. Whatley and A. D. Dale, *J. Chromatogr.*, **171**, 63 (1979).
[282] J. R. Cronin, S. Pizzarello and W. E. Gandy, *Anal. Biochem.*, **93**, 174 (1979).
[283] M. K. Radjai and R. T. Hatch, *J. Chromatogr.*, **196**, 319 (1980).
[284] G. J. Hughes, K. H. Winterhalter, E. Boller and K. J. Wilson, *J. Chromatogr.*, **235**, 417 (1982).
[285] G. R. Barbarash and R. H. Quarles, *Anal. Biochem.*, **119**, 177 (1982).
[286] J. A. M. van der Heyden, K. Venema and J. Korf, *J. Neurochem.*, **32**, 469 (1979).
[287] M. J. Drescher, J. E. Medina and D. G. Drescher, *Anal. Biochem.*, **116**, 280 (1981).
[288] T. Hayashi, H. Tsuchiya and H. Naruse, *J. Chromatogr.*, **274**, 319 (1983).
[289] K. Nakazawa, H. Tanaka and M. Arima, *J. Chromatogr.*, **233**, 313 (1982).
[290] J. Grove, R. G. Alken and P. J. Schechter, *J. Chromatogr.*, **306**, 383 (1984).
[291] N. Seiler, S. Sarhan and B. Knödgen, *Int. J. Dev. Neuroscience.*, **3**, 317 (1985).
[292] A. Schousboe, O. M. Larsson and N. Seiler, *Neurochem. Res.*, **11**, 1497 (1986).
[293] N. Seiler and B. Knödgen, *J. Chromatogr.*, **341**, 11 (1985).
[294] J. R. Benson, *Anal. Biochem.*, **71**, 459 (1976).
[295] E. H. Creaser and G. J. Hughes, *J. Chromatogr.*, **144**, 69 (1979).
[296] T. M. Joys and H. Kim, *Anal. Biochem.*, **94**, 371 (1979).
[297] H. Nakamura, C. L. Zimmerman and J. J. Pisano, *Anal. Biochem.*, **93**, 423 (1979).
[298] G. J. Hughes, K. H. Winterhalter and K. J. Wilson, *FEBS Lett.*, **108**, 81 (1979).
[299] P. R. Carnegie, M. Z. Ilic, M. O. Etheridge and M. G. Collino, *J. Chromatogr.*, **261**, 153 (1983).
[300] K. Deck, S. Uhlhaas and P. Wardenbach, *J. Clin. Chem. Clin. Biochem.*, **18**, 567 (1980).
[301] R. C. Simpson, H. J. Mohammed and H. Veening, *J. Liq. Chromatogr.*, **5**, 245 (1982).
[302] L. J. Marton and P. L. Y. Lee, *Clin. Chem.*, **21**, 1721 (1975).
[303] F. Perini, J. B. Sadow and C. V. Hixon, *Anal. Biochem.*, **94**, 431 (1979).
[304] G. G. Shaw, H. S. Al-Deen and P. M. Elsworthy, *J. Chromatogr. Sci.*, **18**, 166 (1980).
[305] G. Milano, M. Schneider, P. Cambon, J. Boublil, J. Barbe, N. Renee and C. M. Lalanne, *J. Clin. Chem. Clin. Biochem.*, **18**, 157 (1980).
[306] M. Mach, H. Kersten and W. Kersten, *J. Chromatogr.*, **223**, 51 (1981).
[307] P. K. Bondy and Z. N. Canellakis, *J. Chromatogr.*, **224**, 371 (1981).
[308] T. Takagi, T. G. Chung and A. Saito, *J. Chromatogr.*, **272**, 371 (1983).
[309] D. H. Russell, J. D. Ellingson and T. P. Davies, *J. Chromatogr.*, **273**, 263 (1983).
[310] N. Seiler and N. Knödgen, *J. Chromatogr.*, **221**, 227 (1980).
[311] J. Wagner., C. Danzin and P. S. Mamont, *J. Chromatogr.*, **227**, 349 (1982).
[312] S. Ida, Y. Tanaka, S. Ohkuma and K. Kuriyama, *Anal. Biochem.*, **136**, 352 (1984).
[313] C. F. Laureri, E. Gaetani, M. Vitto and F. Bordi, *Farmaco, Ed. Prat.*, **39**, 29 (1984).
[314] P. Kucera and H. Umagat, *J. Chromatogr.*, **255**, 563 (1983).
[315] A. Himuro, H. Nakamura and Z. Tamura, *J. Chromatogr.*, **264**, 423 (1983).
[316] G. Schwedt, *Anal. Chim. Acta*, **92** 337 (1977).
[317] P. M. Froehlich and T. D. Cunningham, *Anal. Chim. Acta*, **97**, 357 (1978).

[318] D. L. Mays, R. J. van Alphendoorn and R. G. Lauback, *J. Chromatogr.*, **120**, 93 (1976).
[319] T. Kawamoto, I. Mashimo, S. Yamaguchi and M. Watanabe, *J. Chromatogr.*, **305**, 373 (1984).
[320] M. E. Rogers, M. W. Adlord, G. Saunders and G. Holt, *J. Chromatogr.*, **257**, 91 (1983).
[321] W. Buchberger, K. Winsauer and F. Nachtmann, *Fresenius' Z. Anal. Chem.*, **315**, 525 (1983).
[322] H. N. Myers and J. V. Rindler, *J. Chromatogr.*, **76**, 103 (1979).
[323] E. E. Stobberingh, A. W. Houben and C. P. A. van Boven, *J. Clin. Microbiol.*, **15**, 795 (1982).
[324] R. T. Krause, *J. Chromatogr. Sci.*, **16**, 281 (1978).
[325] J. C. Hodgin, P. Y. Howard, D. M. Ball, C. Cloete and L. De Jager, *J. Chromatogr. Sci.*, **21**, 503 (1983).
[326] D. C. Turnell and J. D. H. Cooper, *J. Autom. Chem.*, **5**, 36 (1983).
[327] R. Schuster, *Tec. Lab.*, **11**, 47 (1988).
[328] S. S. Simons Jr. and D. F. Johnson, *J. Am. Chem. Soc.*, **98**, 7098 (1976).
[329] S. S. Simons Jr. and D. F. Johnson, *Anal. Biochem.*, **82**, 250 (1977).
[330] S. S. Simons Jr. and D. F. Johnson, *J. Org. Chem.*, **43**, 2886 (1978).
[331] R. C. Simpson, J. E. Spriggle and H. Veening, *J. Chromatogr.*, **261**, 407 (1983).
[332] J. F. Stobbaugh, A. J. Repta, L. A. Sternson and K. W. Garren, *Anal. Biochem.*, **135**, 495 (1983).
[333] G. H. T. Wheler and J. T. Russell, *J. Liq. Chromatogr.*, **4**, 1281 (1981).
[334] M. H. Joseph and P. Davies, *J. Chromatogr.*, **277**, 125 (1983).
[335] D. Linderoth and K. Mopper, *Anal. Chem.*, **51**, 1667 (1979).
[336] R. F. Chen, C. Scott and E. Trepman, *Biochim. Biophys. Acta*, **576**, 440 (1979).
[337] C. R. Krishnamurti, A. M. Heindze and G. Galzy, *J. Chromatogr.*, **315**, 321 (1984).
[338] J. D. H. Cooper, G. Ogden, J. McIntosh and D. C. Turnell, *Anal. Biochem.*, **142**, 98 (1984).
[339] A. M. Felix and G. Terkelson, *Arch. Biochem. Biophys.*, **157**, 177 (1973).
[340] R. H. Buck and K. Krummen, *J. Chromatogr.*, **387**, 225 (1987).
[341] A. Jegorov, J. Triska, T. Trnka and M. Cerny, *J. Chromatogr.*, **434**, 417 (1988).
[342] D. W. Aswad, *Anal. Biochem.*, **137**, 405 (1984).
[343] E. W. Wuis, E. W. J. Beneken Kolmer, L. E. C. van Beisterveldt, R. C. M. Burgers, T. B. Vree and E. van Derkleyn, *J. Chromatogr.*, **415**, 419 (1987).
[344] N. Nimura, K. Iwaki and T. Kinoshita, *J. Chromatogr.*, **402**, 387 (1987).
[345] M. R. Euerby, L. Z. Partridge and P. Rajani, *J. Chromatogr.*, **447**, 392 (1988).
[346] M. O. Fleury and P. V. Ashley, *Anal. Biochem.*, **133**, 330 (1983).
[347] C. Cloete, *J. Liq. Chromatogr.*, **7**, 1979 (1984).
[348] R. L. Smith and K. A. Panico, *J. Liq. Chromatogr.*, **8**, 1793 (1985).
[349] M. Farrant, F. Zia-Gharib and R. A. Webster, *J. Chromatogr.*, **417**, 385 (1987).
[350] D. W. Hill, F. H. Walters, T. D. Wilson and J. D. Stuart, *Anal. Chem.*, **51**, 1338 (1978).
[351] H. Umagat, P. Kucera and L.-F. Wen, *J. Chromatogr.*, **239**, 463 (1982).
[352] B. R. Larsen and F. G. West, *J. Chromatogr. Sci.*, **19**, 259 (1981).
[353] J. P. H. Burbach, A. Prins, J. L. M. Lebouille, J. Verhoef and A. Whitter, *J. Chromatogr.*, **237**, 339 (1982).
[354] M. J. Winespear and A. Oaks, *J. Chromatogr.*, **270**, 378 (1983).
[355] N. B. Jones, S. Pääbo and S. Stein, *J. Liq. Chromatogr.*, **4**, 565 (1981).
[356] S. J. Price, T. Palmer and M. Griffin, *Chromatographia*, **18**, 62 (1984).
[357] J. D. H. Cooper, M. T. Lewes and D. C. Turnell, *J. Chromatogr.*, **285**, 490 (1984).
[358] R. H. Buck and K. Krummen, *J. Chromatogr.*, **303**, 238 (1984).
[359] G. A. Qureshi, L. Fohlin and J. Bergström, *J. Chromatogr.*, **297**, 91 (1984).
[360] S. J. Wassner and B. Li, *J. Chromatogr.*, **227**, 497 (1982).
[361] M. H. Fernstrom and J. D. Fernstrom, *Life Sci.*, **29**, 2119 (1981).
[362] D. L. Hogan, K. L. Kraemer and J. I. Isenberg, *Anal. Biochem.*, **127**, 17 (1982).
[363] B. N. Jones and J. P. Gulligan, *J. Chromatogr.*, **266**, 471 (1983).
[364] H. Godel, T. Graser, P. Földi, P. Pfaender and P. Fürst, *J. Chromatogr.*, **297**, 49 (1984).
[365] H. P. Fiedler and A. Plaga, *J. Chromatogr.*, **386**, 229 (1967).
[366] A. S. Herranz, J. Lerma and R. Martin Del Rio, *J. Chromatogr.*, **309**. 139 (1984).
[367] A. Qureshi, S. van den Berg, A. Gutierrez and J. Bergström, *J. Chromatogr.*, **297**, 93 (1984).
[368] M. L. Patchett, C. R. Monk, R. M. Daniel and H. W. Morgan, *J. Chromatogr.*, **425**, 269 (1988).
[369] E. Mendez, R. Matas and F. Soriano, *J. Chromatogr.*, **323**, 373 (1985).
[370] L. Essers, *J. Chromatogr.*, **305**, 345 (1984).
[371] K. Koshide, R. Tawa, S. Hirose and T. Fujimoto, *Clin. Chem.*, **31**, 1921 (1985).
[372] R. Tawa, K. Koshide, S. Hirose and T. Fujimoto, *J. Chromatogr.*, **425**, 143 (1988).
[373] T. Fujimoto, *J. Chromatogr.*, **425**, 143 (1988).
[374] Y. Tsuruta, I. Kohashi and Y. Ohkuva, *J. Chromatogr.*, **224**, 105 (1981).
[375] J. C. Robert, J. Vatier, B. K. Nguyen Phuoc and S. Boufils, *J. Chromatogr.*, **273**, 275 (1983).
[376] J. L. Devalia, B. D. Sheinman and R. J. Davies, *J. Chromatogr.*, **343**, 407 (1985).
[377] T. P. Davis, *Clin. Chem.*, **24**, 1317 (1978).
[378] D. Mell-Leroy Jr., A. R. Dasler and A. B. Gustafson, *J. Liq. Chromatogr.*, **1**, 261 (1978).
[379] T. P. Davis, C. W. Gehrke Jr., C. H. Williams, C. W.

Gehrke and K. Gerhardt, *J. Chromatogr.*, **228**, 113 (1982).
[380] T. Skaaden and T. Greibrokk, *J. Chromatogr.*, **247**, 111 (1982).
[381] L. A. Sternson, J. F. Stobaugh and A. J. Repta, *Anal. Biochem.*, **144**, 233 (1985).
[382] K. Kostka, *J. Chromatogr.*, **49**, 249 (1970).
[383] K. Kai, M. Yamaguchi and Y. Ohkura, *Anal. Chim. Acta*, **120**, 411 (1980).
[384] M. Kai, T. Miyazaki, M. Yamaguchi and Y. Ohkura, *J. Chromatogr.*, **268**, 417 (1983).
[385] M. Kai, T. Miyazaki and Y. Ohkura, *J. Chromatogr.*, **311**, 257 (1984).
[386] M. Kai, T. Miyazaki, Y. Sakamoto and Y. Ohkura, *J. Chromatogr.*, **322**, 473 (1985).
[387] M. Kai, T. Miuva, J. Ishida, Y. Ohkura, *J. Chromatogr.*, **345**, 259 (1985).
[388] S. S. Chen, A. Y. Kou and H.-H. Y. Chen, *J. Chromatogr.*, **276**, 37 (1983).
[389] N. Nimura, K. Iwaki, K. Takeda and H. Ogura, *Anal. Chem.*, **58**, 2372 (1986).
[390] K. Iwaki, N. Nimura, Y. Hiraga, T. Kinoshita, K. Takeda and H. Ogura, *J. Chromatogr.*, **407**, 273 (1987).
[391] V. E. Barskii, V. B. Ivanov, Y. E. Sklyar and G. I. Mikhailov, *Iz. Akad. Nauk SSSR, Ser. Biol.*, 744 (1968).
[392] R. W. Siegler, *Diss. Abstr. Int. B.*, **47**, 2447 (1986).
[393] R. Siegler and L. A. Sternson, *J. Pharm. Biomed. Anal.*, **6**, 485 (1988).
[394] R. Chayen, R. Dvir, S. Gould and A. Harell, *Anal. Biochem.*, **42**, 283 (1971).
[395] V. Graef, *Z. Klin. Chem. Klin. Biochem.*, **8**, 320 (1970).
[396] T. Kawasaki, M. Maeda and A. Tsuji, *J. Chromatogr.*, **272**, 261 (1983).
[397] G. Avigad, *J. Chromatogr.*, **139**, 153 (1977).
[398] K. Mopper and L. Johnson, *J. Chromatogr.*, **256**, 27 (1983).
[399] T. Kawasaki, M. Maeda and A. Tsuji, *J. Chromatogr.*, **163**, 143 (1979).
[400] T. Kawasaki, M. Maeda and A. Tsuji, *J. Chromatogr.*, **266**, 1 (1981).
[401] T. Kawasaki, M. Maeda and A. Tsuji, *J. Chromatogr.*, **232**, 1 (1982).
[402] W. T. Alpenfels, *Anal. Biochem.*, **114**, 153 (1981).
[403] D. Bion, G. Dulot and M. Pays, *J. Clin. Chem. Biochem.*, **21**, 397 (1983).
[404] G. Gübitz, R. Wintersteiger and R. W. Frei, *J. Liq. Chromatogr.*, **7**, 839 (1984).
[405] S. Mitsutani, Y. Wakuri, N. Yoshida, T. Nakajima and Z. Tamura, *Chem. Pharm. Bull. Jpn.*, **17**, 2340 (1969).
[406] T. Hirata, M. Kai, K. Kohashi and Y. Ohkura, *J. Chromatogr.*, **226**, 25 (1981).
[407] J. F. Gregory III, *Anal. Biochem.*, **102**, 374 (1980).
[408] S. Hase, T. Ikenaka and Y. Matsushima, *Biochem. Biophys. Res. Commun.*, **85**, 257 (1978).
[409] H. Takemoto, S. Hase and T. Ikenaka, *Anal. Biochem.*, **145**, 245 (1985).
[410] H. Weil-Malherbe and A. D. Bone, *Biochem. J.*, **51**, 311 (1952).
[411] H. Nohta, A. Mitsui and Y. Ohkura, *Anal. Chim. Acta*, **165**, 171 (1984).
[412] A. Mitsui, H. Nohta and Y. Ohkura, *J. Chromatogr.*, **344**, 561 (1985).
[413] H. Nohta, A. Mitsui and Y. Ohkura, *J. Chromatogr.*, **380**, 229 (1986).
[414] S. Honda, S. Suzuki, M. Takahashi, K. Kakehi and S. Ganno, *Anal. Biochem.*, **134**, 34 (1983).
[415] S. Honda, M. Takahashi, Y. Araki and K. Kakehi, *J. Chromatogr.*, **274**, 45 (1983).
[416] S. Honda, Y. Matsuda, M. Takahashi, K. Kakehi and S. Ganno, *Anal. Chem.*, **52**, 1079 (1980).
[417] J. E. Spikner and J. C. Towne, *Anal. Chem.*, **34**, 1468 (1962).
[418] T. Hayashi, H. Tsuchiya and H. Naruse, *J. Chromatogr.*, **273**, 245 (1983).
[419] K. Koike and M. Koike, *Anal. Biochem.*, **141**, 481 (1984).
[420] G. A. Qureshi, *J. Chromatogr.*, **400**, 91 (1987).
[421] N. Takeyama, D. Takagi, Y. Kitazawa and T. Tanaka, *J. Chromatogr.*, **424**, 361 (1988).
[422] T. Hayashi, H. Todoriki and H. Naruse, *J. Chromatogr.*, **224**, 197 (1981).
[423] N. E. Hoffman and T. A. Killinger, *Anal. Chem.*, **41**, 162 (1969).
[424] A. Frigerio, P. Martelli and K. B. Baker, *J. Chromatogr.*, **81**, 139 (1973).
[425] U. Langenbeck, H. Hoinowski, K. Mantel and H.-H. Möhring, *J. Chromatogr.*, **143**, 39 (1977).
[426] T. C. Cree, S. M. Hutson and A. E. Harper, *Anal. Biochem.*, **92**, 156 (1979).
[427] H. P. Schwarz, I. E. Karl and D. M. Bier, *Anal. Biochem.*, **108**, 360 (1980).
[428] Z.-J. Wang, K. Zaitsu and Y. Ohkura, *J. Chromatogr.*, **430**, 223 (1988).
[429] M. Nakamura, S. Hara, M. Yamaguchi, Y. Takemori and Y. Ohkura, *Chem. Pharm. Bull.*, **35**, 687 (1987).
[430] W.-F. Chao, M. Kai and Y. Ohkura, *J. Chromatogr.*, **430**, 361 (1988).
[431] M. Kai, J. Ishida and Y. Ohkura, *J. Chromatogr.*, **430**, 271 (1988).
[432] E. Sawicki and R. A. Carnes, *Mikrochim. Acta.*, **I**, 148 (1968).
[433] M. Okamoto, *J. Chromatogr.*, **202**, 55 (1980).
[434] K. Mopper, W. Stahovec and L. Johnson, *J. Chromatogr.*, **256**, 243 (1983).
[435] W. Stahovec and K. Mopper, *J. Chromatogr.*, **298**, 399 (1984).
[436] D. J. Birkett, N. C. Price, G. K. Radda and A. G. Salmon, *FEBS Lett.*, **6**, 346 (1970).
[437] K. Imai, T. Toyo'oka and Y. Watanabe, *Anal. Biochem.*, **128**, 471 (1983).
[438] T. Toyo'oka, H. Miyano and K. Imai, *Proc. 104th Annu. Meet. Pharm. Soc. Jpn.*, Sendai, March 1984, p. 558.
[439] T. Toyo'oka and K. Imai, *J. Chromatogor.*, **282**, 495 (1983).
[440] W. H. Scouten, R. Lubcher and W. Banghman, *Biochim. Biophys. Acta*, **336**, 421 (1974).
[441] E. P. Lankmayr, K. W. Budna, K. Mueller and F. Nachtmann, *Fresenius' Z. Anal. Chem.*, **295**, 371 (1979).

[442] E. P. Lankmayr, K. W. Budna, K. Mueller, F. Nachtmann and F. Rainer, *J. Chromatogr.*, **222**, 249 (1981).
[443] Y. Kanaoka, M. Machida, K. Ando and T. Sekine, *Biochim. Biophys. Acta*, **207**, 269 (1970).
[444] Y. Kanaoka, M. Machida and T. Sekine, *Biochim. Biophys. Acta*, **317**, 563 (1973).
[445] M. Machida, N. Ushijima, M. I. Machida and Y. Kanaoka, *Chem. Pharm. Bull.*, **23**, 1385 (1975).
[446] N. Anzai, T. Kimura, S. Chida, T. Tanaka, H. Takahashi and H. Meguro, *Yakugaku Zasshi*, **101**, 1002 (1981).
[447] B. Kagedal and M. Källberg, *J. Chromatogr.*, **229**, 409 (1982).
[448] H. Takahashi, Y. Nara, H. Meguro and K. Tuzimura, *Agric. Biol. Chem.*, **43**, 1439 (1979).
[449] H. Takahashi, T. Yoshida and H. Meguro, *Bunseki Kagaku*, **30**, 341 (1980).
[450] J. J. Miners, I. Fearnley, K. J. Smith, D. J. Birkett, P. M. Brooks and M. W. Whitehouse, *J. Chromatogr.*, **275**, 89 (1983).
[451] S. Akayima, H. Akimoto, S. Nakatsuji and K. Nakashima, *Bull. Chem. Soc. Jpn.*, **58**, 2129 (1985).
[452] K. Nakashima, C. Umekawa, H. Yoshida, S. Nakatsuji and S. Akiyama, *J. Chromatogr.*, **414**, 11 (1987).
[453] K. Nakashima, C. Umekawa, S. Nakatsuji, S. Akiyama and R. S. Givens, *Biomed. Chromatogr.*, **3**, 39 (1989).
[454] R. C. Fahey, G. L. Newton, R. Dorian and E. M. Kosower, *Anal. Biochem.*, **107**, 1 (1980).
[455] G. L. Newton, R. Dorian and R. C. Fahey, *Anal. Biochem.*, **114**, 383 (1981).
[456] R. C. Fahey, G. L. Newton, R. Dorian and E. M. Kosower, *Anal. Biochem.*, **111**, 357 (1981).
[457] R. C. Fahey and G. L. Newton, *Methods. Enzymol.*, **143**, 85 (1987).
[458] H. S. Kosower, E. M. Kosower, G. L. Newton and H. M. Rauney, *Proc. Natl. Acad. Sci. USA*, **76**, 3382 (1979).
[459] W. Dünges, *Anal. Chem.*, **49**, 442 (1977).
[460] W. Dünges, A. Meyer, K. E. Müller, M. Müller, R. Pietschmann, C. Plachetta, R. Sehr and H. Tuss, *Fresenius' Z. Anal. Chem.*, **288**, 361 (1977).
[461] W. Dünges and N. Seiler, *J. Chromatogr.*, **145**, 483 (1978).
[462] S. Lam and E. Grushka, *J. Chromatogr.*, **158**, 207 (1978).
[463] K. Hayashi, J. Kawase, K. Yoshimura, K. Ara and K. Tsuji, *Anal. Biochem.*, **136**, 314 (1984).
[464] E. Grushka, S. Lam and J. Chassin, *Anal. Chem.*, **50**, 1398 (1978).
[465] J. Turk, S. J. Weiss, J. E. Davis and P. Needleman, *Prostaglandins*, **16**, 291 (1978).
[466] P. Leroy, S. Chakir and A. Nicolas, *J. Chromatogr.*, **351**, 267 (1986).
[467] A. Crozier, J. B. Zaerr and R. O. Morris, *J. Chromatogr.*, **238**, 157 (1982).
[468] M. Iwamoto, S. Yoshida and S. Hirose, *J. Chromatogr.*, **310**, 151 (1984).
[469] H. C. Michaelis, H. Foth and G. F. Kahl, *J. Chromatogr.*, **416**, 176 (1987).
[470] W. Dünges, *Prae-Chromatographische Mikromethoden*, Alfred Huethig, Heidelberg (1979).
[471] M. S. Gandelman and J. W. Birks, *Anal. Chim. Acta*, **155**, 159 (1983).
[472] J. H. Wolf and J. Korf, *J. Chromatogr.*, **436**, 437 (1988).
[473] J. F. B. Lloyd, *J. Chromatogr.*, **178**, 249 (1979).
[474] R. Farinotti, P. Siard, J. Bourson, S. Kirkiacharian, B. Valeur and G. Mahuzier, *J. Chromatogr.*, **269**, 81 (1983).
[475] S. Yoshida, T. Aachi and S. Hirose, *J. Chromatogr.*, **430**, 156 (1988).
[476] H. Tsuchiya, T. Hayashi, H. Naruse and T. Takagi, *J. Chromatogr.*, **234**, 121 (1982).
[477] A. Takadate, T. Tahara, H. Fujino and S. Goya, *Chem. Pharm. Bull.*, **30**, 4120 (1982).
[478] S. Goya, A. Takadate, H. Fujino and T. Tanaka, *Yakugaku Zasshi*, **100**, 744 (1980).
[479] S. Goya, A. Takadate and H. Fujino, *Yakugaku Zasshi*, **102**, 63 (1982).
[480] K. Akasaka, T. Suzuki, H. Ohrni, H. Meguro, Y. Shindo and H. Takahashi, *Anal. Lett.*, **20**, 1581 (1987).
[481] M. Yamaguchi, S. Hara, R. Matsunaga, M. Nakamura and Y. Ohkura, *Anal. Sci.*, **1**, 295 (1985).
[482] M. J. Cooper and M. W. Anders, *Anal. Chem.*, **46**, 1849 (1974).
[483] W. Distler, *J. Chromatogr.*, **192**, 240 (1980).
[484] W. D. Watkins and B. M. Peterson, *Anal. Biochem.*, **125**, 30 (1982).
[485] S. Kamada, M. Maeda and A. Tsuji, *J. Chromatogr.*, **272**, 29 (1983).
[486] W. D. Korte, *J. Chromatogr.*, **243**, 153 (1982).
[487] S. A. Barker, J. A. Monti, S. T. Christian, F. Benington and R. D. Morin, *Anal. Biochem.*, **107**, 116 (1980).
[488] N. Nimura and T. Kinoshita, *Anal. Lett.*, **13**, 191 (1980).
[489] Y. Yamauchi, T. Tomita, M. Senda, A. Hivai, T. Terano, Y. Tamura and S. Yoshida, *J. Chromatogr.*, **357**, 199 (1986).
[490] M. Ikeda, K. Shimada and T. Sakaguchi, *J. Chromatogr.*, **272**, 251 (1983).
[491] J. Goto, N. Goto, A. Hikichi, T. Nishimaki and T. Nambara, *Anal. Chim. Acta*, **120**, 187 (1980).
[492] J. Goto, N. Goto and T. Nambara, *J. Chromatogr.*, **239**, 559 (1982).
[493] H. Lingeman, A. Hülshoff, W. J. M. Underberg, and F. B. J. M. Offermann, *J. Chromatogr.*, **290**, 215 (1984).
[494] S. Goya, A. Takadate and M. Iwai, *Yakugaku Zasshi*, **101**, 1164 (1981).
[495] S. Goya, A. Takadate, H. Fujino and M. Irikura, *Yakugaku Zasshi*, **101**, 1064 (1981).
[496] M. Ikeda, K. Shimada and T. Sakaguchi, *J. Chromatogr.*, **305**, 261 (1984).
[497] J. B.F. Lloyd, *J. Chromatogr.*, **189**, 359 (1980).
[498] S. Goya, A. Takadate, M. Irikura, T. Suchiro and H. Fujimo, *Proc. 104th Annu. Meet. Pharm. Soc. Jpn*, Sendai, March 1984, p. 582.
[499] K. -E. Karlsson, D. Wiesler, I. M. Alasandro and M. Novotny, *Anal. Chem.*, **57**, 229 (1985).
[500] J. Goto, S. Komatsu, N. Goto and T. Nambara, *Chem. Pharm. Bull.*, **29**, 899 (1981).
[501] J. Goto, N. Goto, F. Shamsa, M. Saito, S. Komatsu, K. Suzaki and T. Nambara, *Anal. Chim. Acta*, **147**, 397 (1983).
[502] J. Goto, N. Goto and T. Nambara, *Chem. Pharm. Bull.*, **30**, 4597 (1982).

[503] C. F. Poole, S. Singhawangcha, A. Zlatkis and E. D. Morgan, *J. High Resolut. Chromatogr. Chromatogr. Commun.*, **1**, 96 (1978).
[504] N. K. Kochetkov, V. N. Shibaev and A. A. Kost, *Tetrahedron Lett.*, 1993 (1971).
[505] L. Shugart, *J. Chromatogr.*, **174**, 250 (1979).
[506] M. Yoshioka and Z. Tamura, *J. Chromatogr.*, **123**, 220 (1976).
[507] J. F. Kuttesch, F. C. Schmalstieg and J. A. Nelson, *J. Liq. Chromatogr.*, **1**, 97 (1978).
[508] M. R. Preston, *J. Chromatogr.*, **275**, 178 (1983).
[509] J. Wagner, Y. Hirth, N. Claverie and C. Danzin, *Anal. Biochem.*, **154**, 604 (1986).
[510] M. Yoshioka, K. Nishidate, H. Iizuka, A. Nakamura, M. M. El-Merzabani, Z. Tamura and T. Miyazaki, *J. Chromatogr.*, **309**, 63 (1984).
[511] L. L. Schukorskaya, S. N. Ushakov and N. K. Galania, *Inv. Akad. Nauk SSSR Otd. Khim. Nauk*, 1692 (1962).
[512] J. A. Secrist, III, J. R. Barrio, N. J. Leonard and G. Weber, *Biochemistry*, **11**, 3499 (1972).
[513] M. Kai and Y. Ohkura, *Trends Anal. Chem.*, **6**, 226 (1987).
[514] N. Nakano, M. Kai, M. Ohno and Y. Ohkura, *J. Chromatogr.*, **411**, 305 (1987).

10

Derivatization for Chromatographic Resolution of Optically Active Compounds

M. W. Skidmore
Residue and Environmental Chemistry, ICI Agrochemicals, Jealotts Hill Research Station, Bracknell, Berkshire

1. INTRODUCTION 215
2. INDIRECT VERSUS DIRECT METHODS 217
3. SELECTION OF CHIRAL DERIVATIZING REAGENTS FOR CHROMATOGRAPHIC RESOLUTION OF ENANTIOMERS 218
4. REACTIONS OF INDIVIDUAL FUNCTIONALITIES 219
   4.1 Hydroxyl groups (alcohols and phenols) 219
   4.2 Aliphatic and acyclic carboxylic acids 225
   4.3 Amines 230
   4.4 Aldehydes and ketones 237
   4.5 Lactones 240
   4.6 Amino acids 241
   4.7 Amino alcohols 247
   4.8 Synonyms for some of the reagents described in the text 249
5. CONCLUSIONS AND FUTURE TRENDS 249
6. REFERENCES 249

1. Introduction

Optically active compounds form an integral part of all biological systems and are almost exclusively found in a single enantiomeric form. These systems therefore constitute a chiral environment in which the enantiomers of chiral xenobiotics may exhibit very different properties and behave as two totally different entities; typical examples of the different responses are shown in Table 1.

This type of differential response has led to a considerable amount of research in the 'effects' chemical industries, e.g. pharmaceutical, pesticide, food, flavour and fragrances, where very selective and specific biological responses can be utilized and developed commercially. In general, as demonstrated by the examples in Table 1, while one of the enantiomers is responsible for a certain biological effect, the other form may exhibit a similar activity, be responsible for a very different, unwanted, side effect, or remain totally inactive and be present solely as an impurity. The use of racemic mixtures may therefore be undesirable and could have disastrous implications; e.g. the sedative thalidomide was widely prescribed, as a racemic mixture, to pregnant women during the 1970's; however, it was not initially realized that, although the R enantiomer produced the prescribed sedative effect, the S enantiomer induced severe teratogenic effects. Regulatory bodies are showing increasing interest in determining the relative rates of metabolism and dissipation of the individual enantiomers in biological or environmental systems, and also in product characterization to determine enantiomeric purities. These requirements will undoubtedly increase the amount of academic and industrial research to develop more robust methods for both preparative and analytical separations.

The first enantiomeric separation was credited to Louis Pasteur, who in 1848 manually sorted crystals of sodium ammonium tartrate based on their different crystalline appearances. Advances in technology and separation science have been paralleled by similar advances in enantiomeric separation methods. These have included preferential crystallization, crystallization from optically active solvents, fractional crystallization of diastereo-

Handbook of Derivatives for Chromatography
Edited by K. Blau and J. M. Halket © 1993 John Wiley & Sons Ltd

Table 1 Effects and responses of the enantiomers of some biologically active compounds

Compound	Structure	Enantiomer	Biological response/effect
Tyrosine		D	Sweet taste
		L	Bitter taste
Limonene		R	Odour of oranges [1]
		S	Odour of lemons [1]
Propranolol		R	No defined biological effect
		S	β-Adrenergic blocking agent
Propoxyphene		+	Analgesic
		−	Antitussive; respiratory depressant
Paclobutrazol		+	Fungicidal activity, little or no effect on plant growth
		−	Good plant growth regulation (PGR) and fungicidal activity
Fluazifop-butyl		R	Post-emergent herbicidal activity
		S	No post-emergent activity

meric salts, the selective destruction of one enantiomer, and chromatographic method [2]. Resolution of enantiomers by chromatographic techniques, and in particular high performance liquid chromatography, has many advantages over classical methods, and is suitable for both analytical and preparative separations where both enantiomers can be determined/recovered simultaneously.

Currently, the procedures used for the chromatographic separation of enantiomeric pairs can be classified as either 'indirect' or 'direct' methods. The definitions of these methods are outlined below

(1) Indirect method: enantiomers are converted into diastereomers by reaction with an optically pure reagent and the products are subsequently separated on achiral chromatographic phases.

(2) Direct method: this general definition encompasses two techniques; one in which the enantiomers are separated on chiral stationary phases, and a second in which the separation takes place on an achiral stationary phase by means of chiral additives in the mobile phase. The latter technique relies on the formation of diastereomeric complexes in the mobile phase.

This chapter is primarily concerned with procedures involving the indirect method, and will present a review of the chiral reagents used, including their advantages, shortcomings, and the conditions used for derivatization. Structures of each reagent will also be provided, with a supporting bibliography of their use and synthesis where commercial availability may be uncertain. Although the remit of this chapter is primarily concerned with reagents used for indirect methods, the relative merits and disadvantages of both types of method are reviewed in the next section.

2. Indirect versus direct methods

The first step in developing a separation is to define the exact nature of the problem, the success criteria and limiting factors such as the resources and capabilities of the laboratory. A comprehensive literature review is always recommended to establish whether methods have been developed on the same or similar compounds. If

Table 2   Requirements, merits and drawbacks for direct and indirect methods used in the separation of enantiomers from the racemate

Indirect methods	Direct methods[a]	
	HPLC	GC
Substrate needs to contain a derivatizable function	No universal phase is available, therefore lengthy method development times and several types of column may need to be investigated	Phases are mostly unstable at high temperatures
Can be readily used for both analytical and preparative work	Chiral phases are expensive, or high use rates of mobile phase additives may be required, particularly for preparative work	Offers a wide range of selective detectors and very low limits of determination
Chiral derivatization reagents should be 100% pure or of accurately known purity for analytical applications	Enantiomers have the same detector response	Unsuitable for preparative separations
Resolution is carried out on achiral columns, so a suitable column can readily be found thus reducing development time	No derivatization is needed except where necessary to enhance detectability	Capillary columns can be used to give very high separation efficiencies
By choosing the chirality of the derivatization reagent the elution order can be controlled; this is an advantage for trace analysis of one enantiomer in the presence of an excess of a second, where it is preferred for the minor enantiomer to elute first	Little chance of racemization	Derivatization is usually necessary to enhance volatility
Diastereomers may have different responses on the chromatographic systems	Pure enantiomers are eluted directly from the column	
Recovery of the enantiomers may be difficult and require at least one extra stage		
May be time consuming		

[a]For further details of the direct methods, the reader is referred to several comprehensive reviews, which include gas chromatography [1,3,4], high performance liquid chromatography [1,4,5,6] and thin-layer chromatography [7].

prior art exists, it may be preferable to use these methods to reduce development time.

Indirect and direct methods should be considered as complementary, since neither is a panacea for all enantiomeric separations. The strengths and weaknesses of each method are outlined in Table 2.

3. Selection of chiral derivatizing reagents for chromatographic resolution of enantiomers

With indirect methods, the substrate is derivatized with a chiral reagent to generate two diastereomers, which have different physicochemical properties and can be separated on achiral stationary phases.

The choice of the reagent has a significant effect on the degree of separation, detectability and recovery of the individual enantiomers. A number of researchers have made observations and predictions of the structural features which optimize the reagent in terms of stability and separability of the diastereomers; many of these were described in the previous edition of this handbook [8] and are reiterated, with examples, below.

(1) The diastereomers should have conformational rigidity to maximize their physical differences and so enhance the separation. During the separation of the diastereomeric (+)-neomethyl thioureas of amino acid esters, Nambara et al. [9] showed that the separation factors of the t-butyldimethylsilyl esters were greater than those of the methyl esters. This was believed to be an effect of the increased structural rigidity imparted by the t-butyldimethylsilyl group.

[2] A reagent with a large size difference between the groups attached to the chiral centre will, in general, result in improved separations [10].

(3) The distance between the asymmetric centres should be minimal and ideally less than three bonds [8]. Scott et al. [11] showed that the separation factors of the amides of several acyclic isoprenoid acids decreased as the distance between the chiral centres increased.

(4) The presence of polar or polarizable groups close to the chiral centre can promote hydrogen bonding or π-overlap with the stationary phase and enhance the resolution. This was again shown to be a significant structural feature in the separation of diastereomeric amides of acyclic isoprenoid acids [11]. The introduction of an olefinic group close to the chiral centre increased the separation factors. Similarly, separation factors were decreased as the distance between the olefinic group and the chiral centre increased.

It should be emphasized that these factors are not pre-requisites but merely indicate structural features which can be used to identify the more useful reagents; e.g. thioureas from catecholamines and 2,3,4,6-tetra-O-acetyl-β-D-glucopyranosyl isothiocyanate (GITC) and 2,3,4-tri-O-acetyl-α-D-arabinopyranosyl isothiocyanate (AITC) were separated [12] because of their structural rigidity and hydrophobicity, even though the distance between the chiral centres is greater than 'three bonds'.

In addition to the above structural features, there are several other factors which, although more obvious, are also worthy of consideration.

(a) The reagent should have a high optical purity (preferably 100%). Analytical results are only proportional to the enantiomeric composition of the substrate if the optical purity of the reagent is 100%; any enantiomeric impurity will also react with the substrate and result in an unrepresentative assay. The results of a reaction with an enantiomerically impure reagent are shown in Figure 1; during quantification, III will contribute to the peak area of I, and IV to that of II.

The problem can, to some extent, be overcome by analysing a standard of known enantiomeric purity and applying a correction factor to subsequent analyses.

(b) The structure of the chiral reagent and the conditions necessary for a quantitative reaction should minimize the chances of racemization or degradation of the substrate. The reagent should also be stable during storage; this is particularly important if analytical methods are likely to be carried out over extended periods.

(c) The nature of the available functional groups will define the types of reagent possible. It is also necessary to consider the remaining functional groups on the substrate and their reactivity with the reagent. Multiple reactions, unless quantitative, will have a deleterious effect on the assay.

(d) The nature of the reagent should complement the chromatographic technique; e.g. for gas chromato-

$$(\pm)\text{Substrate X} + (+)\text{Reagent Y} \longrightarrow (+)X(+)Y + (-)X(+)Y$$
$$\phantom{(\pm)\text{Substrate X} + (+)\text{Reagent Y} \longrightarrow (+)X(+)Y}\ \text{I} \ \text{II}$$

Reaction with the optical impurity in reagent Y

$$(\pm)\text{Substrate X} + (-)\text{Reagent Y} \longrightarrow (-)X(-)Y + (+)X(-)Y$$
$$\phantom{(\pm)\text{Substrate X} + (-)\text{Reagent Y} \longrightarrow (-)X(-)Y}\ \text{III} \ \text{IV}$$

Figure 1. Derivatization of substrate X with an impure chiral reagent.

graphy a volatile derivative will decrease the requirement for high column temperatures.

(e) The nature of the reagent should complement the detector and the limit of detection required. This is a similar consideration to that used in achiral quantitative analyses, where reagents are used to introduce electron capturing, UV-absorbing and fluorogenic species.

(f) For analytical purposes it is advantageous to have both forms of the chiral reagent available. Where the determination of a small amount of one enantiomer is required in the presence of a large excess of a second, a more accurate determination can be achieved when the minor component elutes first. Since it is difficult, and sometimes not possible, to predict the elution order, the availability of both enantiomeric forms of the reagent is an advantage.

(g) For preparative work the cost of the reagent and the possibility of recovering it are usually significant factors.

Reagents which meet all the criteria would be ideal; it is, however, up to the reader to choose the procedures and properties of the reagents described in the following sections to meet the criteria for the separation.

4. Reactions of individual functionalities

In this section, the chiral reagents used for each of the major functional groups will be reviewed and the derivatization procedures will be described in outline.

4.1 Hydroxyl groups (alcohols and phenols)

The most commonly used reagents for the resolution of racemic alcohols are chiral acids, isocyanates and chloroformates, which give diastereomeric esters, carbamates and carbonates respectively (Figure 2).

Esters are normally prepared from an alcohol and an 'activated' chiral carboxylic acid, usually in the form of a chloride, anhydride, or imidazole.

Acid chlorides are the most reactive, and are particularly useful for sterically hindered compounds or those which are difficult to acylate. However, care should be exercised when using these reagents, because certain acid chlorides, particularly those containing a proton attached to the chiral carbon, can racemize under extreme acidic or basic conditions or at elevated temperatures. The stability of these reagents should also be investigated to optimize storage and reaction conditions.

The most widely used reagents are the chlorides of 2B-acetoxy-Δ^5-etienic acid (**1**), R(+)-*trans*-chrysanthemic acid (**2**), drimanoic acid (**3**), (−)-menthoxyacetic acid (**4**), (S)-acetoxypropionic acid (**7**) and S(−)-N-(trifluoroacetyl)proline (**11**). (The numbers correspond to the structure in section 4.1.2) Etienic esters have been used for the resolution of asymmetric alcohols by fractional crystallization and, more recently, by gas chromatography [13]. Although acceptable separations are achieved, due in part to the conformational rigidity of the diastereomers, the acid has a high molecular weight and may well significantly decrease the volatility of the diastereomers.

(+)-*Trans*-chrysanthemoyl chloride is the most widely recommended reagent, and has been compared with a range of other reagents [14–16]. The advantages of chrysanthemoyl esters include: a fairly low molecular weight, which results in a derivative having a reasonable volatility; two centres of electron density in the isobutenylcyclopropanoyl moiety which interact with the stationary phase and improve chromatographic separation [14]; the derivatives are readily formed under mild reaction conditions; and the mass spectra of the derivatives have a weak molecular ion and a strong base peak at *m/z* 123, useful for quantitative analysis by single ion monitoring [14, 16].

Fully substituted chiral acetic acids are stable to racemization, even under severe reaction conditions. An example of such a reagent is Mosher's reagent, S(−)-α-methoxy-α-(trifluoromethyl)phenylacetyl chloride (MPTA) (**5**), which yields highly volatile derivatives with good electron capturing properties. Using a fluorine probe, the diastereomeric derivatives can also be quantitatively determined by nuclear magnetic resonance spectroscopy [17]. Michelsen and Odham [18] achieved a detection limit of 0.1 pmol for MPTA diastereomeric derivatives of 1,2-diacylglycerols using single ion monitoring, and an estimated detection limit of 300 fmol for the R-2-phenylselenopropionyl esters (**6**) of R-2-

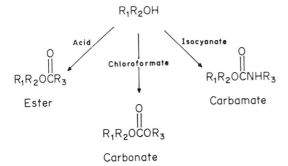

Figure 2. Typical derivatization reactions used to separate the enantiomers of chiral alcohols.

octanol. Doolittle and Heath [19] compared the separation factors of MPTA, S-acetoxypropionyl (AP) (7) and S-tetrahydro-5-oxo-2-furancarboxyl (TOF) (8) diastereomeric esters of a range of alcohols using both gas chromatography (GC) and high performance liquid chromatography (HPLC). The observations and conclusions were similar for both modes of chromatography; AP esters were most suitable for aliphatic alcohols, TOF esters for acetylenic, olefinic and aromatic alcohols, and MPTA esters were recommended for alcoholic lactones. The use of TOF was particularly recommended for preparative separations because of the fairly low cost and low molecular weight of the reagent.

Few examples of derivatizations with other activated forms of the carboxylic acids were noted, but since these can readily be formed from the chiral acid they are worthy of further investigation and may offer several advantages over acid chlorides.

The formation of diastereomeric esters using dicyclohexylcarbodiimide as a coupling reagent was used by Banfield and Rowland [20] in a quantitative assay for warfarin in biological fluids.

Recovery of the individual enantiomeric alcohols from the diastereomeric esters can be achieved by saponification [2, 15, 21] or by reaction with lithium aluminium hydride [2]. Saponification produces a mixture of products that can readily be separated by differential solvent extraction, but the risk of racemization is high. Reaction with lithium aluminium hydride is less likely to cause recemization but results in two alcohols, which need further separation by conventional achiral chromatography.

A novel in vitro glucuronidation of a phenolic substrate was carried out with rat liver microsomes [22]. The method is simple, the reagent is optically pure, only the (+)-glucuronic acid (13) is present, and the individual enantiomers are readily recovered by a simple enzymatic hydrolysis. Glucosidation with acetobromo-α-D-glucose (14) has also been used to separate the enantiomers of menthol [23].

A significant advance in the resolution of asymmetric alcohols has been the formation of carbamates from the reaction with a chiral isocyanate. Isocyanates are highly reactive and are widely used for the derivatization of hydroxyl and amine functional groups. Alcohols react relatively slowly and require heating for several hours, whereas phenols react rapidly, even at room temperature. Reagents used have included $(R(+)/S(-)$ phenylethyl isocyanate (15) and $R(+)/S(-)$-1-(1-naphthyl)ethyl isocyanate (16); the diastereomeric carbamates are stable and no evidence for racemization has been recorded. The individual enantiomers can be recovered by refluxing the dias-tereomeric carbamates in ethanolic sodium ethoxide for 30 minutes [24] or by silanolysis [25].

The formation of diastereomeric carbonates, from (−)-menthyl chloroformate (17) is carried out in alkaline aqueous solution at room temperature, and is particularly recommended for heat/acid-labile alcohols [26]. Base-labile alcohols, e.g. substituted racemic 2-hydroxyacetonitrile, have been separated as chiral hemiacetals [27] using (1R, 4R, 5S)-4-hydroxy-6,6-dimethyl-3-oxabicyclo-[3.1.0]hexan-2-one (18) as the chiral reagent.

4.1.1 Derivatization

4.1.1.1 Preparation of diastereomeric esters from chiral acid chlorides

Procedure [14, 16] for preparation of R(+)-trans-*chrysanthemoyl esters* Heat (+)-*trans*-chrysanthemic acid (2 mg) with freshly distilled thionyl chloride (200 µl) for 1 h at 60 °C. Remove the excess thionyl chloride under a stream of dry nitrogen. Mix the alcohol (1 mg dissolved in 20 µl of dry toluene) with 3 molar equivalents of the acid chloride in a sealed tube and heat at 40 °C for 1–2 h. The temperature used for the reaction is dependent on the nature of the alcohol, and temperature from 40 °C to reflux have been reported for this reaction.

Chromatograhpy The products of the derivatization are readily separated by gas chromatography, usually on non-polar (SE-30), or low polarity (OV-17) columns. Burden et al. [15] separated the diastereomeric chrysanthemoyl esters of several triazolyl fungicides using bonded phase SE-52 and OV-1 columns, and also used TLC to separate the derivative from the unreacted alcohol (silica gel plates eluted with hexane/ethyl acetate, 1:1).

Comments Many reactions involving acid chlorides include a basic catalyst, Burden et al. [15], however, observed some enantioselectivity in incomplete reactions when either 4-dimethylaminopyridine or N-methylimidazole was added to the reaction mixture.

Both chrysanthemoyl and drimanoyl esters produce mass spectra dominated by intense ions at m/z 123 arising from the $(C_9H_{15})^+$ drimanoyl or norchrysanthemoyl moieities, which could be useful for single ion monitoring. Molecular ions have a low abundance, < 10% of the base peak [14].

Procedure [19] for preparation of diastereomeric esters from S-tetrahydro-5-oxo-2-furancarboxylic acid chloride (TOF) Dissolve the acid chloride in an aprotic solvent (2 M in dichloromethane). Mix 100 µl of the acid chloride solution with an ice cold mixture of the chiral alcohol (20 μl) and pyridine (80 μl). Allow the mixture to reach room temperature and stir. Extract the diastereomeric derivatives from an acidified solution.

Chromatography Separation of the TOF diastereomeric esters of a range of secondary aliphatic, olefinic and aromatic alcohols was carried out on a fused-silica capillary column coated with SP2340 (14 m × 0.25 mm ID) and also by HPLC on a Lichrosorb EF54 silica gel column using 20% ethyl acetate in hexane.

The elution order of the diastereomeric esters in GC analysis was reversed in HPLC. Doolittle and Heath [18] suggest that this phenomenon may have been due to the different conformations of the diastereomers in solution and in the gas phase.

Comments The reagent is economical and easy to use, making it suitable for analytical or preparative analyses.

4.1.1.2 Formation of diastereomeric esters using dicyclohexylcarbodiimide as a coupling reagent

Procedure [20] for esterification of warfarin Place a solution containing warfarin in a tapered tube and evaporate to dryness. Add carbobenzyloxy-L-proline (200 mg ml^{-1} in acetonitrile; 10 μl), imidazole (1 mg ml^{-1} in acetonitrile; 10 μl) and dicyclohexylcarbodiimide (200 mg ml^{-1} in acetonitrile; 10 μl) and agitate for approximately 10 sec. Allow the mixture to stand for 2 h at room temperature. Analyse the mixture by HPLC.

Chromatography and comments The diastereomeric esters were separated using a Spherisorb 5 μm column eluted with ethyl acetate/hexane/methanol/acetic acid mixtures (25:74.75:0.25:0.3) and detected by fluorescence following aminolysis of the esters group (excitation wavelength 313 nm).

4.1.1.3 Formation of diastereomeric esters from phenols and D(+)-glucuronic acid using an in vitro glucuronidation with rat liver microsomes.

Procedure [22] Isolation of the microsomes is carried out by recognised methods described in the literature [28] and the protein content of the suspension is determined [29].

Mix the microsomal protein (1 mg equivalent), magnesium chloride (50 mM), Triton X-100 (0.01%, v/v), and uridine-5'-diphosphoglucuronic acid (UDPGA) (6.6 mM) in Tris–HCl buffer (75 mM), pH 7.5, 0.5 ml) in a polyethylene incubation cup. Incubate the mixture in a water bath at 37 °C for 10 min. Add the substrate (300 mM) to the pre-incubated mixture and replace in the water bath for 60 min, with constant shaking.

Terminate the reaction by placing the container in an ice bath, and centrifuge to separate out the microsomal pellet. Analyse the supernatant by HPLC.

Chromatography HPLC separation of the diastereomeric derivatives of the phenols used in this paper was carried out on a Nucleosil 5 C_{18} column (15 cm × 4.6 mm ID) eluted with water/acetonitrile (91:9) containing 0.005 M ammonia.

4.1.1.4 Formation of diastereomeric carbamates from chiral isocyanates

Procedure A [30] Mix the tertiary alcohol, e.g. linalool (10 μl), a chiral isocyanate (20 μl), and dimethylaminopyridine (5% solution in benzene, 20 μl) in a tube, purge with nitrogen and seal. Heat at 60 °C for 6 days and purify the reaction mixture on silica gel prior to analysis.

Procedure B [31] Add 3-O-hexadecylglycerol (0.3 mM) and S-(−)-1-naphthylethylisocyanate (0.6 mM) to a vial containing N,N-dimethylethanolamine in dry toluene (1% w/w, 7 ml). Heat the mixture in a sealed vial at 80 °C for 36 h. Evaporate the solvent and dissolve the residue in the HPLC mobile phase.

Chromatography Gas chromatographic analysis of the diastereomeric carbamates from linalool was carried out on a DB-1 column at temperatures between 150 and 200 °C [30]; R(+)-phenylethylcarbamates of several simple secondary alcohols were separated on a Carbowax 20M column [32].

HPLC separation of S-1-naphthylethyl carbamates of 1,2-diacylglycerols was carried out on a 5 μm normal phase silica column eluted with hexane/ethyl acetate [31].

Comments The reaction with isocyanates works equally well for primary, secondary, and tertiary, allylic and benzylic alcohols; no racemization has been reported.

4.1.1.5 Formation of diastereomeric carbonates from (−)-menthyl chloroformate

Procedure [26] Add the chiral alcohol (1 mM) and pyridine (0.1 ml) to a solution of (−)-menthyl chloroformate in toluene (1.1 mM). Allow the reaction mixture to stand at room temperature for 30 min. Wash the organic phase with water and dry over sodium sulphate. Analyse by gas chromatography.

Comments Annett and Stumpf [33] reported that commercially available reagents supplied as a 0.1 M solution in toluene gave low yields compared with reactions using 'in-house' preparations. The authors recommend the latter to ensure quantitative reactions.

The separation factors of menthyl carbonates and etienic esters of phenylalkylcarbinols were compared using gas chromatography [14]; similar values were obtained for each of the substrates. Since the derivatives from (−)-menthyl chloroformate give good resolution with less rigorous reaction conditions, this reagent has obvious advantages for labile compounds.

4.1.2 Reagents used for the chiral derivatization of hydroxyl groups

Reagent	Availability[a]	Chromatographic method cited	Substrate/bibliography
3β-Acetoxy-Δ⁵-etienic acid (1)	C[b]	GC	Carbinols [13]
R(+)-*trans*-Chrysanthemic acid (2)	C[b]	GC	Various [14]
		GC	Triazole fungicides [15]
		GC	3-Octanol [16]
Drimanoic acid (3)	C[b]	GC	Various [14]
(−)-Menthenyoxyacetic acid (4)	C[b]	HPLC	PAH dihydrodiols [21,34–36]
R(+)/S(−)-α-Methoxy-α-(trifluoromethyl)phenylacetic acid (5) (Mosher's acid)	C[b][17]	GC	1,2-diacylglycerols [18]
		GC/HPLC	Aliphatic, olefinic, acetylenic and aromatic alcohols [19]
		HPLC	PAH dihydrodiols [21,37–39]
R(+)-2-Phenylselenopropionic acid (6)	[40][b]	GC	2-octanol [18]

Reagent	Availability[a]	Chromatographic method cited	Substrate/bibliography
S-Acetoxypropionic acid (7)	[41][b]	GC	2-n-alcohols [41–44]
(S)-Tetrahydro-5-oxo-2-furancarboxylic acid (8)	[19][b]	GC/HPLC	Aliphatic, olefinic, acetylenic and aromatic alcohols [19]
S(+)-Phenylpropionic acid (9)	C[b]	GC	Hydroxy acids [45]
D(−)/L(+)-Mandelic acid (10)	C[b]	GC	2-n-Alkanols [46]
S(−)-N-(trifluoroacetyl)proline (11)	C[47]	GC	3-Octanols [16]
Carbobenzyloxy-L-proline (12)	C	HPLC	Warfarin [20]
(+)-Glucuronic acid (13)	From rat liver	HPLC	Phenolic drugs [22]

(continued)

Reagent	Availability[a]	Chromatographic method cited	Substrate/bibliography
Acetobromo-α-D-glucose (14)	C	GC	Menthol [23,48]
R(+)/S(−)-1-Phenylethyl isocyanate (15)	C[32]	GC	Secondary alcohols [32] Dialkylglycerols [49] Linalool [50]
R(+)/S(−)-1-(1-Naphthyl)ethyl isocyanate (16)	C[24]	GC	Dialkylglycerols [49] 1-Alkyn-3-ols [51] Carboxybicycloheptan/enol [52]
(−)-Menthyl chloroformate (17)	C[26]	GC HPLC	General hydroxy compounds [26] Warfarin [53]
(1R,4R,5S)-4-hydroxy-6,6-dimethyl-3-oxabicyclo[3.1.0]hexan-2-one (18)		LC	2-Hydroxy-2-(3-phenoxyphenyl)acetonitrile [27]

[a] C = commercially available. The associated number refers to the reference in which the preparation of the reagent is described.
[b] As the acid chloride, prepared from the acid.

Note: The names used for the reagents represent those most widely used in the literature and therefore do not conform to any one covention. Other synonyms for the reagents can be found in Section 4.8.

4.2 Aliphatic and acyclic carboxylic acids

Esters have historically been the most commonly used diastereomeric derivatives for the separation of the enantiomers of optically active carboxylic acids, particularly where recovery of the individual enantiomers is required. The reactions used to form diasteromeric esters can be summarized as follows.

(a) Acid-catalysed esterification of the carboxylic acid with the chiral alcohol in the presence of dry hydrogen chloride gas or concentrated sulphuric acid. This reaction is routinely carried out at temperatures of up to 100 °C over several hours [54].

(b) Activation of the carboxylic acid, e.g. to the acid chloride by refluxing the acid with thionyl or oxalyl chloride. The excess reagent is removed *in vacuo* and the chiral alcohol is added in the presence of a proton acceptor, e.g. pyridine [55]. Alternatively, hydrogen chloride gas can be dissolved in the chiral alcohol, which is subsequently reacted with the acid [56].

(c) Formation of an N,N-dialkylisourea adduct between a chiral alcohol and N,N'-dialkylcarbodiimide, followed by reaction with the acids in the presence of a tertiary amine as catalyst [57].

(d) Transesterification has also been demonstrated as an effective method to form diastereomeric esters from chiral acids by Chapman and Harris [58], who determined the enantiomeric ratio of two synthetic pyrethroids extracted from soil. In their method the pyrethroid was hydrolysed, the acid moiety was converted to the benzyl ester, the transesterification was carried out using sodium menthylate.

The nature of the chiral derivatization reagent is critical in achieving optimum resolution of the diastereomers, and for simple carboxylic acid reagents containing a phenyl or other π-donating group close to the chiral centre are recommended to improve the resolution. This effect was particularly well demonstrated by Kaneda [59], who showed that the $S(+)$-2-butanol (**20**) or $R(-)$-2-octanol (**21**) esters of 2-methylbutyric acid were unresolved by gas chromatography but the esters with R-α-methylbenzyl alcohol (**19**) were readily separated. (See Section 4.2.2 for the numbered structures).

The recovery of the individual enantiomers of the acids from the diastereomeric esters has been discussed in Section 4.1.

Diastereomeric amides are, in general, more readily separated than the corresponding esters, particularly on normal phase silica or alumina chromatographic systems, where the amide can hydrogen bond with the stationary phase. Helmchen *et al.* [60] developed a series of concepts to predict the structural features of amides, which would improve separation factors, and also to construct models which correlate elution order with the configuration of the diastereomer.

A wide range of chiral amine reagents is available, including $S(-)/R(+)$-1-phenylethylamine (**23**), L-leucinamide (**24**), L-alanine β-naphthylamide (**25**), $R(-)/S(+)$-amphetamine (**26**), and levo/dextrobase (**27**). The reaction of the carboxylic acid with the reagent is usually carried out after the acid has been activated to the acid chloride or imidazole or the mixed anhydride. The formation of acid chlorides has been discussed earlier in this section, and generally requires the use of elevated temperatures; imidazoles and mixed anhydrides are formed under less vigorous conditions, usually in an aprotic solvent at room temperature. A major advantage of the latter reactions is that the entire derivatization can be carried out as a 'single pot' reaction and is complete within 1 h. However, the excess reagent and by-products of the derivatization remain in the reaction mixture and can cause difficulties in the analysis. The formation of mixed anhydrides from alkyl chloroformates [61] or diphenylphosphinyl chloride (DPPC) [61] (Figure 3) exemplifies these difficulties. Excess chloroformate also reacts with the primary or secondary amine to give alkyl carbamates, lowering the overall yield of the amide and causing chromatographic difficulties. This effect is less pronounced with isobutyl than with methyl chloroformate and is thought to be due to steric effects. Diphenylphosphinyl chloride presents less of a problem, since the major by-product, diphenylphosphinic acid, can be readily removed with an alkaline wash. Secondary amines generally give lower yields when DPPC is used to form the anhydride.

The recovery of the individual enantiomeric acids from the diastereomeric amides has not been carried out to any large extent, because of the inherent stability of the amide bond and the risk of racemization during hydrolysis. Jiang and Soderlund [62] reported the recovery of the individual enantiomers of 3-(2,2-dichlorovinyl)-2,2-dimethylcyclopropanecarboxylic acid, a synthetic pyrethoid precursor, from its $S(-)$-(phenyl)ethylamide diastereomers by refluxing in 6 M hydrochloric acid at

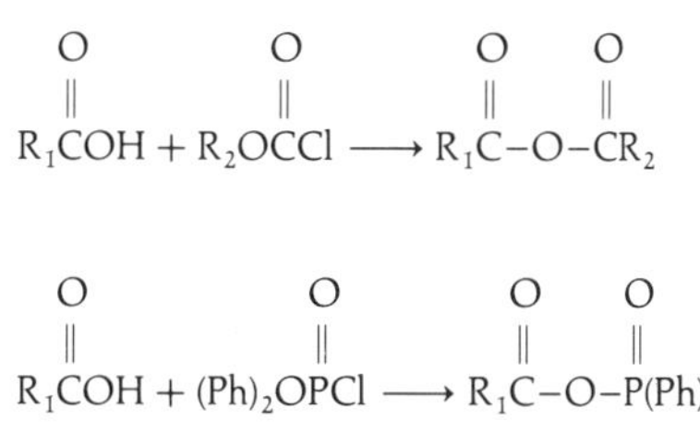

$$R_1\overset{O}{\underset{\|}{C}}OH + R_2O\overset{O}{\underset{\|}{C}}Cl \longrightarrow R_1\overset{O}{\underset{\|}{C}}-O-\overset{O}{\underset{\|}{C}}R_2$$

$$R_1\overset{O}{\underset{\|}{C}}OH + (Ph)_2\overset{O}{\underset{\|}{P}}Cl \longrightarrow R_1\overset{O}{\underset{\|}{C}}-O-\overset{O}{\underset{\|}{P}}(Ph)_2$$

Figure 3. Reactions of racemic acids with alkyl chloroformates and diphenylphosphinyl chloride.

90–100 °C for periods of up to 6h. In other cases, amides have resisted hydrolysis even during reflux in 48% hydrobromic acid/acetic acid [2]. Helmchen et al. [63] reported the directed separation and recovery of acids by neighbouring group assisted hydrolysis. Using amino alcohols as chiral reagents to form amides with the hydroxyl function γ or δ to the carbonyl gave excellent separation factors, and also allowed the hydrolysis to be carried out under somewhat milder conditions, e.g. in 3 M sulphuric acid.

4.2.1 Derivatization

4.2.1.1 Formation of diastereomeric esters from chiral alcohols, e.g S(+)-2-octanol (21) and (−)-menthol (22)

Procedure A [54] Place the acid (1 mg) in a tube fitted with a Teflon-faced screw-capped lid. Add $S(+)$-2-octanol (100 µl), dry toluene (900 µl) and conc. sulphuric acid (5 µl). Purge the vial with nitrogen, seal, and heat at 100 °C for 2h. On cooling, add sodium bicarbonate solution (0.02 M, 1 ml) and ultrasonicate. Dry the toluene layer, containing the ester, over sodium sulphate before analysis.

Procedure B [55] Reflux the acid (5 mg) in freshly distilled thionyl chloride (2 ml) for 30 min. Remove the excess reagent under a stream of dry nitrogen. Boil the acid chloride under reflux (3h) with (−)-menthol (5 mg) in dry toluene in the presence of a few drops of pyridine.

Procedure C [56] Prepare alcoholic HCl by bubbling dry hydrogen chloride gas through the chiral alcohol. Mix the acid (5 mg) and alcoholic HCl (0.5 ml) in a sealed tube and heat for 2h at 100 °C. Excess reagent can be removed under reduced pressure or in a stream of dry nitrogen.

Chromatography Johnson et al. [54] used thin-layer chromatography to monitor the reaction of naproxen with (+)-2-octanol using silica gel plates eluted with hexane tetrahydrofuran acetic acid (50:50:1). In the same paper, the reaction mixture was analysed by liquid chromatography using a microparticle silica column eluted with 0.5% ethyl acetate in hexane. The column elute was monitored at 332 nm because impurities in the batch of octanol eluted at the same retention times as the diastereomers. Lee et al. [64] achieved the separation of the $S(+)$-2-octyl esters of ibuprofen using two 5 µm silica columns connected in series; the mobile phase was 0.05% isopropanol in heptane. Gas chromatography using fused-silica columns coated with DX 1, DB 225 or SP 2100 have been used for the separation of the diastereomeric esters.

Comments The preparation of alcoholic HCl can also be carried out *in situ* by the addition of acetyl chloride to the dry chiral alcohol [8]. Although the conditions used for these reactions are harsh, the authors reported no evidence of racemization during the derivatization of naproxen [54] even after an extended reaction time of 23 h. Lee et al. [64], however, observed a small peak, corresponding to approximately 2%, when enantiomerically pure ibuprofen was derivatized under the conditions described in procedure C; the results could have been attributed to either racemization or the presence of impurities in the substrates or reagents.

4.2.1.2 Preparation of diastereomeric esters from optically active acids and O-(−)menthyl-N,N-diisopropylisourea (28)

Procedure [57] Dissolve the acid (1 mg) in a mixture of tetrahydrofuran and triethylamine (9:1 v/v) and add O-(−)-menthyl-N,N-diisopropylisourea (10 µl). Heat the reaction mixture in a tightly capped vial at 100 °C over-night. Analyse the reaction mixture by capillary gas chromatography.

Comments The reagent was formed by the reaction of a chiral alcohol with N,N'-diisopropylcarbodiimide, and was reported to be stable for up to 4 years [57]. The reaction was quantitative and no racemization was reported.

$$R-N=C=N-R + R'OH \longrightarrow R-NH-C=N-R$$
$$|$$
$$OR'$$

Naproxen

Ibuprofen

4.2.1.3 Formation of diastereomeric amides via mixed anhydrides and chiral amines

(i) Formation of mixed anhydrides from alkyl chloroformates: procedure [65] Dissolve the acid (5–10 µg) in dry acetonitrile containing triethylamine (50 mM) and add ethyl chloroformate (60 mM in acetonitrile, 50 µl). After 2 min add a solution of L-leucinamide and triethylamine in metha-nol (1 M, 50 µl). After a further 2–3 min, acidify the mixture with hydrochloric acid (0.25 M, 0.2 ml) and extract the amide into ethyl acetate (5 ml). Evaporate the solvent and dissolve the residue in a suitable solvent for chromatography.

(ii) Formation of mixed anhydrides from diphenylphosphinyl chloride: procedure [66] Dissolve the acid (1–10 µg) in dry dichloromethane (1 ml) containing triethylamine (0.144 mM, 20 µl) and diphenylphosphinyl chloride (0.105 mM, 20 µl). Agitate the mixture in a vortex mixer for 10 sec add L-leucinamide (0.15 mM, 500 µl) in dry dichloromethane and agitate for a further 10 min.

Comments Prior to derivatization, the amine was converted into the free base by dissolving the salt in sodium hydroxide solution and partitioning with dichloromethane.

Lehr and Damm [66] found that the derivatization of ofloxacin (a bacterial drug) was quantitative only when the anhydrides were formed from DPPC.

Ofloxacin

4.2.1.4 Formation of diastereomeric amides using N,N-dicyclohexylcarbodiimide as a coupling agent

Procedure [67] Place the chiral acid (2 mg) and L-alanine-β-naphthylamide (20 mg) in a vial containing methanol (50 µl) and a chloroform solution of N,N-dicyclohexylcarbodiimide (10 mg in 2 ml). Agitate the solution and leave the mixture to stand for 1 h at room temperature. Analyse the reaction mixture directly by normal phase liquid chromatography.

Comments During standing, N,N-dicyclohexylurea settles out and can be removed by centrifugation. Excess reagent can be decomposed by the addition of glacial acetic acid.

4.2.1.5 Formation of diastereomeric amides from acid chlorides/imidazoles and chiral amines

Procedure A [62] 'Activate' the acid using excess thionyl chloride. The reaction can be quantitative at room temperature or may require heating under reflux, and is dependent on the properties of the acid. Remove excess thionyl chloride *in vacuo*. Add the resulting acid chloride to S(−)-1-phenylethylamine in 1–2 ml of dry toluene containing a small quantity of pyridine (30 µl). Heat the reaction mixture at 70–80 °C for 30 min, cool, allow to stand overnight and analyse by HPLC.

Procedure B [68] Dissolve the acid (15 µg) in dry chloroform (500 µl) and add 1,1'-carbonyldiimidazole (32 mg) as a solution in chloroform. Agitate the reaction mixture and allow to stand at room temperature for 20 min. Add a solution of acetic acid in chloroform (10%, 40 µl) and a solution of S(−)-1-phenylethylamine (10%, 150 µl) in the same solvent. Allow the mixture to react for a further 2 h, evaporate under nitrogen and analyse the HPLC.

Comments 1,1'-Carbonyldiimidazole is an effective reagent and in most cases avoids the problems of racemization, enantioselectivity and other side reactions which can occur when using the acid chloride methods [69].

Chromatography Arylpropionic acids, a group of non-steroidal anti-inflammatory drugs (NSAIDs), are routinely analysed in biological fluids for enantiomeric composition. The derivatives have been separated by GC, HPLC and high performance thin-layer chromatography (HPTLC) and a summary of the methods used for a wide range of 2-arylpropionic acids in biological samples has been prepared by Hutt and Caldwell [70].

Diastereomeric amides from L-leucinamide and arylpropionic acids have been separated by reversed phase liquid chromatography using acetonitrile/phosphate buffer (pH 6.5) as the mobile phase [65]. Some workers also found that the inclusion of triethylamine (0.02%) improved the separation [71]. Rossetti *et al.* [72] used HPTLC to separate the diastereomeric phenylethylamides of several NSAIDs; the solvent systems used were benzene/methanol (93:7) and chloroform/ethyl acetate (15:1), and separation was improved by multiple development in the same solvent system.

Singh *et al.* [73] used a methyl silicone fused silica column to separate the amides formed from NSAIDs and S(−)-1-phenylethylamine or R(+)/S(−)-amphetamine.

4.2.2 Reagents used for the chiral derivatization of carboxylic acids

Reagent	Availability[a]	Chromatographic methods cited	Substrate/ bibliography
R-α-Methylbenzyl alcohol (**19**)	C	GC	2-Methylbutyric acid [59]
R(−)/S(+)-2-butanol (**20**)	C	GC GC	2-Methylbutyric acid [59] Hydroxydicarboxylic acids [56]
R(−)/S(+)-2-octanol (**21**)	C	HPLC	Arylpropionic acids [54, 64] Pyrethroid acid [62]
(−)/(+)-menthol (**22**)	C	GC GC	Isoprenoid acid [55, 74] Hydroxycarboxylic acids [75] Pyrethroid acids [58, 62, 76]
S(−)/R(+)-1-phenylethylamine (**23**)	C	HPLC HPLC, HPTLC, HPLC HPLC GC	Pyrethroid acids [62] Arylpropionic acids [72, 77, 78] Benzoxazoleacetic acids [68] Benzoxazoleacetic acid [79] Chrysanthemic acid [80] 2-Methylbutyric acid [59]
L-Leucinamide (**24**)	C	HPLC HPLC HPLC	Arylpropionic acids [65, 69, 71] Ofloxacin [66] Ketorolac [81]
L-Alanine β-naphthylamide (**25**)	C	HPLC	Arylpropionic acids [67]

Reagent	Availability[a]	Chromatographic methods cited	Substrate/ bibliography
R(−)/S(+)-amphetamine (26)	C	GC	Arylpropionic acids [73]
Levo/dextrobase (27)	C	HPLC, TLC	General acids [82, 83]
O-(−)-Menthyl-N,N'-diisopropylisourea (28)	[57]	GC	General acids [57]
(+)/(−)-1-(4-Dimethylamino-1-naphthyl)ethylamine (29)	[84]	HPLC	Arylpropionic acids [85]
S(−)-1-(1-naphthyl)ethylamine (30)	C	LC	Arylpropionic acid [86, 87]
(R)-α-Methyl-4-nitrobenzylamine (31)	C	HPLC	Isoprenoid acids [11] Citronellic acid [88]

[a] C = commercially available.

Note: The names used for the reagents represent those most widely used in the literature and therefore do not conform to any one convention. Other synonyms for the reagents can be found in Section 4.8.

4.3 Amines

The derivatization of chiral amines to form diastereomeric derivatives has been widely researched and an equally large number of chiral reagents is available. Some of the many reactions found in the literature are illustrated in Figure 4.

Alkylation of a secondary amine was used to form diastereomers from the enantiomers of etilefrine using $R(+)$-tetrahydrofurfuryl-(1S)-camphor-10-sulphonate (**32**) as the chiral reagent. Although this reagent contains two chiral centres, only the tetrahydrofurfuryl group was transferred to the substrate [89].

Etilefrine

The formation of diastereomeric amides from chiral acylating reagents is one of the most widely used reactions, which can be attributed to both the commercial availability of the chiral acids and the separability of the amide. N-Substituted L-prolyl chlorides are versatile

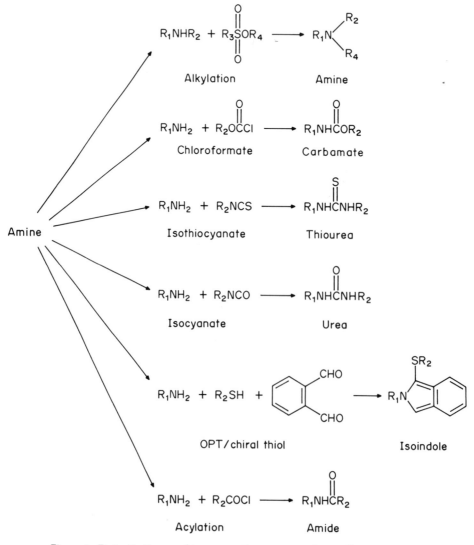

Figure 4. Derivatizations used to prepare diastereomers of optically active amines.

'designer' reagents in which the nature of the substitution can be changed depending on the nature of the final chromatographic analysis; e.g. heptafluorobutyryl (HFB) substitution (**33**) produces a volatile derivative with electron capturing properties, whereas nitrophenylsulphonyl (NPS) substitution (**34**) confers a strong UV chromophore.

A quantitative assay with a limit of detection of 10 ng ml^{-1} was achieved for the enantiomers of tocainide using $S(-)$-α-methoxy-α-(trifluoromethyl)phenylacetyl chloride (**5**) [90]; the limit was up to 2 orders of magnitude lower for amphetamines. Increasing the amount of the reagents did not increase the overall yield (ca. 90%) but increased the level of impurities. The derivative was stable at 4 °C for 48h with evidence of stability for up to two weeks.

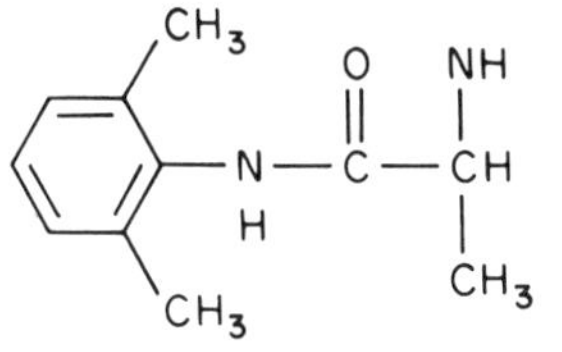

Tocainide

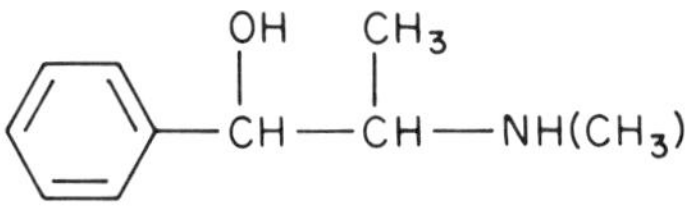

Ephedrine

The N-succinimidyl ester of (S)-2-methoxy-2-phenylacetic acid (**36**) was used by Husain *et al.* [91] for enzyme assays. The reagent is highly stable to hydrolysis, has a long shelf life and is specific for primary amines. The reaction is carried out in aqueous solutions and is suitable for preparative isolation, where the substrate can readily be recovered by acid hydrolysis. The succinimidyl ester of α-methoxy-α-methylnaphthaleneacetic acid (**37**) was demonstrated to have similar properties, and the derivatives had a lower limit of detection [92].

Pirkle and Simmons [93] used aryl-substituted 2-oxazolidones (**35**) as chiral derivatization reagents for the resolution of primary amines as their diastereomeric allophanates. The diastereomers contain a semi-rigid backbone through intramolecular hydrogen bonding. The derivatives were separated by liquid chromatography on silica with separation factors between 1.2 and 4. The reagent also reacts with secondary amines but the derivatives are less well resolved.

2,3,4,6-Tetra-O-acetyl-β-D-glucopyranosyl isothiocyanate (GITC) (**38**) and 2,3,4-tri-O-acetyl-α-D-arabinopyranosyl isothiocyanate (AITC) (**39**) have been widely used in chiral assays of physiologically active amines in forensic and medicinal analyses [94, 95]. The derivatives are readily separated and have a strong UV absorbance near 290 nm, resulting in highly sensitive assays. Miller *et al.* [96] compared the resolution of the diastereomers of phenylethylamine and amphetamine following derivatization with MPTA (**5**), GITC, AITC and $R(+)$-1-phenylethyl isocyanate (PEIC) (**15**) by HPLC. The reactions were carried out in an organic solvent containing a base catalyst at temperatures between 25 and 70 °C. In general, the authors found that the first three reagents resulted in derivatives with the highest separation factors. AITC derivatives eluted in the reverse order to those from the other reagents, a phenomenon previously described by Sedman and Gal [97].

The demand for improved methods with lower limits of detection, particularly in biological fluids, has led to the development of reagents which form fluorescent derivatives. *o*-Phthalaldehyde/N-acetyl-L-cysteine (**40**) is selective for primary amines, and the reactions are often complete within 5 min of adding the reagent. The derivatives are, however, unstable and it has been recommended [98] for analytical purposes that the time between reaction and determination be standardized. Although a significant disadvantage in many instances, the derivatization is ideal for automated analyses. Other more stable fluorescent derivatives have been formed from (+)-1-(9-fluorenyl)ethyl chloroformate (**41**), $S(+)$-benzoxaprofen chloride (**42**) and $S(+)$-flunoxaprofen chloride (**43**). The derivatizations are carried out under mild conditions and are complete in 30 min at room temperature; however, unlike the *o*-phthaldehyde (OPT) reagents they are non-specific and readily react with both secondary amines and hydroxyl groups [99].

The formation of diastereomeric ureides from chloroformate reagents has been used in quantitative assays and for the preparative separation of both secondary and tertiary amines. Quantitative assays for promethazine, a tertiary amine of the phenothiazine group of pharmaceuticals, have been developed. Maibaum [100] carried out a three-stage reaction in which the racemic promethazine was reacted with vinyl chloroformate and the carbamate was subsequently hydrolysed to the secondary amine. The enantiomeric ratio of the amine was then determined as the urea following reaction with a chiral isocyanate. In a later development, Witte *et al.* [101] directly derivatized the same compound with (−)-menthyl chloroformate (**17**) and separated the diastereomers by HPLC. (−)-Menthyl chloroformate has also been used to resolve the enantiomers of nornicotine as

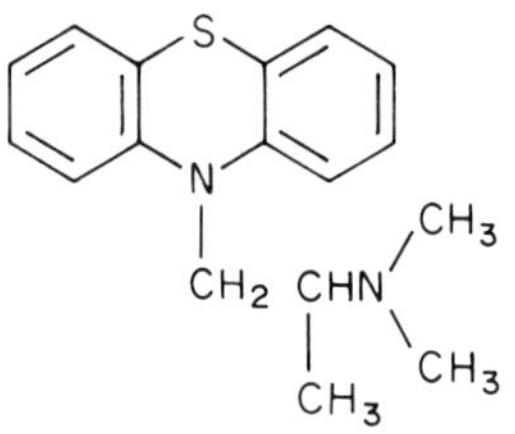

Promethazine

part of an asymmetric synthesis; the enantiomers were recovered by acid-catalysed hydrolysis [102].

4.3.1 Derivatizations

In all the following reactions the amines are first converted into the free base by dissolving the amine salt, or a sample containing the amine salt in sodium hydroxide (0.05 M) or in pH 11 buffer solution and partitioning the amine into an organic solvent. Samples of the organic phase are taken for analysis.

4.3.1.1 Formation of diastereomeric amides from chiral acid chlorides and succinimides

Procedure A [103] Mix tetrahydrofuran (THF, 2 ml), 4-nitrophenylsulphonyl-L-prolyl chloride (0.03 M in THF, 1 ml), the amine (as an 0.006 M aqueous solution, 1 ml) and sodium bicarbonate solution (10% w/w, 0.7 ml) in a Reacti-vial (5 ml) equipped with a Teflon-lined screwcap. Seal the vial and heat at 65 °C for 1 h. Cool, extract the aqueous phase with chloroform, wash the organic phase with water and dry over magnesium sulphate. Evaporate the solvent and dissolve the residue in the HPLC mobile phase.

Comments Although the reactions with N-substituted L-prolyl chlorides are usually carried out in organic media, Lim et al. [104] cited an example of acylation under Schotten–Baumann type conditions. The reagent was added to the substrate as a solution in ice cold aqueous bicarbonate/carbonate buffer (5% w/v 7:3 ratio) and the solution was agitated for 10 min. The mixture was allowed to stand in an ice bath for a further 20 min. Lim et al. also showed that when the reaction was carried out in an organic solvent, in the presence of triethylamine, racemization occurred, even at −75 °C.

Procedure B [99] Dissolve the free amine in a solution of R(−)-benoxaprofen chloride (10% w/v in dichloromethane, 0.5 ml) and keep the solution at room temperature for 30 min. Analyse the reaction mixture directly or remove the solvent and dissolve the residue in the HPLC mobile phase.

Comments Spahn [105] reported the use of S(+)-flunoxaprofen chloride (FLOP) as a reagent with similar properties to those of benoxaprofen chloride. FLOP derivatives are more hydrophilic, which may assist in certain analyses and can give some control of the retention characteristics of the derivatives. In the non-activated form (free acid) both reagents are highly stable. Procedure B is typical of those used for the majority of the acyl chloride reagents, e.g. MPTA, (+)-camphor-10-sulphonyl chloride, drimanoyl chloride and R(+)-trans-chrysanthemoyl chloride. In some cases, however, the reaction may require heating at 40–60 °C for periods of up to 2 h with a threefold molar excess of reagent.

Procedure C [93]: *Formation of diastereomeric allophanates from* (4R, 5S-cis-4,5-diphenyl-2-oxazolidone-3-carbamyl chloride (DOCC) Heat a mixture of DOCC (1 mM), the amine (1.1 mM), triethylamine (1.1 mM) and dry benzene (3 ml) under reflux for approximately 3 h. Cool, dilute with ether and partition with hydrochloric acid (1 M) and finally water. Evaporate the organic phase and dissolve the residue in a solvent suitable for chromatography.

Chromatography Pirkle and Simmons [93] separated the derivatized amines by HPLC on a silica column with 0.5% 2-propanol in hexane as mobile phase.

Procedure D [91] Mix the succinimide ester of (S)-2-methoxy-2-phenylacetic acid (2–10 mg), the amine (1 mg in 1 ml of THF) and water (1 ml) in a vial and heat the mixture on a steam bath for 15 min. Analyse the amides by direct injection, or partition into a suitable organic phase and dissolve in a more appropriate solvent.

Chromatography The resolution of the diastereomeric amides has been carried out using a wide range of chromatographic techniques, including HPLC, GC and TLC, and the reader is referred to the individual papers for further details.

4.3.1.2 Formation of isothioureas from chiral isothiocyanates

Procedure [96] Dissolve the amine (0.05 mg) in dichloromethane (50 μl) in a centrifuge tube. Add a 1.1 molar excess of the isothiocyanate (7.6 μmol ml^{-1} in dichloromethane; 50 μl) containing 0.1% of triethylamine and ultrasonicate. Leave at room temperature for up to 1 h.

Evaporate the reaction mixture to dryness and dissolve the residue in a suitable solvent for chromatography. Alternatively, add hydrochloric acid (1 M), agitate, separate the organic layer, and wash it with sodium hydroxide (1 M, 1 ml). Analyse the organic phase by direct injection GC or dilute with methanol (1 ml) prior to analysis by HPLC.

Chromatography Separations have usually been carried out using reversed phase HPLC with mixtures of THF or acetonitrile and water. Some authors have included acid or used acidic buffers in the mobile phase to maximize resolution.

4.3.1.3 Formation of diastereomeric ureides from chiral chloroformates

Procedure A [106] Dissolve the amine (50 μl) in borate buffer (1 M, adjusted to pH 7.85; 10 μl) and add (+)-1-(9-fluorenyl)ethyl chloroformate (FLEC) (1 mM solution in acetone, 50 μl). Set aside at room temperature for 30 min and then analyse by HPLC.

Procedure B [102] Dissolve the amine (0.101 M) in ether and add triethylamine (0.15 M). Cool to 0 °C in an ice bath and place under nitrogen. Add (−)-menthyl chloroformate (0.127 M) over a period of 10 min, shake for a further 10 min and leave at room temperature for 60 min.

Chromatography Procedure B was part of an asymmetric synthesis using 15 g of nornicotine. Resolution of the diastereomers was achieved on a 10 μm silica column using a mobile phase consisting of hexane/acetone/triethylamine (89:11:3).

Ureides from FLEC were separated on reversed phase silica using acetonitrile/water mixtures.

4.3.1.4 Formation of diastereomeric N-alkyl-2-(S-(N-acetyl-L-cystein))-yl isoindoles from o-phthalaldehyde/N-acetyl-L-cysteine (OPT/AcCys)

Procedure [107] Dissolve the amine in acetic acid (0.1 M) at a concentration of approximately 500 μg ml^{-1}. Add 20 μl of the solution to borate buffer (0.1 M; pH 9.5) and OPT/AcCys/(10 mg ml^{-1} solution in methanol, 20 μl). Set aside at room temperature for 5 min, then analyse by HPLC.

Chromatography Nimura et al. [107] separated the derivatives of norepinephrine and dopamine using a reversed phase HPLC column wih pH 6.5 phosphate buffer/methanol. The eluate was monitored using a fluorescent detector with excitation and emission wavelengths of 340 and 450 nm, respectively.

Comments The reaction is selective for primary amines and can be used in the presence of other functional groups, e.g. phenols. The derivatives are, however, unstable and Nimura et al. [107] showed that extended reaction times result in decreased yields. Detections limits for the compounds investigated were 65–100 pg.

4.3.2 Reagents used for the chiral derivatization of amines

Reagent	Availability[a]	Chromatographic method cited	Substrate/bibliography
R(+)-trans-Chrysanthemic acid (**2**)	C[b]	GC	Amphetamine [14]
Drimanoic acid (**3**)	C[b]	GC	Amphetamine [14]
R(+)/S(−)-α-methoxy-α-(trifluormethyl)phenylacetic acid (**5**)	C[b][17]	HPLC	Amphetamines [96] Tocainide [90]
R(+)/S(−)-1-Phenylethyl isocyanate (**15**)	C	HPLC	Tertiary amines (following N-dealkylation) [100]
		HPLC	Amphetamines [96]
(−)-Menthyl chloroformate (**17**)	C[33]	HPLC	Nornicotine [102] Promethazine [101]

(continued)

Reagent	Availability[a]	Chromatographic method cited	Substrate/bibliography
R(+)-Tetrahydrofurfuryl-(1S)-camphor-10-sulphonate (**32**)		HPLC	Etilefrine [89]
N-Heptafluorobutyryl-L-prolyl chloride (**33**)	[104]	GC	Amphetamines [104, 108]
4-Nitrophenylsulphonyl-L-prolyl chloride (**34**)	[109]	HPLC	Amphetamines [103, 109]
cis-4,5-Diphenyl-2-oxazolidone-3-carbamyl chloride (**35**)	[93]	LC	Primary amines [93]
(S)-2-Methoxy-2-phenylacetic acid N-succinimidyl ester (**36**)	[91]	HPLC	Primary amines [91]

Reagent	Availability[a]	Chromatographic method cited	Substrate/bibliography
1-α-methoxy-α-methylnapthalene acetic acid N-succinimidyl ester (**37**)	[92]	HPLC	Amphetamines [92]
2,3,4,6-Tetra-O-acetyl-β-D-glucopyranosyl isothiocyanate (GITC) (**38**)	C[110]	HPLC HPLC	Amphetamines [96] Propranolol [97]
2,3,4-tri-O-acetyl-α′-D-arabinopyranosyl isothiocyanate (AITC) (**39**)	C	HPLC HPLC	Amphetamines [94–96] Propranolol [97]
o-Phthalaldehyde/N-acetyl-L-cysteine (**40**)	C	HPLC	Catecholamines [107]
(+)/(−)-1-(9-Fluorenyl)ethyl chloroformate (**41**)	C[106]	HPLC	Pimary/secondary amines [106]

(continued)

Reagent	Availability[a]	Chromatographic method cited	Substrate/bibliography
2-(p-Chlorophenyl)-α-methyl-benzoxazole-5-acetic acid (benoxaprofen) (42)	C[b]	HPLC	Amphetamine [99]
S(+)-(p-Fluorophenyl)-α-methyl-benzoxazole-5-acetic acid (flunoxaprofen) (43)	C[b]	HPLC	General amines [105]
R(+)-1-phenylethyl isothiocyanate (44)	C	HPLC	Primary, secondary amines [111]
O-Methylmandelic acid (45)	C[b]	HPLC	Phenylethylamine [112]
(+)-Camphor-10-sulphonic acid (46)	C[b]	HPLC	Primary amines [113]

[a] C = commercially available.
[b] Prepared from the acid usually as the acid chloride.

Note: The names used for the reagents represent those most widely used in the literature and therefore do not conform to any one convention. Other synonyms for the reagents can be found in section 4.8.

4.4 Aldehydes and ketones

Jacques et al. [2] described the three principal types of derivative formed for chiral resolution of compounds containing a carbonyl function as:

Carbonyl

(a) nitrile formation

$$R_1R_2C=O + R_3NH_2 \longrightarrow R_1R_2C=NR_3$$

(b) formation of ketals or thioketals

$$R_1R_2C=O + R_3O(S)H \longrightarrow R_1R_2C\begin{smallmatrix}O(S)R_3\\O(S)R_3\end{smallmatrix}$$

(c) formation of heterocyclic derivatives

$$R_1R_2C=O + R_3CH(OH)CNHCH_3$$
$$\longrightarrow R_1R_2C\begin{smallmatrix}O-C-R_3\\|\\N-C-CH_3\end{smallmatrix}$$

Although a wide variety of derivatizations has been used to facilitate the enantiomeric separation of carbonyl compounds, the final step has inevitably involved crystallization and relatively few have used chromatography.

Halpern [8] described the reaction of cyclic ketones containing an asymmetric carbon α to the carbonyl function with (+)-2,2,2-trifluoro-1-phenylethylhydrazine (**47**) (see Section 4.4.2 for structures of numbered reagents) to form diastereomeric hydrazones, which were resolved by GC. However, some racemization was noted and it was recommended that the reaction should not be considered for ketones containing a hydrogen α to the asymmetric carbon. The enantiomers of cyclic ketones have also been separated as diastereomeric oximes by Pappo [114]; in this reaction the racemic ketone was treated with R-2-amino-oxy-4-methylvaleric acid (**48**) in methanol/pyridine (10:1) at room temperature for 18 h. The diastereomers were separated on silica using chloroform/ethanol (99:1) and the individual enantiomers were recovered with titanium trichloride in aqueous tetrahydrofuran and ammonium acetate.

Resolution of the enantiomers of polygodials has been achieved from diastereomeric pyrroles by reaction with R(−)-amphetamine (**26**). The reaction was carried out under mild conditions in an aprotic solvent at 20 °C and resolution was achieved by GC [115].

The separation of the enantiomers of sugars dominates the literature on indirect methods for carbonyl groups. The first methods, however, were long, multi-step reactions, e.g. oxidation of the aldose to aldonic acid, activation to the acid chloride and esterification with a chiral alcohol. The products were finally acylated or silylated and analysed by gas chromatography [116]. Although effective, this procedure was labour intensive and unsuitable for routine analyses. More recent reports have used one-or two-step reactions forming diastereomers directly with the carbonyl group.

Hara et al. [117] demonstrated an excellent separation of the enantiomers of nine aldoses by GC as their diastereomeric thiazolidine derivatives. The sugars were derivatized with L-cysteine methyl ester (**49**) and subsequently silylated. This simple derivatization, including the acylation, was carried out in pyridine at 60 °C. The rapid reaction of thiols with the carbonyl group of sugars was also exploited by Little [118], who used (+)-1-phenylethanethiol (**50**) to form acyclic dithioacetals.

Diastereomeric oximes with O-(−)-bornylhydroxyammonium chloride (**51**) or O-(−)-menthylhydroxyammonium chloride (**52**) have been formed from several aldoses and ketoses by Schweer [119, 120]. The disadvantage of the reaction, which was carried out in aqueous solution at 60 °C, is that the water needs to be removed before acylation can be carried out. Each enantiomer also produces two peaks corresponding to the *syn* and *anti* oximes. Reductive amination was used by Oshima et al. [121] to form acyclic diastereomers of a series of aldoses by reaction with R(+)-1-phenylethylamine in the presence of sodium cyanoborohydride. The product was acylated and the diastereomers were separated by liquid chromatography. The reagent is particularly recommended for LC analysis, where the introduction of the phenyl group both enhances the UV absorption and assists separability through interaction with the stationary phase.

4.4.1 Derivatization

4.4.1.1 Formation of diastereomeric hydrazones from ketones

Procedure [122]   Dissolve the ketone (0.1 mM) and (+)-2,2,2-trifluoro-1-phenylethylhydrazine in a mixture of ethanol/sodium acetate/acetic acid (100:2:1). Stand the mixture at room temperature for 2 h (for hindered ketones the reaction mixture may require heating under reflux). Analyse by GC.

Chromatography   Pereira et al. [122] separated a range of cyclic ketones and steroids as their diastereomeric hydrazones using an OV-17 column at temperatures between 140 and 280 °C.

4.4.1.2 Formation of diastereomeric methyl-2-(polyhydroxyalkyl)thiazolidine-4(R)-carboxylate TMS ethers (aldoses)

Procedure [117]   Dissolve the sugar (0.04 mM) and L-cysteine methyl ester hydrochloride (0.06 mM) in pyridine (200 µl) and heat at 60 °C for 1 h. Add a suitable silylation reagent, e.g. hexamethyldisilazane/trimethylchlorosilane (HMDS/TMCS, 150 µl) and heat at 60 °C for a further 30 min. Centrifuge, and analyse the supernatant by gas chromatography.

Alternatively, the derivatives can be acylated with acetic anhydride (150 µl) by heating the reaction mixture at 90 °C for 1 h and removing the excess reagent under a flow of nitrogen.

Comments   Hara et al. [117] reported that thiazolidine derivatives can be formed in aqueous or aqueous/pyridine solutions at room temperature, giving a quantitative yield after 3 h. However, the presence of water is unsuitable if further acylation/silylation is required, e.g. for gas chromatographic analysis. The chromatography of the aldose derivatives on an open tubular OV-17 support-coated column gave broad peaks, and although the separation of the individual diastereomers was acceptable mixtures of several different sugars were unresolved, making the method unsuitable for complex mixtures. The authors observed that derivatives of the nine aldoses examined having an R configuration at C-1' eluted before the S enantiomer. The configuration of the remaining carbon atoms was also believed to have a significant influence on the elution order, and the method cannot therefore be used to determine the absolute configuration of the sugars.

4.4.1.3 Formation of diastereomeric methylbenzylaminoalditols from sugars

Procedure [121]   Add an aqueous solution of the sugar (10 mg in 200 µl) to S(−)-1-phenylethylamine (83 µM, 10 mg) and sodium cyanoborohydride (32 µM, 3 mg) in ethanol (200 µl). Set aside overnight at ambient temperature or heat at 40 °C for 3 h. Acidify to pH 3–4 with acetic acid and evaporate to dryness. Dissolve the residue in acetic anhydride/pyridine (1:1, 1 ml) and heat in a sealed tube at 100 °C for 60 min.

Alternatively, the alditol can be silylated. Dry the alditol thoroughly, redissolve in dry acetonitrile, add N,O-bis(trimethylsilyl)acetamide (BSA), allow to stand for 15 min and analyse by GC.

Chromatography   The derivatives can be analysed by several techniques. Liquid chromatography was used to separate acetylated products following extraction from aqueous solutions with chloroform, evaporation of the solvent and dissolution of the residue in the HPLC mobile phase. Separations were achieved by ODS reversed phase liquid chromatography using acetonitrile/water as the eluent or with hexane/ethanol (19:1) on normal phase columns (Develosil 60).

Silylated products were also separated on WCOT gas chromatographic columns (SE-54) or Carbowax 20 M).

Comments   Oshima et al. [121] achieved resolution of 24 monosaccharides, which included ketoses and N-acetylamino sugars, by gas chromatography using the procedure described above.

4.4.1.4 Formation of diastereomeric oximes from carbohydrates

Procedures [123]   Dissolve the carbohydrate (ca. 1 mg) in water (0.2 ml) containing sodium acetate (6 mg) and O-alkylhydroxyammonium chloride (8 mg). Heat at 80 °C for 60 min and remove the water under a stream of nitrogen. Add methanol (0.1 ml) and evaporate. Remove all traces of water as an azeotrope with dichloromethane and add trifluoroacetic anhydride (0.03 ml) in ethyl acetate (0.015 ml). Allow to stand for 2 h at room temperature and analyse by gas chromatography.

4.4.2 Reagents used for the chiral derivatization of carbonyl groups

Reagent	Availability[a]	Chromatographic method cited	Substrate/bibliography
R(−)-Amphetamine (**26**)	C	GC	Drimenedial [115]
S(−)-1-Phenylethylamine (**29**)	C	HPLC, GC	Sugars [121, 124]
(+)-2,2,2-Trifluoro-1-phenylethylhydrazine (**47**)	[8]	GC	Ketones [122]
CF$_3$—CH—C$_6$H$_5$ | NHNH$_2$			
2-Amino-oxy-4-methylvaleric acid (**48**)		LC	Cyclic ketones [114]
(CH$_3$)$_2$CH—CH$_2$CH(COOH)—O—NH$_2$			
L-Cysteine methyl ester (**49**)	C	GC	Sugars (aldoses) [117]
HS—CH$_2$CH(NH$_2$)—COCH$_3$			
(+)-1-Phenylethanethiol (**50**)	[125]	GC	Sugars [118]
CH$_3$CHSH-C$_6$H$_5$			
O-(−)-Bornylhydroxyammonium chloride (**51**)	[126]	GC	Sugars [120]
O-(−)-Menthylhydroxyammonium chloride (**52**)	[126]	GC	Sugars [119, 123]

[a] C = commercially available; reference in which preparation of the reagent is described.

Note: The names used for the reagents represent those most widely used in the literature and therefore do not conform to any one convention. Other synonyms for the reagents can be found in section 4.8.

4.5 Lactones

The enantiodifferentiation of γ and δ-lactones (4 and 5-alkanolides) has received considerable interest in flavour chemistry and, although direct methods using capillary gas chromatography [127] have been developed, applications are so far limited. Indirect methods include derivatization of the lactone with (−)-2,3-butanediol (**53**) (see Section 4.5.2) to form diastereomeric ketals [128], or more usually a two-stage reaction in which the lactone ring is cleaved followed by derivatization with a chiral reagent. Reactions used to cleave the lactone ring include reduction to 1,4- and 1,5-diols or hydrolysis to 4- and 5-hydroxyalkanoic acids. In most cases the hydroxyl attached to the chiral carbon is derivatized with the chiral reagent, e.g. (S)-acetoxypropionic acid (**7**), (S)-tetrahydro-5-oxo-2-furancarboxylic acid (**8**) or R(+)-1-phenylethyl isocyanate (**15**).

Helmchen et al. [63] used S(−)-1-phenylethylamine (**23**) to form amides of racemic lactones, which were separated directly on a silica Lobar column using petroleum ether/ethyl acetate. The enantiomeric lactones were recovered using acid hydrolysis (1 M sulphuric acid at 80 °C overnight).

The procedures described below exemplify the reactions used for ring opening; only brief descriptions are given of the reactions with the chiral reagents, most of which have been covered in previous sections of this chapter.

4.5.1 Derivatization

4.5.1.1 Reduction of γ- and δ-lactones to 1,4- and 1,5-diols

Procedure A [129]   Add lithium aluminium hydride to a solution of the lactone in dry diethyl ether (0.2 ml). Agitate for 15 min at room temperature and acidify the mixture with 1 M hydrochloric acid (1 ml). Partition the resulting diols into chloroform (2 × 2 ml), and wash the organic phase with sodium bicarbonate solution and water. Dry the chloroform over sodium sulphate and evaporate to dryness under nitrogen.

4.5.1.2 Formation of 4-hydroxyisopropyl esters from γ-lactones

Procedure B [130]   Add the lactone to methanolic potassium hydroxide (2.2 mM) and stir for 20 h at room temperature. Remove the solvent and triturate the residue with diethyl ether (3 portions). Filter and evaporate the ether and add dry dimethylformamide (200 μl) followed by 2-bromopropane (75 μl). Stir for 6 days at room temperature in the dark and partition between diethyl ether (2 ml) and brine (5 ml). Dry the organic phase over sodium sulphate and evaporate to dryness.

4.5.1.3 Formation of 4- and 5-hydroxycarboxamides

Procedure C [131]   Add the γ or δ lactone (ca. 4 μl) to butylamine (10 μl) and heat at 80 °C for 5 h in a screw-capped vial. Dissolve the reaction mixture in chloroform and wash with hydrochloric acid (5%), sodium bicarbonate solution (5%) and finally water. Dry the organic phase over sodium sulphate and evaporate to dryness.

Comments   Although the reaction with butylamine was quantitative without the use of a catalyst, Helmchen et al. [63] included a 1-molar equivalent of 2-hydroxypyridine in the derivatization of racemic 2-butyrolactones with S(−)-1-phenylethylamine. Engel et al. [129] carried out carbamoylation of the hydroxy group with R(+)-phenylethylisocyanate in the presence of 4-aminopyridine, which not only catalysed the reaction but also prevented recyclization of the lactone.

4.5.2 Reagents used for the chiral derivatization of lactones

Reagent	Availability[a]	Method	Chromatographic method cited	Bibliography
S-Acetoxypropionic acid (**7**)	[132]	Ring opening (diols)	GC	[132, 133]
		Ring opening (hydroxy alkanoic acid)	GC	[130]
(S)-Tetrahydro-5-oxo-2-furancarboxylic acid (**8**)	C[b]	Ring opening	GC	[134]

Reagent	Availability[a]	Method	Chromatographic method cited	Bibliography
R(+)-1-Phenylethyl-isocyanate (15)	C	Ring opening (hydroxycarboxamides)	GC	[131]
		Ring opening (diols)	GC	[129]
S(−)-1-Phenylethylamine (23)	C	Ring opening	LPLC[e]	[63]
(−)/(+)-2,3-Butanediol (53) OH OH | | CH$_3$CH–CHCH$_3$	C	Direct	LPLC	[128]

[a]C = commercially available.
[b]Prepared from the acid usually as the acid chloride.
[c]LPLC—low pressure liquid chromatography.

Note: The names used for the reagents represent those most widely used in the literature and therefore do not conform to any one convention. Other synonyms for the reagents can be found in Section 4.8.

4.6 Amino acids

Early work on the resolution of amino acids was carried out by GC after esterification with a chiral alcohol and acylation with an achiral reagent; the most widely used combination was R(+)-2-butanol and N-trifluoroacetyl chloride. Pollock [135], however, found that with these reagents the diastereomeric derivatives of isovaline, β-aminoisobutyric acid, N-methylalanine, pipecolic acid and β-aminosobutyric acid were unresolved, and recommended 2-pentyl or 2-hexyl esterification and N-pentafluoropropionyl acylation as the preferred alternatives. Slightly improved resolution was achieved for 2-heptyl and 2-octyl esters but the excessively long retention times proved to be a significant disadvantage, hampering their general application.

Chiral derivatization of the amino group has included the formation of diastereomeric amides, ureas, thioureas and isoindoles, the carboxylic acid group being either protected or left unreacted. Diastereomeric dipeptides have been prepared using a range of N-substituted activated amino acids, e.g. S(−)-N-(trifluoroacetyl)propyl chloride (11), L-leucine-N-carboxyanhydride (45) and t-butoxycarbonyl-L-leucine N-hydroxysuccinimide ester (55). (See Section 4.6.2 for reagent structures).

Iwase and Murai [136–138] used S(−)-N-(trifluoroacetyl)prolyl chloride to derivatize a range of neutral, acidic and basic amino acids in a two-step reaction for gas chromatographic analysis: the esterification was carried out with anhydrous alcohol containing hydrogen chloride gas (ca. 3 N) or with thionyl chloride followed by acylation in methylene chloride. Iwase also confirmed the findings of Pollock [135] that higher separation factors could be achieved when esterifications were carried out using the higher primary alcohols and acylation with reagents having a high degree of fluorine substitution.

L-Amino acid N-carboxyanhydrides were used as chiral reagents by Manning and Moore [139] in a one pot derivatization. The reaction was carried out in aqueous solution, gave minimal racemization and was complete in 2 min; however, the limited availability and the instability of the reagents restricted their wider application. The authors recommended that neutral and acidic amino acids should be coupled with L-leucine and that basic and aromatic amino acids be coupled with L-glutamic acid. L-Leucine adducts of the latter classes were strongly adsorbed on the ion exchange columns used in the amino acid analyser, resulting in excessively long retention times. Mitchell et al. [140] improved the chromatography by using shorter columns and buffer solutions at higher pH. They also proposed the use of t-butoxycarbonyl-L-leucine-N-hydroxysuccinimide (55) as a more stable reagent. The dipeptide was deprotected prior to chromatography. Takaya et al. [141] carried out a comparison of both methods to determine the optical purity of an amino acid using reversed phase columns, and concluded that, although the succinimide ester procedure caused the least racemization, the carboxyanhydride method gave better accu-

Table 3 Separation factors and resolution factors for diastereomeric derivatives of amino acids

Amino acid	Diastereomeric derivative								
	FLEC (i)	AITC (ii)		SINEC (iii)		OPT/BOC (iv)	TFPCI (v)	NEI (vi)	Dipeptide (vii)
	α^a	α	R	α	R	R	α	R	R
Glutamic acid	1.13	1.17	1.14	–	–	–	–	1.00	6.4
Aspartic acid	1.07	1.18	1.43	–	–	–	–	0.90	4.4
Phenylalanine	1.26	1.78	8.43	–	–	–	1.06	2.5	11.5
Tyrosine	1.19	1.48	3.56	1.11	1.67	2.0	–	0.9	6.8
Leucine	1.21	1.73	7.75	1.25	2.92	–	1.09	2.7	11.5
Isoleucine	1.19	1.79	6.94	–	–	4.3	1.10	1.8	–
Tryptophan	1.12	1.50	6.0	1.16	2.33	3.6	–	1.8	4.3
Alanine	1.10	1.53	3.78	1.17	2.71	3.7	1.12	1.4	7.0
Valine	1.18	1.77	4.96	1.30	3.69	5.2	1.14	3.0	13.2
Proline	1.00	1.39	3.00	1.17	2.42	–	1.10	1.1	–

$^a \alpha$ = separation factor = $tr_2 - tr_0/tr_1 - tr_0$.
R = resolution factor = $2(tr_2 - tr_1)/w_1 + w_2$.

Key:

(i) Einarsson et al. [106] (+)-1-(9-fluorenyl)ethyl chloroformate. Liquid chromatograhpy, Spherisorb ODS 3 μm (150 × 4.6 mm). Acetonitrile/THF/acetic acid buffer (1.8 ml of glacial acetic acid in 1/1 of water, adjusted to pH 4.35 with NaOH); 0.8 ml min^{-1}.
(ii) Kinoshita et al. [142] 2,3,4-Tri-O-acetyl-α-D-arabinopyranosyl isothiocyanate. Liquid chromatography, Develosil ODS-5 (150 × 4.6 mm). Methanol 10 mM phosphate buffer (pH 2.8; 0.9 ml min^{-1}).
(iii) Iwaki et al. [143] Succinimido-(−)-1-(1-naphthyl)ethylcarbamate. Liquid chromatography, Develosil ODS-5 (100 × 6.0 mm). Methanol/0.1% phosphoric acid (1 ml min^{-1}).
(iv) Buck and Krummen [144] o-Phthaldialdehyde/BOC-L-cysteine. Liquid chromatography, Hypersil ODS 3(120 × 4 mm). Methanol/0.05 M phosphate buffer (pH 7) + 1% THF.
(v) Iwase [138] S(−)-N-(Trifluoracetyl)prolyl chloride. Gas chromatography, 5% OV-1 on 100–120 mesh Supelcoport, 210 °C.
(vi) Dunlop and Neidle [145] (+)-1-(1-Naphthyl)ethylisocyanate. Liquid chromatography, ODS 100 Polyol (250 × 4.6 mm). Acetonitrile/0.05 M ammonium acetate (pH 6.2), 1 ml min^{-1}. For Glu and Asp 0.1 M acetic acid was added.
(vii) Takaya et al. [141] Liquid chromatography, Nucleosil 5 c_{18} (150 × 4.0 mm) Acetonitrile/0.1 M phosphate buffer (pH 4.5), 1 ml min^{-1}.

racy. Typical resolution factors for dipeptide derivatives are shown in Table 3.

Diastereomeric amides from camphor and camphor-related compounds have been used for the chiral resolution of amino acids by both gas and liquid chromatographic techniques. Nambara et al. [146] recommended the use of L-teresantalinyl chloride (**56**) to form the amides from alkylated amino acids for GC analyses. Although camphorsulphonamide p-nitrobenzyl esters have been used for LC separations [147, 148], Aberhart et al. [149] found that these derivatives were difficult to prepare on a small scale, e.g. for enzyme assays.

In this type of assay, Aberhart used Marfey's reagent (**65**) [150] and achieved a quantitative reaction for L-leucine 1 h at 40 °C. The amides were separated by reversed phase chromatography.

Biochemical and pharmaceutical research has created a need for simple, sensitive chiral assays for amino acids. One of the most successful and widely used reagents is o-phthaldialdehyde (OPT)/chiral thiol, which selectively derivatizes primary amines to form highly fluorescent diastereomeric isoindoles. The reactions are carried out in alkaline solution and are complete in less than 10 min at room temperature, making the reagent ideal for automated analyses. Chiral thiols used in combination with OPT include N-acetyl-L-cysteine (NAC) (**40**), N-acetyl-D-penicillamine (NAP) (**62**) and N-t-butoxycarbonyl (BOC)-L-cysteine (**61**). Buck and Krummen [144] showed that, in general, BOC L-Cys derivatives gave higher separation factors, being particularly recommended for the enantiomers of lysine which were unresolved using the other reagents. It was also suggested that BOC derivatives had disadvantages for chiral assays of lipophilic amino acids where the combined lipophilicity of the reagent and the substrate give unacceptably long retention times. A less lipophilic reagent for these substrates would be N-trimethylacetyl-L-cysteine (**63**) [151]. With BOC derivatives, the L form elutes before the D form for a wide range of amino acids, which suggests that the reagents may be used to define stereochemistry. The elution order of NAC and NAP derivatives was dependent on the nature of the amino acid.

Isoindoles are highly fluorescent and undergo anodic oxidation, making them suitable for both fluorescence and electrochemical detection; limits of detection of 0.2–1.6 pmol [152] and 30–150 fmol [153] respectively have been achieved. The derivatives have the following disadvantages for quantitative analyses.

(1) Fluorescence intensities of the individual diastereomers have been shown to differ by up to 15%, and signals should only be compared with those of the same enantiomeric forms [144, 154].
(2) The derivatives are unstable and rapidly decompose, requiring standardization of the time between reaction and analysis. Buck and Krummen [144] showed that a 10% loss of the fluorescence intensity occurred for BOC derivatives after 30 min.

More stable fluorescent derivatives have been prepared with isocyanates, isothiocyanates and activated carbamate reagents. Fluorescent (+)/(−)-1-(1-naphthyl)-ethylurea derivatives have been prepared from the isocyanate [145] and succinimido-R(+) or S(−)-1-naphthylethylcarbamate [143]. The reactions were carried out in alkaline solution at room temperature, and were complete in approximately 5 min. The reagents can be used to derivatize both primary and secondary amines. Although the reagents rapidly decomposed in alkaline solution, by ensuring the rapid addition of the reagent to the reaction mixture, as a solution in dry organic solvent, yields in excess of 95% were achieved. The carbamoyl derivatives were stable and readily separated by reversed phase HPLC using either UV absorption or fluorescence detection.

Einarsson et al. [106] introduced (+)-1-(9-fluorenyl)-ethyl chloroformate (**41**) as a versatile reagent for the simple chiral derivatization of primary and secondary amino acids to give stable, highly fluorescent products, which are readily separated by reversed phase liquid chromatography, or by GC after methylation with diazomethane. The authors also illustrated a selective method using this reagent for the determination of imino acids in the presence of primary amino acids.

Thioureas of amino acids have been prepared from (−)-1,7-dimethyl-7-norbornyl isothiocyanate (**59**) or (+)-neomenthyl isothiocyanate (**60**) and separated as their methyl and t-butyldimethylsilyl esters by normal phase liquid chromatography [9]. The derivatives were unsuitable for reversed phase liquid chromatography, and no separation was achieved for either the free acid or the methyl ester. Derivatives from 2,3,4,6-tetra-O-acetyl-β-D-glucopyranosyl isothiocyanate (GITC) and 2,3,4-tri-O-acetyl-α-D-arabinopyranosyl isothiocyanate (AITC) can be directly separated by reversed phase chromatography without the need for further derivatization [12, 142]. The reaction with AITC was carried out in 50% aqueous acetonitrile containing 0.4% of triethylamine, and complete in approximately 30 min at room temperature. Typical separation factors and resolution factors for a range of amino acids are shown in Table 3.

4.6.1 Derivatizations

4.6.1.1 Formation of diastereomeric isoindoles from o-phthaldialdehyde and chiral thiols

Procedure [151] Prepare the derivatization reagents fresh each day by dissolving OPT and the chiral thiol (10 mg each) in high purity methanol (1 ml) and store at 4 °C in the dark.

Mix an aqueous solution of the amino acid (1–10 μmol ml^{-1} solution, 20 μl) with the derivatizing reagents (40 μl) and borate buffer (60 μl) (pH 8.4. adjusted with 2 M sodium hydroxide) and incubate for 5 min at ambient temperature in the dark. Analyse by HPLC.

Chromatography Isoindoles have been routinely separated on reversed phase columns using an elution gradient consisting of either acetate or phosphate buffer/methanol. The buffer solutions were adjusted to between pH 6 and 8 and the gradient extended to approximately 20% of methanol.

Comments Although some workers recommend preparing the reagent directly in alkaline buffer, Nimura and Kinoshita [155] reported that the reagent was unstable in basic solution and recommended that the base be added immediately before the reaction.

The reagent is selective for primary amines, so imino acids require pre-treatement prior to analysis; e.g. proline is oxidized with sodium hypochlorite to the primary amine.

4.6.1.2 Formation of diastereomeric ureas from chiral carbamate reagents

Procedure [143] Mix a solution of the amino acid hydrochloride (0.5 mg ml^{-1}, 20 μl) and borate buffer solution (0.3 M, pH 9.5, 30 μl) with a solution (30 μl) of succinimido-R(+)-1-(1-naphthyl)ethylcarbamate (5 mM in acetonitrile) in a Reacti-vial. Leave to stand for 3 min at room temperature. Analyse by direct injection.

Chromatography The derivatives were separated on a Develosil ODS 5 column (100 × 6.0 mm) eluted with methanol/0.1% aqueous phosphoric acid, with detection either by absorbance at 270 nm or by fluorescence (excitation 280 nm, emission 340 nm).

Comments Excess reagent can be removed by the addition of taurine (50 mM solution in water, 20 μl). The resulting derivative elutes more rapidly than those with other amino acids.

An alternative reagent, succinimido-R-(+)-1-phenyl ethylcarbamate, can be used, the derivatives being monitored by UV absorbance at 220 nm.

4.6.1.3 Formation of diastereomeric thioureas from isothiocyanates

Procedure [142] Dissolve the amino acid in aqueous acetonitrile (50% v/v, containing 0.4% of triethylamine). Add an aliquot (50 μl) of the solution to AITC in acetonitrile (50 μl, 0.2% w/v). Allow to stand at room temperature for 30 min and analyse by HPLC.

Chromatography Kinoshita et al. [142] used a Develosil ODS 5 column (150 × 4.6 mm) eluted with a gradient system comprising methanol/phosphate buffer (pH 2.8, 10 mM) to separate a range of protein amino acids.

4.6.1.4 Formation of diastereomeric amides from chiral acylating reagents

Procedure A [136]: formation of S(−)-N-(trifluoroacetyl)-prolyl amino acid esters Esterify the amino acids (5 mg) with thionyl chloride/methanol (1:9)(5 ml) under reflux for 2 h and remove the excess reagent *in vacuo*. Dissolve the residue in S(−)-N(trifluoroacetyl)prolyl chloride (0.017 M in chloroform, 2 ml) and add triethylamine (0.1 ml). Allow to stand for 15 min at room temperature and partition against distilled water. Separate the organic phase, dry over sodium sulphate and analyse by gas chromatography.

Comments The authors tested triethylamine (TEA), pyridine and N,N-dimethylaniline (DMA) as solvent, and although both TEA and DMA gave quantitatively similar yields, DMA catalysed the formation of an unknown component. When hydroxy amino acids are analysed, the reaction mixture is evaporated *in vacuo* and silylated with BSTFA (0.5 ml) in dry acetonitrile (1 ml). The reaction mixture is maintained at room temperature for 15 min, then analysed by gas chromatography.

4.6.1.5 Formation of diastereomeric N-L-teresantalinyl amino acid esters

Procedure B [146] Dissolve the amino acid ester (1 mg) in tetrahydrofuran (0.8 ml) containing pyridine (0.2 ml) and add the chiral acid chloride (4 mg). Analyse by gas chromatography using 1.5% OV-17 or 1.5% SE-30 on Chromosorb W (100–120 mesh).

4.6.1.6 Formation of N-d-10-camphorsulphonyl-p-nitrobenzoates of amino acids

Procedure C [147] Add d-10-camphorsulphonyl chloride (2 mM, 30 ml) in anhydrous ether dropwise to the amino acid (1 mM in diethyl ether, 10 ml), add sodium hydroxide solution (1 M; 20 ml) and stir at 0 °C. Allow to reach room temperature and stir for 3 h. Separate the phases and wash the aqueous phase with diethyl ether, acidify with hydrochloric acid and re-extract with diethyl ether. Dry the ether phase over anhydrous sodium sulphate and evaporate to dryness. Dissolve the residue in N,N-dimethylformamide (10 ml), triethylamine (1 drop) and p-nitrobenzyl bromide (1.1 mM). Heat at 55 °C for 2 h and dilute with chloroform. Wash the organic phase with water and dry over anhydrous sodium sulphate. Analyse HPLC.

Chromatography The separation of the derivatives of several protein amino acids was carried out on a MicroPak-NH₂ column using isooctane/dichloromethane/isopropanol (79:16:5) as mobile phase. Variations on the constituents were used depending on the nature of the amino acids.

Comments The authors advise that the organic phase be washed sequentially with 5% hydrochloric acid, 5% potassium carbonate and finally water (5 times with each) to obtain consistent retention times.

4.6.1.7 Formation of substituted oxycarbonyl amino acid diastereomeric derivatives

Procedure A [47]: L-menthyloxycarbonyl amino acid methyl esters Esterify the amino acid (1 mM) with thionyl chloride in methanol under reflux for about 30 min. Remove the excess solvent under a stream of dry nitrogen. Add L-menthyl chloroformate (1:1 mM solution in toluene, 1.1 ml) and pyridine (0.2 ml) and agitate, then set aside for 30 min at room temperature. Partition against water, dry the organic phase and analyse by gas chromatography.

Procedure B [106]: (+)-1-(9-fluorenyl)ethyloxycarbonyl amino acids Dissolve the amino acid in borate buffer (1 M, pH 6.85, 0.1 ml) and add FLEC reagent (15 mM in acetone/acetonitrile, 3:1). Allow to stand at room temperature for 4 min. Partition twice with pentane to remove the excess reagent and analyse the aqueous phase by HPLC. Alternatively, acidify and extract with diethyl ether. Methylate the dried organic phase with diazomethane and analyse by gas chromatography.

For the selective derivatization of secondary amino acids, dissolve the substance in borate buffer (0.8 M, pH 9.5, 0.1 ml) and mix with OPT/mercaptoethanol reagent (50 mg ml^{-1} of OPT and 25 μl ml^{-1} of mercaptoethanol in acetonitrile; 0.1 ml). Set aside for 30 s. Add iodoacetamide (140 mg ml^{-1} in acetonitrile, 0.1 ml) and, after 30 s, FLEC reagent (5 mM in acetone). After 2 min extract the mixture with pentane and analyse the aqueous phase by HPLC.

Chromatography HPLC analysis of the diastereomeric free acids was carried out on Spherisorb C_{18} columns (150 × 4.6 mm) using gradient elution with acetonitrile and tetrahydrofuran/acetic acid (pH 4.35).

4.6.1.8 Formation of L,D and L,L dipeptides using t-butyloxycarbonyl-L-leucine

Procedure [140] Add sodium bicarbonate solution (0.12 mM), 40 μl to a Reacti-vial containing the free amino acid (270 μm) and a Teflon-coated stirring bar. Stir for a 1 min and add a solution of t-butyloxycarbonyl-L-leucine N-hydroxysuccinimide ester (BOC-L-Leu-OSU) in tetrahydrofuran (40 μm), 0.12 ml). Seal the vial and stir for 1 h at room temperature, then evaporate to dryness at 45–50 °C. Dissolve the residue in trifluoroacetic acid (1 ml) and keep in a sealed tube for 1 h to remove the protecting group. Evaporate the excess TFA *in vacuo* and dissolve the residue in the mobile phase.

Chromatography Separations of the diastereomeric derivatives were carried out on an amino acid analyser using a sulphonated polystyrene column. For the exact conditions the reader is directed to the original reference [140].

Comments The reagent was shown to be stable as a 0.3 M solution in THF for at least 1 week at room temperature and 1 month at 4 °C.

4.6.2 Reagents used for the chiral derivatization of amino acids

Reagent	Availability[a]	Chromatographic method cited	Bibliography
S(−)-N-(trifluoroacetyl)prolyl chloride (**11**)	C[47]	GC (as alkyl esters)	[136–138]
(+)-1-(1-Naphthyl)ethylisocyanate (**16**)	C	LC	[145]
2,3,4,6-Tetra-o-acetyl-β-D-glucopyranosyl isothiocyanate (GITC) (**38**)	[110]	LC	[142, 156]
2,3,4-Tri-o-acetyl-α-D-arabinopyranosyl isothiocyanate (AITC) (**39**)	[110]	LC	[142]
(+)-1-(9-Fluorenyl)ethyl chloroformate (**41**)	[106]	LC	[106]
10-camphorsulphonic acid (**46**)	C[b]	LC (as p-nitrobenzoate)	[147, 148]
L-Leucine-N-carboxyanhydride (**54**)	C	LC (amino acid analyser)	[139, 141]

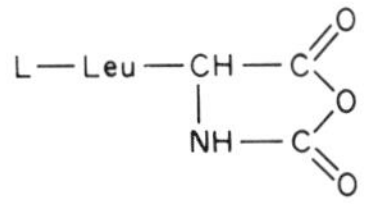

L—Leu—CH—C(=O)
         | \O
        NH—C(=O)

(continued)

Reagent	Availability[a]	Chromatographic method cited	Bibliography
t-Butoxycarbonyl-L-leucine N-hydroxysuccinimide ester (**55**)	[140]	LC	[140]
L-Teresantalinyl chloride (**56**)	[157]	GC (as alkyl ester)	[146]
OPT/BOC-L-cysteine (**61**)	C [151]	LC	[98, 144, 158]
OPT/N-Acetylcysteine (**40**)	C [151]	LC	[151, 152, 154, 158–160]
OPT/N-Acetyl-D-penicillamine (**62**)	C	LC	[144]
OPT/N-Trimethylacetyl-L-cysteine (**63**)	[151]	LC	[151]
OPT/2,3,4,6-Tetra-O-acetyl-1-thio-β-glucopyranoside (**64**)		LC	[161]
Succinimido-R(+)/S(−)-1-phenylethylcarbamate (**57**)	[143]		[143]

Reagent	Availability[a]	Chromatographic method cited	Bibliography
Succinimide-R(+)/S(−)-1-(1-napthlyl)ethylcarbamate (**58**)	[143]	LC	[143]
(−)-1,7-Dimethyl-7-norbornyl isothiocyanate (**59**)	[9]	LC	[9]
(+)-Neomenthyl isothiocyanate (**60**)	[9]	LC	[9]
N^a-(5-Fluoro-2,4-dinitrophenyl)-L-alaninamide (**65**)	C[150]	LC	[149, 150]

[a] C = commercially available. The associated number refers to the reference in which preparation of the reagent in described.
[b] Prepared from the acid usually as the acid chloride.

Note: The names used for the reagents represent those most widely used in the literature and therefore do not conform to any one convention. Other synonyms for the reagents can be found in Section 4.8.

4.7 Amino alcohols

Methods describing the separation of amino alcohols are dominated by analyses of the β-adrenoceptor blocking agents, e.g. betaxolol, widely used in the treatment of hypertension. Although the chiral centre of these compounds is on the secondary alcohols, the majority of the derivatizations are carried out on the secondary amine, presumably because the reactions require less rigorous conditions and the diastereomers are more readily separated.

Reagents used are typical of those described for the derivatization of amino acids and amines.

The formation of esters from the secondary alcohol and disubstituted tartaric acids has been described by Linder *et al.* [162]. The reactions are carried out in an aprotic solvent in the presence of the anhydride and a strong acid, which forms an ion pair with the free amine.

Betaxolol

4.7.1 Derivatizations

4.7.1.1 Formation of diastereomeric tartaric acid monoesters

Procedure [163] Dissolve the amino alcohol (10 mg) in an aprotic solvent, e.g. dichloroethane, THF or acetone (1 ml) and add (R,R)-O,O-dibenzoyltartaric acid anhydride (**66**) (80 mg) and trichloroacetic acid (35 mg). Heat at 50 °C for 4 h in a sealed vial. Analyse by HPLC.

Chromatography Separation of the diastereomeric isomers was carried out on C_8 or C_{18} columns by elution with mixtures of methanol/water (50:50) adjusted to pH 3–4, containing either 2% of acetic acid or 0.1% of triethylamine. The column eluate was monitored using UV absorption at 254 nm.

4.7.2 Reagents used for the separation of amino alcohols

Reagent	Availability[a]	Chromatographic method cited	References
R(−)-1-Naphthylethyl isocyanate (**16**)	C	LC	[164, 166]
S-(+)-benoxaprofen chloride (**42**)	C	LC	[167]
S(−)-N-(Trifluoroacetyl)prolyl chloride (**11**)	C	LC	[168]
2,3,4,6-Tetra-O-acetyl-β-D-glucopyranosyl isothiocyanate (**38**)	110	LC	[166]
(R,R)-O,O-Dibenzoyltartaric acid anhydride (**66**)	169	LC	[162, 163]
t-Butoxycarbonyl-L-leucine anhydride (**67**)	170	LC	[170]

[a] C = commercially available. The associated number refers to the reference in which preparation of the reagent in described.

Note: The names used for the reagents represent those most widely used in the literature and therefore do not conform to any one convention. Other synonyms for the reagents can be found in Section 4.8.

4.8 Synonyms for some of the reagents described in the test

Name used in text	Synonyms
R(+)-*trans*-Chrysanthemic acid	2,2-Dimethyl-3-(2-methylpropenyl)cyclopropane-carboxylic acid
S-Acetoxypropionic acid	2-Hydroxypropanoic acid acetate Acetyllactic acid
Mandelic acid	α-Hydroxybenzeneacetic acid α-Hydroxyphenylacetic acid
Acetobromo-α-D-glucose	α-D-Glucopyranosyl bromide
R-α-Methylbenzyl alcohol	R(+)-sec-phenethyl alcohol
Phenylethylamine	α-Methylbenzylamine
Amphetamine	α-Methylbenzene-ethanamine
Levobase/dextrobase	1-(4-Nitrophenyl)-2-amino-1,3-propanediol

5. Conclusions and future trends

In this chapter I have attempted to review some of the major derivatization reagents used for the chromatographic separation of enantiomers. It was not the intention to produce a glossary of every reagent ever used but merely to demonstrate that, if an optically active compound contains a suitable functional group, it can be derivatized using several different reagents and separated on an achiral column. These diverse methods allow the scientist to design the derivative to produce stable entities which can be readily separated and which have a suitable limit of detection. The technique is therefore ideal for laboratories not routinely carrying out chiral separations or those with limited resources and expertise.

As previously stated, the use of racemic mixtures in biologically active compounds is becoming an increasing concern of many regulatory bodies, e.g. at the time of writing this chapter the Swiss have guidelines for chiral durgs already in use and the German authorities regulate the chirality of food additives. Industry itself is putting an increasing effort into method development for the separation and determination of enantiomers and the investigation of the relative effects of each enantiomeric form. This research must include the development of robust and reproducible methods for use with biological samples, which have a suitably low limit of detection. While much of the current research is in the development of direct methods and the theoretical aspects of the separation mechanisms, there are several notable examples of the development of novel reagents and methods [8, 22, 36, 39, 93, 101].

Perhaps jaundiced by this plethora of references for direct methods, many of the reflections on future trends have been almost a defence of the indirect method based on its advantages. In order to solve any complex scientific problem, an understanding of the principles of a wide range of methods is required; no matter how established the method, it may provide the ideal solution.

To this end of future lies in the inventiveness of the scientist to develop or use reagents which will result in a successful outcome to the problem. My personal view is that this will involve the development of highly stable reagents that give quantitative reactions, particularly in the presence of biological extractives, and produce fluorescent or highly absorbing derivatives.

6. References

[1] S. G. Allenmark, *Chromatographic Enantioseparation: Methods and Applications*, Ellis Horwood, Chichester (1988).
[2] J. Jacques, A. Collet and S. H. Wilen, *Enantiomers, Racemates, and Resolutions*, Wiley-Interscience, New York (1981).

[3] W. A. König, *The Practice of Enantiomer Separation by Capillary Gas Chromatography*, Dr Alfred Hüthig Verlag, Heidelberg (1987).
[4] R. W. Souter, *Chromatographic Separations of Stereoisomers*, CRC Press, Boca Raton (1985).
[5] W. J. Lough (Editor), *Chiral Liquid Chromatography*, Blackie, Glasgow (1989).
[6] W. H. Pirkle and T. C. Pachapsky, *Adv. Chromatogr.*, **27**, 73 (1987).
[7] M. Mack and H. E. Hauck, *J. Planar Chromatogr.*, **2**, 190 (1989).
[8] B. Halpern, *Derivatives for Chromatography*, K. Blau and G. S. King (Editors), Heyden, London (1977), pp. 457–499.
[9] T. Nambara, S. Ikegawa, M. Hasegawa and J. Goto, *Anal. Chim. Acta*, **101**, 111 (1978).
[10] H. C. Rose, R. L. Stern and B. L. Karger, *Anal. Chem.*, **38**, 469 (1966).
[11] C. G. Scott, M. J. Petrin and T. McCorkle, *J. Chromatogr.*, **125**, 157 (1976).
[12] M. Nimura, Y. Kasahara and T. Kinashita, *J. Chromatogr.*, **213**, 327 (1981).
[13] M. W. Anders and M. J. Cooper, *Anal. Chem.*, **43**, 1093 (1971).
[14] C. J. W. Brooks, M. T. Gilbert and J. D. Gilbert, *Anal. Chem.*, **45**, 898 (1972).
[15] R. S. Burden, A. H. B. Deas and T. Clarke, *J. Chromatogr.*, **391**, 273 (1987).
[16] A. B. Attygalle, E. D. Morgan, R. P. Evershed and S. J. Rowland, *J. Chromatogr.*, **260**, 411 (1983).
[17] J. A. Dale, D. L. Dull and H. S. Mosher, *J. Org. Chem.*, **34**, 2543 (1969).
[18] P. Michelsen and G. Odham, *J. Chromatogr.*, **331**, 295 (1985).
[19] R. E. Doolittle and R. R. Heath, *J. Org. Chem.*, **49**, 5041 (1984).
[20] C. Banfield and M. Rowland, *J. Pharm. Sci.*, **73**, 1392 (1984).
[21] C. C. Duke and G. M. Holder, *J. Chromatogr.*, **430**, 53 (1988).
[22] T. K. Gerding, B. F. H. Drenth and R. A. de Zeeuw, *J. High Resolut. Chromatogr. Chromatogr. Commun.*, **10**, 523 (1987).
[23] I. Sakata and H. Iwamura, *Agric. Biol. Chem.*, **43**, 307 (1979).
[24] W. H. Pirkle and M. S. Hoekstra, *J. Org. Chem.*, **39**, 3904 (1974).
[25] W. H. Pirkle and J. R. Hauske, *J. Org. Chem.*, **42**, 2781 (1977).
[26] J. W. Westley and B. Halpern, *J. Org. Chem.*, **33**, 3978 (1968).
[27] J. J. Martel, J. P. Demoute, A. P. Teche and J. R. Tessier, *Pestic. Sci.*, **11**, 188 (1980).
[28] P. Levi and E. Hodgson, *Int. J. Biochem.*, **15**, 349 (1983).
[29] O. H. Lowry, N. J. Rosebrough, A. L. Farr and R. J. Randell, *J. Biol. Chem.*, **193**, 265 (1952).
[30] A. A. Rudmann, and J. R. Aldrich, *J. Chromatogr.*, **407**, 324 (1987).
[31] P. Michelsen, E. Aronsson, G. Odham and B. Akesson, *J. Chromatogr.*, **350**, 417 (1985).
[32] W. Pereira, V. A. Bacon, W. Patton and B. Halpern, *Anal. Lett.*, **3**, 23 (1970).
[33] R. G. Annett and P. K. Stumpf, *Anal. Biochem.*, **47**, 638 (1972).
[34] R. G. Harvey and H. Cho, *Anal. Biochem.*, **80**, 540 (1977).
[35] H. Yagi and D. M. Jerina, *J. Am. Chem. Soc.*, **104**, 4026 (1982).
[36] H. Lee and R. G. Harvey, *J. Org. Chem.*, **49**, 1114 (1984).
[37] P. J. van Bladeren, J. M. Sayer, D. E. Ryan, P. E. Thomas, W. Levin and D. M. Jerina, *J. Biol. Chem.*, **260**, 10226 (1985).
[38] S. K. Balani, P. J. van Bladeren, E. S. Cassidy, D. R. Boyd and D. M. Jerina, *J. Org. Chem.*, **52**, 137 (1987).
[39] R. N. Armstrong, B. Kedzierski, W. Levin and D. M. Jerina, *J. Biol. Chem.*, **256**, 4726 (1981).
[40] P. Michelsen, V. Annby and S. Gronowitz, *Chem. Scr.*, **24**, 253 (1985).
[41] E. Gil-Av, R. Charles-Sigler, G. Fischer and D. Nurok, *J. Gas Chromatogr.*, **4**, 51 (1966).
[42] R. Charles, G. Fischer and E. Gil-Av, *Isr. J. Chem.*, **1**, 234 (1963).
[43] A. Mosandl, M. Gessner, C. Gunther, W. Deger and G. Singer, *J. High Resolut. Chromatogr. Chromatogr. Commun.*, **10**, 67 (1987).
[44] E. Gil-Av and D. Nurok, *Proc. Chem. Soc.*, 146 (1962).
[45] S. Hammarstrom and M. Hamberg, *Anal. Biochem.*, **52**, 169 (1973).
[46] J. M. Cross, B. F. Putney and J. Bernstein, *J. Chromatogr. Sci.*, **8**, 679 (1970).
[47] J. W. Westley and B. Halpern, *J. Org. Chem.*, **33**, 3978 (1968).
[48] I. Sakata and K. Koshimizu, *Agric. Biol. Chem.*, **43**, 411 (1979).
[49] P. E. Sonnet, E. G. Piotrowski and R. T. Boswell, *J. Chromatogr.*, **436**, 205 (1988).
[50] E. M. Gaydou and R. P. Randriamiharisoa, *J. Chromatogr.*, **396**, 378 (1987).
[51] W. H. Pirkle and C. W. Boeder, *J. Org. Chem.*, **43**, 1950 (1978).
[52] Y. Yamazaki and H. Maeda, *Agric. Biol. Chem.*, **50**, 79 (1986).
[53] G. L. Jeyaraf and W. R. Porter, *J. Chromatogr.*, **315**, 378 (1984).
[54] D. M. Johnson, A. Reuter, J. M. Collins and G. F. Thompson, *J. Pharm. Sci.*, **68**, 112 (1979).
[55] R. G. Ackman, R. E. Cox, G. Eglinton, S. N. Hooper and J. R. Maxwell, *J. Chromatogr. Sci.*, **10**, 392 (1972).
[56] J. P. Kamerling, M. Duran, G. R. Gerwig, D. Ketting, L. Bruinvis, J. F. G. Vliegenthart and S. K. Wadman, *J. Chromatogr.*, **222**, 276 (1981).
[57] K. D. Ballard, T. D. Eller and D. R. Knapp, *J. Chromatogr.*, **275**, 161 (1983).
[58] R. A. Chapman and C. R. Harris, *J. Chromatogr.*, **174**, 369 (1979).
[59] T. Kaneda, *J. Chromatogr.*, **366**, 217 (1986).
[60] G. Helmchen, H. Volter and W. Schuhle, *Tetrahedron Lett.*, **16**, 1417 (1977).
[61] S. Bernasconi, A. Comini, A. Corbella, P. Gariboldi and M. Sisti, *Synthesis*, 385 (1980).

[62] M. Jiang and D. M. Soderlund, *J. Chromatogr.*, **248**, 143 (1982).
[63] G. Helmchen, G. Nill, D. Flockerzi and M. S. K. Youssef, *Angew. Chem. Int. Ed. Engl.*, **18**, 62 (1979).
[64] E. J. D. Lee, K. M. Williams, G. G. Graham, R. O. Day and G. D. Champion, *J. Pharm. Sci.*, **73**, 1542 (1984).
[65] H. Spahn, *J. Chromatogr.*, **423**, 334 (1987).
[66] K. H. Lehr and P. Damm, *J. Chromatogr.*, **425**, 153 (1988).
[67] Y. Fujimoto, K. Ishi, H. Nishi, N. Tsumagari, T. Kakimoto and R. Shimizu, *J. Chromatogr.*, **402**, 344 (1987).
[68] S. Pedrazzini, W. Zanoboni-Muciaccia, C. Sacchi and A. Forgione, *J. Chromatogr.*, **415**, 214 (1987).
[69] S. Bjorkman, *J. Chromatogr.*, **414**, 465 (1987).
[70] A. J. Hutt and J. Caldwell, *J. Pharm. Pharmacol.*, **35**, 693 (1983).
[71] R. T. Foster and F. Jamali, *J. Chromatogr.*, **416**, 388 (1987).
[72] V. Rossetti, A. Lombard and M. Buffa, *J. Pharm. Biomed. Anal.*, **4**, 673 (1986).
[73] N. N. Singh, F. M. Pasutto, R. T. Coutts and F. Jamali, *J. Chromatogr.*, **378**, 125 (1986).
[74] S. J. Rowland, A. V. Larcher, R. Alexander and R. I. Kagi, *J. Chromatogr.*, **312**, 395 (1984).
[75] J. P. Kamerling, G. R. Gerwig, J. F. G. Vliegenthart, M. Duran, D. Ketting and S. K. Wadman, *J. Chromatogr.*, **143**, 117 (1977).
[76] M. Horiba, H. Kitahara, K. Takahashi, S. Yamamoto, A. Murano and N. Oi, *Agric. Biol. Chem.*, **43**, 2311 (1979).
[77] B. C. Sallustio, A. Abas, P. J. Hayball, Y. J. Purdie and P. J. Meffin, *J. Chromatogr.*, **374**, 329 (1986).
[78] J. M. Maitre, G. Boss, B. Testa and K. Hostettmann, *J. Chromatogr.*, **356**, 341 (1986).
[79] S. W. McKay, D. N. B. Mallen, P. R. Shrubsall, B. P. Swann and W. R. N. Williamson, *J. Chromatogr.*, **170**, 482 (1979).
[80] F. E. Rickett, *Analyst*, **98**, 687 (1973).
[81] F. Jamali, F. M. Pasutto and C. Lemko, *J. Liq. Chromatogr.*, **12**, 1835 (1989).
[82] L. Ladanyi, I. Sztruhar, P. Slegel and G. Vereczekey-Donath, *Chromatographia*, **24**, 477 (1987).
[83] P. Slegel, G. Vereczekey-Donath, L. Ladanyi and M. Toth-Lauritz, *J. Pharm. Biomed. Anal.*, **5**, 665 (1987).
[84] H. Nagashima, Y. Tanaka, N. Watanabe, R. Hayashi and K. Kawada, *Chem. Pharm. Bull.*, **32**, 251 (1984).
[85] J. Goto, N. Goto and T. Nambara, *J. Chromatogor.*, **239**, 559 (1982).
[86] A. J. Hutt, S. Fournel and J. Caldwell, *J. Chromatogr.*, **378**, 409 (1986).
[87] A. Averginos and A. J. Hutt, *J. Chromatogr.*, **415**, 75 (1987).
[88] D. Valentine, K. K. Chan, C. G. Scott, K. K. Johnson, K. Toth and G. Saucy, *J. Org. Chem.*, **41**, 62 (1976).
[89] H. Knorr, R. Reichl, W. Traunecker, F. Knappen and K. Brandt, *Arzneim.-Forsch.*, **34**, 1709 (1984).
[90] A. J. Sedman and J. Gal, *J. Chromatogr.*, **306**, 155 (1984).
[91] P. A. Husain, J. E. Colbert, S. R. Sirimanne, D. G. VanDerveer, H. H. Herman and S. W. May, *Anal. Biochem.*, **178**, 177 (1989).
[92] J. Goto, N. Goto, A. Hikichi and T. Nambara, *J. Liq. Chromatogr.*, **2**, 1179 (1979).
[93] W. H. Pirkle and K. A. Simmons, *J. Org. Chem.*, **48**, 2520 (1983).
[94] F. T. Noggle and C. R. Clark, *J. Forensic Sci.*, **31**, 732 (1986).
[95] F. T. Noggle, J. DeRuiter and C. R. Clark, *Anal. Chem.*, **58**, 1643 (1986).
[96] K. J. Miller, J. Gal and M. M. Ames, *J. Chromatogr.*, **307**, 335 (1984).
[97] A. J. Sedman and J. Gal, *J. Chromatogr.*, **278**, 199 (1983).
[98] R. H. Buck and K. Krummen, *J. Chromatogr.*, **315**, 279 (1984).
[99] H. Weber, H. Spahn, E. Mutschler and W. Mohrke, *J. Chromatogr.*, **307**, 145 (1984).
[100] J. Maibaum, *J. Chromatogr.*, **436**, 269 (1988).
[101] D. T. Witte, R. A. de Zeeuw and B. F. H. Drenth, *J. High Resolut. Chromatogr. Chromatogr. Commun.*, **13**, 569 (1990).
[102] J. I. Seeman, C. G. Chavdarian and H. V. Secor, *J. Org. Chem.*, **50**, 5419 (1985).
[103] J. M. Barksdale and C. R. Clark, *J. Chromatogr. Sci.*, **23**, 176 (1985).
[104] H. K. Lim, J. W. Hubbard and K. K. Midha, *J. Chromatogr.*, **378**, 109 (1986).
[105] H. Spahn, *J. Chromatogr.*, **427**, 131 (1988).
[106] S. Einarsson, B. Josefsson, P. Moller and D. Sanchez, *Anal. Chem.*, **59**, 1191 (1987).
[107] N. Nimura, K. Iwaki and T. Kinoshita, *J. Chromatogr.*, **402**, 387 (1987).
[108] S. D. Roy and H. K. Lim, *J. Chromatogr.*, **431**, 210 (1988).
[109] C. R. Clark and J. M. Barksdale, *Anal. Chem.*, **56**, 958 (1984).
[110] N. Nimura, H. Ogura and T. Kinoshita, *J. Chromatogr.*, **202**, 375 (1980).
[111] J. Gal and A. J. Sedman, *J. Chromatogr.*, **314**, 275 (1984).
[112] G. Helmchen and W. Strubert, *Chromatographia*, **7**, 713 (1974).
[113] R. W. Souter, *Chromatographia*, **9**, 635 (1976).
[114] R. Pappo, P. Collins and C. Jung, *Tetrahedron Lett.*, **12**, 943 (1973).
[115] C. J. W. Brooks, D. G. Watson and W. J. Cole, *J. Chromatogr.*, **347**, 455 (1985).
[116] G. E. Pollock and D. A. Jermany, *J. Gas Chromatogr.*, **6**, 412 (1968).
[117] S. Hara, H. Okabe and K. Mihashi, *Chem. Pharm. Bull.*, **35**, 501 (1987).
[118] M. R. Little, *Carbohydr. Res.*, **105**, 1 (1982).
[119] H. Schweer, *J. Chromatogr.*, **243**, 149 (1982).
[120] H. Schweer, *J. Chromatogr.*, **259**, 164 (1983).
[121] R. Oshima, Y. Yamauchi and J. Kumanotani, *Carbohydr. Res.*, **107**, 169 (1982).
[122] W. E. Pereira, M. Salomon and B. Halpern, *Aust. J. Chem.*, **24**, 1103 (1971).
[123] H. Schweer, *Carbohyder. Res.*, **116**, 139 (1983).
[124] R. Oshima, J. Kumanotani and C. Watanabe, *J. Chromatogr.*, **259**, 159 (1983).
[125] M. Isola, E. Ciuffarin and L. Sagramora, *Synthesis*, 326 (1976).
[126] W. Theilacker and K. Ebke, *Angew. Chem.*, **68**, 303 (1956).

[127] W. A. König, S. Lutz, C. Colberg, N. Schmidt, G. Wenz, E. Von der Bey, A. Mosandl, C. Gunther and A. Kustermann, *J. High Resolut. Chromatogr. Chromatogr. Commun.*, **11**, 621 (1988).
[128] G. Saucy, R. Borer, D. P. Trullinger, J. B. Jones and K. P. Lok, *J. Org. Chem.*, **42**, 3206 (1977).
[129] K. H. Engel, R. A. Flath, W. Albrecht and R. Tressl, *J. Chromatogr.*, **479**, 176 (1989).
[130] M. Feuerbach, O. Frohlich and P. Schreier, *J. Agric. Food Chem.*, **36**, 1236 (1988).
[131] K. H. Engel, W. Albrecht and J. Heidlas, *J. Agric. Food Chem.*, **38**, 244 (1990).
[132] A. Mosandl, M. Gessner, C. Gunther, W. Deger and G. Singer, *J. High Resolut. Chromatogr. Chromatogr. Commun.*, **10**, 67 (1987).
[133] W. Deger, M. Gessner, C. Gunther, G. Singer and A. Mosandl, *J. Agric. Food Chem.*, **36**, 1260 (1988).
[134] M. Gessner, W. Deger and A. Mosandl, *Int. Food Res. Technol.*, **186**, 417 (1988).
[135] G. E. Pollock, *Anal. Chem.*, **44**, 2368 (1972).
[136] H. Iwase and A. Murai, *Chem. Pharm. Bull.*, **22**, 8 (1974).
[137] H. Iwase and A. Murai, *Chem. Pharm. Bull.* **22**, 1455 (1974).
[138] H. Iwase, *Chem. Pharm. Bull.*, **22**, 2075 (1974).
[139] J. M. Manning and S. Moore, *J. Biol. Chem.*, **243**, 5591 (1968).
[140] A. R. Mitchell, S. B. H. Kent, I. C. Chu and R. B. Merrifield, *Anal. Chem.*, **50**, 637 (1978).
[141] T. Takaya, Y. Kishida and S. Sakakibara, *J. Chromatogr.*, **215**, 279 (1981).
[142] T. Kinoshita, Y. Kasahara and N. Nimura, *J. Chromatogr.*, **210**, 77 (1981).
[143] K. Iwake, S. Yoshida, N. Nimura, T. Kinoshita, K. Takeda and H. Ogura, *Chromatographia*, **23**, 899 (1987).
[144] R. H. Buck and K. Krummen, *J. Chromatogr.*, **387**, 255 (1987).
[145] D. S. Dunlop and A. Neidle, *Anal. Biochem.*, **165**, 38 (1987).
[146] T. Nambara, J. Goto, K. Taguchi and T. Iwata, *J. Chromatogr.*, **100**, 180 (1974).
[147] H. Furukawa, Y. Mori, Y. Takeuchi and K. Ito, *J. Chromatogr.*, **136**, 428 (1977).
[148] H. Furukawa, S. Sakakibara, A. Kamei and K. Ito, *Chem. Pharm. Bull.*, **23**, 1625 (1975).
[149] D. J. Aberhart, J. A. Cotting and H. J. Lin, *Anal. Biochem.*, **151**, 88 (1985).
[150] P. Marfey, *Carlsberg Res. Commun.*, **49**, 591 (1984).
[151] M. R. Euerby, L. Z. Partridge and W. A. Gibbons, *J. Chromatogr.*, **483**, 239 (1989).
[152] T. Takeuchi, T. Niwa and D. Ishii, *J. High Resolut. Chromatogr. Chromatogr. Commun.*, **11**, 343 (1988).
[153] L. A. Allison, G. S. Mayer and R. E. Shoup, *Anal. Chem.*, **56**, 1089 (1984).
[154] D. W. Asward, *Anal. Biochem.*, **137**, 405 (1984).
[155] N. Nimura and T. Kinoshita, *J. Chromatogr.*, **352**, 169 (1986).
[156] N. Nimura, A. Toyama and T. Kinoshita, *J. Chromatogr.*, **316**, 547 (1984).
[157] Y. Asahina and M. Ishidate, *Chem. Ber.*, **66**, 1673 (1933).
[158] M. R. Euerby, L. Z. Partridge and P. Rajani, *J. Chromatogr.*, **447**, 392 (1988).
[159] S. Lam, *J. Chromatogr.*, **355**, 157 (1986).
[160] M. Maurs, F. Trigalo and R. Azerad, *J. Chromatogr.*, **440**, 209 (1988).
[161] S. Einarsson, S. Folestad and B. Josefsson, *J. Liq. Chromatogr.*, **10**, 1589 (1987).
[162] W. Linder, C. Leitner and G. Uray, *J. Chromatogr.*, **316**, 605 (1984).
[163] I. Demain and D. F. Gripshover, *J. Chromatogr.*, **387**, 523 (1987).
[164] A. Darmon and J. P. Thenot, *J. Chromatogr.*, **374**, 321 (1986).
[165] G. Gubitz and S. Mihellyes, *J. Chromatogr.*, **314**, 462 (1984).
[166] T. Walle, D. D. Christ, U. K. Walle and M. J. Wilson, *J. Chromatogr.*, **341**, 213 (1985).
[167] G. Pflugmann, H. Spahn and E. Mutschler, *J. Chromatogr.*, **416**, 331 (1987).
[168] W. Dieterle and J. M. Faigle, *J. Chromatogr.*, **242**, 289 (1982).
[169] F. Zetsche and M. Hubacher, *Helv. Chim. Acta*, **9**, 291 (1926).
[170] J. Hermansson and C. Von Bahr, *J. Chromatogr.*, **227**, 113 (1982).
[171] J. C. Dabrowiak and D. W. Cooke, *Anal. Chem.*, **43**, 791 (1971).

11

Ion-pair Extraction in Chromatography

Bengt-Arne Persson
Bioanalytical Chemistry, Astra Hässle AB, Mölndal, Sweden

and

Göran Schill
Analytical Pharmaceutical Chemistry, Uppsala University, Uppsala, Sweden

1. APPLICATION OF THE ION-PAIR
   CONCEPT   253
2. ION-PAIR EXTRACTION AND SIDE
   REACTIONS   254
   2.1 Ion-pair extraction   254
   2.2 Side reactions   255
   2.3 Other complexing processes   255
3. ION-PAIR DISTRIBUTION IN LIQUID
   CHROMATOGRAPHY   256
   3.1 Normal phase   256
   3.2 Reversed phase   257
   3.3 Enantiomeric separation   260
4. DETECTION BY ION-PAIRING AGENT   260
   4.1 Counter-ion in eluted ion-pairs   260
   4.2 Probe in mobile phase   261
   4.3 Post-column extraction   263
5. ION-PAIR MEDIATED DERIVATIZATION   263
6. THEORETICAL BACKGROUND FOR
   THE DISTRIBUTION OF ION PAIRS   263
7. CONCLUSIONS   264
8. REFERENCES   264

1. Application of the ion-pair concept

Ion-pair extraction is a method for partition of ionic compounds with the aid of counter-ions of opposite charge. Ion-pairs are formed in the organic solvent and their distribution depends on their solvation properties. Initially, the technique was used to extract basic and acidic compounds with counter-ions of different hydrophobicity as extractants. Counter-ion-aided extraction has been applied widely in liquid–liquid extraction, both in batch and column mode [1].

The ion-pair technique has found a great number of applications in normal phase liquid–liquid chromatography, with an aqueous stationary phase supplying the counter-ion and an organic mobile phase [2, 3] and in reversed phase systems [4, 5]. The development of separation science during the last 10 years has established the dominance of reversed phase liquid–solid chromatography on alkyl-bonded silica. This has led to numerous applications of the ion-pair concept in the separation of ionized amines and acids. Other designations of the technique have also been used, such as ion interaction chromatography, dynamic ion exchange and paired-ion chromatography. Irrespective of the 'trade name', the retention of a substance on a column is due to a combined effect of hydrophobic, polar and electrostatic interactions [6, 7].

A characteristic feature of ion-pair extraction is the selection of counter-ions with a high response to specific detectors (such as photometric and fluorimetric). This principle was originally applied in the so-called acid dye extraction for the determination of ammonium compounds with low inherent detectability. Bromothymol blue [8], dipicrylamine [9] and picrate [10], with high molar absorbance and highly fluorescent anions, such as anthracenesulfonate [11] and dimethoxyanthracenesulfonate [12], have been used to attain high detection sensitivity. This ion-pair extraction approach has also been used for post-column detection in liquid chromatography with, e.g., dimethoxyanthracenesulfonate [13] and acridine [14]. Naphthalenesulfonate and picrate were employed as counter-ions in a post-column system for LC–MS to improve the fragmentation pattern [15].

In normal phase liquid chromatography, systems of this kind have been utilized to separate and monitor alkylammonium ions and dipeptides on columns containing naphthalenesulfonate in the stationary phase [3]. For carboxylic and sulfonic acids, dimethylprotriptyline has been used as the counter-ion [16]. Acetylcholine was isolated and measured as the picrate ion-pair in an extract from rat sciatic nerve [17]. In reversed phase liquid chromatography, detection by the ion-pair technique has been performed via the presence of detectable ionic probe in the aqueous mobile phase. More recently this indirect detection method has developed very rapidly for organic ionized compounds, such as carboxylic acids, amines and amino acids [18, 19], and for inorganic anions [20, 21].

The major application of the ion-pair concept is in reversed phase ion-pair chromatography, where the counter-ion in the aqueous mobile phase is used to regulate the retention of ionic compounds. Selection of the nature and concentration of the ion-pairing agent and of the uncharged organic modifiers gives unlimited possibilities for the separation of both hydrophobic and hydrophilic substances. The added species will influence the selectivity [22] and also ensure conditions for efficient chromatographic performance [23]. Enantiomeric separations of optically active anionic and cationic compounds have been made with a chiral counterion in the mobile phase. The chromatographic system usually comprises a hydrophilic adsorbent and an organic mobile phase [24, 25].

Ion-pair-mediated derivatization will also be briefly discussed to illustrate the wide application of the ion-pair concept. A counter-ion in the aqueous phase promotes the derivatization of organic acids in a two-phase reaction system. Phenols [26], carboxylic acids [27] and other compounds with active hydrogen [28] are alkylated as reactive ion-pairs in the organic phase.

2. Ion-pair extraction and side reactions

The process of ion-pair extraction is governed by several parameters, such as the nature and concentration of the ionic species in the two phases, the pH of the aqueous phase, the properties of the organic solvent and the specific interactions of the various ions with one another and with the two phases. A good knowledge of the physicochemical parameters that influence the distribution of the ionic compounds will aid in the successful application of the ion-pair technique [29].

2.1 Ion-pair extraction

A charged compound HA^+, a protonated amine or quaternary ammonium ion, can be distributed between an aqueous and an organic phase with the aid of a counter ion X^-, the anion of a strong or weak acid.

The two oppositely charged ions form ion-pairs in the organic phase as summarized by the formula

$$HA^+_{aq} + X^-_{aq} = HAX_{org} \quad (1)$$

The equilibrium constant $K_{ex(HAX)}$ for this process (the extraction constant) is defined by

$$K_{ex(HAX)} = [HAX]_{org}/[HA^+]_{aq} \cdot [X^-]_{aq} \quad (2)$$

It expresses the combined effect of phase transfer and ion-pair formation. Provided that further reactions can be disregarded, the distribution ratio D_{HA} of HA^+ is given by the relationship

$$D_{HA} = [HAX]_{org}/[HA^+]_{aq} = K_{ex(HAX)} \cdot [X^-]_{aq} \quad (3)$$

The distribution of HA^+ as an ion-pair is determined by the extraction constant and the concentration of the counter-ion in the aqueous phase.

The magnitude of the extraction constant $K_{ex(HAX)}$ depends on the properties of the ions HA^+ and X^- and on the nature of the organic solvent. For a given solvent, the magnitude of the extraction constant can vary within a wide range, as shown for the extraction of the tetrabutylammonium cation into chloroform (Table 1). The extraction constants listed extend over ten orders of magnitude from the low values obtained with hydrophilic inorganic anions to the high ones given by the strongly hydrophobic bromothymol blue and dipicrylamine (hexanitrodiphenylamine). There is thus a high degree of flexibility, a useful feature for the extraction of compounds with widely different properties.

The extraction constant increases with increasing numbers of alkyl and aryl groups, and there is usually within a homologous series a linear relationship between the number of methylene groups and the logarithm of the extraction constant. With chloroform or dichloromethane as organic solvent, $\log K_{ex}$ increases by 0.5–0.6 unit for the addition of one methylene unit. Similar correlations have been found with other organic solvents.

The extraction constant of organic anions increases in the series carboxylate < sulfonate < sulfate, probably because of decreased bonding to the aqueous phase. Substitution by a hydroxyl group in the position ortho to carboxylate, sulfonate or sulfate increases the log extraction constant by approximately 2 units owing to intramolecular hydrogen bonding (compare benzoic acid and salicylic acid in Table 1). Ion-pairs of alkylammonium ions with hydrogen-accepting anions show

Table 1 Extraction constants of tetrabutylammonium ion-pairs; organic phase: chloroform

Class	Anionic compound (X^-)	log $K_{ex(HAX)}$
Inorganic	Cl^-	−0.11
	Br^-	1.29
	NO_3^-	1.39
	I^-	3.01
	ClO_4^-	3.48
Carboxylate	acetate	−2.1
	phenylacetate	0.27
	3-hydroxybenzoate	−1.54
	benzoate	0.39
	salicylate	2.42
Phenolate	phenolate	0.05
	picrate	5.91
	2,4-dinitro-1-naphtholate	6.45
Sulfonate	toluene-4-sulfonate	2.33
	naphthalene-2-sulfonate	3.45
	anthracene-2-sulfonate	5.11
Sulfate	3-phenylpropyl sulfate	4.20
	2-naphthyl sulfate	4.90
'Acid dye'	methyl orange	5.47
	bromothymol blue	8.0
	hexanitrodiphenylamine	9.6

extraction constants increasing in the order quaternary < primary < secondary < tertiary using chloroform or dichloromethane for the extraction. The order may change with hydrogen-accepting solvents or with a different character of the anion.

Hydrophilic substituents, e.g. hydroxy, carboxylic or amino groups, in one of the ion-pair components will decrease the log extraction constant by 1–2 units with a chloroalkane as organic solvent [8], but much less with a strongly hydrogen bonding solvent such as pentanol as extractant [30].

Masking of strongly hydrophilic groups by alkylation or acetylation has a drastic effect on the extraction constant. With CH_2Cl_2 as organic phase, methylation of one hydroxy group in morphine increases the log extraction constant by 2.4 units and acetylation of both hydroxy groups in morphine gives an increase of about four units [8].

2.2 Side reactions

The fundamental equilibrium of ion-pair extraction (1) shows the two oppositely charged ionic species in the aqueous phase and the ion-pair in the organic phase. However, the components may participate in other equilibria which will influence the extraction process significantly. These may be designated as side reactions in the ion-pair extraction.

Anions and cations of weak protolytes, e.g. carboxylic acids or amines, are uncharged at low and high pH respectively, which will counteract the ion-pair extraction. Moreover, they may be distributed in uncharged form to the organic phase. The proportion between the partition in uncharged form and as the ion-pair is determined by pH and the equilibrium constants. If the counter-ion is aprotic, the ion-pair extraction of acids is promoted by high pH and that of bases by low pH. Association in the aqueous phase between the oppositely charged ionic species, and also micelle formation, are processes that will decrease the degree of ion-pair extraction.

Side reactions in the organic phase, on the other hand, will improve the yield of the extraction procedure. These secondary equilibria may just involve the ion-pair, by dissociation to HA^+ and X^- or dimerization to $H_2A_2X_2$ or higher multiples. The occurrence and extent of these processes depend on the nature of the organic solvent and the ionic species and on their concentration.

Ion-pair dissociation in the organic phase is a low-concentration effect influenced by the strength of the binding forces between the ions and by the dielectric properties of the solvent [31]. This side reaction may make a significant contribution to the isolation of compounds in trace analysis, as the distribution ratio increases with decreasing concentration of the analyte ion [11]. Ion-pair dimerization or the formation of higher aggregates in the organic phase can also improve the extraction recovery, particularly on a preparative scale, because it is a high-concentration effect.

2.3 Other complexing processes

Ion-pair formation in the organic phase can be combined with other kinds of complexation. When the solvent has a mainly inert character, the influence of a complex-forming agent can often be expressed as an adduct formation with a fixed stoichiometry [31]:

$$HA^+_{aq} + X^-_{aq} + nS_{org} = HAX \cdot S_{n \, org} \quad (4)$$

If complexes other than $HAX \cdot S_n$ can be disregarded, the distribution ratio of HA^+ is given by the expression

$$D_{HA} = K_{ex(HAX)} \cdot K_n \cdot [X^-]_{aq} \cdot [S]^n_{org} \quad (5)$$

where

$$K_n = [HAX \cdot S_n]_{org}/([HAX]_{org} \cdot [S]^n_{org}) \quad (6)$$

In these systems D_{HA} can be controlled by the concentration of X^- and S and the nature of these species. Lipophilic alcohols, as proton acceptors and donors, are good extractants for ion-pairs. They have been used as complexing agents in studies on ion-pair extraction [29]; in most instances a constant value of n was found.

More selective adduct-forming agents have also been used. Trioctylphosphine oxide (TOPO) has a strong hydrogen-accepting ability. When the concentration of TOPO in the organic phase is increased, the extraction of ion-pairs of hydrogen-donating primary and secondary ammonium ions is enhanced but ion-pairs of tertiary and quaternary ammonium ions are almost unaffected [32]. Cyclic polyethers such as dibenzo-18-crown-6 form 1:1 complexes of high stability with ion-pairs containing primary ammonium ions. The same good fit in the ring structure and possibility of hydrogen bonding are not obtained with ammonium ions of higher degrees of substitution, and the complexing constants are much smaller [33].

An efficient way to increase the extraction of hydrophilic ions such as catecholamines is to use complexing agents, such as bis(2-ethylhexyl)phosphoric acid, which act as both ion-pairing and adduct-forming agents [34]. Another approach suitable for catecholamines is to use diphenylborate, which forms a negatively charged complex extracted into the organic phase as an ion-pair with a hydrophobic ammonium ion [35].

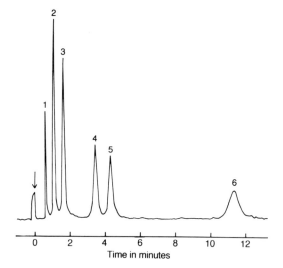

Figure 1. Separation of biogenic ammonium compounds [2]. Stationary phase: 0.2 M $HClO_4$ + 0.8 M $NaClO_4$ on LiChrosorb SI 100 (10 μm). Mobile phase: 1-butanol/dichloromethane (20:80). Peaks: 1, toluene; 2, tryptamine; 3, tryptophan; 4, tyramine; 5, serotonin; 6, dopamine. Reproduced from the *Journal of Chromatographic Science*, **12**, 527 (1974) by permission of Preston Publications, a Division of Preston Industries Inc.

3. Ion-pair distribution in liquid chromatography

Ionic compounds can be transferred with maintained electroneutrality between the mobile and the stationary phase with the participation of a counter-ion. In a liquid–liquid chromatographic system where the influence of the solid phase can be disregarded, the same principles for distribution can be applied as in liquid–liquid extraction system. The ionic compound is moving as an ion-pair in a mobile organic phase in normal phase LC, and is retained as an ion-pair in a stationary organic phase in reversed phase LC. In reversed phase liquid–solid chromatography the ion-pair formation in the stationary phase is less evident. The retention may equally well be due to an exchange of the solute ion for an ion of the same charge, but the prescence of the counter-ion (the ion-pairing agent) is required for electroneutrality.

3.1 Normal phase

Ion-pair distribution in liquid chromatography was initially used in systems where the counter-ion in an aqueous solution was loaded onto a solid phase. In early applications, diatomaceous earth [36] or porous ethanolyzed cellulose [37] was used as support, but the availability of microparticulate silica has improved peak sharpness significantly in liquid–liquid ion-pair chromatography. Perchlorate was used as counter-ion for the separation of biogenic amines by Persson and Karger [2] (Figure 1), and similar systems were used for the separation of tricyclic amines [38] and related compounds [39, 40]. The diastereomeric pairs of N-propylajmaline could be separated in such a liquid–liquid system (Figure 2) [41]. Tetrabutylammonium at neutral pH was the counter-ion for the separation of acidic metabolites of biogenic amines [2]. The same kinds of system were also used for the separation of thyroid hormones and sulfa drugs [42]. Lagerström showed that careful thermostatting and equilibration are necessary for liquid–liquid systems, as demonstrated for the separation of carboxylates as ion-pairs with tetrabutylammonium as counter-ion [43].

In liquid–liquid chromatographic systems, where the solute migrates as an ion-pair in a non-polar mobile phase, the capacity factor of a cationic compound HA^+ with X^- as counter ion is given by

$$k'_{HA} = q/(K_{ex\,(HAX)}[X^-]_{aq}) \qquad (7)$$

where q is the phase volume ratio V_{stat}/V_{mob}.

By analogy with liquid–liquid ion-pair extraction, a more hydrophobic counter-ion X^- (higher $K_{ex\,(HAX)}$) in higher concentration will increase the distribution to

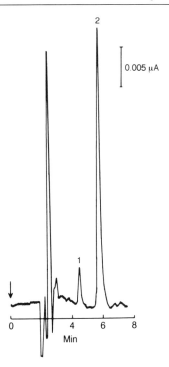

Figure 2. Separation of stereoisomers of N-propylajmaline (1 and 2) in an authentic plasma sample [41]. Stationary phase: 0.2 M HClO$_4$ + 0.8 M NaClO$_4$ on LiChrosorb SI 100 (5 μm). Mobile phase: 1-butanol/1,2-dichloroethane/hexane (15:40:45). Reproduced from Ref. 41, p. 149 by courtesy of Marcel Dekker Inc.

the organic phase and in this case give lower values of k'.

The support will affect the retention to an extent that depends on the properties of its surface and the components in the ion-pair [3, 43]. When silica-based material is used as support, the deviations between the obtained capacity factors and those predicted from batch extraction data are usually significant [44, 45]. Careful thermostatting of the whole chromatographic system is a prerequisite for stable retention.

Lower sensitivity to temperature changes is obtained if the ion-pairing agent is added to the mobile phase as a solution in a polar solvent. Typical examples are systems with silica as the solid phase and aqueous perchloric acid [46, 47] or dodecyl sulfate in methanol [48] added to the hydrophobic organic solvent: dichloromethane or diethyl ether or a mixture of these. The polar components are enriched on the silica; the retention by the stationary phase is probably due to adsorbed anions in a layer of polar solvent. Eriksson and coworkers found that the retention decreased with increasing hydrophobicity of the ion-pair, perchlorate giving lower retention than chloride or methanesulfonate [49].

3.2 Reversed phase

Reversed phase liquid chromatography is the dominant separation technique in which the ion-pair concept is utilized. In liquid–liquid mode rather few studies have been reported. The expression for the capacity ratio is given by

$$k'_{HA} = qK_{ex\,(HAX)}[X^-]_{aq} \qquad (8)$$

Wahlund applied ion-pair principles using alkyl-bonded silica as the support for pentanol as liquid stationary phase [4]. Tetrabutylammonium was used as counter-ion in the mobile aqueous phase for the separation of anionic compounds such as barbiturates, carboxylates, sulfonamides and sulfonates [4, 50]. Hydrophobic amines were separated as ion-pairs with inorganic anions, with long-chain ammonium ions added to the mobile phase to improve peak symmetry [23]. The content of pentanol in the mobile phase had a decisive influence on the retention. The value found for k'_{HA} approached that calculated (eqn. 8) when the mobile phase was saturated with pentanol. At lower concentrations of pentanol, adsorption onto the hydrophobic support had a strong influence on k'_{HA} [51].

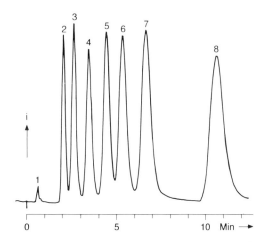

Figure 3. Separation of catecholamines [5]. Stationary phase: tributyl phosphate on LiChrosorb RP-8 (10 μm). Mobile phase: 0.1 M NaClO$_4$ in phosphate buffer, pH 2.1. Peaks: 1, front peak; 2, adrenaline; 3, noradrenaline; 4, dopa; 5, α-methyl-noradrenaline; 6, dopamine, 7, α-methyldopa; 8, α-methyl-dopamine. Reproduced from H. J. L. Janssen et al., *J. Chromatogr.*, **202**, 232 (1980) by permission of Elsevier Science Publishers.

With a strong hydrogen acceptor, tributyl phosphate, as stationary liquid phase, hydrophilic amines have been separated with perchlorate as counter-ion [5]. Owing to the low solubility of tributyl phosphate in the aqueous mobile phase, these systems are stable and have found use in recent applications, as illustrated in Figure 3.

A major part of all ion-pair separations is made in the liquid–solid mode with a hydrophobic adsorbent and an aqueous mobile phase. These systems are compatible with aqueous samples and are highly flexible, which makes them applicable to both hydrophilic and hydrophobic compounds. The solid stationary phase, most often alkyl-bonded silica, has a limited binding capacity, which is of importance for retention and selectivity [22, 23]. Models for the retention can be based on the following statement.

(1) The retention of an ionized solute HA^+ is accompanied by the displacement of a system ion of the same charge or the binding of a counter-ion X^-.
(2) The solute and the mobile phase components may be in competition where the binding capacity of the adsorbing stationary phase is limited.

Assuming a homogeneous solid phase and a Langmuir adsorption, the overall retention of HA^+ can be expressed by

$$k'_{HA} = qK^0 K_{HAX}[X^-]_{aq}/(1 + K_{QX}[Q^+]_{aq}[X^-]_{aq}) \quad (9)$$

where K^0 is the capacity of the adsorbent, K_{HAX} and K_{QX} are distribution constants and Q^+ and X^- are mobile phase components. By analogy with ion-pair liquid–liquid extraction, the retention can be regulated by the concentration and nature of the counter-ion X^-. However, the denominator is also increased by an increase of the hydrophobicity and concentration of the counter-ion, giving a curved relationship as illustrated in Figure 4. A mobile phase ion of the same charge as the solute, i.e. Q^+, will decrease the retention of HA^+ with increased concentration and hydrophobicity (K_{QX}), as seen in Figure 4. The effect of the competing ion Q^+ may be of great importance, particularly for the separation of hydrophobic cationic solutes. The added alkylammonium ion will decrease the retention as well as tailing by eliminating overloading effects in the system [52]. An illustration of this is given in Figure 5 showing the separation of phenoxypropanolamines.

As in other forms of ion-pair distribution, a whole spectrum of counter-ions can be applied in reversed phase ion-pair chromatography. Inorganic anions and hydrophilic cations are suitable for the retention of hydrophobic ionic compounds, whereas such hydrophilic amines as catecholamines are retained as ion-pairs with hydrophobic anions, e.g. hexyl, octyl, decyl and dodecyl sulfates [53]. Hydrophobic quaternary ammonium ions are used for hydrophilic organic and inorganic anions [18]. The influence of the counter-ion is not as straightforward as in liquid–liquid systems, as the limited binding capacity of the solid phase and the presence of competing ions have to be taken into account. Changes in retention of the ion-pair systems can also be made by the addition of uncharged organic modifiers such as acetonitrile,

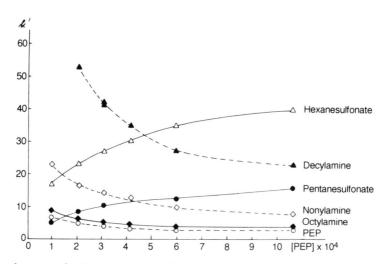

Figure 4. Retention of anionic and cationic compounds with 1-phenethyl-2-picolinium (PEP) in the mobile phase [18]. Stationary phase: μBondapak Phenyl (10 μm). Mobile phase: 1-phenethyl-2-picolinium (PEP) in 0.1 M acetic acid. Reproduced from M. Denkert et al., J. Chromatogr., **218**, 37 (1981) by permission of Elsevier Science Publishers.

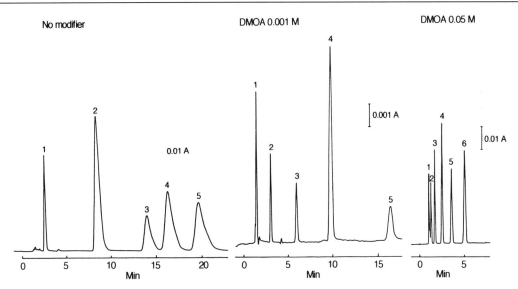

Figure 5. Influence of ionic modifier (dimethyloctylamine, DMOA) on retention and peak shape [52]. Stationary phase: LiChrosorb RP-8. Mobile phase: 0.11 M 1-pentanol and DMOA in phosphate buffer solution (pH 2.2). Sample: 1-phenoxy-3-isopropylamino-2-prapanol derivatives. Reprinted with permission from S.-O. Jansson, *J. Pham. Biomed. Anal.*, **4**, 617, copyright 1986, Pergamon Press PLC.

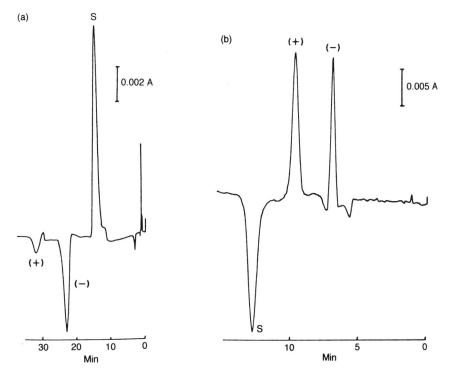

Figure 6. Resolution of camphorsulfonic acid [24]. Stationary phase: LiChrosorb DIOL (5μm). S = system peak. Mobile phase: (a) quinine chloride 3.0×10^{-4} M in dichloromethane/1-pentanol (199:1); (b) quinine 3.5×10^{-4} M and acetic acid 3.5×10^{-4} M in dichloromethane/1-pentanol (99:1). Reproduced from C. Pettersson and K. No, *J. Chromatogr.*, **282**, 674 (1983) by permission of Elsevier Science Publishers.

methanol or other alcohols. In this respect there is no difference compared with other kinds of reversed phase systems.

3.3 Enantiomeric separation

The separation of enantiomeric compounds or optical antipodes cannot be performed in a normal distribution system, but requires interaction with another enantiomeric agent, a chiral selector, which gives diastereomers with different distribution properties with the antipodes of the substrate. The binding can be covalent, but it is often preferable to use coordinative bonding and to separate the enantiomers as diastereomeric complexes. The complexing selector may be incorporated in the stationary phase or added to the mobile phase of the chromatographic system.

Enantiomeric ions can be separated in normal phase liquid–solid systems with one antipode of a chiral counter-ion added to the non-polar mobile phase. The chiral selector should have properties such that several interaction points with the enantiomeric solutes are obtained. Electrostatic interaction, hydrogen bonding and a steric influence from a bulky structure in the vicinity of the asymmetric centre seem to be needed. Pettersson has developed systems for the separation of enantiomers of amino alcohols used as β-adrenoceptor blocking drugs, with (+)-10-camphorsulfonic acid [54] and a dipeptide derivative, N-benzoxycarbonylglycyl-L-proline [25], as selectors, the latter giving higher chiral selectivity.

Quinine, quinidine and cinchonidine, which are amino alcohols with high chiral capability, have been used as selectors for the separation of enantiomers of acids containing a hydrogen bonding function [24]. These chiral selectors have high UV absorbance, providing indirect detection possibilities for solutes without inherent UV absorbance (Figure 6).

4. Detection by ion-pairing agent

High detection sensitivity by absorbance, fluorencence or electrochemical response can be obtained in liquid chromatography provided that the solutes contain structural elements with such activity. If this is not so, detectability can be achieved by pre- or post-column derivatization or, as a third possibility, by ion-pair-mediated indirect detection. In this case, a counter-ion with detector response is added to the chromatographic system or supplied in a post-column extraction system (see Chapter 17).

4.1 Counter-ion in eluted ion-pair

The possibility of measuring ionic compounds without inherent detection properties as ion-pairs with a suitable counter-ion has been utilized in normal phase liquid–liquid chromatography. Picrate was used as counter-ion for acetylcholine [17] and methylguanidine [55] in extracts from biological samples. Aqueous picrate solutions at pH $\approx$ 6.5 were coated on microporous cellulose, and the picrate ion-pairs in the chloroalkane eluent were monitored by a UV detector.

In the high performance mode, with silica as support for the aqueous stationary phase, Crommen et al. used naphthalene-2-sulfonate as counter-ion in the separation of alkylamines, amino acids and dipeptides [3]. A pH of 2.3 was used for dipeptides (Figure 7) and the ion-pairs were detectable with high sensitivity in a UV detector because of the UV-absorbing counter-ion. Careful thermostatting was necessary, and eluents with low background extraction of counter-ion were preferred. The response was directly related to the molar absorp-

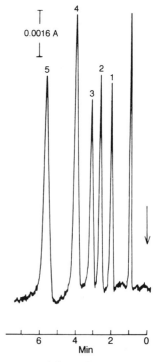

Figure 7. Separation of dipeptides [3]. Stationary phase: 0.1 M naphthalene-2-sulfonate (pH 2.3) on LiChrospher SI 100 (10 μm). Mobile phase: chloroform/1-pentanol (19:1). Detection: 254 nm. Peaks: 1, leucylleucine; 2, phenylalanylvaline; 3, valylphenylalanine; 4, leucylvaline; 5, methionylvaline. Reproduced from J. Crommen et al., J. Chromatogr., **142**, 293 (1977) by permission of Elsevier Science Publishers.

tivity of the ion-pair in the mobile phase, and was independent of the retention.

Hackzell and co-workers utilized dimethylprotriptyline, a tricyclic quaternary ammonium ion, in the stationary aqueous phase for detection of carboxylates and alkylsulfonates [16]. The response was dependent on the retention, which indicated adsorption of the ion-pairs to the solid phase, in this case LiChrosorb DIOL.

4.2 Probe in the mobile phase

The indirect detection technique is used mainly in reversed phase liquid–solid chromatography and it can be applied to charged as well as non-charged analytes. An ion with detectable properties and with affinity for the solid phase is included in the aqueous mobile phase. Analytes that are injected into the system will give rise to equilibrium disturbances and influence the distribution of the detectable ionic component ('the probe'). The response pattern is quite different from that given by direct detection. An example is shown in Figure 8, where a UV-absorbing quaternary ammonium ion, 1-phenethyl-2-picolinium, is added to the mobile aqueous phase as a probe. The solutes are carboxylic acids without inherent response at the detection wavelength. At pH 4.6 they are mainly present in anionic form [18]. There is one peak for each of the analytes and an additional system peak (S) characteristic for the chromatographic system. The response pattern obtained, with negative solute peaks eluted before the system peak and positive after, is typical when the solutes and the probe have opposite charges [56].

A reversed response pattern, with positive solute peaks before the system peak and negative ones after, is obtained when the analytes have the same charge as the probe or are uncharged. The different directions of the peaks of anionic and cationic solutes are demonstrated in the chromatogram in Figure 9, where alkylamines and alkanesulfonates are separated with naphthalene-2-sulfonate in phosphoric acid as an anionic probe [56].

The examples above are typical for systems with one retained component in the mobile phase. The retention of the system peaks is dependent only on the properties of the system, irrespective of the nature of the analytes.

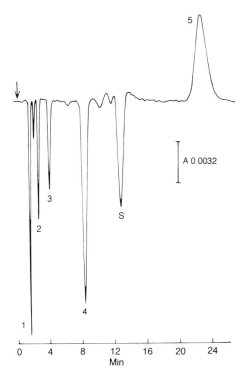

Figure 8. Separation of carboxylic acids [18]. Stationary phase: μBondapak Phenyl (10 μm). Mobile phase: 3×10^{-4} M 1-phenethyl-2-picolinium in acetate buffer (pH 4.6). Detection: 254 nm. Peaks: 1, acetic acid; 2, propionic acid; 3, butyric acid; 4, valeric acid; 5, caproic acid (12 nmole of each); S = system peak. Reproduced from M. Denkert et al., J. Chromatogr., **218**, 34 (1981) by permission of Elsevier Science Publishers.

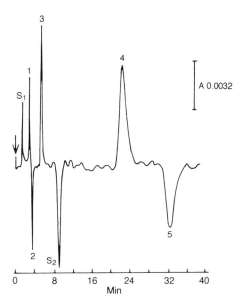

Figure 9. Separation of amines and sulfonates [56]. Stationary phase: μBondapak Phenyl (10 μm). Mobile phase: 4×10^{-4} M naphthalene-2-sulfonate in 0.05 M phosphoric acid. Detection: 254 nm. Peaks: 1, pentanesulfonate; 2, diisopropylamine; 3, hexanesulfonate; 4, heptylamine; 5, octanesulfonate. S_1 and S_2 = system peaks. Reproduced from L. Hackzell and G. Schill, Chromatographia, **15**, 439 (1982) by permission of Vieweg-Publishing.

The direction of a peak representing a solute depends on the charge and retention of the solute relative to the probe, whereas peak size is directly proportional to the amount of sample.

The theoretical basis for detection by ion-pairing probes in reversed phase liquid chromatography can be expressed in a rather simple manner, assuming competition for the binding surface in accordance with a Langmuir model [157]. The relative response of a solute j with a charge opposite to that of the probe k is given by

$$\varepsilon_j^*/\varepsilon_k^* = -(1-\theta_k)\alpha_s/(1-\alpha_s) \quad (10)$$

where $\varepsilon_j^*/\varepsilon_k^*$ is the ratio of the apparent molar absorptivity of the solute peak (ε_j^*) to the known molar absorptivity of the probe (ε_k^*); θ_k is the fractional coverage of the adsorbent by k and $\alpha_s = k_j'/k_k'$, i.e. the retention relative to k. Equation (9) is valid when the hydrophilic buffer components of the mobile phase are present in large excess over the probe, as is most often a prerequisite for good stability of the chromatographic system.

When j and k have the same charge, the expression for the relative response is:

$$\varepsilon_j^*/\varepsilon_k^* = \theta_k\alpha_s/(1-\alpha_s) \quad (11)$$

The relative response of an analyte is influenced by two factors (equations 10 and 11): the relative retention of the solute (α_s) and the fractional coverage of the adsorbent exerted by the probe (θ_k). A system with naphthalene-2-sulfonate as the probe is shown in Figure 10, where the dotted lines give the estimated relative response for cationic and anionic solutes [19, 57]. There is an increase from zero to a maximum value in the α_s range 0–1, and the relative response then decreases and approaches an almost constant level. In order to achieve high detection sensitivity, optimization of the retention is more important than a high absorptivity of the probe.

UV absorbance monitoring has so far predominated but other detection principles have also been utilized [56, 58, 59]. Chromatographic systems of high stability, with different thermostatting, are needed along with high quality detectors, because of the often high background signal [18]. The mobile phase should preferably contain only one retained component, the probe.

It may be kept in mind that this detection principles is non-specific and not suitable for complex biological samples giving numerous peaks. Measurements of impurities in substances and in pharmaceutical products are more typical application areas.

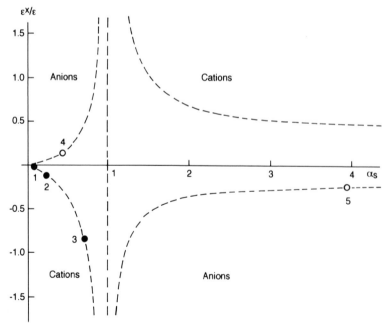

Figure 10. Indirect detection with anionic probe [19]. Stationary phase: μBondpak Phenyl. Mobile phase: 10^{-4} M naphthalene-2-sulfonate in 0.05 M phosphoric acid. Detection: 254 nm. 1, triethylamine; 2, hexylamine; 3, tripropylamine; 4, hexanesulfonate; 5, octanesulfonate. Reproduced from G. Schill and E. Arvidsson, *J. Chromatogr.*, **492**, 307 (1989) by permission of Elsevier Science Publishers.

4.3 Post-column extraction

Detection in liquid chromatography by ion-pair extraction can also be performed post-column by on-line liquid–liquid extraction. A solvent-segmented system is used and the ion-pair extraction is performed in an open-tubular reactor [13]. The eluate from a reversed phase column is mixed with an aqueous solution containing the detectable counter-ion and a non-miscible organic solvent. After extraction in the reaction coil, the aqueous phase containing the excess of the counter-ion is removed and the ion-pair in the organic phase is monitored in the detector. In normal phase liquid chromatography the organic eluent is segmented with an aqueous phase containing the counter-ion, and the ion-pair is monitored in the organic phase after phase separation. In both modes of liquid chromatography it is important to avoid high concentrations of polar modifiers, such as acetonitrile, methanol and other alcohols, since they will promote co-extraction of the ion-pairing agent. A high background will affect the signal-to-noise ratio.

Fluorescent counter-ions such as 9,10-dimethoxyanthracene-2-sulfonate have been used in a number of bioanalytical applications: amines and ammonium compounds used as drugs have been assayed [13]. The detection sensitivity is 10–100 times better than by direct measurement of the inherent absorbance. For alkanesulfonates and related sulfates, the cationic counter-ion, acridine has been employed for determinations in surface water [14]. The ion-pairing agent may be added to the aqueous mobile phase beofre it reaches the column, which simplifies the extraction and detector design and decreases band broadening [60]. In this case only the extraction solvent, generally a chloroalkane, is combined with the eluent. As in all indirect detection methods, careful thermostatting and high system stability are prerequisites for high sensitivity.

5. Ion-pair mediated derivatization

Extractive alkylation processes utilizing ion-pair distribution have been used in organic synthesis [61], as well as in derivatization for gas chromatographic assay. The technique is suitable for organic compounds containing active hydrogen, which are particularly reactive as ion-pairs in non-polar solvents.

Carboxylic acids can be reacted with pentafluorobenzyl bromide after extraction as ion-pairs with symmetrical quaternary ammonium ions at pH 7–9 [62, 63]. The same process can be applied to phenols extracted at more alkaline pH values [26, 27]. Xanthine derivatives [64], barbiturates [65], indomethacin [66] and sulfonamides [28] are other compounds suitable for ion-pair-mediated alkylation. Even very weakly acidic compounds such as the tertiary alcohol chlorthalidone can be reacted as an ion-pair [28].

When the extractive alkylation is carried out from a strongly alkaline aqueous phase, the counter-ion, a quaternary ammonium ion, is extracted into the organic phase as well, as a hydroxide ion-pair. This may improve the reactivtiy of the analyte but sometimes decreases the stability of the alkylated reaction product. The hydroxide will also react with the chloroalkane solvent used [67].

The size and extraction capability of the counter-ion will, as in other kinds of ion-pair extraction, govern the extent of extraction and in many instances the reaction rate and the recovery [65]. This is of particular importance when side reactions occur and the reaction time is critical owing to the limited stability of the alkylating agent [67]. In many instances a low ion-pair distribution ratio is sufficient, since alkylated acid molecules are continuously replaced by the distribution of new ones from the aqueous phase in order to maintain equilibrium.

The technique has also been applied to liquid chromatographic assays. Non-detectable carboxylic acids have been alkylated with the chromophore-containing reagent phenacyl bromide [68]. In a recent paper, 9-bromomethylacridine was used as reagent and the organic solvent was exchanged for a non-ionic surfactant micellar phase [69]. Ion-pairing distribution is also involved in many other processes designed for derivatization for chromatographic separation. An example is the reaction of tertiary amines with chloroformate esters for gas chromatographic analysis [70].

6. Theoretical background for the distribution of ion-pairs

The basic expressions for ion-pair extraction are given by eqns. (1) and (2). The distribution ratio expression (3) derived therefrom is valid if further processes can be disregarded.

However, in many instances the ions or the ion-pair in the organic phase can participate in other equilibria, which have to be taken into consideration in the calculations necessary to arrive at suitable extraction conditions. Side reactions may give increased extraction recovery and selectivity or cause unexpected disturbances in ion-pair extraction processes.

Secondary equilibria such as dissociation and association of the ion-pair in the organic phase are dependent on the concentration of the extracted compound.

Dissociation of the ion-pair increases the degree of

extraction of the cation HA$^+$ into the organic phase [11]:

$$D_{HA} = K_{ex(HAX)}[X^-]_{aq}$$
$$+ (K_{diss(HAX)}K_{ex(HAX)}[X^-]_{aq}/[HA^+]_{aq})^{1/2} \quad (12)$$

where $K_{diss(HAX)}$, the dissociation constant, is defined by

$$K_{diss(HAX)} = [HA^+]_{org}[X^-]_{org}/[HAX]_{org} \quad (13)$$

The expression (11) shows an increases of the distribution ratio with decreasing concentration of the extracted compound HA$^+$.

Association of the ion-pair, such as dimerization and tetramerization, in the organic phase also increases the degree of extraction:

$$D_{HA} = K_{ex(HAX)}[X^-]_{aq} \cdot$$
$$(1 + 2K_{2(HAX)}K_{ex(HAX)}[HA^+]_{aq}[X^-]_{aq}) \quad (14)$$

where the dimerization constant $K_{2\,(HAX)}$ is defined by

$$K_{2\,(HAX)} = [H_2A_2X_2]_{org}/[HAX]^2_{org} \quad (15)$$

In this case the distribution ratio increases with increasing concentration of HA$^+$. The effect is of particular importance in preparative scale work.

The ion-pairing components may be weak protolytes and subject to protolysis in the aqueous phase, and also transferred to the organic phase in uncharged form. An anion of a weak acid will then be protonated and distributed into the organic phase as follows:

$$X^-_{aq} + H^+_{aq} \rightleftharpoons HX_{aq} \xrightarrow{K_{D(HX)}} HX_{org} \quad (16)$$

where K'_{HX} is the acid dissociation constant and $K_{D\,(HX)}$ is the distribution constant. It is obvious that if X$^-$ is the counter-ion the effective concentration of this will decrease owing to side reactions, and so too will the distribution ratio (eqn. 3).

Correspondingly, a cation HA$^+$ of a weak base A may be extracted entirely as the ion-pair HAX at low pH in the aqueous phase, but at higher pH HA$^+$ is protolyzed to A, which form can be extracted into the organic phase. With a simultaneous extraction of the ion-pair HAX and the base A the distribution ratio for A will have the form:

$$D_A = ([A]_{org} + [HAX]_{org})/([A]_{aq} + [HA^+]_{aq}) \quad (17)$$

and

$$D_A = (K_{D(A)}K'_{HA} + K_{ex(HAX)}[X^-]_{aq}a_{H^+})/(K'_{HA} + a_{H^+}) \quad (18)$$

The hydrogen ion activity a_{H^+} will influence the proportions of the species extracted into the organic phase. The extraction conditions will not only determine the distribution ratio but for weak protolytes will also decide whether the ion-pair components are present in equivalent concentrations in the organic phase.

7. Conclusions

Since the previous edition of this handbook, there have been, as we predicted, steady advances in the application of ion-pair chromatographic methods to analytical problems, in the development of new ion-pairing compounds and in the understanding of the underlying physico-chemical mechanisms. These advances are leading to a wider appreciation of the value of ion-pairing techniques among analytical chemists and biochemists, pharmaceutical chemists and, indeed, among chromatographers in general. The attraction lies, of course, in the instantaneous nature of derivatization involving ionic interactions. This widening appreciation and the resulting accounts of successful practical applications will in future years increasingly contribute to the even more widespread use of ion-pair techniques.

8. References

[1] G. Schill, H. Ehrsson, J. Vessman and D. Westerlund, *Separation Methods for Drugs and Related Organic Compounds*, 2nd edn., Swedish Pharmaceutical Press, Stockholm (1984).
[2] B.-A. Persson and B. L. Karger, *J. Chromatogr. Sci.*, **12**, 521 (1974).
[3] J. Crommen, B. Fransson and G. Schill, *J. Chromatogr.*, **112**, 283 (1977).
[4] K. G. Wahlund, *J. Chromatogr.*, **115**, 411 (1975).
[5] H. J. L. Janssen, U. R. Tjaden, H. J. DeJong and K. G. Wahlund, *J. Chromatogr.*, **202**, 223 (1980).
[6] Cs. Horvath, W. R. Melander, I. Molnar and P. Molnar, *Anal. Chem.*, **49**, 2295 (1977).
[7] F. F. Cantwell, *J. Pharm. Biomed. Anal.*, **2**, 153 (1984).
[8] G. Schill, *Acta Pharm. Suec.*, **2**, 13 (1965).
[9] G. Schill and B. Danielsson, *Anal. Chim. Acta*, **21**, 248 (1959).
[10] K. Gustavii and G. Schill, *Acta Pharm. Suec.*, **3**, 259 (1966).
[11] P.-O. Lagerström, K. O. Borg and D. Westerlund, *Acta Pharm. Suec.*, **9**, 53 (1972).
[12] D. Westerlund and K. O. Borg, *Anal. Chim. Acta*, **67**, 89 (1973).
[13] J. F. Lawrence, U. A. Th. Brinkman and R. W. Frei, *J. Chromatogr.*, **203**, 165 (1981).
[14] F. Smedes, J. C. Kraak, C. E. Werkhoven-Goewie, U. A. Th. Brinkman and R. W. Frei, *J. Chromatogr.*, **247**, 123 (1982).
[15] P. Vouros, E. P. Lankmayr, M. J. Hayes, B. L. Karger and J. M. McGuire, *J. Chromatogr.*, **251**, 175 (1982).
[16] L. Hackzell, M. Denkert and G. Schill, *Acta Pharm. Suec.*, **18**, 271 (1981).

[17] B. Ulin, K. Gustavii and B.-A. Persson, *J. Pharm. Pharmacol.*, **28**, 672 (1976).
[18] M. Denkert, L. Hackzell, G. Schill and E. Sjögren, *J. Chromatogr.*, **218**, 31 (1981).
[19] G. Schill and E. Arvidsson, *J. Chromatogr.*, **492**, 299 (1989).
[20] B. A. Bidlingmeyer, C. T. Santasania and F. V. Warren Jr., *Anal. Chem.*, **59**, 1843 (1987).
[21] P. Dorland, M. Tod, E. Postaire and D. Pradeau, *J. Chromatogr.*, **478**, 131 (1989).
[22] A. Tilly-Melin, Y. Askemark, K. G. Wahlund and G. Schill, *Anal. Chem.*, **51**, 976 (1979).
[23] A. Sokolowski and K. G. Wahlund, *J. Chromatogr.*, **189**, 299 (1980).
[24] C. Pettersson and K. No, *J. Chromatogr.*, **282**, 671 (1983).
[25] C. Pettersson and M. Josefsson, *Chromatographia*, **21**, 321 (1986).
[26] H. Brötell, H. Ehrsson and O. Gyllenhaal, *J. Chromatogr.*, **78**, 293 (1973).
[27] A.-M. Tivert and K. Gustavii, *Acta Pharm. Suec.*, **16**, 1 (1979).
[28] M. Ervik and K. Gustavii, *Anal. Chem.*, **46**, 39 (1974).
[29] G. Schill, in *Ion Exchange and Solvent Extraction*, Vol. 6, J. A. Marinsky and Y. Marcus (Editors), Marcel Dekker, New York (1974), pp. 1–54.
[30] R. Modin and S. Bäck, *Acta Pharm. Suec.*, **8**, 585 (1971).
[31] G. Schill, H. Ehrsson, J. Vessman and D. Westerlund, *Separation Methods for Drugs and Related Organic Compounds*, 2nd edn., Swedish Pharmaceutical Press, Stockholm (1984), pp. 1–31.
[32] M. Schröder-Nielsen, *Acta Pharm. Suec.*, **11**, 541 (1974).
[33] M. Schröder-Nielsen, *Acta Pharm. Suec.*, **13**, 145 (1976).
[34] R. Modin and M. Johansson, *Acta Pharm. Suec.*, **8**, 561 (1971).
[35] F. Smedes, J. C. Kraak and H. Poppe, *J. Chromatogr.*, **231**, 25 (1982).
[36] G. Schill, R. Modin and B.-A. Persson, *Acta Pharm. Suec.*, **2**, 119 (1965).
[37] B.-A. Persson, *Acta Pharm. Suec.*, **5**, 343 (1968).
[38] J. H. Knox and J. Jurand, *J. Chromatogr.*, **103**, 311 (1975).
[39] B. A. Persson and P.-O. Lagerström, *J. Chromatogr.*, **122**, 305 (1976).
[40] D. Westerlund, L. B. Nilsson and Y. Jaksch, *J. Liq. Chromatogr.*, **2**, 373 (1979).
[41] I. Grundevik and B.-A. Persson, *J. Liq. Chromatogr.*, **5**, 141 (1982).
[42] B. L. Karger, S. C. Su, S. Marchese and B.-A. Persson, *J. Chromatogr. Sci.*, **12**, 678 (1974).
[43] P.-O. Lagerström, *Acta Pharm. Suec.*, **13**, 213 (1976).
[44] J. Crommen, *Acta Pharm. Suec.*, **16**, 111 (1979).
[45] J. Crommen, *J. Chromatogr.*, **193**, 225 (1980).
[46] P.-O. Lagerström and B.-A. Persson, *J. Chromatogr.*, **149**, 331 (1978).
[47] R. J. Flanagan, C. G. A. Storey and D. W. Holt, *J. Chromatogr.*, **187**, 391 (1980).
[48] P. Haefelfinger, *J. Chromatogr. Sci.*, **17**, 345 (1979).
[49] B.-M. Eriksson, B.-A. Persson and M. Lindberg, *J. Chromatogr.*, **185**, 391 (1980).
[50] B. Fransson, K.-G. Wahlund, I. M. Johansson and G. Schill, *J. Chromatgor.*, **125**, 327 (1976).
[51] K. G. Wahlund and I. Beijersten, *J. Chromatogr.*, **149**, 313 (1978).
[52] S.-O. Jansson, *J. Pharm. Biomed. Anal.*, **4**, 615 (1986).
[53] O. Magnusson, L. B. Nilsson and D. Westerlund, *J. Chromatogr.*, **221**, 237 (1980).
[54] C. Pettersson and G. Schill, *J. Chromatogr.*, **204**, 179 (1981).
[55] S. Eksborg, B.-A. Persson, L.-G. Allgén, J. Bergström. L. Zimmerman and P. Fürst, *Clin. Chim. Acta*, **82**, 141 (1978).
[56] L. Hackzell and G. Schill, *Chromatographia*, **15**, 437 (1982).
[57] J. Crommen, G. Schill, D. Westerlund and L. Hackzell, *Chromatographia.*, **24**, 252 (1987).
[58] J. Ye, R. P. Baldwin and K. Ravichandran, *Anal. Chem.*, **58**, 2337 (1986).
[59] D. R. Bobbitt and E. S. Yeung, *Anal. Chem.*, **56**, 1577 (1984).
[60] J. Paanakker, J. M. S .L. Thio, H. M. van Wildenburg and F. M. Kaspersen, *J. Chromatogr.*, **421**, 327 (1987).
[61] A. Brändström and U. Junggren, *Acta Chem. Scand.*, **23**, 2204 (1969).
[62] H. Ehrsson, *Acta Pharm. Suec.*, **8**, 113 (1971).
[63] J. Vessman and S. Strömberg, *Anal. Chem.*, **49**, 369 (1977).
[64] A. Arbin and P. O. Edlund, *Acta Pharm. Suec.*, **12**, 119 (1975).
[65] H. Ehrsson, *Anal. Chem.*, **46**, 922 (1975).
[66] A. Arbin, *J. Chromatogr.*, **144**, 85 (1977).
[67] H. Brink, R. Modin and J. Vessman, *Acta Pharm. Suec.*, **16**, 247 (1979).
[68] K. Gustavii and A. Furängen, *Acta Pharm. Suec.*, **21**, 295 (1984).
[69] F. A. L. van der Host, M. H. Post, J. J. M. Holthuis and U. A. Th. Brinkman, *Chromatographia*, **28**, 267 (1989).
[70] K. E. Karlsson, *Acta Pharm. Suec.*, **17**, 249 (1980).

12

Derivatization for Fast Atom/Ion Bombardment Mass Spectrometry

Russell C. Spreen
Structural Chemistry Mass Spectrometry Facility, Zeneca Pharmaceuticals Group, Wilmington 19897, Delaware, USA

1. INTRODUCTION 267
2. THE FAB MATRIX: PERFORMANCE CONSIDERATIONS 269
3. DERIVATIVES 270
   3.1 Non-specific methods: sample preparation 270
      3.1.1 Acid–base treatments 270
      3.1.2 Cationization 272
   3.2 Specific derivatization methods: reactions to alter structure 273
      3.2.1 Acetylation 273
      3.2.2 Permethylation 277
      3.2.3 Esterification: methyl esters 278
      3.2.4 Esterification: other esters 280
   3.3 Derivatization for soft ionization MS–MS 280
      3.3.1 Cationization in MS–MS 283
   3.4 Creation of a charged site 284
      3.4.1 ^{18}O exchange reactions 285
      3.4.2 Derivatives for constant neutral loss class screening 285
4. DERIVATIZATION AND CONTINUOUS FLOW FAB: FUTURE PROSPECTS 286
5. CONCLUSIONS 286
6. REFERENCES 286

1. Introduction

Prior to the 1980s, mass spectrometry (MS) was generally limited to the analysis of small volatile compounds or compounds that could be vaporized with moderate heating. In order to obtain mass spectral information on involatile, thermally labile compounds (compounds which decompose prior to vaporization), one could either alter the physical properties of the sample by derivatization, or attempt direct analysis through the use of specialized mass spectral techniques. The rationale behind the direct approach was either to optimize the experimental parameters to improve the vaporization/decomposition ratio, or to eliminate the vaporization requirement altogether by desorbing ions directly from the sample.

Among the early approaches to the vaporization problem was the proposal by Beuhler and co-workers [1] that the vaporization/decomposition ratio could be substantially increased by rapid heating of the sample. Dell *et al.* showed that, by minimizing the distance between the vaporization and ionization sites through the use of a special 'in-beam' probe, significant improvements in the electron ionization (EI) spectra of thermally labile compounds could be obtained [2]. Baldwin and McLafferty observed a similar enhancement when samples were introduced directly into the plasma within a chemical ionization (CI) source [3]. The logical combination of both rapid heating and in-beam techniques quickly followed, and heatable in-beam probes became commercially available for most spectrometers at relatively low cost.

In the intervening period, in-beam techniques rapidly grew in popularity because of their ease of use and relatively good performance for moderately labile compounds of low volatility. Sample loading and probe cleaning were easily accomplished, making the operational aspects attractive for routine MS analysis. The performance of in-beam techniques was somewhat less

Handbook of Derivatives for Chromatography
Edited by K. Blau and J. M. Halket © 1993 John Wiley & Sons Ltd

attractive in the analysis of highly polar involatile compounds. Many compounds of biological interest, especially those with strong hydrogen bonding characteristics and high molecular weights, are fragile, and could not tolerate the conditions necessary for vaporization. In these instances, desorption of ions from the condensed phase appeared to be a requirement for direct mass spectral analysis.

Field desorption (FD), pioneered by Beckey in 1969 [4], was the first and clearly the most successful of the early desorption ionization techniques. In the FD experiment, very high electric fields were used to extract ions from sample-coated thin wire emitters. FD spectra normally contained molecular weight information; however, structural fragments were often absent and signal instability resulted in data acquisition difficulties. As an added complication, FD was experimentally difficult and the method by which ions were formed was not well understood.

Subsequent to the discovery of FD, notable success has been achieved in the use of lasers [5], and ^{252}Cf fission fragments [6] to desorb ions from sample-coated metal surfaces. Similar advances have also been realized in the desorption of ions from liquids using electrohydrodynamic ionization [7]. In all of these desorption ionization techniques, the internal energy imparted to the sample is low ($\ll 1$ eV) and the resulting soft ionization causes minimal fragmentation of the sample ions.

Although a number of innovative desorption ionization methods have been developed, none of these techniques has been commonly employed throughout the mass spectrometry community owing to the nonroutine nature of the experiments and the need for specialized instrumentation and specially trained personnel. In addition, none of the techniques offered the overall performance desired, particularly in terms of information content, spectral reproducibility, and signal lifetime.

When faced with the MS analysis of a moderately labile sample, derivatization followed by conventional MS (EI or CI) often proved to be more practical than the specialized MS techniques. The relatively low cost and subsequent availability of this approach served to bolster the development of sensitive derivatization techniques for use in probe and gas chromatography–mass spectrometry (GC–MS) experiments. The gains in volatility and thermal stability available through derivatization substantially extended the applicability of mass spectrometry. However, as sample molecular weight and thermal instability increased, derivatization was no longer able to provide the increased performance observed at low mass (< 1500 daltons). This problem was further compounded by the net increase in sample molecular weight associated with most derivatization procedures.

A new approach to the analysis of involatile thermally labile compounds emerged in 1981, with the discovery of fast atom bombardement (FAB) by Barber et al [8]. In the FAB experiment, a sample/matrix-coated surface was bombarded with fast atoms (8–10 kV) and both positive and negative ions were sputtered into the gas phase. The matrix consisted of a low volatility liquid such as glycerol or thioglycerol (3-mercapto-1,2-propanediol); this was the key in differentiating the technique from traditional secondary ion mass spectrometry (SIMS), where dry samples and ion rather than neutral bombardment was used. A schematic diagram of the FAB technique is shown in Figure 1.

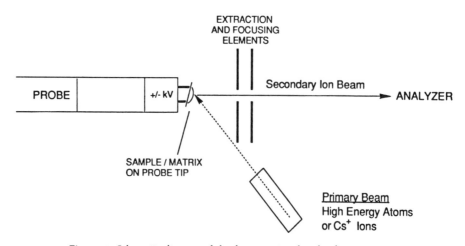

Figure 1. Schematic diagram of the fast atom/ion bombardment process.

In contrast to many of the earlier techniques, FAB capabilities could be easily added to existing instrumentation, operation was routine, and the spectra were normally stable and long lived (minutes). The FAB sputtering process provided a combination of both vaporization and ionization in one step, and was particularly well suited for the analysis of highly polar molecules up to mass 10 000. The technique routinely provided molecular weight and some structural information. Recent studies using very high energy Cs^+ ions (≈ 30 kV) in place or 8 kV fast atoms have allowed extension of the technique to molecules in excess of molecular weight 20 000 [9, 10]. This modification of the SIMS technique is commonly referred to as liquid SIMS (LSIMS) because a liquid matrix is used.

With the advent of FAB, derivatization was usually not required to obtain a molecular ion, $[M + H]^+$ or $[M - H]^-$, from an involatile thermally labile compound. However, the overall information yield in the FAB spectrum of the native compound frequently fell short of what was required for structure identification. Structurally significant fragment ions were often absent from the FAB spectrum, and low abundance fragments from the sample could be difficult to differentiate from the collection of ions originating from the liquid matrix. In addition, overall sensitivity varied widely with compound structure and sample purity. In these circumstances, derivatization has proved to be highly effective method for tailoring the response of a sample compound in order to maximize the yield of structural information.

In FAB experiments, derivatization is used mainly for signal enhancement, molecular weight shifting, or altering the fragmentation properties of a compound. By changing the polarity, hydrophobicity or gas phase basicity of a compound, substantial gains in mass spectral performance may be realized. Derivatization reactions can be specific, and the resulting shifts in the mass spectral pattern can be used for detection and counting of various functional groups.

It may at first appear at variance with the rest of the handbook to discuss derivatization for fast atom/ion bombardment mass spectrometry, where chromatography may be only incidentally rather than directly involved. Nevertheless, it seems appropriate to include this aspect because these techniques are complementary to the electron impact and chemical ionization methods used with gas chromatography–mass spectrometry (GC-MS) as described in Chapter 14, and thus extend the coverage to include derivatization methods for the majority of mass spectrometric techniques.

The primary focus of this chapter will be to illustrate the application of various derivatization techniques in fast atom/ion bombardment mass spectral studies. Rather than restate derivatization procedures which have been thoroughly reviewed in other chapters, only brief procedures will be presented. Emphasis will be placed on derivatization reactions which are commonly used. In general, these are fairly simple procedures which are performed by the mass spectroscopist prior to analyzing a sample. Where convenient, illustrative examples have been included. The well behaved test compound leucine-enkephalin has been selected for demonstration of many of the derivatization methods. The spectra presented are all obtained from reaction mixtures without additional chromatographic clean-up.

The term FAB is generically used throughout the chapter to represent both the FAB MS and the LSIMS technique, which appear to be highly similar in performance below mass 3000. As the major application of desorption techniques is to biological samples, especially peptides, the orientation of the chapter naturally lilts in this direction. It should be noted, however, that many of the derivatizations may serve as models, or be applied equally well in the mass spectral analysis of non-biological samples.

Atom/ion bombardment techniques are relatively new in organic MS, and the understanding of the effects of various experimental parameters is actively being refined. This chapter will serve to outline the current understanding of the technique and provide a basis for the logical application of derivatives in FAB mass spectrometry studies.

2. The FAB matrix: performance considerations

The use of a liquid matrix is perhaps the key discovery in the development of atom/ion bombardment ionization techniques for organic molecules. The matrix provides a medium through which sample concentration at the ionization surface can be replenished as sample ions are sputtered away. Barber's group has shown that, in the absence of a matrix, organic samples produce a transient signal for only 10 ro 15 seconds, compared with several minutes when a liquid matrix is used [8]. Considering that the matrix must be present for an extended period of time in the high vacuum of a mass spectrometer, low volatility, relatively unreactive compounds such as glycerol, thioglycerol, diethanolamine etc. are commonly used. Given the inherent simplicity of the FAB technique, the matrix is one of the few experimental parameters which can easily be changed on an experiment to experiment basis. In general, as the volatility of the matrix increases, the lifetime of the mass spectral signal decreases.

Reviews of various liquid matrix compounds can be found in the literature [11, 12].

One role of the matrix is to act, in essence, as a sample reservoir. The sample must be soluble in the matrix to enable effective diffusion of sample molecules to the surface under analysis. Samples which are insoluble in the matrix normally yeild spectra of marginal quality and substantially reduced intensity. Depending on the degree of insolubility, the resulting spectra may consist almost entirely of ions originating from the matrix itself rather than the sample. In these circumstances, improved solubility and spectral performance can often be obtained by adding acids or bases, substituting another matrix, or derivatizing the sample.

Another important factor in determining the FAB MS performance of a compound is the surface activity. Compounds which display hydrophobic character typically perform well in most of the commonly used matrix compounds (which are generally hydrophilic) because the sample tends to migrate to the surface being analyzed. The degree of hydrophobicity is normally thought to be consistent with surface activity, which is a generalized term describing how efficiently sample molecular ions can be sputtered into the gas phase. Signal suppression effects are commonly observed in mixtures of compounds of varying hydrophobicity because some components of the mixture have a substantially greater surface activity than others and efficiently displace less hydrophobic molecules from the surface. Naylor *et al.* have shown that there is a striking correlation between hydrophobicity (calculated from the Bull and Breese Index [13]) and molecular ion abundance in the analysis of small peptides [14]. Biemann and Martin have noted that gradient reversed phase HPLC serves as a convenient means of splitting a mixture into fractions of similar hydrophobicity [15]. By fractionating the sample prior to mass spectral analysis, suppression effects can be minimized. In a typical gradient reversed phase HPLC system (C_{18}, C_8 or C_4 column, water/acetonitrile/0.1% trifluoroacetic acid solvent system) for separation of a proteolyte digest, the early eluting (almost 100% water) peptides are generally more hydrophilic and are often difficult or impossible to characterize by FAB MS without prior derivatization [16].

Charge-competition effects are also important in determining FAB mass spectral performance. In positive-ion studies, compounds with high gas phase basicities will effectively suppress the signal of compounds with lower gas phase basicities [17]. Therefore, depending on the analysis mode (positive or negative), the matrix must function as either a proton donor or acceptor. In order to achieve the proper balance, relatively basic matrices (e.g. triethanolamine) are often used for negative-ion studies, and mildly acidic matrices (e.g. thioglycerol) are used in the positive-ion mode. By selective derivatization, one may substantially alter the gas phase basicity of a mixture component in order to obtain enhanced sensitivity.

3. Derivatives

The various derivatization methods for desorption MS can be conveniently divided into non-specific methods, which are generally applicable, and specific methods, which are somewhat compound- or class-dependent. Specific methods normally require that the compound be chemically modified through some type of derivatization reaction. A collection of specific derivatization reactions used in FAB MS experiments is briefly summarized in Table 1. In order to utilize sample efficiently, one must have some knowledge concerning the reactivity of the analyte before selecting a specific derivatization method. In contrast, the simplicity and general applicability of non-specific methods facilitate their use in the analysis of unknown samples. In fact these techniques fall on the boundary line between derivatization and routine sample preparation, because of the lack of structural alteration in the condensed phase.

3.1 Non-specific methods: sample preparation

3.1.1 *Acid–base treatments*

The simplest non-specific derivatization reaction for FAB MS involves the addition of acid or base to a sample. In positive-ion experiments, acetic, oxalic, *p*-toluenesulfonic or dilute (0.1 N) hydrochloric acids are often added to improve signal intensity. Similar increases in sensitivity have been observed in negative-ion studies when bases such as sodium, potassium or tetramethylammonium [18] hydroxide are added to the sample. The alteration of sensitivity with pH has been attributed both to increased sample solubility and to enhanced ion production in solution and in the gas phase.

As noted previously, the solubility of a sample in the liquid matrix has been shown to have a major influence in determining FAB sensitivity. If the analyte is not soluble, effective diffusion of sample molecules to the surface of the matrix will not occur, and the resulting spectrum will be dominated by matrix ions. In order to improve solubility in the matrix, the sample is often first dissolved in a mutually compatible co-solvent (e.g. methanol, dimethyl sulfoxide, water, etc.) and acid or base is added. A few microliters of this solution is then

Table 1 Applications of derivatives for fast atom/ion bombardment mass spectrometry

Reactive site	Derivatization	Comments	Reference
Free primary amines e.g. Peptide N-terminus	Acetylation	May be performed on FAB probe tip Commonly used with H_3/D_3 labeling Increases hydrophobicity Delocalizes charge Commonly followed by methyl esterification Useful for counting free amine groups Peracetylation possible with addition of base such as pyridine	23, 43
	Formation of dodecyl Schiff base	Increases hydrophobicity	18
	Dansylation	Increases hydrophobicity Adds chromophore for HPLC	23
	Amidination	Converts N-terminus to highly basic moiety for MS–MS fragmentation steering	52
	β-Guanidinopropionylation	Converts N-terminus to highly basic moiety for MS–MS fragmentation steering	52
	N-methyl quaternization	Generates charged species for enhanced sensitivity Highly directed MS–MS fragmentation	21, 58
	N-acylation with thioalkyl trimethylammonium salt	Generates charged species for enhanced sensitivity Highly directed MS–MS fragmentation Alkyl chain can be adjusted to alter hydrophobicity	51
Lysine ε-amino group	Conversion to 2,4,6-trimethylpyridinium cation	Highly localized charge for fragmentation steering Enhanced sensitivity via pre-formed ion	49, 50
Glutathione N-terminus	Addition of ethoxy or benzyloxycarbonyl	Normally combined with methyl esterification. Produces strong diagnostic (-89 u) loss for MS–MS constant neutral loss scanning Improved HPLC performance	
Free carboxyl groups e.g. peptide C-terminus	Methyl esterification	Commonly used with H_3/D_3 labeling May be performed on FAB probe tip Increases hydrophobicity Useful for counting free carboxyl groups	44, 29
	Isopropyl esterification	Increase hydrophobicity	59, 44
	n-Butyl esterification	Increase hydrophobicity	23
	Hexyl esterification	Increase hydrophobicity	16
	^{18}O-incorporation	Useful in fragmentation and HPLC studies	53
N–H, O–H, S–H groups e.g. peptides and carbohydrates	Permethylation of peptides by modified Hakomori process	Increase hydrophobicity Gives well defined ionic fragments	39, 43
	Permethylation of carbohydrates Cincanu and Kerek process	Increase hydrophobicity Gives well defined fragments Works very well at high mass when doped with NH_4^+	42

(continued)

Table 1 (*continued*)

Reactive site	Derivatization	Comments	Reference
Aldehydes, ketones and ketosteroids (carbonyl moiety)	Formation of immonium salt	Enhanced sensitivity via pre-formed ion	21
	Girard's reagents for formation of hydrazones May be used for selective detection of steroids	Enhanced sensitivity via per-formed ions	58
Sugars	Coupling of Girard's reagent	Enhanced sensitivity via pre-formed ions	60
Triols, sugars, nucleosides, etc.	Reaction with boronic acid to form negatively charged boronate cage compounds	Enhanced sensitivity in the negative-ion mode via pre-formed ions	61

added to the matrix. Acid or base may also be added directly to the sample/matrix mixture on the FAB probe tip.

In addition to solubility, pH also has an important function in other aspects of the FAB ionization process. In SIMS experiments, Benninghoven and Sichtermann [19] observed substantial enhancement of the $[M + H]^+$ signals of amino acids upon acidification, and correspondingly increases in $[M - H]^-$ ion yields with basification. The addition of acid can also be particularly advantageous if the matrix does not contain labile hydrogens for sample protonation [20]. In such analyses, the tendency to form $[M + H]^+$ ions rather than the cationized species $[M + Na]^+$ and $[M + K]^+$ may be optimized through the addition of excess acid.

A number of studies suggest that the role of acid or base in pre-forming ions in solution may be partly responsible for enhanced sensitivity in FAB MS experiments [21]. The association of enhanced sensitivity with the abundance of pre-formed ions originates mainly from early desorption studies, in which derivatives exhibiting strong ionic character showed superior performance relative to the underivatized analogs. The sensitivity enhancement may be due to elimination of the ionization process resulting from the desorption of ions directly from the sample. The relative roles of pre-formed ions versus gas phase ionization processes (ion–molecule reactions) in desorption mass spectrometry is controversial and under study [17, 22].

The spectral changes resulting from the addition of acid to a sample are variable, and depend on the acid used. Naylor and Moneti [23] have suggested that the effect is sensitive to the volatility and gas phase acidity of the acid. Experiments run with sulfuric or perchloric acid doping of the matrix show dramatically increased fragmentation in peptides due to the large excess of protons in the sample during exposure to the atom/ion beam. Under these conditions basic sites are readily protonated and fragmentation is accelarated. The effect of hydrochloric or acetic acid doping on fragmentation is minimal, presumably because the concentration of these acids decreases rapidly when the sample is exposed to high vacuum.

Although addition of acid or base is common practice in FAB MS, one must consider the stability of the analyte in overly acidic or basic solutions. For example, acid could promote hydrolysis of glucuronide or sulfate ester conjugates. The resulting analysis would be erroneous owing to the detection of hydrolysis products rather than the spectrum of the intact conjugate.

3.1.2 Cationization

Alkali metals such as sodium and potassium are common in many samples. Samples isolated from biological fluids, such as urine or bile normally contain high levels of alkali metal salts. In addition, low levels of sodium and potassium can easily be introduced into a sample through exposure to glassware. Alkali ion contamination is generally undesirable because of the potential for interference and suppression of the sample signal. In cases where a sample molecular ion $[M + H]^+$ is observed, satellite peaks corresponding to $[M + Na]^+$ and $[M + K]^+$ are often observed as well, and may actually dominate the spectrum.

In highly polar molecules the proton and cation affinities of a specific functional group or binding site may be different; thus, the sites of proton and action attachment are not necessarily identical. Given these consideration, it may be advantageous to promote cationization through the addition of an alkali metal salt such as LiI, NaCl or KCl in order to alter the fragmentation of the sample molecule [24, 25]. In contrast to protonated $[M + H]^+$ species, where the positive charge is often delocalized, cationized species tend to exhibit charge localization. Röllgen and Schulten have shown that cation attachment usually results in the formation of $[M + \text{cation}]^+$ ions which are more stable than the corresponding $[M + H]^+$ ions [26]. The stability of cationized molecular ions, $[M + \text{cation}]^+$, increases with increasing cation size, and fragmentation requires a larger activation energy relative to the $[M + H]^+$ species because of the necessity of shifting the charge. The increased stability makes the use of alkali metal adducts particularly advantageous for molecular weight determinations. An example of the fragmentation of cationized versus protonated molecular ions is shown in the MS–MS section (Section 3.3) of this chapter.

Cationized molecular ions can easily be formed through the addition of a few microliters of a dilute alkali halide solution to the sample/matrix solution on the FAB probe tip. Addition of lithium, sodium or potassium produces $[M + \text{cation}]^+$ ions with corresponding shifts of 6, 22 and 38 daltons compared with the mass of the protonated molecular ion $[M + H]^+$. In some samples higher mass species are observed, owing to the incorporation of additional cations through cation/hydrogen exchange reactions.

Figure 2 shows an example of the effect of alkali metal ion contamination on a typical FAB mass spectrum. In Figure 2A the molecular ion region of a spectrum obtained from 1 μg of leucine-enkephalin dissolved in glycerol is shown. The base peak at m/z 556 is the protonated parent and the minor fragment ions at m/z 540/541 (not labeled) presumably originate from the loss of CH_3 and CH_4. Figures 2B and 2C were obtained from identical 1 μg samples of leucine-enkephalin, except that 1 μl of 0.1 M NaCl was added in spectrum 2B, and 1 μl of 0.1 M KCl was added in spectrum 2C. Total suppression of the $[M + H]^+$ signal, with an overall reduction in the signal-to-noise level, is evident in each of the spectra where the alkali salt was added. Because of the large molar excess of alkali halide (mole ratio alkali halide/leucine-enkephalin $\approx$ 50:1), both cationization and alkali metal/hydrogen exchange are evident. Spectrum 2B shows a base peak at m/z 600, $[M - H + 2Na]^+$ corresponding to cationization and the exchange of one hydrogen for sodium. The peaks at m/z 578 and 622 correspond with $[M + Na]^+$ and $[M - 2H + 3Na]^+$. The ions at m/z 585 and 607 (not labeled) correlate to the loss of 15 daltons from m/z 600 and 622 respectively, analogous to the fragment ion at m/z 541 in spectrum 2A. Similarly, the base peak in spectrum 2C corresponds to $[M - H + 2K]^+$. The different background levels in spectrum 2B versus 2C suggest different supression efficiencies or surface activities for NaCl versus KCl.

3.2 Specific Derivatization Methods: Reactions to Alter Structure

These derivatization methods tend to be somewhat more traditional, in that they focus on the chemical modification of the analyte. By changing a functional group, or adding a chemical moiety, one may be able to change the nature of the analyte so as to increase the FAB MS signal or alter the fragmentation. Given the ability of the mass spectrometer to differentiate between isotopes, the combination of derivatization and isotopic labelling can provide a wealth of informaion not available in the mass spectrum of the underivatized compound. The most commonly used derivatizations are reviewed in the following text. A more comprehensive list of derivatizations for FAB MS is presented in Table 1, along with appropriate references.

3.2.1 Acetylation

Acetyl and peracetyl derivatives are commonly used in the FAB mass spectral analysis of many classes of compounds, including peptides, oligosaccharides and glycoconjugates. The rationale for this derivatization is to facilitate the desorption of ions from the sample/matrix surface by increasing hydrophobicity and reducing the degree of hydrogen bonding. For this reason acetyl and peracetyl derivatives typically display greater FAB sensitivity than the native compound. For example, Naylor and Moneti have shown that acetylation uniformly increases $[M + H]^+$ and fragment ion yields in a series of three peptides (angiotensin I, bradykinin, and delta sleep-inducing peptide, DSIP [23].

Acetylation is one of the most frequently used derivatization techniques in the mass spectral analysis of peptides. Reaction with freshly distilled acetic anhydride in methanol [27] results in acetylation of α-amino functionalities within one minute. Acetylation at other less reactive sites may be accelerated by the addition of a small amount of base, such as $NaHCO_3$ or triethylamine [28]. Peptide hydroxyl and amino groups can be acetylated by exposure to a 1:1 mixture of acetic anhydride/pyridine for about 40 minutes. The net result of the reaction is a positive shift of 42 daltons for each acetyl group addition. The failure of an acetylation reaction in a peptide analysis may be diagnostic of a blocked N-terminus.

N-Acetylation of a peptide is normally performed by adding a threefold excess of a 3:1 methanol/acetic

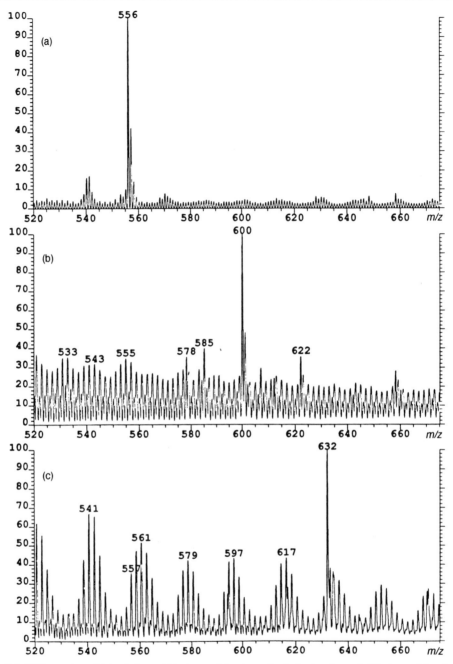

Figure 2. Partial FAB spectra obtained from (A) 1 µg of leucine-enkephalin in glycerol; (B) 1 µg of leucine-enkephalin + 1 µl NaCl (0.1 M) in glycerol; (C) 1 µg of leucine-enkephalin + 1 µl KCl (0.1 M) in glycerol.

anhydride solution to the peptide dissolved in the minimum volume of a compatible solvent such as water. The reaction is allowed to proceed for a few minutes, then the mixture is dried and reconstituted in a suitable solvent (water, methanol, DMSO) for FAB analysis. Micro scale acetylation may be performed on the FAB sampling probe by adding a few microliters of acetic anhydride to a small droplet of sample solution just prior to the FAB MS analysis. N-Acetylation may also be performed directly on a sample dissolved in a liquid matrix without adverse effects.

Consider for example the pentapeptide leucine-enkephalin. A partial FAB mass spectrum of leucine-enkephalin is shown in Figure 3. Upon treatment with acetic anhydride/methanol, the compound undergoes rapid N-acetylation. The spectrum of the N-acetylated product is shown in Figure 4. A single acetylation at the N-terminus results in a positive mass shift of 42 daltons in the molecular weight of the compound to yield the $[M + H]^+$ ion at m/z 598. A corresponding mass shift of 42 daltons is observed in the mass of all fragment ions containing the N-terminus. Fragment ions which do not contain the N-teminus are not shifted in mass. The ions at m/z 620 and 636 correspond to $[M + Na]^+$ and $[M + K]^+$, and originate from trace level contamination of the sample by sodium and potassium.

In order to simplify further the identification of the N-terminus-containing ions, Hurt and Morris have suggested the use of stable isotopes [29]. If the N-acetylation of leucine-enkephalin is performed with a 1:1 mixture of H_6 and D_6 acetic anhydride in methanol, the mass spectrum of the product consists of doublets owing to the equal probability for incorporation of CH_3CO and CD_3CO. The corresponding positive mass shift of 42 or 45 daltons is due to addition of the H_3- or D_3-acetyl group respectively. Where multiple sites for acetylation are possible, overlapping H_3/D_3 doublets would lead to the formation of multiplet patterns in the mass spectrum, with component peaks appearing at 3 dalton intervals. The distribution of ion intensities in the pattern can be used to deduce the number of sites of acetylation (1:1 = one acetylation; 1:2:1 = two acetylations; 1:3:3:1 = three acetylations, and so on). The number of acetylations determined from the isotope pattern should be consistent with the shift in molecular weight.

An example of the FAB mass spectrum of the H_3/D_3 N-acetylation product of leucine-enkephalin is shown in Figure 5. The ions in each doubet contain the N-terminus of the peptide and correspond to analogous ions 42 and 45 daltons lower in the spectrum of the underivatized compound (Figure 3). Comparison of Figures 4 and 5 shows that the relative intensity of N-terminus-containing versus non-N-terminus-containing ions is roughly halved in the labelling experiment because of splitting of the signal between the H_3 and D_3 analogs.

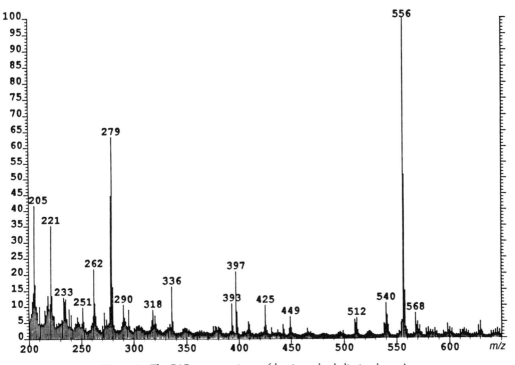

Figure 3. The FAB mass spectrum of leucine-enkephalin in glycerol.

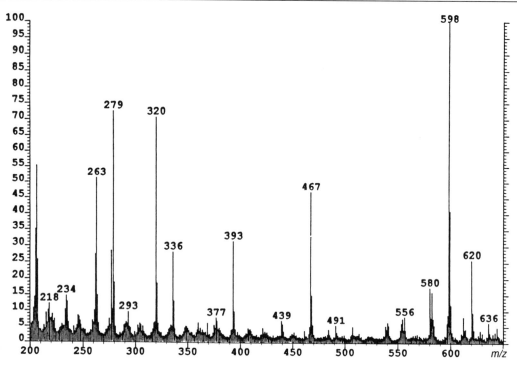

Figure 4. The FAB mass spectrum of N-acetyl leucine-enkephalin in glycerol.

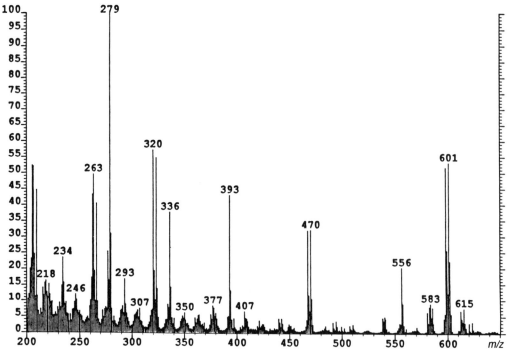

Figure 5. The FAB mass spectrum of 1:1 H_3/D_3 N-acetyl leucine-enkephalin in glycerol + 1 µl of oxalic acid-saturated water.

From these data one may easily locate the various N-terminus-containing and non-N-terminus-containing ions as shown below.

Ions not containing N-terminus
(no shift)
279
336
393

Ions containng N-teminus	
underivatized	H_3/D_3 N-acetylated
221	263/266
278	320/323
397	439/442
425	467/470
556	598/601

Based on these data, the interpretations shown in Figure 6 are proposed. The data are fully consistent with the work previously reported by Gaskell et al. [30].

The spectrum of the H_3/D_3 N-acetylated peptide shows an interesting ion at m/z 556 which is very small in the spectrum of the N-acetylated parent (Figure 4). This ion corresponds to protonated leucine-enkephalin, and probably results from partial N-deacetylation caused by the oxalic acid present only in the H_3/D_3 sample. In the previous edition of this handbook such deacetylations were noted to occur under acidic conditions [31]. Therefore, in addition to selecting the appropriate derivative, one must consider the reactivity of the matrix and matrix additives in the experimental design. A detailed review of chemical reactions accompanying desorption mass spectrometry has been published by Detter et al. [32].

Acetylation is also useful in the negative-ion mode. Peter-Katalinic et al. [33] have reported the advantages of peracetylation on the FAB target in the analysis of sulfated phlorotannins. In this study, micro scale peracetylation of hydroxy groups on aromatic carbons was performed using the direct addition of $2\,\mu l$ each of pyridine and acetic anhydride to the analyte. Derivatization on the FAB probe tip facilitates rapid mass spectral analysis and subsequently minimizes the time available for the decomposition of an unstable derivative.

Dell and co-workers have demonstrated the further application of peracetylation as a means of sample clean-up [34]. Small quantities of oligosaccharides were perfluoroacylated using trifluoroacetic anhydride and acetic acid. The resulting solubility change allowed subsequent desalting of the reaction product by extraction into chloroform followed by washing with water. Dell et al. have also reported peracetylation to be advantageous in the HPLC purification of oligosaccharides prior to FAB MS analysis [35].

3.2.2 Permethylation

Permethylation has been used extensively as a means of increasing the vapor pressure of polar compounds for GC–MS or probe MS analyses. By substituting $-NCH_3$, $-OCH_3$ or $-SCH_3$ for $-NH$, $-OH$ and $-SH$, hydrogen

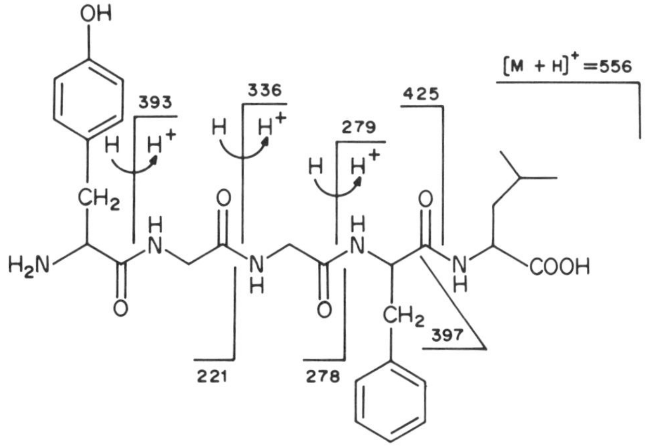

Figure 6. The FAB mass spectral fragmentation of leucine-enkephalin.

bonding is reduced and a net increase in vapor pressure is typically observed. With respect to the desorption ionization process, one would expect the surface activity and subsequent mass spectral sensitivity to increase with permethylation due to increased hydrophobicity. In some cases permethylation enables FAB as well as complementary EI and CI mass spectra to be obtained from an involatile, thermally labile sample.

One common application of permethyl derivatives is in the analysis of glycolipids, oligosacchardes and peptides. A popular derivatization procedure was originally developed by Hakomori [36] and later extened to peptide analysis by Vilkas and Lederer [37] and by Thomas [38]. The reaction involves exposure of the peptide to the methylsulfinyl carbanion base, followed by addition of methyl iodide. Through serial additions of CD_3I and CH_3I, Morris has shown that methylation occurs instantaneously at almost all anioic sites except the carboxylate anion, which reacts slowly [39]. One of the problems with the technique as originally presented was salt formation due to quaternization of histidine; however, Morris has shown that, by limiting reaction time to 1 min, salt formation can be prevented. This modification has the added advantage that accurate balancing of reagent quantities is not critical as long as a substantial excess is present. Should the carboxylate sites not be completely methylated, follow-up esterification with methanolic HCl can be performed.

The modified Hakomori reaction consists of dissolving approximately 1 mg of sample in 100 μl of a 1 M solution of NaH in dimethyl sulfoxide. Approximately 35 μl of methyl iodide is added, the reaction vessel is capped and agitated for 60–75 s, and reaction is immediately quenched by addition of 250 μl of water and 250 μl of chloroform. The chloroform layer is removed, washed repeatedly with water and evaporated to yield the permethyl derivative.

The Hakomori method has also been used extensively in the methylation of carbohydrates; however, yields were typically low and non-sugar products were observed. Subsequent studies by Ciucanu and Kerek [40] led to the conclusion that OH^- and H^- were more effective bases than the methylsulfinyl carbanion. These authors developed a new method for the permethylation of carbohydrates using CH_3I and solid NaOH in DMSO. This reaction is relatively clean, proceeds to completion in about 6–7 min, and gives very high yields ($\approx 98\%$). The derivatization may also be performed with KOH, NaH (excess) and sodium t-butoxide. Ciucanu and Kerek have extensively reviewed the experimental parameters and have published a table of procedures, reagents, reaction times and yeilds for various other versions of the permethylation reaction [40].

The literature shows many applications of permethylated derivatives for both FAB and traditional mass spectral ionization techniques. In the case of oligosaccharides, permethylation serves to direct fragmentation along well defined pathways [41]. Recently Vath, Domon and Costello have reported that micro scale (nmol level) permethylation of gangliosides using the Ciucanu and Kerek procedure resulted in a fiftyfold increase in mass spectral sensitivity relative to the underivatized ganglioside [42]. Permethylation of peptides, especially when combined with acetylation, leads to substantial simplification of the spectra because the resulting derivatives fragment almost exclusively at peptide bonds [43]. Permethylation of peptides without prior acetylation typically results in quaternization of the N-terminus due to reaction of methyl iodide with the tertiary amine. As will be discussed in the MS–MS section, incorporation of a positive charge has a very strong effect in directing fragmentation.

3.2.3 Esterification: Methyl Esters

Simple methyl ester formation can be accomplished by exposure of the sample to methanolic HCl. In the case of peptides all free carboxyl groups (COOH) are converted to methyl esters. As a result, the hydrogen bonding capacity of the compound is substantially reduced and hydrophobicity and surface activity typically increase. In addition to increasing the FAB performance of the compound, the resulting mass shift is diagnostic of the number of free carboxyl groups present in the analyte.

Figure 7 shows the spectrum obtained from the methyl ester of leucine-enkephalin. The ester was prepared by dissolving approximately 300 μg of dry peptide directly in 150 μl of a fresh solution of HCl-saturated methanol (pH $\approx$ 1). The reaction vial was sealed and heated at 42 °C for 3 h, and the mass spectrum was obtained from 1 μl of the reaction mixture ($\approx 2\mu$g of the derivative) dissolved in glycerol.

The spectrum of the leucine-enkephalin methyl ester shows a series of diagnostic 14 dalton shifts for the C-terminus-containing fragment ions. In addition, the molecular ion has shifted positively by 14 daltons, consistent with the single reactive cabonyl group in the molecule. In contrast to the previously shown N-acetylation results, methyl esterification produces shifts in the complementary set of ions as shown below. By using N-acetyl and methyl ester derivatives one can obtain sequence information from both the N- and C-terminus of a peptide, while at the same time improving FAB mass spectral performance.

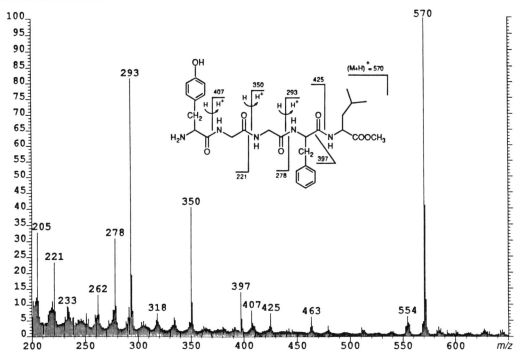

Figure 7. The FAB mass spectrum of the methyl ester of leucine-enkephalin in glycerol.

Ions not containing C-terminus	Ions containing C-terminus	
(no shift)	underivatized	methyl ester
221	297	293
278	336	350
379	393	407
425	556	570

Methyl esterification does not appear significantly to alter the abundance of fragment ions relative to the molecular ion. Comparison of the methyl ester spectrum (Figure 7) with the spectra of underivatized or N-acetylated leucine-enkephalin (Figure 3 and 4) shows that the peak heights of analogous ions are constant within a factor of two. These results are consistent with the general statement by Naylor and Moneti [23] that esterification may result in increased ion abundances relative to backgrounds, but fragmentation of the molecular ion is not substantially increased.

Hunt and Morris [29] have incorporated isotopic labeling into the technique in order to aid in the interpretation of the resulting spectra. By treating the sample with a 1:1 mixture of CH_3OH/CD_3OH plus HCl, all products of the derivatization will now show replacement of the COOH moiety with equal amounts of $COOCH_3$ and $COOCD_3$. In the case of an analyte containing a single carboxyl group, the spectrum of the derivatization will show a doublet separated by 3 daltons and shifted positive by 14 and 17 daltons. When multiple esterification occurs the patterns will build on one another, creating, for example, a 1:2:1 triplet when two carboxyl groups are present and a 1:3:3:1 quartet when three carboxyl groups are present in the analyte. The presence of the proper cluster as well as the mass shift doubly confirms the number of esterification sites.

Interpretation of the mass spectra obtained from samples with multiple esterification sites is often complex because of overlapping clusters and intermediate, partially esterified products. Depending on the number of free carboxyl groups and the extent to which the derivatization reaction goes to completion, the effectiveness of this approach in determining the structure of an unknown can very substantially. Complex spectra of esterified samples can provide valuable information when a complementary spectrum of the underivatized compound is available for comparison. In addition, MS–MS can be used to provide individual spectra of the various mixture components.

3.2.4 Esterification: other esters

Recently Falick and Maltby [16] have reported the use of alkyl and benzyl ester derivatives in the FAB mass spectral analysis of peptides. A series of ester derivatives was evaluated, and the n-hexyl ester derivative was reported as the optimum from the standpoint of sensitivity and ease of preparation. Hexyl ester derivatives were shown to give substantially higher FAB sensitivity than the corresponding methyl esters (about 25 times more for the test peptide Thr-Lys-Pro-Arg), presumably because of the greater hydrophobicity imparted by the C_6 alkyl chain. Hexylation generally increased the hydrophobicity and FAB performance of hydrophilic peptides, but only minor improvements in performance were observed for hydrophobic peptides. This 'leveling effect' serves roughly to equate the hydrophobicity of all of the peptides in a mixture such that surface activity and FAB performance become similar. For some compounds, intact molecular ions of hexylated hydrophilic peptides were observed where no signal was obtained from the underivatized analog.

Esterification is easily accomplished by any of the standard techniques. Falick and Maltby have produced a variety of esters by addition of 5 ml of the dry alcohol, rendered 0.2 M in HCl by addition of acetyl chloride, to the dry peptide and heating the reaction mixture at 45 °C for 1 h. Naylor et al. produced isopropyl esters of peptides by dissolving the peptide in 2-propanol/HCl (1 M) and heating the reaction mixture at 37 °C for 5 h [44].

Figure 8 shows the FAB mass spectrum of an equimolar mixture of three peptides before and after hexylation. The upper spectrum shows clear supression of the peptides TKPR (THr-Lys-Pro-Arg) and ETYSK (Glu-Thr-Tyr-Ser-Lys) by WHWLQL (Trp-His-Trp-Leu-Gln-Leu). After hexylation, all three peptides are clearly observed and exhibit similar FAB sensitivity.

Falick and Maltby have also evaluated benzylation as a potential derivatization reaction for FAB MS of peptides. The benzyl esters show a moderate improvement in sensitivity over methyl esters (about fourfold for Thr-Lys-Pro-Arg). Since the benzyl moiety is a strong chromatophore, one can also obtain a substantial increase in HPLC sensitivity (at 215 nm) in addition to the improved FAB MS performance.

3.3 Derivatization for soft ionization MS–MS

In the last few years the applicability of derivatization for use in soft ionization MS has grown substantially, due in part to the increasing availability of mass spectrometry–mass spectrometry (MS–MS) instrumentation. A detailed discussion of MS–MS methods and instrumentation is beyond the scope of this chapter, but an extensive review of the subject has recently been

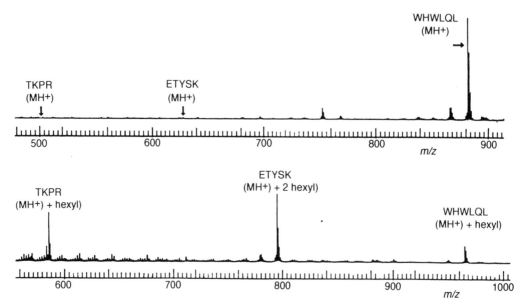

Figure 8. (Top) Mass spectrum of an equimolar mixture of three peptides. The peptides are: TKPR = Thr-Lys-Pro-Arg, ETYSK = Glu-Thr-Tyr-Ser-Lys and WHWLQL = Trp-His-Trp-Leu-Gln-Leu. (Bottom) Spectrum of an equal aliquot of the same mixture after hexylation. (Reprinted with permission from Ref. 16).

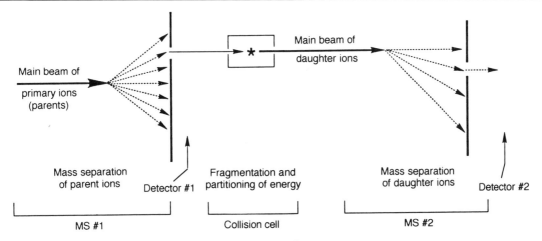

Figure 9. Schematic diagram of MS–MS experiment.

published by Busch, Glish and McLuckey [45]. In simplified form, the typical MS–MS instrument consists of two mass spectrometers (MS1 and MS2) coupled together by means of a gas collision cell (see Figure 9). The combined instrument allows one to select ions of a specific mass with MS1, to induce fragmentation of these ions by collision with a neutral target gas (collisional activation, CA), and to measure the distribution of ionic products with MS2. The combination of derivatization and MS–MS provides a great deal of flexibility in mixture analysis and mass spectral fragmentation studies.

The MS–MS experiment involves multiple stages of analysis (MS1, collison cell, MS2) and approximately 90% of the signal is lost in each stage. In a typical CA MS–MS experiment, the intensity of the various fragment ions represents a small percentage ($<1\%$) of the parent ion beam intensity coming out of MS1. If the parent ion beam is weak, the signal emerging from MS2 may disappear in the background noise and be undetectable.

In order to provide an intense beam of sample ions for MS–MS analysis, soft ionization techniques such as FAB are often used. FAB normally creates a narrower distribution of fragment ions than EI or CI. By bunching the sample ion current into fewer peaks, interpretation of the primary spectrum is simplified and more intense peaks are available for MS–MS analysis. Further increases in primary ion intensity can be realized through derivatization methods such as esterification or permethylation, which are used to increase surface activity without fundamentally altering fragmentation. Using these derivatives, one would expect to see a distribution of fragment ions analogous to those observed in the underivatized analyte.

MS–MS instrumentation is well suited for application in derivatization studies because of the ability to obtain unique daughter mass spectra from each of the ions in the primary spectrum. Derivatization reactions often do not go to completion, and the primary spectrum of a crude derivatization mixture can be highly complex owing to the superimposition of the various component mass spectra. Although chromatographic separations may be used to isolate products prior to the next step in the analysis, with MS–MS techniques an ion of interest (the parent) can be cleanly separated from the mixture by MS1 and fragmented through collisions with neutral atoms, and the daughter mass spectrum can be immediately measured by MS2. Assuming that the primary ions can be manufactured in quantity, the MS–MS analysis can be performed quickly on small amounts of impure sample.

The distribution of fragment ions created in a collisional process can vary substantially from the distribution created by atom or ion bombardment. For example, in MS studies, Naylor and Moneti [23] have shown that underivatized, N-acetylated, dansylated, isopropyl and n-butyl esterified peptides show similar fragment-to-parent, $[M+H]^+$, ion abundance ratios. The net effect of these derivatives is to increase the signal-to-noise ratio but not to alter the fragmentation. In contrast, Gaskell *et al.* have noted an increase in leucine-enkephalin N-terminus containing fragment ions in the MS–MS spectra of N-acetyl leucine-enkephalin, relative to underivatized leucine-enkephalin. The MS–MS fragmentation of N-acetylated (or N-acetylated permethylated) peptides generally proceeds in a better defined fashion, frequently resulting in abundant daughter ions and more easily interpreted spectra. This trend probably results from decreased charge localization following derivatization [30].

A practical example of the utility of MS–MS relates to the H_3/D_3 N-acetylation of leucine-enkephalin (see Figure 5). This derivative was generated from a synthetic sample, and the spectrum of the reaction mixture is composed of at least three components: leucine-enekephalin, H_3 N-acetyl leucine-enkephalin, and D_3 N-acetyl leucine-enkephalin. The spectrum is reasonably complex, and one could envisage far greater complexity if the sample had been extracted from a real biological source such as tissue. In subsequent MS–MS studies, the $[M + H]^+$ ions of H_3 and D_3 N-acetyl leucine-enkephalin (m/z 598 and 601) were individually selected by MS1, collisionally fragmented with argon, and the distribution of fragment ions was measured by MS2. The resulting daughter ion spectra are shown in Figure 10. The spectra are similar but unique because of the H_3 acetyl group in one parent ion and the D_3 acetyl group in the other. Upon comparison of the daughter spectra, the N-terminus-containing ions are readily identified by the 3 Da shift. The peaks in Figure 10A at m/z 320, 439, 467, 580 and 598 shift to 323, 442, 470, 583 and 601 in Figure 10B owing to the incorporation of three deuterium atoms at the N-terminus. The unshifted ions in the spectra, m/z 279, 336 and 393, do not contain the N-terminus. These results are consistent with those previously presented in the acetylation section.

A major feature of the MS–MS spectra is the lack of chemical noise relative to the primary MS spectrum. The ions at m/z 580, 583, 439 and 442 appeared in clusters of small background peaks in the conventional mass spectrum, but they are well defined with a substantially greater signal-to-noise ratio in the MS–MS spectra. No isotopes are observed in the MS–MS daughter spectra because the monoisotopic parent was chosen for each of the MS–MS experiments.

MS–MS is a powerful technique for the characterization of complex mixtures; however, there are circumstances where the daughter ion spectrum does not contain sufficient structural information for fully characterizing a sample. To cope with this problem, a number of studies have been targeted at derivatizing the analyte in order to alter or direct the CA mass spectral fragmentation and increase the infromation content of the daughter ion spectra.

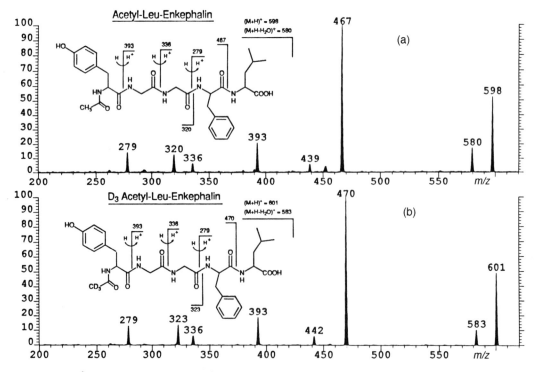

Figure 10. (a) Daughter ion MS–MS spectrum of the molecular ion of N-acetyl-leucine-enkephalin, m/z 598. (B) Daughter ion MS–MS spectrum of the molecular ion of D_3 N-acetyl leucine-enkephalin, m/z 601.

3.3.1 Cationization in MS–MS

One of the simplest methods of altering the CA fragmentation process is through cationization. Depending on the site of cation attachment, some of the fragmentation processes observed in the daughter ion spectrum of the $[M + H]^+$ ion may be disabled. New, highly informative routes of fragmentation may become apparent in the daughter ion spectrum of the cationized molecular ion. These experimental results are often complementary, and the combination of both analyses can provide very strong support for a given interpretation.

Figure 11 shows the MS–MS daughter ion spectrum obtained from the sodium adduct of N-acetyl leucine-enkephalin (MW = 619). This spectrum varies dramatically from the analogous daughter ion spectrum obtained from protonated N-acetyl leucine-enkephalin (Figure 10A). There appears to be no consistency in the

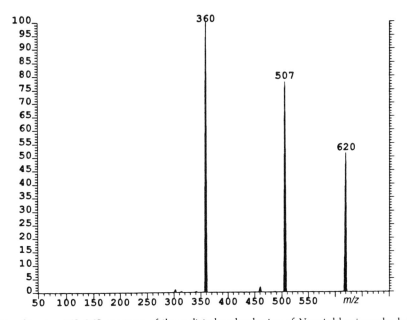

Figure 11. Daughter ion MS–MS spectrum of the sodiated molecular ion of N-acetyl leucine-enkephalin, m/z 620.

fragmentation observed in the two spectra. The daughter ion spectrum of the sodiated molecular ion, m/z 620, does not exhibit characteristic peptide cleavages, presumably because of chelation of the sodium ion across the peptide carbonyl groups. Similar alterations in [M + H]$^+$ versus [M + Na]$^+$ fragmentation have been independently observed by Tang et al. [46] and Kulik et al. [47]. Each of these studies showed that small cationized peptides can decompose by rearrangement of the C-terminal hydroxyl group to the carbonyl of the neighboring peptide bond followed by elimination of the C-terminal amino acid residue. The resulting product ion is a smaller cationized peptide which can also undergo rearrangement and cleavage of the C-terminal residue. Tang et al. have presented results using leucine-enkephalin which are analogous to those presented for sodiated N-acetyl leucine-enkephalin in Figure 11. A generalized fragmentation mechanism (I) is shown below [46].

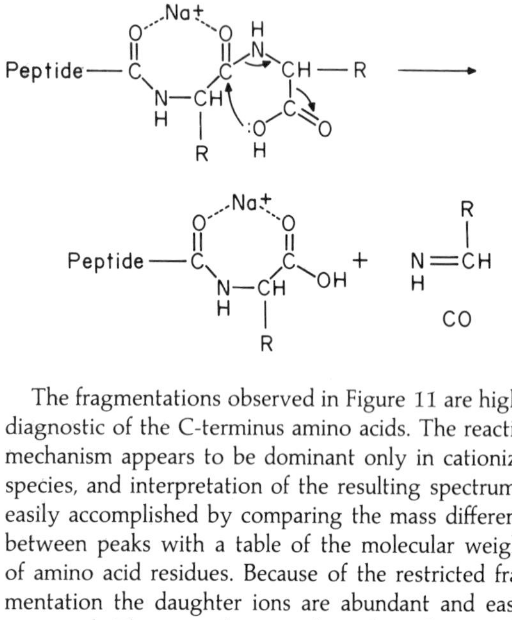

$$\text{Peptide} - \underset{\underset{\text{N}-\text{CH}}{|}}{\overset{\overset{\text{Na}^+}{\cdots \overset{\cdot}{\text{O}}}}{\text{C}}} \cdots \overset{\overset{\text{O}}{\|}}{\underset{\underset{\text{R}}{|}}{\text{C}}} \overset{\text{H}}{\underset{\underset{\text{H}}{|}}{\text{N}}} \text{CH} - \text{R} \longrightarrow$$

$$\text{Peptide} - \underset{\underset{\text{N}-\text{CH}}{|}}{\overset{\overset{\text{Na}^+}{\cdots \overset{\cdot}{\text{O}}}}{\text{C}}} \cdots \overset{\text{O}}{\underset{\underset{\text{R}}{|}}{\text{C}}} \text{OH} + \overset{\text{R}}{\underset{\underset{\text{CO}}{|}}{\text{N}=\text{CH}}} \quad \text{(I)}$$

The fragmentations observed in Figure 11 are highly diagnostic of the C-terminus amino acids. The reaction mechanism appears to be dominant only in cationized species, and interpretation of the resulting spectrum is easily accomplished by comparing the mass difference between peaks with a table of the molecular weights of amino acid residues. Because of the restricted fragmentation the daughter ions are abundant and easily measured. These results provide independent confirmation of the C-terminus amino acid(s) through use of an additional fragmentation process not normally observed in underivatized peptides. Kulik has also shown that the reaction proceeds in many small peptides and that good results can be obtained when the cation is either sodium or lithium.

3.4 Creation of a charged site

The FAB ionization process is thought to encompass two steps in one concerted action. There must be a transfer of sample to the gas phase and ions must be created. In the case of highly polar compounds bearing a fixed charge, such as ammonium or sulfonium salts, the ionization step is no longer required, and cations or anions can be sputtered directly into the vapor phase.

In studies using N-pyridinium derivatives of compounds containing hydroxyl or primary amine groups, Flurer and Busch were able to demonstrate that the vaporization of pre-formed ions is substantially more efficient than the vaporization and ionization of the corresponding underivatized compound [48]. The FAB spectrum of the N-pyridinium derivatives showed an enhanced signal-to-noise level and a corresponding reduction in detection limit. Unlike the non-specific ion attachment derivatives noted previously, these are specific derivatives in that the charge is placed at a well defined site in the sample molecule. A number of investigators have reported the use of fixed charge derivatives in MS–MS studies. Typically, these derivatives exhibit well directed fragmentation because of the localized charge. The interpretation of MS–MS spectra obtained from these derivatives is generally simplified because fewer but more intense peaks are observed.

In order to examine the effect of a fixed positive charge within a peptide, Johnson et al. [49] have investigated a series of lysine-containing peptides in which the primary amine in lysine had been converted to a 2,4,6-trimethylpyridinium cation (Reaction II [50]). Depending on the location of the derivatized lysine in the peptide, the distribution of fragments shifted to favor formation of fragments which retained the charge. Similar fragmentation has been observed in peptides containing basic amino acids. Protonation at the most basic site creates a concentrated area of positive charge; however, the subsequent fragmentation is less well defined owing to the mobility of the proton.

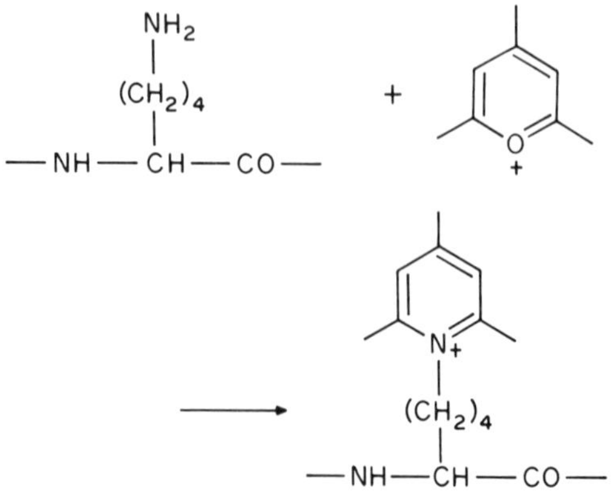

$$-\text{NH}-\underset{\underset{\underset{\underset{\text{NH}_2}{|}}{(\text{CH}_2)_4}}{|}}{\text{CH}}-\text{CO}- \quad + \quad \underset{\text{(trimethylpyridine)}}{} $$

$$\longrightarrow \quad -\text{NH}-\underset{\underset{\underset{\underset{\text{N}^+\text{(trimethylpyridinium)}}{|}}{(\text{CH}_2)_4}}{|}}{\text{CH}}-\text{CO}- \quad \text{(II)}$$

Stults et al. have recently reported a derivatization reactions (III) for incorporation of a quaternary ammonium group at the N-terminus of peptides [51]. The

$$(I-CH_2-CO)_2O + H_2N-CHR-CO\sim$$

$$\downarrow \begin{array}{l} pH\ 6 \\ 0\ °C \\ 3\ min\ (x2) \end{array}$$

$$I-CH_2-CO-NH-CHR-CO\sim + (CH_3)_3\overset{+}{N}-CH_2-CH_2-SH \quad\quad (III)$$

$$\downarrow \begin{array}{l} pH\ 8 \\ 37\ °C \\ 1\ h \end{array}$$

$$(CH_3)_3\overset{+}{N}-CH_2-CH_2-S-CH_2-CO-NH-CHR-CO\sim$$

MS–MS daughter ion spectra of these derivatives are simplified relative to the underivatized compounds and show abundant, almost exclusively N-terminus-containing ions.

Johnson has also investigated a number of derivatives for use in altering the CA fragmentation of peptides [52]. By altering the basicity of the N-terminus of a peptide, a substantial increase in the N-terminus-containing daughter ions can be obtained. In these studies, Johnson has successfully used reactions for the amidination and the coupling of β-guanidinopropionic acid to the peptide N-terminal amino group. In a number of peptides the incorporation of the very basic guanidino group dramatically altered the CA fragmentation in favor of enhanced N-terminus-containing daughter ions. The reader is referred to the literature references listed in Table 1 for information concerning the preparation of these derivatives.

3.4.1 ^{18}O Exchange Reactions

The incorporation of stable isotopes in derivatization processes can be highly advantageous because of the mass spectrometer's ability to differentiate ions of different mass. Previously, examples of the use of deuterated reagents have been shown. Currently, ^{18}O labelling is also becoming a common practice in FAB MS studies. The specific incorporation of ^{18}O at a peptide C-terminus is useful for verifying fragmentation mechanisms and in the production of co-eluting internal standards for HPLC–MS quantitation studies.

Incorporation of ^{18}O can be accomplished using the process reported by Desiderio and Kai [53]. A small amount of peptide is dissolved in methanolic HCl (3 M) and twofold excess of $H_2^{18}O$. The mixture is placed in a sealed vial and heated to 37 °C. The reaction is allowed to proceed for 2–7 days and the mixture is periodically examined mass spectrometrically until a stable level of exchange is obtained. The reaction typically produces a mixture of labeled products which are amenable to subsequent analysis by MS–MS.

3.4.2 Derivatives for constant neutral loss class screening

Constant neutral loss (CNL) scanning is an MS–MS technique where both spectrometers (MS1 and MS2) scan the mass spectrum in parallel, but offset by some selected mass difference. In order to produce a signal at the detector, an ion originating from MS1 must fragment in the collision cell and lose a fragment corresponding to the mass offset of the two spectrometers. The surviving ion, $[M - \text{offset mass}]^+$, is transmitted through MS2 and strikes the detector. Given a complex mixture, a constant neutral loss scan can be used to generate a plot of all primary ions which undergo a selected fragmentation in the collision cell. Since many chemical moieties undergo characteristic CA fragmentations, one can use CNL scanning selectively to screen a mixture for compounds of a specific functionality. Upon locating the precursor ions which exhibit the selected mass loss, one may, without separation, further characterize these ions by measuring the MS–MS daughter ion spectrum. In order to use the CNL technique for screening, one must be able to identify a fragmentation characteristic of the functionality of interest. In many instances a strong fragmentation characteristic of a specific functional group occurs naturally; e.g. O-glucuronide conjugates typically lose 176 daltons by cleavage of the glucuronide moiety.

The role of derivatization in CNL scanning experiments is still very limited, but there are significant advantages to be realized if derivatives can be used to create or enhance a characteristic fragmentation process. One of the best examples of this application is the derivatization of glutathione conjugates. Pearson and co-workers [54, 55] have used a combination of derivatization and constant neutral loss scanning to screen selectively for glutathione conjugates in complex biological matrices. Benzyloxycarbonyl derivatives were formed in high yield through the addition of benzyl chloroformate to a buffered aqueous solution. The derivative was stable and well suited for separation by reversed phase HPLC. Purified samples were reacted with methanolic HCl for 2 h at room temperature in order to form the corresponding dimethyl esters. Upon MS–MS analysis, these derivatives displayed a prominent loss of 89 daltons corresponding to the elimination of glycine methyl ester. The combination of improved HPLC performance with the introduction of a strong characteristic CA fragmentation provides an attractive approach for the characterization of glutahione conjugates in complex biological samples.

4. Derivatization and continuous flow FAB: future prospects

One of the most exciting recent developments in mass spectrometry is a technique known as continuous flow FAB (CF FAB). Caprioli and Fan have shown that it is possible to reduce the liquid matrix concentration and subsequently increase FAB sensitivity by allowing a flow of $2-10\ \mu l\ min^{-1}$ of a sample/matrix solution over the surface of the FAB probe [56]. Because of the constant replenishment of the matrix, it is possible to work with dilute solutions containing 5–30% of matrix. The resulting spectra show increased sample signal-to-noise ratios to the point where some hydrophilic peptides, normally not detected by conventional FAB, can be observed. In addition, the technique does not appear to display suppression effects to the same extent as those observed in normal FAB experiments.

One potential major application of CF FAB is the coupling of liquid chromatography with FAB. Although flow rates are restricted to $3-10\ \mu l\ min^{-1}$, recent developments in micro-bore and capillary liquid chromatography technology appear to offer a near perfect combination of high sensitivity HPLC efficiently coupled to FAB MS with minimal sample and solvent requirements. On-line derivatization appears to be an obvious extension of the technique. Caprioli has already reported the use of CF FAB for continuously monitoring the progress of enzymatic hydrolysis reactions [57]. In addition, on-line HPLC/CF FAB MS systems are currently in use in a number of laboratories. One would expect that appropriate derivatization could be used to provide the optimum combination of good HPLC performance and MS or MS–MS performance in a single on-line experiment. Sequential injection of a series of derivatizing agents into a flowing stream of analyte may provide an attractive method for very quick and efficient characterization of unknowns.

5. Conclusions

The applicability of mass spectrometry to the analysis of involatile, thermally labile compounds has grown substantially in the past ten years, principally because of the discovery and application of FAB. While the need to derivatize to increase volatility and thermal stability has declined, derivatization has grown as a technique for enhancing the yield of structural information in both FAB MS and MS–MS studies. Derivatization is expected to continue to provide an attractive path by which both signal intensity and information content may be optimized.

6. References

[1] R. J. Beuhler, E. Flanigan, L. J. Greene and J. L. Friedman, *J. Am. Chem. Soc.*, **96**, 3990 (1974).
[2] A. Dell, D. H. Williams, H. R. Morris, G. A. Smith, J. Feeney and G. C. K. Roberts, *J. Am. Chem. Soc.*, **97**, 2497 (1975).
[3] M. A. Baldwin and F. M. McLafferty, *Org. Mass Spectrom.*, **7**, 1353 (1973).
[4] H. D. Beckey, *Int. J. Mass Spectrom. Ion Phys.*, **2**, 500 (1969).
[5] P. G. Kistemaker, G. J. Q. van der Peyl and J. Haverkamp, *Soft Ionization Biological Mass Spectrometry*, H. R. Morris (Editor), Heyden & Son, London (1981), p. 120.
[6] R. D. Macfarlane and D. F. Torgerson, *Science*, **191**, 920 (1976).
[7] K. D. Cook, *Mass Spectrom. Rev.*, **5**, 467 (1986).
[8] M. Barber, R. S. Bordoli, R. D. Sedgwick and A. N. Tyler, *J. Chem. Soc. Chem. Commun.*, 325 (1981).
[9] G. Elliott, J. S. Cottrell and S. Evans, *34th Annual Conference of Mass Spectrometry and Allied Topics*, Cincinnati (1986).
[10] M. Barber and B. N. Green, *Rapid Commun. Mass. Spectrom.*, **1**, 80 (1987).
[11] J. L. Gower, *Biomed. Mass. Spectrom.*, **12**, 191 (1985).
[12] E. De Pauw, *Mass Spectrum. Rev.*, **5**, 191 (1986).
[13] H. B. Bull and K. Breese, *Arch. Biochem. Biophys.*, **161**, 665 (1975).
[14] S. Naylor, G. Moneti and S. Guyan, *Biomed. Mass Spectrom.*, **17**, 393 (1988).
[15] K. Biemann and S. A. Martin, *Mass Spectrom. Rev.*, **6**, 1 (1987).
[16] A. M. Falick and D. A. Maltby, *Anal. Biochem.*, **182**, 165 (1989).
[17] J. Sunner, A. Morales and P. Kebarle, *Anal. Chem.*, **59**, 1378 (1987).
[18] W. V. Ligon, *Anal. Chem.*, **58**, 487 (1986).
[19] A. Benninghoven and W. K. Sichtermann, *Anal. Chem.*, **50**, 1181 (1978).
[20] S. A. Martin, C. Costello, K. Biemann, *Anal. Chem.*, **54**, 2362 (1984).

[21] K. L. Busch, S. E. Unger, A. Vincze, R. G. Cooks and T. Keough, *J. Am. Chem. Soc.*, **104**, 1507 (1982).
[22] S. J. Pachuta and R. G. Cooks, *Desorption Mass Spectrometry; Are SIMS and FAB the Same?*, Edit P. A. Lyon (Editor), American Chemical Society, Washington, DC (1985), p. 1.
[23] S. Naylor and G. Moneti, *Biomed. Environ. Mass Spectrom.*, **18**, 405 (1989).
[24] D. H. Russell, E. S. McGlohon and L. Mallis, *Anal. Chem.*, **60**, 1818 (1988).
[25] L. M. Mallis, F. M. Raushel and D. H. Russell, *Anal. Chem.*, **59**, 980 (1987).
[26] F. W. Röllgen and H. R. Schulten, *Org. Mass Spectrom.*, **10**, 660 (1975).
[27] D. W. Thomas, B. C. Das, S. D. Gero and E. Lederer, *Biochem. Biophys. Res. Commun.*, **33**, 519 (1968).
[28] H. R. Morris, *Biochem. Soc. Trans.*, **2**, 806 (1974).
[29] E. Hunt and H. R. Morris, *Biochem. J.*, **135**, 833 (1973).
[30] S. J. Gaskell, M. H. Reilly and C. J. Porter, *Rapid Commun. Mass Spectrom.* **2**, 142 (1988).
[31] K. Blau and G. S. King (Editors) *Handbook of Derivatives for Chromatography*, K. Heyden, London (1977), p. 123.
[32] L. D. Detter, O. W. Hand, R. G. Cooks and R. A. Walton, *Mass Spectrom. Rev.*, **7**, 465 (1988).
[33] J. Peter-Katalinic, H. Egge, B. Deutscher, W. Knoss and K. Glombitza, *Biomed. Environ. Mass Spectrom.*, **15**, 595 (1988).
[34] A. Dell, J. E. Oats and C. E. Ballou, *Proc. 7th Int. Symp. Glycoconjugates*, Lund (1983), p. 127.
[35] A. Dell, J. E. Oats, H. R. Morris and H. G. Egge, *Int. J. Mass Spectrom. Ion Phys.*, **46**, 415 (1983).
[36] S. I. Hakomori, *J. Biochem. (Tokyo)*, **55**, 205 (1964).
[37] E. Vilkas and E. Lederer, *Tetrahedron Lett.*, 3089 (1968).
[38] D. W. Thomas, *Biochem. Biophys. Res. Commun.*, **33**, 483 (1968).
[39] H. R. Morris, *EFBS Lett.*, **22**, 257 (1972).
[40] I. Ciucanu and F. Kerek, *Carbohydr. Res.*, **131**, 209 (1984).
[41] A. Dell and M. Panico, *Mass Spectrometry in Biomedical Research*, S. Gaskell (Editor), Wiley (1986), p. 149.
[42] J. E. Vath, B. Domon and C. E. Costello, *37th ASMS Conference on Mass Spectrometry and Allied Topics*, Miami Beach (1989).
[43] H. R. Morris, *Nature*, **286**, 447 (1980).

[44] S. Naylor, S. G. Ang, D. H. Williams, C. H. More and K. Walsh, *Biomed. Environ. Mass Spectrom.*, **18**, 424 (1989).
[45] K. L. Busch, G. L. Glish and S. A. McLuckey, *Mass Spectrometry/Mass Spectrometry: Techniques and Applications of Tandem Mass Spectrometry*, Verlag Chemie, Weinheim (1988).
[46] X. Tang, W. Ens, K. Standing and J. Westmore, *Anal. Chem.*, **60**, 1791 (1988).
[47] W. Kulik, W. Heerma and J. K. Terlouw, *Rapid Commun. Mass Spectrom.*, **3**, 276 (1989).
[48] R. A. Flurer and K. L. Busch, *Proc. Indiana Acad. Sci.*, **95**, 171 (1986).
[49] R. S. Johnson, S. A. Martin and K. Biemann, *Int. J. Mass Spectrom. Ion Phys.*, **86**, 137 (1988).
[50] M. H. O'Leary and G. A. Samberg, *J. Am. Chem. Soc.*, **93**, 3530 (1971).
[51] J. T. Stults, R. Halualani and R. Wetzel, *37th ASMS Conference on Mass Spectrometry and Allied Topics*, Miami Beach (1988).
[52] R. S. Johnson, Ph. D. Thesis, Massachusetts Institute of Technology, Cambridge (1988).
[53] D. M. Desiderio and M. Kai, *Biomed. Mass Spectrom.*, **10**, 471 (1983).
[54] P. G. Pearson, M. D. Threadgill, W. N. Howald and T. A. Baillie, *Biomed. Environ. Mass Spectrom.*, **16**, 51 (1988).
[55] D. Thomassen, P. G. Pearson and S. D. Nelson, *37th ASMS Conference on Mass Spectrometry and Allied Topics*, Miami Beach (1989).
[56] R. M. Caprioli and T. Fan, *Anal. Chem.*, **58**, 2949 (1986).
[57] R. M. Caprioli, *Mass Spectrometry in Biomedical Research*, S. J. Gaskell (Editor), Wiley, Chichester (1986), p. 41.
[58] D. A. Kidwell, M. M. Ross and R. J. Colton, *Biomed. Mass Spectrom.*, **12**, 254 (1985).
[59] S. Naylor, A. F. Findeis, B. W. Gibson and D. H. Williams, *J. Am. Chem. Soc.*, **108**, 6359 (1986).
[60] R. J. Colton, D. A. Kidwell, G. O. Ramseyer and M. M. Ross, *Desorption Mass Spectrometry: Are SIMS and FAB the Same?*, P. A. Lyon (Editor), American Chemical Society, Washington, DC (1985), p. 160.
[61] M. E. Rose, C. Longstaff and P. D. G. Dean, *Biomed. Mass Spectrom.*, **10**, 512 (1983).
[62] K. Hoffmann and T. A. Baillie, *Biomed. Environ. Mass Spectrom.*, **15**, 637 (1988).

13

Derivatives for Supercritical Fluid Chromatography

Keith D. Bartle, Anthony A. Clifford and **Naila Malak**
School of Chemistry, University of Leeds, Leeds

1. INTRODUCTION 289
2. DERIVATIZATION TO INCREASE SOLUBILITY 290
   2.1 General 290
   2.2 Mono- and polysaccharides 290
   2.3 Amino acids 291
   2.4 Polymers and industrial chemicals 293
3. DERIVATIZATION FOR SELECTIVE AND SENSITIVE DETECTION 293
4. CONCLUSIONS 294
5. ACKNOWLEDGEMENTS 294
6. REFERENCES 294

1. Introduction

Supercritical fluid chromatography (SFC) is being more and more widely used. The principle, as in other kinds of chromatography, is to separate the components of a mixture on a chromatographic column; here, a supercritical fluid is used as mobile phase under conditions of temperature and pressure chosen to maintain it in this state. The advantages are the superior solvent power of supercritical fluids and the low temperatures used, which make many compounds accessible to chromatography which could otherwise not be separated in this way because of high molecular weight or instability to heat. Another benefit is the ready removal of the mobile phase on emergence from the chromatographic system and the consequent ease of recovery of the separated components. A variety of supercritical fluids have been used, but liquid carbon dioxide is by far the most popular. The solvent power of the supercritical fluid can be modified by the addition of small percentages of a modifier such as methanol or trimethylamine.

SFC has important applications in the analysis of involatile, polar and thermally unstable compounds, and has a number of advantages over gas chromatography (GC) and high performance liquid chromatography (HPLC) [1, 2]. Although GC has the highest resolving power, it is limited by the requirements of volatility and thermal stability of the analyte. Polar, unstable and involatile compounds may be analysed by HPLC, but this technique is limited by the lower efficiencies associated with packed columns and by the lack of a simple universal detector for compounds without a chromophore. SFC allows high resolution separations of such compounds at temperatures well below those of thermal decomposition, and can be interfaced with almost any HPLC or GC detection system [1, 2].

Since part of the ethos of SFC is to extend high resolution analysis to high molecular weight, reactive and thermolabile compounds not accessible to GC *without* derivatization, and to use universal detection, applications of derivatization in SFC have been sparse. They fall into two main groups: derivatization to increase solubility in the supercritical mobile phase, and hence to extend further the molecular weight range of SFC; and derivatization to provide selectivity and increase sensitivity in detection.

Handbook of Derivatives for Chromatography
Edited by K. Blau and J. M. Halket © 1993 John Wiley & Sons Ltd

2. Derivatization to increase solubility

2.1 General

Although many polar molecules may be eluted in SFC without derivatization, e.g. carboxylic acids and their hydroxylated derivatives [3], the substitution of polar groups into a molecule tends to reduce the solubility in supercritical CO_2. Figure 1 compares the solubilities of naphthalene and 1-naphthol and is an illustration of this general trend [4]. Addition of a modifier such as an alcohol or ether may be necessary; thus, Raynor et al. found [5] that plant ecdysteroids with up to four hydroxyl substituents were soluble enough in supercritical CO_2 for separation by SFC. Chromatography with CO_2 modified with methanol as mobile phase allowed elution of ecdysteroids with up to eight hydroxyl constituents [5]; of course, flame ionization detection was not possible with use of the modifier. Derivatization improves the SFC of mono- and diglycerides and fatty acids when using certain stationary phases, but is not needed as often as in GC [6, 7].

2.2 Mono- and polysaccharides

Most carbohydrates are insufficiently soluble in CO_2 for successful SFC, and a variety of derivatization procedures have been employed to facilitate elution by increasing the solubility. Calvey et al. [8] converted a mixture of monosaccharides (rhamnose, fucose, arabinose, xylose, mannose, glucose and galactose) to their peracetylated aldonitrile derivatives. GC of the reaction products gave one peak per sugar, but capillary SFC with CO_2 mobile phase showed a range of other products. Combined SFC–FTIR spectrometry and SFC–MS suggested the presence of both aldonitrile acetates and other derivatives in which an acetate moiety is attached to the nitrogen of the oxime.

The glucose oligo- and polysaccharides of corn syrup were successfully chromatographed with pressure programmed CO_2 as the trimethylsilyl derivatives by Chester and Innis [9]. Figure 2 shows the chromatograms on capillary SFC columns. The two peaks for each degree of polymerization correspond to the α and β anomers. The oligomer with 18 glucose units has MW 2934 before derivatization and 6966 after silylation-well beyond the GC range. The baseline resolution achieved in SFC was considered [9] superior to that possible by HPLC [10]. Later work by Pinkston and co-workers [11], verified by SFC–MS [12], showed that both members of each pair of closely eluting compounds below degree of polymerization (DP) 7 have the same mass.

Kuei et al. preferred [13] permethylation to achieve separation by capillary SFC of maltodextrin glucose oligomers with DP up to 15. A single peak for each oligomer derivative was observed. The same procedure was applied to three classes of glycosphingolipids, which are amphoteric conjugates of a long-chain base, a single fatty acid and one or more carbohydrate residues. Compounds with MW up to 3000 were resolved for different carbohydrate moieties, and the method was applied to bovine brain (Figure 3) and human gangliosides.

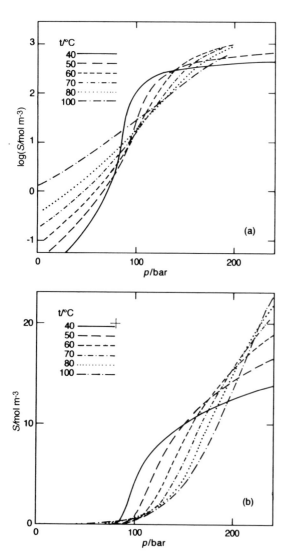

Figure 1. Predicted values of the solubilities S of (A) naphthalene (note logarithmic scale) and (B) 2-naphthol in CO_2 as a function of pressure p at various temperatures (reprinted with permission from Ref 4).

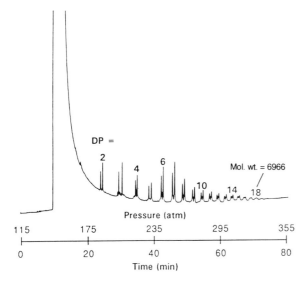

Figure 2. SFC chromatogram of silylated Maltrin 100. Conditions: 10 m × 50 μm ID open-tubular column; poly (5% phenyl (methylsiloxane stationary phase; CO_2 mobile phase; 89 °C; FID. (Reprinted with permission from Ref 9).

Myoinositol is a naturally occurring vitamin with a hexahydroxycyclohexane structure; its phosphorylated forms are readily converted to trimethylsilyl derivatives which, as has been observed for alkyl ethoxy phosphates, can be separated by SFC at temperatures some hundreds of degrees below those necessary for GC of such compounds [14]. Both silylated inositol triphosphate and phytic acid (inositol hexaphosphate) were eluted by density programmed CO_2 [14]. The identities of the derivatives were confirmed [15] by SFC–MS with both ammonia and isobutane chemical ionization, and using electron ionization–CO_2 charge exchange. The latter spectrum of the inositol triphosphate TMS derivative was very similar to the library electron ionization spectrum of TMS-inositol diphosphate.

2.3 Amino acids

The solubility of amino acids in CO_2 modified with methanol, water and methylamine was increased sufficiently for SFC on columns packed with silica by derivatizing with 9-fluorenylmethyl chloroformate (FMOC) [16]. Valine, alanine, phenylalanine, lysine and serine were separated (Figure 4) within 40 minutes. Derivatization with (+)-1-(9-fluorenyl)ethyl chloroformate (FLEC) allowed separation of enantiomers more rapidly than by HPLC but with similar selectivity.

A number of groups have shown how enantiomeric resolution of amino acids derivatized with non-chiral reagents is possible in SFC with chiral stationary phases. N-Acetylamino acid t-butyl ester racemates were rapidly resolved [17] on (N-formyl-L-valylamino)propyl silica with CO_2 modified with methanol, acetonitrile and diethyl ether. A similar stationary phase allowed [18] rapid (<5 min) separation of racemic N-4-nitrobenzoyl-amino acid isopropyl esters with methanol-modified CO_2; the enantioselectivity in SFC was comparable with that in HPLC with isopropanol/n-hexane as mobile phase. Capillary column SFC on polysiloxane stationary phases containing chiral side chains has been employed

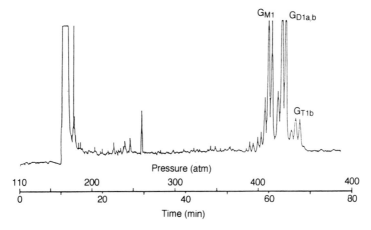

Figure 3. SFC chromatogram of a brain extract consisting of the methylated gangliosides G_{M1}, $G_{D1a,b}$ and G_{T1b}. Conditions: 10 m × 50 μm ID open-tubular column, poly (5% phenyl)methylsiloxane stationary phase; CO_2 mobile phase; 120 °C; pressure programme from 110 to 400 atm at 5 atm min^{-1} after a 5 min isobaric period; FID. (Reprinted with permission from Ref 13).

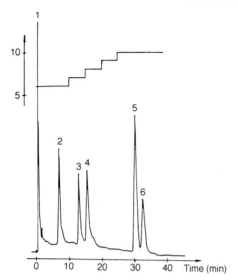

Figure 4. SFC separation of five FMOC-amino acids: 1, acetone (solvent); 2, valine; 3, alanine; 4, phenylalanine; 5, lysine; 6, serine. Conditions: 15 cm × 4.6 mm ID packed column; silica stationary phase; CO_2 mobile phase with methanol/water/methylamine (98.99:1.00:0.01 v/v) modifier gradient indicated on chromatogram; 42 °C; UV detection at 269 nm. (reprinted with permission from Ref. 16).

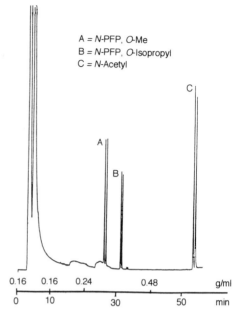

Figure 5. SFC separation of derivatives of proline. Conditions: 10 m × 50 μm ID open-tubular column; CHIRALNEB stationary phase; CO_2 mobile phase; 60 °C; FID. (Reprinted with permission from Ref 20).

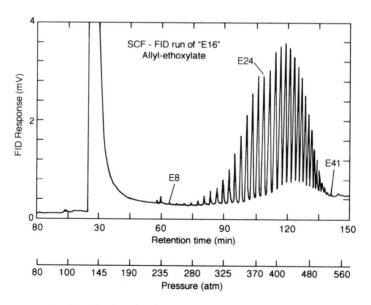

Figure 6. SFC chromatogram of silylated ethoxylated allyl alcohol. Conditions: 10 m × 50 μm ID open-tubular column; 30% biphenyl stationary phase; CO_2 mobile phase; 100 °C; FID. (Reprinted with permission from Ref 22).

[19, 20] to separate amino acid derivatives, e.g. N-pentafluoropropionyl isopropyl esters [20] (Figure 5).

Separation of phenylthiohydantoin (PTH) amino acid derivatives, which are produced in the sequential analysis of proteins by the Edman degradation, is required. Ashraf-Khorassani *et al.* showed [21] how SFC on a cyanopropyl-modified silica column allowed the separation of more than twenty PTH-amino acids if an ion-pairing reagent (tetramethylammonium hydroxide) was present at low concentration in the methanol-modified CO_2 mobile phase.

2.4 Polymers and industrial chemicals

Ethoxylated alcohols, an important class of non-ionic surfactants, and polyethylene glycols (PEG) are sufficiently soluble in CO_2 without derivatization for SFC on capillary columns to be the analytical separation method of choice. The range of polyethylene glycols which may be eluted has been shown [22] to be extended if the samples are trimethylsilylated before analysis. PEG samples containing at least 55 units and ethoxylated allyl alcohol with 41 units have been separated at pressures up to 560 atmospheres by this method (Figure 6).

Poly(acrylic acid) samples were derivatized [23] to increase solubility in supercritical CO_2. t-Butyldimethylsilyl esters showed better chromatographic performance and a more extended MW range than did trimethylsilyl or methyl esters. Individual oligomers up to about $n = 28$ were resolved, and oligomers with MW up to 3300 ($n = 45$) were eluted but not resolved. SFC–MS with ammonia chemical ionization and electron impact ionization showed that the terminal groups of the acids were sulphonate and hydrogen. Trimethylsilylation, on the other hand, proved to be a satisfactory derivatization method for the SFC analysis of oligomers of *m*-cresol Novolak resins [24].

3. Derivatization for selective and sensitive detection

The availability of both GC and HPLC detectors in SFC means that derivatization to enhance selectivity and sensitivity should be an important analytical tool. Thus, David and Novotny showed [25] how nitrogen thermionic detection of quinoxalinol derivatives of α-keto acids was sensitive at the pg level with a response linear over 3–4 orders of magnitude. The derivatives were formed by reaction with *o*-phenylenediamine:

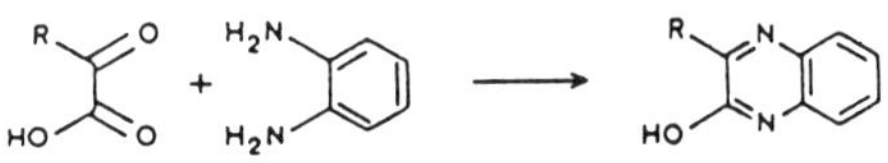

and could be analysed without the additional silylation of hydroxyl that is required for GC analysis. The procedure has been applied to the α-keto acids of human urine (Figure 7).

The thermionic detector was used [26] in phosphorus mode for the capillary SFC analysis of thiophosphinic esters derived from steroids by reaction with dimethylthiophosphinic chloride in the presence of 4-dimethylaminopyridine catalyst.

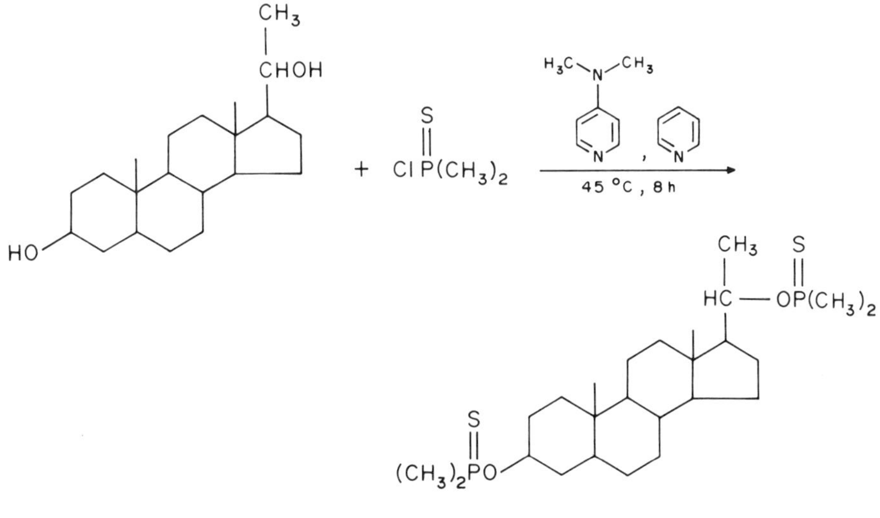

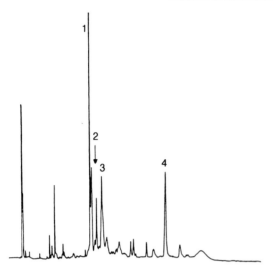

Figure 7. SFC separation of quinoxalinol derivatives of α-keto acids of human urine. Tentative peak assignments: 1, methyl; 2, ethyl; 3, n-propyl; 4, benzyl quinoxalinols. Conditions: 10 m × 50 μm ID open-tubular column; SE-30 stationary phase; N_2O mobile phase; temperature 110–200 °C; thermionic detection. (Reprinted with permission from Ref 25).

A sensitivity of $120 \, fg \, s^{-1}$ was observed for the bis(thiophosphinic) ester of pregnanediol. The SFC analysis of steroids from human urine and plasma as thiophosphinic esters was demonstrated.

Derivatization has also been applied to generate UV-absorbing compounds from amphetamines [27]. The reagent used was FMOC, which reacts with primary and secondary amines to form derivatives which provided selective and sensitive UV detection during packed column SFC with modified CO_2 as mobile phase. David and Sandra noted [28] that CO_2 is able to react with amines at elevated temperatures and pressures to give carbamic acid derivatives insoluble in supercritical CO_2. Formation of the trifluoroacetyl derivatives, however, gave excellent peak shapes in capillary SFC for a variety of C_{16}–C_{21} amines with CO_2 mobile phase and FID.

Caesar et al. [29] used the reagent pentafluorobenzyl aminobenzoate during the analysis of oligosaccharides. The reagent attaches glycosidically to the reducing end of the glycan residues to form fluorescent and UV-absorbing derivatives for SFC. Femtomole detection limits were also possible during SFC with negative-ion chemical ionization mass spectrometric detection; ester cleavage and pentafluorobenzyl elimination gave a single MW-related fragment ion in high abundance.

4. Conclusions

Generally, compounds not accessible to GC (those of high molecular weight or thermally labile) or to HPLC (owing to the absence of a UV chromophore) are amenable to study by SFC without the need for derivatization. However, derivatization is advantageous when an increase in solubility in the mobile phase (supercritical CO_2) is required, and for enhancing the selectivity and sensitivity of detection.

Successful derivatization of several compounds has enabled analysis by SFC under more favourable conditions than GC (i.e. lower temperatures), and resulting in superior resolution to that obtained with HPLC. Increase in the MW after derivatization may, in fact, render the product unsuitable for GC. Analysis of the derivatives by SFC can be more rapid than by HPLC. Attainment of very sensitive detection has also been made possible with a range of detectors.

The actual methods used for derivatizing are similar to those employed in the other chromatographic techniques, the prevailing methods being silylation and the use of FMOC.

A systematic study of common derivatizing agents for SFC has been made [30]. Unique derivatizing reagents, especially those with large blocking groups, have also been investigated as a means of reducing the polarity of functional groups more effectively. The results showed that the use of N-methylbis(trifluoroacetamide) (MBTFA) with pyridine, or 1-(trimethylsilyl)imidazole (TMSI), was successful for almost every functional group studied (–OH, –NH, –NH–CH_3, –SH and –PO_2). These reagents should therefore serve as general-purpose derivatizing agents for SFC.

5. Acknowledgements

The authors thank the Science and Engineering Research Council and SmithKline Beecham for a studentship (N.M.).

6. References

[1] R. M. Smith (Editor), *Supercritical Fluid Chromatography*, Royal Society of Chemistry, London (1988).
[2] M. L. Lee and K. E. Markides (Editors), *Analytical Supercritical Fluid Chromatography*, Chromatography Conferences, Provo, Utah (1990).
[3] M. W. Raynor, K. D. Bartle, A. A. Clifford, J. M. Chalmers, T. Katase, C. A. Rowe, K. E. Markides and M. L. Lee, *J. Chromatogr.*, **505**, 179 (1990).

[4] K. D. Bartle, A. A. Clifford and G. F. Shilstone, *J. Supercrit. Fluids*, **2**, 30 (1989).
[5] M. W. Raynor, J. P. Kithinji, I. K. Barker, K. D. Bartle and I. D. Wilson, *J. Chromatogr.*, **436**, 497 (1988).
[6] T. L. Chester and D. P. Innis, *J. High Resolut. Chromatogr.*, **9**, 178 (1986).
[7] T. Gorner and M. Perrut, *LC-GC*, **7** 503 (1989).
[8] E. M. Calvey, L. T. Taylor and J. K. Palmer, *J. High Resolut. Chromatogr.*, **11**, 739 (1988).
[9] T. L. Chester and D. P. Innis, *J. High. Resolut. Chromatogr.*, **9**, 209 (1986).
[10] W. Praznik and R. H. F. Beck, *J. Chromatogr.*, **303**, 427 (1984).
[11] T. L. Chester, J. D. Pinkston and G. D. Owens, *Carbohydr. Res.*, **194**, 273 (1989).
[12] J. D. Pinkston, G. D. Owens, L. J. Burkes, T. D. Delaney, D. S. Millington and D. A. Maltby, *Anal. Chem.*, **60**, 962 (1988).
[13] J. Kuei, G. R. Her and V. N. Reinhold, *Anal. Biochem.*, **172**, 228 (1988).
[14] T. L. Chester, J. D. Pinkston, D. P. Innis and D. J. Bowling, *J. Microcolumn Sep.*, **1**, 182 (1989).
[15] J. D. Pinkston, D. J. Bowling and T. E. Delaney, *J. Chromatogr.*, **474**, 97 (1989).
[16] J. L. Veuthey, M. Caude and R. Rosset, *Chromatographia*, **27**, 105 (1989).
[17] S. Hara, A. Dobashi, K. Kinoshita, T. Hondo, M. Saito and M. Senda, *J. Chromatogr.*, **371**, 153 (1986).
[18] A. Dobashi, Y. Dobashi, T. Ono, S. Hara, M. Saito, S. Higashidate and Y. Yamauchi, *J. Chromatogr.*, **461**, 121 (1989).
[19] W. Roeder, F. J. Ruffing, G. Schomburg and W. H. Pirkle, *J. High Resolut. Chromatogr.*, **10**, 665 (1987).
[20] J. S. Bradshaw, S. K. Aggarwal, C. A. Rouse, B. J. Tarbet, K. E. Markides and M. L. Lee, *J. Chromatogr.*, **405**, 423 (1987).
[21] M. Ashraf-Khorassani, M. G. Fessahaie, L. T. Taylor, T. A. Berger and J. F. Deye, *J. High Resolut. Chromatogr.*, **11**, 352 (1988).
[22] T. L. Chester, D. J. Bowling, D. P. Innis and J. D. Pinkston, *Anal. Chem.*, **62**, 1299 (1990).
[23] J. D. Pinkston, T. E. Delaney and D. J. Bowling, *J. Microcolumn Sep.*, **2**, 181 (1990).
[24] M. Taguchi and S. Toda, *Bunseki Kagaku*, **38**, 189 (1989).
[25] P. A. David and M. Novotny, *J. Chromatogr.*, **452**, 623 (1988).
[26] P. A. David and M. Novotny, *J. Chromatogr.*, **461**, 111 (1989).
[27] J. L. Veuthey and W. Haerdi, *J. Chromatogr.*, **515**, 385 (1990).
[28] F. David and P. Sandra, *J. High Resolut. Chromatogr.*, **11**, 897 (1988).
[29] J. P. Caesar, D. M. Sheeley and V. N. Reinhold, *Anal. Biochem.*, **191**, 247 (1990).
[30] L. A. Cole, J. G. Dorsey and T. L. Chester, *Analyst*, **116**, 1287 (1991).

14

Derivatives for Gas Chromatography–Mass Spectrometry

John M. Halket
Trace Analysis Unit, Bernhard Baron Memorial Research Laboratories; 339 Goldhawk Road, London

1. INTRODUCTION	297
2. THE MASS SPECTROMETRIC DETECTOR	298
2.1 Electron ionization	298
2.2 Chemical ionization	298
2.3 Scanning and selected ion monitoring	299
2.4 Quantitative analysis and derivatization: sensitivity and specificity	300
3. SILYL DERIVATIVES	303
3.1 Trimethylsilyl	305
3.2 t-Butyldimethylsilyl and similar derivatives	310
3.3 Other silyl derivatives	313
4. PERFLUOROACYLATION	313
5. ESTERIFICATION	317
6. CYCLIZATION	318
7. CONCLUSIONS	321
7.1 Future perspective	321
8. REFERENCES	322

1. Introduction

The development of low cost bench-top mass spectrometers has led to a recent upsurge in their use as detectors in the hyphenated technique of gas chromatography–mass spectrometry (GC–MS). Laboratories accustomed to the application of conventional GC detectors such as FID or ECD are now acquiring additional mass selective detectors or ion trap detectors and taking advantage of the higher degree of specificity and, in many cases, sensitivity which they offer. The trend will certainly continue, and it may be anticipated that mass spectrometric detection will increasingly be a statutory requirement for official testing or for product registration purposes.

The present chapter presents some mass spectrometric properties of those derivatives most commonly employed in GC–MS. Each category of derivative is illustrated by selected examples covering major GC–MS analyte classes. Unfortunately space does not permit a full review, so that many interesting and important references to derivatives employed in GC–MS have had to be omitted. Most emphasis is given to silylation and perfluoroacylation and to the more readily available and routine electron ionization mode of ion production. Detailed derivative preparation instructions are generally given in the specialist chapters of this handbook. Unlike the recent review of Anderegg on derivatization in mass spectrometry [1], which is written more for the mass spectrometrist, the present chapter is intended for the chromatographer, and is limited to the major derivative types with illustrations using the more common analyte classes. The aim is that the newcomer to mass spectrometry may acquire some idea of the principles upon which derivative choice may be made. The derivatizations considered include silylation, acylation, esterification, cyclization and formation of mixed derivatives. The preparation and chromatographic properties of the derivatives are described and discussed in considerable detail in the respective chapters of the handbook.

In addition to the usual advantages for chromatography, including reduction in polarity, protection of labile compounds, alteration of retention times and resolution and enhancement of sensitivity, there are further advantages to be gained from the use of derivatives in GC–MS, and these are covered in the next section,

Handbook of Derivatives for Chromatography
Edited by K. Blau and J. M. Halket © 1993 John Wiley & Sons Ltd

which is devoted to the mass spectrometric detector. Only the GC–MS technique is considered. Direct mass spectrometry, including fast atom bombardment, and the increasingly important liquid chromatography-mass spectrometry are covered in Chapter 12.

2. The mass spectrometric detector

The reader new to mass spectrometry is advised to consult an appropriate introductory text [2–9]. A few mass spectrometric terms will be explained here by way of background and to outline the principles of choosing a derivative. As in the flame ionization detector, ions are produced in the mass spectrometric detector, but the mass spectrometer is able to analyze these ions further according to their molecular weights or rather, mass-to-charge ratios (m/z, see below) to provide a mass spectrum. Different principles are employed to achieve this in a variety of types of mass spectrometer. The instruments most commonly used in GC–MS are known as magnetic sector, quadrupole and ion trap mass spectrometers. Their differences are not further described here. 'Bench-top' systems are of the quadrupole or ion trap type.

The mass spectrometric mode of detection, therefore, has properties related to the actual structure of the analyte and not just the elemental composition. Since chemical derivatization can radically alter the analyte structure, it can have a profound and often advantageous effect on the specificity and sensitivity of detection. Several striking examples of this will be given.

A further strength of the mass spectrometric detector is its ability to distinguish between analytes labelled with stable isotopes, thereby allowing such compounds to be used as convenient internal standards for quantification of a wide variety of biomedical and environmental compounds. Alternatively, stable-isotope-labelled compounds may themselves be selectively detected and measured in connection with metabolic and isotopic enrichment studies.

The two major ionization methods will be considered: electron ionization (EI) and chemical ionization (CI).

2.1 Electron ionization (EI)

Electron ionization, sometimes referred to as electron impact, is the most common ionization technique in GC–MS and the most readily available to the chromatographer. The molecule [M] eluting from the GC column is bombarded by electrons with, normally, an energy of 70 eV. Under these conditions, so-called molecular ions are formed, and most of these are positively charged, $M^{+\bullet}$.

$$M + e^- \longrightarrow M^{+\bullet} + 2e^-$$

Perhaps an energy below about 20 eV is necessary for ionization, and the excess energy imparted to the so-called molecular ion $M^{+\bullet}$ will cause it to fragment until the supplied energy is exhausted:

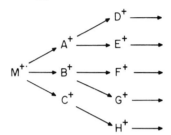

The fragmentation routes are therefore related to the structure of $M^{+\bullet}$. The mass spectrometric detector is capable of measuring the masses and abundances of the ions produced. The mass spectrum or fragmentation pattern is a bar graph showing the mass to charge ratio (m/z, generally the mass m in Daltons where the charge z is unity) along the x axis and the ion abundance on the y axis. In certain cases doubly or even multiply charged ions are produced, but these will not be considered here. It is clear that the fragmentation pattern can be radically altered by chemical derivatization. The EI mass spectrum of n-heptanol is reproduced in Figure 1a, and it is clear that the molecular ion (116 Daltons) has fragmented completely, the highest mass ion being $[M-H_2O]$ at m/z 98 and of very low abundance.

The mass spectrum of the corresponding trimethylsilyl ether derivative is shown in Figure 1b, and an important advantage of this derivatization is immediately apparent in the prominent $[M-CH_3]$ ion at m/z 173. This ion allows the estimation of the molecular weight of the derivative in cases where a molecular ion is absent. In addition, because the fragmentation pattern is characteristic of the molecule, it may be used as a mass spectral 'fingerprint' to confirm the identity of the GC peak, and libraries of reference spectra are readily available. Trimethylsilylation is one of the most common derivatizations in GC–MS. Further details and examples are given later in the section on silylation.

2.2 Chemical ionization (CI)

The other ionization method to be considered here is chemical ionization (CI), which involves the addition of a reactant gas such as methane or ammonia. The technique is of a less routine nature, requiring a more skilled operator than EI, owing to the effects of various

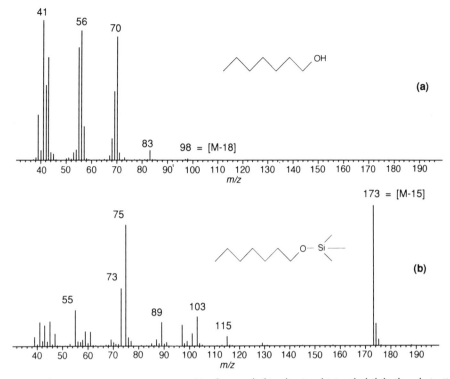

Figure 1. Electron ionization mass spectra: (a) n-heptanol; (b) n-heptanol trimethylsilyl ether derivative.

parameters such as reactant gas pressure on the results. In addition, problems can arise through a more rapid fouling of the ion source, requiring frequent maintenance. The reactant gas is first of all ionized by means of high energy electrons; for methane:

$$CH_4 + e^- \longrightarrow CH_4^{+\bullet} + 2e^-$$

These ions then collide with neutral methane molecules and reactions take place:

$$CH_4^{+\bullet} + CH_4 \longrightarrow CH_5^+ + CH_3^{\bullet}$$

The CH_5^+ ions react with and ionize the analyte M:

$$M + CH_5^+ \longrightarrow [M+H]^+ + CH_4$$

The so-called adduct ions $[M+H]^+$ or $[M+CH_4]^+$ are produced normally with much less fragmentation than in the EI case. Thus, CI is generally more useful for the determination of molecular weights than EI.

Both positively and negatively charged ions can be recorded. In EI, negative ions are generally of very low abundance, but CI, especially with electron capturing derivatives, can increase the ionization efficiency to produce abundant ions together with low chemical background, leading to highly sensitive analyses.

2.3 Scanning and selected ion monitoring

The mass spectrometric detector can be operated in two major modes determined by the nature and objectives of the analysis. These modes are **scanning** and **selected ion monitoring** (SIM).

2.3.1 Scanning

Scanning refers to the detector being scanned over a range of m/z values, not unlike the scanning of a multi-wavelength or diode array detector in HPLC. The time taken is typically 1s to scan a mass range of m/z 50–500, and repetitive scanning at this or faster rates is generally adequate to obtain a required minimum number of sample scans from a capillary column GC peak. Scanning is important when dealing with an unknown compound, in order to determine the molecular weight and to gain valuable structural information, which can be deduced from the fragmentation pattern. Clearly, the choice of a suitable derivative to maximize the abundance of the molecular ion or to influence or direct the fragmentation may be of prime importance. A further important aspect is the ability of the mass spectrometer to detect stable isotopes. The derivatizing reagent can be used to introduce, e.g. deuterium into the molecule. The resulting changes in the fragmentation pattern can assist the deduction of structural features.

Alternatively, the scanned mass spectrum may be compared with a library of reference spectra, and many spectra of derivatives have been published for this purpose. This comparison step may be done by computer or manually using a variety of suitable reference publications. The comparison, when supplemented by the GC retention index, demonstrates the true power of GC–MS for the unambiguous identification of mixture components at trace levels. The choice of derivative may be influenced by that used in specialist libraries of reference spectra, e.g. methyl esters of organic acids.

The scanning mode may also be employed to detect the presence of a specific compound or group of compounds or for quantitative analysis. Mass spectra are recorded repetitively during the GC run, and the computer is employed afterwards to plot individual ion currents, which are sometimes referred to as mass chromatograms. This is analogous to the use of a variable wavelength detector in HPLC where spectra are scanned 'on the fly'. The choice of specific signals available to the analyst then reveals the further power of the GC–MS technique. Indeed, the presence of characteristic compound-specific fragment ions can be of significant diagnostic value and profiles can be plotted using such ions. Such profiles can enable samples to be compared and classified. Derivatization may be employed to alter and 'improve' the fragmentation pattern for this purpose.

2.3.2 Selected ion monitoring (SIM)

The sensitivity of detection can be increased dramatically, 1000-fold in some cases, by allowing the detector to monitor only one, two or several ions, analogous to the use of a diode array detector in HPLC when the instrument is set to monitor a few wavelengths. The sensitivity improvement comes from many more ions of a single m/z value being counted in unit time. The specificity of the analysis can be constantly checked by further monitoring the ratios of abundances of the analyte ions. The ratios should be constant, and any significant variation will indicate the presence of some contaminant(s). This is analogous to the wavelength ratio method for determining peak purity in HPLC. Further ions may be representative of the internal standard, and stable-isotope-labelled compounds are particularly suitable for this purpose, providing a basis for the definitive nature of many GC–MS reference assay methods. This mode is only suitable where the mass spectrum and retention characteristics of the analyte are known.

2.4 Quantitative analysis and derivatization: sensitivity and specificity

Both scanning and selected ion monitoring (SIM) modes may be employed for quantitative analysis. SIM is generally used for lower level measurements. In both modes, the abundances of m/z values specific for the analyte(s) and internal standard(s) are plotted to form selected ion chromatograms. The peak areas or heights are then used in the conventional way to obtain calibration curves and quantitative results. The specific chromatograms obtained from scanning are sometimes referred to as mass chromatograms and those from SIM as mass fragmentograms, but the literature is not consistent in the use of these terms.

In general, the requirement in quantitative analysis, both scanning and SIM, is to have one or more high mass ions of high abundance. Referring to Figure 1a, the fragment ions of low mass in the mass spectrum of n-heptanol are of greater abundance than the ions of higher mass, and this pattern is repeated for most 70 eV EI mass spectra of aliphatic or partly aliphatic compounds. To quantify n-heptanol, the only abundant ions to monitor are at low mass, and any measurement will suffer from interference from the fragment ions of contaminants such as hydrocarbons ('chemical noise'). Such low mass ions lack specificity. Figure 2 shows the partial composite mass spectrum, up to m/z 150, obtained by averaging all 53 995 spectra [10] in the NIST/EPA/MSDC Mass Spectral Database [11].

It is apparent that the lower mass ions in these electron ionization spectra are generally more abundant than those at higher values, i.e. the lower the m/z value, the greater is the chance of contamination. Therefore, a derivative should be prepared to raise the m/z values of molecular or fragment ions as high as possible, without, of course, adversely affecting the GC properties of the analyte. The situation is improved on trimethylsilylation (Figure 1b), as a more suitable ion is produced at m/z 173 ($M - CH_3$) by virtue of the methyl leaving group. Silylation is considered in more detail in Section 3 and in Chapter 4. The column bleed should not be forgotten, as it also has its own mass spectrum depending on the stationary phase. Figure 3 shows a random mass spectrum of the background from a chemically bonded methylsilicone (OV-1) fused-silica capillary column at elevated temperature.

The typical silicon ions at m/z 207, 281 and 355 are evident, and should be avoided for analytical purposes.

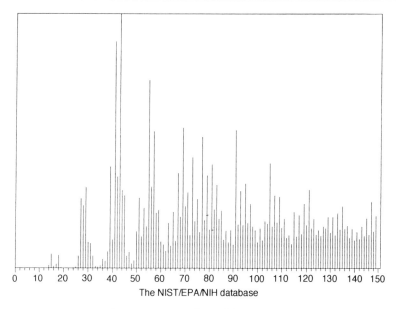

Figure 2. Composite mass spectrum (partial, up to m/z 150) obtained by averaging all 53 995 mass spectra in the NIST/EPA/MSDC Mass Spectral Database [10].

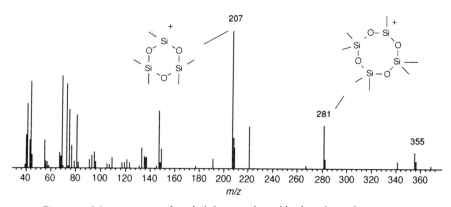

Figure 3. Mass spectrum of methylsilicone column bleed at elevated temperature.

The application of high resolution MS (see below) can often overcome such background problems.

Further sensitivity improvements may be achieved with electron capturing derivatives, usually containing halogen atoms, and negative-ion chemical ionization mass spectrometry. This method is particularly sensitive because the hydrocarbon and other contaminants give only low yields of negative ions and are hardly detected. The resulting improvement of signal-to-noise ratio (analogous to the ECD technique) means that very low levels of detection are possible down to the femtomole, attomole and even zeptomole levels in suitable cases, as illustrated below.

Further improvement in specificity can be achieved by means of a high resolution mass spectrometer, which can separate, measure and monitor accurate m/z values to several decimal places. This means that substances with the same nominal m/z values for molecular ions and fragments but different elemental compositions can be distinguished and measured. Introduction of halogen-containing groups by derivatization can lower the accurate m/z values of halogen-containing ions relative to non-halogen ions, and such a high resolution mass spectrometer can distinguish between them. In this way, the apparent chemical noise (non-derivatized) is reduced and the specificity and sensitivity are improved. High

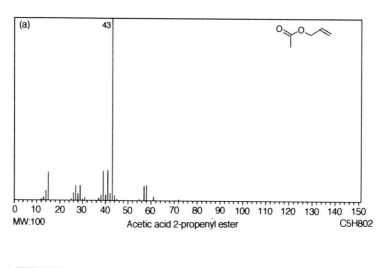
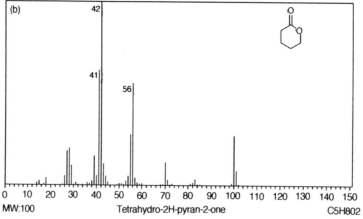
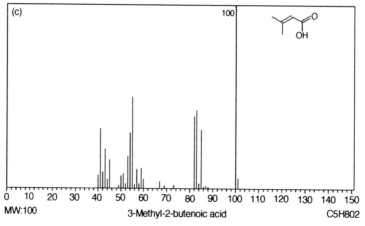

Figure 4. Electron ionization mass spectra [11] of three isomers: (a) acetic acid 2-propenyl ester; (b), tetrahydro-2H-pyran-2-one, and (c) 3-methyl-2-butenoic acid illustrating the variation in mass spectral pattern with change in structural features.

resolution mass spectrometers are very expensive and generally require experienced operators. The reader should refer to a suitable mass spectrometry text for further details [2–9].

The effect of structural change on the mass spectral pattern is illustrated in Figure 4, which compares the 70 eV mass spectra of the three isomers acetic acid 2-propenyl ester, tetrahydro-2H-pyran-2-one and 3-methyl-2-butenoic acid. The spectra were taken from the NIST Mass Spectral Database [11].

Although these are not derivatives, they illustrate that the stability of the molecular ion m/z 100 and the abundance of this ion increase with ring content (Figure 4b) and with resonance stabilization of a conjugated system (Figure 4c). In the same way, derivatives which can effect ring closure and/or produce conjugation may enable ions of higher abundance to be formed. Examples are given in Section 6 (Cyclization), and cyclic derivatives are covered in detail in Chapter 7.

In addition to many chromatographic advantages, derivatives may be prepared in order to:

(1) direct the mass spectral fragmentation in such a way as to facilitate the structure elucidation of the compound, e.g. double bond location;

(2) study the fragmentation process itself and aid the structure elucidation process by inclusion of stable isotopes such as deuterium in the derivative;

(3) increase the abundance of the molecular ion or a related ion to enable the determination of molecular weight;

(4) improve the sensitivity of the analysis by directing the formation of high abundance, high mass ions;

(5) improve the sensitivity of analysis by introducing strongly electron capturing groups for detection by negative-ion chemical ionization;

(6) improve the specificity of the analysis by provision of multiple ions of high abundance whose abundance ratios may be monitored;

(7) introduce halogen or other mass-deficient atoms to facilitate high resolution mass spectrometric detection by lowering the chemical noise level.

Each category of derivative application is illustrated by selected examples covering major GC–MS analyte classes. Unless otherwise stated, EI mass spectra are referred to.

Unfortunately, space does not permit a full review, so that many examples and important applications of derivatives in GC–MS have been omitted. However, some mass spectrometric properties and applications as well as detailed derivative preparation instructions are given in the specialist chapters on Esterification (Chapter 2), Acylation (Chapter 3), Silylation (Chapter 4), Formation of cyclic derivatives (Chapter 7), and also in the respective chapters of the first edition of the handbook [12].

3. Silyl derivatives

The main reactions considered here are formation of the trimethylsilyl (TMS) and t-butyldimethylsilyl (TBDMS) derivatives of hydroxy, amine and carboxylic acid groups.
and, alone or in combination with others, have been applied to a diverse range of analytes. They often have complementary mass spectrometric properties.

TMS derivatives generally produce a fair number of diagnostically useful ions which can assist with the structure elucidation of unknowns or provide detailed 'fingerprints' for comparison with reference libraries.

TBDMS derivative mass spectra are normally dominated by large $[M-57]^+$ ions, which can enable molecular weight determination and are very suitable for quantitative measurements by selected ion monitoring. The mass spectra of cholesterol TMS and TBDMS ether derivatives are compared in Figure 5. The TMS spectrum (Figure 5a) is characterized by several prominent fragment ions, whereas the TBDMS spectrum (Figure 5b) is dominated by the $[M-57]^+$ ion at m/z 443.

Care must be taken that the silylation of multi-functional group compounds does not raise the molecular mass of the analyte or the m/z values of important

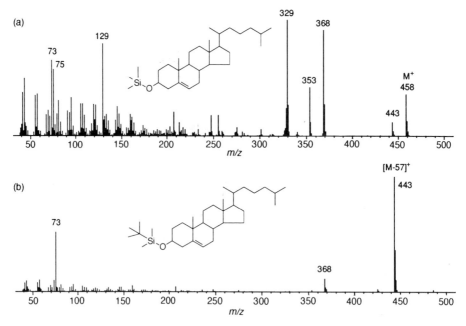

Figure 5. Electron ionization mass spectra of (a) the trimethylsilyl (TMS) ether and (b) the t-butyldimethylsilyl (TBDMS) ether derivatives of cholesterol.

Table 1 Mass increments produced by the major silyl derivatives used in GC–MS

Derivative	Structure	Number of derivatized groups	Increment
TMS		1	72
		2	144
		3	216
	R—Si—	4	288
		5	360
TBDMS		1	114
		2	228
		3	342
	R—Si—	4	456
		5	570

fragment ions above the mass range of the detector. As Table 1 shows for the two major silylations, the mass can increase substantially, and one must bear in mind that the measurement limit of most bench-top instruments is 650 or 800 amu.

The enol problem

A problem can arise with unprotected keto groups owing to the formation of the hydrolytically and thermally unstable enol–TMS ethers, which are difficult to prepare quantitatively (see Chapter 4). However, examples are given below of the application of TMS enol–TMS ether derivatives in the steroid field (see also Section 3.2, TBDMS). In many cases multiple products are formed, which can complicate the analysis.

$$R-\underset{\underset{O}{\|}}{C}-CH_2-\underset{\underset{OH}{|}}{\overset{\overset{R'}{|}}{CH}} \longrightarrow$$

$$R-\underset{\underset{O}{|}}{\overset{\overset{Si}{|}}{C}}=CH-\underset{\underset{O-Si-}{|}}{\overset{\overset{R'}{|}}{CH}}$$

For this reason, such keto groups are normally protected by oxime or alkyloxime formation (see Chapters 4 and 6) prior to silylation:

$$R-CO-CH_2-CH(OH)-R' \longrightarrow R-C(=N-OH)-CH_2-CH(OH)-R'$$
$$\longrightarrow R-C(=N-OTMS)-CH_2-CH(OTMS)-R' \quad \text{(oxime)}$$

or

$$R-CO-CH_2-CH(OH)-R' \longrightarrow R-C(=N-OR'')-CH_2-CH(OH)-R'$$
$$\longrightarrow R-C(=N-OR'')-CH_2-CH(OTMS)-R' \quad \text{(alkyloxime)}$$

Although the oxime-containing derivatives exist in *syn* and *anti* forms, which usually separate on the column, this does not normally present problems in either qualitative or quantitative analyses. However, artefacts can be produced in certain cases, e.g. 3-ketoacids can form 2-isoxazolin-5-ones on reaction with hydroxylamine hydrochloride and cyclization of the resulting oximes after acidification [13], giving rise to multiple peaks.

3.1 Trimethylsilyl

Trimethylsilyl (TMS) derivatives are among the most commonly used in GC–MS and are characterized by ease of formation, excellent GC properties [14] and useful mass spectrometric fragmentation (Figure 5a). In addition, many TMS derivative mass spectra are available as reference spectra in a number of available libraries [11, 15–20].

The molecular ion is often very small or absent, as shown in Figure 1b, but the molecular weight can usually be obtained from the $[M-15]^+$ ion which is normally present. This common ion is formed by loss of a methyl group from the molecular ion:

$$M^{+\cdot}$$

$$\left[R-CH(R')-O-Si\diagup \right]^{+\cdot} \longrightarrow$$

$$[M-15]^+$$

$$\left[R-CH(R')-O-Si\diagup \right]^+$$

For molecular weight determination, the ionization energy can be reduced in cases of doubt to suppress fragmentation. The molecular ion will then usually appear. Nearly every EI mass spectrum of a TMS derivative has an m/z 73 ion, which is often the most abundant peak (referred to as the base peak), together with m/z 75. Their structures are $[(CH_3)_3Si]^+$ and $[HO=Si(CH_3)_2]^+$, respectively. A profile of the trimethylsilylated compounds in a complex mixture can be indicated by a mass chromatogram of m/z 73. The presence of an ion at m/z 147 ($[(CH_3)_2Si=O-Si(CH_3)_3]^+$) indicates two or more TMS groups in the molecule and these can be similarly profiled. Table 2 lists some common TMS mass spectral peaks and their interpretations [21].

In the following sections, some selected examples of TMS and mixed derivatives in GC–MS are briefly described.

3.1.1 Amino acids

The optimum conditions for the pertrimethylsilylation of amino acids have been worked out by Gehrke and co-workers [22, 23], and the mass spectra of over 50 amino acid TMS derivatives are available in the literature [24, 25]. Unfortunately, persilylation is not universally applicable to this group of compounds having such a diverse range of properties: arginine is converted to ornithine, glutamine to glutamic acid and asparagine to aspartic acid during the silylation procedure. Great care must thus be exercised when interpreting the mass spectra of unknowns after derivatization. α-Amino acids may be selectively detected by monitoring the significant fragment at m/z 218 [26].

$$\text{TMS-NH=CH-COOTMS} \quad m/z \; 218$$

Even such high molecular weight compounds as thyroxine may be pertrimethylsilylated [27] for GC–MS analysis, but this derivative shows mainly lower m/z fragment ions. Thus most thyroxine assays tend to be carried out with perfluoroacyl derivatives, which have more abundant higher m/z fragments (see Section 5, below).

3.1.2 Biogenic amines

Although catecholamines can be persilylated [28], most GC–MS work has utilized a variety of perfluoroacyl

Table 2 Characteristic ion fragments of trimethylsilyl derivatives [21]

Ion	Assignment	Comments
43	CH_3Si	
45	CH_3SiH_2	
59	$(CH_3)_2SiH$	
73	$(CH_3)_3Si$	Frequently base peak
74	$(CH_3)_2SiO$	SE-30 and OV-17 liquid phase
75	$(CH_3)_2SiOH$	
84	(i) $(CH_3)_2SiCN$ (ii) $(CH_3)_2SiNC$	
89	$H_2C=O-SiH(CH_3)_2$	
91	$(CH_3)_2SiSH$	
98	$(CH_3)_2SiNH-C\equiv CH$	
99	(i) $(CH_3)_2SiOC\equiv CH$ (ii) $(CH_3)_2SiNHC\equiv N$	
100	(i) $(CH_3)_2Si=N=CH-CH_3$ (ii) $(CH_3)_2SiOC\equiv N$ (iii) $(CH_3)_2Si=N=C=O$	Amino acids Urea
102	$TMS-NH=CH_2$	Primary amines
103	$TMS-O=CH_2$	Intense in primary alcohols but not in polyfunctional molecules, i.e. not present in low molecular weight hydroxy acids.
112	$(CH_3)_2SiNHC\equiv C-CH_3$	
113	$(CH_3)_2SiO-C\equiv C-CH_3$	
115	$(CH_3)_2SiS-C\equiv CH$	
116	$(CH_3)_2SiS-C\equiv N$	
117	(i) $CH_3CH=OTMS$ (ii) $(CH_3)_2SiO-CH_2CHO$ (iii) $TMSO-CH_2CH_2$ (iv) $(CH_3)_2SiO-C(OH)=CH_2$	TMS ethers C_5 and C_6 of hexoses Glycerols Loss of CH_3 from 132 of TMS esters
127	$TMSO=C=C=CH_2$	
129	(i) $TMSO=CH-CH=CH_2$ (ii) $(CH_3)_2Si-S-C=C-CH_3$ (iii) $(CH_3)_2SiO-CO-CH=CH_2$	Carbohydrates, glycerides, esters, and steroids Loss of methane from 145 of aliphatic esters
131	$TMSO-C(CH_3)_2$	Isopropyl alcohol branch
$131 + (14)_n$	$TMSO-CH(CH_2)_nH$	Alkyldiols
132	$TMSO-C(OH)=CH_2$	TMS esters
145	$TMSO-CO-CH_2-CH_2$	Aliphatic esters
147	$(CH_3)_2Si=O-Si(CH_3)_3$	Intense in nearly all poly-TMS derivatives and indicates that more than one site of silylation is present. May be in background from pyrolysis of TMS compounds.
149	$(CH_3)_2Si=O-\overset{\overset{OH}{\mid}}{Si}-(CH_3)_2$	
170	cyclohexenyl–OTMS	Dicarboxylic TMS esters
173	$TMSO-CH=CH-NH-CO-CH_3$	N-acetylhexosamines
174	$CH_2=N(TMS)_2$	
179	phenyl-CH_2/OTMS or phenyl-CHOTMS or methylenedioxyphenyl-$Si(CH_3)_2$	From diTMS catechols

Table 2 (continued)

Ion	Assignment	Comments
179 + R	R—C₆H₄(CH₂)(OTMS) or R—C₆H₄—CHOTMS or H₂C—C₆H₃(R)—O—Si(CH₃)₂—O (methylenedioxy-type)	From substituted diTMS catechols
191	TMSO–CH=OTMS	Carbohydrates
202	3-CH₂-indole, N-TMS	Indoles
202 + R	R-substituted 3-CH₂-indole, N-TMS	Substituted indoles
204	(i) TMSO–CH=CH–OTMS (ii) (TMSO)₂C=CH₂	Carbohydrates TMS esters
205	TMSO–CH₂–CH–OTMS	Alditols, carbohydrates
206	C₆H₅–CO–N(TMS)=CH₂	Hippuric acids
217	(i) TMSO–CH=CH–CH–OTMS (ii) TMSO–C(=OTMS)–CH=CH₂ (iii) TMSO=CH–CH=CH–OTMS	Carbohydrates Esters Alditols
243	HO–P(=O)(OTMS)–OTMS	Phosphate esters
292	TMSO–CH=C(OTMS)(OTMS)	Polyhydroxy acids
299	TMSO–P(=O)(OTMS)–O=Si(CH₃)₂	Phosphate esters
315	TMSO–P(=O)(OH)–OTMS... OTMS	Phosphate esters
319	TMSO–CH=CH–CH(OTMS)–CHOTMS	Carbohydrates
387	(TMSO)₄P	

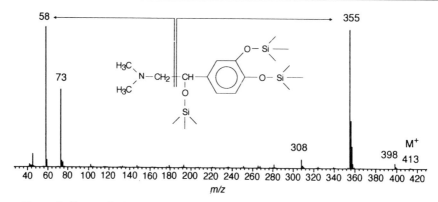

Figure 6. Electron ionization mass spectra of pertrimethylsilylated N-methyladrenaline.

derivatives (see Section 4 and Chapter 3), which tend to give better fragmentation characteristics and have long been used for low level determinations by GC with ECD. In addition, the amine group is relatively difficult to silylate.

An example of the mass spectrum of a catecholamine TMS derivative is shown in Figure 6, the pertrimethylsilylated N-methyladrenaline.

The cited derivative has been employed for the determination of this compound [29]. Although a strikingly abundant ion at m/z 355 is evident, it coincides with the same ion from silicone column bleed (Figure 3) and would therefore be precluded from use in the low level SIM assay of this compound with a silicone-containing stationary phase. A high resolution mass spectrometer would be necessary.

TMS derivatives of primary amines may be detected at the femtomole level by monitoring the ion m/z 174 [30] produced by β-cleavage with charge retention on the nitrogen, which is characterisic of TMS amines:

$$CH_2=N^+{<}^{TMS}_{TMS} \quad m/z\ 174$$

Similarly, the 5-hydroxyindolic group can be specifically detected by monitoring the ion at m/z 290 [30]:

m/z 290

3.1.3 Carbohydrates and sugars

One of the earlier GC–MS applications involved the hexatrimethylsilylation of glucose and its quantification [31]. The method was unsatisfactory owing to the formation of multiple products. Later examples, involving the methoxime–penta-TMS derivative, showed this derivative to be sufficiently stable for application in stable-isotope-dilution reference methods [32, 33] and in the determination of the enrichment of 6,6-d_2-glucose in human plasma [34]. The role of mass spectrometry in carbohydrate analysis and the application of TMS derivatives to sugar analysis, including N, O-acylneuraminic acids (sialic acids) and methyl glycosides, has been reviewed by Kamerling and Vliegenhart [35]. Pertrimethylsilylated derivatives are useful in sequence analysis, but permethylated derivatives are more commonly used.

3.1.4 Drugs

Although most assay methods rely on high m/z ions, as discussed above, it is still possible to utilize low mass ions of high abundance provided that the mass spectrometer is operated at a high enough resolution to 'tune out' interference from the column bleed. In the case of the β-blocker betaxolol, the fragment ion at m/z 72 corresponding to $[CH_2=NH-CH(CH_3)_2]^+$ was employed to measure concentrations down to 10–20 pg per sample with little difficulty [36]. TMS derivatives again find wide application for both qualitative and quantitative analysis of drugs and metabolites. A current example from the forensic field is the low level detection of Δ^9-tetrahydrocannabinol-9-carboxylic acid. The mass spectrum of the TMS derivative exhibits abundant high mass ions, enabling low nanogram level measurements to be achieved on urine by EI MS [37, 38]. In CI mode, levels as low as 0.2 ng ml^{-1} of tetrahydrocannabinol can be measured from 1 ml of plasma using this derivative [39]. The CI mass spectra have abundant [M+H]$^+$ ions, providing sensitivity, but the lack of fragment ions can lead to confirmation difficulties.

3.1.5 Fatty acids and lipids

The mass spectra of unsaturated fatty acids with different double bond positions are very similar, owing to double bond migration during ionization. Thus, mass spectrometry alone cannot be employed to determine double bond positions, and has to be combined with some chemical transformation or derivatization, or both. 'Fixing' of the double bond position(s) by ozonolysis of a diol [40] followed by mass spectrometry has been employed to solve this problem. Many derivatives including TMS ethers have been employed to enhance the diagnostic value of the mass spectrum and facilitate the location of the double bond(s), and this topic has been reviewed in detail [41]. TMS derivatives are also highly suitable for determination of the numbers and positions of the substituents in the polyhydroxylated products of fatty acid methyl esters [42]. Abundant diagnostic ions are formed by cleavage between the pairs of derivatized groups. The TMS derivatives suffer from an incremental molecular weight problem when large numbers of hydroxyl groups are derivatized, i.e. the molecular weight of the derivative can easily exceed the m/z range of the instrument employed, especially if it is of the 'bench-top' type. Pyrrolidide, picolinyl and nicotinylidene derivatives having improved properties for these structure elucidation steps are briefly described below (see Section 5 on esterification). Chemical and mass spectrometric methods for double bond location are also included in the review by Anderegg [1].

The mass spectra of TMS ether derivatives of monoacyl-, monoalkyl- and monoalk-1-enyl-glycerols have been described [43]. Diradylglycerols may also be trimethylsilylated, and the mass spectra have structurally informative features [44–47].

3.1.6 Nucleosides

Derivatives for nucleic acids have been reviewed [48]. The application of TMS and d_9-labelled TMS derivatives is useful in the structure elucidation of newly discovered nucleosides [49]. Recent examples include the GC–MS (SIM) of trimethylsilylated nucleosides in the assessment of DNA damage [50] and the identification of a novel nucleoside, $1,N^6$-dimethyladenosine, in human cancer urine [51].

3.1.7 Organic acids

The profiling of organic acids in urine is usually performed with TMS derivatives [21] or alkyloxime–TMS derivatives; the ethyloximes have improved GC separation characteristics [52, 53], although some laboratories prefer methyl esters (see Section 5, below), a probable consequence of their early introduction and the availability of in-house mass spectral collections. The universal properties of pertrimethylsilylation for organic acid analysis are strikingly demonstrated by the quantitative GC–MS analysis of 130 different compounds, including organic acids, amino acids, sugars and creatinine, in urine [54]. The method uses a commercial target compound analysis programme which allows specific quantification of 106 analytes in 6 minutes following data acquisition.

3.1.8 Peptides

Sequences of peptides may be analysed by the reduction of N-acetylated peptides to N-ethylpolyamino alcohols followed by interpretation of the mass spectra of the O-TMS derivatives [55, 56]. The latter have excellent mass spectrometric properties, but the GC–MS method has been largely supplanted by developments in direct mass spectrometric methods such as fast atom bombardment (see Chapter 12).

3.1.9 Prostaglandins and related compounds

Many GC–MS methods for prostaglandins utilize the more common EI method rather than CI. Both TMS [57–59] and TBDMS ether derivatives (see next section) are commonly employed as mixed derivatives with other types, e.g. oximes and methyl esters. As expected, the TMS derivatives provide multiple fragment ions, which aid specificity as discussed earlier. The TBDMS derivatives are generally chosen for sensitivity, much of the ion current being concentrated on only a few fragment ions. Diagnostic ions for methyloxime–methyl ester–TMS ether derivatives of 11 prostaglandins and thromboxane B_2 have been tabulated by Kelly [60].

The mass spectrometry of leukotrienes, particularly as methyl ester–TMS ether derivatives, has been described by Murphy and Harper [61].

3.1.10 Steroids, sterols and bile acids

Trimethylsilylation in combination with methyloxime (MO) formation is the most popular derivatization method for steroids [62, 63], and the literature contains many examples of corresponding fragmentation pathways and details of the *syn* and *anti* isomers formed with non-sterically-hindered keto groups. A typical steroid MO–TMS derivative mass spectrum is reproduced in Figure 7 for pregnenolone [64], together with an indication of the origins of the major ions.

Diagnostic ions such as m/z 129, also seen in Figure 5a for cholesterol, point to the presence of the Δ^5-3-

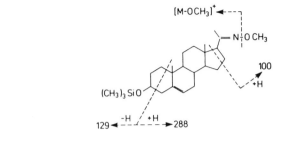

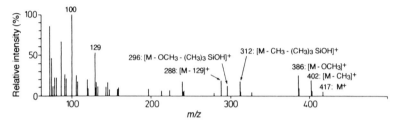

Figure 7. Electron ionization mass spectrum of pregnenolone methoxime–TMS ether. The origins of the major fragment ions are indicated. Reproduced from reference [65] with permission.

hydroxy-group, the [M − 31]$^+$ ion to the methoxime, and fragmentation differences of 90 mass units indicate the TMS groups. Recent applications of GC–MS methods, both EI and CI, with various derivatives have also been reviewed [65].

Enol–TMS derivatives of steroids have been prepared with potassium acetate as catalyst and applied with success [66]. Steroid glucuronides can be analysed as their per-TMS derivatives [67] but low abundance molecular ions are generally obtained.

The most popular method for bile acids is based on the TMS ether–methyl ester mixed derivative, and recent applications have been reviewed [68].

A striking example of the improvement in the mass spectrum of a hydroxylated vitamin D metabolite TMS derivative by formation of a cyclic boronate ester mixed derivative is shown in Section 6 (Cyclization).

Although the TMS derivatives are applicable to a very wide range of analytes, care must be taken to avoid exposure to moisture. They have another major handicap for the GC–MS of compounds with multiple derivatizable functional groups: the molecular weight is increased by 72 daltons for each TMS group added (Table 1). As the m/z range of the majority of bench-top instruments is 0–650 or 0–800, one can quickly obtain derivatives which produce important ions outside this range, so that only partial mass spectra are recorded. In such cases, acylation or methylation (see Sections 4 and 5, below, and Chapters 2, 3 and 5) may be preferred.

3.2 t-Butyldimethylsilyl and similar derivatives

The t-butydimethylsilyl (TBDMS) derivatives have been known for some time as conferring selective protecting groups in synthetic procedures [69, 70]. As illustrated in Figure 5b for cholesterol, the mass spectra of these derivatives are often characterized by abundant [M − 57]$^+$ ions caused by ready loss of the sterically hindered t-butyl moiety:

The leaving group helps to remove much of the excess energy which would otherwise cause further fragmentation of the molecular ion. These very popular derivatives are characterized by ease of formation, hydrolytic stability (see Chapter 4) and the useful mass spectrometric properties outlined above. They are, however, less suitable for compounds with large numbers of derivatizable groups, each TBDMS group adding 114 daltons to the molecular weight (Table 1). A further disadvantage is the difficulty in derivatizing sterically hindered groups, so that in some cases partial derivatives are formed. The abundant [M − 57]$^+$ ions are generally ideal for quantitative analysis by SIM, and the literature contains impressive examples of applications, also in mixed derivative form, to a wide variety of analyte classes. A few of the many examples are outlined in the following sections.

3.2.1 Amino acids

Several reports have appeared on the application of TBDMS derivatives for GC analysis with peak identification using CI MS [71–73]. The method appears to be more universally applicable than TMS derivatization in that arginine and glutamine are also derivatized and not converted to ornithine and glutamic acid as in the TMS case.

In the GC–MS analysis of isotopic enrichment in leucine, the use of the TBDMS derivative gave [M − 57]$^+$ ions of sufficient intensity for EI to be employed for the measurements, which had previously required the more difficult and less readily available CI method [74]. EI with a modern bench-top system and autoinjector allowed complete stable isotope kinetic studies involving hundreds of samples to be reliably completed within days.

TBDMS derivatives have also been applied to the analysis of dipeptides [75].

3.2.2 Biogenic amines

The mass spectrum of isatin (2,3-indoledione), a monoamine oxidase inhibitor, as its oxime–TBDMS derivative

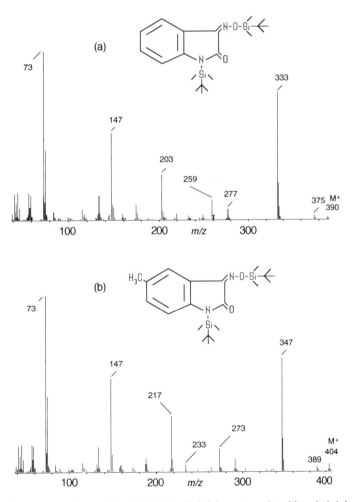

Figure 8. Electron ionization mass spectrum of the N-t-butylmethylsilyl-3-oxime-t-butyldimethylsilyl derivatives of (a) isatin and (b) 5-methylisatin. Reproduced from Reference 76 with permission.

is shown in Figure 8, together with that of 5-methylisatin (internal standard). This mixed derivative was chosen from several, including fluoroacyl derivatives, evaluated for the quantitative assay of isatin in body fluids [76]. The derivative allowed the GC–MS quantification of isatin in urine with minimal sample preparation.

3.2.3 Fatty acids and lipids

The advantages of TBDMS derivatization are amply demonstrated in many applications from the lipid field: alcohols [77], phthiocerols and mycocerosic alcohols [78], deuterium-labelled saturated and unsaturated fatty acids [79], hydroxy fatty acids [80, 81], tuberculostearic acid [82], diacylglycerols [83–85], ceramides [86], 25-hydroxy-vitamin D_3 [87] and vitamin D metabolites [88]. A recent stable-isotope-dilution assay for platelet activating factor (PAF) utilizes hydrolysis to the 2-acetyl-1-O-alkylglycerol. The latter isomerizes to the more stable 3-acetyl form, which is then purified on silica gel and derivatized with t-butyldimethylchlorosilane/imidazole for GC–MS analysis [89]. The derivative is the 3-acetyl-2-TBDMS rather than the 2-acetyl-3-TBDMS, and has a more favourable mass spectrum for selected ion monitoring (76% compared to 31% for m/z 415). Maximum peak areas are obtained at 20 eV ionizing energy. The limit of detection (S/N-3) for $C_{16:0}$ PAF is 20 pg.

The mass spectrometry of lipids has been reviewed in detail by Kuksis and Myher [90], including many examples of the application of derivatives.

3.2.4 Organic acids

TBDMS derivatives have been applied to the analysis of organic acids in urine as oxime–TBDMS [91] and in tissues [92] as methoxime–TBDMS derivatives. The intense $[M-57]^+$ ions found in the spectra of the latter were shown to contain all the original hydrogen atoms. Thus, the derivatization should find application in quantification by isotope dilution and in metabolic studies with deuterium-labelled compounds. TMS and oxime–TMS derivatives, however, still remain the more popular derivatives for general screening, a consequence of the more structurally informative mass spectra, shorter analysis times and the ready availability of reference spectra.

3.2.5 Prostaglandins and related compounds

TBDMS derivatives found early application in prostaglandin and thromboxane analysis [77, 93, 94]. The TBDMS ether/ester derivative of leukotriene B_4 can enable NICI sensitivity to be obtained in the EI mode owing to the presence of a high mass ion of high abundance [95]. A comparative study has been made of the TMS–, TBDMS– and allyldimethylsilyl–methyl ester derivatives of leukotrienes [96].

3.2.6 Steroids, sterols and bile acids

The TBDMS derivatives have been applied to steroids, but have not replaced the TMS derivatives owing to the problems of the molecular weight range (especially corticosteroids) exceeding the normal range of many instruments, long retention times and sterically hindered groups giving partial derivatives which can confuse the profile obtained. As with many other analytes, the TBDMS derivatives are at their best for the quantitative analysis of known compounds [77, 97]. Examples include testosterone [98], testosterone and dehydroepiandrosterone [99], the latter as the methoxime derivative, oestrogens [100] and oestradiol [101]. The GC–MS properties of spirostane TBDMS derivatives have been described and compared with those of TMS, acetate and trifluoroacetate derivatives [102].

Ketosteroids may be converted to enol–TBDMS ether derivatives with t-butyldimethyliodosilane as a catalyst [103]. Steroid enol–TBDMS and mixed TBDMS–TMS derivatives have been applied to the analysis of steroids in rat testes [104]. The enol–TBDMS mass spectra are characterized by the presence of intense M^+ ions.

3.2.7 Phytosterols

The analysis of phytosterols having a single hydroxyl group is greatly facilitated by these derivatives, such that a low level screening may be easily carried out. Figure 9 shows a mass chromatogram of sterols after TBDMS ether formation in an extract of the red alga *Laurencia papillosa* [105]. The powerful combination of specific $[M-57]^+$ ions and retention times provided evidence for the tentative identification of nine sterols. Many more were detected, even at such trace levels that no peaks were clearly visible in the total ion current chromatogram. Retention indices of such sterol TBDMS ethers may be used in a correlation procedure to predict the retention times of the corresponding TMS ethers, which possess more structural information for identification purposes [106].

3.2.8 General

The TBDMS derivatives are much more stable to hydrolysis than the corresponding TMS derivatives. However, the molecular weight increment problem is made worse for GC–MS in some cases by the corresponding 2–3

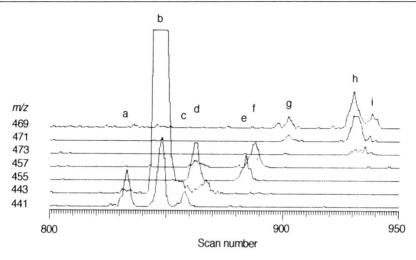

Figure 9. Detection of trace sterols in the red alga *Laurencia papillosa* by mass chromatography of $[M-57]^+$ fragment ions after t-butyldimethylsilyl ether derivatization: (a) m/z 441 (22-dehydrocholesterol), (b) m/z 443 (cholesterol), (d) m/z 455 (brassicasterol), and (e) m/z 457 (campesterol), m/z (h) 469 (stigmasterol), m/z 471 (sitosterol) and m/z 473 (stigmastanol). Reproduced with permission from Reference [105].

times increase in GC retention time. These limitations may preclude the use of this derivative in many GC–MS applications and should be considered when planning experiments. The molecular weight and retention time problems may often be eliminated by the preparation of mixed derivatives, as seen above for the prostaglandins. For quantitative work by SIM these derivatives are hard to beat, and new applications are continually appearing. Alternatives such as the allyldimethylsilyl derivatives (next section) have been suggested, but none of these have come anywhere near to providing useful alternatives.

3.3 Other silyl derivatives

Further alkyldimethylsilyl derivatives have been evaluated in mixed derivatives for the GC–MS analysis of steroids [107–109] and prostaglandins and thromboxane B$_2$ [110]. Although they have favourable properties, they generally provide no significant advantages over the TBDMS derivatives.

3.3.1 Allyldimethylsilyl

The allyldimethylsilyl derivatives were originally made in an attempt to obtain mass spectral characteristics similar to those of TBDMS derivatives, but with ready formation from hindered hydroxy groups. Never really satisfactory, owing to the formation of multiple reaction products and reduced sensitivity in certain cases [96], they have, however, been applied to steroids [111, 112],

steroids and cannabinoids [113, 114], leukotrienes [96, 115] and pyrimidine nucleosides [116]. Their mass spectra are characterized by the presence of prominent $[M-41]^+$ ions analogous to the $[M-57]^+$ ions of the TBDMS derivatives.

4. Perfluoroacylation

Perfluoroacyl derivatives such as trifluoroacetyl (TFA), pentafluoropropionyl (PFP) and heptafluorobutyryl (HFB) are very commonly employed in GC–MS. Much of their popularity arose from their ease of preparation and their useful employment in GC with ECD. Workers then employed them directly in the GC–MS system and in many cases excellent results were obtained—high mass increments which can be conveniently adjusted by choice of derivative (see Table 3) and high abundance fragment ions. All this is in addition to their advantages for negative-ion chemical ionization mass spectrometry, where their electron capturing properties can enable highly sensitive analyses to be carried out.

Once again, care must be taken not to exceed the instrumental mass range with multifunctional group analytes, particularly in the HFB case. Although HFB has such a high mass increment, this can be very useful for the derivatization of compounds such as monohydroxylated steroids, raising the molecular weight well above the region of chemical noise. In the following pages, some selected examples of perfluoroacyl and mixed derivatives in GC–MS are briefly described.

Table 3. Structure and mass increments obtained with the commonly used perfluoroacyl derivatives

Derivative	Structure	Number of derivatized groups	Mass increment
Trifluoroacetyl	R–C(=O)–CF$_3$	1	96
		2	192
		3	288
		4	384
		5	480
Pentafluoropropionyl	R–C(=O)–CF$_2$CF$_3$	1	146
		2	292
		3	438
		4	584
		5	730
Heptafluorobutyryl	R–C(=O)–CF$_2$CF$_2$CF$_3$	1	196
		2	392
		3	588
		4	784
		5	980
Pentafluorobenzoyl	R–C(=O)–C$_6$F$_5$	1	194
		2	388
		3	582
		4	776
		5	970

4.1 Amino acids

The gas chromatography of amino acids commonly employs an esterification step followed by perfluoroacylation. The alcohol is commonly n-butanol, and esterification is followed by trifluoroacylation to give the N-trifluoroacetyl–n-butyl ester (TAB) derivatives. Optimum conditions for this procedure have been worked out in detail by Kaiser et al. [23], and the mass spectra of the TAB derivatives for a large number of amino acids have been published [117, 118]. Further popular derivative combinations employ esterification with alcohols such as isopropanol in combination with PFP or HFB perfluoroacylations.

Serum thyroxine has been precisely measured by selected ion monitoring as its methyl ester–N,O-bis(trifluoroacetyl) [119, 120] and methyl ester–N,O-bis(heptafluorobutyryl) derivatives [121] in connection with the application of isotope dilution GC–MS as a reference method to control the results obtained from less specific immunoassay methods. The fragment ions monitored are in the higher mass range of around 800 amu, with little interference.

4.2 Biogenic amines

Perfluoroacyl derivatives are generally preferred over TMS derivatives for these compounds. In addition to the ready application of ECD for measurement of low levels of biogenic amines and their metabolites, the mass spectra often have more favourable ion abundances for selected ion monitoring and the literature contains an abundance of applications. PFP derivatives found early application to the low level GC–MS (CI) assay of biogenic amines and metabolites in cerebrospinal fluid [122]. The GC–MS properties of the PFP and HFB derivatives of the methyl, trifluoroethyl, pentafluoropropyl and hexafluoroisopropyl esters of twelve acidic metabolites have been compared [123], and the mass spectral behaviour of perfluoroacyl derivatives after acetylation has been investigated [124]. Also, the production of negative ions, assisted by the electron capturing properties of the derivatives, makes them ideal for detection by negative-ion electron capture chemical ionization mass spectrometry. By this technique, sensitivities an order of magnitude greater than those in conventional GC–MS may be obtained. The role of mass spectrometry in the analysis of biogenic amines has recently been reviewed by Blau [125], with 252 literature references.

Further fluorinated derivatives have recently been evaluated for the quantification of biogenic trace amines, e.g. β-phenylethylamine at the pg level in biological samples [126]. The negative-ion CI mass spectra of the PFP, HFB, perfluorosuccinimido, perfluorobenzyl and pentafluorobenzoyl derivatives of phenylethylamine have intense peaks due to loss of HF. Spectra of the 3-(trifluoromethyl)benzoyl, 4-(trifluoromethyl)benzoyl and 3,5-bis(trifluoromethyl)benzoyl derivatives exhibit fragment ions due to loss of H or 2H. The pentafluorobenzoyl derivative has no molecular ion, but acetylation of the fluorobenzoyl derivatives gives primarily molecular ions with increased electron capture cross-sections. The most suitable for phenylethylamine are the N-acetyl-N-pentafluorobenzoyl derivatives, allowing phenylethylamine to be detected at the 75 fg level. The N-acetyl-N-3,5-bis(trifluoromethyl)benzoyl derivative also has satisfactory mass spectral characteristics (Figure 10) [126].

The electron capture method, although very sensitive, does suffer from reproducibility problems, the ionization efficiency and mass spectral patterns being influenced by small changes in operating parameters such as the mass spectrometer ion source temperature.

Sometimes, high sensitivity can be achieved in the more reproducible EI mode by advantageous derivative properties. Application of the spirocyclic derivative

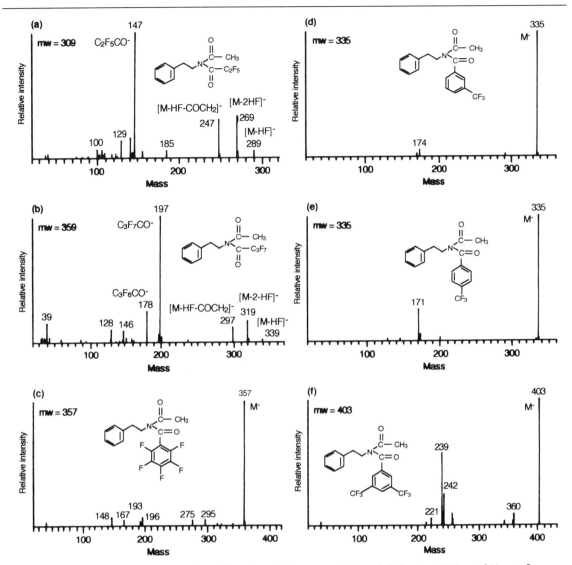

Figure 10. Negative ion mass spectra of acetyl-fluorobenzyl derivatives of phenylethylamine: (a) N-acetyl-N-pentafluoropropionyl; (b) N-acetyl-N-heptafluorobutyryl; (c) N-acetyl-N-pentafluorobenzoyl; (d) N-acetyl-N-(3-trifluoromethyl)benzoyl; (e) N-acetyl-N-(4-trifluoromethyl)benzoyl; (f) N-acetyl-N-[3,5-bis(trifluoromethyl)benzoyl]. Reproduced from Reference 126 with permission.

formed by reaction of pentafluoropropionic anhydride with melatonin is described in Section 6 (Cyclization), below.

4.3 Carbohydrates and sugars

Sugars are normally reduced to their alditols prior to acylation, so that each sugar gives a single derivative. Trifluoroacetates have been used for sugar alcohols in urine [127] by GC–MS and for monosaccharides [128]. However, popular derivatization methods employed in this field are trimethylsilylations, giving highly volatile products, as well as acetylations and methylations. The latter provide lower mass increments when large numbers of derivatizable groups are present.

4.4 Drugs

Pentafluoropropionyl derivatives have been successfully employed for the determination of 11-nor-Δ^9-tetrahy-

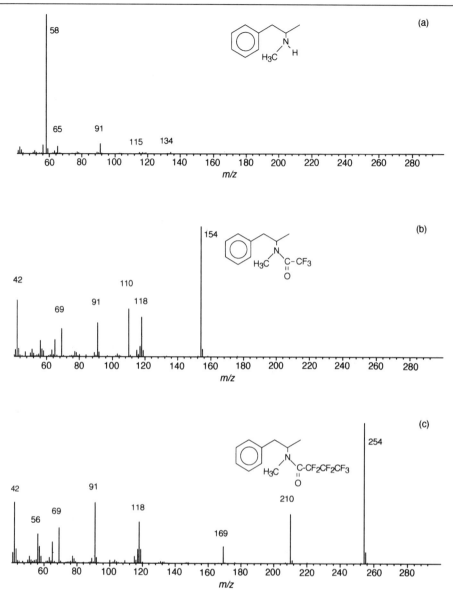

Figure 11. Electron ionization mass spectra of methamphetamine: (a) underivatized, (b) trifluoracetyl and (c) heptafluorobutyryl derivatives.

drocannabinol-9-carboxylic acid, the major urinary metabolite of Δ^9-tetrahydrocannabinol [129, 130]. The mass spectrum of the derivative obtained after a single step reaction with a mixture of PFP anhydride and 2,2,3,3,3-pentafluoropropanol exhibits abundant high mass fragment ions m/z 622 (molecular ion), m/z 607 and m/z 459, enabling measurement down to 5 ng ml^{-1} on a few ml of urine. The same derivative can be employed in the NICI mode [131], a detection limit of 0.7 ng ml^{-1} being obtained.

In the case of amphetamines, the dramatic improvement in mass spectral properties which can be obtained on derivatization is illustrated in Figure 11, which compares the fragmentation patterns of trifluoroacylated and heptafluoroacylated methamphetamine with that of the underivatized compound.

These derivatives are commonly employed for the forensic confirmation of the presence of amphetamines in body fluids. An improved derivatization with heptafluorobutyric anhydride has been described for the analysis of twelve ring- and N-substituted amphetamine drugs in body fluids or seized materials. No heating or standing period is required. Most drugs are distinguishable by their retention times, and all of them by mass spectra [132].

Similar reaction with PFP anhydride can be employed for the derivatization of opiates, such as morphine and codeine [133] and benzoylecgonine [134], an indicator of cocaine abuse. The mass spectra obtained are all highly suitable for confirmation work by GC–MS, leading to the popularity of this derivative type in the forensic laboratory.

4.5 Fatty acids and lipids

A stable isotope dilution reference method for total glycerol utilizes the HFB derivative [135]. Methods are available for the quantification of platelet activating factor by negative CI after hydrolysis and pentafluorobenzoylation [136–138].

4.6 Nucleosides

Trifluoroacetates have been employed for the mass spectrometric analysis of nucleosides and hydrolysates of deoxyribonucleic acid [139], but TMS derivatives are now much more popular.

4.7 Prostaglandins and related compounds

Methyl ester–HFB derivatives have been employed in prostaglandin analysis by GC–MS [140], but most applications employ methyl ester–silyl derivatives (see Section 3) or mixed pentafluorobenzyl ester derivatives for NICI (Section 5, below).

4.8 Steroids, sterols and bile acids

Heptafluorobutyrate derivatives are popular in isotope dilution reference methods for steroids such as testosterone [141], 17β-oestradiol [142–144] and progesterone [145–147]. Even 'difficult' steroids such as aldosterone can be quantified in serum as the 3-enol–HFB ester by stable isotope dilution GC–MS [141] or after a new derivatization procedure with HFB anhydride [148]. Pentafluoropropionyl derivatives have also found application to the quantification of oestriol [141] and urinary progesterone [149].

5. Esterification

Trimethylsilyl esters (see Silylation, Section 3 above) are less stable to hydrolysis than methyl esters. In many cases, the latter predominate for the analysis of acidic compounds such as fatty acids, bile acids and prostaglandins. Some examples of ester derivatives, particularly pentafluorobenzyl esters and mixed derivatives, follow.

5.1 Amino acids

The acidic groups of amino acids are often esterified prior to perfluoroacylation of the amine group(s). The most usual alcohols for this purpose are n-butanol [23], n-propanol [150, 151] or isobutanol [152], all of which provide useful mass spectrometric fragmentation patterns for many amino acids after a further perfluoroacylation step.

5.2 Fatty acids and lipids

The real power of the mass spectrometric detector for structure elucidation is amply illustrated by the application of a variety of ester derivatives to the determination of double bond positions in fatty acids [153]. The double bond may opened by chemical means, e.g. by osmium tetroxide, to give two hydroxy groups at the double bond position. Subsequent trimethylsilylation gives the di-TMS ether–TMS ester, the mass spectrum of which possesses abundant peaks originating from cleavage between the TMS ethers. Multiple double bonds can also be tackled by this means, but the derivative molecular weight increases rapidly (see Table 1) and may soon exceed the mass range of the instrument. An alternative technique is to derivatize the acidic or hydroxy group with a stabilizing group which prevents the double bond migrating during mass spectrometry. Such derivatives are also used to elucidate the positions of branches and of hydroxy and epoxy groups, and include pyrrolidides [154, 155] and picolinyl derivatives for long-chain branched and unsaturated fatty acids [156] and polyenoic acids, hydroxy acids and di-acids [157]. The basic picolinyl nitrogen abstracts a hydrogen from the side chain at the position of branching or double bond. The fragmentation is then directed to the abstraction site, enabling the side chain structure to be determined. The mass spectrum of the picolinyl derivative of 14-methyloctadecanoic acid extracted from guinea-pig Harderian gland is shown in Figure 12 [158].

The branch position is obtained from the 'missing' ion between m/z 304 and 332. Similarly acting derivatives include nicotinates [159, 160], dimethylsilylnicotinates [161] and, more recently, nicotinylidene derivatives, with improved fragmentation characteristics for dihydroxy compounds [162].

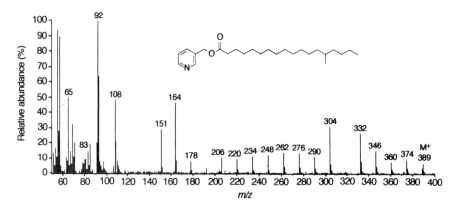

Figure 12. Electron ionization mass spectrum of the picolinyl derivative of 14-methyloctadecanoic acid extracted from the guinea pig Harderian gland, illustrating the determination of branching position. Reproduced from reference [158] with permission.

5.3 Nucleosides

Gas chromatography–electron capture negative-ion mass spectrometry is now possible at the zeptomole level for suitable derivatives. When the GC–MS system is tuned precisely for octafluoronaphthalene and four products derived from DNA adducts, the detection limit for N^1,N^3-bis(pentafluorobenzyl)-5-[(pentafluorobenzyloxy)methyl]uracil is 200 zmol, with response linear to 22 amol [163], a clear indication of the potential sensitivity achievable by GC–MS.

5.4 Organic acids

Esterification with diazomethane is employed for the profiling of carboxylic acids in body fluids [164, 165]. The mass spectra of organic acid methyl esters show a variety of structurally informative base peaks compared with the corresponding TMS esters. However, artefacts are produced on reaction of diazomethane with 2-keto acids and 2,3-unsaturated acids [166, 167]. The methyl esters also have the disadvantage of providing low molecular weight increments, especially for the low molecular weight volatile acids such as lactic acid.

5.5 Prostaglandins and related compounds

As outlined above (Sections 2 and 3), methyl ester formation with diazomethane is still often employed for analysis of these labile compounds, including thromboxanes and leukotrienes. For high sensitivity analyses by negative ion electron capture GC–MS, pentafluorobenzyl esters have proved to be very useful [168–170], with low pg sensitivities being obtained. For 6-keto-prostaglandin $F_{1\alpha}$, the methyloxime–PFB ester gave a mass spectrum with m/z 614 as base peak, allowing 2 pg to be detected.

6. Cyclization

Cyclization is particularly important in derivatization for GC–MS. Cyclic groups, particularly where conjugated, are often more stable to mass spectral fragmentation, leading to higher mass, higher abundance ions particularly suitable for quantitative measurements. Thus, the possibility of preparing a cyclic derivative should always be borne in mind for mass spectrometric detection. The reader is also directed to Chapter 7 on cyclic derivatives and to the corresponding chapter in the previous edition of this handbook [171]. A few examples illustrating these properties follow.

6.1 Amino acids

The mass spectra of oxazolidinones [172, 173] of amino acids have been reported. Recently, the 2-trifluoromethyloxazolinone derivative (Figure 13) has found application in an isotope dilution candidate reference method for phenylalanine [174].

The mass spectrum shows an abundant molecular ion and the derivatization procedure is a simple one-step reaction with trifluoroacetic anhydride/trifluoroacetic acid (1:1). The 2-perfluoroalkyl-3-oxazolin-5-one (FOx) derivatives appear to have significant potential for the low level definitive determination of 'difficult' amino acids and labelled amino acids, as seen in the recent application to plasma arginine [175].

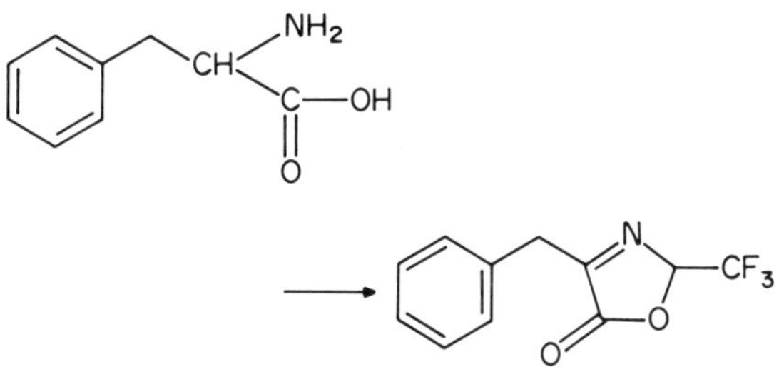

Figure 13. Formation of the 2-trifluoromethyloxazolinone derivative employed in a candidate reference method for the determination of phenylalanine by isotope dilution GC–MS. Reproduced from reference [174] with permission.

6.2 Biogenic amines

Catecholamines saw the early application of boronate derivatization to stabilize these compounds for mass spectrometry [176].

In suitable cases, the formation of cyclic derivatives can provide mass spectra suitable for measurements by EI which would normally require the more expensive and less readily available negative CI technique. The EI, CI and NICI mass spectra obtained from the spirocyclic derivative formed on gas chromatography of the reaction product of melatonin and PFP anhydride are shown in Figure 14. The EI mass spectrum has the molecular ion (m/z 360) as base peak, enabling determinations in plasma to be performed at low levels ($<$10–100 pg ml^{-1}) [177].

The method provides a simpler alternative to the more difficult NICI method [178]. The structure of the spirocyclic derivative has been investigated by Blau et al. [179].

6.3 Carbohydrates and sugars

The presence of neighbouring hydroxy group in most carbohydrate samples makes them ideal candidates for cyclic derivative formation. Examples suitable for GC–MS include butylboronate acetates [180] and several boronate–TMS derivatives [181]. Many other early examples of cyclic derivatives, such as acetals and isopropylidene derivatives, for these compounds have been reviewed by Knapp [182] and by Darbre [183].

6.4 Diols

Butylboronates of diols such as ethylene glycol allow low levels to be measured by GC–MS in plasma [184].

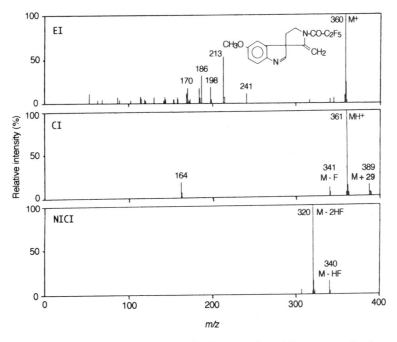

Figure 14. The EI, CI and NICI mass spectra of the spirocyclic derivative formed by reaction of melatonin with pentafluoropropionic anhydride. Reproduced from reference [179] with permission.

Cyclic TBDMS derivatives may be formed by reaction with di-t-butyldichlorosilane [185], and the derivatives possess the expected advantages for GC–MS.

6.5 Fatty acids and lipids

Methylboronates have proved useful for the GC–MS of sphingosines and ceramides [186] and have been applied to phospholipid mixtures after enzymatic hydrolysis [187].

2-Alkenyl-4,4-dimethyloxazoline derivatives formed by condensation of long-chain unsaturated fatty acids with 2-amino-2-methylpropanol are useful for the location of double bonds [188].

6.6 Organic acids

α-Keto (2-oxo) acids may be conveniently derivatized by quinoxalinol formation (Scheme 1) prior to TMS [189] or TBDMS [190] derivatization.

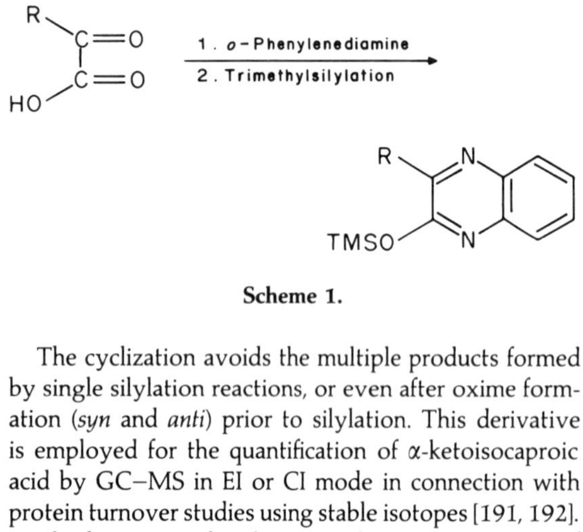

Scheme 1.

The cyclization avoids the multiple products formed by single silylation reactions, or even after oxime formation (*syn* and *anti*) prior to silylation. This derivative is employed for the quantification of α-ketoisocaproic acid by GC–MS in EI or CI mode in connection with protein turnover studies using stable isotopes [191, 192].

The formation of cyclic isoxazolinones on reaction of organic acids with hydroxylamine hydrochloride [13] is utilized for the derivatization of succinylacetone (2,4-dioxoheptanoic acid). After stabilization of the keto groups by isoxazole formation in this way, the remaining carboxylic acid group is esterified with pentafluorobenzyl bromide for GC–MS assay in EI [193] or NICI [194] mode. The latter mode is required to measure the low levels present in amniotic fluid.

6.7 Prostaglandins and related compounds

The ion current obtained from the TMS derivative of prostaglandin F_2 can be concentrated on a single ion by formation of the n-butylboronate mixed derivative [195, 196], allowing the assay sensitivity to be improved. Boronate cyclization can also be conveniently combined with TBDMS derivatization to enable the highly sensitive assay of thromboxane B_2 in human aorta [197].

6.8 Steroids and sterols

Boronate ester derivatives of steroid diols have been prepared and their mass spectrometric properties reported [198, 199]. A striking example of the improvement in mass spectral pattern of the TMS derivative of 24R,25-dihydroxyvitamin D_3 (Figure 15a) brought about by mixed derivatization and cyclization is illustrated in Figures 16a and 16b.

The mass spectrum of the tris-TMS ether of the pyro isomer of this compound (Figure 16a), formed by thermal cyclization in the gas chromatograph, is dominated by a fragment ion at m/z 131 resulting from fission between C-24 and C-25. This ion lies in the lower m/z region, where contamination can be a problem, and is therefore less suitable for low level quantification by SIM or mass chromatography. Formation of a cyclic methylboronate ester (Figure 15b) between the C-24 and C-25 hydroxyls results in a dramatic improvement in the fragmentation

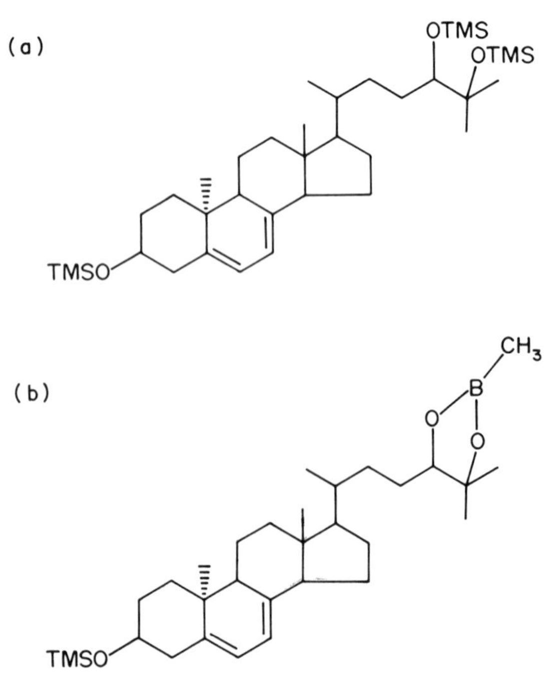

Figure 15. Structures of 24,25-dihydroxyvitamin D_3 (pyro forms cyclized in gas chromatograph) derivatives: (a) tris-trimethylsilyl ether, (b) 3-trimethylsilyl ether–24,25-methylboronate ester.

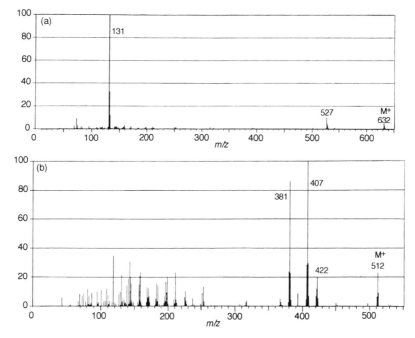

Figure 16. Electron ionization mass spectra of pyro 24,25-dihydroxyvitamin D_3 derivatives: (a) tris-trimethylsilyl ether, (b) 3-trimethylsilyl ether-24,25-methylboronate ester obtained by GC–MS. Reproduced from Reference [200] with permission.

pattern from both a qualitative (specificity) and quantitative (sensitivity) point of view (Figure 16b) [200].

n-Butylboronate derivatives have been employed in routine GC–MS assay procedure for 24,25- and 25,26-dihydroxyvitamin D in human plasma [201].

7. Conclusions

This chapter has concentrated on only the major derivative classes employed in GC–MS, and only a few examples of each could be mentioned. However, the literature, including the other chapters of the present handbook, contains an abundance of references and descriptions of other derivative types, many of which will have hidden mass spectrometric potential for particular applications. Trace analyses have been covered in depth for steroids [202] and cannabinoids [203] and in the application of NICI to drugs [204] and neurotransmitters [205]. A recent volume [206] includes a number of specialist chapters devoted to the application of mass spectrometry in different areas of biomedical research. It can be consulted for more in-depth information of work done with particular analyte classes, and contains a short overview of chemical derivatization for mass spectrometry, including some GC–MS examples [207]. Comprehensive reviews by Evershed, covering developments and new applications of GC–MS, contain many references to derivatization applications in this field [208, 209]. Current awareness publications [210, 211] are further sources of information concerning derivatives and their applications.

There is clearly considerable scope for the choice of derivative to achieve optimization of the assay procedure for an analyte or analyte group. Indeed, there is an excitement to be had from preparation of a new derivative and first sight of the mass spectral fragmentation pattern, particularly if it has the desired features!

7.1 Future perspective

Derivatization is usually a prerequisite for efficient and sensitive GC–MS of compounds possessing functional groups, but also has potential for application in the rapidly developing LC–MS field. Although one of the real advantages of LC–MS is the avoidance of derivatization, early indications are that even here derivatization can lead to significant sensitivity gains in suitable cases. This is particularly true for the particle beam technique, which enables conventional EI and CI mass spectra to be obtained [212]. The LC–MS detection limit for fatty acids can be considerably improved (40 pg) after suitable derivatization [213]. Also, derivatization has potential in tandem mass spectrometry (MS–MS) and GC–MS–MS [1, 207]. Substantial growth should be seen in these areas.

8. References

[1] R. J. Anderegg, *Mass Spectrom. Rev.*, **7**, 395 (1988).
[2] F. W. McLafferty, *Interpretation of Mass Spectra*, 3rd edn., University Science Books, Mill Valley (1980).
[3] M. E. Rose and R. A. W. Johnstone, *Mass Spectrometry for Chemists and Biochemists*, Cambridge University Press, Cambridge (1982).
[4] G. M. Message, *Practical Aspects of GC/MS*, Wiley, New York (1984).
[5] J. R. Chapman, *Practical Organic Mass Spectrometry*, Wiley-Interscience, New York (1985).
[6] J. T. Watson, *Introduction to Mass Spectrometry*, 2nd edn., Raven, New York (1985).
[7] J. M. Halket and M. E. Rose, *Introduction to Bench-top GC/MS*, HD Science, Nottingham (1990).
[8] M. E. Rose, VG Monographs 1 (1990).
[9] B. J. Millard, *Quantitative Mass Spectrometry*, Heyden, London (1978), p. 129.
[10] S. E. Stein, National Institute for Standards and Technology, Gaithersburg, personal communication.
[11] *NIST/EPA/MSDC Mass Spectral Database*, v. 3.0, National Institute for Standards and Technology, Gaithersburg (1990).
[12] K. Blau and G. S. King, *Handbook of Derivatives for Chromatography*, Heyden, London (1977).
[13] R. Libert, J. P. Draye, F. Van Hoof, A. Schanck, J. Ph. Soumillion and E. de Hoffmann, *Biol Mass Spectrom.*, **20**, 75 (1991).
[14] A. E. Pierce, *Silylation of Organic Compounds*, Pierce Chemical Company, Chicago (1968).
[15] S. P. Markey, W. G. Urban and S. P. Levine (Editors), *Mass Spectra of Compounds of Biological Interest*, University of Colorado Medical Center, Denver, 1975.
[16] F. W. McLafferty and D. B. Stauffer (Editors), *The Wiley/NBS Registry of Mass Spectral Data*, Wiley, New York (1989).
[17] *The Important Peak Index of the Wiley/NBS Registry of Mass Spectral Data*, Wiley, New York (1991).
[18] *The GC/MS Companion Series*, HD Science, Nottingham (1990).
[19] *The Eight Peak Index of Mass Spectra*, 4th edn., Royal Society of Chemistry, Cambridge (1992).
[20] K. Pfleger, H. Maurer and A. Weber (Editors), *Mass Spectral and GC Data of Drugs, Poisons and Their Metabolites*, 2nd edn., VCH, Weinheim (1992).
[21] S. F. Markey, *Diagnosis of Organic Acidemias by Gas Chromatography–Mass Spectrometry*, Liss, New York (1981), p. 123.
[22] C. W. Gehrke and K. Leimer, *J. Chromatogr.*, **57**, 219 (1971).
[23] F. E. Kaiser, C. W. Gehrke, R. W. Zumalt and K. C. Kuo, *J. Chromatogr.*, **94**, 113 (1974).
[24] K. R. Leimer, R. H. Rice and C. W. Gehrke, *J. Chromatogr.*, **141**, 355 (1977).
[25] J. M. Halket, K. Blau and S. Down (Editors), *GC/MS Companion Handbook of Amino Acid Derivatives, Part I–Trimethylsilyl*, HD Science, Nottingham (1990).
[26] F. P. Abramson, M. W. McCaman and R. E. McCaman, *Anal. Biochem.*, **57**, 482 (1974).
[27] A. M. Lawson, D. B. Ramsden, P. J. Raw and R. Hoffenberg, *Biomed. Mass Spectrom.*, **1**, 374 (1975).
[28] M. G. Horning, A. M. Moss and E. C. Horning, *Biochim. Biophys. Acta*, **148**, 597 (1967).
[29] E. Gerloh, R. Malfait and A. G. Dupont, *J. Chromatogr.*, **414**, 301 (1987).
[30] F. P. Abramson, M. W. McCaman and R. E. McCaman, *Anal. Biochem.*, **57**, 482 (1974).
[31] C. C. Sweeley, W. H. Elliott, I. Fries and R. Ryhage, *Anal. Chem.*, **38**, 1449 (1966).
[32] I. Björkhem, R. Blomstrand, O. Falk and G. Öhman, *Clin. Chim. Acta*, **72**, 353 (1976).
[33] O. Pelletier and S. Cadieux, *Biomed. Mass Spectrom.*, **10**, 130 (1983).
[34] D. Küry, U. Keller, *J. Chromatogr. Biomed. Appl.*, **572**, 302 (1991).
[35] J. P. Kamerling and J. F. G. Vliegenhart, in *Clinical Biochemistry, Principles, Methods, Applications. Vol. 1. Mass Spectrometry: Applications in Clinical Medicine*, A .M. Lawson (Editor), Walter de Gruyter, Berlin (1989), p. 175.
[36] C. R. Lee, A. C. Coste and J. Allen, *Biomed. Environ. Mass Spectrom.* **16**, 387 (1988).
[37] T. S. Baker, J. V. Harry, J. W. Russell and R. L. Myers, *J. Anal. Toxicol.*, **8**, 255 (1984).
[38] M. C. H. Oon, R. N. Smith and M. J. Whitehouse, *Forensic Sci. Int.*, **46**, 219 (1990).
[39] R. L. Foltz, A. F. Fentiman Jr. and R. B. Foltz, *GC/MS Assays for Abused Drugs in Body Fluids*, National Institute on Drug Abuse Research Monograph 32, Department of Health and Human Services publication number (ADM) 80-1014, US. Government Printing Office, Washington, DC (1980), pp. 66–89.
[40] R. Ryhage and E. Stenhagen, *Ark. Kemi*, **15**, 545 (1960).
[41] D. E. Minniken, *Chem. Phys. Lipids*, **21**, 313 (1978).
[42] V. Dommes, F. Wirtz-Pietz and W. H. Kunau, *J. Chromatogr. Sci.*, **14**, 360 (1976).
[43] J. J. Myher, L. Marai and A. Kuksis, *J. Lipid Res.*, **15**, 586 (1974).
[44] M. G. Horning, G. Casparrini and E. C. Horning, *J. Chromatogr. Sci.*, **7**, 267 (1969).
[45] K. Satouchi and K. Saito, *Biomed. Mass Spectrom.*, **3**, 122 (1976).
[46] T. Curstedt, *Biochim. Biophys. Acta*, **489**, 79 (1977).
[47] K. Satouchi, K. Saito and M. Kates, *Biomed. Mass Spectrom.*, **5**, 87 (1978).
[48] J. A. McCloskey, in *Mass Spectrometry in the Health and Life Sciences*, A. L. Burlingame and N. Castagnoli (Editors), Elsevier, New York (1985), p. 521.
[49] S. P. Dutta, P. F. Crain, J. A. McCloskey and G. B. Chedha, *Life Sci.*, **24**, 1381 (1979).
[50] O. I. Aruoma and B. Halliwell, *Chem. Br.*, **27**, 149 (1991).
[51] W. M. Hammargren, K. H. Schram, K. Nakano and T. Yasaka, *Anal. Chim. Acta*, **247**, 201 (1991).
[52] R. A. Chalmers and A. M. Lawson, *Organic Acids in Man*, Chapman and Hall, London (1982).

[53] R. A. Chalmers, J. M. Halket and G. A. Mills (Editors), *GC/MS Companion, Organic Acid Derivatives*, HD Science, Nottingham (1993).
[54] J. D. Shoemaker and W. H. Elliott, *J. Chromatogr. Biomed. Appl.*, **562**, 125 (1991).
[55] J. A. Kelley, H. Nau, H. J. Förster, and K. Biemann, *Biomed. Mass Spectrom.*, **2**, 313 (1975).
[56] H. Nau, H. J. Förster, J. A. Kelley and K. Biemann, *Biomed. Mass Spectrom.*, **2**, 326 (1975).
[57] F. Vane and M. G. Horning, *Anal. Lett.*, **2**, 257 (1969).
[58] B. S. Middleditch and D. M. Desiderio, *Anal. Biochem.*, **55**, 509 (1973).
[59] J. Rosello, J. Tusell and E. Gelpi, *J. Chromatogr.*, **130**, 65 (1977).
[60] R. W. Kelly, in *Clinical Biochemistry, Principles, Methods, Applications. Vol. 1. Mass Spectrometry: Applications in Clinical Medicine*, A. M. Lawson (Editor), Walter de Gruyter, Berlin (1989), p. 481.
[61] R. C. Murphy and T. W. Harper, in *Mass Spectrometry in Biomedical Research*, S. J. Gaskell (Editor), Wiley, Chichester (1986), p. 11.
[62] J. P. Thenot and E. C. Horning, *Anal. Lett.*, **5**, 801 (1972).
[63] J. Sjövall and M. Axelson, *Vitam. Horm.*, **39**, 31 (1982).
[64] S. J. Gaskell, in *Clinical Biochemistry, Principles, Methods, Applications. Vol. 1, Mass Spectrometry: Applications in Clinical Medicine*, A. M. Lawson (Editor), Walter de Gruyter, Berlin (1989), p. 581.
[65] S. J. Gaskell, in *Clinical Biochemistry, Principles, Methods, Applications. Vol. 1, Mass Spectrometry: Applications in Clinical Medicine*, A. M. Lawson (Editor), Walter de Gruyter, Berlin (1989), p. 571.
[66] E. M. Chambaz, G. Defaye and C. Madani, *Anal. Chem.*, **45**, 1090 (1973).
[67] B. Spiegelhalder, G. Rohle, L. Siekmann and H. Breuer, *J. Steroid Biochem.*, **7**, 749 (1976).
[68] K. D. R. Setchell and A. M. Lawson, in *Clinical Biochemistry, Principles, Methods, Applications. Vol. 1. Mass Spectrometry: Applications in Clinical Medicine*, A. M. Lawson (Editor), Walter de Gruyter, Berlin, (1989), p. 53.
[69] G. Stork and P. F. Hudrlik, *J. Am. Chem. Soc.*, **90**, 4462 (1968).
[70] E. J. Corey and A. Venkateswarlu, *J. Am. Chem. Soc.*, **94**, 6190 (1972).
[71] S. L. Mackenzie, D. Tenaschuk and G. Fortier, *J. Chromatogr.*, **387**, 241 (1987).
[72] W. F. Schwenck, P. J. Berg, B. Beaufrere, J. M. Miles and M. W. Haymond, *Anal. Biochem.*, **141**, 101 (1984).
[73] H. J. Chaves das Neves and A. M. P. Vasconcelos, *J. Chromatogr.*, **392**, 249 (1987).
[74] G. L. Loy, A. N. Quick Jr., C. C. Teng, W. W. Hay Jr. and P. V. Fennessey, *Anal. Biochem.*, **185**, 1 (1990).
[75] M. E. Corbett, C. M. Scrimgeour and P. W. Watt, *J. Chromatogr.*, **419**, 263 (1987).
[76] J. M. Halket, P. J. Watkins, A. Przyborowska, B. L. Goodwin, A. Clow, V. Glover and M. Sandler, *J. Chromatogr. Biomed. Appl.*, **562**, 279 (1991).
[77] R. W. Kelly and P. W. Taylor, *Anal. Chem.*, **48**, 465 (1976).

[78] A. I. Mallet, D. E. Minniken and G. Dobson, *Biomed. Mass Spectrom.*, **11**, 79 (1984).
[79] H. Parsons, E. A. Emken, L. Marai and A. Kuksis, *Lipids*, **21**, 247 (1986).
[80] P. M. Woollard, *Biomed. Mass Spectrom.*, **10**, 143 (1983).
[81] P. M. Woollard and A. I. Mallet, *J. Chromatogr.*, **306**, 1 (1984).
[82] L. Larsson, P.-A. Mardh, G. Odham and G. Westerdahl, *J. Chromatogr. Biomed. Appl.*, **182**, 402 (1980).
[83] J. J. Myher, A. Kuksis, L. Marai and S. K. F. Yeung, *Anal. Chem.*, **50**, 557 (1978).
[84] K. Satouchi and K. Saito, *Biomed. Mass Spectrom.*, **6**, 396 (1979).
[85] B. F. Dickens, C. S. Ramesha and G. A. Thomson, *Anal. Biochem.*, **127**, 37 (1982).
[86] J. Myher and A. Kuksis, *Can. J. Biochem.*, **59**, 626 (1981).
[87] I. Bjorkhem and I. Holmberg, *Methods Enzymol.*, **67**, 385 (1978).
[88] B. P. Lisboa and J. M. Halket, in *Recent Developments in Chromatography and Electrophoresis*, A. Frigerio and L. Renoz (Editors), Elsevier, Amsterdam (1979), p. 141.
[89] A. Triolo, J. Bertini, C. Mannucci, A. Perico and V. Pestellini, *J. Chromatogr. Biomed. Appl.*, **568**, 281 (1991).
[90] A. Kuksis and J. J. Myher, in *Clinical Biochemistry, Principles, Methods, Applications. Vol. 1. Mass Spectrometry: Applications in Clinical Medicine*, A. M. Lawson (Editor), Walter de Gruyter, Berlin (1989), p. 265.
[91] A. P. J. M. de Jong, J. Elema and B. J. T. van de Berg, *Biomed. Mass Spectrom.*, **7**, 359 (1980).
[92] T. Cronholm and C. Norsten, *J. Chromatogr.*, **344**, 1 (1985).
[93] A. G. Smith, W. A. Harland and C. J. W. Brooks, *J. Chromatogr.*, **142**, 533 (1977).
[94] A. C. Bazan and D. R. Knapp, *J. Chromatogr.*, **236**, 201 (1982).
[95] R. C. Murphy, *Prostaglandins*, **28**, 597 (1984).
[96] S. Steffenrud, P. Borgeat, M. J. Evans and M. J. Bertrand, *Biomed. Environ. Mass Spectrom.*, **14**, 313 (1987).
[97] G. Phillipou, D. A. Bigham and R. F. Seamark, *Steroids*, **26**, 516 (1975).
[98] J. F. Sabot, D. Deruaz, P. Bernard and H. Pinatel, *J. Chromatogr.*, **339**, 233 (1985).
[99] S. J. Gaskell, A. W. Pike and K. Griffiths, *Steroids*, **36**, 219 (1980).
[100] L. Dehennin, C. Blacker, A. Reiffsteck and R. Scholler, *J. Steroid Biochem.*, **20**, 564 (1984).
[101] S. J. Gaskell, C. J. Porter and B. N. Green, *Biomed. Mass Spectrum.*, **12**, 139 (1985).
[102] I. Ganschow and B. P. Lisboa, in *Recent Developments in Mass Spectrometry in Biochemistry, Medicine and Environmental Research*, A. Frigerio (Editor), Vol. 8, Elsevier, Amsterdam (1983), pp. 265–268.
[103] M. Donike and J. Zimmermann, *J. Chromatogr.*, **202**, 483 (1980).
[104] S. H. G. Andersson and J. Sjövall, *J. Chromatogr.*, **289**, 195 (1984).
[105] B. P. Lisboa, I. Ganschow, J. M. Halket and F. Pinheiro-Joventino, *Comp. Physiol. Biochem.*, **73B**, 257 (1982).

[106] J. M. Halket and B. P. Lisboa, *J. Chromatogr.*, **189**, 267 (1980).
[107] M. A. Quilliam and J. B. Westmore, *Steroids*, **29**, 579 (1977).
[108] D. J. Harvey, *J. Chromatogr.*, **147**, 291 (1978).
[109] H. Miyazaki, M. Ishibashi and K. Yamashita, *Biomed. Mass Spectrom.*, **6**, 57 (1981).
[110] M. Ishibashi, K. Yamashita, K. Watanabe and H. Miyazaki, *Mass Spectrometry in Biomedical Research*, S. J. Gaskell, (Editor), Wiley, Chichester (1986), p. 423.
[111] G. Phillipou, *J. Chromatogr.*, **129**, 384 (1976).
[112] I. A. Blair and G. Phillipou, *J. Chromatogr. Sci.*, **15**, 478 (1977).
[113] D. J. Harvey, *Biomed. Mass Spectrum.*, **4**, 265 (1977).
[114] D. J. Harvey, *J. Chromatogr.*, **147**, 291 (1978).
[115] S. Steffenrud and P. Borgeat, *Prostaglandins*, **28**, 593 (1984).
[116] J. Boutagy and D. J. Harvey, *J. Chromatogr.*, **156**, 153 (1978).
[117] K. R. Leimer, R. Rice and C. W. Gehrke, *J. Chromatogr.*, **141**, 121 (1977).
[118] J. M. Halket, K. Blau and S. Down (Editors), *GC/MS Companion Handbook of Amino Acid Derivatives, Part II—Trifluoroacetyl, n-butyl ester derivatives*, HD Science, Nottingham (1990).
[119] B. Möller, O. Falk and I. Björkhem, *Clin. Chem.*, **29**, 2106 (1983).
[120] L. Siekmann, *Biomed. Mass Spectrom.*, **14**, 683 (1987).
[121] D. B. Ramsden and M. J. Farmer, *Biomed. Mass Spectrom.*, **8**, 421 (1984).
[122] H. Miyazaki, M. Hashimoto, M. Iwanaga and T. Kubodera, *J. Chromatogr.*, **99**, 575 (1974).
[123] B. A. Davis and D. A. Durden, *Biomed. Mass Spectrum.*, **14**, 197 (1987).
[124] R. T. Coutts, G. B. Baker, F. M. Pasutto, S.-F. Liu, D. F. LeGatt and D. B. Prelusky, *Biomed. Mass Spectrum.*, **11**, 441 (1984).
[125] K. Blau, in *Clinical Biochemistry, Principles, Methods, Applications. Vol. 1. Mass Spectrometry: Applications in Clinical Medicine*, A. M. Lawson (Editor), Walter de Gruyter, Berlin (1989), p. 129.
[126] D. A. Durden, *Biol Mass Spectrum.*, **20**, 367 (1991).
[127] H. Haga, T. Imanari, Z. Tamura and A. Momose, *Chem. Pharm. Bull.*, **20**, 1805 (1972).
[128] W. A. Koenig, H. Bauer, W. Voelter and E. Bayer, *Chem. Ber.*, **106**, 1905 (1973).
[129] T. S. Baker, J. V. Harry, J. W. Russell and R. L. Myers, *J. Anal. Toxicol.*, **8**, 255 (1984).
[130] D. Rosenthal, T. M. Harvey, D. R. Brine and M. E. Wall, *Biomed. Mass Spectrom.*, **5**, 312 (1978).
[131] L. Karlsson, J. Jonsson, K. Aberg and C. Roos, *J. Anal. Toxicol.*, **7**, 198 (1983).
[132] P. Lillsunde and T. Korte, *Forensic Sci. Int.*, **49**, 205 (1991).
[133] N. B. Wu Chen, M. I. Shaffer, R.-L. Lin and R. J. Stein, *J. Anal. Toxicol.*, **6**, 231 (1982).
[134] R. W. Taylor and N. C. Jain, *J. Anal. Toxicol.*, **11**, 233 (1987).

[135] L. Siekmann, A. Schönfelder and A. Siekmann, *Fresenius' Z. Anal. Chem.*, **324**, 280 (1986).
[136] C. S. Ramesha and W. C. Pickett, *Biomed. Environ. Mass Spectrom.*, **13**, 107 (1986).
[137] B. W. Christman, J. C. Gay, J. W. Christman, C. Prakash, I. A. Blair, *Biol. Mass Spectrom.*, **20**, 545 (1991).
[138] M. Balazy, P. Braquet, N. G. Bazan, N. G., *Anal. Biochem.*, **196**, 1 (1991).
[139] W. A. Koenig, L. C. Smith, P. F. Crain and J. A. McCloskey, *Biochemistry*, **190**, 3968 (1971).
[140] B. S. Middleditch and D. M. Desiderio, *Prostaglandins*, **2**, 195 (1972).
[141] L. Siekmann, *J. Steroid Biochem.*, **11**, 117 (1979).
[142] L. Siekmann, A. Siekmann and H. Breuer, *Fresenius' Z. Anal. Chem.*, **290**, 122 (1978).
[143] R. Knuppen, O. Haupt, W. Schramm and H.-O. Hoppen, *J. Steroid Biochem.*, **11**, 153 (1979).
[144] L. Siekmann, *J. Clin. Chem. Clin. Biochem.*, **22**, 551 (1984).
[145] I. Björkhem, R. Blomstrand and O. Lantto, *Clin. Chim. Acta*, **65**, 343 (1975).
[146] W. Schramm, T. Louton, W. Schill and H.-O. Hoppen, *Biomed. Mass Spectrom.*, **7**, 273 (1980).
[147] S. J. Gaskell, B. G. Brownsey and G. V. Groom, *Clin. Chem.*, **30**, 1696 (1984).
[148] D. Stockl, H. Reinauer, L. M. Thienpont and A. P. De Leenheer, *Biol. Mass Spectrom.*, **20**, 657 (1991).
[149] D. W. Johnson, G. Phillipou, M. M. Ralph and R. F. Seamark, *Clin. Chim. Acta*, **94**, 207 (1979).
[150] C. W. Moss, M. A. Lambert and F. J. Diaz, *J. Chromatogr.*, **60**, 134 (1971).
[151] *Hewlett-Packard Application Brief AB85-7, Analysis of Amino Acids*, Hewlett-Packard, Palo Alto, 1985.
[152] J. Desgres, D. Boisson and P. Padieu, *J. Chromatogr.*, **162**, 133 (1979).
[153] D. E. Minniken, *Chem. Phys. Lipids*, **21**, 313 (1978).
[154] B. A. Andersson and R. T. Holman, *Lipids*, **9**, 185 (1974).
[155] W. Vetter, W. Meister and G. Oesterhelt, *Helv. Chim. Acta*, **61**, 1287 (1978).
[156] D. J. Harvey, *Biomed. Mass Spectrom.*, **9**, 33 (1982).
[157] D. J. Harvey, *Biomed. Mass Spectrom.*, **11**, 340 (1984).
[158] D. J. Harvey, *Biol. Mass Spectrom.*, **20**, 61 (1991).
[159] W. Vetter and W. Meister, *Org. Mass Spectrom.*, **16**, 118 (1981).
[160] D. J. Harvey and J. M. Tiffany, *Biomed. Mass Spectrom.*, **11**, 353 (1984).
[161] D. J. Harvey, *Biomed. Environ. Mass Spectrom.*, **14**, 103 (1987).
[162] D. J. Harvey, *Biol. Mass Spectrom.*, **20**, 61 (1991).
[163] S. Abdel-Baky and R. W. Giese, *Anal. Chem.*, **63**, 2986 (1991).
[164] P. Divry, C. Vianey-Liaud and J. Cotte, *Biomed. Environ. Mass Spectrom.*, **14**, 663 (1987).
[165] E. Jellum, E. A. Kvittingen and O. Stokke, *Biomed. Environ. Mass Spectrom.*, **16**, 57 (1988).
[166] S. Bauer, M. Neupert and G. Spiteller, *J. Chromatogr.*, **309**, 243 (1984).

[167] H. M. Liebich and C. Först, *J. Chromatogr.*, **309**, 225 (1984).
[168] K. W. Waddell, I. A. Blair and J. Wellby, *Biomed. Mass Spectrom.*, **10**, 83 (1983).
[169] K. A. Waddell, S. E. Barrow, C. Robinson, M. A. Orchard, C. T. Dollery and I. A. Blair. *Biomed. Mass Spectrom.*, **11**, 68 (1984).
[170] S. E. Barrow, D. K. Heavey, M. Ennis, C. Chappell, I. A. Blair and C. T. Dollery, *Prostaglandins*, **28**, 743 (1984).
[171] A. Darbre, in *Handbook of Derivatives for Chromatography*, K. Blau and G. S. King (Editors), Heyden, London (1977), p. 262.
[172] R. Liardon, U. Otto-Kuhn and P. Husek, *Biomed. Mass Spectrom.*, **6**, 381 (1979).
[173] V. Ferrito, R. Borg, J. Eagles and G. R. Fenwick, *Biomed. Mass Spectrom.*, **6**, 499 (1979).
[174] G. W. Lynes and M. Hjelm, *J. Chromatogr.*, **562**, 213 (1991).
[175] S. K. Branch, C. D. Warner, A. M. Ajami and V. R. Young, *Proc. 39th Annual Conference on Mass Spectrometry and Allied Topics*, Nashville, (1991), p. 7.
[176] C. J. W. Brooks, G. M. Anthony and B. S. Middleditch, *Org. Mass Spectrom.*, **2**, 1023 (1969).
[177] C. R. Lee and H. Esnaud, *Biomed. Environ. Mass Spectrom.*, **15**, 249 (1988).
[178] A. J. Lewy and S. P. Markey, *Science*, **201**, 741 (1978).
[179] K. Blau, G. S. King and M. Sandler, *Biomed. Mass Spectrom.*, **7**, 232 (1977).
[180] J. Wiecko and W. R. Sherman, *J. Am. Chem. Soc.*, **98**, 7631 (1976).
[181] V. N. Reinhold, F. Wirtz-Peitz and K. Biemann, *Carbohydr. Res.*, **37**, 203 (1974).
[182] D. R. Knapp, *Handbook of Analytical Derivatization Reactions*, Wiley, Chichester (1979), p. 539.
[183] A. Darbre, in *Handbook of Derivatives for Chromatography*, K. Blau and G. S. King (Editors), Heyden, London (1977), p. 262.
[184] C. Giachetti, G. Zanolo, A. Assandri and P. Poletti, *Biomed. Environ. Mass Spectrum.*, **18**, 592 (1989).
[185] C. J. W. Brooks, W. J. Cole and G. M. Barrett, *J. Chromatogr.*, **315**, 119 (1984).
[186] S. J. Gaskell, C. G. Edmonds and C. J. W. Brooks, *J. Chromatogr.*, **126**, 5912 (1976).
[187] S. J. Gaskell and C. J. W. Brooks, *J. Chromatogr.*, **142**, 469 (1977).
[188] J. Y. Zhang, Q. T. Yu, B. N. Liu and Z. H. Huang, *Biomed. Environ. Mass Spectrum.*, **15**, 33 (1988).
[189] U. Langenbeck, A. Mench-Hoinowski, K.-P. Dieckmann, H.-U. Möhring and M. Petersen, *J. Chromatogr.*, **145**, 185 (1978).
[190] U. Langenbeck, H. Luthe and G. Schaper, *Biomed. Mass Spectrom.*, **12**, 507 (1980).

[191] D. E. Matthews, H. P. Schwarz, R. D. Yang, K. J. Motil, V. R. Young and D. M. Bier, *Metabolism*, **31**, 1105 (1982).
[192] G. C. Ford, K. N. Cheng and D. Halliday, *Biomed. Mass Spectrom.*, **12**, 432 (1985).
[193] B. R. Pettit, F. MacKenzie, G. S. King and J. V. Leonard, *J. Inherited Metab. Dis.*, **7** (Suppl. 2), 135 (1984).
[194] C. Jakobs, L. Dorland, B. Wikkerink, R. M. Kok, A. P. de Jong and S. K. Wadman, *Clin. Chim. Acta*, **171**, 223 (1988).
[195] C. Pace-Asciak and L. S. Wolfe, *J. Chromatogr.*, **56**, 129 (1971).
[196] R. W. Kelly and P. L. Taylor, *Anal. Chem.*, **45**, 2079 (1973).
[197] A. G. Smith, W. A. Harland and C. J. W. Brooks, *J. Chromatogr.*, **142**, 533 (1977).
[198] T. A. Baillie, C. J. W. Brooks and B. S. Middleditch, *Anal. Chem.*, **44**, 30 (1972).
[199] S. J. Gaskell and C. J. W. Brooks, *J. Chromatogr.*, **158**, 331 (1978).
[200] J. M. Halket, I. Ganschow and B. P. Lisboa, *J. Chromatogr.*, **192**, 434 (1980).
[201] R. D. Coldwell, D. J. H. Trafford, M. J. Varley, H. L. J. Makin and D. N. Kirk, *Biomed. Environ. Mass Spectrom.*, **16**, 81 (1988).
[202] S. J. Gaskell, V. J. Gould and H. M. Leith, in *Massitor Spectrometry in Biomedical Reserch*, S. J. Gaskell (Editor), Wiley, Chichester (1986), p. 347.
[203] D. J. Harvey, in *Mass Spectrometry in Biomedical Research*, S. J. Gaskell (Editor), Wiley, Chichester, (1986), p. 363.
[204] W. A. Garland and M. P. Barbalas, in *Mass Spectrometry in Biomedical Research*, S. J. Gaskell (Editor), Wiley, Chichester, (1986), p. 379.
[205] K. F. Faull and O. Beck, in *Mass Spectrometry in Biomedical Research*, S. J. Gaskell (Editor), Wiley, Chichester, (1986), p. 403.
[206] J. A. McCloskey (Editors), *Methods in Enzymology* Vol, 193, Academic Press, New York (1990).
[207] D. R. Knapp, in *Methods in Enzymology*, 193, J. A. McCloskey (Editor), Academic Press, (1990), p. 314.
[208] R. P. Evershed, in *Mass Spectrometry*, M. E. Rose (Editor), Specialist Periodical Reports Vol. 9, The Royal Society of Chemistry, London (1987), p. 196.
[209] R. P. Evershed, in *Mass Spectrometry*, M. E. Rose, (Editor), Specialist Periodical Reports Vol. 10, The Royal Society of Chemistry, Cambridge (1989), p. 181.
[210] *CA Selects, Mass Spectrometry*, American Chemical Society, Chemical Abstracts Service, Columbus, OH.
[211] *GC/MS Update, Part A—Environmental; Part B—Biomedical, Clinical, Drugs*, HD Science, Nottingham.
[212] V. Raverdino, *J. Chromatogr.*, **554**, 125 (1991).
[213] A. P. Tinke, A. P., R. A. M. van der Hoeven, W. M. A. Niessen, U. R. Tjaden and J. van der Greef, *J. Chromatogr.*, **554**, 119 (1991).

15

Post-chromatographic Derivatization

Karl Blau
Department of Chemical Pathology, Bernhard Baron Memorial Research Laboratories, Queen Charlotte's and Chelsea Hospital, London

1. THE RATIONALE AND SCOPE OF POST-CHROMATOGRAPHIC DERIVATIZATION ... 327
2. REVELATION OF SPOTS ON PAPER AND THIN-LAYER CHROMATOGRAMS ... 329
3. POST-CHROMATOGRAPHIC DERIVATIZATION IN LIQUID CHROMATOGRAPHY ... 329
   - 3.1 General considerations ... 329
   - 3.2 Post-chromatographic derivatization techniques ... 330
   - 3.3 Mixers and liquid reactors in manifolds ... 331
   - 3.4 Packed-bed reactors ... 332
   - 3.5 Immobilized enzyme reactors ... 333
   - 3.6 Photochemical derivatization ... 334
4. POST-CHROMATOGRAPHIC DERIVATIZATION IN GAS CHROMATOGRAPHY ... 336
   - 4.1 Pyrolysis ... 336
   - 4.2 Transmodulators ... 336
   - 4.3 Post-column concentrators ... 337
   - 4.4 Post-column reactors ... 337
   - 4.5 Photochemical reactor ... 338
5. RADIOACTIVE DERIVATIZATION ... 338
5. MISCELLANEOUS APPLICATIONS ... 338
   - 6.1 Amines ... 338
   - 6.2 Catecholamines ... 339
   - 6.3 Amino acids and peptides ... 339
   - 6.4 Drugs and pharmaceuticals ... 339
   - 6.5 Carbohydrates ... 340
   - 6.6 Steroids ... 341
   - 6.7 Vitamins ... 341
   - 6.8 Phospholipids ... 341
   - 6.9 Guanidino and urea compounds ... 341
   - 6.10 Acids ... 342
   - 6.11 Miscellaneous compounds ... 343
7. CONCLUSIONS: FUTURE DEVELOPMENTS ... 343
8. REFERENCES ... 344

1. The rationale and scope of post-chromatographic derivatization

Originally, when chromatography was literally the separation of coloured compounds, the separation of the mixture into its individual components could be followed by eye, or in specific instances under ultraviolet light. But when 'chromatography' was extended to colourless substances, and most chemical compounds are colourless rather than coloured, the separated components had to be revealed in some way, and this was the beginning of post-chromatographic derivatization. At first, stains and dyes and general colour reactions were used, and these included specifically designed chromophoric derivatization reagents or reagents adapted from spot-tests [1, 2]; see also Chapter 8. The earliest, classic example is the ninhydrin reaction with amino acids after paper [3], thin-layer [4] or column chromatography [5–7], and its use perfectly illustrates all the main objectives of post-chromatographic derivatization listed in Table 1.

The choice of detection reagents depends more than anything on the chemical properties of the selected compounds of interest, and chemical modifications can be made not only by the use of reagents but also by

Handbook of Derivatives for Chromatography
Edited by K. Blau and J. M. Halket © 1993 John Wiley & Sons Ltd

Table 1 Objectives of post-chromatographic derivatization

To detect the separated components
To provide specificity
To discriminate in favour of selected compounds of interest
To amplify the sensitivity of detection
To achieve quantitation

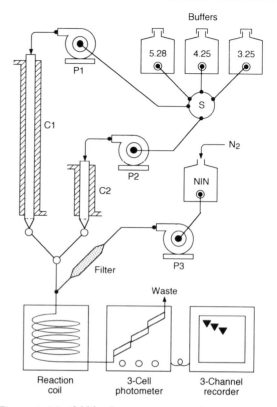

Figure 1. Manifold for the continuous post-chromatographic detection of amino acids by the classical colorimetric ninhydrin (NIN) reaction (greatly simplified from Ref. 7). C1: main 150 cm ion-exchange resin column; C2: short 15 cm column for basic amino acids; P1–P3: pumps; S: solenoid-controlled stopcock.

the action of heat or light, or electrochemically. Special chapters in this handbook are devoted to coloured and UV-absorbing derivatives (Chapter 8) and fluorescent derivatives (Chapter 9). The chromatographic process used to separate the compounds in which the analyst is interested also affects the means used to detect them. In paper and thin-layer chromatography (and indeed in electrophoresis), most post-chromatographic reactions are designed to give coloured or fluorescent derivatives on the surface of the chromatographic medium (see Chapters 8 and 9), and the selected reagent is applied to the chromatograms by spraying or dipping. Occasionally reagents are applied with overlays, and this is generally done to avoid diffusion or smearing and to preserve the separation already achieved. The avoidance of smearing and the preservation of resolution are recurring themes in the design or choice of post-chromatographic procedures.

In liquid chromatographic methods (column chromatography, LC, HPLC or gel permeation chromatography) the separated components emerge one by one from the chromatographic system, and reagents are added to the column effluent either discontinuously following fraction collection, as in Stein and Moore's first classical amino acid separations [5, 6], or continuously via a derivatization manifold, as in the classical automated ninhydrin method of Spackman, Stein and Moore [7], and this is the preferred way nowadays; see Figure 1.

After derivatization has occurred in some kind of reaction coil or reactor, the products are determined by colorimetric or fluorimetric methods, by electrochemical detection or by some other specific property. One of the advantages of post-chromatographic derivatization is precisely that the reaction(s) do not need to go to completion, nor do the reagents necessarily have to be separated from the products; nor indeed does each eluting substance have to yield a single derivative. Even side reactions need not be a problem: everything resulting from the reaction is swept into the detection system in what one tries to ensure will be a narrow band, or at least one that is broadened as little as possible from the original width of the chromatographically separated peak. As long as detection of the products takes place in a different way (at different wavelengths, for example)

from that for the starting materials, satisfactory analyses will be possible.

In gas chromatography, derivatization is most often pre-chromatographic, associated with procedures for making various compounds susceptible to analysis in the vapour phase. This is why a considerable proportion of the derivatization recipes in this handbook was devised for gas chromatographic purposes. In gas chromatography the detector achieves the most important objectives of sensitive detection and quantitation, and some detectors provide a degree of specificity as well, at least to the extent of reacting to certain chemical groupings and not to others, e.g. the electron capture detector. However, even in gas chromatography, there may in some circumstances be scope for post-chromatographic derivatization procedures additional to and preceding instrumental detection. This is particularly true if the derivative moiety is likely to dominate the chromatographic properties too much for clean separation, for instance where it adds too much to the molecular weight and thus tends to diminish the effects of differences in size between the molecules to the separated. In such circumstances it may be better to separate the compounds of interest without prior derivatization, or at least with only minimal derivatization, after which post-chromatographic modification of the eluted components may be best. Such modification may be discontinuous, involving addition of derivatization reagents to selected fractions trapped at the column outlet, or less commonly continuous, involving addition of derivatization reagents in the vapour phase to the gas stream leaving the column, or by pyrolysis or other reactions matched to a particular detection system. Thus, in GC–MS a reagent gas may be added at a chemical ionization (CI) source to yield CI mass spectra for identification and quantitation (See chapter 14).

These principles have been applied very widely, as will be described, and examples will be given of how specific problems might be solved. This overview is meant to be illustrative rather exhaustive: it may suggest possible solutions to problems that readers may have and point to potential fresh approaches.

2. Revelation of spots on paper and thin-layer chromatograms

Substances separated on paper and thin-layer chromatograms are revealed or detected by derivatization to coloured or fluorescent products, and these are interpreted visually, evaluated semi-quantitatively with reference to standards run under identical conditions, or quantitated by elution of the coloured or fluorescent spots and instrumental determination by colorimetry or fluorimetry respectively. In paper chromatography the spots can be cut out with scissors or removed with a paper punch; thin layers are usually scraped off the support and analysed in some appropriate way. A neat application is to analyse TLC spots directly by FAB MS [8], and this potentially provides identification as well as quantitation. It is one of the advantages of paper chromatographic and TLC methods that several samples and standards can be run under identical conditions, but the accuracy and precision of quantitation by these means are usually not nearly as good as with liquid or gas–liquid chromatographic separations.

Substances on paper chromatograms are generally revealed by spraying or dipping. It is important to make sure that the reagents used are dissolved in a solvent which will not smear or diffuse the developed spots and so spoil the separation achieved by chromatography. The nature of the solvent may also affect the development of the colours. For many reagents acetone is the most useful solvent, because it is a good general solvent and it dries so quickly. Thin-layer chromatograms, being less flexible than paper, are more often sprayed than dipped; nevertheless, dipping methods usually give more uniform and reproducible results than spraying. Both kinds of chromatogram may also be treated in the vapour phase (e.g. with iodine vapour, where colour formation is reversible, or chlorine followed by potassium iodide, where the reaction is usually not reversible), with or without a subsequent colorimetric or fluorimetric reaction. Reaction with fluorescein/bromine vapour is also a fairly general visualization technique [9]. Silica thin-layer plates are often subjected to charring as a convenient non-specific but almost universal method of showing up separated zones. A useful variation is to heat the TLC plates in the presence of ammonia vapour, which produces fluorescent spots with many compounds [10].

Many commercial thin-layer media incorporate a fluorescent indicator, and under ultraviolet light the separated spots may show up dark against the brightly fluorescent background, but a host of substances will *not* show up under ultraviolet light on these fluorescent layers.

There must be hundreds of visualization recipes, some of them highly ingenious, particularly where sequences of several colorimetric or fluorimetric methods have been elaborated to provide highly specific identification of certain compounds. These visualization methods have been gathered into various handbooks on paper and thin-layer chromatography [4, 11–14], and the wealth of applications they contain is far more abundant than can be included here.

3. Post-chromatographic derivatization in liquid chromatography

3.1 General considerations

In liquid chromatography, including HPLC, Schlabach and Weinberger [15] classify analytes as co-operative if they lend themselves to methods of detection such as UV or visible absorption, fluorescence, changes in refractive index, electrochemical detection or other instrumental procedures. Analytes are classed as non-co-operative if such detection methods are not applicable or lack the necessary sensitivity, and in general it is such analytes that are the prime candidates for post-chromatographic derivatization. The scope of post-chromatographic procedures is summarized in Table 2.

Table 2 Post-chromatographic derivatization in liquid chromatography

Addition of reagents in solution by a liquid manifold
Post-chromatographic photochemical derivatization
Solid-phase reactors
Electrochemical methods

3.2 Post-chromatographic derivatization techniques

Liquid chromatographic systems are particularly suited to post-chromatographic derivatization, because the vast majority of chemical reactions are done in solution or in the liquid phase. Another reason in that liquid chromatographic separations can very often be achieved without any need for derivatization before the chromatographic step, whereas derivatization may be needed for detection or quantitation (or both) afterwards.

Generally speaking, derivatization of the components separated by the column is achieved by the addition of reagent solutions to the column effluent, which is straightforward because the pressure of the effluent at the end of the column is much lower than the inlet pressure. In order to meet the objectives of Table 1, there are a number of other requirements, which are listed in Table 3.

The arrangement by which reagents and column effluent are allowed to interact is called a 'manifold' and a great variety of manifolds has been devised; many were adapted or modified from available continuous-flow systems, which lend themselves well to these applications. Two main types of highly developed technology are available from which to elaborate post-chromatographic derivatization manifolds. These are the AutoAnalyzer, developed by Skeggs [16], which by and large involves flows of between 0.1 and 2.5 ml min^{-1}

Table 3 Requirements for post-chromatographic derivatization in liquid chromatographic systems

Reagent solution must be miscible with column effluent
Reagent must be compatible with column effluent, i.e. no precipitation or phase separation
Reagent must be sufficiently stable
Mixing of reagent with effluent must be rapid
Mixing must be efficient to prevent generation of 'noise'
Reagent and effluent flow rates must be matched
The background signal must be minimal
Reaction must be fast enough to minimize loss of resolution
The design of the reactor should minimize dispersion

for individual reagent-metering pump tubes; and methods depending on flow injection analysis (FIA) [17, 18], which use much narrower bore tubing and lower flow rates. The difference between these two types is characterized by the general use of air segmentation in AutoAnalyzer methods, a technique where the air bubble acts as a means of isolating segments of fluid from one another (to minimize what is called 'carry-over', i.e. the merging of successive separated zones) and of reducing longitudinal diffusion (band broadening or 'smearing') by 'scrubbing' the walls of the tubing. On the other hand, in FIA-based methods band broadening, now dubbed 'dispersion', is accepted as part of the methodology. The distinction between these two related continuous-flow techniques involves significant differences in flow chracteristics such as the Reynolds number, and whether there is turbulent or laminar flow within the system, but from a practical viewpoint the analyst selects whichever procedure best ensures that the compounds emerging from the folumn can be detected, recorded and quantitated as efficiently as possible. Alternatively, where the reaction is a straightforward addition of reagent followed by a rapid reaction with the separated components as they leave the column one by one, a simple one-pump arrangement for reagent metering followed by an efficient mixing arrangement may be all that is necessary (Figure 2).

Derivatization manifolds make use of a variety of fittings which are connected together between the

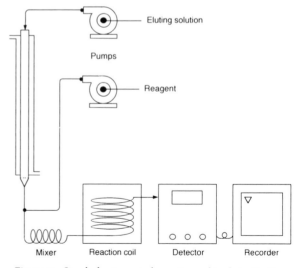

Figure 2. Simple basic post-chromatographic derivatization manifold. The detector may be a photometer, fluorimeter, electrochemical detector or any other appropriate type of instrument.

column and the detection system. These include unions, junctions (tees or 'cacti'), mixers, reaction coils, solid-phase reactors, debubblers (in AutoAnalyzer manifolds) and colorimeter or fluorimeter flow cells. The design of all of these has become more sophisticated in a continuing search for ways of minimizing degradation of the signal. This is done in two main ways: by preventing the development of a 'noisy' baseline, which impairs the signal-to-noise ratio and so reduces the sensitivity of detection by raising the minimum detectable amount; and by avoiding band broadening as much as possible, because a loss of chromatographic resolution obviously leads to deterioration of analytical performance.

Attention must also be given to the speed of the reactions that occur in the manifolds after addition of the reagents and mixing: 'fast' reactions are completed in a matter of seconds, but 'slow' ones may take minutes. The speed of a reaction influences the time delay needed between mixing and detection: for fast reactions the reactor may be quite short, because a long dwell time is not needed, but slow reactions will need a much longer time delay before detection to allow a sufficient proportion of the reaction to go to completion to provide the necessary sensitivity. Obviously increased temperatures favour shorter reaction times, but may also promote diffusion and lead to some band broadening and increased signal noise. As a rough guide, 'fast' or 'short' manifolds favour flow injection techniques, whereas 'slow' manifolds will probably benefit from air segmentation.

3.3 Mixers and liquid reactors in manifolds

Attention has been concentrated particularly on the design of mixers and reactors. The geometries of a variety of mixing devices have been evaluated in order to optimize the additon of reagents to the flowing stream.

Consideration must also be given to the materials of which the manifold is made [19], because glass and the various types of polymeric tubing have different wetting characteristics (contact angles), and these have profound effects on flow characteristics such as skin friction and pressure drop [19, 20].

Creating local turbulence and keeping the volume of the mixing chamber as low as possible are the main objectives of good mixer design [21, 22]. In liquid reactor design the main aim has been to find ways of keeping band broadening to a minimum. Band broadening is mainly produced by friction of the flowing stream against the walls of the tubing, which results in progressively severe tailing, and is worst in conditions of laminar flow such as obtain in FIA-based manifolds [18]. In AutoAnalyzer manifolds, turbulent flow is more likely to obtain, usually as a result of the comination of the flow along the tubing and skin friction, and gas segmentation is used to control carry-over and longitudinal smearing. In post-chromatographic non-segmented open tube reactors much ingenuity has been used to overcome dispersion [23–25]. Experience has shown coiling to be the best answer to the dispersion caused by laminar flow and skin friction, and in practice the most effective

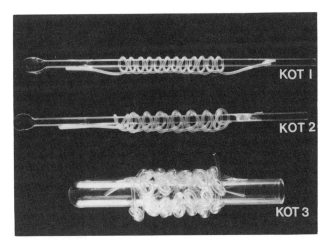

Figure 3. Knotted open tube (KOT) reactors. KOT1 consists of a series of alternating clove hitches. KOT2 is started similarly, but a second loop in the half-hitches is made by pulling the tubing through the knot again. KOT3 is made by tying alternating left- and right-handed overhand knots closely together, and then wrapping the knotted tubing round a 21 mm diameter spool (constructed according to Ref. 27).

configurations involve bending the tubing in more than one dimension [26]. The ultimate in three-dimensional tube deformation is the knotted open tube (KOT) reactor [27], as shown in Figure 3, but these are also knitted reactors [26, 27], which have the advantages of taking up the minimum of space and of providing an increased surface for heat exchange and for photochemical irradiation (see below) and stitched reactors (Figure 4). Of these procedures knitting is the fastest way of producing a suitable reactor.

In stitched, knotted and, even more, in knitted reactors dispersion is kept low, because the introduction of eddies at every turn is in practice equivalent to turbulent flow, and there are many turns. Each separated zone thus travels through the reactor as a bolus with almost the same flow velocity in the lumen of the tubing as at the wall, but unfortunately these advantages are obtained at the cost of increased back-pressures. A reactor recently devised to overcome this pressure problem is the 'single bead string reactor', which is intermediate between the open tube and a packed bed. It was found to work with much lower dispersion, and consists of a tube of internal diameter ca. 0.6 mm filled with somewhat smaller diameter beads [28]. The fluid stream passes round the beads, which take up a regular staggered packing pattern. In operation the reactor is analogous to a 'jet mixer' in non-segmented flowing streams such as in FIA (a comparison of these in shown in Figure 5).

3.4 Packed bed reactors

The alternative to the open tube reactor is the packed bed or solid-phase reactor, essentially a tube filled with some solid, either inert or non-inert, inorganic or poly-

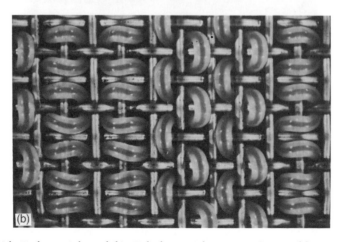

Figure 4. Structure of (a) knitted open tube and (b) stitched open tube reactors. Reprinted by permission of the author and John Wiley & Sons Ltd from H. Engelhardt, *Eur. Chromatogr. News*, **2**, 21 (1988).

Figure 5. Comparison of single bead string reactor and jet mixer. In both reactors turbulence is created in the flowing stream by alternating wide and narrow flow paths.

meric, pure or with some surface pre-treatment. Here the development of back-pressure sets practical limits to the dimensions, which therefore rarely exceed a bore of 5–6 mm and a length of 25–30 cm. As reactions in an open-tube reactor occur in a homogeneous liquid medium, and because there is a time delay between the entry of the reactants and the exit of the products, relatively slow reactants can be accommodated. In a packed bed reactor, by contrast, the chemistry is heterogeneous, and the rate of reaction depends on the surface area presented to the liquid phase, so that packed bed reactors are suitable for fast reactions, and the packings may be deliberately selected or modified so that the surface will have catalytic properties. The great advantage of packed bed reactors is that they provide a high local reagent concentration but do not require removal of excess reagent.

Little et al. were among the first to give careful consideration to the design aspects of the geometry of a packed bed reactor and a suitable mixer system [29] with special emphasis on minimizing band broadening, but reactor design has to be a compromise between efficiency and the practical limits set by the pressures that can be used [30]. There are a number of factors that lead to improved efficiency, including the use of small particles (though this will inevitably lead to an increase in back-pressure) and increased temperatures [31]. The actual process of the reaction may also contribute to band broadening and signal noise. These factors have been discussed in several reviews [32–34].

To give some examples, an ion-exchange resin in the permanganate form has been used as packing in an oxidation reactor [35], and the same group has also developed a reductive borohydride reactor [36, 37]. A reductive reactor was used by Lankmayer et al. to measure thyroid hormones by reducing them to iodide in a post-chromatographic packed bed reactor containing powdered zinc [38]. A reductive reactor was also developed by Studebaker to detect disulphides: he used a Sepharose with thiol groups. He also had a resin which incorporated a chromophore (N-dinitrophenylcysteine) or fluorophore (N,N'-didansylcystine) to render his reduced thiols detectable [39]. Ion-exchange resins lend themselves to such applications: Nondek et al. used a basic resin to hydrolyse N-methylcarbamates, which were then converted into fluorescent derivatives with the OPA reaction [40], and a strongly acidic resin was used to hydrolyse non-reducing oligosacharides to reducing monosaccharides for colorimetric analysis [41]. In fact, post-column reactors have found their widest application in carbohydrate analysis. Jolley's group at Oak Ridge developed a post-column glass bead reactor in which the sugars reacted with sulphuric acid and phenol to give coloured products for quantitative determination [42, 43]. Finally, an interesting reactor was made from the protein–dye combination azocoll, which was used to detect the enzyme α-chymotrypsin [39]. The packing in the reactor is gradually depleted as the enzyme splits off the detectable azo-dye.

3.5 Immobilized enzyme reactors

The covalent linking of an enzyme to a coil of tubing makes a useful reactor for the specific determination of the corresponding substrates, and this was first exploited in continuous-flow automated systems designed to measure a single substrate in many successive samples. However, it is equally applicable to substances emerging from a chromatographic column, and a variety of such systems has been described [44–47]. Some of these use classical coil reactors, others link the enzyme to a bead matrix packed into a solid phase reactor; in either case the reaction is, of course, heterogeneous.

Okuyama et al. used immobilized 3α-hydroxysteroid dehydrogenase to oxidize bile acids in a reaction coupled to the reduction of the cofactor nicotinamide adenine dinucleotide to a fluorescent species [48], which gave good sensitivity. Immobilized cholesterol oxidase was used to convert cholesterol to the more conjugated cholestenone, with a shift in absorption maximum and a sevenfold increase in detectability [49]. A more complex system was the use of immobilized xanthine oxidase, which has hydrogen peroxide as a product; this was detected in a manifold where a second enzyme, peroxidase, produced a fluorescent product for sensitive measurement [50, 51]. Dalgaard et al. separated glycosides chromatographically and used immobilized β-glucuronidase to yield detectable species [52], and Meek and Eva used immobilized choline oxidase, with or without acetylcholinesterase, to detect choline or acetylcholine. Here the product of the oxidase reaction, again hydrogen peroxide, was measured in an electrochemical detector [53].

The use of immobilized enzymes for post-chromatographic derivatization reactions is likely to grow with their commercial availability, and with it experience in how to maintain enzyme activity, avoid reversible or irreversible inhibition ('poisoning') and optimize the detectability of the products. As an example, one might cite a recently advertised post-column reactor developed for HPLC, consisting of a cartridge of one of a selection of oxidases which yields hydrogen peroxide for sensing in the linked electrochemcial detector. A variety of cartridges permits analysis of purines, sugars, lactate, oxalate and alcohols, depending on the particular HPLC separation and the appropriate oxidase being used.

3.6 Photochemical derivatization

Chemical compounds which can absorb light energy may become activated sufficiently to undergo a great variety of chemical reactions. As one might expect, many of these compounds have aromatic or conjugated structures and absorb light energy in the UV region of the spectrum, where the quanta are more energetic. Although photochemical reactions have been studied extensively, their application to liquid chromatography, and particularly to HPLC, has opened up a whole new analytical field, because photochemical activation and derivative formation is another useful way of making peaks emerging from a chromatographic column detectable, or of improving their detectability [54, 55].

Table 4 gives a list of the types of reactions which may be initiated by photochemical activation, and indi-

Table 4 Types of photochemically activated reactions

Oxidations and reductions
Photolysis and photo-hydrolysis reactions
Photo-ionization reactions
Molecular rearrangements
Addition and elimination reactions
Cyclization reactions
Dimerization and polymerization reactions
Other miscellaneous photochemical reactions

cates the wide scope of the process, although of course the applicability of any one of these reactions is limited to those compounds that have the right molecular structure to absorb light of sufficient energy and also to react appropriately.

Such reactions are of practical use only if the products are more readily detectable than the starting materials, although in some cases detectability has been achieved by demonstrating the disappearance of peaks on irradiation. In general, it is essential first to establish that a photochemical reaction does occur, and to work out the optimal conditions, such as wavelength of irradiation and mode of detection. One can then go on to employ a photochemical derivatization system for a post-chromatographic application: when the process works, photochemical detection is capable of dramatic successes.

Light must be the cheapest derivatizing agent, and photochemical derivatization is very straightforward, requiring little extra instrumentation; it is also flexible and clean. In operation the system consists merely of a lamp for irradiating the effluent from the chromatographic column before it passes to the detector, which may be a UV/VIS spectrophotometer, a fluorimeter, an electrochemical detector or even in specific instances simply a photomultiplier to detect luminescence (Figure 6).

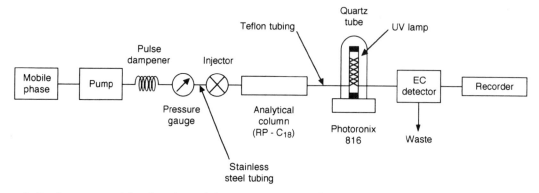

Figure 6. Simple arrangement for photochemical derivatization. Reprinted with permission from X.-D. Ding and I. S. Krull, *J. Agric. Food Chem.*, **32**, 622 (1984). Copyright (1984) American Chemical Society.

Table 5 Factors contributing to successful photochemical derivatization

---
Absorption at a suitable wavelength
Transparency of the chromatographic effluent at the wavelength used
High light flux at the wavelength chosen
Formation of detectable products (UV, visible, fluorescent, electrochemical)
Product(s) preferably detectable at wavelength(s) different from the wavelength of the activating light source
Sufficiently rapid rate of reaction
Products stable long enough for sensitive detectability

It may, however, be necessary to add reagents to the manifold downstream from the photochemical reactor in order to produce a detectable species by a subsequent chemical reaction. The lamp is usually a high-intensity UV source, although lasers will no doubt increasingly be used [55]. In order to achieve a high degree of photochemical conversion of the compounds eluting from the column, it is clearly essential for the column effluent to intercept as much light energy as possible: to do this a number of different conditions should be met (Table 5).

The insertion of a photochemical derivatization system between the chromatographic column and the detector carries the risk of band broadening. To minimize the danger of this various different designs have been tried out [54, 55], and the latest, a commercially available photochemical derivatization reaction module, consists of Teflon tubing formed into a KOT reactor wrapped round the UV lamp (Figure 7).

Some of the earliest work on the photochemical irradiation of flowing streams was by Fitzgerald and co-workers [56–58], who established the main features as already listed in Tables 4 and 5. In particular, they appreciated that constant flow conditions, as in FIA or AutoAnalyzer methods, would lead to consistent and reproducible kinetic reaction parameters, and thus to the possibility of quantitation, without the necessity of reactions going to completion or even to equilibrium.

Although a variety of applications have been described, drugs and pharmaceutical substances have received particular attention, not only for the obvious practical reason that they are among the most analysed classes of compounds, but also because many of them have aromatic structures which absorb strongly in the UV/VIS region and are therefore more likely to exhibit photochemical reactivity [59, 60]. Probably the most studied compounds are tamoxifen and its metabolites, and presumably because they are particularly suited to this approach they were some of the earliest to be analysed by photochemical reaction, but derivatization to fluorescent phenanthrene derivatives was initially *pre*-column, followed by ion-pair chromatography [61]. Subsequently, however, post-column photochemical derivatization has been the rule [62–64]. Other drugs analysed by post-column photochemical reactions include cannabinoids [65], phenothiazines [66], benzodiazepines [60, 66], barbiturates [67], clomiphene [68], reserpine [69], methotrexate [70], and penicillins and cephalosporins [71], among others.

For the rest, there have been reports concerning alcohols, aldehydes and ethers [72–74] and carbohydrates [73, 75], diethylstilboestrol [76], pesticides and herbicides

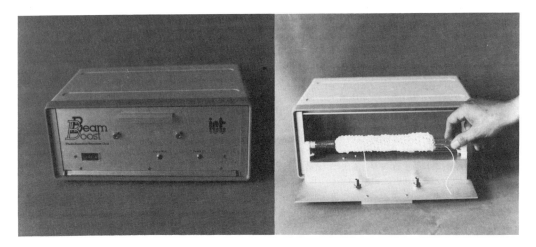

Figure 7. 'Beam Boost' system for use in HPLC (courtesy of Franz Morgenbesser and ICT Handels GmbH, Frankfurt, Germany).

[77, 78], quinones [79], alkaloids [80], nitro compounds [81] and nitrosamines [82, 83], and again this is only an illustrative list.

The interest in this form of post-column derivatization is likely to lead to more applications of photochemical reactions, particularly now that practical instrumentation is commercially available, and there are already signs of considerable activity in this area [55].

4. Post-chromatographic derivatization in gas chromatography

4.1 Pyrolysis

Pyrolysis generally precedes gas chromatography: the chromatographic step then resolves the mixture of pyrolysis products. However, the usefulness of pyrolysis as an aid to chemical identification has also been applied to separated gas chromatographic peaks by Levy and Paul [84], who switched the peaks from the first gas chromatograph to a heated coiled quartz reactor and analysed the products of this thermo-cracking on a second gas chromatograph. There is an obvious similarity here to GC–MS, and indeed pyrolysis has been called 'the poor man's mass spectrometer' [85], although the two techniques are complementary rather than competitive [86]. The quartz coil was replaced first with a non-catalytic gold coil reactor, and eventually by a modular two-stage oxidation–reduction reactor, to yield the 'Pyrochrom' a commercial analyser which, like some earlier schemes [87, 88], gave elemental C, H, O and N analyses of separated gas chromatographic peaks [89]. More recently Franc and Pour have worked out methods for elemental analyses of nitrogenous compounds [90] and for identification of substances containing carbon–carbon single bonds [91] by the use of reaction gas chromatography. Commercial instruments are now available which will provide a continuous instrumental display of specific elemental composition of a gas chromatographic separation, so that the elemental composition of eluted gas chromatographic peaks can be determined from a combination of several such elemental traces.

Although the idea of the Pyrochrom was to sidestep the lack of a GC–MS instrument, not enough development and application works was done before it went commercial. Uden et al. built a Pyrochrom, together with a multi-purpose thermal analyser, mass chromatograph and infrared spectrophotometer, into a comprehensive and versatile computerized GC peak identification system [92]. However, the 'bench-top' type of GC–MS apparatus has proved a much more satisfactory and much simpler solution to the problem, especially now that it is no longer a 'rich man's toy'. It also seems to solve many of the problems for which the multiple elemental detectors were developed, because identification of a gas chromatographic peak includes knowledge of its elemental composition. The multiple elemental analysis mode is thus limited to specialized applications.

A form of pyrolysis called 'electron pyrolysis' has been investigated by Schildknecht's group, who used tritiated water as a source of β-particles for the electron fragmentation of selected chemical compounds followed by chromatographic separation. Although the process was too slow to be used post-chromatographically, Schildknecht suggests such a possibility with sources of higher specific activity [93].

4.2 Transmodulators

Lovelock and co-workers [94, 95] developed the palladium transmodulator, based on the well known ability of heated metallic palladium to allow hydrogen but no other gas to diffuse through. They developed this device because their thermal conductivity detector was flow-sensitive. The transmodulator was used to remove hydrogen from the carrier gas stream and to replace it with a scavenger gas, here helium, the flow rate of which was kept strictly constant. In their system they were able to use flow programming with hydrogen as carrier gas to optimize their separations, but by adding a constant flow of helium downstream of the column and getting rid of the hydrogen by causing it to diffuse away through three heated palladium coils in parallel (Figure 8) they were able to achieve a 40-fold increase in sensitivity.

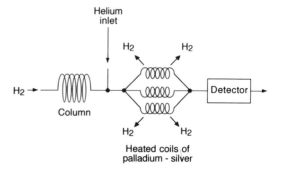

Figure 8. The palladium transmodulator. Hydrogen diffuses out of the system via the heated palladium–silver coils (three in parallel, to handle the volume of hydrogen involved). Helium replaces it and carries the separated peaks to the detector, whose response is kept consistent through maintaining a constant flow of helium. The sensitivity of detection can be increased by reducing the helium flow.

A feature of the heated palladium system is that it is catalytically active in proportion to its hydrogen permeability, so reducible compounds will be completely hydrogenated on passing through; hydrogenation proved to be specific to compounds with conjugated unsaturated bonds such as dienes, α,β-unsaturated aldehydes, ketones and nitriles, but did not occur otherwise.

A useful extension of the palladium transmodulator was the hydrogen transfer system described by Thompson [96], which is marketed commercially [97]. The hydrogen in the column effluent was kept in a jacket sealed round the heated palladium tube, and carried to a second detector in a stream of nitrogen (Figure 9). This is a neat method for the specific quantitative analysis of hydrogen in a gas mixture: sensitivity ranged linearly 'from a few parts per million to 100%' [96].

A different type of transfer between flowing streams, this time one gas and the other liquid, was developed by Chow and Karmen, who analysed primary amines and ammonia separated by a gas chromatograph by bubbling the GC effluent through a solution which was then removed via a short capillary tube and reacted in a manifold with o-phthaldialdehyde (OPA) and ethanethiol to yield fluorescent derivatives, which were recorded fluorimetrically [98]. These analyses gave good sensitivity with only slight peak broadening, but extraction recovery dropped in parallel with water solubility with increasing chain length. This approach echoes the original work of James and Martin, who scrubbed out lower molecular weight fatty acids into a water-filled colorimetric cell for continuous automatic titration [99]. Bubbling the gas stream emerging from a gas chromatograph through specific colour reagents responding to specific molecular groupings was also used by Walsh and Merritt as a way of classifying compounds eluting from gas chromatographic columns [100, 101].

A specific kind of post-column functional group analysis was done by Casu and Cavallotti by moving a paper strip wetted with nitrochromic acid past the column outlet at the same rate as the recorder chart. Oxygenated compounds gave coloured spots on the strip which could be accurately correlated with the peaks of the recorded GC profile [102].

4.3 Post-column concentrators

A variety of post-column concentrators has been described, mainly for gas chromatography–mass spectrometry, but these can also be used for improving the signal-to-noise ratio, as usefully reviewed by Freedman [103]. The devices fall into four main groups: the jet separators [104–106]; the fritted glass separators [107, 108]; the porous membrane type [109–112]; and those dependent on polymeric membranes such as silicone or heated Teflon [112–114].

4.4 Post-column reactors

One of the earliest gas chromatographic, post-column reactors was developed by Zlatkis and his group, and consisted of a micro-reactor containing a nickel–kieselguhr catalyst heated at 425 °C through which the carrier gas, hydrogen, was passed, so that any peaks contained in it were hydrocracked to methane and water. The latter was removed by a Drierite column and the methane was determined in a thermal conductivity cell operated at room temperature for maximum sensitivity [115, 116]. One of the applications of this hydrogen reactor was for the analysis of aldehydes, which were themselves generated from a mixture of amino acids in a *pre*-column reactor packed with ninhydrin [117].

Coulson passed a gas chromatographic effluent

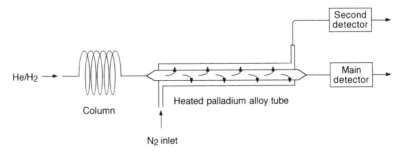

Figure 9. The hydrogen transfer system. Hydrogen diffuses out of the main carrier gas stream through the heated palladium tube and is swept to a second detector by a stream of nitrogen and determined there, while the remaining components are determined at the first detector. In the commercial instrument the two sides of a single thermal conductivity detector operate independently as if they were two separate detectors.

through a hydrogen reactor to reduce halogen- or nitrogen-containing peaks to halogen acids or ammonia. The reaction products were detected by coulometry or conductimetry [118]. A post-column hydrogen reactor was also used by Need et al. for specific detection of nitrogen-containing substances in gas chromatographic peaks [119]. The ammonia produced was scrubbed from the reactor effluent into a suitable manifold and detected fluorimetrically by reaction with OPA and ethanethiol as described earlier [119, 98].

Post-column derivatization was achieved by Aue et al. using a plasma reactor which could be switched between two gas chromatographic columns, so that a given peak could be diverted from the end of the first column, reacted and separated on the second column [120]. The reactor consisted of an ionization chamber lined with a ^{63}Ni-coated gold foil containing parallel tungsten electrodes to provide the high voltage for producing the plasma. It was found that plasma processing was 'typically, based on a weak β-stimulated argon discharge, and a variety of doping gases can be added to provide different chemical environments'. One can envisage that, as an alternative to the hydrogen described in their paper, one might use a halogen such as chlorine or bromine and place an electron capture detector at the end of the second column to analyse the reaction products.

The whole subject of identification techniques in gas chromatography has been extensively reviewed [121, 122], but reaction methods tend to be relatively insensitive and rather complicated. GC–MS is by far the most elegant solution to the chemical identification of gas chromatographic peaks, and is now relatively affordable, so that this will increasingly become the preferred way of identifying unknown chromatographic peaks of confirming the chemical structures of known compounds emerging from a column.

4.5 Photochemical reactor

Although one might wonder whether a gas stream leaving a gas chromatographic column is likely to occupy a photochemical reactor for long enough to undergo photolytic reactions, such a reactor was designed and built by Aue and Aigner-Held, who made a quartz tube which fitted over a powerful light source. The tube was double-walled, and the annular space had a narrow gap between the inner and outer walls, with an inlet connected to the column end and an exit leading to the detector. In operation it functioned just like the analogous reactor in liquid chromatography, although its scope is probably not so wide [123, 124].

5. Radioactive derivatization

Karmen and Longo analysed radioactive substances by gas chromatography and burned the column effluent in a gas train to yield radioactive CO_2, which was trapped by bubbling into a stream of alkali solution flowing through a short capillary tube. The CO_2 was quantitatively scrubbed out and peak broadening was minimal, the original chromatographic separation being preserved [125].

A method for radiochemical amplification also using a moving-band reactor to apply radioactive reagents to peaks eluting from the outlet of a gas chromatographic column was described by Bächmann et al. [126], who found that the sensitivity depended on the half-life of the radionuclide applied. Analysis of the products could be achieved by scanning the strip, or by cutting it into small pieces and using, e.g., liquid scintillation counting.

6. Miscellaneous Applications

6.1 Amines

Amines, because of their polar character, do not chromatograph well because of non-specific adsorption to column packings, which leads to tailing, distorted peaks and poor recoveries. Amines are therefore usually derivatized before chromatography to produce derivatives with better chromatographic properties. An interesting example of post-chromatographic derivatization to a fluorescent derivative has already been mentioned, however [98]. A number of papers describe the application of amino acid analysers to amine analysis with post-column ninhydrin colorimetric determination [127–129]. An interesting variant of this was the use of a ninhydrin-containing solvent system for the HPLC of amines (and amino acids) with post-column heating to produce coloured zones from the peaks eluting from the column [130]. Post-column fluorimetric reactions have also been used after ion exchange chromatography of histamine [131] or HPLC of primary [132] or secondary amines [133], biogenic amines [134] and 5-hydroxy- and 5-methoxyindoles [135, 136]. Tertiary amines have often proved an analytical problem, and of course many pharmaceutical substances and pesticides etc. are tertiary amines. Although volatile tertiary amines may be analysed directly by gas chromatography, their polar nature usually demands pre-chromatographic derivatization (see Chapter 3). One way of analysing tertiary amines is by HPLC with post-chromatographic derivatization using the Palumbo reaction, a specific reaction of tertiary amines with acetic anhydride and citric acid [137–139].

6.2 Catecholamines

Catecholamines, although polar in character, can be chromatographed without derivatization. Nowadays they are generally determined by HPLC and quantitated by electrochemical detection without further reaction. However, they can also be converted to fluorescent derivatives by post-column reaction with a fluorophore such as borate [140], cyanoacetate [141] or glycylglycine [142, 143]. Glycylglycine was similarly used by the same group to determine the catecholamine–acetaldehyde condensation product and alkaloid salsolinol [144], which may be implicated in alcohol addiction. A most interesting report describes the analysis of fluorescamine-labelled catecholamines by HPLC followed by a sensitive post-column reaction in which the fluorescence of the derivatized catecholamines was elicited by a chemical excitation reaction between hydrogen peroxide and 2,4,6-trichlorophenyl oxalate [145]. The catecholamine metabolite vanilmandelic acid was determined by post-column periodate oxidation to vanillin, which was measured by UV spectrophotometry [146].

6.3 Amino acids and peptides

Apart from paper chromatography and TLC in one or two dimensions, amino acids are traditionally separated by ion-exchange chromatography followed by the classical post-column reaction with ninhydrin [5–7]. As a variant, trinitrobenzenesulphonic acid has been used [147]. Ion-exchange chromatography is quite demanding, requiring careful attention to the composition and programming of the buffer sequence and regeneration of the column after each run, not to mention the need for scarce technically trained manpower, and so HPLC methods for amino acids are becoming popular. Almost all of these involve derivatization prior to separation, as much to render the amino acids susceptible to separation by HPLC as to make them more detectable afterwards. Numerous recipes have been given for the ninhydrin reagent used in post-column colorimetric reactions, with the aim of increasing colour yield, reducing background and improving stability [5–7, 148, 149]. The ninhydrin reaction produces a purple colour with an absorption maximum of around 570 nm with most amino acids, but the imino acids proline and hydroxyproline give a brownish product which absorbs at 405–440 nm, so that at least two colorimeters are required.

A significant advance in post-chromatographic derivatization to fluorescent products came with the introduction of fluorescamine by Udenfriend's group [150], and this was applied to a post-column system by Stein et al. [151], leading to very much greater sensitivity of detection. Post-column fluorescamine was also used in the analysis of small peptides [152]. Because of difficulties with the stability of the fluorescamine reagent, there has been a shift towards using the 'OPA plus a thiol' post-column derivatization reaction [149–159]. Again, the imino acids need special treatment, such as prior reaction with N-chlorosuccinimide [150, 160] or hypochlorite [161]. Alternatively, both amino and imino acids may be derivatized with 4-fluoro-7-nitrobenzofurazan (NBD-F) as the fluorogenic reagent [162], as shown in Figure 10.

An interesting development is the application of electrochemical detection methods for improved sensitivity. Originally metal electrodes were used, without derivatization, [163–165], but sensitivity was fairly poor and varied a good deal between the amino acids. For these reasons OPA derivatization was tried [166, 167], and the latest derivatization schemes use p-dimethylaminophenyl isothiocyanate [168] or phenyl isothiocyanate [169] for making detectable derivatives.

6.4 Drugs and pharmaceuticals

As already mentioned, this is a popular area, and all types of chromatographic determination are used. Most of the drugs which are determined by gas chromatography are derivatized prior to chromatographic separation. Drugs separated by paper or thin-layer chromatography are of course revealed by specific colour or UV reagents [4, 11–14].

By far the majority of drugs are determined by HPLC, almost invariably by UV or visible spectrophotometric methods; in addition to the procedures mentioned earlier which involve photochemical reactions, scattered references describe other post-column reaction procedures, practically all of which involved derivatization to yield fluorescent products. Thus, amoxycillin was derivatized with fluorescamine for fluorimetry [170], and so were cefatrizine [171], felypressin [172] and sulfapyridine [173]. On the other hand, OPA procedures were used to derivatize aminoglycoside antibiotics [174–176], β-lactam antibiotics [177], fludalanine [178] and phenylpropanolamine [179]. Some drugs were hydrolysed to fluorescent derivatives: alkaline hydrolysis to give a fluorescent product from indomethacin [180] and acid hydrolysis for digitalis glycosides [181] and also for digoxin [182], where a PVC pressure vessel was used for hydrochloric acid hydrolysis followed by reaction with hydrogen peroxide and dehydroascorbic acid to give a fluorescent product. Oxidation was used to produce fluorescent derivatives from morphine and some morphine analogues, but not from heroin or codeine [183, 184], and also from thioridazine [185]. Reserpine was converted to 3,4-dehydroreserpine with sodium

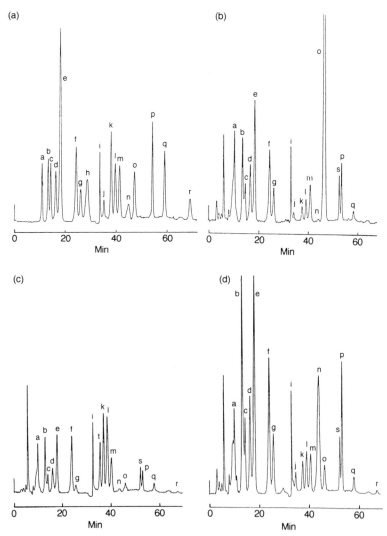

Figure 10. HPLC of amino and imino acids after post-column derivatization with NBD-F (Ref. 162). Abbreviations: a = Asp; b = Thr; C = Ser; d = Glu; e = Pro; f = Gly; g = Ala; h = Cys; i = Val; j = Met; k = Ile; l = Leu; m = Nle (internal standard); n = Tyr; o = Phe; p = Lys; q = His; r = Arg; s = Orn; t = *allo*-Ile. Profile A is of a standard mixture. Profile B is of amino acids eluted from a filter-paper blood spot of a patient with phenylketonuria, showing a normal profile but with an abnormally large phenylalanine peak. Profiles C and D are from patinets with branched chain ketoaciduria (maple syrup urine disease) and tyrosinaemia respectively. Reprinted with permission from Y. Watanabe and K. Imai, *Anal. Chem.*, **55**, 1786 (1983). Copyright (1983) American Chemical Society.

nitrite for fluorimetry [186]. Penicillins were also determined by a post-column degradation reaction carried out in a hollow-fibre membrane reactor, with subsequent UV spectrophotometry [187].

6.5 Carbohydrates

Although one tends to think of carbohydrates as a group, as one thinks of amino acids as a group, carbohydrates are no more a homogeneous group than amino acids. They consist of monosaccharides, oligo- and polysaccharides, reducing and non-reducing sugars, amino sugars and uronic acids, and these all require different methods to detect them. Unfortunately many of the detection reagents used for carbohydrates contain strong acids, and, what is worse, many of these reactions are not very rapid. For post-column work, the drawbacks are that the manifolds have to be made of materials that can stand up to the reagents, and slow reactions either require long dwell times in the reactors or elevated temperatures—or both. The penalties are band broadening and unfavourable effects on the signal-to-noise ratio. Nevertheless, these difficulties can be overcome by good design and attention to detail to provide acceptable HPLC with post-chromatographic carbohydrate analysis, such as the phenol/sulphuric acid system and its variants developed at Oak Ridge National Laboratory [188, 189]. Non-corrosive reagents were developed using tetrazolium salts, to avoid the problems of strong acid [190]. Reducing carbohydrates can of course be detected by their ability to reduce cupric ions, which may be detected colorimetrically after reaction with bicinchoninic acid [191–193]. Alternatively, reduction of cerate produced fluorescent Ce^{3+}, which could be detected at 350 nm [194]. The slowness of these reactions led to the use of cuprammonium ions [195], the reaction of which is not confined only to reducing sugars. Another way of dealing with reducing sugars is reaction with cyanoacetamide followed by electrochemical detection [196]. The presence of non-reducing sugars such as sucrose or maltose has also been allowed for by the use of post-column enzymatic hydrolysis [193, 197]. These procedures used enzyme solutions rather than immobilized enzymes. However, a galactose oxidase reactor has been described for post-column analysis of saccharides. The products were converted to fluorescent derivatives by addition of purified p-hydroxyphenylacetic acid at pH 7.5 to give a fluorescent product [198]. A variety of post-chromatographic methods was developed for amino sugars, starting as early as 1953 with post-column reaction with the Elson–Morgan reagent (alkaline acetylacetone) followed by Ehrlich reagent [199, 200]. More recently, fluorimetric determination has been achieved following post-column reactions with 2,4-pentanedione and formaldehyde [201] or cyanoacetamide [202]. The latter reagent has also been applied to post-chromatographic derivatization for the colorimetric and fluorimetric determination of uronic acids [203]. A considerable literature exists concerning the chromatography of acidic glycosaminoglycans (mucopolysaccharides) in connection with a number of rare inherited lysosomal enzyme deficiencies in which these substances accumulate. Almost all of these analyses, although post-chromatographic, are done 'off-line' following fraction collection and hydrolysis of the fractions that contain individual peaks. Another possibility is to determine the peaks leaving the column by nephelometry following continuous addition of Pennock's reagent, cetylpyridinium chloride. A fluorimeter was used as detector at the end of the manifold [204].

6.6 Steroids

Fluorimetric reaction detectors were developed by Reh and Schwedt and by Verbeke and Vanhee [205, 206], and fluorimetric detection has been the rule in post-column derivatization reactions of steroids, including the use of such fluorophores as glycinamide [207], benzamidine [208] and 3-chloroformyl-7-methoxycoumarin [209]. Dansylated steroids were separated by reversed phase HPLC and the pre-column derivatization was complemented by a post-column treatment with peroxyoxalate and detection by chemiluminescence [210].

6.7 Vitamins

Vitamins, like pharmaceutical substances, can often be determined by UV absorption or electrochemical detection without derivatization. However, B_6 vitamins were determined in a semi-automated manifold by post-column reaction with a coupling reaction with diazotized 5-chloroaniline-2,4-disulphonyl chloride and continuous colorimetric determination of the orange products at 440 nm [211]. Post-column derivatization schemes for thiamine and its phosphate esters have also been described [212–214].

6.8 Phospholipids

An early procedure for sphingosine and sphingolipids used labelling with fluorescamine [215], and another labelling reagent was 2-naphthoxyacetate [216]. More recent procedures used 1,6-diphenyl-1,3,5-hexatriene (DPH) for fluorescence enhancement [217, 218]; see Figure 11. The mechanism for this is not clear, but seems to involve some kind of intercalation of the DPH within the phospholipid molecules or micelles rather than the formation of true derivatives.

6.9 Guanidino and urea compounds

Durzan determined guanidino compounds on an amino acid analyser using a manifold with the classic Sakaguchi reaction (N-bromosuccinimide as a source of bromine, followed by 8-hydroxyquinoline) [219]. Ninhydrin can be used [220, 221], but of course the presence of amino acids in the samples would confuse the interpretation of the chromatographic profiles. More recently fluorimetric post-chromatographic reagents have been used, including phenanthroquinone [222], benzoin [223, 224] and naphthalene-2,3-dicarboxaldehyde [225].

Urea and related compounds were separated by HPLC, and the peaks emerging from the column were converted to the N-chloramines with hypochlorite. Excess reagent

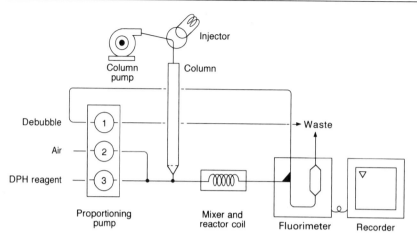

Figure 11. Manifold for quantitation of phospholipids separated by HPLC by addition of diphenylhexatriene (DPH) to enhance fluorescence (after Ref. 218).

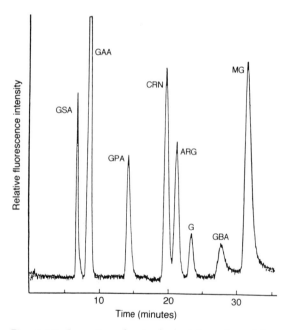

Figure 12. Separation of a standard mixture of guanidines by HPLC followed by post-column derivatization with phenanthraquinone and fluorescence detection. GAS = guanidinosuccinic acid; GAA = guanidinoacetic acid; GPA = guanidinopropionic acid; CRN = creatinine; ARG = arginine; G = guanidine; GBA = guanidinobutyric acid; MG = methylguanidine. Reprinted with permission from M. D. Baker et al., Anal. Chem., **53**, 1658 (1981). Copyright (1981) American Chemical Society.

was selectively removed with sodium nitrite. The chloramines reacted with potassium iodide to give the triiodide ion, which was monitored spectrophotometrically at 370 nm [226].

6.10 Acids

Chromatographic analysis of the simpler organic acids is generally done by GC after conversion to a more volatile and less polar derivative. However, many acids do not lend themselves to this approach, and various HPLC methods have been used, usually detecting the acids by fluorescence [227–230]. Hydroxy acids were separated by anion exchange chromatography and detected by the classical hydroxamate–ferric ion procedure in a post-column manifold [231]. Samuelson's group developed a five-channel post-column manifold: one channel went to a fraction collector, leaving open the options for further analysis; another used cleavage with periodate, followed by measurement of periodate consumption. The third used periodate cleavage followed by reaction of the formaldehyde formed with 2,4-pentanedione; a fourth channel used chromic acid oxidation, and the last channel used the Dische carbazole reaction [232–234]. The results were displayed on a multi-pen recorder, and the profiles therefore also gave some information about the reactivity of the eluted acids. Carboxylic acids were also determined in an electrochemical detector following post-column reaction with benzoquinone to yield the detectable species hydroquinone [235].

6.11 Miscellaneous compounds

Prostaglandins and other arachidonic acid metabolites, previously determined after HPLC by UV detection, were measured by oxidation with bromine generated in a post-column electrochemical cell, and the bromide formed was measured electrochemically [236]. An even more high technology solution to this difficult analytical problem was to use HPLC–thermospray MS after post-chromatographic methylation with trimethylanilinium hydroxide [237]. Cannabinoids were complexed with Fast blue B to give a dye complex that was determined colorimetrically at 490 nm [238]. Carbamate insecticides were converted to methylamine by post-column alkaline hydrolysis, and the methylamine was determined fluorimetrically after reaction with OPA and mercaptoethanol [239]. Since the OPA reaction for amino compounds requires a thiol to produce the characteristic fluorescent isoindole derivatives, it can also be used to analyse thiols in the presence of an amino compound, and this was the basis of the analysis of thiols carried out with taurine as the amino compound [240, 241]. Although the measurement of the activity of enzymes after their chromatographic separation is not strictly speaking derivatization, this topic has been reviewed by Torens and Vacik [242]. Chloroanilines were determined fluorometrically after HPLC and reaction with fluorescamine [243]; aflatoxins after reaction with iodine [244]; and biopterins after nitrite oxidation [245]. Nitrosamines were separated by reversed phase HPLC, and the peaks leaving the column were hydrolysed with sulphuric acid at 80 °C. The products were oxidized with Ce^{4+} to give the fluorescent Ce^{3+} [246]. Kynurenine was chromatographed with hydrogen peroxide actually in the mobile phase. After the analyte had emerged from the column, ethanol was added, and UV irradiation was used to induce fluorescence [247]. Finally, thymidine hydroperoxides were selectively detected after addition of Fe^{2+}–xylenol orange reagent [248].

7. Conclusions: future developments

This brief account, taken together with applications mentioned in earlier sections of the chapter, is by no means exhaustive, but reflects the extent and scope of post-chromatographic derivatizations. However, in spite of its kaleidoscopic character some principles do stand out. The general objective is to perform some reaction that will produce derivatives that can be determined by the most sensitive detection methods available, and for this reason fluorimetry and electrochemical detection are the most popular; they may increasingly be joined by photochemical detection. Mass spectrometric detection may in some circumstances also require post-chromatographic derivatization for improved sensitivity and identification (see Chapters 12 and 14). For the rest, in any particular analytical problem some of the approaches and principles outlined may well suggest possible ways in which post-chromatographic derivatization or modification reactions may be applicable.

As will be abundantly clear from the rest of this handbook, derivatization for chromatography is still most usually done *before* the chromatographic separation step, because the choice of derivative includes as one of its aims conversion to compounds with improved chromatographic properties. However, this chapter describes many situations where post-chromatographic derivatization has been found necessary or desirable, not only as an alternative to pre-chromatographic derivatization, but also as a complementary additional procedure.

Doing things with a chromatographic effluent is growing in popularity, particularly in liquid chromatographic procedures, because, as previously mentioned, most chemical reactions are done in liquid solution, and because, unlike pre-column derivatization, there is no absolute requirement for derivatization reactions to go to completion, which at once makes the analytical problem less demanding. The major advances in post-chromatographic derivatization are therefore likely to be made in liquid chromatography, especially HPLC. Enough illustrations have been given to indicate some of the direction these advances could take.

The improved commercial availability of solid-phase reaction systems, immobilized enzyme reactors and photochemical derivatization instruments is one such direction; for a brief outline of some of these see [34]. There will undoubtedly be new developments in the application of analytical methods, particularly in liquid chromatography, that will exploit the practical advantages of these devices when they are linked to the most sensitive detection modes, particularly the fluorescence and electrochemical types of detector.

One area that looks distinctly promising is the further application of immunochemical methods in fast protein liquid chromatography. The ingenuity already shown in the development of immunochemical assay procedures, not just with proteins but more generally, suggests at least one likely growth area for a more widespread application in post-chromatographic derivatization.

8. References

[1] F. Feigl, *Spot Tests in Organic Analysis*, 7th edn., Elsevier, Amsterdam (1966).
[2] F. Feigl, *Spot Tests in Inorganic Analysis*, 6th edn., Elsevier, Amsterdam (1972).
[3] R. Consden, A. H. Gordon and A. J. P. Martin, *Biochem. J.*, **38**, 224 (1944).
[4] E. Stahl, *Thin-Layer Chromatography, a Laboratory Handbook*, 2nd fully revised and expanded edition, Springer, Berlin (1969).
[5] W. H. Stein and S. Moore, *J. Biol. Chem.*, **176**, 337 (1948).
[6] S. Moore and W. H. Stein, *J. Biol. Chem.*, **178**, 53 (1949).
[7] D. H. Spackman, W. H. Stein and S. Moore, *Anal. Chem.*, **30**, 1190 (1958).
[8] T. T. Chang, J. O. Lay and R. J. Francel, *Anal. Chem.*, **56**, 109 (1984).
[9] G. D. Barrett, *Adv. Chromatogr.*, **11**, 145 (1974).
[10] R. Segura and A. M. Gotto, *J. Chromatogr.*, **99**, 643 (1974).
[11] I. Smith, *Chromatographic and Electrophoretic Techniques*, 4th edn., Heinemann, London (1976).
[12] J. G. Kirchner, *Thin-Layer Chromatography*, 2nd edn., Wiley, New York (1978).
[13] J. C. Touchstone and M. F. Dobbins, *Practice of Thin-Layer Chromatography*, Wiley, New York (1978).
[14] Anon., *Dyeing Reagents for Thin Layer and Paper Chromatography*, E. Merck, Darmstadt (1980).
[15] T. D. Schlabach and R. Weinberger, in *Reaction Detection in Liquid Chromatography*, I. S. Krull (Editor), Marcel Dekker, New York (1986), pp. 63–127.
[16] L. T. Skeggs, *Am. J. Clin. Pathol.*, **28**, 311 (1957).
[17] J. Růzicka and E. H. Hansen, *Anal. Chim. Acta*, **78**, 145 (1975).
[18] K. K. Stewart, G. R. Beecher and P. E. Hare, *Anal. Biochem.*, **70**, 167 (1976).
[19] A. M. Schwartz, C. A. Rader and E. Huey, *Adv. Chem. Ser.*, (**43**), 250 (1964).
[20] A. L. Chaney, *Automation in Analytical Chemistry: Technicon Symposium 1967*, Vol. I, Mediad, New York (1968), pp. 115–117.
[21] B. Lillig and H. Engelhardt, in *Reaction Detection in Liquid Chromatography*, I. S. Krull (Editor), Marcel Dekker, New York (1986), pp. 1–61.
[22] S.-I. Kobayashi and K. Imai, *Anal. Chem.*, **52**, 1548 (1980).
[23] A. H. M. T. Scholten, U. A. T. Brinkman and R. W. Frei, *Anal. Chem.*, **54**, 1932 (1982).
[24] M. Margoshes, *Anal. Chem.*, **49**, 17 (1977).
[25] W. E. van der Linden, *Trends Anal. Chem.*, **1**, 188 (1982).
[26] H. Engelhardt, *Eur. Chromatogr. News.*, **2**, 20 (1988).
[27] C. M. Selavka, K.-S. Jiao and I. S. Krull, *Anal. Chem.*, **59**, 2221 (1987).
[28] J. M. Reijn, W. E. van der Linden and H. Poppe, *Anal. Chim. Acta*, **126**, 1 (1981).
[29] C. J. Little, J. A. Whatley and A. D. Dale, *J. Chromatogr.*, **171**, 63 (1979).
[30] R. S. Deelder, M. G. F. Kroll, A. J. B. Beeren and J. H. M. Van Den Berg, *J. Chromatogr.*, **149**, 669 (1978).
[31] L. Nondek, U. A. T. Brinkman and R. W. Frei, *Anal. Chem.*, **55**, 1466 (1983).
[32] R. W. Frei and J. F. Lawrence, *Chemical Derivatization in Analytical Chemistry*, Vol. 1; Chromatography, Plenum, New York (1981).
[33] C. F. Poole and S. Schuette, *Contemporary Practice of Chromatography*, Elsevier, Amsterdam (1984).
[34] S. T. Colgan and I. S. Krull, in *Reaction Detection in Liquid Chromatography*, I. S. Krull (Editor), Marcel Dekker, New York (1986), pp. 227–258.
[35] K.-H. Xie, C. T. Santasania, I. S. Krull, U. Neue, B. Bidlingmeyer and A. Newhart, *J. Liq. Chromatogr.*, **6**, 2109 (1983).
[36] I. S. Krull, K.-H. Xie, S. Colgan, T. Izod, U. Neue, R. King and B. Bidlingmeyer, *J. Liq. Chromatogr.*, **6**, 605 (1983).
[37] I. S. Krull, S. Colgan, K.-H. Xie, U. Neue, R. King and B. Bidlingmeyer, *J. Liq. Chromatogr.*, **6**, 1015 (1983).
[38] E. P. Lankmayr, B. Maichin, G. Knapp and F. Nachtmann, *J. Chromatogr.*, **224**, 239 (1981).
[39] J. F. Studebaker, *J. Chromatogr.*, **185**, 497 (1979).
[40] L. Nondek, R. W. Frei and U. A. T. Brinkman, *J. Chromatogr.*, **282**, 141 (1983).
[41] P. Vratny, J. Ouhrabkova and J. Copikova, *J. Chromatogr.*, **191**, 313 (1980).
[42] R. L. Jolley and M. L. Freeman, *Clin. Chem.*, **14**, 538 (1968).
[43] C. D. Scott, R. L. Jolley, W. W. Pitt, Jr, and W. F. Johnson, *Am. J. Clin. Pathol.*, **53**, 701 (1970).
[44] L. Martinek and I. V. Berizin, *J. Solid Phase Biochem.*, **2**, 343 (1977).
[45] T. D. Schlabach and F. E. Regnier, *J. Chromatogr.*, **158**, 349 (1978).
[46] P. W. Carr and L. D. Bowers, in *Immobilized Enzymes in Analytical and Clinical Chemistry*, Wiley, New York (1980).
[47] L. D. Bowers, in *Reaction Detection in Liquid Chromatography*, I. S. Krull (Editor), Marcel Dekker, New York (1986), pp. 195–225.
[48] S. Okuyama, K. Kokubun, S. Higashidate, D. Uemura and Y. Hirata, *Chem. Lett.*, **12**, 1443 (1979).
[49] L. Ogren, I. Cisky, L. Risinger and G. Johansson, *Anal. Chim. Acta*, **117**, 71 (1980).
[50] R. Tawa, M. Kito, S. Hirose and K. Asachi, *Chem. Pharm. Bull.*, **30**, 615 (1982).
[51] M. Kito, R. Tawa, S. Takeshima and S. Hirose, *J. Chromatogr.*, **278**, 32 (1983).
[52] L. Dalgaard, L. Nordholm and L. Brimer, *J. Chromatogr.*, **265**, 83 (1983).
[53] J. L. Meek and C. Eva, *J. Chromatogr.*, **317**, 343 (1984).
[54] I. S. Krull and W. R. LaCourse, in *Reaction Detection in Liquid Chromatography*, I. S. Krull (Editor), Marcel Dekker, New York (1986), pp. 303–352.
[55] I. S. Krull, C. M. Selavka, M. Lookabaugh and W. R. Childress, *LC–GC Int.*, **2**, 28 (1989).
[56] H. D. Drew and J. M. Fitzgerald, *Anal. Chem.*, **41**, 974 (1969).
[57] V. R. White and J. M. Fitzgerald, *Anal. Chem.*, **44**, 1267 (1972).
[58] V. R. White and J. M. Fitzgerald, *Anal. Chem.*, **47**, 903 (1975).

[59] A. H. M. T. Scholten, U. A. T. Brinkman and R. W. Frei, *Anal. Chim. Acta*, **114**, 137 (1980).
[60] C. M. Selavka, I. S. Krull and I. S. Lurie, *J. Chromatogr. Sci.*, **23**, 499 (1985).
[61] Y. Colander and L. A. Sternson, *J. Chromatogr.*, **149**, 683 (1978).
[62] R. R. Brown, R. Bain and V. C. Jordan, *J. Chromatogr.*, **272**, 351 (1983).
[63] M. Nieder and H. Jaeger, *J. Chromatogr.*, **413**, 207 (1987).
[64] C. Kikuta and R. Schmid, *J. Pharm. Biomed. Anal.*, **7**, 329 (1988).
[65] P. J. Twitchett, P. L. Williams and A. C. Moffat, *J. Chromatogr.*, **149**, 683 (1978).
[66] U. A. T. Brinkman, P. L. M. Welling, G. de Vries, A. H. M. T. Scholten and R. W. Frei, *J. Chromatogr.*, **217**, 463 (1981).
[67] R. W. Schmid and C. Wolf, *Proc. 3rd Int. Symp. Drug Analysis, Antwerp* (1988).
[68] P. J. Harman and G. I. Lackman, *J. Chromatogr.*, **225**, 131 (1981).
[69] J. R. Lang, J. T. Stewart and I. L. Honigberg, *J. Chromatogr.*, **264**, 144 (1983).
[70] J. Salamoun, M. Smrz, F. Kiss and A. Salamounova, *J. Chromatogr.*, **419**, 213 (1987).
[71] C. M. Selavka, I. S. Krull and K. Bratin, *J. Pharm. Biomed. Anal.*, **4**, 83 (1986).
[72] M. S. Gandelman and J. W. Birks, *Anal. Chem.*, **54**, 2131 (1982).
[73] M. S. Gandelman and J. W. Birks, *J. Chromatogr.*, **242**, 21 (1982).
[74] W. LaCourse and I. S. Krull, *Anal. Chem.*, **59**, 49 (1987).
[75] M. S. Gandelman, J. W. Birks, U. A. T. Brinkman and R. W. Frei, *J. Chromatogr.*, **282**, 193 (1983).
[76] A. T. Rhys Williams, S. A. Winfield and R. C. Belloli, *J. Chromatogr.*, **235**, 461 (1982).
[77] R. G. Luchtefeld, *J. Chromatogr. Sci.*, **23**, 516 (1985).
[78] C. J. Miles and H. A. Moye, *Anal. Chem.*, **60**, 220 (1988).
[79] M. F. Lefevre, R. W. Frei, A. H. M. T. Scholten and U. A. T. Brinkman, *Chromatographia*, **15**, 459 (1982).
[80] A. H. M. T. Scholten and R. W. Frei, *J. Chromatogr.*, **176**, 349 (1979).
[81] I. S. Krull, X. D. Ding, C. Selavka, K. Bratin and G. Forcier, in *Drug Determination in Therapeutic and Forensic Extracts*, E. Reid and I. Wilson (Editors), Series *Methodological Surveys in Biochemistry and Analysis*, Vol. 14, Plenum Press, London, 1984, pp. 365–366.
[82] D. E. G. Shuker and S. R. Tannenbaum, *Anal. Chem.*, **55**, 2152 (1983).
[83] M. Righezza, M. H. Murello and A. M. Siouffi, *J. Chromatogr.*, **410**, 145 (1987).
[84] E. J. Levy and D. G. Paul, *J. Gas Chromatogr.*, **5**, 136 (1967).
[85] C. A. M. G. Kramer and A. I. M. Keulemans, in *Instrumentation in Gas Chromatography*, J. Krugers (Editor), Centrex Publishing Co., Amsterdam (1968), p. 71.
[86] J. D. Kelley and C. J. Wolf, *J. Gas Chromatogr.*, **8**, 583 (1970).
[87] V. G. Berezhkin and V. S. Tartarinski, *Zh. Anal. Khim,*, **25**, 398 (1971).

[88] V. Real, B. Kaplanova and J. Janák, *J. Chromatogr.*, **65**, 47 (1972).
[89] S. A. Liebman, D. H. Ahlstrom, C. D. Nauman, R. Averitt, J. L. Walker and E. J. Levy, *Anal. Chem.*, **45**, 1360 (1973).
[90] J. Franc and J. Pour, *J. Chromatogr.*, **131**, 285 (1977).
[91] J. Franc and J. Pour, *J. Chromatogr.*, **131**, 291 (1977).
[92] P. C. Uden, D. E. Henderson and R. J. Lloyd, *J. Chromatogr.*, **126**, 225 (1976).
[93] H. Schildknecht, *Angew. Chem. Int. Ed. Engl.*, **5**, 751 (1966).
[94] J. E. Lovelock, K. W. Charlton and P. G. Simmonds, *Anal. Chem.*, **41**, 1048 (1969).
[95] P. G. Simmonds, G. R. Shoemake and J. E. Lovelock, *Anal. Chem.*, **42**, 881 (1970).
[96] B. Thompson, *Fundamentals of Gas Analysis by Gas Chromatography*, Varian Associates, Palo Alto (1977).
[97] Carle Instrument Corp., *Curr. Peaks*, **7** (2) (1974).
[98] F. Chow and A. Karmen, *Clin. Chem.*, **26**, 1480 (1980).
[99] A. J. P. Martin and R. L. M. Synge, *Biochem. J.*, **50**, 679 (1952).
[100] J. T. Walsh and C. Merritt, *Anal. Chem.*, **32**, 1378 (1960).
[101] R. Ikan, *Chromatography in Organic Microanalysis*, Academic Press, New York (1982).
[102] B. Casu and L. Cavallotti, *Anal. Chem.*, **34**, 1514 (1968).
[103] A. N. Freedman, *Anal. Chim. Acta*, **59**, 19 (1972).
[104] E. W. Becker, in *Separation of Isotopes*, H. London (Editor), Newnes, London (1961), p. 360.
[105] R. Ryhage, *Anal. Chem.*, **36**, 759 (1964).
[106] R. Ryhage, *Ark. Kemi*, **26**, 305 (1967).
[107] K. Biemann and J. T. Watson, *Anal. Chem.*, **36**, 1134 (1964).
[108] K. Biemann and J. T. Watson, *Anal. Chem.*, **37**, 844 (1965).
[109] M. B. Morin, *Methodes Phys. Anal.*, 157 (1967).
[110] P. M. Krueger and J. A. McCloskey, *Anal. Chem.*, **41**, 1930 (1969).
[111] A. Copet and J. Evans, *Org. Mass Spectrom.*, **3**, 1457 (1970).
[112] M. A. Grayson and C. J. Wolf, *Anal. Chem.*, **42**, 426 (1970).
[113] S. R. Lipsky, C. G. Horvath and W. J. MacMurray, *Anal. Chem.*, **38**, 1585 (1986).
[114] J. E. Hawes, R. Mallaby and V. P. Williams, *J. Chromatogr. Sci.*, **7**, 690 (1969).
[115] A. Zlatkis and J. A. Ridgway, *Nature*, **182**, 130 (1958).
[116] A. Zlatkis and J. F. Oro, *Anal. Chem.*, **30**, 1156 (1958).
[117] A. Zlatkis, J. F. Oro and A. P. Kimball, *Anal. Chem.*, **32**, 162 (1960).
[118] D. M. Coulson, *J. Gas Chromatogr.*, **4**, 285 (1966).
[119] A. Need, C. Karmen, S. Sivakoff and A. Karmen, *J. Chromatogr.*, **158**, 153 (1978).
[120] W. A. Aue, V. Paramasigamani and S. Kapila, *Mikrochim. Acta*, **I**, 193 (1978).
[121] D. A. Leathard and B. C. Shurlock, *Identification Techniques in Gas Chromatography*, Wiley-Interscience, London (1970).
[122] M. N. Inscoe, G. S. King and K. Blau, in *Handbook of Derivatives for Chromatography*, K. Blau and G. S. King (Editors), 1st edn., Heyden, London (1977), pp. 317–345.

[123] W. Aue and R. Aigner-Held, *J. Chromatogr.*, **189**, 119 (1980).
[124] R. Aigner-Held and W. A. Aue, *J. Chromatogr.*, **189**, 127 (1980).
[125] A. Karmen and N. S. Longo, *J. Chromatogr.*, **112**, 637 (1975).
[126] K. Bächmann, K. Büttner and J. Rudolph, *Fresenius' Z. Anal. Chem.*, **282**, 189 (1976).
[127] M. Yoshioka, A. Ohara, H. Kondo and H. Kanazawa, *Chem. Pharm. Bull.*, **17**, 1276 (1969).
[128] H. Hatano, K. Sumizu, S. Rokushika and F. Murakami, *Anal. Biochem.*, **35**, 377 (1970).
[129] S. Rokushika, S. Funakoshi, F. Murakami and H. Hatano, *J. Chromatogr.*, **56**, 137 (1971).
[130] J. N. LePage and E. M. Roicha, *Anal. Chem.*, **55**, 1360 (1983).
[131] Y. Yamatodani, H. Fukuda, H. Wada, T. Iwaeda and T. Watanabe, *J. Chromatogr.*, **344**, 115 (1985).
[132] J. C. Gfeller, G. Frey, J. M. Huen and J. P. Thevenin, *J. Chromatogr.*, **172**, 141 (1979).
[133] H. Himuro, H. Nakamura and Z. Tamura, *J. Chromatogr.*, **264**, 423 (1983).
[134] G. Schwedt, *Anal. Chim. Acta*, **92**, 337 (1977).
[135] T. Hojo, H. Nakamura and Z. Tamura, *J. Chromatogr.*, **247**, 157 (1982).
[136] F. Engback and I. Magnusson, *Clin. Chem.*, **24**, 376 (1978).
[137] M. Palumbo, *Farmaco, Ed. Sci.*, **3**, 675 (1948).
[138] A. B. Groth and G. Wallerberg, *Acta Chem. Scand.*, **20**, 2628 (1966).
[139] M. Kudoh, I. Matoh and S. Fudano, *J. Chromatogr.*, **261**, 293 (1983).
[140] N. Nimura, K. Ishida and T. Kinoshita, *J. Chromatogr.*, **221**, 249 (1980).
[141] S. Honda, M. Takahashi, Y. Araki and K. Kakehi, *J. Chromatogr.*, **274**, 45 (1983).
[142] T. Seki and Y. Yamaguchi, *J. Chromatogr.*, **287**, 407 (1984).
[143] T. Seki and Y. Yamaguchi, *J. Chromatogr.*, **332**, 9 (1985).
[144] Y. Yanagihara and K. Noguchi, *J. Chromatogr.*, **459**, 245 (1988).
[145] S. Kobayashi and K. Imai, *Anal. Chem.*, **52**, 424 (1980).
[146] J. G. Flood, M. Granger and R. B. McComb, *Clin. Chem.*, **25**, 1234 (1979).
[147] A. C. C. Spadaro, W. Draghetta, S. N. Del Lama, A. C. M. Camargo and L. J. Greene, *Anal. Biochem.*, **96**, 317 (1979).
[148] R. L. Niece, *J. Chromatogr.*, **103**, 25 (1975).
[149] S. Takahashi, *J. Biochem. (Tokyo)*, **83**, 57 (1978).
[150] M. Weigele, S. L. DeBernardo, J. P. Tengi and W. Leimgruber, *J. Am. Chem. Soc.*, **94**, 5927 (1972).
[151] S. Stein, P. Böhlen, J. Stone, W. Dairman and S. Udenfriend, *Arch. Biochem. Biophys.*, **155**, 203 (1973).
[152] H. Mabuchi and H. Nakahashi, *J. Chromatogr.*, **228**, 292 (1982).
[153] M. Roth and A. Hampad, *J. Chromatogr.*, **83**, 353 (1973).
[154] J. R. Benson and P. E. Hare, *Proc. Natl. Acad. Sci. U.S.A.*, **72**, 619 (1975).
[155] J. R. Cronin and P. E. Hare, *Anal. Biochem.*, **81**, 151 (1977).

[156] K. S. Lee and D. G. Drescher, *Int. J. Biochem.*, **9**, 457 (1978).
[157] H.-M. Lee, D. Forder, M. C. Lee and D. J. Bucher, *Anal. Biochem.*, **96**, 298 (1979).
[158] R. L. Cunico and T. Schlabach, *J. Chromatogr.*, **266**, 461 (1983).
[159] H. Sista, *J. Chromatogr.*, **359**, 231 (1986).
[160] A. M. Felix and G. Terkelson, *Arch. Biochem. Biophys.*, **157**, 177 (1973).
[161] A. J. Thomas, in *Amino Acid Analysis*, J. M. Rattenbury (Editor), Ellis Horwood, Chichester, (1981), pp. 37–47.
[162] Y. Watanabe and K. Imai, *Anal. Chem.*, **55**, 1786 (1983).
[163] P. W. Alexander and C. Maltra, *Anal. Chem.*, **53**, 1590 (1981).
[164] J. A. Polta and D. C. Johnson, *J. Liq. Chromatogr.*, **6**, 1727 (1983).
[165] W. T. Kok, U. A. T. Brinkman and R. W. Frei, *J. Chromatogr.*, **256**, 17 (1983).
[166] M. H. Joseph and P. Davies, *J. Chromatogr.*, **277**, 125 (1983).
[167] L. A. Allison, G. S. Mayer and R. E. Shoup, *Anal. Chem.*, **56**, 1089 (1984).
[168] T. J. Mahachi, R. M. Carlsson and D. P. Poe, *J. Chromatogr.*, **298**, 279 (1984).
[169] E. Kuzniar, *EDT Digest*, EDT Analytical, London (1989).
[170] T. L. Lee, L. D'Arconte and M. A. Brooks, *J. Pharm. Sci.*, **68**, 555 (1979).
[171] E. Crombez, G. van der Weken, W. van den Bossche and P. de Moerloose, *J. Chromatogr.*, **177**, 323 (1979).
[172] M. Svensson and K. Gröningsson, *J. Chromatogr.*, **521**, 141 (1990).
[173] H. S. Sista, D. M. Dye and J. Leonard, *J. Chromatogr.*, **273**, 464 (1983).
[174] D. L. Mays, R. J. van Appeldoorn and R. G. Lauback, *J. Chromatogr.*, **120**, 93 (1976).
[175] J. P. Anhalt and S. D. Brown, *Clin. Chem.*, **24**, 1940 (1978).
[176] H. N. Myers and J. V. Rindler, *J. Chromatogr.*, **176**, 103 (1979).
[177] M. E. Rogers, M. W. Adlard, G. Saunders and G. Holt, *J. Chromatogr.*, **257**, 91 (1983).
[178] D. G. Musson, S. M. Maglietto and W. F. Bayne, *J. Chromatogr.*, **338**, 357 (1985).
[179] D. Dye and T. East, *J. Chromatogr.*, **284**, 457 (1984).
[180] E. F. Bayne, T. East and D. Dye, *J. Pharm. Sci.*, **70**, 458 (1981).
[181] J. C. Gfeller, G. Frey and R. W. Frei, *J. Chromatogr.*, **142**, 271 (1977).
[182] L. Embree and K. M. McErlane, *J. Chromatogr.*, **496**, 321 (1989).
[183] P. E. Nelson, S. L. Logan and K. R. Bedford, *J. Chromatogr.*, **234**, 407 (1982).
[184] P. E. Nelson, *J. Chromatogr.*, **298**, 59 (1984).
[185] C. E. Wells, E. C. Juenge and W. B. Furman, *J. Pharm. Sci.*, **72**, 622 (1983).
[186] J. R. Lang, I. L. Honigberg and J. T. Stewart, *J. Chromatogr.*, **252**, 288 (1982).
[187] J. Haginaka, J. Waka, Y. Nishimura and H. Yasuda, *J. Chromatogr.*, **447**, 365 (1988).

[188] S. Katz, S. R. Dinsmore and W. W. Pitt, Jr., *Clin. Chem.*, **17**, 731 (1971).
[189] S. Katz and L. H. Thacker, *J. Chromatogr.*, **64**, 247 (1972).
[190] K. Mopper and E. T. Degens, *Anal. Biochem.*, **43**, 147 (1972).
[191] K. Mopper and E. M. Gindler, *Anal. Biochem.*, **56**, 440 (1973).
[192] M. D'Amboise, T. Hanai and D. Noel, *Clin. Chem.*, **24**, 1384 (1980).
[193] R. S. Ersser, *Med. Lab. World* (August) 27 (1983).
[194] S. Katz, W. W. Pitt, Jr., J. E. Mrochek and S. Dinsmore, *J. Chromatogr.*, **101**, 193 (1974).
[195] G. K. Grimble, H. M. Barker and R. H. Taylor, *Anal. Biochem.*, **128**, 422 (1983).
[196] S. Honda, T. Konishi and S. Suzuki, *J. Chromatogr.*, **299**, 245 (1984).
[197] K. Omichi and T. Ikenaka, *J. Chromatogr.*, **428**, 415 (1988).
[198] N. Kiba, K. Shitara and M. Furusawa, *J. Chromatogr.*, **463**, 183 (1989).
[199] L. E. Elson and W. T. J. Morgan, *Biochem. J.*, **33**, 1824 (1933).
[200] S. Gardell, *Acta Chem. Scand.*, **7**, 206 (1953).
[201] S. Honda, T. Konishi, S. Suzuki, K. Kakehi and S. Ganno, *J. Chromatogr.*, **281**, 340 (1983).
[202] S. Honda, T. Konishi, S. Suzuki, M. Takahashi, K. Kakehi and S. Ganno, *Anal. Biochem.*, **134**, 483 (1983).
[203] S. Honda, S Suzuki, M. Takahashi, K. Kakehi and S. Ganno, *Anal. Biochem.*, **134**, 34 (1983).
[204] K. Blau and C. S. Dodge, *Anal. Biochem.*, **58**, 650 (1974).
[205] E. Reh and G. Schwedt, *Fresenius' Z. Anal. Chem.*, **303**, 117 (1980).
[206] R. Verbeke and P. Vanhee, *J. Chromatogr.*, **265**, 239 (1983).
[207] T. Seki and Y. Yamaguchi, *J. Liq. Chromatogr.*, **6**, 1131 (1983).
[208] T. Seki and Y. Yamaguchi, *J. Chromatogr.*, **305**, 188 (1983).
[209] C. Hamada and M. Iwasaki, *J. Chromatogr.*, **341**, 426 (1985).
[210] T. Koziol, M. L. Grayeski and R. Weinberger, *J. Chromatogr.*, **317**, 355 (1984).
[211] K. Yasumoto, K. Tadera, H. Tsuji and H. Mitsuda, *J. Nutr. Sci. Vitaminol.*, **21**, 117 (1975).
[212] M. Kimura and Y. Itokawa, *J. Chromatogr.*, **332**, 181 (1981).
[213] M. Kimura and Y. Itokawa, *Clin. Chem.*, **29**, 2073 (1983).
[214] H. Ohta, T. Baba and Y. Suzuki, *J. Chromatogr.*, **284**, 281 (1984).
[215] M. Naoi, Y. C. Lee and S. Roseman, *Anal. Biochem.*, **58**, 571 (1974).
[216] S. S.-H. Chen, A. Y. Kou and H.-H. Y. Chen, *J. Chromatogr.*, **276**, 37 (1983).
[217] E. London and G. W. Feigenson, *Anal. Biochem.*, **88**, 203 (1978).
[218] D. O. E. Gebhardt, W. Soederhuizen and J. H. M. Feyen, *Ann. Clin. Biochem.*, **22**, 321 (1985).
[219] D. J. Durzan, *Can. J. Biochem.*, **47**, 657 (1969).
[220] Y. Hiraga and T. Kinoshita, *J. Chromatogr.*, **226**, 43 (1981).
[221] Y. Kobayashi, H. Kubo and T. Kinoshita, *J. Chromatogr.*, **400**, 113 (1987).
[222] M. D. Baker, H. Y. Mohammed and H. Veening, *Anal. Chem.*, **53**, 1658 (1981).
[223] Y.-L. Hung, M. Kai, H. Nohta and Y. Ohkura, *J. Chromatogr.*, **305**, 281 (1984).
[224] M. Ohna, M. Kai and Y. Ohkura, *J. Chromatogr.*, **392**, 309 (1987).
[225] T. Miura, M. Kashiwamura and M. Kimura, *Anal. Biochem.*, **139**, 432 (1984).
[226] J. Kawase, H. Ueno, A. Nakae and K. Tsuji, *J. Chromatogr.*, **252**, 209 (1982).
[227] H. Tsuchiya, T. Hayashi, H. Naruse and N. Takagi, *J. Chromatogr.*, **234**, 121 (1982).
[228] R. Tarinotti, P. Siard, J. Bourson, S. Kirkiacharian, B. Valeur and G. Mahuzier, *J. Chromatogr.*, **269**, 81 (1983).
[229] M. Ikeda, K. Shimada and T. Sakaguchi, *J. Chromatogr.*, **272**, 251 (1983).
[230] N. Ichinose, K. Nakamura, C. Shimizu, H. Kurokura and K. Okamoto, *J. Chromatogr.*, **295**, 463 (1984).
[231] Y. Kasai, T. Tanimura and Z. Tamura, *Anal. Chem.*, **49**, 655 (1977).
[232] S. Johnson and O. Samuelson, *Anal. Chim. Acta*, **36**, 1 (1966).
[233] B. Carlsson, T. Isaaksson and O. Samuelson, *Anal. Chim. Acta*, **43**, 47 (1970).
[234] B. Carlsson and O. Samuelson, *Anal. Chim. Acta*, **49**, 247 (1970).
[235] Y. Takata and G. Muto, *Anal. Chem.*, **45**, 1864 (1973).
[236] W. T. King and P. T. Kissinger, *Clin. Chem.*, **26**, 1484 (1980).
[237] R. D. Voyksner and E. D. Bush, *Biomed. Environ. Mass Spectrom*, **14**, 213 (1987).
[238] H. K. Borys and R. Karler, *J. Chromatogr.*, **205**, 303 (1981).
[239] R. T. Krause, *J. Chromatogr. Sci.*, **16**, 281 (1978).
[240] H. Nakamura and Z. Tamura, *Anal. Chem.*, **53**, 2190 (1981).
[241] H. Nakamura and Z. Tamura, *Anal. Chem.*, **54**, 1951 (1982).
[242] E. C. Tornes, Jr. and D. N. Vacik, *Anal. Chim. Acta*, **152**, 1 (1983).
[243] A. H. M. T. Scholten, U. A. T. Brinkman and R. W. Frei, *J. Chromatogr.*, **218**, 3 (1981).
[244] L. G. M. T. Tuinstra and W. Haasnoot, *J. Chromatogr.*, **282**, 457 (1983).
[245] K. Yazawa and Z. Tamura, *J. Chromatogr.*, **254**, 327 (1983).
[246] S. H. Lee and L. R. Field, *J. Chromatogr.*, **386**, 137 (1987).
[247] K.-I. Mawatari, F. Iinuma and M. Watanabe, *J. Chromatogr.*, **488**, 349 (1989).
[248] J. R. Wagner, M. Berger, J. Cadet and J. E. van Lier, *J. Chromatogr.*, **504**, 191 (1990).
[249] X.-D. Ding and I. S. Krull, *J. Agric. Food Chem.*, **32**, 622 (1984).

16

Practical Considerations

Karl Blau
Department of Chemical Pathology, Bernhard Baron Memorial Research Laboratories, Queen Charlotte's and Chelsea Hospital, London

1. INTRODUCTION 349
2. APPARATUS AND EQUIPMENT 349
   2.1 Reaction vessels 349
   2.2 Reagent metering 351
   2.3 Care of microlitre syringes 351
   2.4 How to manage various manipulations 352
3. SOLVENTS AND REAGENTS 353
   3.1 Solvents 353
   3.2 Reagents 353
4. SAFETY 353
5. CONCLUSIONS 354
6. REFERENCES 354

1. Introduction

Anyone looking through this handbook will be struck by how easy derivatization seems: and they will not be wrong. The beauty of it is that it not only looks easy, for the most part it really is not hard. The main reason for this is that the conditions are chosen to make the reactions work well and to achieve what, in the rest of organic chemistry, would be regarded as exceptional yields. This is possible for two reasons: the first is that chromatographic methods are so sensitive, and require only small samples, the other is that often there is no need to isolate the products completely and so we can use very high reagent ratios. When one is analysing only a few microlitres, and when one does not have to worry about getting rid of excess reagent, one can work on the micro scale, and working at this level, as much as anything, is what makes derivatization chemistry easy and successful.

A word of caution is necessary, however. The very sensitivity which makes chromatography such an attractive group of techniques also poses the risk of interference from a variety of sources, such as the environment, the equipment, the reagents and solvents, and even the operator. *It is essential that this potential risk should be evaluated by running appropriate blank determinations whenever demanding and highly sensitive analyses are involved.* In most cases the blank runs will show a reassuring absence of interference, but there are scattered references to problems here and there, such as absorption, contamination (with plasticizers, for example), artefacts, and side reactions in particular cases (such as unwanted esterification occurring in alcohol extracts). One enthusiast has even gathered these pitfalls within the covers of a single book—and it is a fascinating one [1]!

It must again be emphasized how important isolation procedures are, and how getting them right may take a considerable effort. Fortunately the development of solid-phase extraction columns and filters has made isolation of many compounds very much easier [2]. It is a rapidly developing area, and for this reason anyone embarking on any demanding isolations should consult the latest catalogues of the various companies which have specialized in solid-phase extraction to discover the most appropriate adsorbents.

This book is primarily concerned with what happens *after* isolation, though it is worth making the point that the chemical nature of the substances to be isolated and the matrix from which they have to be extracted govern the details of the initial clean-up procedure. However,

Handbook of Derivatives for Chromatography
Edited by K. Blau and J. M. Halket © 1993 John Wiley & Sons Ltd

if the nature of the material to be analysed is known beforehand, it pays to sort out the chromatographic analysis on pure authentic samples first, and this is where this handbook comes in. Once the analytical scheme has been finalized, then one can readily use the chromatography to monitor various different work-up approaches in developing the best way to isolate the substances of interest.

2. Apparatus and equipment

2.1 Reaction vessels

Derivatization reactions are most conveniently done in disposable glass vials, typically with a volume of less than 2 ml. One type of vial is sealed with a 'serum cap' type of closure held in place by a crimped aluminium top. This is designed to be pierced with a syringe needle for sampling. The other type of vial is closed with a plastic screw-cap, and this kind of closure may have to be unscrewed for sampling the products after the reaction is complete, although many have an open top to the screw-cap and are closed with a septum-type liner, again for sampling with a syringe.

While in conventional-scale chemistry the relatively large volumes that are used require reflux condensation of solvents and reagents, in small vials heating does not produce large volumes of vaporized solvents or reagents, and the pressure produced is readily contained and also promotes the reaction. Small vials are also easier to heat up externally quickly and evenly, and are equally quick to cool down when heat is removed.

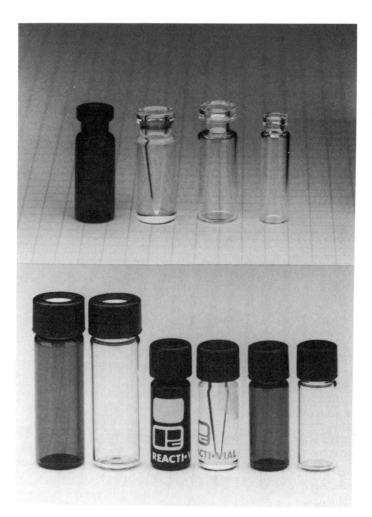

Figure 1. A selection of reaction vials. (By courtesy of Tim Warriner and Pierce Chemicals).

There is also no risk of exothermic reactions getting out of hand, because cooling is so effective.

A great variety of reaction vials is commercially available. Some are specifically designed for derivatization, are made of heavy-walled glass and have tapering construction, so that even small volumes may easily be sampled with a microlitre syringe. Many are designed to fit commercial automatic samplers or injection robots. This enables derivatization reactions to be done simultaneously on many samples, which are then analysed sequentially in programmed analytical schemes, so that the analytical chromatography or GC–MS can be mechanized for continuous operation without human intervention. On the other hand, much cheaper vials can be used where derivatization conditions are less demanding (e.g. room temperature); provided that there are no problems with the materials used for the vials and closures, ordinary screw-capped so called 'half dram' vials, each holding about 2 ml, are very convenient single-use derivatization vessels.

Glass is the best all-round material because of its relative inertness and resistance to chemical attack. However, it is a silicate with some silanol groups on the surface, and for extremely demanding derivatization applications the surface can conveniently be deactivated by silanization. This is done with a 2% solution of dimethyldichlorosilane in toluene or dichloromethane. After half an hour the vial is rinsed with pure solvent, treated with methanol to methylate any remaining chloro groups, and finally oven dried. Silanized glass vials are also available commercially. For some applications (e.g. cannabinoid metabolites) deactivation is essential. Glass may be either clear, or amber coloured to protect light-sensitive compounds.

Various plastics may be used for derivatization re action vials in circumstances where they will not react or interfere. Vials made of polyethylene, polypropylene or PTFE (prices increase in that order) are available.

The caps which are used to seal reaction vials need to be considered for their suitability in any given application. Some vials have a rubber or silicone cap, and often these are perfectly satisfactory—but this cannot be taken for granted and should be established by determining a blank value. Some vials have foil-lined caps, and these *may* be acceptable, but metal is often contraindicated, especially where corrosive reagents or products are involved. PTFE is also widely used to line seals in reaction vials, because it is both chemically inert and resistant to moderately high temperatures (up to 250 °C). However, plastics need to be used cautiously: they must obviously not interfere in any way with the derivatization reaction, they must not contribute contaminants such as plasticizers, and they must not physically absorb or extract reaction products and so diminish yields. Incidentally, where reaction products have been sampled by piercing a liner with a syringe needle, if the sample is then to be stored the liner should be replaced with an unpierced one. If necessary, the air may be replaced by an inert gas at the same time.

2.2 Reagent metering

Small-scale handling of solvents and of liquid derivatization reagents is generally done with microlitre syringes. These are usually glass with either stainless steel or PTFE-tipped plungers and stainless steel needles. They are graduated volumetrically and are quite accurate. From a practical viewpoint microlitre syringes up to 50 μl are most satisfactory. Above that volume it becomes difficult to maintain leak-free operation because of plunger wear. From a practical standpoint, syringes with reinforced tops to the plungers or extensions to the barrel that support the plunger, or those that have flexible plungers (Figure 2), will have the longest working life. Small volume pipettors, especially positive displacement ones using PTFE-tipped plungers in glass capillaries, are useful above 50 μl. Air displacement pipettors and plastic disposable tips are acceptable provided that the possibility of plasticizers leaching out of the tips is borne in mind and is not going to cause problems. For the most sensitive analyses, plastics must be ruled out because of plasticizers leaching, particularly when organic solvents are being used.

2.3 Care of microlitre syringes

Microlitre syringes are also widely used for spotting samples onto thin-layer plates and for the injection of samples into chromatographs. There is a strong tendency for microlitre syringes that hold 10 μl or less to clog up. Either the needle blocks up or the plunger gets hopelessly kinked or seizes up or gets covered with sticky material, which may contribute contaminants. It is essential to keep syringes scrupulously clean. In the first place this may best be done by drawing a couple of microlitres of pure solvent up before the main sample. When the sample is injected the pure solvent then rinses it into the chromatograph or onto the thin-layer plate and leaves the syringe clean. Second, the syringe can be rinsed with pure solvent a number of times after each injection. Finally, the plunger can be taken out and cleaned separately; the barrel can be flushed with solvent or other cleaning solution from the top of the barrel while suction is applied to the needle end (Figure 3).

More expensive versions of this simple arrangement are available commercially, and these additionally heat

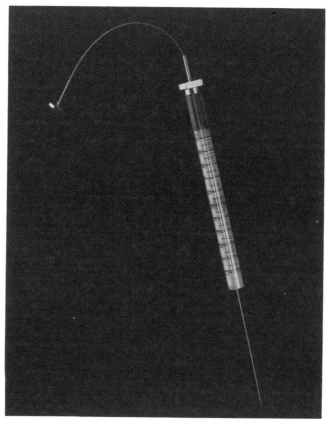

Figure 2. The new type of microlitre syringe with a flexible non-kinking plunger. (By courtesy of Mr P. Knight and Scientific Glass Engineering (UK) Ltd).

the needle, but there seems little advantage in doing so, and these devices risk baking carbonaceous residues onto the inside of the needle.

The most popular 10 μl and 5 μl syringes will eventually have metal from the plunger deposited round the top of the barrel. Although this may not cause problems, it is easily removed by soaking the syringe barrel with the top centimetre or so dipping into strong HCl or nitric acid in a test tube, and after the metal has been dissolved washing the syringe with water. Syringes with flexible plungers look set to become popular: with them, a kinked plunger is a thing of the past.

Blocked needles are a problem. The only effective way of unblocking them is with fine stainless steel wire, provided for the purpose by some manufacturers. This is fed in from the tip end with a pair of forceps, and needs great patience and some skill. The moment one pushes too hard, the wire kinks and has to be discarded. One has to proceed a millimetre or two at a time. However, with the price of microlitre syringes, it is well worth while persisting with efforts to clear a blocked needle, otherwise the syringe is useless. But these problems do not arise with syringes that have removable needles: with those a fresh needle is all that is required. However, prevention is the best policy: keeping the syringe clean and rinsing it frequently, especially with suction at the sharp end, will repay the extra work—and avoid contamination as well.

2.4 How to manage various manipulations

Although the dedicated microchemist may use some of the small-scale procedures so ingeniously developed by Dünges [3], which are well worth considering, many analysts wishing to carry out derivatization reactions may find the simple procedures outlined here quite satisfactory. Heating a reaction mixture is done either in a conventional oven or in a heating block. Various commercial models are available, and consist of a solid metal block with appropriate sized holes drilled for the

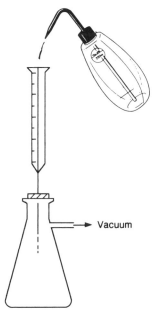

Figure 3. An easy way of cleaning microlitre syringes.

reaction vials. Different blocks can be obtained, each the same size but drilled with a different sized set of holes for selected reaction vials, and in use the blocks are kept at the chosen temperature in a thermostatted heater. As suggested in the preface, using a microwave oven is another way of supplying heat to a derivatization reaction.

Stirring on a very small scale is scarcely possible, so other ways of agitating the reaction mixture are used. The easiest is to shake the whole vial during the reaction. Another very good way is to put the vial into an ultrasonic bath, because the energy pumped into the vial may also promote the reaction. Mild agitation can be achieved in open vials by gently bubbling nitrogen through the reaction mixture via a fine needle.

Many recipes call for evaporation of the solvent or excess reagents on completion of the derivatizaton. The most popular method is to use a stream of nitrogen gas passed through a narrow bore tube such as a Pasteur pipette or a fairly large bore syringe needle. Manifolds for achieving evaporation in many tubes simultaneously are commercially available (Figure 4), and similar manifolds are easily assembled: one way is to use the inexpensive plastic valves obtainable from aquarium shops for bubbling air into fish-tanks. Several of these valves may be assembled in series.

The reaction vials may, if necessary be kept in the heating block while the contents are being evaporated to dryness. Although nitrogen is almost universally used, in applications where air oxidation is unlikely air is just as good, and may be supplied not only from cylinders or the house air line but even from something as simple and cheap as an aquarium pump.

Alternatively, a vacuum may be used. The easiest way of doing this is to put the vials into a vacuum

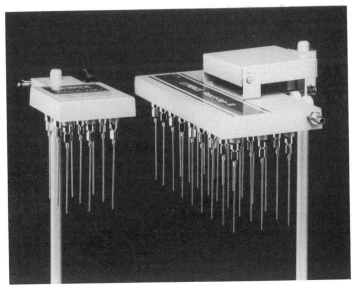

Figure 4. A typical manifold for the simultaneous evaporation of several samples. (By courtesy of Tim Warriner and Pierce Chemicals).

Figure 5. A typical centrifugal vacuum evaporator. (By courtesy of Claire Wittekind and V. A. Howe & Co. Ltd).

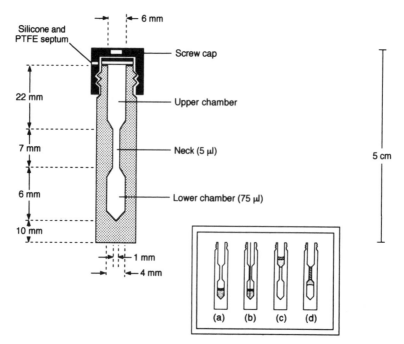

Figure 6. The 'Keele' reaction vial [4]. The two immiscible liquids are mixed in the lower chamber (by courtesy of Ms Crystal Ridgeway and Wheaton Science Products Inc., on p. 261 of their catalogue). The inset shows how the liquids can be moved to the upper chamber with a syringe for separation, and then into the most advantageous place in the narrow section by sucking air out of the lower chamber with a syringe. Each layer may then be sampled with a syringe.

desiccator and to pump it down, but care is needed to avoid frothing or 'bumping' sudden explosive evaporation of supercooled solvents, which causes losses. This problem is overcome by using a centrifugal evaporator, where the vials are in a spinning carousel which holds their contents down by centrifugal force while vacuum is applied (Figure 5).

Recipes often inovlve washing or extraction with an immiscible solvent. Mixing on a vortex mixer is the usual mode of agitation, but an efficient commercial device called a 'Mixxor' is available for volumes down to 2 ml. Where layers have to be separated on the micro scale, in a small vial, this is achieved by centrifugation and sampling of the appropriate layer with a microlitre syringe. It is a good idea to pull up a little air into the syringe before putting the tip of the needle into either layer, and to expel the air before sucking up the liquid. A very neat device for separating immiscible layers is the 'Keele vial' (Figure 6), where the desired layer can be manipulated into a narrowed section of the vial for efficient removal with a microlitre syringe [4].

Although very small centrifugal filters are available commercially, filteration too is generally replaced by centrifugation and withdrawal of the supernatant, again with a microlitre syringe.

3. Solvents and reagents

3.1 Solvents

The highest grade of solvents should be used to avoid the risk of impurities contaminating the reaction products. Since the volumes used are small the extra investment is not prohibitive. There are commercial companies that specialize in high-purity solvents, as do many suppliers of materials and reagents specially selected for chromatography. A solvent handbook may also be helpful; for example Ref. [5]. Unfortunately, manufacturers sometimes have to add preservatives to certain solvents to prevent chemical changes, e.g. air oxidation, on storage. In other words, it is a paradox that without the addition of contaminating preservatives some solvents may not *stay* pure. For this reason, many suppliers to the chromatography trade prefer to put up highly purified solvents and reagents in small sealed vessels, often under nitrogen, to avoid the need for contaminating them with preservatives. We must repeat our earlier warning: check for contamination, impurities, unexpected peaks etc.; *run the analysis blank* by doing the whole derivatization reaction without the starting material. This will only be necessary once or twice and may avoid a nasty surprise later on.

For people who want to, or may have to, redistill their own solvents, it is not to use too large a scale. A 100 ml distillation flask, possibly incorporating a small Vigreux-type fractionating column, will distill enough of a solvent to last some time for purposes of derivatization A valuable dodge is to add 1% of a long-chain hydrocarbon or ordinary paraffin wax to the solvent before distillation: this is very effective in reducing the residual trace impurities in the distillate [6]. Procedures for purification of solvents and reagents may be found in several places, including the first edition of the handbook [7]. There is a lot to be said for keeping pure reagents and solvents in small glass-stoppered vessels— possibly even under nitrogen.

Some solvents are reaction-promoting for selected derivatizations [8]. These include acetonitrile, pyridine, N-methylimidazole, dimethylformamide (DMF) and dimethyl sulphoxide (DMSO). Provided that one of these solvents is acceptable, chemically speaking, its use could substantially speed up or increase the yield of a derivatization reaction, or even both.

3.2 Reagents

When it comes to reagents for derivatization, commercial suppliers have responded magnificently to the needs of chromatographers. All the most important reagents are widely available in high purity, and even the more esoteric ones can usually be found by a diligent hunt through catalogues, particularly those of the suppliers who have specialized in meeting the requirements of chromatographers. Those catalogues are a valuable resource and should be part of the chromatographer's library, because they are the source of reagents specially packaged for chromatography: small lots, sealed glass ampoules, often inert atmospheres, and frequently a range of summarized instructions for the successful application of the reagents and much incidental information relating to derivatization and chromatography. It is really no longer appropriate or necessary to include instructions for the purification and storage of solvents and reagents when all the information is so freely available; for further authoritative information see also Ref. 9 and 10.

4. Safety

Any extensive chemistry involves powerful reagents, some of which are corrosive, toxic, unstable or unpleasant. In derivatization one is cushioned against the worst effects of such hazards by the minuscule amounts of these chemicals that one needs to use. Nevertheless one should not ignore the potential hazards [11], nor treat the reagents lightly; even a tiny drop of a reagent as powerful as sulphuric acid, to take an obvious example, if left on the skin would cause a nasty burn.

A few obvious safety measures are worth mentioning. A good fume hood or fume cupboard is essential. It should be large enough to work with comfortably, and the air flow should meet modern regulations. Perforated downdraught extraction benches have recently become available as an alternative, and they would seem ideal for derivatization chemistry and far less claustrophobic.

Some reagents may be unstable: diazomethane is an example. Although on a small scale, and with the appropriate generating kit, diazomethane and its homologues are perfectly safe, it is sensible to wear protective goggles or to work behind a plastic or reinforced-glass screen when handling these reagents.

Some people may be sensitive to certain substances. It is a general safety principle to wear surgical gloves for procedures in which highly reactive reagents might otherwise get on the fingers.

Nowadays good labelling practice ensures that chemical hazards are brought to the user's notice. Nevertheless, even small scale hazards should be kept in mind and the chromatographer should know what to do when things go wrong. Each laboratory should have a spill kit, including absorbents that will contain and neutralize spilled hazardous reagents. Finally, although one hopes they will not be needed, each laboratory should also have an eye-washing facility and a first aid kit [12], and a fire extinguisher by the exit.

5. Conclusions

In practice derivatization is an enjoyable branch of chemistry. The small scale makes light work of the procedures and requires only limited bench space while taking most of the unpleasantness and danger out of chemistry that may include some vigorous reactions. And beyond all that, the high success rate gives one a sense of satisfaction and a feeling of achievement.

6. References

[1] B. S. Middleditch, *Analytical Artifacts: GC, MS, HPLC, TLC and PC*, Elsevier Amsterdam (1989).
[2] K. C. Van Horne (Editor), *Sorbent Extraction Technology, A Handbook*, Analytichem International, Harbor City (1985).
[3] W. Dünges, *Prä-chromatographische Mikromethoden: Techniken für die biomedizinische Spurenanalye*, Hüthig, Heidelberg (1979).
[4] A. B. Attygale and E. D. Morgan, *Anal. Chem.*, **58**, 3054 (1986).
[5] J. A. Riddick, W. B. Bunger and T. K. Sakano, *Techniques of Chemistry, Volume 2, Organic Solvents*, 4th edn., Wiley, New York (1986).
[6] N. Fisher and R. M. Cooper, *Chem. Ind. (London)*, 619 (1968).
[7] P. Kováč and D. Anderle, in *Handbook of Derivatives for Chromatography*, K. Blau and G. S. King (Editors), Heyden, London (1977), pp. 227–229.
[8] C. Reichardt, *Solvents and Solvent Effects in Organic Syntheses*, 2nd edn, VCH, Weinheim (1988).
[9] *Reagent Chemicals*, 7th edn., American Chemical Society, Washington, DC (1986).
[10] D. D. Perrin and W. L. F. Armarego, *Purification of Laboratory Chemicals*, 3rd edn., Pergamon, New York (1988).
[11] National Research Council, *Prudent Practices for Handling Hazardous Chemicals in laboratories*, National Academy Press, Washington, DC (1981).
[12] M. J. LeFevre, *First Aid Manual for Chemical Accidents*, 2nd edn., Van Nostrand, New York (1989).

Index

The following abbreviations, additional to those on p. xv, are used.

abs	absorbing, absorbance
chrom	chromatography, chromatographic
cpd, cpds	compound, compounds
deriv	derivative, derivatization of
prep	prepared, preparation of
react	reacted with, reaction of
sep	separated, separation of

Derivatizations are indexed under classes of compounds (alcohols, amines, amino acids, sugars, steroids, etc., see Chapter 1, Table 1). Individual compounds are not indexed unless they undergo a unique derivatization. Thus for derivatizations applying to, for example, alanine see under amino acids; or to cortisone see under steroids; and so on. The entry '(table)' indicates the location of a table giving further details and references.

A

Acetal formation 4, 142
Acetobromo-α-D-glucose
    for chiral deriv of –OH cpds
    (table) 224
Acetolysis
    of polysaccharides 38
    of sulphate ester conjugates 38
Acetonides (isopropylidenes) 143
Acetate, as ion-pair (table) 225
2B-Acetoxy-Δ^5-etienic acid,
    for chiral deriv of –OH cpds 219
    (table) 222
S-Acetoxypropionic acid,
    for chiral deriv of lactones 240
    for chiral deriv of –OH cpds 220
    synonym 249
    (tables) 223, 240
Acetylacetone,
    deriv of aldehydes to fluorescent

lutidines 199
Acetylation 36
    with acetic anhydride 36
    with acetic anhydride and
        pyridine 36
    with acetic anhydride and
        sodium acetate 36
    acid-catalysed 38
    with acetyl chloride 39
    in aqueous solution 38
    for FAB–MS (table) 271
    deriv 273–7, 281
    with ketene 40
    with N-methylimidazole 38
    with simultaneous silylation 38
2-Acetylbenzaldehyde (OAB),
    analogue of OPA, for fluorescent
        deriv 195
N-Acetylcysteine (NAC),
    for chiral deriv of amino acids 242

(table) 246
N-Acetyl-D-penicillamine,
    for chiral deriv of amino acids 242
    (table) 246
O-Acetylsalicyloyl deriv for HPLC 36
Acids (both fatty acids and organic
    acids generally),
    chiral sep of optically active
        acids 225–8
    deriv for FAB–MS (table) 271
    esterification (see Chapter 2) 2–30
    fluorescent esters 201–3
        coumarin deriv 201–2
    ion-pair chrom 261
    post-chrom deriv 342–3
    (table) 2
    TBDMS deriv for GC–MS 76–8,
        81–2, 312
    TMS deriv 68–70
        for GC–MS 309, 317–8

N-(9-Acridinyl)maleimide for
    thiols 200–1
Activated carboxylic acids for
    fluorescent deriv 202–3
    carbodiimides and other activating
        (coupling) agents 203
Acylation (Chapter 3) 31–50
    of amides 43
    detector-oriented 34, 35, 36
    with ethyl chloroformate 47
    for FAB–MS (table) 271
    with MBTFA 44
    with methyl chloroformate 47
    with pentafluorobenzyl chloroformate
        for ECD 47
    with perfluoroacyl anhydrides 40–3
    with perfluoroacyl imidazoles
        43, 44
    for sep of chiral amines 230
    with simultaneous
        silylation 38, 43, 44
    with simultaneous esterification 43
    (table) 6
    with thioalkyl trimethylammonium
        salt for FAB–MS (table) 271
    with trichloroethyl chloroformate for
        ECD 47
Acyl chlorides, prep from acids to make
        amides 167
Acyl glycerols,
    TBDMS deriv 81–2
    TMS deriv 8–70
    see also lipids
Acyl imidazoles, reactivity 33
Aflatoxins,
    post-chrom deriv 343
    (table) 2
Alanine-β-naphthylamide for chiral deriv
        of acids 225, 227
    (table) 228
Alcohols,
    chloroformates 186
    CMDMS deriv 95–6
    chiral sep of optically active 219
    deriv for FAB–MS (table) 271
    fluorescent labelling 203
    various photochemical deriv 335
    silyl deriv, comparative studies 87
Aldehydes,
    alkylation 113
    benzyloxime and nitrobenzyloxime
        deriv 137
    chiral deriv 237–9
    condensation with diketones to
        fluorescent lutidines 199
    deriv for FAB–MS (table) 271
    DNP-hydrazones 136
    flash exchange for GC 132

photochemical deriv 335
    Schiff bases (Chapter 6) 131–4, 136,
        137–8
Alditols, (from monosaccharides),
    acetyl deriv 37
    TFA deriv 42
    for GC–MS 315
    see also polyols
Aldononitrile acetates (from
        monosaccharides) for SFC 290
Alkaloids,
    Dns deriv 184
    photochemical deriv 336
Alkylation (Chapter 5) 109–129
    with alkyl fluorosulphonates 120
    with alkyl p-tolyltriazenes 113, 120
    with alkyl trichloroacetimidates 114
    with barium oxide or hydroxide 112
    chiral deriv of amines 230
    with diazoalkanes 112, 113
    with DMF–dialkyl acetates 113
    with DMF–DMA 120
    with silver oxide 111, 112
    with sodium hydride 112
    (table) 4
    with TMAH 113, 120
    deriv of sterically hindered functional
        groups 117
    with trialkyloxonium fluoro-
        borates 113
Alkyl fluorosulphonates 120
Alkyl oximes,
    of carbonyl cpds 2, 5
    in GC–MS 305
Alkylthio-2-alkylisoindoles, prep by
        OPA/thiol react 194
Alkyl trichloroacetimidates 114, 126
Allydimethylsilyl (ADMS),
    deriv for GC–MS 312
    general 88
    (table) 56
Amides,
    perfluoroacylation of 43
    prep via acyl chlorides 167
    (table) 7
Amidination for FAB–MS 271
Amines (amines generally and biogenic
        amines),
    p-anisoyl deriv 158
    benzopyrone deriv 195–6
    benzoyl deriv 158
    p-t-butylbenzoyl deriv 158
    chiral deriv 230–6
    chloroformates 186
    chlorophenylmethylbenzoxazole
        acetyl deriv 187
    DTAF deriv, from secondary
        amines 196

Dns deriv 177, 178
deriv for FAB MS (table) 271
fluorescent deriv with fluorene- and
    pyrenecarboxaldehyde 193
formyl deriv 39
formylhydroxyacetophenone
    deriv 195
ion-pair chrom 256, 261, 262
isothiocyanatoacridine deriv 190
p-methoxybenzoyl deriv 158
8-methoxyquinolinesulphonyl
    deriv 186
naphthylenebenzimidazole sulphonyl
    deriv 186
NBD deriv 187
m and p-nitrobenzoyl deriv 158
OPA/thiol deriv 193
pentafluorobenzaldehyde deriv 137
perfluoroacyl deriv 41
deriv with PITC and NITC 163
post-chrom deriv 338
(tables) 2, 6
TBDMS deriv 311–12
TMS deriv 305–8
m-toluyl 158
Amino acids,
    alkylation 112, 114
    automated analysis 328
        with detection by succini-
            midylnaphthyl carbamate 196
    cyclic deriv 150, 151
    DABS deriv 160–1
        for automated amino acid
            analysis 160
    Dns deriv 177, 79, 182
    esters for GC–MS 318
    FLEC deriv for SFC 291–3
    fluorenymethyloxycarbonyl (FMOC)
        deriv 186
        for SFC 291–3
    fluorescamine deriv 192
    fluorescent thiohydantoins 190–1
    methyl esters 15
    8-methoxyquinoline sulphonyl
        deriv 186
    NBD deriv (table and properties)
        188
    OPA/thiol deriv 193
    perfluoroacyl deriv 40–1
    post-chrom deriv 339
    N-propionyl deriv 39
    n-propyl esters 37, 38
    PTH deriv 150
    pyridoxyl deriv 193
    (tables) 2, 7
    TBDMS deriv 83–5
    TFA-n-butyl esters (TAB) 314
        for GC–MS 26, 40–1, 317

TMS deriv 74
trifluoromethyl oxazolinones for
  GC–MS 318
Amino alcohols,
  chiral deriv for sep of chiral amino
    alcohols 247–8
  cyclic boronates 147, 150
  cyclization with phosgene 150
  OPA/thiol deriv 193
Aminoglycoside antibiotics,
  DNP deriv with FDNB 161
  TNP deriv with TNBS 162
  (table) 2
R-2-Amino-oxy-4-methylvaleric acid,
  chiral deriv of carbonyl cpds 237
  (table) 239
Aminophenols, Dns deriv 180
2-Aminopyridine, deriv with
  sugars 198
$R(-)/S(+)$-Amphetamine for chiral sep
  of acids 225
  (table) 229
Anhydrides,
  mixed, for activation of
    acids 225, 227
Anils (Schiff bases from aromatic
  amines) 133
Anthracene-9-carbonyl chloride 169
  UV-abs deriv with alcohols 169
Anthracene sulphonate as ion-pair 253
  (table) 255
1- and 9-Anthroylnitrile, fluorescent
  deriv with alcohols 203
p-(Anthryloxy)-phenacyl bromide
  (panacyl bromide) 202
  fluorescent deriv for acids 202
  UV–abs deriv of eicosanoids 166
  UV–abs deriv of
    prostaglandins 166
9-Anthryldiazomethane, fluorescent
  deriv for acids 202
Aromatic amines,
  react with acyl chlorides to give
    amides 167
  anils (Schiff bases) with carbonyl
    cpds 133
AutoAnalyzer manifolds for post-chrom
  deriv 330–4
Azibenzyl, for prep of
  phenyldiazomethane 18

B

Bansyl (Bns), see di-n-butylamino-
  naphthalene sulphonyl 183
Barbiturates,
  alkylation with TMAH and
    on-column pyrolysis 113, 121
  Dns deriv 179
  react with pentafluorobenzyl
    bromide 114
  UV-abs deriv with CIMP, CIMPI
    and CIMIS 166
$R(-)$-Benoxaprofen chloride,
  for chiral deriv of amines 232
  for chiral deriv of amino alcohols
    248
  (table) 236, 248
Benzeneboronates
  see also phenylboronates 147
Benzenesulphonyl deriv 46
  of gentamicin 160
Benzoate as ion-pair (table) 255
Benzoin, deriv with guanidines 196
Benzo-γ-pyrone 195
N-[(p-2-Benzoxazolyl)-phenyl]-
  maleimide for thiols 200–1
Benzoyl deriv,
  with benzoyl chloride 4, 6, 46, 159
  of –OH cpds 168
  (table) 2
Benzyldimethylsilyl (BZDMS) deriv
  (table) 56
Benzyl esters 20
  with benzyl-3-p-tolyltriazenes 27
  with benzyl-N-nitroso-p-toluene
    sulphonamide 18
  with dicyclohexyl-O-benzylurea
    24
  for FAB–MS 280
  with phenyldiazomethane 18
Benzyl ethers,
  of catechols 122
  deriv with phenyldiazomethane
    119
  of tritylated carbohydrates 117
Benzylidene deriv,
  of alditols 114
  of sugars 114
O-Benzyloximes of steroids 132
Benzyloxycarbonyl (carbobenzoxy)
  deriv for FAB–MS (table) 271
Bidentate ligands for cyclization
  react 142
Bile acids
  dinitrophenylhydrazine
    deriv 171
  EDMS deriv 86
  TBDMS deriv 79–81
  for GC-MS 312
  TMS deriv 63–7
  UV-abs deriv with p-NBDI 167
Bimanes, fluorophoric reagents for
  thiols 201
N,O-Bis(trimethylsilyl)acetamide
  (BSA) general silyl donor 58, 60
O-(–)-Bornylhydroxyammonium
  chloride,
  chiral deriv of carbonyl
    cpds 237–8
  (table) 239
Boronates (cyclic boronates, see
  under individual reagents),
  of amino alcohols 147
  of catecholamines 147
  of diamines 147
    for GC–MS 320
  for MHPG 147
  of steroids 146
    for GC–MS 320
  (tables) 2, 5
Bromoacetaldehyde, fluorescent
  reagent for adenines 203–4
2-Bromoacetonaphthane
  (naphthacyl bromide) 202
  fluorescent deriv for acids 202
1-Bromoacetylpyrene, fluorescent
  deriv of acids 202
p-Bromobenzeneboronates 147
p-Bromobenzyl esters 20, 35
4-Bromomethyl-7-acetoxycoumarin
  (B-Mac) deriv of acids 202
9-Bromomethylacridine,
  fluorescent deriv for acids 202
  ion-pairing distribution 263
4-Bromomethyl-6,7-dimethoxy-
  coumarin (Br-Mdmc),
  fluorescent quantum yield 202
  deriv of fluorouracil 202
  deriv of imides 202
3-Bromomethyl-6,7-dimethoxy-1-
  methyl-2(1H)-quinoxalinone
  fluorescent deriv of acids 202
Bromomethyldimethylsilyl
  (BMDMS) 22, 35
  deriv 95
  (table) 57
4-Bromomethyl-7-
  methoxycoumarin (Br-Mmc),
  deriv of acids 201
  esters of biotin and deriv 165
  properties of Mmc esters 202
p-Bromophenacyl esters 20, 35
  (table) 2
  UV–abs esters 165–6
Bromothymol blue, as ion-pair 253
  (table) 255
Butaneboronates 146
  of diols 147
    for GC–MS 319
  of D vitamins for GC–MS 321
  of sugars 146
$S(+)$-2-Butanol for chiral deriv of
  acids 225
  (table) 228

2,3-Butanediol for chiral deriv of
  lactones (table) 241
Butoxydimethylsilyl (table) 56
Butaneboronates, butylboronates
  146
  of sugars 146
  of diols 147
t-Butyldimethylsilyl (TBDMS) 75
  of amines 311–12
  of amino acids 311
  of bile acids 79–81
  chloride (TBDMCS) 75, 76–85
  deriv for GC–MS 310–13
  ethers 304
  imidazole (TBDSIM) 76
  of steroids 79–81
    for GC-MS 312
  (table) 56
N-t-Butyldimethylsilyl-N-methyl
  trifluoroacetamide (MTBSTFA),
  use 76–85
t-Butyldiphenylsilyl (TBDPS)
  (table) 56
Butyl esters 20
  for FAB–MS (table) 271
  for MS–MS 281
  from triglycerides by ester
    interchange 26
t-Butylmethoxyphenylsilyl
  (TBMPS) (table) 56
t-Butylmethoxyphenylbromosilane,
  chiral deriv of chiral –OH
    cpds 92–3
t-Butyloxycarbonyl-L-leucine
  N-hydroxysuccinimide ester,
    for chiral deriv of amino
    acids 241, 245
  (tables) 242, 246
  for chiral deriv of amino
    alcohols 248
  (table) 248

C

(+)-Camphor-10-sulphonic acid,
  for chiral deriv of amines
    (tables) 236, 245
  for chiral deriv of amino
    acids 244
  chiral sep by ion-pair chrom 259
Cannabinoids,
  ADMS deriv for GC–MS 313,
    315–16
  CMDMS deriv 95–6
  Dns deriv 184
  post-chrom deriv 343
  silyl deriv, comparisons 87
Carbamate formation (table) 6, 7

Carbobenzoxy-L-proline,
  chiral deriv of –OH cpds
    (table) 223
Carbohydrates,
  (table) 2
  see under sugars
  also polyols, alditols
N,N'-Carbonyldiimidazole (CDI),
  esterification reagent 23
Catecholamines,
  cyclic boronate deriv 147
    for GC–MS 319
  Dns deriv 184
  DEHS deriv 86
  DMNPS deriv 86
  fluorescent deriv with DPE 198
  sep by ion-pair chrom 257
  isothiocyanates (mustard oils) with
    CS_2 134–6
  perfluoroacyl deriv 40, 42
  post-chrom deriv 339
  Schiff bases with
    pentafluorobenzaldehyde 133,
    137–8
Catechols, benzyl ethers 122
Cationization,
  in FAB–MS 272
  in MS–MS 283–4
Chemical ionization (CI) mass
  spectrometry 288
  in GC–MS 298
Chiral deriv reagents, selection
  of 218
Chiral sep: a. indirect; b. direct
  216–8
Chloroacetaldehyde: fluorescence
  labelling of adenines 203–4
p-Chlorobenzyl esters 35
7-(Chlorocarbonylmethoxy)-4-
  methylcoumarin,
  fluorescent reagent for
    alcohols 203
6-Chloro-p-chlorobenzene-
  sulphonylbenzotriazole (CCBBT),
  in esterification 23, 24
Chlorodifluoroacetic anhydride 41
2-Chloroethyl esters 15, 16, 35
Chloroformates 47
  alcohols and amines, react with 186
  chiral deriv of amines 230, 231
  ion-pairing distribution in react
    with tert amines 263
  menthyl, for chiral sep of –OH
    cpds 220
  mixed anhydrides of acids, prep 227
9-Chloromethylanthracene,
  fluorescent deriv of acids 202
Chloromethyldimethylsilyl

  (CMDMS) 22, 35
  deriv 95
  (table) 57
N-Chloromethylisatin (CIMIS) 166
  UV–abs deriv of acids 166
  UV–abs deriv of barbiturates 166
N-Chloromethyl-4-nitrophthalimide
  (CIMNPI) 166
  UV–abs esters with 166
  UV–abs deriv of barbiturates 166
N-Chloromethylphthalimide
  (CIMPI) 166
  UV-abs deriv of barbiturates 166
  UV-abs deriv of mono- and
    dicarboxylic acids 166
4-Chloro-7-nitro-2,1,3-benzoxadiazole
  (NBD-Cl) 187
[R,S]-2(p-Chlorophenyl)-α-methyl-5-
  benzoxazoleacetyl chloride 187
2-p-(Chlorosulphophenyl)-3-
  phenylindole (Dis-Cl) 185
Coloured deriv (Chapter 8) 157–174
R(+)trans-Chrysanthemic acid, also
  chloride 219
  chiral deriv of –OH cpds 219
  synonym 249
  (tables) 222, 233
Cinchonidine for enantiomeric sep 260
Collisional activation (CA) in
  MS–MS 281–6
Coloured and UV–abs deriv
  (Chapter 8) 142–174
Constant Neutral Loss (CNL) in
  MS–MS 285
Continuous flow fast atom
  bombardment (CF–FAB) 286
  combined with HPLC 286
Counter-ions in ion-pair
  formation 253–4
  tetramethylammonium, as
    counter-ion 167
Crown ethers in esterification 19, 21,
  164
  in automatic sample prep 201–2
  dibenzo-18-crown-6, adduct-forming
    agent 256
  hexaoxacyclooctadecane
    (18-crown-6) 20–1, 164–5
  pentaoxapentadecane
    (15-crown-5) 21
Cyanoethyldimethylsilyl (CEDMS)
  deriv 96–7
  use with NPD detector 96–7
  (table) 57
Cyclic boronates, see under
  individual boronates
Cyclic carbonates of diols 149
Cyclic derivatives, cyclization

(Chapter 7) 141–153
for GC–MS 318–321
(tables) 4, 5, 6, 7
Cyclohexane-1,3-dione (CHD, dihydroresorcinol)
condensation with aldehydes to fluorescent lutidines 199
Cyclohexyl boronate 147
Cyclohexylidene deriv of sugars 144
Cyclohexylidene–ethylene acetal deriv of sugars 144
Cyclopentylidene deriv of sugars 144
Cyclotetramethylene-t-butylsilyl (TMTBS) (table) 56
Cycloptetramethyleneisopropylsilyl (TMIPS) (table) 56
L-Cysteine methyl ester, chiral deriv of sugars 237
(table) 239

D

Dansyl, Dns, *see under* dimethylaminonaphthalene sulphonyl
2-Dns aminoethanol, fluorescent alcohol for deriv of acids 203
Dns aziridine (Dns-A), fluorescent reagent for thiols 200
2-Dns ethyl chloroformate, fluorescent reagent for –OH cpds 187
Dns hydrazine (Dns-H),
applications 197
fluorescent reagent for steroids 197
fluorescent reagent for sugars 197
Schiff base formation 132
Dextrobase: 1-(4-nitrophenyl)-2-amino-propane-1,3-diol,
for chiral deriv of acids (table) 229
synonym 249
Dialkyl acetals in esterification 25
Diamines, cyclic boronate deriv 147
1,2-Diamino-4,5-methylenedioxybenzene (DMB), for α-keto acids 199
9,10-Diaminophenanthrene (DAP) for fluorescent deriv of acids 203
'Diazald', *see under* methylnitrosotoluenesulphonamide
Diazoalkanes 17, 112, 117
Lewis acid-catalysed alkylation with 118
Diazomethane 17
esterification of acids for GC–MS 318
4-Diazomethyl-7-methoxycoumarin, for fluorescent deriv of acids 202

Dibenzo-18-crown-6: adduct-forming agent 256
(R,R)-O,O-Dibenzoyltartaric anhydride, for chiral deriv of aminoalcohols (*also* table) 248
Dibromobimane, deriv for thiols 201
5-Di-n-butylaminonaphthalene-1-sulphonyl (Bns) 183
chloride 183
fragmentation of deriv on electron impact in MS 183–4
Di-t-butylsilylene (DTBS) 90
deriv of methylsalicylic acid for GC 90
(table) 57
2,4-Dichlorobenzeneboronate 147
Dichlorotetrafluoroacetone
for preparation of cyclic deriv 150, 151
5-(4,6-Dichloro-1,3,5-triazene-2-yl)-aminofluorescein (DTAF), deriv for secondary amines 196
Dicyclohexyl carbodiimide (DCCI) 22, 23
for coupling of acids with nitrophenylhydrazine 167
Chiral deriv, coupling in 221, 227
cyclization with 151
in esterification 22, 23
2-(N,N-Diethylamino)-ethoxydimethylsilyl (table) 56
Diethylsilylene (DES),
prep of diethylhydrogensilyl (DEHS) deriv with –OH cpds 57
deriv of steroids 90–2
(table) 57
Di-isopropylmethylsilyl (MDIPS) (table) 57
1,3-Diketones for fluorescent lutidines from aldehydes 199
Dimethoxyanthracene as ion-pair 253
1,2-Dimethoxyethane, diazoalkane alkylations in 118
2,2-Dimethoxypropane, water scavenger 14, 147
N,N-Dimethylacetamide, alkylation in 117
N,N-Dimethylamino acids 126
Dimethylaminoazobenzene sulphonyl (DABS),
chloride (DABS1) 160
deriv of amino acids 160
Dimethylaminonaphthalene sulphonyl (Dns) 2, 4, 164, 179–184
chloride (Dns-Cl) 164, 176, 177
purification 177
double isotope techniques in MS

of 183
fragmentation under EI–MS 183
MS of deriv 182, 183
of N-terminal amino acids in proteins and peptides 184
quantitation 182
radioactive 182
sep of deriv 181
spectra of deriv 180
4-N,N'-Dimethylaminoazobenzene-4-isothiocyanate,
for HPLC of amines 163
N-[4-(6-Dimethylamino-2-benzofuranyl)-phenyl]-maleimide (DBPM),
for deriv of thiols 200–1
N-(7-Dimethylamino-4-methylcoumarinyl)-maleimide,
for deriv of thiols 200–1
4-Dimethylamino-1-naphthoylnitrile,
fluorescent deriv for –OH cpds 203
1-(4-Dimethylamino-1-naphthyl)-ethylamine (DANE),
for fluorescent deriv of acids 203
1-(4-Dimethylamino-1-naphthyl)-ethylamine,
for chiral deriv of acids 229
(table) 229
4-Dimethylamino-1-naphthylisocyanate 191
Dimethylchlorosilane (DMCS) 145, 146
deriv of steroids 146
5,5-Dimethyl-1,3-cyclohexanedione (dimedone),
deriv with aldehydes to fluorescent lutidines 199
Dimethyldiacetoxysilane (DMDAS) 145
deriv of steroids 146
N,N-Dimethylformamide dimethylacetal (DMF-DMA),
alkylation with 25, 119
deriv of sterically hindered acids 120
1,7-Dimethyl-7-bornyl isothiocyanate,
for chiral deriv of amino acids 243
(table) 246
Dimethyloctylamine (DMOA),
for ion-pair chromatography 259
Dimethylprotriptylene as ion-pair 254, 261
Dimethylsilyl (DMS) (table) 56
Dimethylsilylene (table) 57

Dimethyl sulphate for
    methylation 20
Dimethyl sulphide, enhancement of
    O-alkylation 111
2,4-Dinitrobenzene sulphonyl
    deriv 35
3,5-Dinitrobenzoyl deriv 46
  of —OH cpds 169
Dinitrofluorobenzene (DNFB),
    (fluorodinitrobenzene, FDNB),
  of amines, amino acids and
    aminoglycosides 161–2
  for polypeptide end-group studies
    (historical) 176
  (tables) 2, 4
2,4-Dinitrophenyl (DNP)
    deriv 6, 7, 35
2,4-Dinitrophenylhydrazine deriv of
    carbonyl cpds 131–2, 170
Diols,
  acetals and ketals 142
  boronates 147, 203
    GC–MS 319
  cyclic carbonates 149
  3,5-dinitrobenzoyl deriv 169
Dipeptides,
  enantiomeric deriv of amino acids
    (table) 242
  sep by ion-pair chrom 260
Diphenylphosphinyl chloride
    (DPPC),
  for activation of acids as mixed
    anhydrides 225, 227
1,2-Diphenylethylenediamine
    (DPE) 198
cis-4,5-Diphenyl-2-oxazolidone-3-
    carbamyl chloride (DOCC),
  for chiral deriv of amines 232
  (table) 236
Dipicrylamine
    (hexanitrodiphenylamine),
  as ion-pair 253
  (table) 255
Dodecyl sulphate, as ion-pair 257
Drimanoic acid for chiral deriv of
    —OH cpds (tables) 222, 233
Drugs,
  deriv with chloroformates 187
  chiral deriv with fluorenylmethyl
    chloroformate 231
  chiral deriv with glucuronic
    acid 223
  fluorescent deriv with
    fluorescamine 192
  perfluoroacyl deriv for
    GC–MS 315–16
  photochemical deriv 335
  post-chrom deriv 339–40

TMS deriv for GC–MS 308

E

Edman degradation 152
  fluorescent deriv in 191
Electron capture detector (ECD) for
    GC 34
  p-bromophenacyl esters 20
  deriv for high sensitivity
    analyses 35, 132
  DNPH deriv, sensitivity
    compared with NICI–MS 133
  flophemesyl deriv 93
  halocarbondimethylsilyl
    deriv 95–6
  halogenated silyl esters 22
  HFB 42, 44
  pentafluorobenzyloxycarbonyl
    deriv of tert. amines 47, 124–5
  perfluoroacyl deriv 40–1
  PFB esters 20–1
  PFBzO deriv 45
  specificity 328
  trichloroethyloxycarbonyl deriv
    of tertiary amines 47
Electron impact (EI) MS 268
  in GC–MS 298
  (table) 306–7
Electron pyrolysis, post chrom 336
Enantiomeric sep
    (Chapter 10) 215–249
  by ion-pair chrom 260
  direct, by SFC with chiral
    phases 291
Eneamines (Schiff bases from amines
    and a ketone) 137
Enols,
  problems with silylation 58, 64,
    304–5
  silyl deriv of ketosteroids 64–5
Episulphides (dithiocarbonates)
    of steroids 149
Epoxides (oxiranes) 149
Esterification (Chapter 2) 11–30
  with alcoholic HCl 14
  catalysis 13
  with CCBBT 23
  with CDI 23
  use of crown ethers 19
  with DCCI 22, 23
  diazomethane and analogues 17
  formation of higher esters 15
  for GC–MS 317
  KBr methods 19
  reaction mechanisms 12
  silver salt methods 19
  with simultaneous
    perfluoroacylation 43

thionyl chloride 16
water removal 13
Ethanethiol in OPA/thiol
    deriv 194
Ethers (deriv not generally needed)
    photochemical deriv 335
Ethoxy deriv for FAB–MS
    (table) 271
Ethoxydimethylsilyl (table) 56
Ethylchloroformate 47
Ethyldimethylsilyl (EDMS)
    (table) 56
Ethyl esters with diazoethane 18
bis(2-Ethylhexyl)-phosphoric acid,
    as ion-pairing and adduct-
    forming agent 256
O-Ethyloximes (ethoximes) of
    ketones 132
Extractive alkylation with
    pentafluorobenzyl bromide 21,
    263

F

FAB-ionization in MS, creation of a
    charged site 284–5
FAB-matrix in MS 269
    deriv 270
    non-specific 270
    (table) 271
Fast atom ion bombardment (FAB)
    MS (Chapter 12) 267–286
    analytical applications 268
Fatty acids (see under acids),
    methyl esters (FAME) 15
    (table) 2
Field desorption (FD) MS 268
Filtration, on micro-scale;
    centrifugal filters 355
'Fingerprints' in MS 298
Fischer's glycosidation
    method 122–4
Fischer–Speier esterification
    (alcohol + HCl) 14
Flash exchange of carbonyl cpds for
    GC 132
Flow-injection analysis (FIA) for
    post-chrom manifolds 330–4
(S(+)-Flunoxaprofen chloride for
    chiral deriv of amines 232
    (table) 236
2-Fluorenecarboxaldehyde 193
(+)-1-(9-Fluorenyl)ethyl
    chloroformate (FLEC),
  for chiral deriv of amino
    acids 243
  for chiral deriv of drugs 231, 233
  for SFC of amino acids 291–3
  (tables) 235, 242, 245

9-Fluorenylmethyloxycarbonyl
    (FMOC) deriv   186–7
  FMOC-Cl   186
  for SFC of amino acids   291–3
  for SFC of amphetamines   294
Fluorescamine   191–2
  fluorescent deriv of amines, amino
    acids and peptides   192
  quantum yields   192
  post-chrom deriv   339
Fluorescein isothiocyanate
    (FITC)   190–1
Fluorescence polarization   176
Fluorescent deriv (Chapter 9)
    175–213
1-Fluoro-2,4-dinitrobenzene,
    see under Dinitrofluorobenzene
N^{α}-(5-Fluoro-2,4-dinitrophenyl)-L-
    alaninamide,
  for chiral deriv of amino acids
    (table)   247
4-Fluoro-7-nitrobenzofurazan
    (NBD-F)   189
4-Fluoro-3-nitrobenzotrifluoride
    (FNBT)   162
4-Fluoro-7-sulphamoylbenzofurazan
    for thiols   199–200
7-Fluorobenzofurazan-4-sulphonate
    (SBD-F) for thiols   199–200
Formyl deriv, of amines and of
    steroids   39
ω-Formyl-o-hydroxyacetophenone
    deriv   195
Fragmentation routes in MS,
  of silyl deriv of steroids   85
  of peptides   277, 282–3, 284, 285
  in EI–MS   298
  (table)   306–7

G

Gas chromatography
  deriv for making cpds volatile for
    GC are too many to
    list individually. See particularly
    Chapters 2–5, 7
Gas chromatography linked to mass
    spectrometry (GC–MS)   3, 268
  see also Chapter 14   297–321
D-(+)-Glucuronic acid chiral deriv
    of phenols,
  using rat liver microsomal
    coupling   221
  (table)   223
  glucuronide cleavage with
    simultaneous
    perfluoroacylation   43
Glycosidation, Fischer's
    method   122–4

Guanidino cpds,
  deriv with benzoin   196
  deriv with
    hexafluoroacetylacetone   148
  cyclization with
    malondialdehyde   148
  post-chrom deriv   341–2
Guanidinopropionylation for
    FAB–MS (table)   271

H

Heptafluorobutyryl (HFB) see also
    perfluoroacyl deriv   16, 35
  for GC–MS   313–8
HFB-L-prolyl chloride for chiral deriv of
    amines (table)   234
Heptafluoropentyldimethylsilyl
    (table)   57
Hexafluoroacetylacetone for deriv of
    guanidines   148, 149
Hexamethyldisilazane (HMDS),
  selective silyl donor   59
  solvent power   59
  use with TMCS   59
Hexanitrodiphenylamine (dipicrylamine)
    as ion-pair   253
  (table)   255
Hexosamines, acetylation   38
Hexyl esters for FAB–MS   271
  deriv   280
High performance liquid
    chromatography HPLC,   3
  (as with GC, derivatization is used
    extensively for HPLC applications and
    appears throughout the handbook as
    well as in this arbitrary selection of
    pages)
  automated sample prep, in   193, 202
  combined with CA–FAB–MS   286
  of Dns deriv   181, 182
  of dimethylaminoazobenzene
    isothiocyanate deriv of
    amines   163
  of FMOC deriv   186–7
  of NBD deriv   189
  of diol naphthaleneboronates   203
  of OPA/thiol deriv   195
  of diol phenanthreneboronates   203
  post-chrom deriv in   286
    'Beam-Boost' system   335
  photochemical deriv and
    detection   334–5
  requirements and techniques   330
4-Hydrazino-7-nitrobenzofurazan
    (NBD-H)   197
4'-Hydrazino-2-stilbazole (4H2S)   197
Hydrazone formation   4
  for FAB–MS (table)   272

3-Hydroxybenzoate as ion-pair
    (table)   255
4-Hydroxy-6,6-dimethyl-3-
    oxabiscyclohexan-2-one,
  for chiral deriv of –OH cpds
    (table)   224
Hydroxylamine for carbonyl cpd
    deriv   131
9-Hydroxymethylanthracene,
  fluorescent alcohol for chiral deriv of
    acids   203
4-Hydroxymethyl-7-methoxycoumarin
    (HO-Mmc),
  fluorescent alcohol for deriv of
    acids   203

I

Imidazoles
  acyl, reactivity   33
  Dns deriv   180
  perfluoroacyl   43
  silylated   59, 60, 76
Iminohydantoins   152
Immobilized enzyme reactor for post-
    chrom analysis   333–4
Immonium salt formation for FAB–MS
    (table)   272
Indoleamines and indole alcohols
  perfluoroacylation with
    perfluoracylated imidazoles   44
Intramolecular rearrangements   142
Iodomethyldimethylsilyl (IMDMS)   35
  deriv   96
  (table)   57
Ion-pair extraction in chrom
    (Chapter 11)   253–265
  complexation   255–6
  extraction constant   254
  ion-pair dimerization   255
  ion-pair dissociation   255
  side-reactions   255
  theory   254–6, 263–4
Ion-pair formation   5, 6, 7
  ion chiral deriv of chiral amino
    alcohols   247
  combined with complexation   255–6
  concept   253
  for enantiomeric sep   260
  side-reactions   255
Ion-pairing probes   262
Ion-pair-mediated deriv   254
  indirect detection   260, 263
Ion-pair sep   258–262
Ion trap detectors for GC–MS   297
Isocyanates   190
  chiral deriv for optical resolution of
    amines   230
Isopropyldimethylsilyl (IPDMS),

Isopropyldimethylsilyl (cont.)
  deriv 88
  (table) 56
Isopropyl esters 26
  deriv 280, 281
  for FAB–MS (table) 271
  from triglyceride fats by ester interchange 26
Isopropylidenes (acetonides) 143
  of diols 143
  of glycerol and glyceryl ethers 143
  of steroids 144, 145
  of sugars 143, 144
Isothiocyanates (mustard oils) 190–1
  of amines 134, 138
  chiral deriv for optical resolution of amines 230
  GC 135
  MS with EI 135
  stability 135
9-Isothiocyanatoacridine 190

J

Jet mixer in post-chrom manifolds 333

K

'Keele' reaction vial 354–5
Ketal deriv 4, 142, 237
Ketene, acetylation with 39
α-Keto acids,
  Br-Mmc esters 201
  fluorescent hydrazinostilbazole deriv 196
  N-methylquinoxalones 61
  quinoxalinol deriv with o-phenylene diamine 61
    fluorescent 198–9
    for GC–MS 320
    for SFC 293
    silylated 61
Ketones,
  benzyloxime deriv 137
  chiral deriv 237–9
  dinitrophenylhydrazones 136
  deriv for FAB–MS (table) 272
  flash exchange for GC 132
  fluorescent reagents for 196
  ketone-base deriv to Schiff bases (Chapter 6) 131–140
  methoximes 132
  p-nitrobenzyloxime deriv 137
  pentafluorobenzyloxime deriv 114
  phenylmethylpyrazolone deriv 171
  (table) 4
Knitted and knotted tube reactors for post-chrom manifolds 331–2

L

Lactones, chiral deriv for sep of chiral lactones 240–1
L-Leucinamide for chiral deriv of chiral acids 225
  (table) 228
L-Leucine-N-carboxy anhydride (Leuchs anhydride),
  for chiral deriv of amino acids 241
  (tables) 242, 245
Levobase, 1-(4-nitrophenyl)-2-aminopropan-1,3-diol,
  for chiral deriv of chiral acids 229
  synonyms 249
  (table) 229
Lipids,
  TMS deriv 68
  transesterification of triglyceride fats 14, 25–6
  see also acyl glycerols
Lipopolysaccharides, alkylation 112
Liquid–liquid ion-pair chrom 256
Liquid reactors in post-chrom manifolds 331–2
Lutidines, fluorescent deriv from aldehydes 199

M

Maleimide deriv for thiols 200
Malondialdehyde, reagents for guanidines 148
D(−)/L(+)-Mandelic acid,
  for chiral deriv of −OH cpds 223
  synonym 249
  (table) 223
Mass increments after prep of perfluoroacyl deriv (table) 314
Mass spectrometers,
  bench-top 297
  ion trap 298
  MS–MS 280
  mass-selective detectors 297
  quadrupole 298
(−)-Menthenoxyacetic acid for chiral deriv of −OH cpds (table) 222
O-(−)-Menthyl-N,N-diisopropylurea for chiral deriv of acids 226
  (table) 229
(−)-Menthylchloroformate,
  for chiral deriv of alcohols 220–2
  for chiral deriv of amino acids 244
  (tables) 224, 231
Menthylhydroxyammonium chloride, chiral deriv of carbonyl cpds 237–8
  (table) 239
Mercaptoethanol, in OPA/thiol deriv 194
3-Mercapto-1-propanol, in OPA/thiol deriv 194
Mercaptopropionic acid, in OPA/thiol deriv 194
p-Methoxybenzoyl deriv 157–8
  of alcohols 169
  of amines 158
7-Methoxycoumarin-3 or 4-carbonyl azide (3-MCCA or 4-MCCA), fluorescent reagent for −OH cpds 203
Methoxydiethylsilyl and methoxydimethylsilyl (table) 56
Methoxyethylmethylsilyl (table) 56
3-Methoxy-4-hydroxyphenylethanol, perfluoroacylation 42
3-Methoxy-4-hydroxyphenethylene glycol (MHPG),
  cyclic boronate deriv 147
  perfluoroacylation 42
$S(−)$-α-Methoxy-α-(trifluoromethyl)-phenylacetyl chloride (MPTA), (Mosher's reagent)
  for chiral deriv of −OH cpds 219–20
  amphetamine deriv 231
  tocainamide deriv 231
  (tables) 222, 233
6-N-Methylanilinonaphthalene-2-sulphonyl (Mns) deriv 184–5
Methylation procedures,
  Purdie and Irvine 110
  Hakomori 110, 115, 278
2-Methyl-1,1'-binaphthalene-2'-carbonitrile,
  fluorescent deriv for −OH cpds 203
N-Methylbis(trifluoroacetamide) (MBTFA) 44
  acylation with 44
  on-column deriv 54
  for SFC 294
Methylboronates 147
  of steroids for GC–MS 320
  silylated 66
2-Methylbutoxydimethylsilyl (table) 56
Methyl chloroformate 47
Methylenedioxy deriv of sugars and steroids 143
Methyl esters,
  of amino acids 15
  with BF_3–methanol 15
  with dimethyl sulphate 20
  of fatty acids (FAME) 15
  by pyrolysis with quaternary ammonium salts 25
  silver salt method 20
  with TMS–diazomethane 18
Methyl ethers (table) 112–14

N-Methylimidazole (catalyst for
  acylations) 38
O-Methylmandelic acids for chiral deriv
  of amines (table) 236
4-Methyl-7-methoxycoumarin
  esters 21
R-α-Methyl-4-nitrobenzylamine for
  chiral deriv of acids (table) 229
Methylnitronitrosoguanidine (MNNG)
  for prep of diazomethane 17
Methylnitrosotoluenesulphonamide for
  prep of diazomethane 17
Methyl orange as ion-pair (table) 255
O-Methyloximes (methoximes) of
  ketones 132
N-Methyl quaternization for FAB–MS
  (table) 271
2-Methylquinoxanol deriv of lactic
  acid 167–8
Methylsulphinyl carbanion 110, 115,
  116
N-Methyl-*N*-(trimethylsilyl)-
  trifluoroacetamide (MSTFA),
  general silyl donor 54
  silylations with 58
Microlitre syringes 351–2
  blocked needles, unblocking 352
  care of 351–2
  cleaning and maintenance 352
  flexible, non-kink plungers 352
Micro-scale manipulations 353
  evaporation 353–4
  filtration 355
  heating and cooling 352–3
  'Keele' reaction vial 354
Mixing in post-chrom manifolds 331–2
Monobromobimane, deriv with
  thiols 201
Monobromotrimethylammoniobimane,
  deriv with thiols 201
Mosher's reagent, *S*(−)-α-Methoxy-α-
  (trifluoromethyl)-phenacetyl
  chloride (MPTA),
  for chiral deriv 219–220
  (tables) 223, 233
MS–MS (two mass spectrometers in
  tandem),
  cationization 283
  deriv for soft ionization 280–5
Mustard oils (isothiocyanates) 134
  of amines 134–6, 138
  EI–MS of 135
  GC 135
  stability 135
Mycotoxins (table) 2

N

Naphthacyl bromide (2-bromonaphthane),

fluorescent deriv of acids 202
  UV-abs deriv of acids 164–6
Naphthaleneboronic acid, fluorescent
  deriv of diols 203
Naphthalenesulphonate, as ion-pair 253
  for deriv of amines 261–2
  chloride 176
  ion-pair chrom of dipeptides 260
  (table) 255
1-Naphthylamine, for fluorescent deriv
  of acids 203
2-Naphthyl chloroformate
  (NCF–Cl) 187
1,2-Naphthylbenzimidazole-6-sulphonyl
  deriv 186
S(−)-1-(1-Naphthyl)-ethylamine for
  chiral deriv of acids (table) 229
1-(1-Naphthyl)-ethyl isocyanate for
  chiral deriv of –OH cpds 220–2
  for chiral deriv of amino acids 243
  (tables) 224, 242, 246
Naphthyl isocyanate (NIC)
  deriv 162–3
Naphthyl isothiocyanate, deriv of
  amines 163
NBD deriv (see also nitrobenzoxadiazole
  deriv) 187–9
  properties 189
  post-chrom deriv of amino acids 339
  solvolysis to NBD–OCH$_3$ 189
Negative ion chemical ionization (NICI)
  MS 133
  cleavage of β-phenylalkyl
  isothiocyanates 136
  MS, sensitivity compared with
  ECD 133
(+)-Neomenthyl isothiocyanate for
  chiral deriv of acids 243
  (table) 247
Nicotine deriv for GC–MS of
  unsaturated fatty acids 317
Ninhydrin react with amino
  acids 164, 327
  post-chrom 328
m- and *p*-Nitrobenzoyl deriv,
  of amines 158
  of amino acids, direct sep by SFC on
  chiral phase 291
  of –OH cpds 169
Nitrobenzylimines for visualizing amines
  in TLC 133
p-Nitrobenzyl deriv,
  bromide 163
  esters 20–35
  of tertiary amines 163
o- and *p*-Nitrobenzyl-*N*,*N*′-diisopropylurea
  (NBD–I),
  UV–abs deriv of bile acids 167

p-Nitrobenzylhydroxylamine, react with
  prostaglandins 171
Nitro cpds,
  photochemical deriv 336
  (table) 8
Nitrogen–phosphorus detector (NPD),
  acyl deriv 36
  with CDMS deriv 96–7
  for thiophosphinic esters of
  steroids 293
p-Nitrophenacyl esters 21
  UV–abs 165–6
o- and *p*-Nitrophenylhydrazines,
  deriv of acids, using coupling with
  DCCI 167
  deriv of carbonyl cpds 131–2
4-Nitrophenylsulphonyl-L-prolyl
  chloride,
  for chiral deriv of amines 232
  (table) 234
Nitrosamines,
  photochemical deriv 336
  post-chrom deriv 343
  (table) 2
Nonafluorohexyldimethylsilyl
  (DMNFHS),
  deriv 95
  (table) 57
Nucleic acid bases, nucleosides and
  nucleotides,
  ADMS deriv 313
  fluorescence labelling with bromo-
  and chloroacetaldehyde 203–4
  fluorescence labelling with
  Mdmc 202
  pentafluorobenzyl deriv 318
  perfluoroacylation 41
  TMS deriv 73
  for GC–MS 309
  TBDMS deriv 82–3
  (table) 2

O

Oestrogens, Dns deriv 184
OPA, see under *o*-Phthaldialdehyde
Optical resolution
  (Chapter 10) 215–252
  with Dns deriv 184
  of 2-methyl-1,1′-binaphthalene-
  carbonyl nitrile deriv 203
  following OPA/thiol react with chiral
  thiols 195
  see also enantiomeric sep
Optimization of deriv formation 3, 7,
  9, 43
  of isolation 350
  TLC for use in 3

Osazones, formation from hydrazine and diketones 131–2
Oxazolidinones from amino acids 150
  for GC–MS 318
Oximes from carbonyl cpds (tables) 2, 4
  in GC–MS 305
^{18}Oxygen incorporation for FAB–MS 285
  (table) 271

P

Packed-bed reactors in post-chrom manifolds 332–3
  ion-exchange resins 333
  OPA/thiol 333
  reductive 333
Palumbo reaction (heating with citric acid and Ac$_2$O),
  colour react for tertiary amines 163, 338
Panacyl bromide, fluorescent deriv for acids 202
Pasteur; first enantiomeric sep (manual) 215
Pentafluorobenzaldehyde deriv 35, 125, 132–4
  for catecholamines 133–4
Pentafluorobenzoyl (PFBzO) deriv 35, 44, 45
Pentafluorobenzyl (PFB),
  aminobenzoate for deriv of oligosaccharides for SFC 294
  bromide in extractive alkylation 21
  chloroformate for react with tertiary amines 35, 47, 114, 124
  esters 21–35
    in GC–MS of nucleosides and of prostaglandins 318
  ethers of –OH cpds 125
  ion-pair-mediated alkylation 263
  deriv of sulphonamides 125
O-PFB-hydroxylamine deriv of carbonyl cpds 114, 125, 132–4
  TMS deriv 132
Pentafluorophenyl-t-butylmethylsilyl (table) 57
Pentafluorophenylchloromethylsilyl (table) 57
Pentafluorophenyldiazoalkanes 19
Pentafluorophenyldimethylsilyl (flophemesyl) 35
  analogues 94
  deriv 93
  (table) 57
Pentafluorophenylethyldimethylsilyl (table) 57
Pentafluoropropyl 16

esters 35
Pentafluoropropionyl (PFP) 16, 35
  for GC–MS 313–8
  melatonin deriv for GC–MS 319
  see also Perfluoroacyl
Pentoxydimethylsilyl (table) 56
Peptides,
  alkylation 112
  amino acid terminal residue studies 184
    see also tyrosyl residues 151
  Dns deriv 178
  8-methoxyquinoline sulphonyl deriv 186
  OPA/thiol deriv 193
  post-chrom deriv 229
  sequencing studies 151–3
  structure determination by FAB–MS 273–286
  TMS deriv for GC–MS 309
  (table) 2
Peptide and protein end-group determination 184, 185
  by FAB–MS (table) 271
Peracetic acid in epoxide prep 149
Perchlorate as counter-ion 256
Perfluoroacyl (TFA, PFP, HFB, etc.) deriv 40
  acylation with anhydrides alone 40
    with a basic catalyst 41, 42
    with perfluoroacyl imidazoles 43
    with MBTFA 44
    for GC–MS 313–8
    in a solvent 41
  O-perfluoroacyl deriv and the ECD 40
Perfluoroacyl imidazoles, see under Imidazoles
Permethylation (Hakomori procedure) 15
  for FAB–MS (table) 271
  methods 277–8
Phenacyl bromide, for extractive alkylation 263
Phenacyl esters, UV-abs deriv for acids 164–6
Phenanthreneboronic acid, fluorescent deriv of diols 203
1-Phenethyl-2-picolinium (PEP) in ion-pair chrom 258, 261
Phenolic acids, perfluoroacyl deriv 42
Phenolic alkaloids, alkylation with TMAH and pyrolysis 113
Phenols,
  chiral deriv with glucuronic acid 221
  chiral sep of optically active 219
  alkylation with p-tolyltriazenes 113
  TMS deriv 72–3

Phenyldiazomethane for benzyl esters 18, 119
Phenyldimethylsilyl (PhDMS), chloride, for deriv of alditols 170
  (table) 56
o-Phenylenediamine,
  analogues 199
  deriv of formate to benzimidazole 167
  quinoxalinol deriv from α-keto acids 147–8, 198–9
  for GC–MS 320
  for SFC 293
(+)-1-Phenylethanethiol chiral deriv for carbonyl cpds 237
  (table) 239
(+)/(−)-1-Phenylethylamine,
  for chiral deriv of acids 225–7
  of aldoses 236, 238
  of lactones 240
  of mandelic acid 236
  (tables) 228, 240–1
β-Phenylethylamine
  Dns deriv 180, 184
  mustard oils (isothiocyanates) 134
  NICI–MS of acetyl perfluoroacyl deriv 315
  PFBzO deriv 37
R(+)/S(−)-Phenylethyl isocyanate, for chiral deriv of –OH cpds 220–1, 231
  of lactones 241
  (tables) 224, 233, 241
R(+)-1-Phenylethyl isothiocyanate, chiral deriv of amines (table) 236
Phenylhydrazine for deriv of carbonyl cpds 131–2
Phenyl isocyanate (PIC) deriv of amines 162–3
  of –OH cpds 169–170
  (table) 2
Phenyl isothiocyanate (PITC) in Edman degradation 152, 163
1-Phenyl-3-methyl-5-pyrazolone (PMP) deriv of ketones 171
Phenylosazones from sugars and phenylhydrazine 132
S(+)-Phenylpropionic acid for chiral deriv of –OH cpds (table) 223
Phenylpropyl sulphate as ion-pair (table) 225
R(+)-2-Phenylselenopropionic acid for chiral deriv of –OH cpds, (table) 222
4-Phenylsemicarbazide for deriv of carbonyl cpds 131
1-Phenyl-3-p-sulphophenylisobenzofuran deriv 185–6

Phenylthiohydantoin (PTH) of amino
    acids,
  in peptide sequencing 150
  direct SFC sep on chiral phase 293
Phosgene for cyclization react 150
Phospholipids, post-chrom deriv 341
Photochemical deriv in HPLC 334–6
  in GC 338
o-Phthaldialdehyde with an alkyl thiol
    (OPA/thiol),
  deriv of –NH$_2$ cpds to yield
    fluorescent products 164, 193–5
  applications 195
  use of chiral thiol for chiral deriv of
    amines 230–1, 233
    (table) 235
  in a packed bed reactor 333
  in post-chrom deriv of amino
    acids 339
  scope of the reaction 195
  spectra of glycine deriv 194
    (tables) 2, 242, 246
Picolinyl deriv of unsaturated fatty acids
    for GC 317
Picolinyldimethylsily (PICSI) 97
  deriv with –OH cpds 97–9
  MS 98–9
  (table) 56
Picrate as ion-pair 253
  (table) 255
Pivaloyl deriv 46
Polyamines,
  benzoyl deriv 159–60
  Dns deriv 184
  fluorescamine deriv 192
  nitrotrifluorophenyl deriv 162
  OPA/thiol deriv 193
  (table) 2
Polychlorinated biphenyls (PCB)
    (table) 2
Polyethylene glycols (PEG) sep by
    SFC 293
Polycyclic aromatic hydrocarbons (PAH)
    (table) 2
Polyfunctional cpds, cyclic deriv
    formation (Chapter 7) 142–155
Polymers, sep by SFC 293
Polyols,
  acetal and ketal deriv 142
  acetyl deriv 37
  alkylboronate deriv 147
  benzylidine deriv 144
  TFA deriv 42
    for GC–MS 315
  TMS deriv 70–1
  UV–abs PhDMS deriv 170
  see also Alditols
Polysaccharides (mainly indexed under
    sugars),
  acetolysis 38
  alkylation with diazoalkanes 113
Polytetrafluoroethylene (PTFE, 'Teflon'),
  adverse effect in acetylation 37
  affinity for perfluoro cpds 36
Porphyrins,
  TBDMS deriv 85
  TMS deriv 74
  (table) 2
Post-chrom deriv (Chapter 15) 327–47
  coloured and UV–abs deriv 158
  manifolds 330–3
  objectives 328
  OPA/thiol 193
Post-column extraction in HPLC 263
  reactors 337–8
Practical considerations of deriv
    (Chapter 16) 349–56
Propionyl deriv 39
Propoxydimethylsilyl (table) 56
Propoxyethylmethylsilyl (table) 56
n-Propyldimethylsilyl (n-PDMS)
    (table) 56
n-Propyl esters 15, 37
n-Propyl ethers, side reaction of
    diazomethane in n-propanol 117
Prostaglandins,
  ADMS deriv 89–90
  butylboronate deriv 147
  7-(chlorocarbonylmethoxy)-
    methylcoumarin deriv 203
  IPDMS deriv 88
  Mmc deriv 201
  p-nitrobenzyloximes 171
  oxidation to 15-oxo deriv 167
  post-chrom deriv 343
  silylation 85–6
  TBDMS deriv 78–9
    for GC–MS 312
  TMS deriv 61–3
    for GC–MS 63, 309
  UV–abs anthryloxyphenacyl
    deriv 166
  (table) 2
Proximal reactive groups, candidates for
    cyclization 142
Pyrazoles from diketones and hydrazine
    reagents 132
1-Pyrene carboxaldehyde 193
N-(1-Pyrenyl)-maleimide for
    thiols 200–1
Pyridine, artefact formation with acid
    anhydrides 37, 42
  acylation in 42
  as reaction-promoting solvent and
    basic catalyst 46
Pyridoxal and pyridoxal-5-phosphate
    (PLP) 192
  deriv 192–3
  fluorescent deriv of PLP with
    4H2S 197
  properties of deriv 193
Pyrolysis in deriv with quaternary
    ammonium salts 24, 25
Pyrrolidides of unsaturated fatty acids
    for GC–MS 317

Q

Quaternary ammonium salts (table) 7
Quinidine deriv for enantiomeric
    sep 260
Quinine chloride for ion-pair
    chrom 259
  for enantiomeric sep 260
Quinoxalinol deriv from α-keto acids and
    O-phenylenediamine
  as fluorescent deriv 198
  prep 147
  deriv for GC–MS 320
  deriv for SFC 293
  silylated 61
  (table) 2

R

Racemization, at various places from
    p. 217 in Chapter 10
Reaction GC, post chrom 336
Reaction vessels 350
  'Keele' reaction vial 354
  mixing in, on a small scale 355
Reagent metering at the microlitre
    level 351
  deriv reagents 355
  microlitre syringes 351
  pipettors and pipettor tips 351
Resonance stabilization 303
Retinoic acid methyl ester 25
Reversed phase in ion-pair chrom 254

S

Safety considerations and
    precautions 355–6
Salicylate as ion-pair (table) 255
Scanning, in GC–MS 299
Schiff base formation (Chapter 6)
    131–40
  from catecholamines and
    pentafluorobenzaldehyde 137–8
  for FAB–MS (table) 271
  hydrazine in the atmosphere with
    acetone 138
  with pyridoxal and PLP 192
Schotten–Baumann acylation 38, 46
Selected ion monitoring (SIM) in
    GC–MS 299

Semicarbazide,
  deriv of carbonyl cpds 131, 132
  fluorescent deriv with PLP 197
Siliconides of steroids 145
Silylation (Chapter 4) 51–100
  alkoxyalkyl-, (or aryl)silyl deriv 92
  other alkylsilyl or arylsilyl
    deriv 85–8
  GC of silyl deriv 54–5
  with mixtures of silyl donors 60
  selective 59
  with simultaneous acylation
    38, 43, 44
  for stability enhancement 75
  substitution, effect on react
    velocity 85
  theory and mechanism 52–3
  (tables) 4, 5, 6, 56–7
  see also TMS, TBDMS, etc.
Single bead string reactor in post-chrom
  manifolds 33–4
Solvent modifiers in SFC 290
Solvents for deriv, also purification 355
Sphingosines, methylboronate
  deriv 147
Stable-isotope-lebelling in GC–MS 300
Steroids,
  ADMS deriv 88–9
  benzoyl deriv 169
  O-benzyloximes 132, 137
  CMDMS deriv 95–6
  cyclic siliconides with DMCS 145
    with DMDAS 146
  DEHS deriv 90–2
  dimethoxymethylsilyl ethers 92
  dimethylhydrazones 136
  DNPH deriv 171
  DTBS deriv 90–1
  episulphides (dithiocarbonates) 149
  formyl deriv 39
  fluorescent deriv with coumarin
    cpds 203
    with Dns–H 197
  flophemesyl deriv 94
    MS 94
  IDMS deriv 96
  IPDMS deriv 88
  isopropylidene deriv 145
  methylboronate esters for
    GC–MS 320
  methylenedioxy deriv 144
  O-methoxime/TMS deriv 137
  PFB-hydroxylamine deriv 114
  perfluoroacyl deriv 341
  silyl deriv, various, comparison
    studies 87
  TBDMS deriv 79–81
    for GC–MS 31

TMS deriv 63–67
  for GC–MS 309–10
  (table) 2
Stitched open tube reactors in
  post-chrom manifolds 332
Structure elucidation in MS 303
N-Succinimidyl-α-methoxy-α-
  methylnaphthalene acetic acid,
  for sensitive chiral deriv of
    amines 231
  (table) 235
N-Succinimidyl-(S)-2-methoxy-2-
  phenylacetic acid,
  for sensitive chiral deriv of
    amines 231
  (table) 234
N-Succinimidyl-2-naphthoxyacetate,
  post-chrom deriv with
    aminophospholipids 196
Succinimido-(–)-1-(1-naphthyl)-ethyl-N-
  carbamate,
  for chiral deriv of amino acids 243–4
  (tables) 242, 247
N-Succinimidyl-1-naphthyl carbamate,
  for automated amino acid
    analysis 196
Succinimidyl-R(+)/S(–)-1-phenylethyl
  carbamate,
  for chiral deriv of amino acids 246
  (table) 246
Sugars,
  acetylation 36
    of hexosamines 38
  acylation 168
  alkylation 112
  benzylidene deriv 144
  butaneboronate deriv 146
  chiral deriv (thialidines and
    oximes) 237
  cyclic carbonates 150
  cyclic deriv for GC–MS 319
  cyclohexylidene deriv 144
    ethylene acetal deriv 144
  cyclopentylidene deriv 150
  deriv for FAB-MS (table) 272
  fluorescent deriv with
    aminopyridine 198
    with Dns–H 197
  methylenedioxy deriv 143
  p-nitrobenzoyl deriv 169
  permethylation 117
  per-O-benzyl deriv 117
  phenylmethylpyrazolone deriv 172
  photochemical deriv 335
  post-chrom deriv 340–1
  deriv for SFC 290–1
  TFA deriv 40, 41, 42
    for GC–MS 315

TMS deriv 70–1
  for GC–MS 308
  multiple peaks 70
  (table) 2
Sulphate, esters,
  deriv by acetolysis 38
  simultaneous hydrolysis and
    perfluoroacylation 43
Sulphhydryl (thiol) reagents for
  fluorescent deriv 199–201
Supercritical fluid chrom (SFC)
  (Chapter 13) 289–94
  ease of recovery 289
  high resolving power 289
  intermediate between GC and
    HPLC 289, 294
  solvent modifiers 290
  superior solvent power 289
  for unstable or high MW cpds 289

T

N-L-Teresantalinyl deriv for chiral sep of
  amino acids 244
  (table) 246
Tertiary amines,
  catalysts for esterification with crown
    ethers 164
  ion-pairing distribution in reaction
    with chloroformates 203
  deriv with naphthyl chloroformate
    (NCF–Cl) 187
  Palumbo reaction with citric acid and
    Ac$_2$O 163, 338
  quaternization with p-nitrobenzyl
    bromide 163
  react with PFB-chloroformate
    47, 114
  react with trichloroethyl
    chloroformate 47
  (table) 7
2,3,4,6-Tetra-O-acetylglucopyranosyl
  isothiocyanate (GITC),
  for chiral deriv of amines 231
    of amino acids 243
    of amino alcohols 248
  (tables) 235, 242, 245, 248
Tetrabutylammonium as ion-pair 254
  as counter-ion 256, 257
  (table) 255
R(+)-Tetrahydrofurfuryl-(–1S)-
  camphor-10-sulphonate,
  for chiral deriv of amines 230
  (table) 234
S-Tetrahydro-5-oxo-2-furancarboxyl
  (TOF),
  for chiral deriv of alcohols and
    lactones 220–1
  (tables) 223, 240

1,1,3,3-Tetramethyldisilazane deriv of
 steroids 146
Thin-layer chrom (TLC) 3
 Bns deriv on polyamide layers 183
 of benzopyrone deriv 196
 charring to reveal separated
  zones 3, 329
 Dns deriv 181
 deriv of diols 203
 formylhydroxyacetophenone
  deriv 196
 NBD deriv 189
 naphthalene- and
  phenanthreneboronate deriv of
  diols 203
 optimization of deriv, monitoring
  by 3
 nitrated benzylimines for amines 133
 revelation of spots in 329
Thiazolidines, chiral deriv of sugars 237
Thiohydantoins in peptide sequencing
 152
 fluorescent 190
Thiols, fluorescent deriv 199–200
 deriv for FAB–MS (table) 271
Thionyl chloride in esterification 16
 for prep of acyl chlorides 167
Thiosemicarbazide for deriv of
 carbonyl cpds 131–2
 metal deriv of resulting
  thiosemicarbazones 132
Thromboxanes,
 fluorescent Mmc deriv 201
 IPDMS deriv 88
 TMS deriv 62
 (table) 2
p-Toluenesulphonate as ion-pair
 (table) 255
 tosyl deriv of amines 160
m-Toluyl deriv. of amines 159
p-Tolyltriazenes for benzyl esters 27
 alkylation with 120
Transesterification (interesterification,
 ester interchange) 14
 with aluminium chloride 26
 of amino acid esters 26
 with methanolic HCl 26
 with sodium methoxide 26
 with trifluoromethyl
  trimethylammonium hydroxide 26
 of triglycerides 25, 26
Transmodulators, post-chrom 336–7

2,3,4-Tri-O-acetyl-α-D-arabinopyranosyl
 isothiocyanate (AITC),
 for chiral deriv of amines 231
 of amino acids 243
 (tables) 235, 245
Trialkyloxonium fluoroborate 113
 prep of methyl esters 27, 121, 122
Tributyl phosphate as ion-pair 258
Tri-n-butylsilyl (TnBS) (table) 56
Trichloroacetyl deriv 35
2,2,2-Trichloroethanol and trichloroethyl
 esters 16, 35
 chloroformate and deriv 35, 47
Triethylsilyl (TES) deriv (table) 56
Trifluoroacetyl (TFA) (see also
 perfluoroacyl deriv) 35
 deriv for GC–MS 313–8
S(−)-N-(Trifluoroacetyl)-prolyl,
 for chiral deriv of amino
 acids 241, 244
 of amino alcohols 247
 of −OH cpds 248
 (tables) 223, 242, 245
2,2,2-Trifluoroethoxydimethylsilyl
 (table) 56
3,5-Trifluoromethyl benzeneboronate
 deriv 147
 oxazolinone deriv of amino acids for
 GC–MS 319
2,2,2-Trifluoro-1-phenylethylhydrazine
 for chiral deriv of carbonyl
 cpds 237, 238
 (table) 239
3,3,3-Trifluoropropyldimethylsilyl
 (table) 57
Tri-n-hexylsilyl (TnHS) (table) 56
Trimethylanilinium hydroxide (TMAH),
 on-column pyrolysis to methyl
 deriv 25, 113, 120, 121
Trimethylchlorosilane (TMCS),
 on-column silylation 53–4
 as silylation catalyst 57, 59, and
 generally
Trimethylpyridinium cation for
 FAB–MS (table) 271
Trimethylsilyl (TMS) deriv 22
 applications 60–75
 deriv for GC 55–75
 for GC–MS 305–13
 diethylamine (TMSDEA) as silyl
 donor 59
 high volatility of 59

esters of GC–MS 317–8
 imidazole (TMSIM) 44
 deriv for −OH cpds 59
 for general deriv in SFC 294
 non-deriv of enols 59
 problems with enols 304
 removal of excess 59
 selective, non-acidic silyl donor 59
 as silylation catalyst 58
 in vapour phase silylation 53
 O,N-bis(trimethylsilyl)-
  trifluoroacetamide (BSTFA),
  most general silyl
  donor 57 onwards
 on-column silylation 54
 (tables) 2, 4, 56
 see also Silylation
Trinitrobenzenesulphonic acid
 (TNBS) and deriv 162
 for post-chrom deriv of amino
 acids 339
Trinitrophenyl deriv of aminoglycosides
 and amino acids 162
Trioctylphosphine oxide (TOPO)
 adduct-forming agent 256
Triols, acetal and ketal deriv 142
Tri-n-propylsilyl (TnPS) (table) 56
Tropylium ion in MS 142
Tyrosine, dimethyloxy benzimidazole
 deriv 151

U

Ultraviolet (UV)-abs deriv
 (Chapter 8) 142–174
Uronic acids,
 alkylation 112
 base-promoted β-elimination 115
 ester deriv 115

V

Vitamins
 acetylation of α-tocopherol 37
 butyl- and methylboronates of D
  vitamins for GC–MS 321
 perfluoroacylation of B_6 vitamers 42
 post-chrom deriv of B vitamins 341
 TMS deriv for GC–MS 320–1
 UV–abs deriv of biotin and related
  cpds 165

W

Water removal 13, 14